Thomson™ NOW!

What do you need to learn NOW?

Social Problems

Table of Contents

Part I: SOCIOLOGY AND THE STUDY OF SOCIAL PROBLEMS
1. Thinking About Social Problems

Part II: PROBLEMS OF WELL-BEING
2. Illness and the Health Care Crisis
3. Alcohol and Other Drugs
4. Crime and Social Control
5. Family Problems

Part III: PROBLEMS OF INEQUALITY
6. Poverty and Economic Inequality
7. Work and Unemployment
8. Issues in Education
9. Race, Ethnicity, and Immigration
10. Gender Inequality
11. Issues in Sexual Orientation
12. Problems of Youth and Aging

Part IV: PROBLEMS OF GLOBALIZATION
13. Population Growth and Urbanization
14. Environmental Problems
15. Science and Technology
16. Conflict, War, and Terrorism

Take charge of your learning with **ThomsonNOW™**, (formerly SociologyNow) the first assessment-centered student learning tool for social problems.

This new online diagnostic tool could be your key to success in your social problems course!

ThomsonNOW™ will help you:

◇ create a personalized study plan for each chapter of your social problems text
◇ understand key concepts in the course
◇ better prepare for exams—and increase your chances of success

Lift the page for more information.

THOMSON

WADSWORTH™

How can you access ThomsonNOW?

◆ **Your instructor may have chosen to package the access code card with your new text.** In this case, you'll find the access card within this text, which contains your free four-month pass code, allowing you anytime access to **ThomsonNOW.**

◆ **If your instructor did not order the free access code card to be packaged with your text—or if you have a used copy of the text— you can still obtain an access code for a nominal fee.** Just visit the Thomson Wadsworth E-Commerce site at www.thomsonedu.com/sociology, where easy-to-follow instructions help you purchase your access code.

Are you ready to get started?
Help is just a click away with ThomsonNOW.

Thomson™ NOW!

The exciting program that is firmly grounded in sociology and lets you take diagnostic quizzes, review chapter content, conduct online research, think critically about social problems, watch videos of well-known sociologists as they discuss important concepts, and ultimately improve your success in the course, all on one, easy to use web-based program.

www.thomsonedu.com/thomsonnow

Log on today!

THOMSON
✦
WADSWORTH

Thomson Wadsworth ◆ P.O. Box 6904 ◆ Florence, KY 41022
800-423-0563 ◆ Fax 859-647-5020 ◆ Email: review@kdc.com

Source code 7TPSOM01

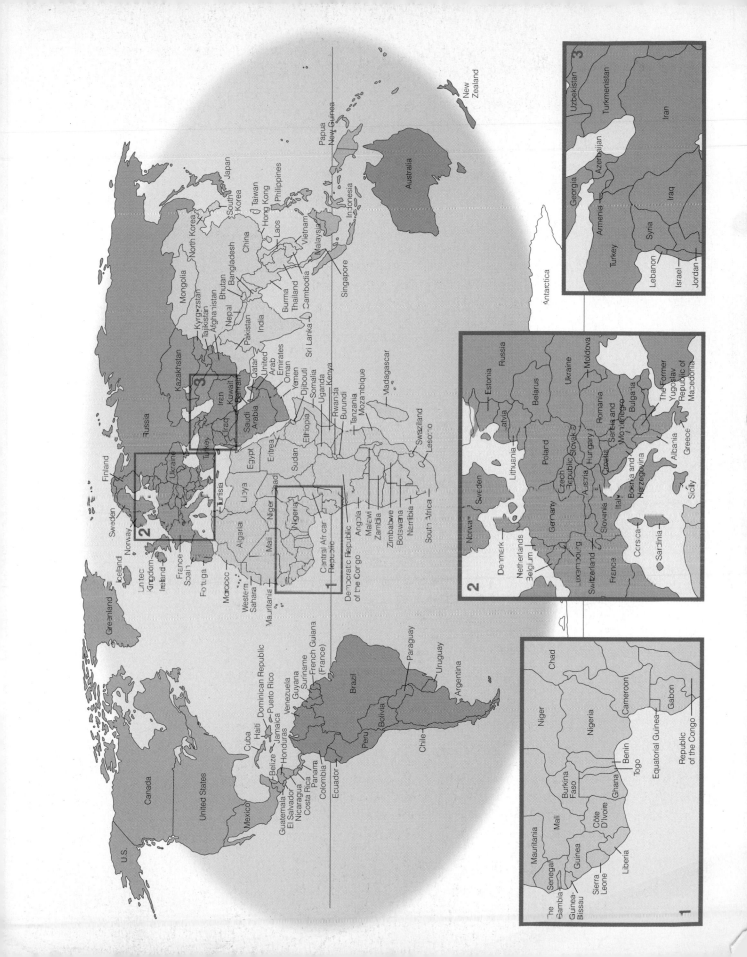

Understanding Social Problems

Fifth Edition

Linda A. Mooney

David Knox

Caroline Schacht

East Carolina University

THOMSON

WADSWORTH

Australia • Brazil • Canada • Mexico • Singapore
Spain • United Kingdom • United States

THOMSON

WADSWORTH

Understanding Social Problems, **Fifth Edition**
Linda A. Mooney, David Knox, and Caroline Schacht

Senior Acquisitions Editor, Sociology: Robert Jucha
Assistant Editor: Kristin Marrs
Editorial Assistant: Katia Krukowski
Technology Project Manager: Dee Dee Zobian
Marketing Manager: Michelle Williams
Marketing Assistant: Jaren Boland
Marketing Communications Manager: Linda Yip
Project Manager, Editorial Production: Cheri Palmer
Creative Director: Rob Hugel
Print Buyer: Judy Inouye
Permissions Editor: Bob Kauser
Production Service: G&S Book Services

Text Designer: Jeanne Calabrese
Photo Researcher: Myrna Engler
Copy Editor: Mimi Braverman
Cover Designer: Larry Didona
Cover Images (top to bottom): Doug Menuez/Photodisc Green/ Getty Images; © Photodisc/Getty Images; © Royalty-Free/Corbis; © Viviane Moos/Corbis; © Don Farrall/Photodisc Green/Getty Images; © Paul Fusco/Magnum Photos; © David Hume Kennerly/ Reportage/Getty Images; © Peter Hendrie/The Image Bank/Getty Images
Compositor: G&S Book Services
Printer: Quebecor World/Taunton

Thomson Higher Education
10 Davis Drive
Belmont, CA 94002-3098
USA

For more information about our products, contact us at:
Thomson Learning Academic Resource Center
1-800-423-0563

For permission to use material from this text or product, submit a request online at **http://www.thomsonrights.com**.

Any additional questions about permissions can be submitted by e-mail to **thomsonrights@thomson.com**.

Library of Congress Control Number: 2005934718

ISBN-13: 978-0-495-09158-5
ISBN-10: 0-495-09158-8

To Carol L. Jenkins—
writer, mentor, and friend.
And to Isabelle Troadec, in appreciation for her
dedication to social justice and for the hope she has
brought to communities in New Orleans.

Brief Contents

Contents

Part 1 : Sociology and the Study of Social Problems

Thinking About Social Problems 1

Illness and the Health Care Crisis   28

Alcohol and Other Drugs 72

Part 3 : Problems of Inequality

Race, Ethnicity, and Immigration   289

Gender Inequality   331

Problems of Youth and Aging 408

Environmental Problems   472

Science and Technology   512

Conflict, War, and Terrorism 545

Preface

Understanding Social Problems is intended for use in a college-level sociology course. We recognize that many students enrolled in undergraduate sociology classes are not sociology majors. Thus, we have designed our text with the aim of inspiring students—*no matter what their academic major or future life path may be*—to care about the social problems affecting people throughout the world. In addition to providing a sound theoretical and research basis for sociology majors, *Understanding Social Problems* also speaks to students who are headed for careers in business, psychology, health care, social work, criminal justice, and the nonprofit sector, as well as to those pursuing degrees in education, fine arts, the humanities, and to those who are "undecided." Social problems, after all, affect each and every one of us, directly or indirectly. And everyone, whether a leader in business or politics, a stay-at-home parent, or a student, can become more mindful of how his or her actions (or inactions) perpetuate or alleviate social problems. We hope that *Understanding Social Problems* not only informs, but also inspires, planting seeds of social awareness that will grow no matter what academic, occupational, and life path students choose.

New to This Edition

The Fifth Edition of *Understanding Social Problems* includes an exciting new pedagogical feature called **ThomsonNOW,** which enables students to take an online diagnostic quiz on each chapter of the text to identify the areas that the student has mastered as well as those areas that require further study. The chapters in the Fifth Edition have been reorganized so that topics of poverty and economic inequality, work and unemployment, and problems of education are covered earlier in the text. The chapter previously called "Race and Ethnic Relations" has been renamed "Race, Ethnicity, and Immigration" to reflect the substantial expansion of U.S. immigration issues covered in that chapter. Most of the chapter opening vignettes are new, and many of the chapter features (The Human Side, Focus on Technology, Social Problems Research Up Close, and Self and Society) have been changed. Finally, each chapter has new and updated figures and tables, as well as new and revised material, detailed below.

    Chapter 1 (Thinking About Social Problems) now includes a discussion of social change and social movements in the Human Side feature, which also includes new examples of college student activism. The Self and Society feature has been updated with 2004 freshman norms from the UCLA's Higher Education Research Institute, and the opening vignette now includes 2005 data on Americans' level of satisfaction on a variety of social issues.

Chapter 2 (Illness and the Health Care Crisis) contains new information on the global HIV/AIDS crisis, the problem of inadequate health insurance coverage, and the effects of international free-trade agreements on health. A new section on the growing problem of obesity has also been added, as well as a new Focus on Technology feature that looks at Americans' use of the Internet to find health information, and a new Taking a Stand feature that addresses whether college and universities should require students to have health insurance coverage.

Chapter 3 (Alcohol and Other Drugs) contains an expanded discussion of tobacco and advertising as it relates to marketing cigarettes, particularly of flavored cigarettes, to minors. Alcohol and drug use statistics have been updated using 2004 Monitoring the Future and 2003 National Survey on Drug Use and Health data. There is new information on fetal alcohol syndrome, binge drinking, and the increased dangers of "crystal meth." This revised chapter also features a new Human Side box containing excerpts from the 2005 autobiography by Koren Zailckas, titled *Smashed: The Story of a Drunken Girlhood.*

Chapter 4 (Crime and Social Control) presents recent crime statistics available from the *National Crime Victimization Survey* and the *Uniform Crime Reports.* It also contains a new opening vignette on Brian Nichols, and expanded sections on white-collar crime, identity theft and internet fraud, child pornography, international incarceration rates, and serial killers. The Focus on Technology feature on DNA evidence has also been updated.

In **Chapter 5** (Family Problems), information on changing structures and patterns in U.S. families and households has been updated, and the "marital decline perspective" and "marital resiliency perspective" have been added to the discussion of the opposing views on the state of the family. Other new information in this chapter includes a new section on the role of women in the family, new research on why some adults stay in abusive relationships, a discussion of bans on corporal punishment in other countries, and the current status of sex education programs in the United States. A new Human Side contains an excerpt from the book *We're Still Family: What Grown Children Have to Say About Their Parents' Divorce* (by Ahrons, 2004), and the new Taking a Stand feature addresses whether or not pharmacists should have the right to refuse to fill a prescription for birth control.

Chapter 6 (Poverty and Economic Inequality) includes a discussion of how poor populations are most vulnerable to the effects of natural disasters, such as the Asian tsunami of 2004 and Hurricane Katrina of 2005. This chapter also has new information about a proposed universal living wage and about efforts to alleviate poverty through the Faith-based and Community Initiative policy of the G. W. Bush administration. Statistics on poverty, economic inequality, and welfare programs have been updated, with an increased emphasis on the growing economic inequality in the United States and throughout the world. The new Self and Society feature presents a "Food Insecurity Scale," allowing students to assess their own level of food security.

In **Chapter 7** (Work and Unemployment) there is increased attention to **corporatocracy**—the system of government that involves ties between government and corporations and that serves the interests of corporations. A new section on "McDonaldization" has been added, and information has been updated on economic globalization and free trade agreements, global and U.S. unemployment, and work-family issues. New material in this chapter also includes a discussion of long-term unemployment, recent examples of corporate misbehavior, a new Self and Society feature ("How Do You Define the American Dream?"), and the results of the *Annual Survey of Violations of Trade Union Rights* (in the Social Problems Research Up Close feature).

Chapter 8 (Problems in Education) has new sections on bullying, Head Start assessment, total immersion programs, and violence against teachers. A new Human Side feature titled "The Pendulum of Change" describes one teacher's frustration with federal mandates and state-imposed accountability guidelines. This revised chapter also includes updated information on distance education; expanded coverage of President Bush's *No Child Left Behind* initiative; and new sections on the Individuals with Disabilities Education Act (IDEA), special education programs, and school system compliance.

Chapter 9, previously titled "Race and Ethnic Relations," has been renamed "Race, Ethnicity and Immigration" to reflect the new coverage of a topic of growing concern: U.S. immigration. A new Human Side feature describes the experience of an immigrant day laborer in Georgia who was victimized by a violent hate-crime attack, the new Self and Society feature is on attitudes toward U.S. immigrants and immigration, and the new Taking a Stand feature addresses the issue of whether undocumented immigrants should qualify for in-state tuition. This chapter also includes the effects of Hurricane Katrina and its aftermath on U.S. race relations, as well as new data on attitudes toward affirmative action.

Chapter 10 (Gender Inequality) includes a new section on gender and religion, and an increased focus on global gender inequality. For example, Table 10.1 includes scores and rankings of the ten countries with the smallest gender gaps and the ten countries with the largest gender gaps. This revised chapter also includes increased coverage of gender issues related to boys and men, including an expanded section on the men's movement, as well as a lengthier discussion of gender as it exists on a continuum. The new Focus on Technology feature discusses sex selection technology and the future of "designer babies." Finally, a new Social Problems Research Up Close discusses "Gender and the Quantity and Quality of Free Time."

In **Chapter 11** (Issues in Sexual Orientation), we provide updated information on the legal status of homosexuality and same-sex relationships globally and in the United States. This revised chapter includes updated survey data on same-sex marriage, attitudes toward homosexuality, and discrimination against gays and lesbians. A new section on police mistreatment of sexual orientation minorities has been added. New information also has been added to the sections on reparative therapy ("therapy" that purports to change homosexuals' sexual orientation), religion and sexual orientation, gays and lesbians in the media, discrimination against sexual orientation minorities, and laws and policies that protect gays and lesbians from discrimination. The Human Side feature addresses the effects of a federal marriage amendment on gay and lesbian families. A new Taking a Stand feature asks if religious organizations that receive federal funds to provide social services under the Faith-Based Initiative should be allowed to discriminate in hiring.

In **Chapter 12** (Problems of Youth and Aging), we have expanded the section on child prostitution and present new information on child obesity and on children and the death penalty. The revised chapter also includes new sections on adoption and foster care and on phased retirement. The chapter incorporates recent statistics from the Administration on Aging (AOA) and presents new data on income and marital status of the elderly. A new Social Problems Research Up Close feature focuses on television portrayals of the elderly.

Chapter 13 (Population Growth and Urbanization) presents new statistics on population growth and discusses a Population Institute report that identifies rapid population growth as a contributing factor to global insecurity, war, and terrorism. New information is presented on population density, the youth bulge, and the concept of the environmental footprint. This revised chapter contains new sections on the involvement of men in family planning and the role of community development corporations in urban renewal efforts. A new Self and Society feature focuses on attitudes toward walking and on creating better walking communities, and the new Social Problems Research Up Close feature looks at the relationship between urban sprawl, physical activity, obesity, and morbidity.

Chapter 14 (Environmental Problems) includes new data on a wide range of topics, including the effects of free trade agreements on the environment, the 2005 Millennium Ecosystem Assessment, global warming and climate change, the ozone hole, nuclear power, pesticide use, environmental injustice, disappearing species, environmental destruction by the U.S. military, and growing concern over China's increasing income and consumption. New sections on "green energy" have been added, specifically on biofuels and hydrogen power, as well as new sections on green building and the role of corporations in the environmental movement. A new Self and Society feature focuses on attitudes toward energy and the environment, and the revised Social Problems Research Up Close feature presents the Centers for Disease Control and Prevention's *Third National Report on Human Exposure to Environmental Chemicals.*

Chapter 15 (Science and Technology) contains expanded sections on privacy and security issues, online activities, and Internet addiction. This revised chapter also includes a new section on SCNT (somatic cell nuclear transfer). The new Taking a Stand feature addresses the issue of federal funding of stem cell research and the Focus on Technology feature has been updated with new technological innovations.

Chapter 16 (Conflict, War, and Terrorism) begins with a new vignette that traces the evolution of an Iraqi suicide bomber. This revised chapter highlights the cost of continued military presence in Iraq, including information on military deaths, civilian causalities, and the financial cost of U.S. involvement. Additionally, there is a new section on military abuses and mistreatment of prisoners. This revised chapter also has a new section on social psychological variables associated with being a terrorist, and a new Social Problems Research Up Close feature on attitudes of Generation Xers toward the military.

Features and Pedagogical Aids

We have integrated a number of features and pedagogical aids into the text to help students learn to think about social problems from a sociological perspective. It is our mission to help students not only apply sociological concepts to observed situations in their everyday lives and think critically about social problems and their implications, but also learn to assess how social problems relate to their lives on a personal level.

Study Aids in Each Chapter
That Are New to This Edition

Placement of the Taking a Stand Feature In the Fourth Edition of *Understanding Social Problems,* we added the Taking a Stand exercise at the end of each chapter to help students hone critical thinking skills. In the revised Fifth Edition, we have instead integrated the Taking a Stand feature into the text of each chapter.

Exercises and Boxed Features

Self and Society Each chapter includes a social survey designed to help students assess their own attitudes, beliefs, knowledge, or behavior regarding some aspect of the social problem under discussion. In Chapter 5 (Family Problems), for example, the "Abusive Behavior Inventory" invites students to assess the frequency of various abusive behaviors in their own relationships.

The Human Side In addition to the Self and Society boxed features, each chapter includes a boxed feature that further personalizes the social problems under discussion by describing personal experiences of individuals who have been affected by them. The Human Side feature in Chapter 9 (Race, Ethnicity and Immigration), for example, describes the experience of an immigrant day laborer who was victimized by a violent hate crime.

Social Problems Research Up Close In every chapter, these features present examples of social science research. These features demonstrate for students the sociological enterprise, from theory and data collection to findings and conclusions, thus exposing students to various studies and research methods.

Focus on Technology Because technology is a pervasive part of our society, each chapter contains a Focus on Technology feature, in addition to an entire chapter on the topic (Chapter 15, Science and Technology). In any given chapter, the Focus on Technology features present information on how technology may contribute to that social problem and its solution. In Chapter 6 (Poverty and Economic Inequality), for example, the Focus on Technology feature examines the controversy surrounding the use of agricultural biotechnology, weighing its potential benefits (such as increasing food supplies and nutrition for the poor) against the potential risks (such as adverse effects on human and environmental health).

In-Text Learning Aids

Vignettes Each chapter begins with a vignette designed to engage students and draw them into the chapter by illustrating the current relevance of the topic under discussion. Chapter 2 (Illness and the Health Care Crisis), for instance, begins with the story of Nkosi Johnson, a South African boy who was born HIV-positive to a 17-year-old mother who later died of AIDS. At age 11, while suffering from full-blown AIDS, Nkosi stood up before the 13th International AIDS Conference in Durban, South Africa, and delivered a speech he wrote about AIDS. About one year later, at age 12, Nkosi died.

Critical Thinking Each chapter ends with a brief section called "Critical Thinking," which raises several questions related to the chapter topic. These questions invite students to apply their critical thinking skills to the information discussed in the chapter.

Key Terms Important terms and concepts are highlighted in the text where they first appear. To re-emphasize the importance of these words, they are listed at the end of every chapter and included in the glossary at the end of the text.

Glossary All key terms are defined in the end-of-text glossary.

Understanding [specific social problem] Sections All too often students, faced with contradictory theories and study results, walk away from social problems courses without any real understanding of their causes and consequences. To address this problem, chapter sections titled "Understanding . . . [specific social problem]" cap the body of each chapter just before the chapter summaries. Unlike the chapter summaries, these sections synthesize the material presented in the chapter, summing up the present state of knowledge and theory on the chapter topic.

Supplements

The Fifth Edition of *Understanding Social Problems* comes with a full complement of supplements designed with both faculty and students in mind.

Supplements for the Instructor

Instructor's Resource Manual with Test Bank This supplement offers the instructor learning objectives, key terms, lecture outlines, student projects, classroom activities, and Internet and InfoTrac® College Edition exercises. Test items include 60–80 multiple-choice and 10–15 true-false questions with answers and page references, as well as 5 short-answer and 10 essay questions for each chapter. Each multiple-choice item has the question type (factual, applied, or conceptual) indicated. All questions are labeled as new, modified, or pickup, so instructors know if the question is new to this edition of the test bank, modified but picked up from the previous edition of the test bank, or picked up straight from the previous edition of the test bank. Concise user guides for InfoTrac College Edition and WebTutor™ are included as appendixes.

ExamView® Computerized Testing Create, deliver, and customize tests and study guides (both print and online) in minutes with this easy-to-use assessment and tutorial system. ExamView offers both a quick test wizard and an online test wizard that guide you step by step through the process of creating tests. The test appears on screen exactly as it will print or display online. Using ExamView's complete word-processing capabilities, you can enter an unlimited number of new questions or edit existing questions included with ExamView (it contains all the test bank questions).

Multimedia Manager Instructor Resource CD: A 2006 Microsoft® PowerPoint® Link Tool With this one-stop digital library and presentation tool, instructors can assemble, edit, and present custom lectures with ease. The Multimedia Manager contains figures, tables, graphs, and maps from this text; pre-assembled Microsoft PowerPoint lecture slides; video clips from DALLAS TeleLearning; ShowCase presentational software; tips for teaching; the instructor's manual; and more.

Videos Adopters of *Understanding Social Problems* have several different video options available with the text.

CNN Today: Social Problems/Issues and Solutions Video

Social Problems: Volume II: 0-534-54124-0

Social Problems: Volume III: 0-534-61898-7

Social Problems: Volume IV: 0-534-61933-9

Integrate the up-to-the-minute programming power of CNN and its affiliate networks right into your classroom! Updated yearly, *CNN Today* videos are course-specific videos that can help you launch a lecture, spark a discussion, or demonstrate an application—using the top-notch business, science, consumer, and political reporting of the CNN networks. Produced by Turner Learning, Inc., these 45-minute videos show your students how the principles they learn in the classroom apply to the stories they see on television. Special adoption conditions apply.

AIDS in Africa DVD Southern Africa has been overcome by a pandemic of unparalleled proportions. This documentary series focuses on Namibia, a new democracy, and the many actions that are being taken to control HIV/AIDS there. Included in this series are four documentary films created by the Project Pericles scholars at Elon University.

Wadsworth Sociology Video Library This large selection of thought-provoking films is available to adopters based on adoption size.

Supplements for the Student

Study Guide The study guide includes learning objectives, brief and detailed chapter outlines, key terms, Internet activities, InfoTrac College Edition exercises, student projects and classroom activities, practice tests consisting of 20–25 multiple-choice and 10–15 true-false questions with answers and page references, as well as 5 short-answer questions and 5 essay questions with page references to enhance and test student understanding of chapter concepts.

Online Resources

ThomsonNOW. This online tool provides students with a customized study plan based on a diagnostic "pretest" that they take after reading each chapter. The study plan provides interactive exercises, videos, and other resources to help students master the material. After the study plan has been reviewed, students can then take a "posttest" to monitor their progress in mastering the chapter concepts. Instructors may bundle this product for their students with each new copy of the text for free! If your instructor did not order the free access code card to be packaged with your text—or, if you have a used copy of the text—you can still obtain an access code for a nominal fee. Just visit the Thomson Wadsworth E-Commerce site at http://www.thomsonedu.com, where easy-to-follow instructions help you purchase your access code.

Extension: Wadsworth's Sociology Reader Collection Create your own customized reader for your sociology class, drawing from dozens of classic and contemporary articles found on the exclusive Thomson Wadsworth TextChoice database. Using the TextChoice website (http://www.TextChoice.com), you can preview articles, select your content, and add your own original material. TextChoice will then produce your materials as a printed supplementary reader for your class.

JoinIn™ on TurningPoint® Thomson Wadsworth is now pleased to offer you book-specific JoinIn content for Response Systems tailored to *Understanding Social Problems,* Fifth Edition, allowing you to transform your classroom and assess your students' progress with instant in-class quizzes and polls. Our exclusive agreement to offer TurningPoint software lets you pose book-specific questions and display students' answers seamlessly within the Microsoft PowerPoint slides of your own lecture, in conjunction with the "clicker" hardware of your choice. Enhance how your students interact with you, your lecture, and each other.

Turnitin™. This proven online plagiarism-prevention software promotes fairness in the classroom by helping students learn to correctly cite sources and allowing instructors to check for originality before reading and grading papers. Turnitin quickly checks student papers against billions of pages of Internet content, millions of published works, and millions of student papers and within seconds generates a comprehensive originality report.

Wadsworth's Sociology Home Page (http://sociology.wadsworth.com) Here you will find a wealth of sociology resources such as *Census 2000: A Student Guide for Sociology, Breaking News in Sociology, Guide to Researching Sociology on the Internet, Sociology in Action,* and much more. Contained on the home page is the companion website for *Understanding Social Problems,* Fifth Edition.

Mooney/Knox/Schacht's *Understanding Social Problems* **Companion Website (http://sociology.wadsworth.com/mooney_knox_schacht5e)** This site provides access to useful learning resources for each chapter of the book. Instructors can also access password-protected instructor's manuals, PowerPoint lectures, and important sociology links. Click on the companion website to find useful learning resources for each chapter of the book. Some of these resources include

- Tutorial Practice Quizzes that can be scored and e-mailed to the instructor
- Web Links
- Internet Exercises
- Video Exercises
- InfoTrac College Edition Exercises
- Flash cards of the text's glossary
- Crossword Puzzles
- Essay Questions
- Learning Objectives
- Virtual Explorations

And much more!

WebTutor™ Toolbox for WebCT or Blackboard Preloaded with content and available free via access code when packaged with this text, WebTutor Toolbox pairs all the content of this text's rich book companion website with all the sophisticated course management functionality of a WebCT or Blackboard product. You can assign materials (including online quizzes) and have the results flow automatically to your grade book. Toolbox is ready to use as soon as you log on—or you can customize its preloaded content by uploading images and other resources, adding web links, or creating your own practice materials. Students have access only to student resources on the website. Instructors can enter a pincode for access to password-protected instructor resources.

InfoTrac College Edition Four months' access to this online database—featuring reliable, full-length articles from thousands of academic journals and periodicals—is available with this text at no additional charge! This fully searchable database now features stable, topically bookmarked InfoMarks® URLs to assist in research, plus InfoWrite critical thinking and writing tools. The database also offers 20 years' worth of full-text articles from almost 5000 diverse sources, such as academic journals, newsletters, and up-to-the-minute periodicals, including *Time, Newsweek, Science, Forbes,* and *USA Today.* This incredible depth and breadth of material—available 24 hours a day from any computer with Internet access—makes conducting research so easy that your students will want to use it to enhance their work in every course!

Opposing Viewpoints Resource Center (OVRC) Newly available from Wadsworth, this online center presents varying perspectives on today's most compelling issues. OVRC draws on Greenhaven Press's acclaimed Social Issues series, as well as core reference content from other Gale and Macmillan Reference USA sources. The result is a dynamic online library of current events topics—the facts as well as the arguments of each topic's proponents and detractors. Special sections focus on critical thinking—walking students through the steps involved in critically evaluating point-counterpoint arguments—and researching and writing papers.

Acknowledgments

This text reflects the work of many people. We would like to thank the following for their contributions to the development of this text: Bob Jucha, Acquisitions Editor; Kristin Marrs, Assistant Editor; Elise Smith, Assistant Editor; and Gretchen Otto. We would also like to acknowledge the support and assistance of Carol Jenkins (thanks, CJ), Marieke Van Willigen, and Lee Maril. To each we send our heartfelt thanks. Additionally, we are indebted to those who read the manuscript in its various drafts and provided valuable insights and suggestions, many of which have been incorporated into the final manuscript:

Bob Cordell
West Virginia University at Parkersburg

Furjen Deng
Sam Houston State University

Katherine Dietrich
Blinn College

Bill Feigelman
Nassau Community College

Gay Moore
Chattanooga State Technical Community College

Carl Riden
Longwood University

Brenda Wilhelm
Mesa State College

We are also grateful to the reviewers of the first, second, and third and fourth editions: David Allen, *University of New Orleans;* Patricia Atchison, *Colorado State University;* Wendy Beck, *Eastern Washington University;* Walter Carroll, *Bridgewater State College;* Deanna Chang, *Indiana University of Pennsylvania;* Roland Chilton, *University of Massachusetts;* Verghese Chirayath, *John Carroll University;* Margaret Chok, *Pellissippi State Technical Community College;* Kimberly Clark, *DeKalb College–Central Campus;* Anna M. Cognetto, *Dutchess Community College;* Robert R. Cordell, *West Virginia University at Parkersburg;* Barbara Costello, *Mississippi State University;* William Cross, *Illinois College;* Kim Davies, *Augusta State University;* Doug Degher, *Northern Arizona University;* Katherine Dietrich, *Blinn College;* Jane Ely, *State University of New York–Stony Brook;* William Feigelman, *Nassau Community College;* Joan Ferrante, *Northern Kentucky University;* Robert Gliner, *San Jose State University;* Roberta Goldberg, *Trinity College;* Roger Guy, *Texas Lutheran University;* Julia Hall, *Drexel University;* Millie Harmon, *Chemeketa Community College;* Sylvia Jones, *Jefferson Community College;* Nancy Kleniewski, *University of Massachusetts, Lowell;* Daniel Klenow, *North Dakota State University;* Sandra Krell-Andre, *Southeastern Community College;* Pui-Yan Lam, *Eastern Washington University;* Mary Ann Lamanna, *University of Nebraska;* Phyllis Langton, *George Washington University;* Cooper Lansing, *Erie Community College;* Tunga Lergo, *Santa Fe Community College, Main Campus;* Dale Lund, *University of Utah;* Lionel Maldonado, *California State University, San Marcos;* Judith Mayo, *Arizona State University;* Peter Meiksins, *Cleveland State University;* Madonna Harrington-Meyer, *University of Illinois;* JoAnn Miller, *Purdue University;* Clifford Mottaz, *University of Wisconsin–River Falls;* Lynda D. Nyce, *Bluffton College;* Frank J. Page, *University of Utah,* James Peacock, *University of North Carolina;* Barbara Perry, *Northern Arizona University;* Ed Ponczek, *William Rainey Harper College;* Donna Provenza, *California State University at Sacramento;* Cynthia Reynaud, *Louisiana State University;* Carl Marie Rider, *Longwood University;* Jeffrey W. Riemer, *Tennessee Technological University,* Cherylon Robinson, *University of Texas at San Antonio;* Rita Sakitt, *Suffolk County Community College;* Mareleyn Schneider, *Yeshiva University;* Paula Snyder, *Columbus State Community College;* Lawrence Stern, *Collin County Community College;* John Stratton, *University of Iowa;* D. Paul Sullins, *The Catholic University of America,* Joseph Trumino, *St. Vincent's College of St. John's University;* Robert Turley, *Crafton Hills College;* Alice Van Ommeren, *San Joaquin Delta College;* Joseph Vielbig, *Arizona Western University;* Harry L. Vogel, *Kansas State University;* Robert Weaver, *Youngstown State University;* Rose Weitz, *Arizona State University;* Bob Weyer, *County College of Morris;* Oscar Williams, *Diablo Valley College;* Mark Winton, *University of Central Florida;* Diane Zablotsky, *University of North Carolina.*

Finally, we are interested in ways to improve the text, and invite your feedback and suggestions for new ideas and material to be included in subsequent editions. You can contact us at mooneyl@ecu.edu, knoxd@ecu.edu, or schachtc@ecu.edu.

About the Authors

Linda A. Mooney, Ph.D., is an associate professor of sociology at East Carolina University in Greenville, North Carolina. In addition to social problems, her specialties include law, criminology, and juvenile delinquency. She has published over thirty professional articles in such journals as *Social Forces, Sociological Inquiry, Sex Roles, Sociological Quarterly,* and *Teaching Sociology.* She has won numerous teaching awards, including the University of North Carolina Board of Governor's Distinguished Professor for Teaching Award.

David Knox, Ph.D., is professor of sociology at East Carolina University. He has taught Social Problems, Introduction to Sociology, and Sociology of Marriage Problems. He is the author or co-author of ten books and more than sixty professional articles. His research interests include various aspects of college student relationships, sexual values, and behavior.

Caroline Schacht, M.A., is an instructor of sociology at East Carolina University. She has taught Introduction to Sociology, Deviant Behavior, Sociology of Education, Individuals in Society, and Courtship and Marriage. She has co-authored several textbooks in the areas of social problems, introductory sociology, courtship and marriage, and human sexuality. Her areas of interest include mediation and conflict resolution, alternative education, social inequality, and environmental problems.

> " Unless someone like you cares a whole awful lot, nothing is going to get better. It's not. " *Dr. Seuss, The Lorax*

Thinking About Social Problems

Since 2001 the Gallup Organization has conducted a survey asking Americans to rate their satisfaction with a variety of social issues. In 2005 Americans' net satisfaction (percentage satisfied minus percentage dissatisfied) was highest in reference to the position of women in the U.S. (41%), military strength and preparedness (35%), and "the opportunity for a person to get ahead by working hard" (33%). Americans were much less satisfied with the availability of affordable health care (−48%), efforts to deal with poverty and homelessness (−41%), the Social Security and Medicare systems (−34%), and the quality of public education in the nation (−18%) (Newport 2005). Moreover, survey results indicate that, overall, only 38% of Americans were satisfied "with the way things are going in the country today" (Gallup 2005).

A global perspective on social problems is also troubling. In 1990 the United Nations Development Programme published its first annual *Human Development Report,* which measured the well-being of populations around the world according to a "human development index." This index measures three basic dimensions of human development: longevity, as measured by life expectancy at birth; knowledge (i.e., literacy, educational attainment); and a decent standard of living. The most recent report concludes that "unless people who are poor and marginalized—who more often than not are members of religious or ethnic or migrant groups—can influence political action at local and national levels, they are unlikely to get equitable access to jobs, schools, hospitals, justice, security, and other basic services" (Human Development Report 2004, p. 1).

Problems related to poverty and malnutrition, inadequate education, acquired immunodeficiency syndrome (AIDS) and other sexually transmitted diseases (STDs), inadequate health care, crime, conflict, oppression of minorities, environmental destruction, and other social issues are both national and international concerns. Such problems present both a threat and a challenge to our national and global society. The primary goal of this textbook is to facilitate increased awareness and understanding of problematic social conditions in U.S. society and throughout the world. Although the topics covered in this book vary widely, all chapters share common objectives: to explain how social problems are created and maintained; to indicate how they affect individuals, social groups, and societies as a whole; and to examine programs and policies for change. We begin by looking at the nature of social problems.

Only relatively recently have suicide bombers been considered a social problem to the U.S. public. More specifically, since the horror of September 11, 2001, terrorism in the United States has taken on new meaning. Here airport security guards inspect vehicles approaching the terminals.

AP/Wide World Photos

What Is a Social Problem?

There is no universal, constant, or absolute definition of what constitutes a social problem. Rather, social problems are defined by a combination of objective and subjective criteria that vary across societies, among individuals and groups within a society, and across historical time periods.

Objective and Subjective Elements of Social Problems

Although social problems take many forms, they all share two important elements: an objective social condition and a subjective interpretation of that social condition. The **objective element** of a social problem refers to the existence of a social condition. We become aware of social conditions through our own life experience, through the media, and through education. We see the homeless, hear gunfire in the streets, and see battered women in hospital emergency rooms. We read about employees losing their jobs as businesses downsize and factories close. In television news reports we see the anguished faces of parents whose children have been killed by violent youths.

The **subjective element** of a social problem refers to the belief that a particular social condition is harmful to society or to a segment of society and that it should and can be changed. We know that crime, drug addiction, poverty, racism, violence, and pollution exist. These social conditions are not considered social problems, however, unless at least a segment of society believes that these conditions diminish the quality of human life.

By combining these objective and subjective elements, we arrive at the following definition: A **social problem** is a social condition that a segment of society views as harmful to members of society and in need of remedy.

Variability in Definitions of Social Problems

Individuals and groups frequently disagree about what constitutes a social problem. For example, some Americans view the availability of abortion as a social problem, whereas others view restrictions on abortion as a social problem. Similarly, some Americans view homosexuality as a social problem, whereas others view prejudice and discrimination against homosexuals as a social problem. Such variations in what is considered a social problem are due to differences in values, beliefs, and life experiences.

Definitions of social problems vary not only within societies but also across societies and historical time periods. For example, before the 19th century it was a husband's legal right and marital obligation to discipline and control his wife through the use of physical force. Today, the use of physical force is regarded as a social problem rather than a marital right.

Tea drinking is another example of how what is considered a social problem can change over time. In 17th- and 18th-century England tea drinking was regarded as a "base Indian practice" that was "pernicious to health, obscuring industry, and impoverishing the nation" (Ukers 1935, cited by Troyer & Markle 1984). Today, the English are known for their tradition of drinking tea in the afternoon.

Because social problems can be highly complex, it is helpful to have a framework within which to view them. Sociology provides such a framework. Using a sociological perspective to examine social problems requires a knowledge of the

Whereas some individuals view homosexual behavior as a social problem, others view homophobia as a social problem. Here, participants carry a giant rainbow flag during a gay pride parade in Toronto, Canada. The 2006 Canadian census has recently been revamped to include "same-sex married spouse" as a response option (Beeby 2005).

AP/Wide World Photos

basic concepts and tools of sociology. In the remainder of this chapter we discuss some of these concepts and tools: social structure, culture, the "sociological imagination," major theoretical perspectives, and types of research methods.

Elements of Social Structure and Culture

Although society surrounds us and permeates our lives, it is difficult to "see" society. By thinking of society in terms of a picture or image, however, we can visualize society and therefore better understand it. Imagine that society is a coin with two sides: On one side is the structure of society, and on the other is the culture of society. Although each side is distinct, both are inseparable from the whole. By looking at the various elements of social structure and culture, we can better understand the root causes of social problems.

Elements of Social Structure

The *structure* of a society refers to the way society is organized. Society is organized into different parts: institutions, social groups, statuses, and roles.

Institution. An **institution** is an established and enduring pattern of social relationships. The five traditional institutions are family, religion, politics, economics, and education, but some sociologists argue that other social institutions, such as science and technology, mass media, medicine, sports, and the military, also play important roles in modern society. Many social problems are generated by inadequacies in various institutions. For example, unemployment may be influenced by the educational institution's failure to prepare individuals for the job market and by alterations in the structure of the economic institution.

Social Groups. Institutions are made up of social groups. A **social group** is defined as two or more people who have a common identity, interact, and form a social relationship. For example, the family in which you were reared is a social group that is part of the family institution. The religious association to which you may belong is a social group that is part of the religious institution.

Social groups can be categorized as primary or secondary. **Primary groups,** which tend to involve small numbers of individuals, are characterized by intimate and informal interaction. Families and friends are examples of primary groups. **Secondary groups,** which may involve small or large numbers of individuals, are task oriented and are characterized by impersonal and formal interaction. Examples of secondary groups include employers and their employees, and clerks and their customers.

Statuses. Just as institutions consist of social groups, social groups consist of statuses. A **status** is a position that a person occupies within a social group. The statuses we occupy largely define our social identity. The statuses in a family may consist of mother, father, stepmother, stepfather, wife, husband, child, and so on. Statuses can be either ascribed or achieved. An **ascribed status** is one that society assigns to an individual on the basis of factors over which the individual has no control. For example, we have no control over the sex, race, ethnic background, and socioeconomic status into which we are born. Similarly, we are assigned the status of child, teenager, adult, or senior citizen on the basis of our age—something we do not choose or control.

An **achieved status** is assigned on the basis of some characteristic or behavior over which the individual has some control. Whether you achieve the status of college graduate, spouse, parent, bank president, or prison inmate depends largely on your own efforts, behavior, and choices. One's ascribed statuses may affect the likelihood of achieving other statuses, however. For example, if you are born into a poor socioeconomic status, you may find it more difficult to achieve the status of college graduate because of the high cost of a college education.

Every individual has numerous statuses simultaneously. You may be a student, parent, tutor, volunteer fund-raiser, female, and Hispanic. A person's **master status** is the status that is considered the most significant in a person's social identity. Typically, a person's occupational status is regarded as his or her master status. If you are a full-time student, your master status is likely to be student.

Roles. Every status is associated with many **roles,** or the set of rights, obligations, and expectations associated with a status. Roles guide our behavior and allow us to predict the behavior of others. As a student, you are expected to attend class, listen and take notes, study for tests, and complete assignments. Because you know what the role of teacher involves, you can predict that your teacher will lecture, give exams, and assign grades based on your performance on tests.

A single status involves more than one role. For example, the status of prison inmate includes one role for interacting with prison guards and another role for interacting with other prison inmates. Similarly, the status of nurse involves different roles for interacting with physicians and with patients.

ThomsonNOW™

Learn more about **Social Groups** by going through the How Social Groups Shape Our Actions Data Experiment.

"When I fulfill my obligations as a brother, husband, or citizen, when I execute contracts, I perform duties that are defined externally to myself. . . . Even if I conform in my own sentiments and feel their reality subjectively, such reality is still objective, for I did not create them; I merely inherited them."

Emile Durkheim
Sociologist

ThomsonNOW™

Learn more about **Roles and Status** by going through the Roles and Status Learning Module.

Elements of Culture

Whereas social structure refers to the organization of society, culture refers to the meanings and ways of life that characterize a society. The elements of culture include beliefs, values, norms, sanctions, and symbols.

Beliefs. Beliefs refer to definitions and explanations about what is assumed to be true. The beliefs of an individual or group influence whether that individual or group views a particular social condition as a social problem. Does secondhand smoke harm nonsmokers? Are nuclear power plants safe? Does violence in movies and on television lead to increased aggression in children? Our beliefs regarding these issues influence whether we view the issues as social problems. Beliefs influence not only how a social condition is interpreted but also the existence of the condition itself. For example, police officers' beliefs about their supervisors' priorities affected officers' problem-solving behavior and the time devoted to it (Engel & Worden 2003). The *Self and Society* feature in this chapter allows you to assess your own beliefs about various social issues and to compare your beliefs with a national sample of first-year college students.

Values. Values are social agreements about what is considered good and bad, right and wrong, desirable and undesirable. Frequently, social conditions are viewed as social problems when the conditions are incompatible with or contradict closely held values. For example, poverty and homelessness violate the value of human welfare; crime contradicts the values of honesty, private property, and nonviolence; racism, sexism, and heterosexism violate the values of equality and fairness.

Values play an important role not only in the interpretation of a condition as a social problem but also in the development of the social condition itself. Sylvia Ann Hewlett (1992) explains how the American values of freedom and individualism are at the root of many of our social problems:

> There are two sides to the coin of freedom. On the one hand, there is enormous potential for prosperity and personal fulfillment; on the other are all the hazards of untrammeled opportunity and unfettered choice. Free markets can produce grinding poverty as well as spectacular wealth; unregulated industry can create dangerous levels of pollution as well as rapid rates of growth; and an unfettered drive for personal fulfillment can have disastrous effects on families and children. Rampant individualism does not bring with it sweet freedom; rather, it explodes in our faces and limits life's potential. (pp. 350–351)

Absent or weak values may contribute to some social problems. For example, many industries do not value protection of the environment and thus contribute to environmental pollution.

Norms and Sanctions. Norms are socially defined rules of behavior. Norms serve as guidelines for our behavior and for our expectations of the behavior of others.

There are three types of norms: folkways, laws, and mores. **Folkways** refer to the customs and manners of society. In many segments of our society it is customary to shake hands when being introduced to a new acquaintance, to say "excuse me" after sneezing, and to give presents to family and friends on their birthdays. Although no laws require us to do these things, we are expected to do them because they are part of the cultural traditions, or folkways, of the society in which we live.

Laws are norms that are formalized and backed by political authority. It is normative for a Muslim woman to wear a veil. However, in the United States failure to

Personal Beliefs About Various Social Problems

Indicate whether you agree or disagree with each of the following statements:

Statement	Agree	Disagree
1. Federal military spending should be increased.	_____	_____
2. Colleges should prohibit racist/sexist speech on campus.	_____	_____
3. There is too much concern in the courts for the rights of criminals.	_____	_____
4. Abortion should be legal.	_____	_____
5. The death penalty should be abolished.	_____	_____
6. The activities of married women are best confined to the home and family.	_____	_____
7. Marijuana should be legalized.	_____	_____
8. It is important to have laws prohibiting homosexual relationships.	_____	_____
9. Colleges have the right to ban extreme speakers.	_____	_____
10. The federal government should do more to control the sale of handguns.	_____	_____
11. Racial discrimination is no longer a major problem in America.	_____	_____
12. Realistically, an individual can do little to bring about changes in our society.	_____	_____
13. Wealthy people should pay a larger share of taxes than they do now.	_____	_____
14. Affirmative action in college admissions should be abolished.	_____	_____
15. Same-sex couples should have the right to legal marital status.	_____	_____

Percentage of First-Year College Students Agreeing with Belief Statements*

Statement Number	Percentage Agreeing in 2004		
	Total	Women	Men
1. Military spending	35	32	40
2. Prohibit speech on campus	59	62	55
3. Too much concern for criminals' rights	58	56	61
4. Abortion rights	54	53	55
5. Abolishment of death penalty	33	36	30
6. Women's activities confined to home	21	16	27
7. Legalization of marijuana	37	33	43
8. Laws prohibiting gay relationships	30	23	38
9. Ban speakers	44	42	46
10. Federal control of handgun sales	78	86	70
11. Racial discrimination not a problem	23	19	28
12. Individuals can't influence social change	27	23	31
13. Wealthy should pay higher taxes	56	56	55
14. Affirmative action abolished in college	50	46	56
15. Legal right of same-sex couples to marry	57	64	48

*Percentages are rounded.

Source: *The American Freshman; National Norms for Fall 2004* (2005). Copyright © 2004 by the Regents of the University of California. Used by permission.

Table 1.1

Types and Examples of Sanctions

	Positive	Negative
Informal	Being praised by one's neighbors for organizing a neighborhood recycling program.	Being criticized by one's neighbors for refusing to participate in the neighborhood recycling program.
Formal	Being granted a citizen's award for organizing a neighborhood recycling program.	Being fined by the city for failing to dispose of trash properly.

remove the veil for a driver's license photo is grounds for revoking the permit. Such is the case of a Florida woman who has brought suit against the state, claiming that her religious rights are being violated because she is required to remove her veil for the driver's license photo (Canedy 2003).

Some norms, called **mores,** have a moral basis. Violations of mores may produce shock, horror, and moral indignation. Both littering and child sexual abuse are violations of law, but child sexual abuse is also a violation of our mores because we view such behavior as immoral.

All norms are associated with **sanctions,** or social consequences for conforming to or violating norms. When we conform to a social norm, we may be rewarded by a positive sanction. These may range from an approving smile to a public ceremony in our honor. When we violate a social norm, we may be punished by a negative sanction, which may range from a disapproving look to the death penalty or life in prison. Most sanctions are spontaneous expressions of approval or disapproval by groups or individuals—these are referred to as informal sanctions. Sanctions that are carried out according to some recognized or formal procedure are referred to as formal sanctions. Types of sanctions, then, include positive informal sanctions, positive formal sanctions, negative informal sanctions, and negative formal sanctions (see Table 1.1).

Symbols. A **symbol** is something that represents something else. Without symbols, we could not communicate with each other or live as social beings.

The symbols of a culture include language, gestures, and objects whose meaning is commonly understood by the members of a society. In our society a red ribbon tied around a car antenna symbolizes Mothers Against Drunk Driving, a peace sign symbolizes the value of nonviolence, and a white hooded robe symbolizes the Ku Klux Klan. Sometimes people attach different meanings to the same symbol. The Confederate flag is a symbol of southern pride to some, a symbol of racial bigotry to others.

The elements of the social structure and culture just discussed play a central role in the creation, maintenance, and social response to various social problems. One of the goals of taking a course in social problems is to develop an awareness of how the elements of social structure and culture contribute to social problems. Sociologists refer to this awareness as the "sociological imagination."

The Sociological Imagination

The **sociological imagination,** a term developed by C. Wright Mills (1959), refers to the ability to see the connections between our personal lives and the social world in which we live. When we use our sociological imagination, we are able to distinguish between "private troubles" and "public issues" and to see connections between the events and conditions of our lives and the social and historical context in which we live.

For example, that one person is unemployed constitutes a private trouble. That millions of people are unemployed in the United States constitutes a public issue. Once we understand that personal troubles such as HIV infection, criminal victimization, and poverty are shared by other segments of society, we can look for the elements of social structure and culture that contribute to these public issues and private troubles. If the various elements of social structure and culture contribute to private troubles and public issues, then society's social structure and culture must be changed if these concerns are to be resolved.

Rather than viewing the private trouble of being unemployed as a result of an individual's faulty character or lack of job skills, we may understand unemployment as a public issue that results from the failure of the economic and political institutions of society to provide job opportunities to all citizens. Technological innovations emerging from the Industrial Revolution led to individual workers being replaced by machines. During the economic recession of the 1980s, employers fired employees so the firm could stay in business. Thus, in both these cases, social forces rather than individual skills largely determined whether a person was employed.

Theoretical Perspectives

Theories in sociology provide us with different perspectives with which to view our social world. A perspective is simply a way of looking at the world. A theory is a set of interrelated propositions or principles designed to answer a question or explain a particular phenomenon; it provides us with a perspective. Sociological theories help us to explain and predict the social world in which we live.

Sociology includes three major theoretical perspectives: the structural-functionalist perspective, the conflict perspective, and the symbolic interactionist perspective. Each perspective offers a variety of explanations about the causes of and possible solutions to social problems.

Structural-Functionalist Perspective

The structural-functionalist perspective is based largely on the works of Herbert Spencer, Emile Durkheim, Talcott Parsons, and Robert Merton. According to **structural functionalism,** society is a system of interconnected parts that work together in harmony to maintain a state of balance and social equilibrium for the whole. For example, each of the social institutions contributes important functions for society: Family provides a context for reproducing, nurturing, and socializing children; education offers a way to transmit a society's skills, knowledge, and culture to its youth; politics provides a means of governing members of society; economics provides for the production, distribution, and consumption of goods and services; and religion provides moral guidance and an outlet for worship of a higher power.

The structural-functionalist perspective emphasizes the interconnectedness of society by focusing on how each part influences and is influenced by other parts. For example, the increase in single-parent and dual-earner families has contributed to the number of children who are failing in school because parents have become less available to supervise their children's homework. As a result of changes in technology, colleges are offering more technical programs, and many adults are returning to school to learn new skills that are required in the workplace. The increasing number of women in the workforce has contributed to the formulation of policies against sexual harassment and job discrimination.

Structural functionalists use the terms *functional* and *dysfunctional* to describe the effects of social elements on society. Elements of society are functional if they contribute to social stability and dysfunctional if they disrupt social stability. Some aspects of society can be both functional and dysfunctional. For example, crime is dysfunctional in that it is associated with physical violence, loss of property, and fear. But according to Durkheim and other functionalists, crime is also functional for society because it leads to heightened awareness of shared moral bonds and increased social cohesion.

Sociologists have identified two types of functions: manifest and latent (Merton 1968). **Manifest functions** are consequences that are intended and commonly recognized. **Latent functions** are consequences that are unintended and often hidden. For example, the manifest function of education is to transmit knowledge and skills to society's youth. But public elementary schools also serve as babysitters for employed parents, and colleges offer a place for young adults to meet potential mates. The baby-sitting and mate-selection functions are not the intended or commonly recognized functions of education; hence they are latent functions.

Structural-Functionalist Theories of Social Problems

Two dominant theories of social problems grew out of the structural-functionalist perspective: social pathology and social disorganization.

Social Pathology. According to the social pathology model, social problems result from some "sickness" in society. Just as the human body becomes ill when our systems, organs, and cells do not function normally, society becomes "ill" when its parts (i.e., elements of the structure and culture) no longer perform properly. For example, problems such as crime, violence, poverty, and juvenile delinquency are often attributed to the breakdown of the family institution, the decline of the religious institution, and inadequacies in our economic, educational, and political institutions.

Social "illness" also results when members of a society are not adequately socialized to adopt its norms and values. People who do not value honesty, for example, are prone to dishonesties of all sorts. Early theorists attributed the failure in socialization to "sick" people who could not be socialized. Later theorists recognized that failure in the socialization process stemmed from "sick" social conditions, not "sick" people. To prevent or solve social problems, members of society must receive proper socialization and moral education, which may be accomplished in the family, schools, churches, workplace, and/or through the media.

Social Disorganization. According to the social disorganization view of social problems, rapid social change disrupts the norms in a society. When norms become weak or are in conflict with each other, society is in a state of **anomie,** or norm-lessness. Hence people may steal, physically abuse their spouses or children, abuse drugs, commit rape, or engage in other deviant behavior because the norms regarding these behaviors are weak or conflicting. According to this view, the solution to social problems lies in slowing the pace of social change and strengthening social norms. For example, although the use of alcohol by teenagers is considered a violation of a social norm in our society, this norm is weak. The media portray young people drinking alcohol, teenagers teach each other to drink alcohol and buy fake identification cards (IDs) to purchase alcohol, and parents model drinking behavior by having a few drinks after work or at a social event. Solutions to teenage drinking may involve strengthening norms against it through public education, restricting media depictions of youth and alcohol, imposing stronger sanctions against the use of fake IDs to purchase alcohol, and educating parents to model moderate and responsible drinking behavior.

Conflict Perspective

The structural-functionalist perspective views society as composed of different parts working together. In contrast, the **conflict perspective** views society as composed of different groups and interests competing for power and resources. The conflict perspective explains various aspects of our social world by looking at which groups have power and benefit from a particular social arrangement. For example, feminist theory argues that we live in a patriarchal society—a hierarchical system of organization controlled by men. Although there are many varieties of feminist theory, most would hold that feminism "demands that existing economic, political, and social structures be changed" (Weir and Faulkner 2004, p. xii).

The origins of the conflict perspective can be traced to the classic works of Karl Marx. Marx suggested that all societies go through stages of economic development. As societies evolve from agricultural to industrial, concern over meeting survival needs is replaced by concern over making a profit, the hallmark of a capitalist system. Industrialization leads to the development of two classes of people: the bourgeoisie, or the owners of the means of production (e.g., factories, farms, businesses); and the proletariat, or the workers who earn wages.

The division of society into two broad classes of people—the "haves" and the "have-nots"—is beneficial to the owners of the means of production. The workers, who may earn only subsistence wages, are denied access to the many resources available to the wealthy owners. According to Marx, the bourgeoisie use their power to control the institutions of society to their advantage. For example, Marx suggested that religion serves as an "opiate of the masses" in that it soothes the distress and suffering associated with the working-class lifestyle and focuses the workers' attention on spirituality, God, and the afterlife rather than on such worldly concerns as living conditions. In essence, religion diverts the workers so that they concentrate on being rewarded in heaven for living a moral life rather than on questioning their exploitation.

"Underlying virtually all social problems are conditions caused in whole or in part by social injustice."

Pamela Ann Roby
Sociologist

Conflict Theories of Social Problems

There are two general types of conflict theories of social problems: Marxist and non-Marxist. Marxist theories focus on social conflict that results from economic inequalities; non-Marxist theories focus on social conflict that results from competing values and interests among social groups.

Marxist Conflict Theories. According to contemporary Marxist theorists, social problems result from class inequality inherent in a capitalistic system. A system of haves and have-nots may be beneficial to the haves but often translates into poverty for the have-nots. As we will explore later in this textbook, many social problems, including physical and mental illness, low educational achievement, and crime, are linked to poverty.

In addition to creating an impoverished class of people, capitalism also encourages "corporate violence." Corporate violence can be defined as actual harm and/or risk of harm inflicted on consumers, workers, and the general public as a result of decisions by corporate executives or managers. Corporate violence can also result from corporate negligence, the quest for profits at any cost, and willful violations of health, safety, and environmental laws (Reiman 2004). Our profit-motivated economy encourages individuals who are otherwise good, kind, and law-abiding to knowingly participate in the manufacturing and marketing of defective brakes on American jets, fuel tanks on automobiles, and contraceptive devices (e.g., intrauterine devices [IUDs]). The profit motive has also caused individuals to sell defective medical devices, toxic pesticides, and contaminated foods to developing countries. As Eitzen and Baca Zinn noted, the "goal of profit is so central to capitalistic enterprises that many corporate decisions are made without consideration for the consequences" (Eitzen & Baca Zinn 2000, p. 483).

Marxist conflict theories also focus on the problem of **alienation,** or powerlessness and meaninglessness in people's lives. In industrialized societies workers often have little power or control over their jobs, a condition that fosters in them a sense of powerlessness in their lives. The specialized nature of work requires workers to perform limited and repetitive tasks; as a result, the workers may come to feel that their lives are meaningless.

Alienation is bred not only in the workplace but also in the classroom. Students have little power over their education and often find that the curriculum is not meaningful to their lives. Like poverty, alienation is linked to other social problems, such as low educational achievement, violence, and suicide.

Marxist explanations of social problems imply that the solution lies in eliminating inequality among classes of people by creating a classless society. The nature of work must also change to avoid alienation. Finally, stronger controls must be applied to corporations to ensure that corporate decisions and practices are based on safety rather than on profit considerations.

Non-Marxist Conflict Theories. Non-Marxist conflict theorists, such as Ralf Dahrendorf, are concerned with conflict that arises when groups have opposing values and interests. For example, antiabortion activists value the life of unborn embryos and fetuses; pro-choice activists value the right of women to control their own bodies and reproductive decisions. These different value positions reflect different subjective interpretations of what constitutes a social problem. For antiabortionists the availability of abortion is the social problem; for pro-choice advocates restrictions on abortion are the social problem. Sometimes the social problem is not the conflict itself but rather the way that conflict is expressed. Even most pro-life advocates agree that shooting doctors who perform abortions and blowing up abortion clinics constitute unnecessary violence and lack of respect for life. Value conflicts may occur between diverse categories of people, including nonwhites versus whites, heterosexuals versus homosexuals, young versus old, Democrats versus Republicans, and environmentalists versus industrialists.

Solving the problems that are generated by competing values may involve ensuring that conflicting groups understand each other's views, resolving differences through negotiation or mediation, or agreeing to disagree. Ideally, solutions should be win-win, with both conflicting groups satisfied with the solution. However, outcomes of value conflicts are often influenced by power; the group with the most power may use its position to influence the outcome of value conflicts. For example, when Congress could not get all states to voluntarily increase the legal drinking age to 21, it threatened to withdraw federal highway funds from those that would not comply.

Symbolic Interactionist Perspective

Both the structural-functionalist and the conflict perspectives are concerned with how broad aspects of society, such as institutions and large social groups, influence the social world. This level of sociological analysis is called **macro sociology**: It looks at the big picture of society and suggests how social problems are affected at the institutional level.

Micro sociology, another level of sociological analysis, is concerned with the social psychological dynamics of individuals interacting in small groups. Symbolic interactionism reflects the micro-sociological perspective and was largely influenced by the work of early sociologists and philosophers such as Max Weber, George Simmel, Charles Horton Cooley, G. H. Mead, W. I. Thomas, Erving Goffman, and Howard Becker. **Symbolic interactionism** emphasizes that human behavior is influenced by definitions and meanings that are created and maintained through symbolic interaction with others.

Sociologist W. I. Thomas (1966) emphasized the importance of definitions and meanings in social behavior and its consequences. He suggested that humans respond to their definition of a situation rather than to the objective situation itself. Hence Thomas noted that situations that we define as real become real in their consequences.

Symbolic interactionism also suggests that our identity or sense of self is shaped by social interaction. We develop our self-concept by observing how others interact with us and label us. By observing how others view us, we see a reflection of ourselves that Cooley calls the "looking glass self."

Last, the symbolic interactionist perspective has important implications for how social scientists conduct research. The German sociologist Max Weber argued that to understand individual and group behavior, social scientists must see the world through the eyes of that individual or group. Weber called this approach *Verstehen,* which in German means "empathy." *Verstehen* implies that in conducting research, social scientists must try to understand others' view of reality and the subjective aspects of their experiences, including their symbols, values, attitudes, and beliefs.

> "Each to each a looking glass, Reflects the other that doth pass."
>
> **Charles Horton Cooley**
> **Sociologist**

Symbolic Interactionist Theories of Social Problems

A basic premise of symbolic interactionist theories of social problems is that a condition must be defined or recognized as a social problem for it to be a social problem. Based on this premise, Herbert Blumer (1971) suggested that social problems develop in stages. First, social problems pass through the stage of *societal recognition*—the process by which a social problem, for example, drunk driving, is "born." Second, *social legitimation* takes place when the social problem achieves recognition by the larger community, including the media, schools, and churches. As the visibility of traffic fatalities associated with alcohol increased, so did the legitimation of drunk driving as a social problem. The next stage in the development of a social problem involves *mobilization for action,* which occurs when individuals and groups, such as Mothers Against Drunk Driving, become concerned about how to respond to the social condition. This mobilization leads to the *development and implementation of an official plan* for dealing with the problem, involving, for example, highway checkpoints, lower legal blood-alcohol levels, and tougher drunk-driving regulations.

Blumer's stage-development view of social problems is helpful in tracing the development of social problems. For example, although sexual harassment and date rape occurred throughout the 20th century, these issues did not begin to receive recognition as social problems until the 1970s. Social legitimation of these problems was achieved when high schools, colleges, churches, employers, and the media recognized their existence. Organized social groups mobilized to develop and implement plans to deal with these problems. For example, groups successfully lobbied for the enactment of laws against sexual harassment and the enforcement of sanctions against violators of these laws. Groups also mobilized to provide educational seminars on date rape for high school and college students and to offer support services to victims of date rape.

Some disagree with the symbolic interactionist view that social problems exist only if they are recognized. According to this view, individuals who were victims of date rape in the 1960s may be considered victims of a problem, even though date rape was not recognized at that time as a social problem.

Labeling theory, a major symbolic interactionist theory of social problems, suggests that a social condition or group is viewed as problematic if it is labeled as such. According to labeling theory, resolving social problems sometimes involves changing the meanings and definitions that are attributed to people and situations. For example, so long as teenagers define drinking alcohol as "cool" and "fun," they will continue to abuse alcohol. So long as our society defines providing sex education and contraceptives to teenagers as inappropriate or immoral, the teenage pregnancy rate in the United States will continue to be higher than in other industrialized nations.

Social constructionism is another symbolic interactionist theory of social problems. Similar to labeling theorists and symbolic interactionism in general, social constructionists argue that reality is socially constructed by individuals who interpret the social world around them. Society, therefore, is a social creation rather than an objective given. As such, social constructionists often question the origin and evolution of social problems. For example, most Americans define "drug abuse" as a social problem in the United States but rarely include alcohol or cigarettes in their discussion. A social constructionist would point to the historical roots of alcohol and tobacco use as a means of understanding their legal status. Central to this idea of the social construction of social problems are the media, universities, research institutes, and government agencies that are often responsible for the public's initial "take" on the problem under discussion.

Table 1.2 summarizes and compares the major theoretical perspectives, their criticisms, and social policy recommendations as they relate to social problems.

Table 1.2

Comparison of Theoretical Perspectives

	Structural Functionalism	Conflict Theory	Symbolic Interactionism
Representative theorists	Emile Durkheim Talcott Parsons Robert Merton	Karl Marx Ralf Dahrendorf	George H. Mead Charles Cooley Erving Goffman
Society	Society is a set of interrelated parts; cultural consensus exists and leads to social order; natural state of society—balance and harmony.	Society is marked by power struggles over scarce resources; inequities result in conflict; social change is inevitable; natural state of society—imbalance.	Society is a network of interlocking roles; social order is constructed through interaction as individuals, through shared meaning, make sense out of their social world.
Individuals	Individuals are socialized by society's institutions; socialization is the process by which social control is exerted; people need society and its institutions.	People are inherently good but are corrupted by society and its economic structure; institutions are controlled by groups with power; "order" is part of the illusion.	Humans are interpretative and interactive; they are constantly changing as their "social beings" emerge and are molded by changing circumstances.
Cause of social problems?	Rapid social change: social disorganization that disrupts the harmony and balance; inadequate socialization and/or weak institutions.	Inequality; the dominance of groups of people over other groups of people; oppression and exploitation; competition between groups.	Different interpretations of roles; labeling of individuals, groups, or behaviors as deviant; definition of an objective condition as a social problem.
Social policy/solutions	Repair weak institutions; assure proper socialization; cultivate a strong collective sense of right and wrong.	Minimize competition; create an equitable system for the distribution of resources.	Reduce impact of labeling and associated stigmatization; alter definitions of what is defined as a social problem.
Criticisms	Called "sunshine sociology"; supports the maintenance of the status quo; needs to ask "functional for whom?" Does not deal with issues of power and conflict; incorrectly assumes a consensus.	Utopian model; Marxist states have failed; denies existence of cooperation and equitable exchange. Can't explain cohesion and harmony.	Concentrates on micro issues only; fails to link micro issues to macro-level concerns; too psychological in its approach; assumes label amplified problem.

The study of social problems is based on research as well as on theory, however. Indeed, research and theory are intricately related. As Wilson (1983) stated:

> Most of us think of theorizing as quite divorced from the business of gathering facts. It seems to require an abstractness of thought remote from the practical activity of empirical research. But theory building is not a separate activity within sociology. Without theory, the empirical researcher would find it impossible to decide what to observe, how to observe it, or what to make of the observations. (p. 1)

Should Sociologists Be Required by Law to Reveal Their Sources?

In a free society there must be freedom of information. That is why journalists' sources are protected by the U.S. Constitution and, more specifically, the First Amendment. If journalists are compelled to reveal their sources, their sources may be unwilling to share information, and this would jeopardize the public's right to know. A journalist cannot reveal information given in confidence without permission from the source or a court order. Sociologists, in some circumstances, have been given the same legal rights as journalists. However, not all states protect sociologists' confidential information.

It Is Your Turn Now to Take a Stand!

Do you think sociologists should be granted the same protections as journalists? Why or why not?

Social Problems Research

Most students taking a course in social problems will not become researchers or conduct research on social problems. Nevertheless, we are all consumers of research that is reported in the media. Politicians, social activist groups, and organizations attempt to justify their decisions, actions, and positions by citing research results. As consumers of research, we need to understand that our personal experiences and casual observations are less reliable than generalizations based on systematic research. One strength of scientific research is that it is subjected to critical examination by other researchers (see this chapter's *Social Problems Research Up Close* feature). The more you understand how research is done, the better able you will be to critically examine and question research rather than to passively consume research findings. In the remainder of this section we discuss the stages of conducting a research study and the various methods of research used by sociologists.

Stages of Conducting a Research Study

Sociologists progress through various stages in conducting research on a social problem. In this section we describe the first four stages: formulating a research question, reviewing the literature, defining variables, and formulating a hypothesis.

Formulating a Research Question. A research study usually begins with a research question. Where do research questions originate? How does a particular researcher come to ask a particular research question? In some cases, researchers have a personal interest in a specific topic because of their own life experiences. For example, a researcher who has experienced spouse abuse may wish to do research on such questions as "What factors are associated with domestic violence?" and "How helpful are battered women's shelters in helping abused women break the cycle of abuse in their lives?" Other researchers may ask a particular research question because of their personal values—their concern for humanity and the desire to improve human life. Researchers who are concerned about the spread of human immunodeficiency virus (HIV) infection and AIDS may conduct research on such questions as "How does the use of alcohol influence condom use?" and "What educational strategies are effective for increasing safer sex behavior?" Researchers may also want to test a particular sociological theory, or some aspect of it, to establish its validity or conduct studies to evaluate the effect of a social policy or program. Research questions may also be formulated by the concerns of community groups and social activist organizations in collaboration with academic researchers. Government and industry also hire researchers to answer questions such as "How many children are victimized by episodes of violence at school?" and "What types of computer technologies can protect children against being exposed to pornography on the Internet?"

Reviewing the Literature. After a research question is formulated, the researcher reviews the published material on the topic to find out what is already known about it. Reviewing the literature also provides researchers with ideas about how to conduct their research and helps them formulate new research questions. A literature review serves as an evaluation tool, allowing a comparison of research findings and other sources of information, such as expert opinions, political claims, and journalistic reports.

Defining Variables. A **variable** is any measurable event, characteristic, or property that varies or is subject to change. Researchers must operationally define the variables they study. An **operational definition** specifies how a variable is to be measured. For example, an operational definition of the variable "religiosity" might be the number of times the respondent reports going to church or synagogue. Another operational definition of "religiosity" might be the respondent's answer to the question, "How important is religion in your life?" (1, not important; 2, somewhat important; 3, very important).

Operational definitions are particularly important for defining variables that cannot be directly observed. For example, researchers cannot directly observe concepts such as "mental illness," "sexual harassment," "child neglect," "job satisfaction," and "drug abuse." Nor can researchers directly observe perceptions, values, and attitudes.

Formulating a Hypothesis. After defining the research variables, researchers may formulate a **hypothesis,** which is a prediction or educated guess about how one variable is related to another variable. The **dependent variable** is the variable that the researcher wants to explain; that is, it is the variable of interest. The **independent variable** is the variable that is expected to explain change in the dependent variable. In formulating a hypothesis, the researcher predicts how the independent variable affects the dependent variable. For example, Kmec (2003) investigated the impact of segregated work environments on minority wages, concluding that "minority concentration in different jobs, occupations, and establishments than whites is a considerable social problem because it perpetuates racial wage inequality" (p. 55). In this example the independent variable is workplace segregation and the dependent variable is wages.

In studying social problems, researchers often assess the effects of several independent variables on one or more dependent variables. Jekielek (1998) examined the impact of parental conflict and marital disruption (two independent variables)

> "In science (as in everyday life) things must be believed in order to be seen as well as seen in order to be believed."
>
> Walter L. Wallace
> Social scientist

ThomsonNOW

Learn more about **Variables** by going through the Independent and Dependent Variables Animation.

The Sociological Enterprise

Each chapter in this book contains a *Social Problems Research Up Close* box that describes a research report or journal article that examines some sociologically significant topic. Some examples of the more prestigious journals in sociology include the *American Sociological Review,* the *American Journal of Sociology,* and *Social Forces.* Journal articles are the primary means by which sociologists, as well as other scientists, exchange ideas and information. Most journal articles begin with *an introduction and review of the literature.* It is here that the investigator examines previous research on the topic, identifies specific research areas, and otherwise "sets the stage" for the reader. It is often in this section that research hypotheses, if applicable, are set forth. A researcher, for example, might hypothesize that the sexual behavior of adolescents has changed over the years as a consequence of increased fear of sexually transmitted diseases and that such changes vary on the basis of sex.

The next major section of a journal article is *sample and methods.* In this section the investigator describes the characteristics of the sample, if any, and the details of the type of research conducted. The type of data analysis used is also presented in this section (see Appendix A). Using the sample research question, a sociologist might obtain data from the Youth Risk Behavior Surveillance Survey collected by the Centers for

Disease Control and Prevention. This self-administered questionnaire is distributed biennially to more than 10,000 high school students across the United States.

The final section of a journal article includes the *findings and conclusions.* The findings of a study describe the results, that is, what the researcher found as a result of the investigation. Findings are then discussed within the context of the hypotheses and the conclusions that can be drawn. Often research results are presented in tabular form. Reading tables carefully is an important part of drawing accurate conclusions about the research hypotheses. In reading a table, you should follow the steps listed here (see table on next page):

1. *Read the title of the table and make sure that you understand what the table contains.* The title of the table indicates the unit of analysis (high school students), the dependent variable (sexual risk behaviors), the independent variables (sex and year), and what the numbers represent (percentages).
2. *Read the information contained at the bottom of the table, including the source and any other explanatory information.* For example, the information at the bottom of this table indicates that the data are from the Centers for Disease Control, that "sexually active" was defined as having intercourse in the last three

months, and that data on condom use were only from those students who were defined as currently sexually active.
3. *Examine the row and column headings.* This table looks at the percentage of males and females, over four years, that reported ever having sexual intercourse, having four or more sex partners in a lifetime, being currently sexually active, and using condoms during the last sexual intercourse.
4. *Thoroughly examine the data in the table carefully, looking for patterns between variables.* As indicated in the table, with the exception of 1999, the first three columns suggest that "risky" sexual behaviors of males have remained fairly constant over the years surveyed. For females, however, risky sexual behaviors have, in general, gone down, although there were increases from 2001 to 2003 for ever having sexual intercourse or being currently sexually active. In addition, Column 4 indicates that condom use among males and females has increased over the four years surveyed— the largest increases being between 2001 and 2003.
5. *Use the information you have gathered in Step 4 to address the hypotheses.* Clearly, sexual practices, as hypothesized, have changed over time. For example, both males and females have a

on the emotional well-being of children (the dependent variable). Her research found that both parental conflict and marital disruption (separation or divorce) negatively affect children's emotional well-being. However, children in high-conflict intact families exhibit lower levels of well-being than children who have experienced high levels of parental conflict but whose parents divorce or separate.

general increase in condom use during sexual intercourse. In addition, from a comparison of the data from 1997 to 2003, it is clear that changes in risky sexual behaviors are greater for females than for males.

6. *Draw conclusions consistent with the information presented.* From the table can we conclude that sexual practices have changed over time? The answer is probably yes, although the limitations of the survey, the sample, and the measurement techniques used always should be considered. Can we conclude, however, that the observed changes are a consequence of the fear of sexually transmitted diseases? Although the data may imply it, having no measure of fear of sexually transmitted diseases over the time period studied, we would be premature to come to such a conclusion. More information, from a variety of sources, is needed. The use of multiple methods and approaches to study a social phenomenon is called **triangulation.**

Percentage of High School Students Reporting Sexual Risk Behaviors, by Sex and Survey Year

Survey Year	Ever Had Sexual Intercourse	Four or More Sex Partners During Lifetime	Currently Sexually Active[a]	Condom Used During Last Intercourse[b]
Male				
1997	48.8	17.6	33.4	62.5
1999	52.2	19.3	36.2	65.5
2001	48.5	17.2	33.4	65.1
2003	48.0	17.5	33.8	68.8
Female				
1997	47.7	14.1	36.5	50.8
1999	47.7	13.1	36.3	50.7
2001	42.9	11.4	33.4	51.3
2003	45.3	11.2	34.6	57.4

[a]Sexual intercourse during the three months preceding the survey.

[b]Among currently sexually active students.

Sources: Centers for Disease Control and Prevention (1998, 1999, 2002, 2004).

Methods of Data Collection

After identifying a research topic, reviewing the literature, defining the variables, and developing hypotheses, researchers decide which method of data collection to use. Alternatives include experiments, surveys, field research, and secondary data.

Experiments. Experiments involve manipulating the independent variable to determine how it affects the dependent variable. Experiments require one or more experimental groups that are exposed to the experimental treatment(s) and a control group that is not exposed. After the researcher randomly assigns participants to either an experimental group or a control group, she or he measures the dependent variable. After the experimental groups are exposed to the treatment, the researcher measures the dependent variable again. If participants have been randomly assigned to the different groups, the researcher may conclude that any difference in the dependent variable among the groups is due to the effect of the independent variable.

An example of a "social problems" experiment on poverty would be to provide welfare payments to one group of unemployed single mothers (experimental group) and no such payments to another group of unemployed single mothers (control group). The independent variable would be welfare payments; the dependent variable would be employment. The researcher's hypothesis would be that mothers in the experimental group would be less likely to have a job after 12 months than mothers in the control group.

The major strength of the experimental method is that it provides evidence for causal relationships, that is, how one variable affects another. A primary weakness is that experiments are often conducted on small samples, usually in artificial laboratory settings; thus the findings may not be generalized to other people in natural settings.

Surveys. Survey research involves eliciting information from respondents through questions. An important part of survey research is selecting a sample of those to be questioned. A **sample** is a portion of the population, selected to be representative so that the information from the sample can be generalized to a larger population. For example, instead of asking all abused spouses about their experience, the researcher could ask a representative sample of them and assume that those who were not questioned would give similar responses. After selecting a representative sample, survey researchers either interview people, ask them to complete written questionnaires, or elicit responses to research questions through computers.

Interviews. In interview survey research, trained interviewers ask respondents a series of questions and make written notes about or tape-record the respondents' answers. Interviews may be conducted over the telephone or face to face. A recent Gallup Poll (Newport 2003) involved telephone interviews with a randomly selected national sample of more than 1,000 U.S. adults. One of the questions the interviewers asked was for respondents to "name the most important problem facing the country." The top three responses were fear of war (see Chapter 16), the economy (see Chapter 6), and unemployment (see Chapter 7) (Newport 2003).

One advantage of interview research is that researchers are able to clarify questions for the respondent and follow up on answers to particular questions. Researchers often conduct face-to-face interviews with groups of individuals who might otherwise be inaccessible. For example, some AIDS-related research attempts to assess the degree to which individuals engage in behavior that places them at high risk for transmitting or contracting HIV. Street youth and intravenous drug users, both high-risk groups for HIV infection, may not have a telephone or address because of their transient lifestyle. These groups may be accessible, however, if the researcher locates their hangouts and conducts face-to-face interviews. Research on drug addicts may also require a face-to-face interview survey design (Jacobs 2

ThomsonNOW™

Learn more about **Experiments** by going through the Classical Experiment Learning Module.

"My latest survey shows that people don't believe in surveys."

Laurence Peter
Humorist

ThomsonNOW™

Learn more about **Interviews** by going through the Role of the Survey Interviewer Animation.

The most serious disadvantages of interview research are cost and the lack of privacy and anonymity. Respondents may feel embarrassed or threatened when asked questions that relate to personal issues such as drug use, domestic violence, and sexual behavior. As a result, some respondents may choose not to participate in interview research on sensitive topics. Those who do participate may conceal or alter information or give socially desirable answers to the interviewer's questions (e.g., "No, I do not use drugs.").

Questionnaire. Instead of conducting personal or phone interviews, researchers may develop questionnaires that they either mail or give to a sample of respondents. Questionnaire research offers the advantages of being less expensive and time-consuming than face-to-face or telephone surveys. In addition, questionnaire research provides privacy and anonymity to the research participants. This reduces the likelihood that they will feel threatened or embarrassed when asked personal questions and increases the likelihood that they will provide answers that are not intentionally inaccurate or distorted.

The major disadvantage of mail questionnaires is that it is difficult to obtain an adequate response rate. Many people do not want to take the time or make the effort to complete and mail a questionnaire. Others may be unable to read and understand the questionnaire.

"Talking" Computers. A new method of conducting survey research is asking respondents to provide answers to a computer that "talks." Romer et al. (1997) found that respondents rated computer interviews about sexual issues more favorably than face-to-face interviews and that the computer interviews were more reliable. Such increased reliability may be particularly valuable when conducting research on drug use, deviant sexual behavior, and sexual orientation because respondents reported that the privacy of computers was a major advantage.

Field Research. **Field research** involves observing and studying social behavior in settings in which it occurs naturally. Two types of field research are participant observation and nonparticipant observation.

In participant observation research the researcher participates in the phenomenon being studied so as to obtain an insider's perspective on the people and/or behavior being observed. Palacios and Fenwick (2003), two criminologists, attended dozens of raves over a 15-month period to investigate the South Florida drug culture. In nonparticipant observation research the researcher observes the phenomenon being studied without actively participating in the group or the activity. For example, Dordick (1997) studied homelessness by observing and talking with homeless individuals in a variety of settings, but she did not live as a homeless person as part of her research.

Sometimes sociologists conduct in-depth detailed analyses or case studies of an individual, group, or event. For example, Fleming (2003) conducted a case study of young auto thieves in British Columbia. He found that unlike professional thieves, the teenagers' behavior was primarily motivated by thrill seeking—driving fast, the rush of a possible police pursuit, and the prospect of getting caught.

The main advantage of field research on social problems is that it provides detailed information about the values, rituals, norms, behaviors, symbols, beliefs, and emotions of those being studied. A potential problem with field research is that the researcher's observations may be biased (e.g., the researcher becomes too involved

ThomsonNOW

Learn more about **Questionnaires** by going through the Questionnaire Construction Coached Problem.

"When I was younger I could remember anything— whether it happened or not."

Mark Twain
American humorist and writer

ThomsonNOW™

Learn more about **Field Research** by going through the Qualitative Field Research Learning Module.

"Feminists in all disciplines have demonstrated that objectivity has about as much substance as the emperor's new clothes."

Connie Miller
Feminist scholar

in the group to be objective). In addition, because field research is usually based on small samples, the findings may not be generalizable.

Secondary Data Research. Sometimes researchers analyze secondary data, which are data that have already been collected by other researchers or government agencies or that exist in forms such as historical documents, police reports, school records, and official records of marriages, births, and deaths. Caldas and Bankston (1999) used information from Louisiana's 1990 Graduation Exit Examination to assess the relationship between school achievement and television-viewing habits of more than 40,000 tenth graders. The researchers found that, in general, television viewing is inversely related to academic achievement for whites but has little or no effect on school achievement for African Americans. A major advantage of using secondary data in studying social problems is that the data are readily accessible, so researchers avoid the time and expense of collecting their own data. Secondary data are also often based on large representative samples. The disadvantage of secondary data is that the researcher is limited to the data already collected.

Goals of the Textbook

This textbook approaches the study of social problems with several goals in mind.

1. *Providing an integrated theoretical background.* The book reflects an integrative theoretical approach to the study of social problems. More than one theoretical perspective can be used to explain a social problem because social problems usually have more than one cause. For example, youth crime is linked to (1) an increased number of youths living in inner-city neighborhoods with little or no parental supervision (social disorganization), (2) young people having no legitimate means of acquiring material wealth (anomie theory), (3) youths being angry and frustrated at the inequality and racism in our society (conflict theory), and (4) teachers regarding youths as "no good" and treating them accordingly (labeling theory).

2. *Encouraging the development of a sociological imagination.* A major insight of the sociological perspective is that various structural and cultural elements of society have far-reaching effects on individual lives and social well-being. This insight, known as the sociological imagination, enables us to understand how social forces underlie personal misfortunes and failures and contribute to personal successes and achievements. Each chapter in this textbook emphasizes how structural and cultural factors contribute to social problems. This emphasis encourages you to develop your sociological imagination by recognizing how structural and cultural factors influence private troubles and public issues.

3. *Providing global coverage of social problems.* The modern world is often referred to as a "global village." The Internet and fax machines connect individuals around the world, economies are interconnected, environmental destruction in one region of the world affects other regions of the world, and diseases cross national boundaries. Understanding social problems requires an awareness of how global trends and policies affect social problems. Many social problems call for collective action involving countries around the world; efforts to end poverty, protect the environment, control population growth, and reduce the spread of HIV are some of the social problems that have been addressed at the global level. Each chapter in this book includes coverage of global aspects of social problems. We hope that at-

tention to the global aspects of social problems will broaden students' awareness of pressing world issues.

4. *Providing an opportunity to assess personal beliefs and attitudes.* Each chapter in this textbook contains a section called *Self and Society,* which offers you an opportunity to assess your attitudes and beliefs regarding some aspect of the social problem discussed. Earlier in this chapter, the *Self and Society* feature allowed you to assess your beliefs about a number of social problems and to compare your beliefs with a national sample of first-year college students.

5. *Emphasizing the human side of social problems.* Each chapter contains a feature called *The Human Side,* which presents personal stories of how social problems have affected individual lives. By conveying the private pain and personal triumphs associated with social problems, we hope to elicit a level of understanding and compassion that may not be attained through the academic study of social problems alone. This chapter's *Human Side* feature presents stories about how college students, disturbed by various social conditions, have participated in social activism.

6. *Highlighting social problems research.* In every chapter there are boxes called *Social Problems Research Up Close,* which present examples of social science research. These boxes demonstrate for students the sociological enterprise from theory and data collection to findings and conclusions. Examples of research topics covered include "Perceptions of Marriage Among Low Income Single Mothers," "The Social Construction of the Hacking Community," and The National Health-Risk Behavior Survey.

7. *Focusing on technology.* Boxes called *Focus on Technology* also appear in every chapter. These boxes present information on how technology can contribute to social problems and their solutions. Hate on the Web, the use of DNA evidence in criminal investigations, and telework—the new workplace of the 21st century—are a few examples.

8. *Challenging students to "take a stand."* A relatively new feature, *Taking a Stand,* challenges students to take and defend a position on a current issue.

"Activism pays the rent on being alive and being here on the planet. . . . If I weren't active politically, I would feel as if I were sitting back eating at the banquet without washing the dishes or preparing the food. It wouldn't feel right."

Alice Walker
Novelist

One way to effect social change is through demonstrations. A U.S. survey of first-year college students revealed that 49% reported having participated in organized demonstrations in the last year (Sax et al. 2004). Here, students march against the war in Iraq.

© Paul Fusco/Magnum Photos

Social Change and College Student Activism

Both structural functionalism and conflict theory address the nature of social change, although in different ways. Durkheim, a structural functionalist, argued that social change, if rapid, was disruptive to society and that the needs of society should take precedence over the desires of individuals; that is, social change should be slow and methodical regardless of popular opinion. When it is slow and methodical, social change can be responsive to the needs of society and can contribute to the preferred state of society—that of equilibrium.

To conflict theorists, social change is a result of the struggle for power by different groups. Specifically, Marx argued that social change was a consequence of the struggle between different economic classes as each strove for supremacy. Marx envisioned social change as primarily a revolutionary process ultimately leading to a utopian society.

Social movements are one means by which social change is realized. A **social movement** is an organized group of individuals with a common purpose to either promote or resist social change through collective action. Some people believe that to promote social change, one must be in a position of political power and/or have large financial resources. However, the most important prerequisite for becoming actively involved in improving levels of social well-being may be genuine concern and dedication to a social "cause." The following vignettes provide a sampler of college student activism—college students making a difference in the world:

- Sean Sellers, a recent graduate of the University of Texas at Austin successfully led a "Boot the Bell" campaign in which "22 colleges and high schools either managed to remove a Taco Bell franchise from the campus or prevent one from being built" as part of a general protest against sweatshops in the field (Berkowitz 2005, p. 1). Yum Brands Inc., which owns Taco Bell as well as KFC, Long John Silver, and Pizza Hut, has agreed to increase the pay of tomato pickers and to improve working conditions of farm workers in general.

- In May 1989, hundreds of Chinese college students protested in Tiananmen Square in Beijing, China, because Chinese government officials would not meet with them to hear their pleas for a democratic government. These students boycotted classes and started a hunger strike. On June 4, 1989, thousands of students and other protesters were massacred or arrested in Tiananmen Square.

- Students at several colleges are petitioning their university administrations to buy "fair trade coffee"—coffee that is certified by monitors to have come from farmers who were paid a fair price for their beans. Many of these students are members of Students for Fair Trade. As one student said, "This is easy activism." Students make their voices heard by buying coffee with a fair-trade certified label or by not buying coffee at all (Batsell 2002).

- Students at the University of California, Berkeley, came together to protest cutbacks in the campus's Ethnic Studies

In doing so, students are encouraged to use reason, scientific evidence, and logic rather than emotionality and anecdotal evidence in making a cohesive argument.
9. *Encouraging students to take pro-social action.* Individuals who understand the factors that contribute to social problems may be better able to formulate interventions to remedy those problems. Recognizing the personal pain and public costs associated with social problems encourages some to initiate social intervention.

Individuals can make a difference in society by the choices they make. Individuals may choose to vote for one candidate over another, demand the right to reproductive choice or protest government policies that permit it, drive drunk or stop a friend from driving drunk, repeat a racist or sexist joke or chastise the person who tells it, and practice safe sex or risk the transmission of sexually transmitted

"In a certain sense, every single human soul has more meaning and value than the whole of history."

Nicholas Berdyaev
Philosopher

Department. The protest lasted more than a month with over 100 arrests and 6 hunger strikes. As a consequence of the students' activism, the university administration agreed to reopen a multicultural student center, hire eight tenure-track ethnic studies faculty over the next five years, and invest $100,000 in an Ethnic Research Center (Alvarado 2000).

- Students at Grinnell College in Iowa, in response to accusations of human rights violations of union workers in Coca-Cola bottling plants in Colombia (South America), formed an anti-Coke campaign. Using the official boycotting policy of the college, in November 2004 the student initiative passed a boycott on all Coca Cola products. Because Coca-Cola had an exclusive contract with Grinnell, Coke and Coke products continued to be sold on campus. However, wherever they are sold, there are now signs reading, "The Grinnell College student body has voted to boycott Coca-Cola products. This is a Coca-Cola product" (*Killer Coke* 2005).

- While a student at George Washington University, Ross Misher started an organization called Students Against Handgun Violence. When Ross was 13, his father was shot and killed by a co-worker who had purchased a handgun during his lunch hour and returned to shoot Ross's father before killing himself (Lewis 1991).
- While a zoology major at the University of Colorado, Jeff Galus began the Animal Rights Student Group (website at http://www.colorado.edu/StudentsGroups/animalrights/links.html). This organization focuses on informing the public about how animals are treated in research and which corporations use animals in testing their products.
- Students at more than 150 campuses are members of the anti-sweatshop movement, many belonging to the Worker's Rights Consortium (WRC). The WRC is a student run watchdog organization that inspects factories worldwide, monitoring the monitors, as part of the anti-sweatshop movement. The WRC

requires that member schools agree to closely scrutinize manufacturers of collegiate apparel. The WRC also mandates "the protection of workers' health and safety, compliance with local labor laws, protection of women's rights, and prohibition of child labor, forced labor, and forced overtime" (Boston College 2003).

Students who are interested in becoming involved in student activism, or who are already involved, might explore the website for the Center for Campus Organizing (http://www.organizenow.net/cco/)—a national organization that supports social justice activism and investigative journalism on campuses nationwide. The organization was founded on the premise that students and faculty have played critical roles in larger social movements for social justice in our society, including the Civil Rights movement, the anti–Vietnam War movement, the Anti-Apartheid movement, the women's rights movement, and the environmental movement. In 2004 almost half of all college students participated in an organized demonstration (Sax et al. 2004).

diseases. Individuals can also make a difference by addressing social concerns in their occupational role as well as through volunteer work.

Although individual choices make an important impact, collective social action often has a more pervasive effect. For example, although individual parents discourage their teenage children from driving under the influence of alcohol, Mothers Against Drunk Driving contributed to the enactment of national legislation that potentially will influence every U.S. citizen's decision about whether to use alcohol and drive. Schwalbe (1998) reminded us that we do not have to join a group or organize a protest to make changes in the world.

> We can change a small part of the social world single-handedly. If we treat others with more respect and compassion, if we refuse to participate in recreating inequalities, even in little ways, if we raise questions about official representation of reality, if we refuse to work in destructive industries, then we are making change. (p. 206)

Understanding Social Problems

At the end of each chapter we offer a section titled "Understanding" in which we reemphasize the social origin of the problem being discussed, the consequences, and the alternative social solutions. It is our hope that the reader will end each chapter with a "sociological imagination" view of the problem and with an idea of how, as a society, we might approach a solution.

Sociologists have been studying social problems since the Industrial Revolution. Industrialization brought about massive social changes: The influence of religion declined and families became smaller and moved from traditional, rural communities to urban settings. These and other changes have been associated with increases in crime, pollution, divorce, and juvenile delinquency. As these social problems became more widespread, the need to understand their origins and possible solutions became more urgent. The field of sociology developed in response to this urgency. Social problems provided the initial impetus for the development of the field of sociology and continue to be a major focus of sociology.

There is no single agreed-on definition of what constitutes a social problem. Most sociologists agree, however, that all social problems share two important elements: an objective social condition and a subjective interpretation of that condition. Each of the three major theoretical perspectives in sociology—structural-functionalist, conflict, and symbolic interactionist—has its own notion of the causes, consequences, and solutions of social problems.

Chapter Review

ThomsonNOW™

Reviewing is as easy as ① ② ③

1. Before you do your final review, take the ThomsonNOW diagnostic quiz to help you identify the areas on which you should concentrate. You will find information on ThomsonNOW and instructions on how to access all of its great resources on the foldout at the beginning of the text.

2. As you review, take advantage of ThomsonNOW's study videos and interactive Map the Stats exercises to help you master the chapter topics.

3. When you are finished with your review, take ThomsonNOW's posttest to confirm you are ready to move on to the next chapter.

- **What is a social problem?**

Social problems are defined by a combination of objective and subjective criteria. The objective element of a social problem refers to the existence of a social condition; the subjective element of a social problem refers to the belief that a particular social condition is harmful to society or to a segment of society and that it should and can be changed. By combining these objective and subjective elements, we arrive at the following definition: A social problem is a social condition that a segment of society views as harmful to members of society and in need of remedy.

- **What is meant by the structure of society?**

The structure of a society refers to the way society is organized.

- **What are the components of the structure of society?**

The components are institutions, social groups, statuses, and roles. Institutions are an established and enduring pattern of social relationships and include family, religion, politics, economics, and education. Social groups are defined as two or more people who have a common identity, interact, and form a social relationship. A status is a position that a person occupies within a social group and that can be achieved or ascribed. Every status is associated with many roles, or the set of rights, obligations, and expectations associated with a status.

- **What is meant by the culture of society?**

Whereas social structure refers to the organization of society, culture refers to the meanings and ways of life that characterize a society.

- **What are the components of the culture of society?**

The components are beliefs, values, norms, and symbols. Beliefs refer to definitions and explanations about what is assumed to be true. Values are social agreements about what is considered good and bad, right and wrong, desirable and undesirable. Norms are socially defined rules of behavior. Norms serve as guidelines for our behavior and for our expectations of the behavior of others. Finally, a symbol is something that represents something else.

- **What is the sociological imagination, and why is it important?**

The sociological imagination, a term developed by C. Wright Mills (1959), refers to the ability to see the connections between our personal lives and the social world in which we live. It is important because when we use our sociological imagination, we are able to distinguish between "private troubles" and "public issues" and to see connections between the events and conditions of our lives and the social and historical context in which we live.

- **What are the differences between the three sociological perspectives?**

According to structural functionalism, society is a system of interconnected parts that work together in harmony to maintain a state of balance and social equilibrium for the whole. The conflict perspective views society as composed of different groups and interests competing for power and resources. Symbolic interactionism reflects the micro-sociological perspective and emphasizes that human behavior is influenced by definitions and meanings that are created and maintained through symbolic interaction with others.

- **What are the first four stages of a research study?**

The first four stages of a research study are formulating a research question, reviewing the literature, defining variables, and formulating a hypothesis.

- **How do the various research methods differ from one another?**

Experiments involve manipulating the independent variable to determine how it affects the dependent variable. Survey research involves eliciting information from respondents through questions. Field research involves observing and studying social behavior in settings in which it occurs naturally. Secondary data are data that have already been collected by other researchers or government agencies or that exist in forms such as historical documents, police reports, school records, and official records of marriages, births, and deaths.

- **What is a social movement?**

Social movements are one means by which social change is realized. A social movement is an organized group of individuals with a common purpose to either promote or resist social change through collective action.

Critical Thinking

1. People increasingly are using information technologies to get their daily news. As a matter of fact, some research indicates that news on the Internet is beginning to replace television news as the primary source of information among computer users (see Chapter 15). What role do the media play in our awareness of social problems, and will definitions of social problems change as sources of information change?
2. Each of you occupies several social statuses, each one carrying an expectation of role performance—that is, what you should and shouldn't do given your position. List five statuses you occupy, the expectations of their accompanying roles, and any role conflict that may result. What types of social problems are affected by role conflict?
3. Definitions of social problems change over time. Identify a social condition, now widely accepted, that might be viewed as a social problem in the future.

Key Terms

achieved status	objective element
alienation	operational definition
anomie	primary group
ascribed status	role
beliefs	sample
conflict perspective	sanction
dependent variable	secondary group
experiment	social constructionism
field research	social group
folkway	social movement
hypothesis	social problem
independent variable	sociological imagination
institution	status
labeling theory	structural functionalism
latent function	subjective element
law	survey research
macro sociology	symbol
manifest function	symbolic interactionism
master status	triangulation
micro sociology	values
mores	variable
norm	

 ## Media Resources

The Companion Website for
Understanding Social Problems,
Fifth Edition

http://sociology.wadsworth.com/mooney_knox_schacht5e

Supplement your review of this chapter by going to the companion website to take one of the Tutorial Quizzes, use the flash cards to master key terms, and check out the many other study aids you'll find there. You'll also find special features such as *Wadsworth's Sociology Online Resources and Writing Companion,* GSS Data, and Census 2000 information, data, and resources at your fingertips to help you complete that special project or do some research on your own.

> " The defense this nation seeks involves a great deal more than building airplanes, ships, guns and bombs. We cannot be a strong nation unless we are a healthy nation. " *U.S. president Franklin Roosevelt, 1940*

Illness and the Health Care Crisis

The Global Context: Effects of Globalization on Health

Societal Measures of Health and Illness

Sociological Theories of Illness and Health Care

HIV/AIDS: A Global Health Concern

The Growing Problem of Obesity

Mental Illness: The Hidden Epidemic

Lifestyle Behaviors and Social Factors Associated with Health and Illness

Problems in U.S. Health Care

Strategies for Action: Improving Health and Health Care

Understanding Illness and the Health Care Crisis

Chapter Review

nkosi Johnson, a South African boy, was born HIV-positive to a 17-year-old mother who died of AIDS, leaving Nkosi orphaned at the age of 8. At age 11, while suffering from full-blown AIDS, Nkosi stood up before the 13th International AIDS Conference in Durban, South Africa, and delivered a speech he wrote himself. "I hate having AIDS," he said, "because I get very sick, and I get very sad when I think of all the other children and babies that are sick with AIDS. . . . I just wish that the government can start giving AZT to pregnant HIV mothers to help stop the virus being passed on to their babies" (quoted by Wooten 2004, p. 205). Nkosi concluded his speech with a plea on behalf of all those infected with HIV: "Care for us and accept us. We are all human beings . . . and we have needs just like everyone else. Don't be afraid of us. We are all the same" (quoted by Wooten 2004, p. 206). About one year later, at age 12, Nkosi died. In South Africa alone more than 70,000 children are born HIV-positive each year, and half the population under age 15 will die of AIDS over the next decade.

HIV/AIDS is one of the world's most challenging health problems. In this chapter we address HIV/AIDS as well as other problems of illness and health care throughout the world and in the United States. Taking a sociological look at health issues, we examine how social forces affect and are affected by health and illness and why some social groups suffer more illness than others.

Although significant gains in global health have been made over the last century, there have also been serious setbacks. Old diseases thought to be under control, including cholera, tuberculosis, and malaria, have increased in the number of cases or in geographic spread. And new diseases and health threats have emerged in recent years, such as the emergence of a drug-resistant strain of HIV (first discovered in a New York City patient in 2005), growing resistance to antibiotics, the threat of anthrax and other bioterrorist attacks, and the emergence of severe acute respiratory syndrome (SARS) and Asian bird flu. One of the most significant social forces influencing health is the increasing globalization of the world.

The Global Context: Effects of Globalization on Health

Globalization, broadly defined as the growing economic, political, and social interconnectedness among societies throughout the world, has eroded the boundaries that separate societies, creating a "global village." Globalization has had both positive and negative effects on health. On the positive side, globalized communications technology enhances the capacity to monitor and report on outbreaks of disease, disseminate guidelines for controlling and treating disease, and share scientific knowledge and research findings (Lee 2003). Globalization also provides opportunities for establishing international health programs and agreements. In 2003, for example, the World Health Organization (made up of 192 member countries) voted unanimously to adopt the Framework Convention on Tobacco Control, which urges countries to eliminate tobacco advertising, establish stronger warning labels, raise cigarette prices, and adopt smoke-free workplace laws around the world (SmokeFree Educational Services 2003). On the negative side, features of globalization, including the growth of travel and information technologies and the expansion of trade and transnational corporations, have been linked to a number of health problems.

Effects of Increased Travel and Information Technology on Health

Modern information technology has useful applications in health and medicine. For example, the Program for Monitoring Infectious Diseases, an Internet-based global alert system with more than 30,000 subscribers in 150 countries, gathers and disseminates information about disease outbreaks worldwide (Pirages 2005). But information technology is also linked to a number of health problems. After the terrorist attacks of September 11, 2001, the world was reminded that global communications systems and international travel enable terrorist groups to form well-organized networks that can move around the globe. The terrorist attacks of September 11 not only resulted in direct deaths and injuries but also raised awareness of the prospect of biological and chemical weapons attacks as a major public health threat.

Increased business travel and tourism have encouraged the spread of disease, such as the potentially fatal West Nile virus, which first appeared in the United States in 1999, and the spread of SARS, which was first diagnosed in Asia in 2002 and within months spread to 30 countries, infecting thousands of individuals and killing more than 800.

Effects of Increased Trade and Transnational Corporations on Health

Increased international trade, another feature of globalization, has expanded the range of goods available to consumers, but at a cost to global health. The increased transportation of goods by air, sea, and land contributes to the pollution caused by the burning of fossil fuels. The expansion of international trade of harmful products such as tobacco, alcohol, and processed or "fast" foods are associated with a worldwide rise in cancer, heart disease, stroke, and diabetes (World Health Organization 2002).

Expanding trade has also facilitated the growth of transnational corporations that set up shop in developing countries to take advantage of lower labor costs and lax environmental and labor regulations (see also Chapter 7). Because of lax labor and human rights regulations, factory workers in transnational corporations—typically low-status uneducated women—are often exposed to harmful working conditions that increase the risk of illness, injury, and mental anguish (Hippert 2002). These workers often suffer exposure to toxic substances, lack safety equipment such as gloves and goggles, are denied bathroom breaks (which leads to bladder infections), and are assaulted at the workplace, because "physical brutality is frequently used as a mechanism of control on the production floor of the factory" (Hippert 2002, p. 863).

As a result of weak environmental laws in developing countries, transnational corporations are responsible for high levels of pollution and environmental degradation, which negatively affect the health of entire populations. Finally, the

Nkosi Johnson, age 11, gave a speech at the 13th International AIDS Conference in Durban, South Africa, in July 2000. Although he died of AIDS a year later at age 12, his legacy of AIDS activism continues to help support the global fight against HIV/AIDS.

movement of factories out of the United States to other countries has resulted in significant losses of U.S. jobs in manufacturing and the textile and apparel industry, which is significant because of the relationship between physical and mental illness and unemployment (Bartley et al. 2001).

Effects of International Free Trade Agreements on Health

The World Trade Organization (WTO) (which had 148 member countries in 2005) and regional trade agreements such as NAFTA (North American Free Trade Agreement) establish rules aimed at increasing international trade. These trade rules supersede member countries' laws and regulations, including those governing public health, if those laws or regulations represent a barrier to trade. For example, the Methanex Corporation of Canada produces methanol, a component of methyl tertiary butyl ether (MTBE), a gas additive. When the state of California banned the use of MTBE because of its link to cancer, the Methanex Corporation initiated an approximately $1 billion lawsuit against the United States, claiming that California's ban of MTBE violates Chapter 11 of NAFTA. After a five-year legal battle a closed tribunal sided with California and ruled against the Methanex Corporation in 2005, a major victory for environmentalists and Californians. Supporters of NAFTA say that the tribunal's decision demonstrates that U.S. trade agreements do not encroach on governments' right to enforce health and environmental regulations. However, other companies have succeeded in suing governments under NAFTA rules. In 2000 the U.S. Metaclad Company sued Mexico for $16 million because Mexico stopped the company from reopening a toxic waste dump that would contaminate people and the environment (Shaffer et al. 2005).

Another trade rule, the Agreement on Trade-Related Aspects of Intellectual Property Rights (TRIPS), mandates that all WTO member countries implement intellectual property rules that provide 20-year monopoly control over patented items, including medications. TRIPS limits the availability of generic drugs, thus contributing to higher drug costs. TRIPS also affects access to medications for life-threatening diseases in low-income countries. In 2005 India approved legislation that ends the decades-old practice of allowing drug companies to make low-cost generic drugs. This legislation, which resulted from the WTO's requirement that India enforce stricter patent rules for its pharmaceutical industry, is expected to curb the supply of affordable HIV/AIDS medications to impoverished nations (Mahapatra 2005). Half of HIV patients in Africa, Asia, and Latin America have relied on low-cost drugs from India. One month's dose of AIDS drugs cost US$30, or 5% of the price of nongeneric drugs sold by American or European companies.

Public health experts warn that "processes that link global trade and health often occur silently, with little attention or representation by legislators, the public media, and health professionals" (Shaffer et al. 2005, p. 34). However, organizations such as the Center for Policy Analysis on Trade and Health (http://www.cpath.org) are working to raise awareness of the public health effects of global trade.

Societal Measures of Health and Illness

Measures of health and illness reveal striking disparities among countries and regions. In the following sections we describe various measures that provide indicators of the health of populations, including measures of morbidity, life expectancy, and mortality.

Table 2.1

Measures of Longevity and Mortality by Region: 2003

Region	Life Expectancy	Infant Mortality Rate	Under-5 Mortality Rate
World	63	54	80
Industrialized countries	78	5	6
United States	77	7	8
Developing countries	62	60	87
Least developed countries	49	98	155
Sub-Saharan Africa	46	104	175

Source: UNICEF (2004).

Morbidity

Morbidity refers to illnesses, symptoms, and the impairments they produce. Measures of morbidity are often expressed in terms of the incidence and prevalence of specific health problems. *Incidence* refers to the number of new cases of a specific health problem in a given population during a specified time period. *Prevalence* refers to the total number of cases of a specific health problem in a population that exists at a given time. For example, the *incidence* of HIV infection worldwide was 5 million in 2004; this means that there were 5 million people newly infected with HIV in 2004. In the same year the worldwide *prevalence* of HIV was nearly 40 million, meaning that nearly 40 million people worldwide were living with HIV infection in 2004 (Joint United Nations Programme on HIV/AIDS 2004).

Rates of morbidity in a population provide one measure of the health of that population. As we discuss later in this chapter, patterns of morbidity vary according to social factors, such as social class, education, sex, and race. Morbidity patterns also vary according to the level of development of a society and the age structure of the population. In the less developed countries the major health threats are infectious and parasitic diseases, such as HIV disease, tuberculosis, diarrheal diseases (caused by bacteria, viruses, or parasites), measles, and malaria (Weitz 2004). In the industrialized world, where infectious and parasitic diseases have been largely controlled by advances in sanitation, immunizations, and antibiotics, noninfectious diseases have emerged as major sources of morbidity. These include chronic and degenerative conditions such as heart disease, cancer, stroke, arthritis, diabetes, mental disorders, and respiratory diseases.

Patterns of Longevity

ThomsonNOW™

Learn more about **Life Expectancy** by going through the Average Life Expectancy Map Exercise.

Another measure of the health of a population is the average number of years that individuals born in a given year can expect to live, referred to as **life expectancy.** Wide disparities in life expectancy exist between regions of the world (see Table 2.1). In 2003 Japan had the longest life expectancy (82 years), whereas 23 countries (primarily in Africa) had life expectancies of less than 50 years (UNICEF 2004).

Table 2.2
Top Three Causes of Death by Selected Age Groups: United States, 2003

| Age (Years) | Leading Causes of Death | | |
	First	Second	Third
1–4	Unintentional injuries	Congenital/chromosomal abnormalities	Cancer
5–14	Unintentional injuries	Cancer	Congenital/chromosomal abnormalities
15–24	Unintentional injuries	Homicide	Suicide
25–44	Unintentional injuries	Cancer	Heart disease
45–64	Cancer	Heart disease	Unintentional injuries
65 years and older	Heart disease	Cancer	Stroke

Source: Hoyert et al. (2005, Table 7).

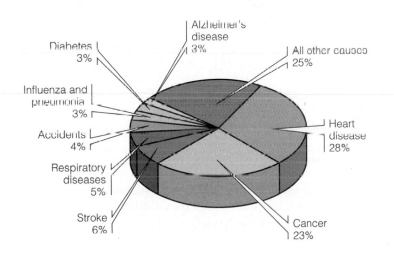

Figure 2.1
Leading causes of death in the United States: 2003.*
Source: Hoyert et al. (2005).
*Rounded to the nearest percent.

Patterns of Mortality

Rates of **mortality,** or death—especially those of infants, children, and women—provide sensitive indicators of the health of a population. Worldwide, the leading cause of death is infectious and parasitic diseases (Weitz 2004). In the United States the three leading causes of death for both women and men are heart disease, cancer, and stroke (see Figure 2.1). Later, we discuss how patterns of mortality are related to social factors, such as social class, sex, race or ethnicity, and education. Mortality patterns also vary by age, as shown in Table 2.2.

Table 2.3

Trained Childbirth Assistance and Lifetime Chance of Maternal Mortality by Region

	Percentage of Births Attended by Skilled Personnel	Lifetime Chance of Dying from Maternal Mortality
More developed countries	99	1 in 2,800
Less developed countries	57	1 in 61
Sub-Saharan Africa	41	1 in 16

Source: Ashford & Clifton (2005).

Infant and Childhood Mortality Rates. The infant mortality rate, the number of deaths of live-born infants under 1 year of age per 1,000 live births (in any given year), provides an important measure of the health of a population. In 2003 infant mortality rates ranged from 104 in sub-Saharan Africa to an average of 5 in industrialized countries (UNICEF 2004). The U.S. infant mortality rate was 6.9, and 36 countries had infant mortality rates that were lower than that of the United States (UNICEF 2004). Diarrhea, resulting from poor water quality and sanitation, is one of the predominant causes of infant death worldwide (Millennium Ecosystem Assessment 2005). The major cause of U.S. infant deaths is disorders related to premature birth and low birth weight (Hoyert et al. 2005).

The *under-5 mortality rate,* another useful measure of child health, refers to the rate of deaths of children under age 5. Under-5 mortality rates range from 175 in sub-Saharan Africa to 6 in industrialized countries (UNICEF 2004.) Most deaths of infants and children under age 5 occur in developing countries, where underweight is a contributing factor in 60% of childhood deaths (World Health Organization 2002).

> "The state of the world's children is the best measure of human well-being, and the health of children is the best measure of the health of our planet."
>
> **Carol Bellamy**
> **Director, UNICEF**

Maternal Mortality Rates. The **maternal mortality rate,** a measure of deaths that result from complications associated with pregnancy, childbirth, and unsafe abortion, also provides a sensitive indicator of the health status of a population. Maternal mortality is the leading cause of death and disability for women ages 15 to 49 in developing countries. The three most common causes of maternal death are hemorrhage (severe loss of blood), infection, and complications related to unsafe abortion.

Rates of maternal mortality show a greater disparity between rich and poor countries than any of the other societal health measures. Of the 529,000 annual maternal deaths worldwide, including 68,000 deaths from unsafe abortion, only 1% occur in high-income countries (World Health Organization 2005). Women's lifetime risk of dying from pregnancy or childbirth is highest in sub-Saharan Africa, where 1 in 16 women dies of pregnancy-related causes, compared to 1 in 2,800 women in developed countries (Ashford & Clifton 2005). High maternal mortality rates in less developed countries are related to poor-quality and inaccessible health care; most women give birth without the assistance of trained personnel (see Table 2.3). High maternal mortality rates are also linked to malnutrition and poor sanitation, and higher rates of pregnancy and childbearing at early ages. Women in many countries also lack access to family planning services and/or do not have the support of their

> "It is not uncommon for women in Africa, when about to give birth, to bid their older children farewell."
>
> **United Nations Population Fund**

male partners to use contraceptive methods such as condoms. Consequently, many women resort to abortion to limit their childbearing, even in countries where abortion is illegal and unsafe.

The Epidemiological Transition

Life expectancy and rates and causes of mortality and morbidity vary dramatically between developing and developed nations. As societies develop and increase the standard of living for their members, life expectancy increases and birthrates decrease. At the same time, the causes of death and disability shift from infectious disease and maternal and infant mortality and morbidity to chronic, noninfectious causes. This shift is referred to as the **epidemiological transition,** whereby low life expectancy and predominance of parasitic and infectious diseases shift to high life expectancy and predominance of chronic and degenerative diseases. As societies make the epidemiological transition, birthrates decline and life expectancy increases, so diseases that need time to develop, such as cancer, heart disease, Alzheimer's disease, arthritis, and osteoporosis, become more common, and childhood illnesses, typically caused by infectious and parasitic diseases, become less common, as do pregnancy-related deaths and health problems.

In the *World Health Report 2002* the World Health Organization (2002) noted that changes in patterns of consumption, particularly of food, alcohol, and tobacco, around the world are creating a "risk transition." Changes in food production and processing and in agricultural and trade policies have led to increased consumption of alcohol, tobacco, salt, sugar, and fat and subsequent increases in noninfectious diseases related to these substances, such as cancer, respiratory disease, heart disease, and diabetes. For low- and middle-income countries that are still dealing with high rates of infectious diseases and malnutrition, this creates a "double burden."

Patterns of Burden of Disease

A new approach to measuring the health status of a population provides an indicator of the overall burden of disease on a population through a single unit of measurement that combines not only the number of deaths but also the impact of premature death and disability on a population (Murray & Lopez 1996). This comprehensive unit of measurement, called the *disability-adjusted life year* (DALY), reflects years of life lost to premature death and years lived with a disability. More simply, 1 DALY is equal to 1 lost year of healthy life.

Worldwide, tobacco is the leading cause of burden of disease (World Health Organization 2002). Hence tobacco has been called "the world's most lethal weapon of mass destruction" (SmokeFree Educational Services 2003, p. 1). The top 10 risk factors that contribute to the global burden of disease are underweight; unsafe sex; high blood pressure; tobacco; alcohol; unsafe water, sanitation, and hygiene; high cholesterol; indoor smoke from solid fuels; iron deficiency; and overweight (World Health Organization 2002).

Sociological Theories of Illness and Health Care

The sociological approach to the study of illness, health, and health care differs from medical, biological, and psychological approaches to these topics. Next, we discuss how the three major sociological theories—structural functionalism, conflict theory, and symbolic interactionism—contribute to our understanding of illness and health care.

Structural-Functionalist Perspective

The structural-functionalist perspective is concerned with how illness, health, and health care affect and are affected by other aspects of social life. For example, rather than look at individual reasons for suicide, the structural-functionalist approach looks for social patterns that may help to explain suicide rates. Durkheim (1951 [1897]) conducted one of the first scientific sociological research studies that found that suicide rates were higher in countries characterized by lower levels of social integration and regulation—a finding that has been replicated in recent studies (Stockard & O'Brien 2002).

The structural-functionalist perspective draws attention to how changes in society affect health. For example, increased modernization and industrialization throughout the world has resulted in environmental pollution—a major health concern (see Chapter 14).

Just as social change affects health, health concerns may lead to social change. The emergence of HIV and AIDS in the U.S. gay male population was a force that helped unite and mobilize gay rights activists. Concern over the effects of exposure to tobacco smoke—the greatest cause of disease and death in the United States and other developed countries—has led to legislation banning smoking in workplaces, restaurants, and bars in at least five states (California, Delaware, New York, Connecticut, and Maine) (SmokeFree Educational Services 2003).

According to the structural-functionalist perspective, health care is a social institution that functions to maintain the well-being of societal members and, consequently, of the social system as a whole. Illness is dysfunctional in that it interferes with people performing needed social roles. To cope with nonfunctioning members and to control the negative effects of illness, society assigns a temporary and unique role to those who are ill—the sick role (Parsons 1951). This role ensures that societal members receive needed care and compassion, yet at the same time, it carries with it an expectation that the person who is ill will seek competent medical advice, adhere to the prescribed regimen, and return as soon as possible to normal role obligations.

Finally, the structural-functionalist perspective draws attention to latent dysfunctions, or unintended and often unrecognized negative consequences of social patterns or behavior. For example, a latent dysfunction of widespread use of some drugs is the emergence of drug resistance, which occurs when drugs kill the weaker disease-causing germs while allowing variants resistant to the drugs to flourish. For generations the drug chloroquine was added to table salt to prevent malaria. But overuse led to drug-resistant strains of malaria, and now chloroquine is useless in preventing malaria (McGinn 2003). Another health-related example of a latent dysfunction is the unintended consequences of highly active antiretroviral therapy (HAART), which reduces the viral loads of HIV-positive patients and delays their progression to AIDS. One study found that HIV-positive young people are engaging in more unprotected sex, have more sexual partners, and are more likely to use illicit drugs than HIV-positive young people did before the availability of HAART (Lightfoot et al. 2005). The researchers explained that many HIV-infected individuals who are on antiretroviral therapy believe that unprotected sex is less risky because they have lower viral load levels. And, because this medication prolongs the lives of HIV-infected people, they might have more opportunities to transmit the virus to others. Engaging in more risky behavior and having more opportunities to infect others with HIV is an unintended negative consequence of antiretroviral therapy.

Conflict Perspective

The conflict perspective focuses on how wealth, status, power, and the profit motive influence illness and health care. Worldwide, populations living in poverty, with little power and status, experience more health problems and have less access to quality medical care (Feachum 2000).

The conflict perspective criticizes the pharmaceutical and health care industry for placing profits above people. For example, pharmaceutical companies' research and development budgets are spent not according to public health needs but rather according to calculations about maximizing profits. Because the masses of people in developing countries lack the resources to pay high prices for medication, pharmaceutical companies do not see the development of drugs for diseases of poor countries as a profitable investment. This explains why 90% of the $70 billion invested annually in health research and development by pharmaceutical companies and Western governments focuses on the health problems of the 10% of the global population living in developed industrialized countries (Thomas 2003). This allocation of health-related funding clearly benefits the wealthier citizens of the world and neglects the needs of poor populations. Consider, for example, malaria—a tropical disease that kills 3 million people annually, killing more people than AIDS. Malaria remains one of the world's leading health threats, yet it is rarely covered in the news and is a relatively low public health priority. McGinn (2003) explained, "The reality is that malaria is a disease of poor countries. If it were a constant threat in industrial countries, the story would be completely different" (p. 63).

Conflict theorists also point to the ways in which health care and research are influenced by male domination and bias. When the male erectile dysfunction drug Viagra made its debut in 1998, women across the United States were outraged by the fact that some insurance policies covered Viagra (or were considering covering it), even though female contraceptives were not covered. The male-dominated medical research community has also been criticized for neglecting women's health issues and excluding women from major health research studies (Johnson & Fee 1997).

The conflict perspective points to ways in which powerful groups and wealthy corporations influence health-related policies and laws through financial contributions to politicians and political candidates and other means. Before the 1970s the American Medical Association used its power to campaign against a national health plan, fearing that the adoption of such a plan would lead to government control of the medical profession. In recent decades the most vehement opponents of national health insurance in the United States have been health insurance companies, who have used their power "to convince politicians that the public opposed health care reform" (Quadagno 2004, p. 39).

Conflict theorists argue that the high cost of medical care in the United States is a result of a capitalistic system in which health care is a commodity rather than a right. The conflict perspective views power and concern for profits as the primary obstacles to U.S. health care reform. Insurance companies fear that health care reform could mean federal regulation of the insurance industry. In an effort to buy political influence to maintain profits, the insurance industry has contributed millions of dollars to congressional candidates.

Symbolic Interactionist Perspective

Symbolic interactionists focus on (1) how meanings, definitions, and labels influence health, illness, and health care and (2) how such meanings are learned through interaction with others and through media messages and portrayals. According to the symbolic interactionist perspective of illness, "there are no illnesses or diseases in nature. There are only conditions that society, or groups within it, have come to define as illness or disease" (Goldstein 1999, p. 31). Psychiatrist Thomas Szasz (1970) argued that what we call "mental illness" is no more than a label conferred on those individuals who are "different," that is, those who do not conform to society's definitions of appropriate behavior.

Defining or labeling behaviors and conditions as medical problems is part of a trend known as **medicalization.** Initially, medicalization was viewed as occurring when a particular behavior or condition deemed immoral (e.g., alcoholism, masturbation, homosexuality) was transformed from a legal problem into a medical problem that required medical treatment. The concept of medicalization has expanded to include (1) any new phenomena defined as medical problems in need of medical intervention, such as post-traumatic stress disorder, premenstrual syndrome, and attention-deficit/hyperactivity disorder and (2) "normal" biological events or conditions that have come to be defined as medical problems in need of medical intervention, including childbirth, menopause, and death.

Conflict theorists view medicalization as resulting from the medical profession's domination and pursuit of profits. A symbolic interactionist perspective suggests that medicalization results from the efforts of sufferers to "translate their individual experiences of distress into shared experiences of illness" (Barker 2002, p. 295). In her study of women with fibromyalgia (a pain disorder that has no identifiable biological cause), Barker (2002) suggested that the medicalization of symptoms and distress through a diagnosis of fibromyalgia gives sufferers a framework for understanding and validating their experience of distress.

Recent theorists have observed a shift from medicalization to *biomedicalization.* Spurred by technoscientific innovation, **biomedicalization** refers to the view that medicine can not only control particular conditions but also *transform* bodies and lives. Examples of biomedical transformations include receiving an organ transplant, becoming pregnant through reproductive technology, and receiving artificial limbs (Clarke et al. 2003).

The concepts of medicalization and biomedicalization suggest that conceptions of health and illness are socially constructed. It follows, then, that definitions of health and illness vary over time and from society to society. In some countries being fat is a sign of health and wellness; in others it is an indication of mental illness or a lack of self-control. Among some cultural groups, perceiving visions or voices of religious figures is considered normal religious experience, whereas such "hallucinations" would be indicative of mental illness in other cultures. In 18th- and 19th-century America masturbation was considered an unhealthy act that caused a range of physical and mental health problems. Individuals caught masturbating were often locked up in asylums, treated with drugs (such as sedatives and poisons), or subjected to a range of interventions designed to prevent masturbation by stimulating the genitals in painful ways, preventing genital sensation, or deadening it. These physician-prescribed interventions included putting ice on the genitals; blistering and scalding the penis, vulva, inner thighs, or perineum; inserting electrodes into the rectum and urethra; cauterizing the clitoris by applying pure carbolic acid; circumcising the penis; and surgically removing the clitoris, ovaries, and testicles

(Allen 2000). Today, most health professionals agree that masturbation is a normal, healthy aspect of sexual expression.

Symbolic interactionists also focus on the stigmatizing effects of being labeled "ill." A **stigma** refers to any personal characteristic associated with social disgrace, rejection, or discrediting. (Originally, the word *stigma* referred to a mark burned into the skin of a criminal or slave.) Individuals with mental illnesses, drug addictions, physical deformities and impairments, and HIV and AIDS are particularly prone to being stigmatized. Stigmatization may lead to prejudice and discrimination and even violence against individuals with illnesses or impairments.

Finally, symbolic interactionism draws attention to the effects that meanings and labels have on behaviors that affect our health. For example, as tobacco sales have declined in developed countries, transnational tobacco companies have looked for markets in developing countries, using advertising strategies that depict smoking as "an inexpensive way to buy into glamorous lifestyles of the upper or successful social class" (Egwu 2002, p. 44).

Meanings and labels also affect social policy. Dennis Altman (2003) suggested that we conceptualize HIV/AIDS as a global security issue rather than as a health issue, "because how we conceptualize the pandemic will impact on the extent of political commitment governments bring to dealing with it. . . . Redefining the disease to encompass security issues almost inevitably pushes it higher on government agendas" (pp. 40–41). We examine HIV/AIDS in the next section.

HIV/AIDS: A Global Health Concern

One of the most urgent worldwide public health concerns is the spread of the human immunodeficiency virus (HIV), which causes acquired immunodeficiency syndrome (AIDS). HIV/AIDS is the leading cause of death among adults ages 15–59 years worldwide. HIV/AIDS has killed more than 20 million people, and in 2004 nearly 40 million people worldwide were living with HIV infection (Joint United Nations Programme on HIV/AIDS 2004; World Health Organization 2004).

HIV is transmitted through sexual intercourse, through sharing unclean intravenous needles, through perinatal transmission (from infected mother to fetus or newborn), through blood transfusions or blood products, and, rarely, through breast milk. Worldwide, the predominant mode of HIV transmission is through heterosexual contact (World Health Organization 2004).

ThomsonNOW

Learn more about **AIDS** by going through the AIDS Map Exercise.

HIV/AIDS in Africa and Other Regions

In 2003 Africa was home to two-thirds of the world's people living with HIV/AIDS. About 1 in 12 African adults has HIV/AIDS, and as many as 9 out of 10 HIV-infected people in sub-Saharan Africa do not know that they are infected (World Health Organization 2004). But HIV/AIDS also affects millions of people living in India and hundreds of thousands of people in China, the Mediterranean region, Western Europe, and Latin America. Eastern European countries and Central Asia are experiencing increasing rates of HIV infection, mainly from drug-injecting behavior and to a lesser extent by unsafe sex. The second highest HIV prevalence rate (after sub-Saharan Africa) is the Caribbean, where 2–3% of adults are infected with HIV (World Health Organization 2004).

The high rates of HIV in developing countries, particularly sub-Saharan Africa, are having alarming and devastating effects on societies. HIV/AIDS has reversed the gains in life expectancy made in sub-Saharan Africa, which peaked at 49 years in the late 1980s and fell to 46 years in 2005 (World Health Organization 2004). The HIV/AIDS epidemic creates an enormous burden on the limited health care resources of poor countries. Economic development is threatened by the HIV epidemic, which diverts national funds to health-related needs and reduces the size of a nation's workforce. AIDS deaths have left 14 million orphans in the world, and by 2010 a projected 25 million children will be orphaned by HIV/AIDS (World Health Organization 2004). Some scholars fear that AIDS-affected countries could become vulnerable to political instability as the growing number of orphans increases the proportion of dependent people and exacerbates poverty. In addition, "these children could become a source of future urban discontent, criminal activity, and recruits for insurgencies" (Mastny & Cincotta 2005).

HIV/AIDS in the United States

An estimated 850,000 to 950,000 U.S. residents are infected with HIV, about one-fourth of whom are unaware of their infection (Centers for Disease Control and Prevention 2004b). About 70% of new HIV infections in the United States each year are among men, with the remainder occurring among women, and half of new infections are in people younger than 25 years of age. Of new infections among U.S. men, 60% were infected through same-sex sexual contact, 25% through injection drug use, and 15% through heterosexual sex. Of new infections among women, 75% were infected through heterosexual sex and 25% through injection drug use (Centers for Disease Control and Prevention 2004b).

Rates of HIV infection among men who have sex with men are alarming. In a study of five U.S. cities, 25% of men who have sex with men were infected with HIV; nearly half of these infected men (48%) were unaware of their infection (Centers for Disease Control and Prevention 2005b).

The African American population is being especially hard hit by HIV/AIDS. A recent fact sheet on HIV/AIDS among African Americans revealed alarming statistics (Centers for Disease Control and Prevention 2005a):

- Although African Americans make up about 12% of the U.S. population, in 2003 African Americans accounted for 42% of all people in the United States living with HIV/AIDS.
- In 2001 HIV/AIDS was the number one cause of death for African American women ages 25–34.
- During 2000–2003, the HIV/AIDS rate for African American females was 19 times the rate for white females and 5 times the rate for Hispanic females; the rate for African American females also exceeded the rates for males of all races and ethnicities other than African Americans. The rate for African American males was 7 times the rate for white males and 3 times the rate for Hispanic males.

Higher rates of HIV/AIDS among African Americans is partly due to the link between higher AIDS incidence and poverty. The nearly 1 in 4 African Americans who live in poverty experience limited access to high-quality health care and HIV prevention education. A recent study of HIV transmission among African American women in North Carolina found that women with HIV infection were more likely than noninfected women to be unemployed, to receive public assistance, to have had 20 or more lifetime sexual partners, or to have traded sex for drugs, money, or shelter (reported by Centers for Disease Control and Prevention 2005a).

Despite the widespread concern about HIV, many Americans—especially adolescents and young adults—engage in high-risk behavior. Although a national survey of U.S. teens found that most teens (81%) say that AIDS is a serious problem for people their age (Kaiser Family Foundation 2000), a national study of U.S. youth in grades 9–12 found that among the sexually active students, only 63% reported that either they or their partner had used a condom during their last sexual intercourse (Grunbaum et al. 2004).

The Growing Problem of Obesity

Obesity, which can lead to heart disease, hypertension, diabetes, and other health problems, is the second biggest cause of preventable deaths in the United States (second only to tobacco use) (Stein & Connolly 2004). In 2005 an alarming report was published in the *New England Journal of Medicine* that suggested that over the next 50 years obesity will shorten the average U.S. life expectancy of 77.6 years by at least 2 to 5 years, reversing the mostly steady increase in life expectancy that has occurred over the past two centuries (Olshansky et al. 2005).

The following statistics reflect the degree to which Americans are overweight or obese (as defined by body mass index) (National Center for Health Statistics 2005; National Center for Health Statistics 2004b; Centers for Disease Control and Prevention 2004a).

- About two-thirds (64%) of U.S. adults are either overweight or obese.
- One-fourth of U.S. adults age 20 or older are obese.
- The percentage of U.S. adults who are obese increased from 19.4% in 1997 to 24.5% in 2004.
- The highest rate of obesity is among non-Hispanic black women (39.3%).
- Among youth ages 6–19, 16% are overweight.
- The percentage of children and adolescents who are overweight has more than doubled since 1970.

Although genetics and certain medical conditions contribute to many cases of overweight and obesity, two major social and lifestyle factors that play a major role in the obesity epidemic are patterns of food consumption and physical activity level. For example, Americans are increasingly eating out at fast food and other restaurants where foods tend to contain more sugars and fats than foods consumed at home. In 1970 Americans spent one-third of their food dollars on food away from home; this amount grew to 39% in 1980, 45% in 1990, and 47% in 2001 (Sturm 2005). A study of more than 6,000 U.S. children and adolescents found that on a typical day, 30% reported eating fast food (Bowman et al. 2004). Research has found that fast food consumption is strongly associated with weight gain and insulin resistance, suggesting that fast food increased the risk of obesity and type 2 diabetes (Pereira et al. 2005). Consumption of snack

Digital Vision./Getty Images

One-fourth of U.S. adults are obese.

foods and sugary soft drinks has also increased. Among children ages 6 to 11 years, consumption of chips, crackers, popcorn, and/or pretzels tripled from the mid-1970s to the mid-1990s. Consumption of soft drinks doubled during the same period (Sturm 2005).

A national survey found that in 2004 less than one-third (30.9%) of U.S. adults (age 18 or older) engaged in regular leisure-time physical activity (defined as moderate activity for 30 minutes or more at least 5 times a week or vigorous activity for 20 minutes at least 3 times a week) (National Center for Health Statistics 2005). According to the National Center for Chronic Disease Prevention and Health Promotion (2004), more than one-third of youth in grades 9–12 do not engage in regular vigorous physical activity, and only one-third of high school students participate in daily physical education classes at school.

Mental Illness: The Hidden Epidemic

What it means to be mentally healthy varies across and within cultures. **Mental health** has nevertheless been defined as the successful performance of mental function, resulting in productive activities, fulfilling relationships with other people, and the ability to adapt to change and to cope with adversity (U.S. Department of Health and Human Services 2001). **Mental illness** refers collectively to all mental disorders, which are health conditions that are characterized by alterations in thinking, mood, and/or behavior associated with distress and/or impaired functioning and that meet specific criteria (such as level of intensity and duration) specified in the classification manual used to diagnose mental disorders, *The Diagnostic and Statistical Manual of Mental Disorders* (American Psychiatric Association 2000) (see Table 2.4).

Mental illness is a "hidden epidemic" because the shame and embarrassment associated with mental problems discourage people from acknowledging and talking about them. Because male gender expectations associate masculinity with emotional strength, men are particularly prone to deny or ignore mental problems. In an effort to raise awareness that depression in men is a major public health problem, the National Institute of Mental Health (NIMH) launched a public health education campaign called "Real Men, Real Depression," which includes print, television, and radio public service announcements (see this chapter's *Human Side* feature).

Extent and Impact of Mental Illness

Although transnational estimates of the prevalence of mental disorders vary, one study found a 40% lifetime prevalence of any mental disorder in the Netherlands and the United States, a 12% lifetime prevalence in Turkey, and a 20% lifetime prevalence in Mexico (WHO International Consortium in Psychiatric Epidemiology 2000). Annually, about 1 in 5 (21%) U.S. adults experience mental illness, and more than 1 in 10 (14%) school-age children (ages 5 to 17) have mental health problems (Center for Mental Health Services 2004; U.S. Department of Health and Human Services 2001). One out of six Americans admit that poor mental health or emotional well-being kept them from doing their usual activities at least once during the last month. And 12% of Americans have visited a mental health professional, such as a psychologist, psychiatrist, or therapist, in the past 12 months (Newport 2004).

Untreated mental disorders can lead to poor educational achievement, lost productivity, unsuccessful relationships, significant distress, violence and abuse, incarceration, and poverty. Half of students identified with emotional disturbances

Table 2.4

Mental Disorders Classified by the American Psychiatric Association

Classification	Description
Anxiety disorders	Disorders characterized by anxiety that is manifest in phobias, panic attacks, or obsessive-compulsive disorder
Dissociative disorders	Problems involving a splitting or dissociation of normal consciousness, such as amnesia and multiple personality
Disorders first evident in infancy, childhood, or adolescence	Disorders including mental retardation, attention-deficit/hyperactivity, and stuttering
Eating or sleeping disorders	Disorders including anorexia, bulimia, and insomnia
Impulse control disorders	Problems involving the inability to control undesirable impulses, such as kleptomania, pyromania, and pathological gambling
Mood disorders	Emotional disorders such as major depression and bipolar (manic-depressive) disorder
Organic mental disorders	Psychological or behavioral disorders associated with dysfunctions of the brain caused by aging, disease, or brain damage (such as Alzheimer's disease)
Personality disorders	Maladaptive personality traits that are generally resistant to treatment, such as paranoid and antisocial personality types
Schizophrenia and other psychotic disorders	Disorders with symptoms such as delusions or hallucinations
Somatoform disorders	Psychological problems that present themselves as symptoms of physical disease, such as hypochondria
Substance-related disorders	Disorders resulting from abuse of alcohol and/or drugs, such as barbiturates, cocaine, or amphetamines

drop out of high school (Gruttadaro 2005). On any given day 150,000 people with severe mental illness are homeless, living on the streets or in public shelters (National Council on Disability 2002). As many as 1 in 5 adults in U.S. prisons and as many as 70% of youth incarcerated in juvenile justice facilities are mentally ill (Honberg 2005; Human Rights Watch 2003).

Mental disorders also contribute to mortality, with suicide being the fourth leading cause of death worldwide among 15- to 44-year-olds (Mercy et al. 2003). In the United States suicide is the third leading cause of death among people ages 15 to 24 (Hoyert et al. 2005). Most suicides in the United States (60–90%) are committed by individuals with mental illness (Ezzell 2003). Suicides outnumber homicides two to one every year in the United States (Ezzell 2003). In 2000 about half of the 1.7 million violent deaths that occurred in the world were the result of suicide, about one-third resulted from homicide, and one-fifth were from war injuries (Mercy et al. 2003).

Causes of Mental Disorders

Biological and social factors are linked to mental illness. Some mental illnesses are caused by genetic or neurological pathological conditions. However, social and environmental influences, such as poverty, history of abuse, or other severe emotional

Men Talk about Depression

As part of the National Institute of Mental Health campaign "Real Men, Real Depression," the following men went public with their experiences of depression. In the following interview excerpts, these men reveal their personal struggles with depression (National Institute of Mental Health 2003).

Jimmy Brown, Firefighter

"CAN YOU DESCRIBE A TYPICAL DAY?"
Many days . . . I just didn't wanna get out of bed. Honestly, the only reason I got out of bed . . . was because the dog had to get walked and my wife had to go to work. So I walked the dog, take her to work, and come back. Some days I'd get back in bed, some days I'd just sit on the couch and wonder what I was going to do next.

"DO YOU BELIEVE IT WILL EVER END?"
No . . . you just don't know if it's gonna end. . . . You don't know if you're ever gonna be the person that you were before. There were days when I thought I'd never be myself again.

"WAS THERE ANYTHING YOU LIKED TO DO?"
I pretty much lost interest in just about everything. Every aspect of your life the interest level just goes. You're just kinda there.

"CAN YOU PULL OUT OF IT ON YOUR OWN?"
They think I'm a big, tough fireman. I'm supposed to be able to deal with anything, I'm supposed to be able to just pick up, carry on. . . . It's not that easy. . . . I don't know if I'd be a firefighter today if I didn't get help.

Patrick McCathern, Retired First Sergeant, U.S. Air Force

"HOW IS DEPRESSION DIFFERENT FROM THE BLUES?"
Everybody gets the blues. I call depression the super blues. . . . When you have the super blues, you can't find your way back cause you've gotten so far in. It's like a hole that closes up behind you and you just get lost in your own mind. You literally get lost.

"WHY DIDN'T YOU TALK TO PEOPLE ABOUT YOUR DEPRESSION?"
Here I am in the Air Force and I'm one of the senior leaders in the enlisted ranks. And that would be a sign that, well, maybe I'm not a leader. And then my career's derailed or maybe I'll lose my security clearance. I can't let anybody know. I've got to gut it out, I've got to fake my way through it. . . . You don't want to be perceived as weak, you finally get to a point where . . . you don't care how you're perceived, because you are barely breathing, you're barely getting up. . . .

"WHAT DID YOU DO TO RELIEVE THE PAIN?"
I'd drink and I'd just get numb. . . . We're talking many, many beers to get to that state where you could shut your head off, but then you wake up the next day and it's still there. Because you have to deal with it, it doesn't just go away.

trauma, also affect individuals' vulnerability to mental illness and mental health problems. For example, iodine deficiency, common in poor countries, is believed to be the single most common preventable cause of mental retardation and brain damage (World Health Organization 2002). War within and between countries also contributes to mental illness. For example, experts predict that 16% of service members serving in Iraq and Afghanistan will develop post-traumatic stress disorder (Miller 2005) (see also Chapter 16 for a discussion of combat-related post-traumatic stress disorder). Garfinkel and Goldbloom (2000) explain, "The radical shifts in society towards technology, changes in family and societal supports and networks and the commercialization of existence . . . may account for the current epidemic of depression and other psychiatric disorders" (p. 503). It may be safe to conclude that the causes of most mental disorders lie in some combination of genetic, biological, and environmental factors (U.S. Department of Health and Human Services 2001).

NIMH

Rene Ruballo, Retired Police Officer

"CAN YOU DESCRIBE HOW IT STARTED?"
It started with my loss of interest in basically everything that I like doing. . . . I just didn't really feel like doing anything any more. . . . Sometimes I didn't even want to get out of bed.

"WHAT DID YOU THINK WAS HAPPENING TO YOU?"
I am thinking there's got to be something wrong because . . . I feel like nothing matters. My children, my family, nothing matters.

"WHY DIDN'T YOU SEEK HELP RIGHT AWAY?"
I was hoping that it would just go away . . . but it didn't. It just got worse. . . . Every day was a struggle, just to do minor things.

"HOW DO THE CHILDREN FEEL NOW THAT YOU ARE BETTER?
Well, they feel they got Daddy back the way he used to be. I'm doing more things with them and . . . taking more interest in things and school.

NIMH

Rodolf Palma-Lulion, Recent College Graduate

"WHAT WERE THE FIRST SYMPTOMS OF DEPRESSION?"
I just felt terrible and I didn't know why. . . . I didn't want to face anyone. I didn't want to talk to anyone. I didn't really want to do anything for myself because I felt . . . like I was such an awful person that there was no real reason for me to do anything for myself.

"DESCRIBE HOW YOU FELT."
I just didn't feel any emotions. I just couldn't feel. My real feeling was just pure numbness. . . .

"HOW DID DEPRESSION AFFECT YOU AT SCHOOL?"
I didn't read a book. I barely went to class. I just couldn't wake up in time for class. If I had class at two, I'd sleep till three. . . .

"DID BEING LATINO MAKE A DIFFERENCE?"
Yeah, I totally think that being a Latino made it harder. . . . My little brother went through depression before me . . . and we never even really talked about it because . . . there's just things you don't talk about. . . . When I told my parents I had depression . . . my mom was like you're not depressed! . . . You're gonna get over it. You just got to be strong. You just gotta finish school. . . .

Lifestyle Behaviors and Social Factors Associated with Health and Illness

Public health education campaigns, articles in popular magazines, college-level health courses, and health professionals emphasize that to be healthy, we must adopt a healthy lifestyle. Despite these efforts, many individuals incorporate high-risk health behaviors into their lifestyle, often during their youth (see this chapter's *Social Problems Research Up Close* feature). However, health and illness are affected by more than personal lifestyle choices. In the following sections we examine how social factors such as social class, poverty, education, race, and gender affect health and illness.

The National Health-Risk Behavior Survey

The Youth Risk Behavior Surveillance System (YRBSS), conducted biennially since 1991, monitors a number of health-risk behaviors among U.S. youth and young adults, including behaviors that contribute to unintentional injuries and violence, tobacco use, alcohol and other drug use, sexual behaviors that contribute to unintended pregnancy and sexually transmitted diseases (STDs), unhealthy dietary behaviors, and physical inactivity and overweight.

Methods and Sample

The YRBSS includes a national school-based survey conducted by the Centers for Disease Control and Prevention (CDC) as well as state and local school-based surveys conducted by education and health agencies. The findings reported here are from the national survey, 32 state surveys, and 18 local surveys conducted among students in grades 9–12 during 2003. The sampling frame for the national YRBSS consisted of a nationally representative sample of public and private school students in grades 9–12. For the national YRBSS, 15,240 questionnaires were com-

pleted in 158 schools. The school response rate was 81%, and the student response rate was 83%. The 2003 State and Local Youth Risk Behavior Surveys used representative samples of students in grades 9–12 in their jurisdiction. The student sample sizes for the state and local YRBSS ranged from 968 to 9,320. School response rates ranged from 67% to 100%; student response rates ranged from 60% to 94%.

Findings and Conclusions[1]

- *Seat belt use.* Nationwide, 18% of students had rarely or never worn seat belts when riding in a car driven by someone else. Overall, the prevalence of having rarely or never worn seat belts was higher among male (22%) than among female (15%) students.
- *Riding with a driver who had been drinking alcohol.* During the 30 days preceding the survey, 30% of students nationwide had ridden in a car or other vehicle one or more times with a driver who had been drinking alcohol. Overall, the prevalence of having ridden with a driver who had been drinking alcohol was

higher among Hispanic students (36%) than among white (29%) and black (31%) students.
- *Driving after drinking alcohol.* During the 30 days preceding the survey, 12% of students nationwide had driven a car or other vehicle one or more times after drinking alcohol. Overall, the prevalence of having driven after drinking alcohol was higher among male (15%) than among female (9%) students.
- *Serious consideration of attempting suicide.* Nationwide, 17% of students had seriously considered attempting suicide during the 12 months preceding the survey. Overall, the prevalence of having considered attempting suicide was higher among female (21%) than among male (13%) students and was higher among white (17%) and Hispanic (18%) students than among black (13%) students.
- *Attempting suicide.* Nationwide, 9% of students had actually attempted suicide one or more times during the 12 months preceding the survey. Overall, the prevalence of having attempted suicide was

Social Class and Poverty

In an address to the 2001 World Health Assembly, UN secretary-general Kofi Annan stated that the biggest enemy of health in the developing world is poverty (United Nations Population Fund 2002). Poverty is associated with malnutrition, indoor air pollution, hazardous working conditions, lack of access to medical care, and unsafe water and sanitation (see also Chapter 6). Half of the urban population in Africa, Asia, Latin America, and the Caribbean suffers from one or more diseases linked to inadequate water and sanitation (Millennium Ecosystem Assessment 2005).

In the United States low socioeconomic status is associated with higher incidence and prevalence of health problems, disease, and death (Malatu & Schooler 2002). The percentage of Americans reporting fair or poor health is more than three times as high for people living below the poverty line as for those with family income at least twice the poverty threshold (National Center for Health Statistics 2004a). In

higher among female (12%) than among male (5%) students.

- *Current tobacco use.* Nationwide, 28% of students had reported current cigarette use, current smokeless tobacco use, or current cigar use on more than 1 of the 30 days preceding the survey. Overall, the prevalence of current tobacco use was higher among male (30%) than among female (25%) students.

- *Episodic heavy drinking.* Nationwide, 28% of students had had more than 5 drinks of alcohol in a row (i.e., within a couple of hours) on more than 1 of the 30 days preceding the survey.

- *Sexual behavior and condom use.* Nationwide, 47% of students had had sexual intercourse during their lifetime. Nationwide, 14% of students had had sexual intercourse with more than 4 sex partners during their lifetime. Among the 34% of currently sexually active students nationwide, 63% reported that either they or their partner had used a condom during their last sexual intercourse.

- *Insufficient amount of physical activity.* Nationwide, 33% of students had not participated in sufficient vigorous or moderate physical activity during the 7 days preceding the survey. Overall, the prevalence of having participated in an insufficient amount of physical activity was higher among female (40%) than among male (27%) students.

- *Dietary behaviors.* Approximately one-fifth (22%) of students nationwide had eaten fruits and vegetables five or more times a day during the 7 days preceding the survey.

- *Weight.* Nationwide, 12% of students were overweight (based on objective criteria). Overall, the prevalence of being overweight was higher among male (17%) than among female (9%) students. However, 30% of students described themselves as slightly or very overweight. Overall, the prevalence of describing themselves as overweight was higher among female (36%) than among male (24%) students. Nationwide, 13% of students had gone without eating for more than 24 hours to lose

weight or to keep from gaining weight during the 30 days preceding the survey. Overall, the prevalence of having gone without eating for more than 24 hours to lose weight or to keep from gaining weight was higher among female (18%) than among male (9%) students. Also, 6% of students had vomited or taken laxatives to lose weight or to keep from gaining weight during the 30 days preceding the survey. Overall, the prevalence of having vomited or taken laxatives to lose weight or to keep from gaining weight was higher among female (8%) than among male (4%) students.

Discussion

The results of the YRBSS indicate that many U.S. teens engage in behaviors that place them at risk for serious health problems. The survey results provide important baseline data for school administrations and health officials to use in reducing health-risk behaviors among U.S. teens.

[1] Percentages are rounded.

the United States poverty is associated with higher rates of health-risk behaviors such as smoking, alcohol drinking, being overweight, and being physically inactive. The poor are also exposed to more environmental health hazards and have unequal access to and use of medical care (Lantz et al. 1998). In addition, the lower class tends to experience high levels of stress and has few resources to cope with it (Cockerham 2005). Stress has been linked to a variety of physical and mental health problems, including high blood pressure, cancer, chronic fatigue, and substance abuse.

Just as poverty contributes to health problems, health problems contribute to poverty. Health problems can limit one's ability to pursue education or vocational training and to find or keep employment. The high cost of health care not only deepens the poverty of people who are already barely getting by but also can financially devastate middle-class families. One study found that in about half of all U.S. bankruptcies, medical bills were a contributing factor (Himmelstein et al. 2005). The average out-of-pocket costs for those bankruptcy filers since the onset of illness or injury was $11,854.

Poverty and Mental Health. Low socioeconomic status is also associated with increased risk of a broad range of psychiatric conditions. People living below the U.S. poverty line are three times as likely as those with incomes twice the poverty line to have serious psychological distress (8% versus 25%) (National Center for Health Statistics 2004a). Two explanations for the link between social class and mental illness are the social causation explanation and the social selection explanation (Cockerham 2005). The *social selection* explanation suggests that mentally ill individuals have difficulty achieving educational and occupational success and thus tend to drift to the lower class, whereas the mentally healthy are upwardly mobile. The *social causation* explanation suggests that lower-class individuals experience greater adversity and stress as a result of their deprived and difficult living conditions, and this stress can reach the point where the individual can no longer cope with daily living. Recent research that tested these two explanations found more support for the social causation explanation (Hudson 2005).

Education

Although economic resources are important, "the strongest single predictor of good health appears to be education" (Cockerham 2004, p. 62). Education level and family income are closely tied, for example, to the likelihood of being uninsured (Nelson 2003). Individuals with low levels of education are more likely to engage in health-risk behaviors such as smoking and heavy drinking. Women with less education are less likely to seek prenatal care and are more likely to smoke during pregnancy, which helps explain why low birth weight and infant mortality are more common among children of less educated mothers (Children's Defense Fund 2000). In some cases lack of education means that individuals do not know about health risks or how to avoid them. A national survey in India found that only 18% of illiterate women had heard of AIDS, compared to 92% of women who had completed high school (Ninan 2003).

Family and Household Factors

The household in which one lives has consequences for one's physical and mental health. A study of adults in their 50s found that married people who live only with their spouse or with spouse and children had the best physical and mental health, whereas single women living with children had the lowest measures of health (Hughes & Waite 2002). Other research findings concur that married adults are healthier and have lower levels of depression and anxiety compared to adults who are single, divorced, cohabiting, or widowed (Mirowsky & Ross 2003; Schoenborn 2004). Better health among married individuals results from the economic advantages of marriage and from the emotional support provided by most marriages—the sense of being cared about, loved, and valued (Mirowsky & Ross 2003).

For children, living in a two-parent household is associated with better health outcomes. A Swedish study found that children living with only one parent have a higher risk of death, mental illness, and injury than those in two-parent families, even when their socioeconomic disadvantage is taken into account (Hollander 2003). For both children and adults psychosocial stress involved in living in a single-parent household appears to have a negative effect on health.

Gender

Gender affects the health of both women and men. Gender discrimination and violence against women produce adverse health effects in girls and women worldwide. Violence against women is a major public health concern: At least one in three women has been beaten, coerced into sex, or abused in some way—most often by someone she knows (Alan Guttmacher Institute 2004). "Although neither health care workers nor the general public typically thinks of battering as a health problem, woman battering is a major cause of injury, disability, and death among American women, as among women worldwide" (Weitz 2004, p. 56).

Sexual violence and gender inequality contribute to growing rates of HIV among girls and women. In 2004 nearly half of all adults living with HIV were women—up from 35% in 1985. In sub-Saharan Africa girls and young women make up 75% of those individuals ages 15 to 24 who are HIV-positive, in part because African women do not have the social power to refuse sexual intercourse and/or to demand that their male partners use condoms (Lalasz 2004). Women are also more susceptible to HIV infection for physical and biological reasons.

As noted earlier, women in developing countries suffer high rates of mortality and morbidity as a result of the high rates of complications associated with pregnancy and childbirth. The low status of women in many less developed countries results in their being nutritionally deprived and having less access to medical care than men have (Murphy 2003).

In the United States before the 20th century, the life expectancy of U.S. women was shorter than that of men because of the high rate of maternal mortality that resulted from complications of pregnancy and childbirth. In the United States today life expectancy of women (79.9 years) is greater than that of men (74.5 years) (Hoyert et al. 2005). Lower life expectancy for U.S. men is due to a number of factors. Men tend to work in more dangerous jobs than women, such as agriculture and construction. In addition, "beliefs about masculinity and manhood that are deeply rooted in culture . . . play a role in shaping the behavioral patterns of men in ways that have consequences for their health" (Williams 2003, p. 726). Men are socialized to be strong, independent, competitive, and aggressive and to avoid expressions of emotion or vulnerability that could be construed as weakness. These male gender expectations can lead men to take actions that harm themselves or to refrain from engaging in health-protective behaviors. For example, socialization to be aggressive and competitive leads to risky behaviors (such as dangerous sports, fast driving, and violence) that contribute to men's higher risk of injuries and accidents. Men are more likely than women to smoke cigarettes and to abuse alcohol and drugs but are less likely than women to visit a doctor and to adhere to medical regimens (Williams 2003).

Gender and Mental Health. A review of recent research suggests that the prevalence of mental illness is higher among U.S. women than among U.S. men (Cockerham 2005). In 2004, U.S. women were more likely than men (3.7% versus 2.4%) to have experienced serious psychological distress during the past 30 days (National Center for Health Statistics 2005). Women and men differ in the types of mental illness they experience; rates of mood and anxiety disorders are higher among women, and rates of personality and substance-related disorders are more common among men. Although women are more likely to attempt suicide, men are more likely to succeed at it because they use deadlier methods.

Table 2.5

Life Expectancy [a] in the United States by Race and Sex

All Races		Black		White	
Female	**Male**	**Female**	**Male**	**Female**	**Male**
79.9	74.5	76.1	69.2	80.5	75.4

[a]For individuals born in 2003.

Source: Hoyert et al. (2005, Table A).

Biological factors may account for some of the gender differences in mental health. Hormonal changes during menstruation and menopause, for example, may predispose women to depression and anxiety, although evidence to support this explanation is "insufficient at present" (Cockerham, 2005, p. 166). High testosterone and androgen levels in males may be linked to the greater prevalence of personality disorders in men, but again, research is not conclusive. Other explanations for gender differences in mental health focus on ways in which gender roles contribute to different types of mental disorders. For example, the unequal status of women and the strain of doing the majority of housework and child care may predispose women to experiencing greater psychological distress. Women may also be more likely to experience depression when their children leave home, because women are socialized to invest more in their parental role than men are.

Racial and Ethnic Minority Status

In the United States racial and ethnic minorities tend to suffer higher rates of mortality and morbidity. Black U.S. residents, especially black men, have a lower life expectancy than whites (see Table 2.5), and they also have the highest rate of infant mortality of all racial and ethnic groups, largely because of higher rates of prematurity and low birth weight (Hoyert et al. 2005). Black Americans are more likely than white Americans to die from stroke, heart disease, cancer, HIV infection, unintentional injuries, diabetes, cirrhosis, and homicide.

Compared with white Americans, Native Americans have higher death rates from motor vehicle injuries, diabetes, and cirrhosis of the liver (caused by alcoholism). Compared with non-Hispanic whites, Hispanics have more diabetes, high blood pressure, and lung cancer and are at higher risk of dying from violence, alcoholism, and drug use (Weitz 2004). Because Asian Americans have the highest levels of income and education of any racial or ethnic U.S. minority group, they typically have high levels of health.

Socioeconomic differences between racial and ethnic groups are largely responsible for racial and ethnic differences in health status (Cockerham 2004; Weitz 2004). One effect of the lower socioeconomic status of U.S. minorities is that they are less likely to have health insurance. In 2004 Hispanic individuals were most likely to be uninsured (32.3%), followed by non-Hispanic blacks (15.9%) and non-Hispanic whites (10.4%) (National Center for Health Statistics 2005). As we note elsewhere, compared with the insured, the uninsured are less likely to get timely and routine care and are more likely to be hospitalized for preventable conditions. As we discuss in Chapter 14, minorities are also more likely than whites to live in environments where they are exposed to hazards such as toxic chemicals, dust, and fumes. In addition, discrimination contributes to poorer health among oppressed racial and ethnic populations by restricting access to the quantity and quality of public education, housing, and health care. For example, a study of heart attack patients at 658 U.S. hospitals found that black patients were much less likely than white patients to get basic diagnostic tests, clot-busting drugs, or angioplasties (Vaccarino et al. 2005). In another study researchers looked at racial disparities in the rates of undergoing nine different surgical procedures. In the early 1990s whites had higher rates than blacks for all nine surgical procedures. By 2001 the difference between the rates among whites and blacks narrowed significantly for only one of the nine procedures, remained unchanged for three of the procedures, and *increased* significantly for five of the nine procedures (Jha et al. 2005).

Race, Ethnicity, and Mental Health. Medical sociologist William Cockerham (2005) reported that "almost all of the data and research that currently record differences in mental disorder between races show there is little or no significant difference in general between whites and members of racial minority groups" (p. 194). Differences that do exist are often associated more with social class than with race or ethnicity. However, other studies suggest that minorities are at higher risk for mental disorders, such as anxiety and depression, in part because of racism and discrimination, which adversely affect physical and mental health (U.S. Department of Health and Human Services 2001). In 2004, Hispanics (3.9%) were more likely to have experienced serious psychological distress during the past 30 days than Blacks (3.3%) and whites (2.9%) (National Center for Health Statistics 2005). Minorities also have less access to mental health services, are less likely to receive needed mental health services, often receive lower quality mental health care, and are underrepresented in mental health research (U.S. Department of Health and Human Services 2001).

Problems in U.S. Health Care

The United States boasts of having the best physicians, hospitals, and advanced medical technology in the world, yet problems in U.S. health care remain a major concern on the national agenda. A World Health Organization report on the first-ever analysis of the world's health systems noted that, although the United States spends a higher portion of its gross domestic product on health care than any other country, it ranks 37 out of 191 countries according to its performance (World Health Organization 2000). The report concluded that France provides the best overall health care among major countries, followed by Italy, Spain, Oman, Austria, and Japan.

After presenting a brief overview of U.S. health care, we address some of the major health care problems in the United States—inadequate health insurance coverage, the high cost of medical care and insurance, the managed care crisis, and inadequate mental health care.

U.S. Health Care: An Overview

In the United States there is no one health care system; rather, health care is offered through various private and public means. In 2004, 27% of Americans were covered by government health insurance plans (Medicare, Medicaid, and military insurance) and 68% were covered by private insurance, most often employment-based (DeNavas-Walt et al. 2005).

In traditional health insurance plans the insured choose their health care provider, who is reimbursed by the insurance company on a fee-for-service basis. The insured individual typically must pay an out-of-pocket "deductible" (usually ranging from a few hundred to a thousand dollars per year per person or per family) and then is often required to pay a percentage of medical expenses (e.g., 20%) until a maximum out-of-pocket expense amount is reached (after which insurance will cover 100% of medical costs up to a limit).

Health maintenance organizations (HMOs) are prepaid group plans in which a person pays a monthly premium for comprehensive health care services. HMOs attempt to minimize hospitalization costs by emphasizing preventive health care. **Preferred provider organizations (PPOs)** are health care organizations in which employers who purchase group health insurance agree to send their employees to certain health care providers or hospitals in return for cost discounts. In this arrangement health care providers obtain more patients but charge lower fees to buyers of group insurance.

Managed care refers to any medical insurance plan that controls costs through monitoring and controlling the decisions of health care providers. In many plans doctors must call a utilization review office to receive approval before they can hospitalize a patient, perform surgery, or order an expensive diagnostic test. Although the terms *HMO* and *managed care* are often used interchangeably, HMOs are only one form of managed care. Recipients of Medicaid and Medicare may also belong to a managed care plan.

Publically Funded Health Programs. Medicare, Medicaid, and the State Children's Health Insurance Program (SCHIP) are the major publicly funded health programs. **Medicare** is funded by the federal government and reimburses the elderly and the disabled for their health care (see also Chapter 12). Medicare consists of two separate programs: a hospital insurance program and a supplementary medical insurance program. The hospital insurance program is free, but enrollees may pay a deductible and a copayment, and coverage of home health nursing and hospice care is limited. Medicare's medical insurance program is not free; enrollees must pay a monthly premium as well as a copayment for services.

Medicare does not cover long-term nursing home care, dental care, eyeglasses, and other types of services, which is why many individuals who receive Medicare also purchase supplementary private insurance, known as *medigap* policies. The Medicare Prescription Drug, Improvement, and Modernization Act of 2003, signed into law by President G. W. Bush, contains a prescription drug benefit for the disabled and seniors; the benefits, available in 2006, will be offered by private insurers and health plans under contract with the government. Because the program is voluntary, beneficiaries pay a monthly premium (about $35) and meet an annual deductible of $275, as well as pay co-insurance of 25% for prescriptions up to a limit of $2,200. This co-insurance drops to 5% once an enrollee reaches an annual limit of $3,600. Critics of the Medicare prescription bill argue that the drug coverage is inadequate and complicated, denies beneficiaries the right to purchase supplemental coverage, and fails to lower the cost of prescription drugs. The Medicare prescription drug legislation provides billions of dollars in subsidies to HMOs and other managed care plans, paying them much more than it costs regular Medicare to provide the same services. These private plans can elect to cover a limited number of drugs and deny coverage for other drugs.

Medicaid, which provides health care coverage for the poor, is jointly funded by the federal and state governments (see also Chapter 6). Contrary to the belief that Medicaid covers all poor people, it does not. Eligibility rules and benefits vary from state to state, and in many states Medicaid provides health care only for the very poor who are well below the federal poverty level. In 2003 more than one-third (36%) of those living beneath the federal poverty level lacked health insurance (Kaiser Family Foundation 2005b).

In 1997 the **State Children's Health Insurance Program** (SCHIP) was created to expand health coverage to uninsured children, many of whom come from families with incomes too high to qualify for Medicaid but too low to afford private health insurance. Under this initiative states receive matching federal funds to provide medical insurance to uninsured children.

Inadequate Health Insurance Coverage

Virtually all elderly Americans have Medicare coverage, and more than half (59.8%) of nonelderly Americans receive health insurance coverage through their employers. But in 2004, 15.7% of the U.S. population—45.8 million Americans—lacked health insurance (DeNavas-Walt et al. 2005). The number of uninsured Americans is expected to grow to 56 million by 2013 (Gilmore & Kronick 2005). The main reason for this increase in the number of uninsured individuals is reductions in employer-sponsored health insurance coverage (Ku et al. 2005).

High rates of uninsurance are especially problematic among racial and ethnic minorities and among young adults. As noted earlier in this chapter, in 2004 one-third of Hispanic people (32.7%) were uninsured, followed by blacks (19.7%), and Asians (18.8%). The rate of uninsurance among non-Hispanic whites was 11.3%. Of all age groups, young adults ages 18 to 24 are the least likely to have health insurance. In 2004 nearly 1 in 3 young adults (31.4%) were uninsured (DeNavas-Walt et al. 2005).

Individuals who lack health insurance are less likely to receive preventive care, are more likely to be hospitalized for avoidable health problems, and are more likely to be diagnosed in the late stages of disease (Kaiser Commission on Medicaid and the Uninsured 2004). As shown in Figure 2.2, individuals without insurance experience more barriers to health care than individuals with insurance. An estimated 18,000 deaths per year in the United States are attributable to lack of health insurance (Institute of Medicine 2004).

Medicaid—the public health insurance program for the poor—is also inadequate. Eligibility levels are set so low that many low-income adults are not eligible. Widespread state budget deficits have led most states to adopt policies to restrict Medicaid or SCHIP enrollment, and politicians have called for substantial federal budget cuts in Medicaid and/or the capping of federal Medicaid funding for states, further threatening the health care needs of low-income Americans. Some states have waiting lists for Medicaid. The *Philadelphia Inquirer* reported that Pennsylvania's state-subsidized health insurance for low-income adults, AdultBasic, had a waiting list of 100,000 in 2004, with the average wait being 16 months (reported by Skala 2005). Another problem is that because of the low reimbursement from Medicaid, many health care providers do not accept Medicaid patients.

It is worth noting that having health insurance does not guarantee that one is protected against financial devastation resulting from illness or injury. According to Families USA (2005), "the middle class, those with college degrees, decent jobs, health insurance—the group of people who feel secure and well-protected—are at

"Of all the forms of inequality, injustice in health care is the most shocking and inhumane."

Martin Luther King Jr.
Civil Rights Leader

Figure 2.2
Barriers to health care by insurance status, 2003.*

Source: Kaiser Family Foundation (2005b).

*Experienced by respondent or member of his or her family.

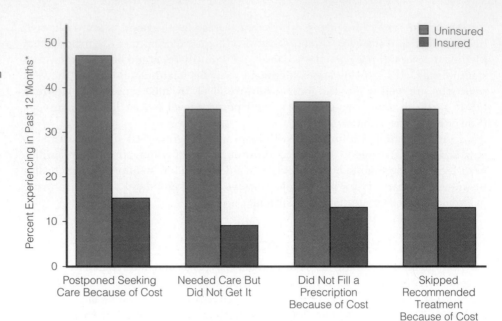

ThomsonNOW™

Learn more about **Inadequate Health Insurance Coverage** by going through the Health Insurance Map Exercise.

high, and often highest, risk of being left penniless when serious illness hits" (p. 1). In a 2005 study that found that half of U.S. bankruptcies are attributable, at least in part, to medical bills, most of those bankrupted (75%) were insured when they first became sick or injured (Himmelstein et al. 2005). One cause of the high rate of medical bankruptcy among the insured is the high cost of copayments, deductibles, and exclusions. In addition, the link between coverage and employment means that insurance is often lost when it is needed the most—when workers lose their jobs because of medical problems. Although the COBRA law allows people to continue their insurance coverage when they lose a job, the premiums for continued coverage are unaffordable (often $10,000 a year or more) (PNHP 2005).

The High Cost of Health Care

Larry Causey, age 57, called the FBI and told them he was going to rob the post office in West Monroe, Louisiana. Then he went to the post office and handed a note to a teller demanding money. He left empty-handed and sat in his car until officers arrested him. Larry had no intention of committing robbery. Larry had cancer and could not afford cancer treatment, so he staged a robbery in order to get arrested and be put in jail, where he would receive medical treatment for his cancer ("Access to Free Health Care," 2001). Larry Causey's story is not uncommon. "Sheriffs nationwide say they're also arresting people willing to trade their freedom for a free visit to the doctor" ("Access to Free Health Care," 2001).

In 2004 about three in four Americans said that they were very concerned about being unable to afford necessary health care when a family member gets sick (Kaiser Family Foundation 2005b). When a national sample of U.S. adults was asked to identify the most important problem in health or health care for the government to address, the most frequent response was the high cost of health care (Kaiser Family Foundation 2005a). Per capita health care expenditures in the United States increased from $2,738 in 1990 to $6,420 in 2005 (Baker & Rosnick 2005; Kaiser Family Foundation 2005b). The United States spends more than twice

Should Universities Require Students to Have Health Insurance?

Some college students have health insurance under either their parent's plan or their employer's plan or have purchased health insurance coverage through their college or university. But many college students are uninsured. Each year hundreds of college students withdraw from school because of their inability to pay medical bills from accidents or unexpected illnesses. About 25% of public universities require students to have health insurance compared to about 90% of private universities (Keith 2005). Although mandatory health insurance may keep students from dropping out, it also adds to college bills and may prevent some individuals who cannot afford health insurance from enrolling in college.

It Is Your Turn Now to Take a Stand!

Do you think that universities should require students to have health insurance? Why or why not?

as much per person for health care as other wealthy countries. Yet virtually every other wealthy nation has better health outcomes, as measured by life expectancy and infant mortality.

Several factors have contributed to escalating medical costs. These include the growing elderly population and the high costs of health care administration, drugs, doctors' fees, hospital services, medical technology, and health insurance.

The Growing Elderly Population. As discussed in Chapter 12, the elderly population in the United States is growing. Because people older than age 65 use medical services more than younger individuals do, the growing segment of the elderly population means that more money is spent on medical care. For example, a national sample of U.S. adults found that 84% of seniors age 65 and older take prescription medicine on a daily basis, compared to only 42% of adults ages 18 to 64 (Kaiser Family Foundation 2005a).

Cost of Health Care Administration. Health care administrative expenses are higher in the United States than in any other nation. According to the World Health Organization, 15% of the money paid to private health insurance companies for premiums goes to administrative expenses, compared to only 4% of the budgets of public insurance companies (i.e., Medicare and Medicaid) (reported by Krugman 2005). Insurance companies and for-profit HMOs spend between 20% and 30% of their budgets to cover the costs of stockholder dividends, lobbyists, huge executive salaries, marketing, and wasteful paperwork (Conyers 2003).

Cost of Drugs. The high cost of drugs also contributes to health care costs. The United States pays 81% more for patented brand-name prescription drugs than Canada and six Western European nations (Sager & Socolar 2004). The high prices that Americans pay for prescription drugs partly explain why the pharmaceutical industry was the most profitable industry in the United States from 1995 to 2002.

"Health care is one of our most basic needs. Making it available is perhaps our greatest test of humanity."

Linda Peeno
Physician

ThomsonNOW

Learn more about the **Growing Elderly Population** by going through the Aging Population Learning Module.

Problems in U.S. Health Care **55**

In 2003 drug companies were the third most profitable industry (following mining/crude oil production and banking) (Kaiser Family Foundation 2005b).

For consumers high drug costs have consequences not only for people's wallets but also for their health. In a national sample of U.S. adults more than one-third said that in the past year they did not fill a prescription because of the cost, cut pills in half, or skipped doses in order to make a medication last (Kaiser Family Foundation 2005a).

Cost of Hospital Services, Doctors' Fees, and Medical Technology. High hospital costs and doctors' fees are factors in the rising costs of health care. For example, the average visit to the emergency room costs a little more than $1,000 (Ehrenreich 2005). The use of expensive medical technology, unavailable just decades ago, also contributes to high medical bills.

Cost of Health Insurance. High costs of public and private insurance have also contributed to escalating health care expenditures. The average monthly cost of a family health plan in 2004 was $829 ($9,950 annually); for an individual the cost was $308 per month ($3,695 annually) (Kaiser Family Foundation 2004b). In 2004 workers who had health insurance provided by their employer paid 16% of the premium costs for an individual ($47 monthly) and 28% for family coverage ($222 monthly). With the rising cost of medical insurance, companies are increasing the employees' share of the cost, decreasing the benefits, or not providing insurance at all. The cost of health care to businesses also affects the prices that consumers pay for goods and services. For example, the cost of health benefits for employees at General Motors accounts for $1,500 of the sticker price of each new vehicle (Ehrenreich 2005).

At the time of this writing, the United States was focused on how to reform Social Security so that this program, which provides income to the elderly and the disabled, would not run out of money (see Chapter 12). Yet the amount of tax increase necessary to keep Social Security solvent is dwarfed by the impact of rising health care costs on the living standards of U.S. workers (Baker & Rosnick 2005).

In 2003, 18,000 General Electric workers in 23 states went on strike to protest an increase in health insurance copayments.

© AP/Wide World Photos

If the prospect of increased Social Security taxes is seen as posing a crisis, then, according to Baker and Rosnick (2005),

> the burden of rising health care costs must be viewed as an even more cataclysmic event. Politicians and commentators who claim to be concerned about the living standards of future generations of workers seem to be misdirecting their energy by focusing on the comparatively minor problem of Social Security. Clearly the inefficiency of the U.S. health care system poses a far larger and more immediate danger to the livings standards of our children and grandchildren. (p. 6)

The Managed Care Crisis

Increasingly, Americans are concerned about the reduced quality of health care resulting from the emphasis on cost containment in managed care. Surveys done each year from 1997 to 2004 found that more people have said that managed care plans do a "bad job" than a "good job" in serving customers (Kaiser Health Poll Report 2004). In a survey of physicians' views on the effects of managed care, the majority responded that managed care has negative effects on the quality of patient care because of limitations on diagnostic tests, length of hospital stay, and choice of specialists (Feldman et al. 1998). One former director of an HMO described the situation:

> I've seen from the inside how managed care works. I've been pressured to deny care, even when it was necessary. I have seen the bonus checks given to nurses and doctors for their denials. I have seen the medical policies that keep patients from getting care they need . . . and the inadequate appeal procedures. (Peeno 2000, p. 20)

Inadequate Mental Health Care

Since the 1960s, U.S. mental health policy has focused on reducing costly and often neglectful institutional care and on providing more humane services in the community. This movement, known as **deinstitutionalization,** had good intentions but has largely failed to live up to its promises. Only one in five U.S. children with mental illness is identified and receives treatment, and fewer than half of adults with a serious mental illness received treatment or counseling for a mental health problem during the past year (Gruttadaro 2005; Substance Abuse and Mental Health Services Administration 2003).

Reasons for not seeking treatment include the stigma associated with mental illness (and thus with treatment for mental illness), fear and mistrust of treatment, cost of care (which is often not covered by health insurance), and lack of access to services. In 2004, although most workers with health insurance had inpatient and outpatient mental health benefits, they typically faced limits on the number of mental health visits and inpatient mental health days covered by their health plans (typically no more than 30 visits for outpatient visits and 21–30 days for inpatient stays) (Kaiser Family Foundation 2005b). Families of children with mental disorders often face the heart-wrenching choice of not receiving adequate mental health services for their children or relinquishing custody of the children in order to qualify for Medicaid (Lehmann 2003).

The mental health system is also plagued by inadequate funding. Medicaid provides more than half of funding for public mental health services, and every state has cut or plans to cut Medicaid funds. Inadequate federal and state funding of public mental health centers results in rationing care to those most in need. Thus people must "hit bottom" before they can receive services.

Mental health services are often inaccessible, especially in rural areas. In most states services are available from "9 to 5"; the system is "closed" in the evenings and on weekends when many people with mental illness experience the greatest need. Across the nation people with severe mental illness end up in jails and prisons, homeless shelters, and hospital emergency rooms. Many children with untreated mental disorders drop out of school or end up in foster care or the juvenile justice system. As much as 70% of youths incarcerated in juvenile justice facilities suffer from mental disorders (Honberg 2005). Given the increasing growth of minority populations, another deficit in the mental health system is the inadequate number of mental health clinicians who speak the client's language and who have awareness of cultural norms and values of minority populations (U.S. Department of Health and Human Services 2001).

ThomsonNOW™

Learn more about **U.S. Health Care: An Overview** by going through the Health Care in the United States Learning Module.

Strategies for Action: Improving Health and Health Care

There are two broad approaches to improving the health of populations: selective primary health care and comprehensive primary health care (Sanders & Chopra 2003). **Selective primary health care** uses technocratic solutions to target a specific health problem, such as using immunization and oral rehydration therapy to promote child survival. In contrast, **comprehensive primary health care** focuses on the broader social determinants of health, such as poverty and economic inequality, gender inequality, environment, and community development.

In recent decades the health agenda has focused on selective primary health care approaches, such as child immunization, which increased from 20% (of children under age 1) in 1980 to 80% by 1990 (Sanders & Chopra 2003). However, more recent declines in immunization coverage in some regions along with rising infant mortality rates provide evidence that selective primary health care produces "short term but unsustainable results" (Sanders & Chopra 2003, p. 107).

Targeting specific health problems may be necessary, but not sufficient, for achieving long-term health gains. Sanders and Chopra (2003) emphasize that "only where health interventions are embedded within a comprehensive health care approach, including attention to social equity, health systems and human capacity development, can real and sustainable improvements in health status be seen" (p. 108).

As you read the following sections on improving maternal and infant health, preventing and alleviating HIV/AIDS, and fighting obesity, see whether you can identify which strategies represent selective primary health care approaches and which strategies are comprehensive. Also, be mindful that strategies to alleviate social problems discussed in subsequent chapters of this textbook are also important elements to a comprehensive primary health care approach.

Although we focus on governmental, organizational, and cultural strategies for improving health and health care, another strategy is for individuals to take a proactive role in their health and health care by making careful lifestyle choices and by being informed about health and medical issues. This chapter's *Focus on Technology* feature looks at the increasing trend among Americans to use the Internet to seek health and medical information and support.

Improving Maternal and Infant Health

As discussed earlier, pregnancy and childbirth are major causes of mortality and morbidity among women of reproductive age in the developing world. In 1987 the Safe Motherhood Initiative was launched to improve maternal health. This global initiative involves a partnership of governments, nongovernmental organizations, agencies, donors, and women's health advocates working to protect women's health and lives, especially during pregnancy and childbirth.

In many developing countries women's lack of power and status means that they have little say over their reproductive health (Murphy 2003). Men make the decisions about whether or when their wives (or partners) will have sexual relations, use contraception, or use health services. Thus improving the status and power of women is an important strategy in improving their health. Promoting women's education increases the status and power of women to control their reproductive lives, exposes women to information about health issues, and also delays marriage and childbearing. Around the world women carry disproportionate responsibility for reproductive health and family size. And although women receive the bulk of reproductive health education, including family planning information, gender dynamics can render women powerless to make decisions. Men often hold decision-making power over matters as basic as sexual relations and when and whether to have a child or even seek health care. Although most reproductive health programs focus exclusively on women, some programs also reach out to men with services and education that enable them to share in the responsibility for reproductive health.

Access to family planning services, skilled birth attendants, affordable methods of contraception, and safe abortion services are important determinants of the well-being of mothers and their children (Save the Children 2002). Family planning reduces maternal mortality simply by reducing the number of unintended pregnancies. Spacing births two to three years apart decreases infant mortality significantly (Murphy 2003). Since 1960 contraception use among married couples in developing countries has increased from 10–15% to 60% (United Nations Population Fund 2004), but there are still millions of women who do not have access to contraception.

Improving maternal and infant health globally requires funding, but the cost is not prohibitive. The price of providing basic health services for mothers and infants in low-income countries is only $3 per person (Oxfam GB 2004). The question is, Do the rich countries of the world have the political will to support efforts to protect the health and lives of women and infants in the developing world?

ThomsonNOW™

Learn more about **Improving Maternal and Infant Health** by going through the Contraceptives Map Exercise.

"Children are the future of society, and their mothers are the guardians of that future."

World Health Report 2005 (World Health Organization 2005)

HIV/AIDS Prevention and Alleviation Strategies

As of this writing, there is no vaccine to prevent HIV infection. As researchers continue to work on developing such a vaccine, a number of other strategies are available to help prevent and treat HIV/AIDS.

HIV/AIDS Education and Access to Condoms. HIV/AIDS prevention efforts include educating populations about how HIV is transmitted and how to protect against HIV transmission and providing access to condoms as a means of preventing HIV transmission. Many people throughout the world remain uninformed or misinformed about HIV/AIDS. At least 30% of young people in a survey of 22 countries had never heard of AIDS, and in 17 countries surveyed, more than half of adolescents could not name a single method of protecting themselves against HIV infection (United Nations Population fund 2002). A survey of U.S. adults found that significant numbers say that they do not know how HIV is transmitted or that it is possible to transmit HIV through kissing (38%), sharing a drinking glass (18%), and touching a toilet seat (18%) (Kaiser Family Foundation 2004b). This same survey also found that 12% of U.S. adults did not know that there are drugs that can lengthen the lives of those with HIV and that more than half (57%) did not know

Americans' Use of the Internet to Find Health Information

The Internet has revolutionized the way in which Americans participate in their health care. Two national surveys of U.S. adults found that 80% of adult Internet users, or half the total adult U.S. population, have used the Internet to search for information about health or medical care (Fox 2005; Fox & Fallows 2003). Although most Internet users have conducted health searches, most do so infrequently—every few months or less. More than half of those who recently conducted searches did so on behalf of someone else—a spouse, child, friend, or other loved one—not for themselves (Fox & Fallows 2003). The most common health topics searched by Internet users include the following (Fox 2005):

- specific disease or medical problem (66%)
- certain medical treatment or procedure (51%)
- diet, nutrition, vitamins, or nutritional supplements (51%)
- exercise or fitness (42%)
- prescription or over-the-counter drugs (40%)
- health insurance (31%)
- alternative treatments or medicines (30%)

In addition to information searches, Internet users are increasingly going to disease- or situation-specific support sites and using e-mail to discuss health issues with family and friends. A small number (7%) of survey respondents reported using e-mail to communicate with their doctors regarding appointments, prescriptions, and basic questions (Fox & Fallow 2003). Indeed, many insurance companies are beginning to cover e-mail consultations between physicians and patients (Freudenheim 2005).

Women are more likely than men to seek health care and health information, so it is no surprise that more women than men have searched for health-related information on the Internet. College graduates are also more likely than less educated adults to have searched online for health information.

Three-quarters of Americans who seek online health information say that the Internet has improved the health information and services they receive. Online survey respondents who used the Internet to search for health information reported the following (Fox & Fallows 2003):

- "Information available on the Internet takes the mystery out of illness and gives the patient a sense of power over his/her condition" (p. 7).
- "As the parent of a child with a very rare neurological syndrome, the Internet was vital to putting the pieces of a puzzle together. It saved my son months of struggle when I found a diagnosis prior to the neurologist he was seeing, who openly admitted she had only heard of the syndrome but never treated a child with [Landau-Kleffner Syndrome]" (p. 7).
- A survey respondent living with depression and borderline personality disorder wrote, "Doctors do not spend enough time explaining or consoling. Reading and learning exactly what my conditions are helps me to cope" (p. 9).
- "The doctors do not have the time to remain current in the information about every disorder, so I give the doc the cutting edge information on mine so that I can benefit from new thoughts and therapies" (p. 9).
- "I live in a manufactured home purchased new in 1997. . . . I was diagnosed [with porphyria] in 1998. It was only through the Internet that I was able to learn that the house was making me very ill. . . . It was through these sources that I was able to learn what can be done to make the house safer for me" (pp. 9–10). (Porphyria is a group of disorders characterized by the accumulation of natural chemicals called "porphyrins" or "porphyrin precursors." Environmental factors can affect the severity of symptoms, which show up in the nervous system or on skin.)
- Most patients who are online use the Internet to search for health information

that a pregnant woman with HIV can take medication to reduce the risk of her baby being born with HIV infection.

HIV/AIDS education occurs in a variety of ways, including through media and public service announcements, faith-based groups, health care providers, and schools. Nearly 100% of U.S. parents of junior or senior high school students believe that HIV/AIDS is an appropriate topic for school sexuality programs (SIECUS 2004).

when they are concerned about a specific symptom or diagnosis. One survey respondent wrote, "My mother-in-law suddenly began bruising very badly. Medical personnel simply said it was because she was old. I was not satisfied with this answer because the onset was so sudden. Turning to the Internet, I found it was because her aspirin dosage was too high" (p. 11). After consulting with a doctor, this caregiver lowered the aspirin dosage and gave her mother-in-law vitamin C tablets, and the bruising cleared up within weeks.

- Individuals with various health conditions, and their caregivers, often use e-mail to connect to emotional support. One survey respondent wrote, "The Internet put me in contact with others that KNOW what it was and is like living with a disease that can be disabling with many odd symptoms" (p. 12). And a mother wrote, "Emailing with other mothers of special needs children has been a great way for me to feel linked with women who know what I'm going through" (p. 12).

- People also use the Internet as a way to give as well as receive health information and support. One respondent wrote, "As an old timer, I'll help another MS patient with questions that I might know the answer to" (p. 13).

Using the Internet to access health information and support has been helpful and empowering to countless individuals. But Internet users would be well advised to follow certain guidelines when searching for health information on the Internet. Some of these guidelines are listed here (Fox & Rainie 2002):

1. The Medical Library Association suggests that, when you use a search engine such as Google or Yahoo, you take advantage of the health subsets of these services and learn how to use the advanced searching features of the sites so that you can combine terms to narrow your search. For example, entering the terms *cancer* and *chemotherapy* linked together is more powerful and precise than trying to read through all the hits found by simply entering the general terms.

2. Consider starting your search by using credible general health information finding tools such as MedlinePlus (http://www.medlineplus.gov), produced by the National Library of Medicine, or Healthfinder (http://www.healthfinder.gov), from the U.S. Department of Health and Human Services. The Medical Library Association's "Top Ten" list (http://www.mlanet.org/resources/medspeak/topten.html) is a resource that lists quality health information sites.

3. Pay attention to the sponsor of the websites you visit. Sponsorship is important because it helps establish the site as respected and dependable. Commercial sites (with Web addresses that end with ".com") may represent a specific company or be sponsored by a company using the Web to sell products.

4. Check the date of the information on the website. Health information changes constantly as new information is learned about diseases and treatments. Make sure you visit websites that reflect the most up-to-date information. The date of the latest revision usually appears at the bottom of the page.

5. Make sure the information on the website is factual (not opinion) and can be verified from another source, such as the professional literature, abstracts, or links to other Web pages.

6. The California Health Care Foundation recommends that consumers take ample time to search for health advice, visit four to six websites, and discuss the information with a health care provider before making a treatment decision.

7. Advocates for privacy, such as the Center for Democracy and Technology, recommend reading a website's privacy policy carefully.

With the HIV infection rate growing among the over-50 population, HIV/AIDS education also takes place in some senior centers (Goldberg 2005).

Some HIV/AIDS education is based on the ABC approach—a prevention strategy that involves three elements (Halperin et al. 2004):

A = Abstain. Young people who have not started sexual activity should be encouraged to abstain from or delay sexual activity to prevent HIV and other sexually transmissible infections as well as unwanted pregnancy.

B = Be faithful/reduce partners. After individuals become sexually active, returning to abstinence or remaining faithful to an uninfected partner are the most effective ways to avoid HIV infection.

C = Use condoms. People who have a sexual partner of unknown HIV status should be encouraged to practice correct and consistent use of condoms.

Providing education that advocates condom use and providing youth with access to condoms are controversial topics. Many conservatives believe that promoting condoms sends the "wrong message" that sex outside marriage is OK. Consequently, under the Bush administration federal support for "abstinence-only" education programs, which promote abstinence from sexual activity without teaching basic facts about contraception or providing access to contraception, has expanded rapidly. But abstinence-only programs are criticized for failing to provide youth with potentially life-saving information. A report released by Representative Henry Waxman found that more than 80% of the abstinence-only curricula contains false, misleading, or distorted information about reproductive health (Waxman 2004). For example, the report found that

> many of the curricula misrepresent the effectiveness of condoms in preventing sexually transmitted diseases and pregnancy. One curriculum says that "the popular claim that 'condoms help prevent the spread of STDs,' is not supported by the data"; another states that "[i]n heterosexual sex, condoms fail to prevent HIV approximately 31% of the time". . . . These erroneous statements are presented as proven scientific facts. (Waxman 2004, p. i).

Another controversy involves the question of whether to provide condoms to prison inmates. Vermont and Mississippi allow condom distribution in prisons, as do Canada, most of Western Europe, and parts of Latin America. One deputy at the Los Angeles Sheriff's Department, which allows only homosexual inmates to receive condoms provided by a local nonprofit organization, said, "We're not promoting sex; we're promoting health" (Sanders 2005).

HIV Testing. Another strategy to curb the spread of HIV involves encouraging individuals to get tested for HIV so that they can modify their behavior (to avoid transmitting the virus to others) and so that they can receive early medical intervention, which can slow or prevent the onset of AIDS. An estimated one-fourth to one-third of HIV-infected Americans do not know that they are infected (Kaiser Family Foundation 2004a). About half (48%) of Americans report ever having been tested for HIV, including 20% who say they have been tested in the last year (Kaiser Family Foundation 2004c). Unfortunately, many individuals who have been diagnosed with HIV infection continue to engage in risky behaviors, such as unprotected anal, genital, or oral sex and needle sharing (Diamond & Buskin 2000; Hollander 2005).

The Fight Against HIV/AIDS Stigma and Discrimination. HIV/AIDS-related stigma stems from societal views that people with HIV/AIDS are immoral and shameful. The stigma associated with HIV/AIDS results in discrimination that can lead to loss of employment and housing, social ostracism and rejection, and lack of access to medical care. A survey of 1,000 physicians and nurses in Nigeria found that 1 in 10 admitted to refusing care for an HIV/AIDS patient or had denied HIV/AIDS patients admission to a hospital, and 20% believed that people living with HIV/AIDS have behaved immorally and deserved their fate (*HIV and AIDS Stigma and Discrimination,* 2004). Some societies practice compulsory HIV testing with prior consent or protection of confidentiality. And the stigma surrounding

HIV/AIDS has also led to acts of violence against people perceived to be infected with HIV.

HIV/AIDS stigma and discrimination can deter people from getting tested for the disease, can make them less likely to acknowledge their risk of infection, and can discourage those who are HIV-positive from discussing their HIV status with their sexual and needle-sharing partners. Combating the stigma and discrimination against people who are affected by HIV/AIDS is crucial to improving care, quality of life, and emotional health for people living with HIV and AIDS and to reducing the number of new HIV infections.

Former South African president Nelson Mandela's announcement in 2005 that his son, Makgatho Mandela, 54, had died of AIDS was a public attempt to fight the stigma associated with HIV/AIDS (Timberg 2005). The Miss HIV Stigma Free beauty pageant, first held in 2002 in Botswana, where more than one-third of adults are infected with HIV, combats stigma by showing that HIV-infected individuals need not be ashamed and that with treatment they can lead productive lives (Goering 2005).

Per-Anders Pettersson/Getty Images

Cynthia Leshomo is the 2005 winner of the Miss HIV Stigma Free beauty pageant for HIV-positive women. First held in 2002, the Miss HIV pageant is a way of showing that HIV-positive individuals need not be ashamed and that with treatment, they can look good and lead productive lives.

Needle Exchange Programs. Injection drug use accounts for most HIV cases in China, Russia, Iran, Afghanistan, Nepal, the Baltic states, and all of Central Asia as well as much of Southeast Asia and South America (Human Rights Watch 2005). To reduce transmission of HIV among injection drug users, their sex partners, and their children, some countries and U.S. communities have established **needle exchange programs** (also known as syringe exchange programs), which provide new, sterile syringes in exchange for used, contaminated syringes. Many needle exchange programs also provide other social and health services, such as referrals to drug counseling and treatment, HIV testing and screening for other sexually transmissible diseases, hepatitis vaccinations, and condoms. Needle exchange has been endorsed as an effective means of HIV prevention by the American Medical Association, the American Public Health Association, and the World Health Organization. Needle exchange programs also protect public health by providing safe disposal of potentially infectious syringes.

In Canada sterile injection equipment is available to drug users in pharmacies and through numerous needle exchange programs. In contrast, most U.S. states prohibit the sale or possession of sterile needles or syringes without a medical prescription. In 2004, 184 needle exchange programs were operating in 36 states. Less than half of these programs receive public funding from local and/or state government (Centers for Disease Control and Prevention 2005c). The United States is the only country in the world to explicitly ban the use of federal funds for needle exchange (Human Rights Watch 2005).

Financial and Medical Aid to Developing Countries. In the first four years after antiretroviral medication was introduced in Europe and North America, death rates for HIV/AIDS in those regions fell by 80%, and under Brazil's program to provide universal access to antiretroviral therapy, the average survival time of people with AIDS rose from less than 6 months to at least 5 years (World Health Organization

2004). But life-extending treatment for individuals infected with HIV is not affordable for many people in the developing world. In 2003, 6 million people in developing countries needed antiretroviral medication to fight HIV infection, but only about 15% of them received it (World Health Organization 2004). The risk of HIV transmission to infants can be reduced by more than 50% in mothers receiving antiretroviral therapy. Without this needed medication, not only do HIV-infected mothers die sooner, but they also risk transmitting the virus to their infants.

Developing countries—those hardest hit by HIV/AIDS—depend on aid from wealthier countries to help provide medications, HIV/AIDS education programs, and condoms. In 2002 the United Nations helped to create the Global Fund to Fight AIDS, Tuberculosis, and Malaria to help poor countries fight these diseases. But the biggest obstacle to fighting AIDS in Africa may not be lack of money but lack of health care personnel. At an Ethiopian hospital that serves the bulk of the country's patients on AIDS medication, two doctors and two nurses care for roughly 2,000 people. By contrast, the United States has 15 nurses for the same number of patients (Rosenberg 2005). The shortage of doctors and nurses in developing countries is partly due to the mass emigration of health professionals to wealthy countries. Reducing emigration by improving working conditions and wages for health workers in developing countries is an important piece of the fight against HIV/AIDS.

Commenting on HIV/AIDS prevention, Altman (2003) noted, "The great irony is that we know how to prevent HIV transmission and it is neither technically difficult nor expensive. Most HIV transmission can be stopped by the widespread use of condoms and clean needles" (p. 41). Implementing these strategies, however, conflicts with religious and cultural beliefs and threatens the political power structure. Altman explained that "effective HIV prevention requires governments to acknowledge a whole set of behaviours—drug use, 'promiscuity,' homosexuality, commercial sex work—which they would often rather ignore, and a willingness to support, and indeed empower, groups practicing such behaviours" (p. 42).

Fighting the Growing Problem of Obesity

In general, reducing and preventing obesity requires encouraging people to (1) eat a diet with sensible portions, with lots of high-fiber fruits and vegetables and with minimal sugar and fat, and (2) engage in regular physical activity. Some of the programs and strategies that attempt to achieve these goals include the following:

- *School nutrition programs.* Observe students in U.S. school cafeterias across the country, and you will likely see many students eating french fries, chips, sugary soft drinks, and other high-fat, high-sugar foods. Some schools are implementing nutrition programs that restrict the availability of "junk food" and provide nutrition education to students. In 2004 Arkansas became the first state to pass legislation banning vending machines (which sell soft drinks and snacks) from elementary schools (Mui 2005). One study of 5,200 fifth graders found that students from schools that participated in school-based healthy eating programs exhibited significantly lower rates of overweight and obesity, had healthier diets, and reported more physical activities than students from schools without nutrition programs (Veugelers & Fitzgerald 2005).

- *Interventions to treat obesity.* Interventions to treat obesity include weight-loss or fitness clubs, nutrition and weight-loss counseling, weight-loss medications, and surgeries. In 2004 the U.S. Department of Health and Human Services announced that the Centers for Medicare and Medicaid Services would remove language in Medicare's coverage manual that states that obesity is not an illness (Stein & Connolly 2004). Classifying obesity as an illness means that treatment for obesity, ranging from joining weight-loss or fitness clubs to surgeries and counseling, can be covered by Medicare. Increasingly, private insurers are also covering the cost of weight-loss treatment for obese individuals.

• *Legislation to reduce obesity.* Some politicians are pushing for legislation that would promote exercise in schools, require nutritional labeling on chain restaurant menus, and give the Federal Trade Commission power to restrict food advertising that targets children (Stein & Connolly 2004).

U.S. Health Care Reform

The United States is the only country in the industrialized world that does not have any mechanism for guaranteeing health care to its citizens. Other countries, such as Canada, Great Britain, Sweden, Germany, and Italy, have national health insurance systems, also referred to as *socialized medicine* and *universal health care.* **Socialized medicine,** or **universal health care,** refers to a state-supported system of health care delivery in which health care is purchased by the government and sold to the consumer at little or no additional cost. Despite differences in how socialized medicine works in various countries, in all systems of socialized medicine the government (1) directly controls the financing and organization of health services, (2) directly pays providers, (3) owns most of the medical facilities (Canada is an exception), (4) guarantees equal access to health care, and (5) allows private care for individuals who are willing to pay for their medical expenses (Cockerham 2004). Most countries with national health insurance allow or encourage private insurance as an upgrade to a higher class of service and a fuller range of services (Quadagno 2004). To the extent that health care is rationed in countries with national health insurance, rationing is done on the basis of medical need, not ability to pay.

Since 1912, when Theodore Roosevelt first proposed a national health insurance plan, the idea of health care for all Americans has been advocated by the Truman, Nixon, Carter, and Clinton administrations. A 2002 Gallup poll found that nearly two-thirds (62%) of U.S. adults believe that the federal government is responsible for guaranteeing health care coverage for all Americans (Gallup Poll Social Series 2002). Currently, advocates are promoting a national health insurance bill that would create an improved and expanded Medicare for All program. This program would create a single-payer system in which a single tax-financed public insurance program replaces private insurance companies. Under this plan every U.S. resident would be issued a national health insurance card, would receive all medically necessary services (including prescription drugs and long-term care), would have no copayments or deductibles, and would see the doctor of his or her choice (Conyers 2003). If this plan is adopted, it is estimated to save approximately $100 billion per year on administrative costs—enough to provide coverage for all the uninsured and to substantially help the underinsured. The proposed single-payer program would use the already existing Medicare program by expanding it to cover all U.S. residents, and it would provide comprehensive coverage (including dentistry, eye care, mental health services, substance abuse treatment, prescription drugs, and long-term care) with no copayments or deductibles.

The insurance industry, not surprisingly, opposes the adoption of such a system because the private health insurance industry would be virtually eliminated. The health insurance industry's opposition to a single-payer universal health plan is matched only by the persistent efforts of those who advocate such a plan. Bills for single-payer health programs have been introduced in a number of states, including California, Hawaii, Minnesota, New Mexico, Delaware, Massachusetts, Wisconsin, and Maine. At the federal level Representative Pete Stark (D-California) proposed an amendment to the U.S. Constitution to guarantee health care as a right for every American. The proposed amendment would say that "all persons shall enjoy the right to health care of equal high quality." Stark argued that "the health of every American is vital to their unalienable rights of 'life, liberty, and the pursuit of happiness' . . . To ensure these rights are fully enjoyed, we must be certain that every American can access quality health care—regardless of their income, race, education or job status" ("Stark Introduces Constitutional Amendment," 2005).

Strategies to Improve Mental Health Care

In 1999 the first White House conference on mental health and the first surgeon general's report on mental health established mental health care as a national health priority. Two areas for improving mental health care in the United States are eliminating the stigma associated with mental illness and improving health insurance coverage for treating mental disorders. Efforts in these areas will, it is hoped, promote the delivery of treatment to individuals suffering from mental illness. Although most mental disorders can be successfully treated with medications and/or psychotherapy or counseling, nearly half of all Americans who have a severe mental illness do not seek treatment (U.S. Department of Health and Human Services 1999).

Eliminating the Stigma of Mental Illness. The first White House conference on mental health called for a national campaign to eliminate the stigma associated with mental illness. Fearing the negative label of "mental illness" and the social rejection and stigmatization associated with mental illness, individuals are reluctant to seek psychological services (Komiya et al. 2000). In addition, "stigma deters the public from wanting to pay for care and, thus, reduces consumers' access to resources and opportunities for treatment and social services" (U.S. Department of Health and Human Services 1999, p. viii). To assess your attitudes toward seeking professional psychological help, see this chapter's *Self and Society* feature.

Reducing stigma associated with mental illness might be achieved through encouraging individuals to seek treatment and making treatment accessible and affordable. The surgeon general's report on mental health explains that

> effective treatment for mental disorders promises to be the most effective antidote to stigma. Effective interventions help people to understand that mental disorders are not character flaws but are legitimate illnesses that respond to specific treatments, just as other health conditions respond to medical interventions. (U.S. Department of Health and Human Services 1999, p. viii)

The National Alliance for the Mentally Ill (NAMI) has a StigmaBusters campaign, whereby the public submits instances of media content that stigmatize individuals with mental illness to StigmaBusters, which then investigates and conveys their concerns to media organizations and corporations, urging them to avoid stigmatizing portrayals of mental illness. For example, when McDonalds aired a radio commercial in 2004 stating, "I am hearing voices about the food at McDonald's. I am hearing voices telling me to go to McDonald's. I better listen to them. I am not crazy!" StigmaBusters wrote to McDonalds expressing disapproval of the ad (Stigma Busters Alert 2004).

"Americans assign high priority to preventing disease and promoting personal well-being and public health; so too must we assign priority to the task of promoting mental health and preventing mental disorders."

Mental Health: A Report of the Surgeon General (U.S. Department of Health and Human Services 1999)

Attitudes Toward Seeking Professional Psychological Help

For each of the following statements, indicate your level of agreement according to the following scale: agree, partly agree, partly disagree, or disagree.

1. If I believed I was having a mental breakdown, my first inclination would be to get professional attention. (S) _____

2. The idea of talking about problems with a psychologist strikes me as a poor way to get rid of emotional conflicts. (R) _____

3. If I were experiencing a serious emotional crisis at this point in my life, I would be confident that I could find relief in psychotherapy. (S) _____

4. There is something admirable in the attitude of a person who is willing to cope with his or her conflicts and fears *without* resorting to professional help. (S) _____

5. I would want to get psychological help if I were worried or upset for a long period of time. (S) _____

6. I might want to have psychological counseling in the future. (S) _____

7. A person with an emotional problem is not likely to solve it alone; he or she *is* likely to solve it with professional help. (S) _____

8. Considering the time and expense involved in psychotherapy, it would have doubtful value for a person like me. (R) _____

9. A person should work out his or her own problems; getting psychological counseling would be a last resort. (R) _____

10. Personal and emotional troubles, like many things, tend to work out by themselves. (R) _____

Scoring: Straight items (S) are scored 3, 2, 1, 0, and reversal items (R) are scored 0, 1, 2, 3, respectively, for the response alternatives *agree, partly agree, partly disagree,* and *disagree.* Sum the responses for each item to find your total score, which will range from 0 to 30. The higher your score, the more positive your attitudes toward seeking professional psychological help; the lower your score, the more negative your attitudes toward seeking professional psychological help.

In a study of 389 undergraduate students (primarily 18-year-old freshmen), the average score on this scale was 17.45 (Fischer & Farina 1995). Women were more likely to have higher scores, reflecting more positive attitudes. For women in the study the average score was 19.08, compared to 15.46 for men (Fischer & Farina 1995).

Reprinted by permission of the American College Personnel Association.

Another tool used to reduce the stigma of mental illness is Breaking the Silence, a curriculum for elementary, middle, and high schools available through NAMI. Using true stories, activities, a board game, and posters, Breaking the Silence debunks myths about mental illness and sensitizes students to the pain that words such as *psycho* and *schizo* and frightening or comic media images of mentally ill people can cause (Harrison 2002).

Eliminating Inequalities in Health Care Coverage for Mental Disorders. Another priority on the agenda to improve the nation's mental health care system involves eliminating the inequalities in health care coverage for mental disorders. The Mental Health Parity Act of 1996 was an important step in ending health care discrimination against individuals with mental illnesses by requiring equality between mental health care insurance coverage and other health care coverage—a concept known as **parity.** The act requires that insurers that offer mental health coverage provide benefits equivalent to other health issues, but the act does *not* require group health plans to include mental health coverage in their benefits package.

Some states have enacted mental health parity laws, which vary in their scope and application, but many of these laws do not address substance abuse, are limited to the more serious mental illnesses, or apply only to government employees.

Understanding Illness and the Health Care Crisis

Although human health has probably improved more over the past half century than over the previous three millennia, the gap in health between rich and poor remains wide and the very poor suffer appallingly (Feachum 2000). Poor countries need economic and material assistance to alleviate such problems as HIV/AIDS, high maternal and infant mortality rates, and malaria. The wealthy countries of the world do have resources to make a difference in the health of the world. When a tsunami took the lives of thousands of people in December 2004, media attention to the tragedy elicited an outpouring of aid throughout the world to help affected regions. Meanwhile, malaria, which has been referred to as "a silent tsunami," takes the lives of more than 150,000 African children each month, which is about the same as the death toll of the South Asian tsunami disaster (Sachs 2005). Although deaths from malaria dwarf the deaths resulting from the 2004 tsunami, malaria continues to receive comparatively scant public attention. Sachs (2005) noted that malaria, unlike a tsunami, is largely preventable and treatable. "A $5 mosquito bed net . . . could dramatically lower the rate of malaria illness and death. . . . Medicines, at roughly $1 per dose, could treat the cases that slip by the bed nets. Yet Africa's poverty is so extreme that . . . families . . . can't afford the few dollars they would cost" (Sachs 2005, p. A17).

Although poverty may be the most powerful social factor affecting health, other social factors that affect health include globalization, aging of the population, family structure, gender, education, and race or ethnicity. But U.S. cultural values and beliefs view health and illness as determined by individual behavior and lifestyle choices rather than by social, economic, and political forces. We agree that an individual's health is affected by the choices he or she makes—choices such as

In 2005, the California State Assembly and the state senate approved legislation banning the sale of soda and junk food in public schools. This legislation, the first of its kind nationwide, is an example of a social action designed to curb the public health problem of obesity.

Bill Aron/PhotoEdit, Inc.

whether to smoke, exercise, engage in risky sexual activity, use condoms, wear a seatbelt, and so on. However, the choices that individuals make are influenced by social, economic, and political forces that must be taken into account if the goal is to improve the health not only of individuals but also of entire populations.

By focusing on individual behaviors that affect health and illness, we often overlook social causes of health problems (Link & Phelan 2001). One of the most pressing social causes of health problems is environmental pollution and degradation. The Millennium Ecosystem Assessment (2005) synthesis report warns that pollution and depletion of the earth's natural resources are eroding ecosystems and could lead to an increase in existing diseases such as malaria and cholera as well as to an increased risk of new diseases emerging (see Chapter 14 for more on the health effects of environmental problems).

A sociological view of health and illness not only emphasizes the social causes of health problems but also looks for social solutions. For example, the consumption of trans fats (a component of partially hydrogenated vegetable oils) increases the risk of heart disease, the leading cause of death in the United States. Social solutions for fighting heart disease include (1) Denmark's ban on all processed foods containing more than 2% trans fat for every 100 grams of fat, (2) the U.S. Food and Drug Administration's requirement (as of January 1, 2006) that all food companies include trans fat levels in package labeling information, and (3) the New York City health department's decision to urge all city restaurants to stop serving food containing trans fats (Santora 2005).

Improving health worldwide calls for cooperation in the international community. At a 2005 meeting of the Commission on Human Rights in Geneva, 52 countries voted in favor of a resolution urging countries not only to commit to realizing the universal right to the highest attainable standards of physical and mental health but also to assist developing countries achieve the highest attainable standard of physical and mental health through financial and technical support and through training of personnel (UNOG 2005). The United States was the only country to vote against this resolution.

Improvements in health care systems and health care delivery are also necessary to improve the health of populations around the world. Although certain changes in medical practices and policies may help to improve world health, "the health sector should be seen as an important, but not the sole, force in the movement toward global health" (Lerer et al. 1990, p. 10). Improving the health of a society requires addressing diverse issues, including poverty and economic inequality, gender inequality, population growth, environmental issues, education, housing, energy, water and sanitation, agriculture, and workplace safety.

Improving the health of the world also means seeking nonmilitary solutions to international conflicts. In addition to the deaths, injuries, and illnesses that result from combat, war diverts economic resources from health programs, leads to hunger and disease caused by the destruction of infrastructure, causes psychological trauma, and contributes to environmental pollution (Sidel & Levy 2002). Thus "the prevention of war . . . is surely one of the most critical steps mankind can make to protect public health" (White 2003, p. 228).

The World Health Organization (1946) defines **health** as "a state of complete physical, mental, and social well-being" (p. 3). Based on this definition, we conclude this chapter with the suggestion that the study of social problems *is,* essentially, the study of health problems, as each social problem is concerned with the physical, mental, and social well-being of humans and the social groups of which they are a part. As you read the remaining chapters in this book, consider how the problems in each chapter affect the health of individuals, families, populations, and nations.

Chapter Review

Reviewing is as easy as ① ② ③

1. Before you do your final review, take the ThomsonNOW diagnostic quiz to help you identify the areas on which you should concentrate. You will find information on ThomsonNOW and instructions on how to access all of its great resources on the foldout at the beginning of the text.

2. As you review, take advantage of ThomsonNOW's study videos and interactive Map the Stats exercises to help you master the chapter topics.

3. When you are finished with your review, take ThomsonNOW's posttest to confirm you are ready to move on to the next chapter.

• **What features of globalization have contributed to health problems?**

Features of globalization that have been linked to problems in health are increased transportation and information technology, the expansion of trade and transnational corporations, and free trade agreements.

• **What are three measures that serve as indicators of the health of populations? Which health measure reveals the greatest disparity between developed and developing countries?**

Measures of health that serve as indicators of the health of populations include morbidity (often expressed as incidence and prevalence rates), life expectancy, and mortality rates (including infant and under-5 childhood mortality rates and maternal mortality rates). Maternal mortality rates reveal the greatest disparity between developed and developing countries.

• **Which theoretical perspective criticizes the pharmaceutical and health care industry for placing profits above people?**

The conflict perspective criticizes the pharmaceutical and health care industry for placing profits above people. For example, pharmaceutical companies' research and development budgets are spent not according to public health needs but rather according to calculations about maximizing profits. Because the masses of people in developing countries lack the resources to pay high prices for medication, pharmaceutical companies do not see the development of drugs for diseases of poor countries as a profitable investment.

• **Worldwide, what is the most common means by which HIV is transmitted?**

Most cases of HIV worldwide are transmitted through heterosexual contact.

• **What is the second biggest cause of preventable deaths in the United States (second only to tobacco)?**

Obesity, which can lead to heart disease, hypertension, diabetes, and other health problems, is the second biggest cause of preventable deaths in the United States.

• **Why is mental illness referred to as a "hidden epidemic"?**

Mental illness is a "hidden epidemic" because the shame and embarrassment associated with mental problems discourage people from acknowledging and talking about mental illness. Because male gender expectations associate masculinity with emotional strength, men are particularly prone to deny or ignore mental problems.

• **What, according to UN secretary-general Kofi Annan, is the "biggest enemy of health in the developing world"?**

In an address to the 2001 World Health Assembly, UN secretary-general Kofi Annan remarked, "The biggest enemy of health in the developing world is poverty." Approximately one-fifth of the world's population live on less than US$1 per day and nearly one-half live on less than US$2 per day. Poverty is associated with malnutrition, unsafe water and sanitation, indoor air pollution, hazardous working conditions, and lack of access to medical care.

• **According to a World Health Organization analysis of the world's health systems, which country provides the best overall health care?**

The World Health Organization found that France provides the best overall health care among major countries, followed by Italy, Spain, Oman, Austria, and Japan. The United States ranked 37 out of 191 countries, despite the fact that the United States spends a higher portion of its gross domestic product on health care than any other country.

• **What is the difference between selective primary health care and comprehensive primary health care?**

Selective primary health care uses technocratic solutions to target specific health problems, such as using immunization to promote child survival or condoms to prevent the spread of HIV. In contrast, comprehensive primary health care focuses on the broader social determinants of health, such as poverty and economic inequality, gender inequality, environment, and community development.

• **How does the World Health Organization define health?**

Health, according to the World Health Organization is "a state of complete physical, mental, and social well-being." Based on this definition, we suggest that the study of social problems *is,* essentially, the study of health problems, because each social problem is concerned with the physical, mental, and social well-being of humans and the social groups of which they are a part.

Critical Thinking

1. For nearly a century, campaigns to bring about universal health care in the United States have failed. Do you envision universal health care in the United States in your lifetime? Why or why not?
2. Why do you think the American Psychiatric Association (2000) avoids the use of such expressions as "a schizophrenic" or "an alcoholic" and instead uses the expressions "an individual with schizophrenia" or "an individual with alcohol dependence"?
3. As noted in this chapter, suicides outnumber homicides two to one every year in the United States (Ezzell 2003). Why do you think media coverage of homicides is much more common than media coverage of suicides?

Key Terms

biomedicalization
comprehensive primary
    health care
deinstitutionalization
epidemiological
    transition
globalization
health

health maintenance
    organizations (HMOs)
life expectancy
managed care
maternal mortality rate
Medicaid
medicalization
Medicare

mental health
mental illness
morbidity
mortality
needle exchange programs
parity
preferred provider
    organizations (PPOs)

selective primary health
    care
socialized medicine
State Children's Health
    Insurance Program
    (SCHIP)
stigma
universal health care

Media Resources

The Companion Website for
Understanding Social Problems,
Fifth Edition

http://sociology.wadsworth.com/mooney_knox_schacht5e

Supplement your review of this chapter by going to the companion website to take one of the Tutorial Quizzes, use the flash cards to master key terms, and check out the many other study aids you'll find there. You'll also find special features such as *Wadsworth's Sociology Online Resources and Writing Companion,* GSS Data, and Census 2000 information, data, and resources at your fingertips to help you complete that special project or do some research on your own.

> " Substance abuse, the nation's number one preventable health problem, places an enormous burden on American society, harming health, family life, the economy, and public safety, and threatening many other aspects of life. "
>
> *Robert Wood Johnson Foundation, Institute for Health Policy, Brandeis University*

Alcohol and Other Drugs

The Global Context: Drug Use and Abuse

Sociological Theories of Drug Use and Abuse

Frequently Used Legal and Illegal Drugs

Societal Consequences of Drug Use and Abuse

Treatment Alternatives

Strategies for Action: America Responds

Understanding Alcohol and Other Drugs

Chapter Review

S cott Krueger was athletic, intelligent, and handsome and what you'd call an all-around "nice guy." A freshman at the Massachusetts Institute of Technology, he was a three-letter athlete and one of the top 10 students in his high school graduating class of over 300. He was a "giver," not a "taker," tutoring other students in math after school while studying second-year calculus so he could pursue his own career in engineering. While at MIT he rushed a fraternity and celebrated his official acceptance into the brotherhood. The night he celebrated he was found in his room, unconscious, and after three days in an alcoholic coma he died. He was 18 years old (Moore 1997). In September 2000, MIT agreed to pay Scott's parents, Bob and Darlene Krueger, $4.75 million in a settlement over the death of their son and to establish a scholarship in his name. In 2002 an out-of-court settlement that paid the Kruegers $1.75 million was reached with Phi Gamma Delta and the fraternity officers. The defendants also agreed to produce an educational video (*Tell Me Something I Don't Know*) on the dangers of drinking. This video is being shown on high school and college campuses across the United States (Crittenden 2002; Singley 2004). In 2005 the University of California at Berkeley banned alcohol at all fraternity and sorority events (CNN 2005b).

Drug-induced death is just one of the many negative consequences that can result from alcohol and drug abuse. The abuse of alcohol and other drugs is a social problem when it interferes with the well-being of individuals and/or the societies in which they live—when it jeopardizes health, safety, work and academic success, family, and friends. But managing the drug problem is a difficult undertaking. In dealing with drugs, a society must balance individual rights and civil liberties against the personal and social harm that drugs promote—crack babies, suicide, drunk driving, industrial accidents, mental illness, unemployment, and teenage addiction. When to regulate, what to regulate, and who should regulate are complex social issues. Our discussion begins by looking at how drugs are used and regulated in other societies.

The Global Context: Drug Use and Abuse

Pharmacologically, a **drug** is any substance other than food that alters the structure or functioning of a living organism when it enters the bloodstream. Using this definition, everything from vitamins to aspirin is a drug. Sociologically, the term *drug* refers to any chemical substance that (1) has a direct effect on the user's physical, psychological, and/or intellectual functioning, (2) has the potential to be abused, and (3) has adverse consequences for the individual and/or society. Societies vary in how they define and respond to drug use. Thus drug use is influenced by the social context of the particular society in which it occurs.

Drug Use and Abuse Around the World

Globally, 3.0% of the world's population—185 million people—reported using at least one illicit drug in the previous year (World Drug Report 2004). According to estimates, however, the prevalence of drug use around the world varies dramatically. For example, the lifetime prevalence of illicit drug use varies from 46% of the adult population in the United States, to 36% in England, 26% in Italy, 18% in Poland, and 9% in Sweden (ODCCP 2003). In Europe lifetime prevalence of illegal

U.S. citizens visiting the Netherlands may be shocked or surprised to find people smoking marijuana and hashish openly in public. Pictured here is a tourist using a water pipe to smoke marijuana in a coffee shop.

AP / Wide World Photos

drug use *excluding* cannabis ranges from 12% in England to 2% in Finland. Moreover, 23% of European 15- to 16-year-olds have used marijuana at least once in their lifetime, compared to 35% of 15- to 16-year-olds in Canada, 43% in Australia, and 41% in the United States.

In England illegal drug use continues to spread, particularly among those younger than age 25 (BBC 2003). In a recent survey 10% of the 11- to 15-year-old population reported smoking at least one cigarette a week, 18% reported regularly taking drugs, and 24% reported having an alcohol drink in the previous week. Furthermore, the average age of a caller to the British government drug information hotline fell from 29 to 20 in the last year. Statistics also indicate that, compared with nonusers, drug users in the United Kingdom are more likely to be male, unemployed, and living in or around London.

Some of the differences in international drug use can be attributed to variations in drug policies. The Netherlands, for example, has had an official government policy of treating the use of such drugs as marijuana, hashish, and heroin as a health issue rather than a crime issue since the mid 1970s. In the first decade of the policy, drug use did not appear to increase. However, increases in marijuana use were reported in the early 1990s with the advent of "cannabis cafes." These coffee shops sell small amounts of marijuana for personal use and, presumably, prevent casual marijuana users from coming into contact with drug dealers (MacCoun & Reuter 2001; Drug Policy Alliance 2003). Some evidence suggests that marijuana use among Dutch youth is decreasing (Sheldon 2000), although it is increasing among youth

U.S. Social Policy

Indicate whether you agree or disagree with the following policy-oriented statements and then compare your answers to those from a national sample of U.S. adults age 18 and older.

	Agree	Disagree	Results of a National Survey, 2001	
				Percentage in Support
1. Restrict drinking on city streets.	____	____	1. Restrict drinking on city streets.	93
2. Ban youth-oriented packaging.	____	____	2. Ban youth-oriented packaging.	70
3. Ban keg sales to individuals.	____	____	3. Ban keg sales to individuals.	31
4. Ban home delivery of alcohol.	____	____	4. Ban home delivery of alcohol.	64
5. Restrict drinking in parks.	____	____	5. Restrict drinking in parks.	91
6. Ban beer and wine ads on TV.	____	____	6. Ban beer/wine ads on TV.	59
7. Require server training.	____	____	7. Require server training.	90
8. Ban alcohol marketing with athletes.	____	____	8. Ban alcohol marketing with athletes.	62
9. Restrict drinking at sports stadiums.	____	____	9. Restrict drinking at sports stadiums.	74
10. Ban happy hours.	____	____	10. Ban happy hours.	38
11. Restrict drinking on college campuses.	____	____	11. Restrict drinking on college campuses.	89
12. Require legal age for alcohol servers.	____	____	12. Require legal age for alcohol servers.	78
13. Check everyone's ID.	____	____	13. Check everyone's ID.	80
14. Require beer keg registration.	____	____	14. Require beer keg registration.	62
15. Punish adult providers.	____	____	15. Punish adult providers	87

Source: Alcohol Epidemiology Program (2002).

in South America, Western and Eastern Europe, Asia, and Africa (World Drug Report 2004).

Great Britain has also adopted a "medical model," particularly in regard to heroin and cocaine. As early as the 1960s, English doctors prescribed opiates and cocaine for their patients who were unlikely to quit using drugs on their own and for the treatment of withdrawal symptoms. By the 1970s, however, British laws had become more restrictive, making it difficult for either physicians or users to obtain drugs legally. Today, British government policy (this chapter's *Self and Society* feature deals with different U.S. policies) provides for limited distribution of drugs by licensed drug-treatment specialists to addicts who might otherwise resort to crime to support their habits (Abadinsky 2004). Furthermore, a 2002 report from the Advisory Council on the Misuse of Illicit Drugs recommended that marijuana be downgraded to a Class C drug, the lowest ranking possible, noting that it is less dangerous to the individual or to society than tobacco or alcohol (Drug Policy Alliance 2005a).

In stark contrast to such health-based policies, other countries execute drug users and/or dealers or subject them to corporal punishment, which may include whipping, stoning, beating, and torture. Such policies are found primarily in less developed nations, such as Malaysia, where religious and cultural prohibitions condemn the use of any type of drug, including alcohol and tobacco.

Table 3.1

Percentages Reporting Lifetime, Past Year, and Past Month Use of Illicit and Licit Drugs Among Individuals Age 12 or Older: 2003

Drug	Time Period		
	Lifetime	Past Year	Past Month
Any illicit drug [a]	46.4	14.7	8.2
Marijuana and hashish	40.6	10.6	6.2
Cocaine	14.7	2.5	1.0
Crack	3.3	.6	.3
Heroin	1.6	.1	.1
Hallucinogens	14.5	1.7	.4
LSD	10.3	.2	.1
PCP	3.0	.1	0.0
Ecstasy	4.6	.9	.2
Inhalants	9.7	.9	.2
Nonmedical use of any psychotherapeutic [b]	20.1	6.3	2.7
Pain relievers	13.1	4.9	2.0
Tranquilizers	8.5	2.1	.8
Stimulants	8.8	1.2	.5
Methamphetamine	5.2	.6	.3
Sedatives	4.0	.3	.1
Cigarettes	68.7	35.1	29.8
Alcohol	83.1	65.0	50.1

[a] "Any illicit drug" includes marijuana/hashish, cocaine (including crack), heroin, hallucinogens, inhalants, or any prescription-type psychotherapeutic used nonmedically.
[b] "Nonmedical use" of any prescription-type pain reliever, tranqulizer, stimulant, or sedative; does not include over-the-counter drugs.

Source: U.S. Department of Health and Human Services (2004).

Drug Use and Abuse in the United States

According to government officials, there is a drug crisis in the United States—a crisis so serious that it warrants a multibillion-dollar-a-year "war on drugs." Americans' concern with drugs, however, has varied over the years. Ironically, in the 1970s, when drug use was at its highest, concern over drugs was relatively low. When a sample of Americans were recently asked whether they worried about drug use "a great deal," "a fair amount," "only a little," or "not at all," 42% reported worrying a great deal and 23% reported worrying a fair amount (Lyons 2005).

As Table 3.1 indicates, use of illicit drugs in one's lifetime is fairly common. Of people 12 years old and older, 46.4% reported using an illicit drug sometime in their life. Marijuana and hashish had the highest occurrence in lifetime use

(40.6%), with heroin having the lowest use (1.6%). Despite these relatively high numbers, particularly for marijuana and hashish, use of alcohol and tobacco are much more widespread than use of illicit drugs. Half of Americans age 12 and older report being *current* alcohol drinkers and an estimated 71 million Americans over the age of 12 reported *current* use of a tobacco product (U.S. Department of Health and Human Services 2004). In the United States cultural definitions of drug use are contradictory—condemning it on the one hand (e.g., heroin), yet encouraging and tolerating it on the other (e.g., alcohol). At various times in U.S. history many drugs that are illegal today were legal and readily available. In the 1800s and the early 1900s opium was routinely used in medicines as a pain reliever, and morphine was taken as a treatment for dysentery and fatigue. Amphetamine-based inhalers were legally available until 1949, and cocaine was an active ingredient in Coca-Cola until 1906, when it was replaced with another drug—caffeine (Witters et al. 1992; Abadinsky 2004).

Sociological Theories of Drug Use and Abuse

Most theories of drug use and abuse concentrate on what are called psychoactive drugs. These drugs alter the functioning of the brain, affecting the moods, emotions, and perceptions of the user. Such drugs include alcohol, cocaine, heroin, and marijuana. **Drug abuse** occurs when acceptable social standards of drug use are violated, resulting in adverse physiological, psychological, and/or social consequences. For example, when an individual's drug use leads to hospitalization, arrest, or divorce, such use is usually considered abusive. Drug abuse, however, does not always entail drug addiction. **Drug addiction**, or **chemical dependency**, refers to a condition in which drug use is compulsive—users are unable to stop because of their dependency. The dependency may be psychological (the individual needs the drug to achieve a feeling of well-being) and/or physical (withdrawal symptoms occur when the individual stops taking the drug). For example, withdrawal from marijuana includes depression, anger, decreased appetite, and restlessness (Zickler 2003). In 2003 more than 20 million Americans, 9.1% of the population, were defined as dependent on or abusers of a drug. Of that number, 14.8 million (74%) were dependent on or abused alcohol only, 3.8 million (19%) were dependent on or abused illicit drugs but not alcohol, and 3.1 million were dependent on or abused both illicit drugs and alcohol (U.S. Department of Health and Human Services 2004).

Various theories provide explanations for why some people use and abuse drugs. Drug use is not simply a matter of individual choice. Theories of drug use explain how structural and cultural forces as well as biological factors influence drug use and society's responses to it.

Structural-Functionalist Perspective

Structural functionalists argue that drug abuse is a response to the weakening of norms in society. As society becomes more complex and as rapid social change occurs, norms and values become unclear and ambiguous, resulting in **anomie**—a state of normlessness. Anomie may exist at the societal level, resulting in social strains and inconsistencies that lead to drug use. For example, research indicates that increased alcohol consumption in the 1830s and the 1960s was a response to rapid social change and the resulting stress (Rorabaugh 1979). Anomie produces

This poster from the Office of National Drug Control Policy's National Youth Anti-Drug Media Campaign emphasizes the importance of a close relationship between parent and child in the fight against drug use by youths.

inconsistencies in cultural norms regarding drug use. For example, although public health officials and health care professionals warn of the dangers of alcohol and tobacco use, advertisers glorify the use of alcohol and tobacco and the U.S. government subsidizes alcohol and tobacco industries. Furthermore, cultural traditions, such as giving away cigars to celebrate the birth of a child and toasting a bride and groom with champagne, persist.

Anomie may also exist at the individual level, as when a person suffers feelings of estrangement, isolation, and turmoil over appropriate and inappropriate behavior. An adolescent whose parents are experiencing a divorce, who is separated from friends and family as a consequence of moving, or who lacks parental supervision and discipline may be more vulnerable to drug use because of such conditions. Thus, from a structural-functionalist perspective, drug use is a response to the absence of a perceived bond between the individual and society and to the weakening of a consensus regarding what is considered acceptable. Consistent with this perspective, a national poll of Americans age 18 years or older found that peer pressure and lack of parental supervision were the two most common responses given for why teenagers take drugs (Pew 2002). Interestingly, in a recent survey of British youth, 20% responded that they had friends who "faked" drug use in order to "fit in" with their peers (BBC 2004b).

Conflict Perspective

The conflict perspective emphasizes the importance of power differentials in influencing drug use behavior and societal values concerning drug use. From a conflict perspective drug use occurs as a response to the inequality perpetuated by a capitalist system. Societal members, alienated from work, friends, and family as well as from society and its institutions, turn to drugs as a means of escaping the oppression and frustration caused by the inequality they experience. Furthermore, conflict theorists emphasize that the most powerful members of society influence the definitions of which drugs are illegal and the penalties associated with illegal drug production, sales, and use.

For example, alcohol is legal because it is often consumed by those who have the power and influence to define its acceptability—white males (U.S. Department of Health and Human Services 2004). This group also disproportionately profits from the sale and distribution of alcohol and can afford powerful lobbying groups in Washington to guard the alcohol industry's interests. Because tobacco and caffeine are also commonly used by this group, societal definitions of these substances are also relatively accepting.

Conversely, crack cocaine and heroin are disproportionately used by minority group members, specifically, blacks and Hispanics (U.S. Department of Health and Human Services 2004). Consequently, the stigma and criminal consequences associated with the use of these drugs are severe. The use of opium by Chinese immigrants in the 1800s provides a historical example. The Chinese, who had been brought to the United States to work on the railroads, regularly smoked opium as part of their cultural tradition. As unemployment among white workers increased,

"There are but three ways for the populace to escape its wretched lot. The first two are by route of the wine-shop or the church; the third is by that of the social revolution."

Mikhail A. Bakunin
Anarchist and revolutionary

however, so did resentment of Chinese laborers. Attacking the use of opium became a convenient means of attacking the Chinese, and in 1877 Nevada became the first of many states to prohibit opium use. As Morgan (1978) observed:

> The first opium laws in California were not the result of a moral crusade against the drug itself. Instead, it represented a coercive action directed against a vice that was merely an appendage of the real menace—the Chinese—and not the Chinese per se, but the laboring "Chinamen" who threatened the economic security of the white working class. (p. 59)

The criminalization of other drugs, including cocaine, heroin, and marijuana, follows similar patterns of social control of the powerless, political opponents, and/or minorities. In the 1940s marijuana was used primarily by minority group members and carried with it severe criminal penalties. But after white middle-class college students began to use marijuana in the 1970s, the government reduced the penalties associated with its use. Although the nature and pharmacological properties of the drug had not changed, the population of users was now connected to power and influence. Thus conflict theorists regard the regulation of certain drugs, as well as drug use itself, as a reflection of differences in the political, economic, and social power of various interest groups.

Symbolic Interactionist Perspective

Symbolic interactionism, which emphasizes the importance of definitions and labeling, concentrates on the social meanings associated with drug use. If the initial drug use experience is defined as pleasurable, it is likely to recur, and over time the individual may earn the label of "drug user." If this definition is internalized so that the individual assumes an identity of a drug user, the behavior will likely continue and may even escalate.

Drug use is also learned through symbolic interaction in small groups. In a study of binge drinking, researchers found that students who believed that their friends were binge drinking were more likely to binge-drink themselves (Weitzman et al. 2003). First-time users learn not only the motivations for drug use and its techniques but also what to experience. Becker (1966) explained how marijuana users learn to ingest the drug. A novice being coached by a regular user reports the experience:

> I was smoking like I did an ordinary cigarette. He said, "No, don't do it like that." He said, "Suck it, you know, draw in and hold it in your lungs . . . for a period of time." I said, "Is there any limit of time to hold it?" He said, "No, just till you feel that you want to let it out, let it out." So I did that three or four times. (Becker 1966, p. 47)

Marijuana users not only learn how to ingest the smoke but also learn to label the experience positively. When certain drugs, behaviors, and experiences are defined by peers as not only acceptable but also pleasurable, drug use is likely to continue.

> Because they (first-time users) think they're going to keep going up, up, up till they lose their minds or begin doing weird things or something. You have to like reassure them, explain to them that they're not really flipping or anything, that they're gonna be all right. You have to just talk them out of being afraid. (Becker 1966, p. 55)

Interactionists also emphasize that symbols can be manipulated and used for political and economic agendas. The popular DARE (Drug Abuse Resistance

Education) program, with its antidrug emphasis fostered by local schools and police, carries a powerful symbolic value that politicians want the public to identify with. "Thus, ameliorative programs which are imbued with these potent symbolic qualities (like DARE's links to schools and police) are virtually assured widespread public acceptance (regardless of actual effectiveness) which in turn advances the interests of political leaders who benefit from being associated with highly visible, popular symbolic programs" (Wysong et al. 1994, p. 461).

Biological and Psychological Theories

Drug use and addiction are likely the result of a complex interplay of social, psychological, and biological forces. Biological research has primarily concentrated on the role of genetics in predisposing an individual to drug use. Research indicates that severe, early-onset alcoholism may be genetically predisposed, with some men having 10 times the risk for addiction as those without a genetic predisposition. Interestingly, other problems such as depression, chronic anxiety, and attention-deficit disorder are also linked to the likelihood of addiction. Nonetheless, researchers warn, "Nobody is predestined to be an alcoholic" (Firshein 2003).

Biological theories of drug use hypothesize that some individuals are physiologically predisposed to experience more pleasure from drugs than others and, consequently, are more likely to be drug users. According to these theories, the central nervous system, which is composed primarily of the brain and spinal cord, processes drugs through neurotransmitters in a way that produces an unusually euphoric experience. Individuals not so physiologically inclined report less pleasant experiences and are less likely to continue use (Jarvik 1990; National Institute on Alcohol Abuse and Alcoholism 2000).

Psychological explanations focus on the tendency of certain personality types to be more susceptible to drug use. Individuals who are particularly prone to anxiety may be more likely to use drugs as a way to relax, gain self-confidence, or ease tension. For example, research indicates that female adolescents who have been sexually abused or who have poor relationships with their parents are more likely to have severe drug problems (NIDA 2000). Psychological theories of drug abuse also emphasize that drug use may be maintained by positive and negative reinforcement.

Frequently Used Legal and Illegal Drugs

More than 19.5 million people in the United States are illicit drug users, which represents 8.2% of the population age 12 and older. Users of illegal drugs, although varying by type of drug used, are more likely to live in urban areas, to be male, to be young, and to be a member of a minority group (U.S. Department of Health and Human Services 2004). Social definitions regarding which drugs are legal or illegal, however, have varied over time, circumstance, and societal forces. In the United States two of the most dangerous and widely abused drugs, alcohol and tobacco, are legal.

Alcohol

Americans' attitudes toward alcohol have had a long and varied history. Although alcohol was a common beverage in early America, by 1920 the federal government had prohibited its manufacture, sale, and distribution through the passage of the

Eighteenth Amendment to the U.S. Constitution. Many have argued that Prohibition, like the opium regulations of the late 1800s, was in fact a "moral crusade" (Gusfield 1963) against immigrant groups who were more likely to use alcohol. The amendment had little popular support and was repealed in 1933. Today, the United States is experiencing a resurgence of concern about alcohol. What has been called a "new temperance" has manifested itself in federally mandated 21-year-old drinking-age laws, warning labels on alcohol bottles, increased concern over fetal alcohol syndrome and underage drinking, and stricter enforcement of drinking and driving regulations (e.g., checkpoint traffic stops). Such practices may have had an effect on drinking norms. For example, there has been a significant decline in "all measures of drinking" for eighth, tenth and twelfth graders since 2001 (Johnston et al. 2004).

Nonetheless, alcohol remains the most widely used and abused drug in America. Although most people who drink alcohol do so moderately and experience few negative effects, alcoholics are psychologically and physically addicted to alcohol and suffer various degrees of physical, economic, psychological, and personal harm.

The National Survey on Drug Use and Health, conducted by the U.S. Department of Health and Human Services, reports that 119 million Americans age 12 and older consumed alcohol at least once in the month preceding the survey; that is, they were current users in 2003 (U.S. Department of Health and Human Services 2004). Of this number, 6.8% reported being heavy drinkers (defined as drinking five or more drinks per occasion on five or more days in the survey month) and 22.6% were binge drinkers (defined as drinking five or more drinks on at least one occasion during the survey month).

Even more troubling were the 10.9 million current users of alcohol who were 12 to 20 years old, more than half of whom reported being heavy or binge drinkers (U.S. Department of Health and Human Services 2004). Binge drinking in college has attracted media attention and thus the public's attention. The alcohol consumed by binge drinkers represents 70% of all alcohol consumed by college students (Schemo 2002). Furthermore, the money spent annually by college students on alcohol, $5.5 billion, is more than what they spend on milk, soft drinks, coffee, tea, and books combined (MADD 2003a). Many binge drinkers began in high school, with almost one-third having their first drink before age 13. Research indicates that the younger the age of onset, the higher the probability that an individual will develop a drinking disorder at some time in his or her life. For example, an individual's chance of becoming dependent on alcohol, as defined by the National Survey on Drug Use and Health, is 40% if the person's drinking began before the age of 13. The chances of being alcohol dependent also increase if an individual's parents (1) are alcoholics, (2) drink, (3) have a positive attitude about drinking, or (4) use discipline sporadically (Affects Child 2003). Heavy teenage drinkers, as with their adult counterparts, are more likely to be white, non-Hispanic, and male (U.S. Department of Health and Human Services 2004).

Additional results from the National Survey on Drug Use and Health (U.S. Department of Health and Human Services 2004) include the following:

- The highest levels of both heavy and binge drinking are among 18- to 25-year-olds, peaking at age 21.
- Rates of alcohol use are higher among the employed than the unemployed; however, patterns of heavy or binge drinking are highest among the unemployed.
- College graduates are less likely to be binge drinkers than high school graduates but more likely to report alcohol use in the past month.

An Excerpt from *Smashed: Story of a Drunken Girlhood,* by Koren Zailckas

Still, I am not an alcoholic. As far as I can tell, I have no family history of alcoholism. I am not physically addicted to drinking, and I don't have the genetically based reaction to alcohol that addiction counselors call "a disease." In the nine years that I drank, I never hid bottles or drank alone, and I never spent a night in a holding cell awaiting DUI charges. Today, one glass of wine would not propel me into the type of bender where I'd wind up drinking whole bottles. While I have been to AA meetings, I don't *go* to them.

I am a girl who abused alcohol, meaning I drank for the explicit purpose of getting drunk, getting brave, or medicating my moods. In college, that abuse often took the form of binge drinking, which for women, means drinking four or more drinks in a row at least once during a span of two weeks. But frequently, before college and during it, more time would pass between rounds, and two or three drinks could get me wholly obliterated.

I wrote this book knowing that my alcohol abuse, though dangerous, was not unprecedented. Nor were the after effects I experienced as a result of it. Mine are ordinary experiences among girls and young women in both the United States and abroad, and I believe that very commonness makes them noteworthy.

In the past decade alone, girls have closed the gender gap in terms of drinking. I wrote this book because girls are drinking as much, and as early, as boys for the first time in history, because there has been a threefold increase in the number of women who get drunk at least ten times a month, and because a 2001 study showed 40 percent of college girls binge drink. When you factor in increased rates of depression, suicide, alcohol poisoning, and sexual assault, plus emerging research that suggests women who drink have greater chances of liver disease, reproductive disorders, and brain abnormalities, the consequences of alcohol abuse are far heavier for girls than boys.

I also wrote this book because I wanted to quash the misconceptions about girls and drinking: that girls who abuse alcohol are either masculine, sloppy, sexually available, or all of the above, that girls are drinking more and more often in an effort to compete with men, and that alcohol abuse is a life-stage behavior, a youthful excess that is not as damaging as other drugs. (pp. xv–xvi)

For many girls, alcohol abuse may be a stage that tapers off after the quarter-life mark. Many will be spared arrests, accidents, alcoholism, overdoses, and sexual assaults. A whole lot of them will have close calls, incidents they will recount with self-mocking at dinner parties some fifteen years later. Some of them will have darker stories, memories or half memories or full-out blackouts, that they will store in the farthest corners of their mental histories and never disclose to their families or lovers. But I fear that women, even those women who escape the physical consequences of drinking, won't escape the emotional ones. I fear some sliver of panic, sadness, or self-loathing will always stay with us. (p. xvii)

Nine years after I took my first drink, it occurs to me that I haven't grown up. I am missing so much of the equipment that adults should have, like the ability to sustain eye contact without flinching or letting my gaze roll slantwise to the floor. At this point in time, I should be able to hear my own unwavering voice rise in public without feeling my heart flutter like it's trying to take flight. I should be able to locate a point of conversation with the people I deeply long to know as my friends, like my memoirist neighbor or the woman in my reading group who carries the same tattered paperbacks that I do and wears the same foot-less tights. I should be able to stop self-censoring and smile when I feel like it. I should recognize happiness when I feel it expand in my gut. (pp. xvii–xviii).

In the end, I quit drinking because I didn't want to waste any more time picking up the pieces. I decided *smashed,* when it's used as a synonym for drunk, is a self-fulfilling prophecy. (p. ix)

Zailckas (2005). Reprinted with permission.

- Binge drinking was least likely to be reported by Asians and most likely to be reported by Native Hawaiians or other Pacific Islanders.
- More males than females age 12 to 20 report binge drinking.

Despite the fact that males are more likely to abuse alcohol than females, some evidence suggests that female drinking is on the rise (Armstrong and McCarroll 2004). This chapter's *Human Side* feature describes *Smashed* (2005), the story of a drunken girlhood.

Many alcohol users report using other controlled substances, but the more frequently a student binges, the higher the probability of reporting other drug use. Some evidence suggests that certain combinations of drugs—for example, alcohol and cocaine—heighten the negative effects of either drug separately; that is, there is a negative drug interaction.

Tobacco

Although nicotine is an addictive psychoactive drug and although tobacco smoke has been classified by the Environmental Protection Agency as a Group A carcinogen, tobacco continues to be among the most widely used drugs in the United States. According to a U.S. Department of Health and Human Services survey, 70.8 million Americans continue to smoke cigarettes—30% of the population age 12 and older. In 2003, 14.4% of the 12- to 17-year-old population reported smoking, a slight decrease from the previous year (U.S. Department of Health and Human Services 2004). Interestingly, among 12- to 17-year-olds, three brands account for more than 50% of the tobacco market—Marlboro, Newport, and Camel. Use of all tobacco products, including smokeless tobacco (7.3 million users), cigars (12.1 million users), pipe tobacco (2.3 million users), and cigarettes, is higher for high school graduates than for college graduates, males, and Native Americans and Alaska Natives (NHSDA 2003; U.S. Department of Health and Human Services 2004).

In 2003, 3.6 million youths between the ages of 12 and 17 reported past-month use of a tobacco product (U.S. Department of Health and Human Services 2004). Advertising of tobacco products, despite a 1998 legal settlement between the tobacco companies and the states that prohibited any tobacco company from taking "action, directly or indirectly, to target youth . . . in the advertising, promotion or marketing of tobacco products," is often to blame. Recent marketing of flavored tobacco products (e.g., a pineapple-and-coconut-flavored cigarette called Kauai Kilada) is a case in point (Campaign for Tobacco-Free Kids 2004). There is also evidence that cigarette advertisers target minorities. "Studies have shown a higher concentration of tobacco advertising in magazines aimed at African Americans, such as *Jet* and *Ebony*, than in similar magazines aimed at a broader audiences, such as *Time* and *People*. . . . From 1992 to 2000 smoking rates increased among African American 12th graders from 8.7% to 14.2%" (American Heart Association 2003a).

The Campaign for Tobacco-Free Kids calls the introduction of candy-flavored cigarettes and smokeless tobacco an "outrageous" tactic to lure youth into using tobacco products. Note the appeal to African American youth and women in some of the packaging.

Images of Alcohol and Tobacco Use in Children's Animated Films

The impact of media on drug and alcohol use is likely to be recursive—media images affect drug use while societal drug use helps define media presentations. Previous research has documented the rate of tobacco and alcohol use in print media, advertising, and Hollywood movies. In a recent study, Goldstein and colleagues (1999) used content analysis to investigate the prevalence of tobacco and alcohol use in children's animated films as one step in assessing the growing concern with media influence on children's smoking and drinking behavior.

Sample and Methods
The researchers examined all G-rated animated films released between 1937 (*Snow White and the Seven Dwarfs*) and 1997 (*Hercules, Anastasia, Pippi Longstocking,* and *Cats Don't Dance*). Criteria for sample inclusion included that the film was at least 60 minutes in length and that it had been released to theaters before video distribution. The resulting sample included all of Disney's animated children's films produced during the target years with the exception of three that were unavailable on videocassette. The remaining films included all children's animated films produced by MGM/United Artists, Universal, 20th Century Fox, and Warner Brothers since 1982. Variables included (1) the presence of alcohol or tobacco use, (2) the length of time of use on screen, (3) the number of characters using alcohol or tobacco, (4) the value of the character using tobacco or alcohol (i.e., good, neutral, or bad), (5) any implied messages about the drug use, and (6) the type of tobacco or alcohol being used.

Findings and Conclusions
Of the 50 films analyzed, at least one episode of alcohol and/or tobacco use was portrayed in 34 films (68%) with tobacco use ($N = 28$) slightly exceeding portrayals of alcohol use ($N = 25$). Tobacco was used by 76 different characters with an onscreen time of 45 minutes—an average of 1.62 minutes per movie. Characters were most likely to use cigars, followed by cigarettes and pipes. Of the 76 characters using tobacco, 28 (37%) were classified as good. Surprisingly, the use of tobacco products by "good" characters has increased rather than decreased over time.

Sixty-two characters, averaging 2.5 per film, were shown using alcohol, with a total duration of 27 minutes across all films. Characters were most likely to consume wine, followed by beer, spirits, and champagne. The number of good characters using alcohol was similar to the number of characters classified as bad. In 19 of the 25 films in which alcohol use was portrayed, tobacco use was also pictured. Although several films portrayed the physical consequences of smoking ($N = 10$) (e.g., coughing) or drinking ($N = 7$) (e.g., passing out), no film verbally referred to the health hazards of either drug.

One particularly interesting finding of the research concerned the use of alcohol and tobacco as a visual prop in character development. For example, although cigar smokers were portrayed as tough and powerful (e.g., Sykes in *Oliver and Company*), pipe smokers were most often older, kindly, and wise (e.g., Geppetto in *Pinocchio*), and cigarette smokers were independent, witty, and intelligent (e.g., the Genie in *Aladdin*). There was also a tendency for alcohol and tobacco use to be portrayed together. When one, the other, or both are associated with positively defined characters, the impact may be detrimental to the lifestyle choices of viewers.

Although Goldstein and colleagues' (1999) study cannot assess the "impact question," advertising campaigns such as Joe Camel and the Budweiser frogs have been linked to detrimental results. Although in each of these cases the motivation for the use of such appealing characters is clear, the presentation of "good" characters using alcohol and tobacco products in children's animated films remains unexplained. Interpretation of the results is further complicated by the lack of change over time; that is, as our knowledge of the harmful effects of these products increased, their presence in children's films did not, as expected, decrease. In light of these results, Goldstein and colleagues (1999) call for an end to the portrayal of alcohol and tobacco use in all children's animated films and associated products (e.g., posters, books, games).

Similarly, the tobacco industry is accused of developing advertising campaigns that target Hispanics and women. Advertising is but one venue criticized for the positive portrayal of tobacco use. In this chapter's *Social Problems Research Up Close* feature, images of tobacco and alcohol use in children's animated films are examined.

Tobacco was first cultivated by Native Americans, who introduced it to the European settlers in the 1500s. The Europeans believed tobacco had medicinal properties, and its use spread throughout Europe, ensuring the economic success of the colonies in the New World. Tobacco was initially used primarily through chewing and snuffing, but in time smoking became more popular, even though scientific evidence that linked tobacco smoking to lung cancer existed as early as 1859 (Feagin & Feagin 1994). However, it was not until 1989 that the U.S. surgeon general concluded that tobacco products are addictive and that it is nicotine that causes the dependency. Today, the hazards of tobacco use are well documented and have resulted in federal laws that require warning labels on cigarette packages and prohibit cigarette advertising on radio and television. By the year 2030 tobacco-related diseases will be the number one cause of death worldwide, killing one of every six people. Eighty percent of the deaths will take place in poor nations, where many smokers are unaware of the health hazards associated with their behavior (Mayell 1999).

"Everybody knew it's addictive. Everybody knew it causes cancer. We were all in it for the money."

Victor Crawford
Former tobacco lobbyist and smoker who developed lung cancer

Marijuana

Marijuana is the most commonly used and most heavily trafficked illicit drug in the world. When just the top of the plant is sold, it is called hashish. Hashish is much more potent than marijuana, which comes from the entire plant. Globally, there are 146 million marijuana users, representing 2.3% of the world's population and 3.7% of the global population 15 to 64 years of age. Marijuana use, in general, has increased in Europe, Africa, Oceania, and the Americas. It has decreased in South and Southwest Asia (ODCCP 2003; World Drug Report 2004). Use of marijuana is higher among males in most countries, as high as 90% in traditional Asian cultures. The gender difference in the United States is much narrower, with 56% of marijuana users being male (ODCCP 2003).

Marijuana's active ingredient is THC (δ-9-tetrahydrocannabinol), which in varying amounts can act as a sedative or a hallucinogen. Marijuana use dates back to 2737 B.C. in China and has a long tradition of use in India, the Middle East, and Europe. In North America hemp, as it was then called, was used to make rope and as a treatment for various ailments. Nevertheless, in 1937 Congress passed the Marijuana Tax Act, which restricted the use of marijuana; the law was passed as a result of a media campaign that portrayed marijuana users as "dope fiends" and, as conflict theorists note, was enacted at a time of growing sentiment against Mexican immigrants (Witters et al. 1992, pp. 357–359).

There are more than 14.6 million current marijuana users, representing 6.2% of the U.S. population age 12 and older (U.S. Department of Health and Human Services 2004). According to the 2004 Monitoring the Future survey, 45.7% of twelfth graders have used marijuana or hashish at least once in their lifetime, 34.3% used it in the last year, and 19.9% used it in the last month. This is despite the fact that 54.6% of the twelfth graders responded that smoking marijuana regularly is harmful. The study further shows that 11.2% of eighth graders and 19.3% of tenth graders reported use of an illicit drug in the past month, with marijuana being the most commonly reported (Monitoring the Future 2004).

Not surprisingly, teenage marijuana users who report positive affects from use, such as "feeling happy," "getting really high," and "laughing a lot," were more likely to be addicted to marijuana. Almost 40% of young people between the ages of 16 and 21 who reported five or more positive responses were addicted to marijuana compared to 5.2% of those who had no positive responses (Eisner 2005).

Although the effects of alcohol and tobacco are, in large part, well known, the long-term psychological and physiological effects of marijuana are less understood. According to the Office of National Drug Control Policy (ONDCP 2005d), marijuana is associated with many negative health effects, including impaired memory, anxiety, panic attacks, and increased heart rate. Another important concern is that marijuana may be a **gateway drug** that causes progression to other drugs such as cocaine and heroin.

More likely, however, is that people who experiment with one drug are more likely to experiment with another. Indeed, most drug users are polydrug users. Research has indicated that there is a strong "contemporaneous relationship between smoking cigarettes, drinking alcohol, smoking marijuana, and using cocaine" (Lee and Abdel-Ghany 2004, p. 454).

Cocaine

Cocaine is classified as a stimulant and, as such, produces feelings of excitation, alertness, and euphoria. Although such prescription stimulants as methamphetamine and dextroamphetamine are commonly abused, over the last 10 to 20 years societal concern over drug abuse has focused on cocaine. Such concerns have been fueled by its increased use, addictive qualities, physiological effects, and worldwide distribution. More than any other single substance, cocaine led to the present "war on drugs."

Cocaine, which is made from the coca plant, has been used for thousands of years, but anticocaine sentiment in the United States did not emerge until the early 1900s, when it was primarily a response to cocaine's heavy use among urban blacks, poor whites, and criminals (Witters et al. 1992, p. 260; Thio 2004). Cocaine was outlawed in 1914 by the Harrison Narcotics Act, but its use and effects continued to be misunderstood. For example, a 1982 *Scientific American* article suggested that cocaine was no more habit forming than potato chips (Van Dyck & Byck 1982). As demand and then supply increased, prices fell from $100 a dose to $10 a dose, and "from 1978 to 1987 the United States experienced the largest cocaine epidemic in history" (Witters et al. 1992, pp. 256 and 261).

Cocaine use in recent years has decreased in the United States, although it is increasing in South America, Africa, and Europe (ODCCP 2003). According to the National Survey on Drug Use and Health, 34.9 million people—14.7% of the U.S. population age 12 and older—have tried cocaine once in their lifetime (U.S. Department of Health and Human Services 2004). Forty-two percent of high school seniors surveyed in the 2004 Monitoring the Future study indicated that getting powdered cocaine is "fairly easy" or "very easy." Almost half the high school seniors surveyed reported that use of powdered cocaine or crack cocaine once or twice is a "great risk" (Monitoring the Future 2004).

Crack is a crystallized product made by boiling a mixture of baking soda, water, and cocaine. The result, also called rock, base, and gravel, is relatively inexpensive and was not popular until the mid-1980s. Crack is one of the most dangerous drugs to surface in recent years. Crack dealers often give drug users their first few "hits" free, knowing the drug's intense high and addictive qualities are likely to produce returning customers. An addiction to crack can take six to ten weeks; to pure cocaine, three to four years (Thio 2004).

According to the National Survey on Drug Use and Health, 3.3% of Americans over the age of 12 (7.9 million people) have tried crack cocaine once in their life-

"Crack is a drug peddler's dream: it is cheap, easily concealed, and provides a short-duration high that invariably leaves the user craving more."

Tom Morganthau
Journalist

time. Just over 604,000 are current users, compared to 406,000 in 2002. Results from the Monitoring the Future survey indicate that 3.9% of high school seniors reported using crack at least once in their lifetime. This occurred despite the fact that half of these survey respondents reported that using crack once or twice is a "great risk" (Monitoring the Future 2004). In 2003, 3.1% of college students reported using crack cocaine once in their lifetime (ONDCP 2005b).

Other Drugs

Other drugs abused in the United States include "club drugs" (e.g., LSD, Ecstasy), heroin, prescription drugs (e.g., tranquilizers, amphetamines), and inhalants (e.g., glue).

Club Drugs. Club drugs is a general term for illicit, often synthetic drugs commonly used at nightclubs or all-night dances called raves. Club drugs include Ecstasy, ketamine ("Special K"), LSD ("acid"), GHB ("liquid ecstasy"), and Rohypnol ("roofies").

Ecstasy, manufactured in and trafficked from Europe, is the most popular of the club drugs. It ranges in price from $20 to $30 a dose (Leshner 2003). Use of Ecstasy is growing throughout Europe (UNODC 2004) and in some segments of the United States. In 2003, 4.6% of the American population age 12 and older (10.9 million) had used Ecstasy at least once in their lifetime (U.S. Department of Health and Human Services 2004). In 2001 the comparable percentage was 3.6% (8.1 million) (U.S. Department of Health and Human Services 2002). Ecstasy is associated with feelings of euphoria and inner peace, yet critics argue that as the "new cocaine," it can produce both long-term (e.g., permanent brain damage) and short-term (e.g., hyperthermia) negative effects (Cloud 2000; DEA 2000; ONDCP 2003c; Thio 2004). According to the Monitoring the Future survey (2004), many students are well aware of the dangers of Ecstasy. Approximately 58% of high school seniors reported that using Ecstasy once or twice is a "great risk." Not surprisingly, the use of Ecstasy by high school seniors, as well as by eighth and tenth graders, has decreased over the last several years (Johnston et al. 2004).

Ketamine and LSD (lysergic acid diethylamide) both produce visual effects when ingested. Use of ketamine, an animal tranquilizer, can also cause loss of long-term memory, respiratory problems, and cognitive difficulties. Ketamine use by eighth, tenth, and twelfth graders has decreased slowly but steadily over the last several years (Monitoring the Future 2004). LSD is a synthetic hallucinogen, although many other hallucinogens are produced naturally (e.g., peyote). In 2003, 24.4 million Americans, 10.3% of the population age 12 and older, reported lifetime use of LSD. Although twelfth graders have a higher past-month use of LSD (0.7%) than the general population age 12 and over (0.1%), LSD use among secondary school students has declined since 1996 and remains at "historically low levels" (U.S. Department of Health and Human Services 2004; Monitoring the Future 2004; Johnston et al. 2004, p. 4).

GHB (gamma hydroxybutyrate) and Rohypnol (flunitrazepam) are often called **date-rape drugs** because of their use in rendering victims incapable of resisting sexual assaults. GHB, a central nervous system depressant, was banned by the Food and Drug Administration in 1990, although kits containing all the necessary ingredients to manufacture the drug continued to be available on the Internet (ONDCP 2003c).

On February 18, 2000, President Clinton signed a bill that made GHB a controlled substance and thus illegal to manufacture, possess, or sell. Nonetheless, 2.0% of twelfth graders reported past-year use of GHB (Monitoring the Future 2004).

Rohypnol, presently illegal in the United States, is lawfully sold by prescription in more than 70 countries for the short-term treatment of insomnia (Abadinsky 2004; ONDCP 2005a). It belongs to a class of drugs known as benzodiazepines, which also includes such common prescription drugs as Valium, Halcion, and Xanax. Rohypnol is tasteless and odorless. The effects of Rohypnol begin within 15 to 20 minutes; 1 milligram of the drug can incapacitate a victim for up to 12 hours (NIDA 2000; DEA 2000; ONDCP 2003e). "Roofies" are popular with high school and college students, costing only $5 a tablet.

Heroin. Heroin is an analgesic—that is, a painkiller—and is the most commonly abused of the opiates. Most heroin comes from the poppy fields of Laos, Burma, Thailand, Afghanistan, Pakistan, Iran, Mexico, and Colombia (Thio 2004, p. 291). Overall use of heroin is higher in Europe than in the United States (ODCCP 2003). According to the 2003 National Survey on Drug Use and Health, 3.7 million Americans age 12 and older reported lifetime use of heroin (U.S. Department of Health and Human Services 2004). The number of Americans who are heroin dependent is, of course, much lower, ranging from 850,000 to 1 million addicts. The highest use of heroin is among 18- to 25-year-olds, with 1.6% lifetime heroin use (ONDCP 2003d). Heroin is a highly addictive drug that is increasingly popular among school-aged youth. It can be injected, snorted, or smoked. If intravenous injection is used, the effects are felt within 7 to 8 seconds; if snorted or smoked, the effects are felt within 10 to 15 minutes. Although crack cocaine has become less fashionable among youthful offenders, heroin has increased in acceptability to the point of being glamorized in recent motion pictures and song lyrics (Heroin Drug Conference 1997; NIDA 2000). In 2004 the rate of heroin use among eighth, tenth, and twelfth graders remained the same from the previous year, although the rate was substantially reduced from the record highs of the 1990s (Johnston et al. 2004). In addition to experiencing the negative repercussions that heroin shares with all other drugs, heroin users are subjected to the risks of HIV/AIDS if they take the drug intravenously (U.S. Department of Health and Human Services 2004).

In recent years methadone, a synthetic drug often used in the treatment of heroin addiction, has been responsible for an increasing number of deaths. In Florida, for example, there was a 71% increase in methadone-related deaths between 2000 and 2001, prompting one Florida official to announce that methadone "is the fastest rising killer drug" (UF News 2002; Belluck 2003, p. 1). The increase in methadone abuse is somewhat puzzling, given its sedating effect and delayed reaction, sometimes up to two hours. The increase in methadone use may be linked to the growing abuse of heroin and OxyContin, a prescription painkiller (Belluck 2003).

Psychotherapeutic Drugs. Estimates of lifetime use of psychotherapeutic drugs —that is, nonmedical use of any prescription pain reliever, stimulant, sedative, or tranquilizer—did not increase between 2002 and 2003 (U.S. Department of Health and Human Services 2004). However, more than 6 million people, 2.7% of the population over the age of 12, report current use of a psychotherapeutic drug for nonmedical reasons. Of these users, 4.7 million use pain relievers, 1.8 million use tranquilizers, 1.2 million use stimulants, and 300,000 use sedatives.

"The only livin' thing that counts is the fix. . . . Like I would steal off anybody—anybody, at all, my own mother gladly included."

Heroin Addict

Methamphetamine, a stimulant, is one example of a popular psychotherapeutic drug. Although occasionally prescribed for legitimate medical reasons, "crystal meth" is often made in clandestine laboratories in Mexico and the United States. Recent increases in the use of methamphetamine have alarmed international authorities, with use in some areas of the world rivaling that of cocaine. It is notable that such increases are not just in industrialized nations but also in many developing countries. For example, in Thailand methamphetamine has replaced heroin as the most commonly used drug. In the United States use has remained stable over recent years, with lifetime users over the age of 12 representing 5.2% of the population in 2003. Nonetheless, when a sample of respondents were asked, "How concerned are you about the use or sale of 'crystal meth' in your local community? Very concerned, somewhat concerned, not too concerned, or not concerned at all?" Sixty-five percent of Americans reported they were very concerned (Carroll 2005). Furthermore, in a survey of 500 law enforcement agencies, 58% responded that "meth" is their biggest drug problem (Jefferson 2005). Methamphetamine is linked to violent behavior and is often used in combination with other drugs.

Recently, there has been an increase in several types of psychotherapeutic drugs, particularly painkillers (U.S. Department of Health and Human Services 2004). One such drug is OxyContin. The pharmacological characteristics of OxyContin make it a suitable heroin substitute. As a prescription painkiller, OxyContin is often covered by health insurance plans, which contributes to its appeal. When insurance will no longer pay, it is not uncommon for people to switch to heroin, which is less expensive on the street than OxyContin. High school senior annual use has increased from 4.0% in 2002 to 5.0% in 2004. In response to this increase, a spokesperson for Monitoring the Future commented that "considering the addictive potential of this drug, which is a powerful synthetic narcotic used to control pain, we think that there are disturbingly high rates of involvement by American young people" (Johnston et al. 2004, p. 6).

Inhalants. Common inhalants include adhesives (e.g., rubber cement), food products (e.g., vegetable cooking spray), aerosols (e.g., hair spray, air fresheners), anesthetics (ether), and cleaning agents (e.g., spot remover). In total, more than 1,000 household products are currently abused (UNDCP 2005c). More than 23 million people (9.7%) over the age of 12 have reported trying inhalants at least once in their lifetime (U.S. Department of Health and Human Services 2004). Youth, however, are particularly prone to inhalant use. According to Monitoring the Future, 17.3% of eighth graders, 12.4% of tenth graders, and 10.9% of twelfth graders reported lifetime use of an inhalant (Monitoring the Future 2004). Young people often use inhalants in the belief that they are harmless or that prolonged use is necessary for any harm to result. In fact, inhalants are dangerous because of their toxicity and can result in what is called sudden sniff death syndrome.

Societal Consequences of Drug Use and Abuse

Drugs are a social problem, not only because of their adverse effects on individuals but also because of the negative consequences their use has for society as a whole. Everyone is a victim of drug abuse. Drugs contribute to problems within the family and to crime rates, and the economic costs of drug abuse are enormous. Drug abuse also has serious consequences for health at both the individual and the societal level.

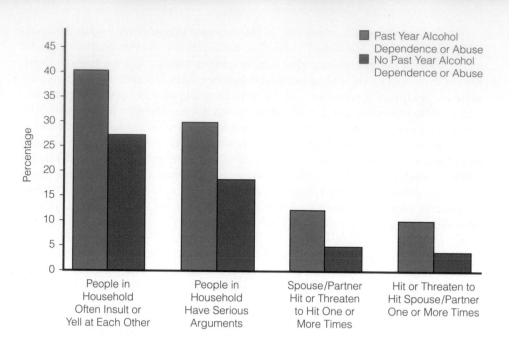

Figure 3.1

Percentages of parents, age 18 or older, reporting household turbulence in the past year, by past-year alcohol dependence or abuse: 2002.

Source: NSDUH (2004a).

Family Costs

ThomsonNOW

Learn more about **Family Costs** by going through the Characteristics of Strong Families Learning Module.

The cost to families of drug use is incalculable. When one or both parents use or abuse drugs, needed family funds may be diverted to purchase drugs rather than necessities. Children raised in such homes have a higher probability of neglect, behavioral disorders, and absenteeism from school as well as lower self-concepts and increased risk of drug abuse (Easley and Epstein 1991; Associated Press 1999; Affects Child 2003).

Nearly 5 million adults who abuse alcohol have children under the age of 18 living with them. Sixty percent of these parents are fathers; 38% are mothers. Parents who report abusing alcohol in the past year are also more likely to report cigarette and illicit drug use than parents who did not report alcohol abuse in the previous year. They were also more likely to report "household turbulence" (see Figure 3.1), including yelling, serious arguments, and violence (NSDUH 2004a).

ThomsonNOW

Learn more about **Family Costs** by going through the Characteristics of Perpetrators Animation.

Alcohol abuse is the single most common trait associated with wife abuse (Charon 2002). The more violent the interaction, the more likely it is that the husband has been drinking excessively. Furthermore, serious violence in the first year of marriage is twice as high among heavy drinkers than among social drinkers (Johnson 2003). In a study of 320 men who were married or living with someone, twice as many reported hitting their partner only after they had been drinking, compared with those who reported the same behavior while sober (Leonard and Blane 1992).

Crime Costs

The drug behavior of individuals arrested, incarcerated, and in drug-treatment programs provides evidence of a link between drugs and crime. Drug users commit a disproportionate number of crimes. For example, at the time of arrest 63.3% of males and 63.9% of females tested positive for cocaine, opiates (e.g., heroin), marijuana, methamphetamine, and/or PCP (ONDCP 2003b). Furthermore, most crimes that take place on college campuses—more than 90%—are alcohol related (Thio 2004).

The relationship between crime and drug use, however, is complex. Sociologists disagree as to whether drugs actually "cause" crime or whether, instead, criminal activity leads to drug involvement. Alternatively, as Siegel (2006, p. 472) noted, criminal involvement and drug use can occur at the same time; that is, someone can take drugs and commit crimes out of the desire to engage in risk-taking behaviors. Furthermore, because both crime and drug use are associated with low socioeconomic status, poverty may actually be the more powerful explanatory variable. After extensive study of the assumed drug-crime link, Gentry (1995) concluded that "the assumption that drugs and crime are causally related weakens when more representative or affluent subjects are considered" (p. 491).

The National Survey on Drug Use and Health documents a relationship between delinquency and drug use. Youths were asked to report the number of times they had engaged in six categories of delinquency, including theft, assault, and carrying a firearm. Youths who reported current drug involvement were more likely to have been involved in any of the six behaviors measured (NSDUH 2005).

In addition to the hypothesized crime−drug use link, some criminal offenses are drug defined: possession, cultivation, production, and sale of controlled substances; public intoxication; drunk and disorderly conduct; and driving while intoxicated. Driving while intoxicated is one of the most common drug-related crimes.

> Nationwide, drunk drivers account for 14 percent of all probationers, 7 percent of local jail inmates, and 2 percent of state prisoners. Of drunk driving offenders, most (89 percent) were on probation; only 11 percent were sentenced to jail time. The average length of incarceration for those who did serve time was 11 months, and nearly half were sentenced to at least six months in jail. About two-thirds of those incarcerated for drunk driving were repeat offenders. (State Legislatures 2000)

In 2004 the National Highway Traffic Safety Administration reported a 2.1% decrease in alcohol-related traffic fatalities between 2003 and 2004. The preliminary report estimated that 16,654 people died in preventable alcohol-related traffic crashes, compared to 17,013 alcohol-related traffic fatalities in 2003 (MADD 2005a). However, automobile accidents remain the leading cause of death of young people between the ages of 16 and 20. Based on survey data from the 2003 National Drug Use and Health Survey, 21% of 16- to 20-year-olds reported driving while under the influence of alcohol or some other substance. Less than 4% reported getting arrested and charged for driving while under the influence (NSDUH 2004b).

Economic Costs

According to a 2004 report of the Office of National Drug Control Policy, the total economic cost of drug use is high, more than $180 billion in 2002, most of which comes from loss of productivity, health care costs, and criminal justice costs (e.g., correctional facilities) (ONDCP 2004). Billions of dollars in insurance are also lost. For example, alcohol-related car accidents account for 18% of the more than $100 billion in U.S. auto insurance payments (MADD 2003b).

Also, revenue is lost to the underground economy because Americans spend an estimated $36 billion on cocaine, $11 billion on marijuana, $10 billion on heroin, and $5.4 billion on methamphetamine (ONDCP 2003b). Concern that on-the-job drug use may impair performance and/or cause fatal accidents has led to drug testing in many industries. For many employees such tests are routine, both as a condition of employment and as a requirement for employees to keep their jobs. This chapter's *Focus on Technology* feature reviews some of the issues related to technologies, privacy rights, and drug testing.

Other economic costs of drug abuse include the cost of homelessness and the cost of implementing and maintaining educational and rehabilitation programs. Also, the cost of fighting the "war on drugs" is likely to increase as organized crime develops new patterns of involvement in the illicit drug trade.

Health Costs

Some consumption of alcohol has been shown to be beneficial in that "moderate drinkers are generally healthier, live longer, and have lower death rates than abstainers" (Thio 2004, p. 324). However, the physical health consequences of abusing alcohol, tobacco, and other drugs are tremendous: shortened life expectancy; higher morbidity (e.g., cirrhosis of the liver, lung cancer); exposure to HIV infection, hepatitis, and other diseases through shared needles; a weakened immune system; birth defects, such as fetal alcohol syndrome; drug addiction in children; and higher death rates.

Cigarette smoking is the leading preventable cause of disease and deaths in the United States. Of the more than 2.4 million U.S. deaths annually, over 440,000 are attributable to cigarette smoking alone. Smoking increases the risk of high blood pressure, blood clots, strokes, lung cancer, chronic obstructive pulmonary disease, and atherosclerosis (American Heart Association 2005). Worldwide, it is estimated that by the year 2020 more than 10 million tobacco-related deaths will occur annually.

Alcohol use is considered the third leading cause of preventable death in the United States, responsible for 75,766 deaths and as many as 2.3 million years of lost life (Jernigan 2005). Furthermore, maternal prenatal alcohol use is one of the leading preventable causes of birth defects and developmental disabilities in children. According to the Centers for Disease Control and Prevention, one of the most extreme effects of drinking while pregnant is **fetal alcohol syndrome,** a syndrome characterized by serious physical and mental handicaps, including low birth weight, facial deformities, mental retardation, and hearing and vision problems (Centers for Disease Control and Prevention 2004).

Heavy alcohol and drug use are also associated with negative consequences for an individual's mental health. Data on both male and female adults have shown that drug users are more likely to suffer from serious mental disorders, including anxiety disorders (e.g., phobias), depression, and antisocial personalities (U.S. Department of Health and Human Services 2004; ONDCP 2005h). Marijuana, the drug most commonly used by adolescents, is also linked to short-term memory loss, learning disabilities, motivational deficits, and retarded emotional development. Twenty-eight percent of suicides by children 12 to 16 years of age are alcohol related (Affects Child 2003).

The societal costs of drug-related health concerns are also extraordinary—an estimated $15.8 billion annually (ONDCP 2004). Health costs include medical services for drug users, the cost of disability insurance, the effects of secondhand smoke, the spread of AIDS, and the medical costs of accident and crime victims, as well as unhealthy infants and children. For example, cocaine use in pregnant women may lead to low-birth-weight babies, increased risk of spontaneous abortions, and abnormal placental functioning (Klutt 2000). In addition, women who smoke while using oral contraceptives are at a greater risk of coronary heart disease and stroke than nonsmoking women who are taking oral contraceptives (American Heart Association 2003b).

The Question of Drug Testing

The technology available to detect whether a person has taken drugs was used during the 1970s by crime laboratories, drug treatment centers, and the military. Today, employers in private industry have turned to chemical laboratories for help in making decisions on employment and retention, and parents and school officials use commercial testing devices to detect the presence of drugs. An individual's drug use can be assessed through the analysis of hair, blood, or urine. New technologies include portable breath (or saliva) alcohol testers, THC detection strips, passive alcohol sensors, interlock vehicle ignition systems, and fingerprint screening devices. Countertechnologies have even been developed—for example, shampoos that rid hair of toxins and "Urine Luck," a urine additive that is advertised to speed the breakdown of unwanted chemicals. A recent law in New Mexico requires that everyone convicted of driving while intoxicated have an ignition interlocking device installed in their vehicle (CNN 2005a).

In 1986 the President's Commission on Organized Crime recommended that all employees of private companies that have contracts with the federal government be regularly subjected to urine testing for drugs as a condition of employment. This recommendation was based on the belief that if employees, such as air traffic controllers, airline pilots, and railroad operators, are using drugs, human lives may be in jeopardy as a result of impaired job performance. In 1987 an Amtrak passenger train crashed outside Baltimore, killing 16 and injuring hundreds. There was evidence of drug use by those responsible for the train's safety. As a result, the Supreme Court ruled in 1989 (by a vote of 7 to 2) that it is constitutional for the Federal Railroad Administration to administer a drug test to railroad crews if they are involved in an accident. Testing those in "sensitive" jobs for drug use may save lives.

An alternative perspective is that drug testing may be harmful. One concern is the accuracy of the tests and the possible effect of false-positives. An innocent person, for example, could lose his or her job. Concern with accuracy of drug tests has led the U.S. Department of Health and Human Services to investigate all federally certified drug-testing laboratories (Brannigan 2000).

A second issue concerns the constitutionality of drug testing. At the heart of the debate is the Fourth Amendment, which states that "the right of the people to be secure in their persons . . . against unreasonable searches and seizures, shall not be violated." Specifically at issue is the definition of "special needs"—an exception to the Fourth Amendment. The special needs exception argues that when a circumstance arises (e.g., drug use among student athletes) that requires action (e.g., controlling drug use), an exception to the Fourth Amendment based on "special needs" (e.g., requiring drug testing as a condition of eligibility) may be made. Thus in 1995 the Supreme Court ruled that random drug testing of student athletes in public schools, where a pattern of drug use had been established and where athletes voluntarily submitted to physical examinations, is not unconstitutional. In 2002 the U.S. Supreme Court held that students in any competitive activity, including cheerleading and choir, may be subject to random drug testing even if no pattern of drug use has been established (Greenhouse 2002).

The issue continues to grow in complexity as drug testing spreads to other venues. For example, a federal lawsuit has been filed that contests a random drug-testing policy that would screen all employees of the Missouri Department of Mental Health. The suit asks the court to "permanently bar implementation of the policy" (Columbia Tribune 2005). Furthermore, several school districts around the United States, although there is no way to know how many, have incorporated sobriety tests into the school routine (Healy 2005).

Finally, the 2006 federal budget has allocated over $20 million to fund student drug testing as part of the National Drug Control Strategy. This initiative "provides competitive grants to support schools in the design and implementation of programs to randomly screen selected students and to intervene with assessment, referral, and intervention" (ONDCP 2005g).

The question in a complex and increasingly technologically dependent society is how to balance the rights of an individual with the needs of society as a whole. The question is not one unique to the United States. In 2004 Prime Minister Tony Blair recommended the use of drug testing, "sniffer dogs," and police patrols in the British school system (BBC 2004a). Ironically, student drug testing has not been found to significantly reduce student drug use (Yamaguchi et al. 2003).

Brown and decaying teeth, called "meth mouth," are just one of the many negative side effects from the use of the drug methamphetamine. Increasingly, methamphetamine is being defined as one of the most dangerous drugs in the United States.

Photo by Dennis Oda, June 2005

Treatment Alternatives

Prevention is always preferable to treatment. Prevention techniques fall into one of two categories (Hanson 2002). First are what may be called "risk and protective" strategies. Here, factors known to be associated with drug use (e.g., child abuse) are targeted and factors that help insulate a person from drug use (e.g., stable family) are encouraged. The second group of prevention techniques, rather than dealing with reducing the vulnerability of an individual, focuses on "the dynamics of the situations, beliefs, motives, reasoning and reactions that enter into the choice to abuse or not abuse drugs" (Hanson 2002, p. 3).

Treatment of drug users has become increasingly important in part as a response to the greater need for treatment programs (U.S. Department of Health and Human Services 2004). In 2003 the number of people needing treatment for an alcohol or illicit drug problem reached 22.2 million, or 9.3% of the population age 12 and older; however, 20 million people also needed but did not receive treatment (see Figure 3.2) (U.S. Department of Health and Human Services 2004).

Individuals who are interested in overcoming chemical dependency have a number of treatment alternatives from which to choose. Some options include family therapy, counseling, private and state treatment facilities, community care programs, pharmacotherapy (i.e., use of treatment medications), behavior modification, drug maintenance programs, and employee assistance programs. Two commonly used techniques are inpatient or outpatient treatment and peer support groups.

Inpatient and Outpatient Treatment

Inpatient treatment refers "to the treatment of drug dependence in a hospital and includes medical supervision of detoxification" (McCaffrey 1998, p. 2). Most inpatient programs last between 30 and 90 days and target individuals whose

"Every $1 invested in addiction treatment programs yields a return of between $4 and $7 in reduced drug-related crime."

Principles of Drug Addiction Treatment, **National Institute of Drug Abuse**

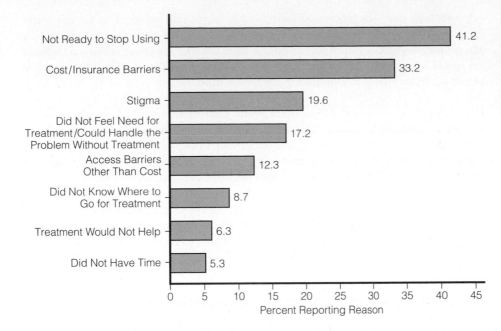

Figure 3.2
Reasons for not receiving treatment among individuals, age 12 or older, who needed but did not receive treatment and felt they needed treatment: 2003.

Source: U.S. Department of Health and Human Services (2004).

withdrawal symptoms require close monitoring (e.g., alcoholics, cocaine addicts). Some drug-dependent patients, however, can be safely treated as outpatients. Outpatient treatment allows individuals to remain in their home and work environments and is often less expensive. In outpatient treatment the patient is under the care of a physician who evaluates the patient's progress regularly, prescribes needed medication, and watches for signs of a relapse.

The longer a patient stays in treatment, the greater the likelihood of a successful recovery (NIDA 2003). Variables that predict success include the user's motivation to change, support of family and friends, criminal justice or employer intervention, a positive relationship with therapeutic staff, and a program of recovery that addresses many of the needs of the patient.

Peer Support Groups

Twelve-Step Programs. Both Alcoholics Anonymous (AA) and Narcotics Anonymous (NA) are voluntary associations whose only membership requirement is the desire to stop drinking or taking drugs. AA and NA are self-help groups in that they are operated by nonprofessionals, offer "sponsors" to each new member, and proceed along a continuum of 12 steps to recovery. Members are immediately immersed in a fellowship of caring individuals with whom they meet daily or weekly to affirm their commitment. Some have argued that AA and NA members trade their addiction to drugs for feelings of interpersonal connectedness by bonding with other group members. In a survey of recovering addicts, more than 50% reported using a self-help program such as AA in their recovery (Willing 2002). AA boasts a membership of over 2 million (Alcoholics Anonymous 2005).

Symbolic interactionists emphasize that AA and NA provide social contexts in which people develop new meanings. Abusers are surrounded by others who offer positive labels, encouragement, and social support for sobriety. Sponsors tell the new members that they can be successful in controlling alcohol and/or drugs "one day at a time" and provide regular interpersonal reinforcement for doing so.

People who smoke are more likely to be heavy drinkers and current illicit drug users. Some evidence suggests that giving up smoking leads to a reduction in alcohol consumption.

AA members may also take medications to help prevent relapses, although some think that the medications are a crutch. In a study of 222 AA members, Rychtarik and colleagues (2000) found that although more than half of those surveyed thought the use of relapse-preventing medication was or might be a good idea, 29% reported pressures from others to stop taking the medication.

Therapeutic Communities. In **therapeutic communities**, which house between 35 and 500 people for up to 15 months, participants abstain from drugs, develop marketable skills, and receive counseling. Synanon, which was established in 1958, was the first therapeutic community for alcoholics and was later expanded to include other drug users. More than 400 residential treatment centers are now in existence, including Daytop Village and Phoenix House, the largest therapeutic community in the country (Abadinsky 2004). The longer a person stays at such a facility, the greater the chance he or she has of overcoming dependency. Symbolic interactionists argue that behavioral changes appear to be a consequence of revised self-definition and the positive expectations of others.

Strategies for Action: America Responds

Drug use is a complex social issue that is exacerbated by the structural and cultural forces of society that contribute to its existence. Although the structure of society perpetuates a system of inequality, creating in some the need to escape, the culture of society, through the media and normative contradictions, sends mixed messages about the acceptability of drug use. Thus developing programs, laws, or initiatives that are likely to end drug use may be unrealistic. Nevertheless, numerous social policies have been implemented or proposed to help control drug use and its negative consequences.

Government Regulations

The largest social policy attempt to control drug use in the United States was Prohibition. Although this effort was a failure by most indicators, the government continues to develop programs and initiatives designed to combat drug use. In the 1980s the federal government declared a "war on drugs," which was based on the belief that controlling drug availability would limit drug use and, in turn, drug-related problems. In contrast to a **harm reduction** position, which focuses on minimizing the costs of drug use for both user and society (e.g., distributing clean syringes to decrease the risk of HIV infection), this "zero-tolerance" approach advocates get-tough law enforcement policies. However, Yale law professor Steven Duke and coauthor Albert C. Gross, in their book *America's Longest War* (1994), argued that the war on drugs, much like Prohibition, has only intensified other social problems: drug-related gang violence and turf wars, the creation of syndicate-controlled black markets, unemployment, the spread of AIDS, overcrowded prisons, corrupt law enforcement officials, and the diversion of police from other serious crimes.

> In 2005, the European Parliament officially acknowledged that the war on drugs was a failure: . . . despite the policies carried out to date at international, European and national levels, the production, consumption and sale of illicit substances . . . have reached extremely high levels in all the Member states, and faced with this failure it is essential that the EU [European Union] revise its general strategy on narcotic substances. (Drug Policy Alliance 2005b, p. 1)

Table 3.2

National Priorities in the Fight Against Drugs: 2006

Initiative	Description
Stopping drug use before it starts	
Student drug testing	Provides federal funds to local schools that develop and implement random drug testing of selected students (see *Focus on Technology* feature)
Research grant funds to local education agencies	Provides support for school safety and prevention programs that have been "demonstrated to be effective in reducing youth drug use and violence"
Healing America's drug users	
Access to Recovery	Provides treatment vouchers that can be used at "a range of appropriate community based services" to those in need of recovery support
Screening, Brief Intervention, Referral, and Treatment (SBIRT)	A program designed to intervene with "nondependent users and stop drug use before it leads to dependence" through the improvement of treatment delivery
Disrupting the drug markets	
Andean Counter-Drug Initiative	Provides support for various counterdrug activities, including illicit crop reduction in Colombia, Bolivia, Peru, and the Andean region; also includes humanitarian efforts in this region, for example, support for vulnerable groups
Afghanistan Initiative	Program designed to provide support for counternarcotics programs, including poppy crop eradication campaigns

Source: ONDCP (2005g).

Consistent with conflict theory, still others argue that the war on drugs unfairly targets minorities (see Chapter 4). U.S. Department of Justice data, analyzed by Human Rights Watch, an international human rights organization, indicates that, in general, black men are 13 times more likely to be incarcerated in state prisons for drug charges than white men (Fletcher 2000, p. A10).

Despite concerns, the war on drugs continues at an astronomical societal cost— a projected $12.4 billion in 2006. The National Control Drug Strategy has three priorities: (1) "stopping drug use before it starts," which focuses on young people, education, and communities; (2) "healing drug users," which targets the more than 6 million citizens who are in need of drug treatment; and (3) "disrupting drug markets" in transit and producer countries, including the United States (see Table 3.2) (ONDCP 2005g).

Many of the countries in which drug trafficking occurs are characterized by government corruption and crime, military coups, and political instability. This is the case with Colombia, which supplies more than 90% of the cocaine in the United States. Aerial spraying of the coca plants "is a major component of Colombia's strategy for fighting the drug trade and is the program with the single greatest potential for disrupting the production of cocaine before it enters the supply train to

Figure 3.3
Federal drug control spending
by function, fiscal year 2006.
Source: ONDCP (2005f).

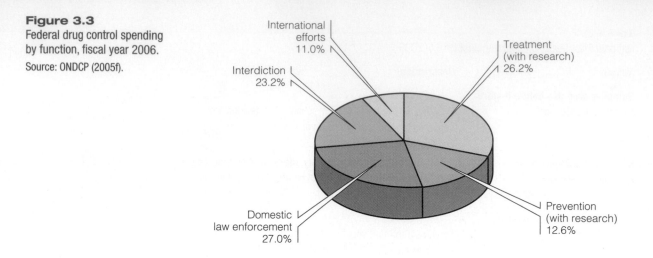

the United States" (ONDCP 2003a, p. 35). To that end, the United States has channeled more than $100 million in support of counterdrug activities in Colombia (See Figure 3.3). When adult Americans in a national poll were asked about financial assistance to foreign countries to fight trafficking, 42% responded that the United States was providing too much assistance (Pew 2002). Rather than foreign aid and military assistance, many argue that trade sanctions should be imposed in addition to crop eradication programs and interdiction efforts. Others, however, noting the relative failure of such programs in reducing the supply of illegal drugs entering the United States, argue that the war on drugs should be abandoned and that deregulation is preferable to the side effects of regulation.

Deregulation or Legalization: The Debate

Americans have mixed feelings about drugs and drug use. For example, only 34% believe that marijuana should be legal; however, 72% believe that recreational marijuana use should result in nothing more than a fine, and 47% have tried the drug (Time/CNN 2003). Given these results, it is not surprising that some advocate alternatives to the punitive emphasis of the last several decades.

Deregulation is the reduction of government control over certain drugs. For example, although individuals must be 21 years old to purchase alcohol and 18 to purchase cigarettes, both substances are legal and purchased freely. Furthermore, in some states possession of marijuana in small amounts is now a misdemeanor rather than a felony, and in other states marijuana is lawfully used for medical purposes. In 1996 both Arizona and California passed acts known as "marijuana medical bills" that made the use and cultivation of marijuana, under a physician's orders, legal. Medical use of marijuana has also been approved in Alaska, Colorado, Maine, Nevada, Oregon, and Washington. Deregulation is popular in other countries as well. Canada has legalized the medical use of marijuana and has approved a cannabis spray that is administered through the mouth for treatment of pain in patients with multiple sclerosis (Grant 2005). In addition, in the United States the drug MDMA, or "Ecstasy," is being tested as a possible aid in dealing with such

Should Drugs Be Legalized for Those 21 Years Old and Older?

Many argue that the right of an adult to make an informed decision includes deciding to use illegal drugs. Similar to arguments about the legalization of prostitution and gambling, drug use is considered a victimless crime by many. In addition, it is argued that any violence associated with drug use is a consequence of drugs being illegal, which drives their prices up and forces users to turn to crime as a means of supporting their drug use. Alternatively, there is the argument that the state not only has the right but the obligation to protect its citizens. Seatbelt laws, traffic laws, and many administrative agency regulations (e.g., U.S. Food and Drug Administration regulations) are designed to do just that. Moreover, if drugs were made legal, there would be an increase in drug use and the associated costs described in this chapter.

It Is Your Turn Now to Take a Stand!

Should drugs be legal for those 21 years and older? What about 18 years and older? What do you think?

mental problems as moderate to severe anxiety and post-traumatic stress disorder (Conant 2005).

Proponents for the **legalization of drugs** affirm the right of adults to make an informed choice. They also argue that the tremendous revenues realized from drug taxes could be used to benefit all citizens, that purity and safety controls could be implemented, and that legalization would expand the number of distributors, thereby increasing competition and reducing prices. Drugs would thus be safer, drug-related crimes would be reduced, and production and distribution of previously controlled substances would be taken out of the hands of the underworld.

Those in favor of legalization also suggest that the greater availability of drugs would not increase demand, pointing to countries where some drugs have already been decriminalized. **Decriminalization,** or the removing of penalties for certain drugs, would promote a medical rather than criminal approach to drug use that would encourage users to seek treatment and adopt preventive practices. For example, making it a criminal offense to sell or possess hypodermic needles without a prescription encourages the use of nonsterile needles that spread infections such as HIV and hepatitis.

Opponents of legalization argue that it would be construed as government approval of drug use and, as a consequence, drug experimentation and abuse would increase. Furthermore, although the legalization of drugs would result in substantial revenues for the government, drug trafficking and black markets would still flourish because all drugs would not be decriminalized (e.g., crack). Legalization would also require an extensive and costly bureaucracy to regulate the manufacture, sale, and distribution of drugs. Finally, the position that drug use is an individual's right cannot guarantee that others will not be harmed. It is illogical to assume that a greater availability of drugs will translate into a safer society.

Collective Action

Social action groups, such as Mothers Against Drunk Driving (**MADD**), have successfully lobbied legislators to raise the drinking age to 21 and to provide harsher penalties for driving while impaired. MADD, with 3.5 million members and 600 chapters, has also put pressure on alcohol establishments to stop "two for one" offers and has pushed for laws that hold the bartender personally liable if a served person is later involved in an alcohol-related accident. Even hosts in private homes can now be held liable if they allow a guest who became impaired while drinking at their house to drive.

"If some drunk gets out and kills your kid, you'd probably be a little crazy about it, too.**"**

Kathy Prescott
Former MADD president

MADD also has several national programs designed to increase public awareness of the problems associated with drinking and driving. For example, the Tie One On For Safety ribbon campaign encourages drivers to place a red ribbon on their vehicle as an indication of their commitment to driving safely. This program is in its 18th year of continuous operation. In addition, the popular "designate a driver" initiative, which focuses on the four deadliest holidays—Labor Day, the holiday season (November 24 through January 2), St Patrick's Day, and Memorial Day—encourages motorists to designate nondrinking drivers at holiday times (MADD 2005b).

Collective action is also being taken against tobacco companies by smokers, ex-smokers, and the families of smoking victims. They charge that tobacco executives knew more than 50 years ago that tobacco was addictive and that they concealed this fact from both the public and the government. Furthermore, the groups charge that tobacco companies manipulate nicotine levels in cigarettes with the intention of causing addiction. In a class action suit by more than 300,000 Florida smokers, a jury ordered the top five cigarette producers to pay $145 billion—the largest settlement to date—to the plaintiffs. In 2004 a New York jury "awarded $20 million in punitive damages to the widow of a long time smoker who died of lung cancer at the age of 57" (Brayton and Purcell Law Firm 2005, p. 1).

Finally, several initiatives have resulted in statewide referendums concerning the cost-effectiveness of government policies. For example, as a result of the passage of Proposition 36, California (as well as many other states) will now require that nonviolent first- and second-time minor drug offenders receive treatment, including job training, therapy, literacy education, and family counseling rather than jail time. The initiative provides more than $100 million a year over the next five and a half years for community-based treatment facilities. The program is predicted to divert 30,000 state and county prisoners to treatment programs, resulting in a net savings to California taxpayers of $1.5 billion over the course of the program. Evaluation of a similar strategy in Arizona boasts a treatment success rate of 71% (Mann 2000; Thompson 2000). In addition, over the past decade voters and state governments have enacted 150 significant drug policy reforms. For example, Connecticut passed significant overdose prevention legislation, following in the footsteps of New Mexico. Texas and Kansas passed legislation providing for treatment instead of incarceration for first-time drug offenders, and Colorado reduced sentences for these offenders as well. Illinois passed legislation allowing for the sale of sterile syringes without a prescription and also mandated the gathering of race-based information during police traffic stops (Drug Policy Alliance 2005c).

Understanding Alcohol and Other Drugs

In summary, substance abuse—that is, drugs and their use—is socially defined. As the structure of society changes, the acceptability of one drug or another changes as well. As conflict theorists assert, the status of a drug as legal or illegal is intricately linked to those who have the power to define acceptable and unacceptable drug use. There is also little doubt that rapid social change, anomie, alienation, and inequality further drug use and abuse. Symbolic interactionism also plays a significant role in the process: If people are labeled "drug users" and are expected to behave accordingly, then drug use is likely to continue. If people experience positive reinforcement of such behaviors and/or have a biological predisposition to use drugs, the probability of their drug involvement is even higher. Thus the theories of drug use complement rather than contradict one another.

Drug use must also be conceptualized within the social context in which it occurs. Many youths who are at high risk for drug use have been "failed by society"—they are living in poverty, victims of abuse, dependents of addicted and neglectful parents, alienated from school. Despite the social origins of drug use, many treatment alternatives, emanating from a clinical model of drug use, assume that the origin of the problem lies within the individual rather than in the structure and culture of society. Although admittedly the problem may lie within the individual when treatment occurs, policies that address the social causes of drug abuse provide a better means of dealing with the drug problem in the United States.

As stated earlier, prevention is preferable to intervention, and given the social portrait of hard-drug users—young, male, minority—prevention must entail dealing with the social conditions that foster drug use. Some data suggest that inner city adolescents are particularly vulnerable to drug involvement because of their lack of legitimate alternatives (Van Kammen and Loeber 1994):

> Illegal drug use may be a way to escape the strains of the severe urban conditions, and dealing illegal drugs may be one of the few, if not the only, way to provide for material needs. Intervention and treatment programs, therefore, should include efforts to find alternate ways to deal with the limiting circumstances of inner-city life, as well as create opportunities for youngsters to find more conventional ways of earning a living. (p. 22)

Social policies that deal with drug use have been predominantly punitive rather than preventive. Recently, however, there appears to be some movement toward educating the public and changing the culture of drugs. For example, a new media campaign by the Office of National Drug Control Policy features real teens "sharing their anti-drug attitudes and commitments." The new program debuted on Fox's *American Idol* in an advertisement in which a 17-year-old girl read her original poetry about the dangers of drug use (ONDCP 2005e).

In this country and throughout the world, millions of people depend on legal drugs for the treatment of a variety of conditions, including pain, anxiety and nervousness, insomnia, overeating, and fatigue. Although drugs for these purposes are relatively harmless, the cultural message "better living through chemistry" contributes to alcohol and drug use and its consequences. But these and other drugs are embedded in a political and economic context that determines who defines what drugs, in what amounts, are licit or illicit and what programs are developed in reference to them.

"A child who reaches 21 without smoking, abusing alcohol or using drugs is virtually certain never to do so."

Joseph A. Califano Jr.
President, Center on Addiction and Substance Abuse

Chapter Review

ThomsonNOW™

Reviewing is as easy as ① ② ③

1. Before you do your final review, take the ThomsonNOW diagnostic quiz to help you identify the areas on which you should concentrate. You will find information on ThomsonNOW and instructions on how to access all of its great resources on the foldout at the beginning of the text.

2. As you review, take advantage of ThomsonNOW's study videos and interactive Map the Stats exercises to help you master the chapter topics.

3. When you are finished with your review, take ThomsonNOW's posttest to confirm you are ready to move on to the next chapter.

- **What is a drug, and what is meant by drug abuse?**

Sociologically, the term *drug* refers to any chemical substance that (1) has a direct effect on the user's physical, psychological, and/or intellectual functioning, (2) has the potential to be abused, and (3) has adverse consequences for the individual and/or society. Drug abuse occurs when acceptable social standards of drug use are violated, resulting in adverse physiological, psychological, and/or social consequences.

- **How do the three sociological theories of society explain drug use?**

Structural functionalists argue that drug abuse is a response to the weakening of norms in society, leading to a condition known as anomie or normlessness. From a conflict perspective drug use occurs as a response to the inequality perpetuated by a capitalist system as societal members respond to alienation from their work, family, and friends. Symbolic interactionism concentrates on the social meanings associated with drug use. If the initial drug use experience is defined as pleasurable, it is likely to recur, and over time the individual may earn the label of "drug user."

- **What are the most frequently used legal and illegal drugs?**

Alcohol is the most commonly used and abused legal drug in America. The use of tobacco products is also very high, with 30% of Americans reporting that they currently smoke cigarettes. Marijuana is the most commonly used illicit drug, with 146 million marijuana users, representing 2.3% of the world's population.

- **What are the consequences of drug use?**

The consequences of drug use are fourfold. First is the cost to the family, often manifesting itself in higher rates of divorce, spouse abuse, child abuse, and child neglect. Second is the relationship between drugs and crime. Those arrested have disproportionately higher rates of drug use. Although drug users commit more crimes, sociologists disagree as to whether drugs actually "cause" crime or whether, instead, criminal activity leads to drug involvement. Third are the economic costs, which are estimated to be more than $180 billion. Finally are the health costs of abusing drugs, including shortened life expectancy; higher morbidity (e.g., cirrhosis of the liver, lung cancer); exposure to HIV infection, hepatitis, and other diseases through shared needles; a weakened immune system; birth defects, such as fetal alcohol syndrome; drug addiction in children; and higher death rates.

- **What treatment alternatives are available for drug users?**

Although there are many ways to treat drug abuse, two methods stand out. The inpatient-outpatient model entails medical supervision of detoxification and may or may not include hospitalization. Twelve-step programs such as Alcoholics Anonymous (AA) and Narcotics Anonymous (NA) are particularly popular, as are therapeutic communities. Therapeutic communities are residential facilities where drug users learn to redefine themselves and their behavior as a response to the expectations of others and self-definition.

- **What can be done about the drug problem?**

First, there are government regulations limiting the use (e.g., the law establishing the 21-year-old drinking age) and distribution (e.g., prohibitions about importing drugs) of legal and illegal drugs. The government also imposes sanctions on those who violate drug regulations and provides treatment facilities for other offenders. Second, there are collective action groups—for example, Mothers Against Drunk Driving. Finally, there are local and statewide initiatives geared toward holding companies responsible for the consequences for their product—for example, class action suits against tobacco producers.

Critical Thinking

1. Are alcoholism and other drug addictions a consequence of nature or nurture? If nurture, what environmental factors contribute to such problems? Which of the three sociological theories best explains drug addiction?

2. Measuring alcohol and drug use is often difficult. This is particularly true given the tendency for respondents to acquiesce, that is, to respond in a way they believe is socially desirable. Consider this and other problems in doing research on alcohol and other drugs, and propose how such problems would be remedied.

3. Jeffrey Reiman (2003, p. 37) noted that "on the basis of available scientific evidence, there is every reason to suspect that we do our bodies more damage, more irreversible damage, by smoking cigarettes and drinking liquor than by using heroin." How would a social constructionist interpret this statement?

Key Terms

anomie
chemical dependency
club drugs
crack
date-rape drugs
decriminalization
deregulation
drug
drug abuse
drug addiction
fetal alcohol syndrome
gateway drug
harm reduction
legalization
MADD
therapeutic communities

Media Resources

The Companion Website for
Understanding Social Problems,
Fifth Edition

http://sociology.wadsworth.com/mooney_knox_schacht5e

Supplement your review of this chapter by going to the companion website to take one of the Tutorial Quizzes, use the flash cards to master key terms, and check out the many other study aids you'll find there. You'll also find special features such as *Wadsworth's Sociology Online Resources and Writing Companion,* GSS Data, and Census 2000 information, data, and resources at your fingertips to help you complete that special project or do some research on your own.

> " Unjust social arrangements are themselves a kind of extortion, even violence. " *John Rawls, A Theory of Justice*

Crime and Social Control

born in Baltimore, Maryland, on December 10, 1971, Brian Nichols by all observers had a happy childhood. He grew up surrounded by friends and family in a middle-class neighborhood; wanting for little, he learned to play the piano on a baby grand and prided himself on his martial arts prowess. Later, at over 6 feet tall and weighing 200 pounds, 16-year-old Brian attended a private high school, played football, and seemingly had the world by the tail. After high school he headed to Kutztown University in Pennsylvania, but would eventually drop out after getting into trouble with the law. At first, he engaged in only minor offenses—loud parties, buying alcohol, and the like. But then, there were more serious offenses—more troublesome—such as disorderly conduct and harassment (Tucker & Kiehl 2005).

Nonetheless, no one would have predicted what Brian Nichols would do in Atlanta's Fulton County Court House on March 11, 2005. A defendant on trial for rape, Brian Nichols, now age 33, overpowered a guard and disarmed her. Officials say he then went on a shooting rampage that left a state judge, a sheriff deputy, a court reporter, and a U.S. customs agent dead. After fleeing the courthouse, Nichols kept the Atlanta metro area of 4 million people in fear for 26 hours. He was eventually captured at an apartment where he had taken a woman prisoner (Copeland 2005, p. 2). The woman, 26-year-old Ashley Smart, had talked of redemption and service to others and pleaded with him not to kill her. "I told him if he hurt me, my little girl won't have a Mommy or a Daddy" (Johnson 2005, p. 29).

Clearly, what happened in this case is not representative of the thousands of court cases that take place every year. In reality, few prisoners escape while being escorted to the courtroom. In addition, when asked to rank the most important problem in the United States, less than 1% of a random sample of Americans responded that the courts were a significant concern (Gallup Poll 2005a). The courts are just one part of a massive bureaucracy called the criminal justice system, a system that often comes under public scrutiny, particularly in high profile cases, such as the kidnapping of Elizabeth Smart, the murder of Laci Peterson, the trial of Michael Jackson, and the arrest and conviction of Martha Stewart. In this chapter we examine the criminal justice system as well as theories, types, and demographic patterns of criminal behavior. The economic, social, and psychological costs of crime are also examined. The chapter concludes with a discussion of social control, including policies and prevention programs designed to reduce crime in the United States.

ThomsonNOW™

Reviewing is as easy as ① ② ③

Use ThomsonNOW to help you make the grade on your next exam. When you are finished reading this chapter, go to the chapter review for instructions on how to make ThomsonNOW work for you.

The Global Context: International Crime and Social Control

Several facts about crime are true throughout the world. First, crime is ubiquitous; that is, there is no country where crime does not exist. Second, most countries have the same components in their criminal justice systems: police, courts, and prisons. Third, worldwide, adult males make up the largest category of crime suspects; and fourth, in all countries theft is the most common crime committed, whereas violent crime is a relatively rare event.

Even so, dramatic differences do exist in international crime rates, although comparisons are made difficult by the paltry amount of research and variations in measurement and crime definitions. However, as Stephens notes (2006), on the basis of the most reliable data available, we can make certain statements about the U.S. crime rate vis-á-vis other nations. The overall U.S. serious crime rate, particularly property crime, was the highest in the world in 1980. By 2000 the overall crime rate had decreased below that of Finland, England and Wales, and Denmark. In terms of violent crime, among industrialized nations the United States has one of the highest homicide rates in the world and a rape rate second only to Canada. The lowest rates of rape are in Asia and the Middle East. In general, "the United States has experienced a downward trend in crime while other Western nations, and even industrialized non-Western nations, are witnessing higher numbers" (Stephens 2006, p. 17).

In 2005 delegates from 150 countries met to deal with, among other things, transnational crime (Associated Press 2005b). As defined by the U.S. Department of Justice, **transnational crime** is "organized criminal activity across one or more national borders" (U.S. Department of Justice 2003). For example, Chinese Triads operate in large cities worldwide, netting billions of dollars a year from prostitution, drugs, and other organized crime activities; Colombian cocaine cartels flourish and have spread to sub-Saharan countries with needy economies; Ecstasy is manufactured in the Netherlands and is sold in the United States by Israeli-organized crime members; and an estimated 50,000 women and children are trafficked into the United States annually for use in pornography rings and prostitution (United Nations 1997; Interpol 1998; Finckenauer 2000; Dobriansky 2001; U.S. Department of Justice 2003). Transnational crime is facilitated by recent trends in globalization, including enhanced transportation and communication technologies. For example, the Internet has led to an explosive growth in child pornography. In 2005 a United Nations expert on the subject told the 53-nation UN Commission on Human Rights that governments must act now to curb the proliferation of child pornography. Children, particularly in developing countries, are "being lured, often with parental consent, to pose naked or semi-naked for illegal photo agencies who then post the material on web sites" (Klapper 2005, p. 1). One other type of transnational crime, terrorism, will be discussed in Chapter 16.

Sources of Crime Statistics

The U.S. government spends millions of dollars annually to compile and analyze crime statistics. A **crime** is a violation of a federal, state, or local criminal law. For a violation to be a crime, however, the offender must have acted voluntarily and with intent and have no legally acceptable excuse (e.g., insanity) or justification (e.g., self-defense) for the behavior. The three major types of statistics used to measure crime are official statistics, victimization surveys, and self-report offender surveys.

Official Statistics

Local sheriffs' departments and police departments throughout the United States collect information on the number of reported crimes and arrests and voluntarily report them to the Federal Bureau of Investigation (FBI). The FBI then compiles these statistics annually and publishes them, in summary form, in the Uniform Crime Reports (UCR). The UCR lists **crime rates** or the number of crimes committed per 100,000 population, the actual number of crimes and the percentage of change over time, and clearance rates. **Clearance rates** measure the percentage of cases in which an arrest and official charge have been made and the case has been turned over to the courts.

These statistics have several shortcomings. For example, many incidents of crime go unreported. It is estimated that between 1992 and 2000 only 31% of rapes

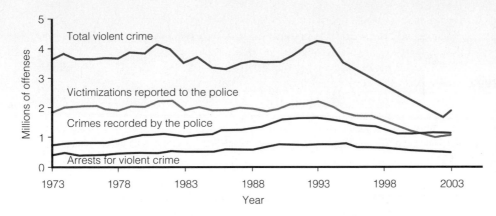

Figure 4.1
Four measures of serious violent crime.

Note: The serious violent crimes included are rape, robbery, aggravated assault, and homicide.

Source: U.S. Bureau of Justice Statistics. Available at http://www.ojp.usdoj.gov/bjs/glance/cv2.htm

and sexual assaults, 55% of aggravated assaults, and 57% of robberies were actually reported to the police (U.S. Bureau of the Census 2004). Even if a crime is reported, it may not be recorded by the police (see Figure 4.1). Alternatively, some rates may be exaggerated. Motivation for such distortions may come from the public (e.g., demanding that something be done), from political officials (e.g., election of a sheriff), and/or from organizational pressures (e.g., budget requests). For example, a police department may "crack down" on drug-related crimes in an election year. The result is an increase in the recorded number of these offenses for that year. Such an increase reflects a change in the behavior of law enforcement personnel, not a change in the number of drug violations. Thus official crime statistics may be a better indicator of what police are doing rather than of what criminals are doing.

Victimization Surveys

Victimization surveys ask people whether they have been victims of crime. The National Crime Victimization Survey, conducted annually by the U.S. Department of Justice, interviews nearly 100,000 people about their experiences as victims of crime. Interviewers collect a variety of information, including the victim's background (e.g., age, race and ethnicity, sex, marital status, education, and area of residence), relationship to offender (stranger or nonstranger), and the extent to which the victim was harmed. For example, in 2003 males were more likely to be victimized by a stranger and females were more likely to be victimized by a friend, intimate, or acquaintance (BJS 2005b). Although victimization surveys provide detailed information about crime victims, they provide less reliable data on offenders.

Self-Report Offender Surveys

Self-report surveys ask offenders about their criminal behavior. The sample may consist of a population with known police records, such as a prison population, or it may include respondents from the general population, such as college students. Self-report data compensate for many of the problems associated with official statistics but are still subject to exaggerations and concealment. The Criminal Activities Survey in this chapter's *Self and Society* feature asks you to indicate whether you have engaged in a variety of illegal activities.

Self-report surveys reveal that virtually every adult has engaged in some type of criminal activity. Why then is only a fraction of the population labeled criminal?

Criminal Activities Survey

Read each of the following questions. If, since the age of 16, you have ever engaged in the behavior described, place a "1" in the space provided. If you have not engaged in the behavior, put a "0" in the space provided. After completing the survey, read the section on interpretation to see what your answers mean.

Questions

1. Have you ever been in possession of drug paraphernalia? ___

2. Have you ever lied about your age or about anything else when making application to rent an automobile? ___

3. Have you ever intentionally destroyed or erased someone else's phone messages? ___

4. Have you ever tampered with a coin-operated vending machine or parking meter? ___

5. Have you ever shared, given, or shown pornographic material to someone under 18? ___

6. Have you ever begun and/or participated in a basketball, baseball, or football pool? ___

7. Have you ever used "filthy, obscene, annoying, or offensive" language while on the telephone? ___

8. Have you ever given or sold a beer to someone under the age of 21? ___

9. Have you ever been on someone else's property (land, house, boat, structure, etc.) without that person's permission? ___

10. Have you ever forwarded a chain letter with the intent to profit from it? ___

11. Have you ever improperly gained access to someone else's e-mail or other computer account? ___

12. Have you ever written a check when you knew it was bad? ___

Interpretation

Each of the activities described in these questions represents criminal behavior that was subject to fines, imprisonment, or both under the laws of Florida in 2004. For each activity the following table lists the maximum prison sentence and/or fine for a first-time offender. To calculate your "prison time" and/or fines, sum the numbers corresponding to each activity you have engaged in.

Offense	Maximum Prison Sentence	Maximum Fine
1. Possession of drug paraphernalia	1 year	$1,000
2. Fraud	5 years	$5,000
3. Unlawful interference with telecommunications	2 months	$500
4. Fraud	2 months	$500
5. Protection of minors from obscenity	5 years	$5,000
6. Illegal gambling	2 months	$500
7. Harassing/obscene telecommunications	2 months	$500
8. Illegal distribution of alcohol	2 months	$500
9. Trespassing	1 year	$1,000
10. Illegal gambling	1 year	$1,000
11. Illegal misappropriation of cybercommunication	5 years	$5,000
12. Worthless check	1 year	$1,000

Source: *Florida Criminal Code* (2004).

Like a funnel, which is large at one end and small at the other, only a small proportion of the total population of law violators is ever convicted of a crime. For an individual to be officially labeled a criminal, his or her behavior (1) must become known to have occurred and (2) must come to the attention of the police who then file a report, conduct an investigation, and make an arrest; and finally, (3) the arrestee must go through a preliminary hearing, an arraignment, and a trial and may or may not be convicted. At every stage of the process an offender may be "funneled" out. As Figure 4.1 indicates, the measures of crime used at various points in time lead to different results.

Sociological Theories of Crime

Some explanations of crime focus on psychological aspects of the offender, such as psychopathic personalities, unhealthy relationships with parents, and mental illness. Other crime theories focus on the role of biological variables, such as central nervous system malfunctioning, stress hormones, vitamin or mineral deficiencies, chromosomal abnormalities, and a genetic predisposition toward aggression. Sociological theories of crime and violence emphasize the role of social factors in criminal behavior and societal responses to it.

Structural-Functionalist Perspective

According to Durkheim and other structural functionalists, crime is functional for society. One of the functions of crime and other deviant behavior is that it strengthens group cohesion: "The deviant individual violates rules of conduct that the rest of the community holds in high respect; and when these people come together to express their outrage over the offense . . . they develop a tighter bond of solidarity than existed earlier" (Erikson 1966, p. 4).

Crime can also lead to social change. For example, an episode of local violence may "achieve broad improvements in city services . . . [and] be a catalyst for making public agencies more effective and responsive, for strengthening families and social institutions, and for creating public-private partnerships" (National Research Council 1994, pp. 9–10).

Although structural functionalism as a theoretical perspective deals directly with some aspects of crime, it is not a theory of crime per se. Three major theories of crime have developed from structural functionalism, however. The first, called **strain theory**, was developed by Robert Merton (1957) and uses Durkheim's concept of *anomie,* or normlessness. Merton argued that when legitimate means (e.g., a job) of acquiring culturally defined goals (e.g., money) are limited by the structure of society, the resulting strain may lead to crime.

Individuals, then, must adapt to the inconsistency between means and goals in a society that socializes everyone into wanting the same thing but provides opportunities for only some (see Table 4.1). Conformity occurs when individuals accept the culturally defined goals and the socially legitimate means of achieving them. Merton suggested that most individuals, even those who do not have easy access to the means and goals, remain conformists. Innovation occurs when an individual accepts the goals of society but rejects or lacks the socially legitimate means of achieving them. Innovation, the mode of adaptation most associated with criminal behavior, explains the high rate of crime committed by uneducated and poor

Table 4.1

Merton's Strain Theory

Mode of Adaptation	Seeks Culturally Defined Goals?	Uses Structurally Defined Means to Achieve Them?
1. Conformity	Yes	Yes
2. Innovation	Yes	No
3. Ritualism	No	Yes
4. Retreatism	No	No
5. Rebellion	No—seeks to replace	No—seeks to replace

individuals who do not have access to legitimate means of achieving the social goals of wealth and power.

Another adaptation is ritualism, in which the individual accepts a lifestyle of hard work but rejects the cultural goal of monetary rewards. The ritualist goes through the motions of getting an education and working hard, yet he or she is not committed to the goal of accumulating wealth or power. Retreatism involves rejecting both the cultural goal of success and the socially legitimate means of achieving it. The retreatist withdraws or retreats from society and may become an alcoholic, drug addict, or vagrant. Finally, rebellion occurs when an individual rejects both culturally defined goals and means and substitutes new goals and means. For example, rebels may use social or political activism to replace the goal of personal wealth with the goal of social justice and equality.

Whereas strain theory explains criminal behavior as a result of blocked opportunities, **subcultural theories** argue that certain groups or subcultures in society have values and attitudes that are conducive to crime and violence. Members of these groups and subcultures, as well as other individuals who interact with them, may adopt the crime-promoting attitudes and values of the group. For example, Kubrin and Weitzer (2003) found that retaliatory homicide is a response to subcultural norms of violence that exist in some neighborhoods.

But if blocked opportunities and subcultural values are responsible for crime, why don't all members of the affected groups become criminals? **Control theory** may answer that question. Hirschi (1969), consistent with Durkheim's emphasis on social solidarity, suggests that a strong social bond between individuals and the social order constrains some individuals from violating social norms. Hirschi identified four elements of the social bond: attachment to significant others, commitment to conventional goals, involvement in conventional activities, and belief in the moral standards of society. Several empirical tests of Hirschi's theory support the notion that the higher the attachment, commitment, involvement, and belief, the higher the social bond and the lower the probability of criminal behavior. For example, Laub and colleagues (1998) found that a good marriage contributes to the cessation of a criminal career. In addition, Warner and Rountree (1997) reported that local community ties, although varying by neighborhood and offense, decrease the probability that crimes will occur. On the other hand, Vander Ven and colleagues (2001) reported that maternal employment is unrelated to behavioral problems of children.

Conflict Perspective

Conflict theories of crime suggest that deviance is inevitable whenever two groups have differing degrees of power; in addition, the more inequality in a society, the greater the crime rate in that society. Social inequality leads individuals to commit crimes such as larceny and burglary as a means of economic survival. Other individuals, who are angry and frustrated by their low position in the socioeconomic hierarchy, express their rage and frustration through crimes such as drug use, assault, and homicide. In Argentina, for example, the soaring violent crime rate is hypothesized to be "a product of the enormous imbalance in income distribution . . . between the rich and the poor" (Pertossi 2000).

According to the conflict perspective, those in power define what is criminal and what is not, and these definitions reflect the interests of the ruling class. Laws against vagrancy, for example, penalize individuals who do not contribute to the capitalist system of work and consumerism. Furthermore, D'Alessio and Stolzenberg (2002, p. 178) found that "in cities with high unemployment, unemployed defendants have a substantially higher probability of pretrial detention" than employed defendants. Rather than viewing law as a mechanism that protects all members of society, conflict theorists focus on how laws are created by those in power to protect the ruling class. For example, wealthy corporations contribute money to campaigns to influence politicians to enact tax laws that serve corporate interests (Reiman 2004).

In addition, conflict theorists argue that law enforcement is applied differentially, penalizing those without power and benefiting those with power. For example, although the race of a victim should not matter, blacks are more likely to be arrested when involved in black-on-white crime than when involved in black-on-black crime (Eitle et al. 2002). Moreover, female prostitutes are more likely to be arrested than are the men who seek their services. Unlike street criminals, corporate criminals are often punished by fines rather than by lengthy prison terms, and rape laws originated to serve the interests of husbands and fathers who wanted to protect their property—wives and unmarried daughters.

© Michael Newman/PhotoEdit

To Marxists the cultural definition of women as property contributes to the high rates of female criminality and, specifically, involvement in prostitution, drug abuse, and petty theft. In 2003 there were 79,733 arrests for prostitution and commercial vice in the United States.

Societal beliefs also reflect power differentials. For example, "rape myths" are perpetuated by the male-dominated culture to foster the belief that women are to blame for their own victimization, thereby, in the minds of many, exonerating the offender. Such myths include the notion that when a woman says no she means yes, that "good girls" don't get raped, that appearance indicates willingness, and that women secretly want to be raped. Not surprisingly, in societies where women and men have greater equality, there is less rape.

Symbolic Interactionist Perspective

Two important theories of crime emanate from the symbolic interactionist perspective. The first, **labeling theory**, focuses on two questions: How do crime and deviance come to be defined as such, and what are the effects of being labeled criminal or deviant? According to Howard Becker (1963):

> Social groups create deviance by making rules whose infractions constitute deviance, and by applying those rules to particular people and labeling them as outsiders. From this point of view, deviance is not a quality of the act a person commits, but rather a consequence of the application by others of rules and sanctions to an "offender." The deviant is one to whom the label has successfully been applied; deviant behavior is behavior that people so label. (p. 238)

Labeling theorists make a distinction between **primary deviance**, which is deviant behavior committed before a person is caught and labeled an offender, and **secondary deviance**, which is deviance that results from being caught and labeled. After a person violates the law and is apprehended, that person is stigmatized as a criminal. This deviant label often dominates the social identity of the person to whom it is applied and becomes the person's "master status," that is, the primary basis on which the person is defined by others.

Being labeled as deviant often leads to further deviant behavior because (1) the person who is labeled as deviant is often denied opportunities for engaging in nondeviant behavior, and (2) the labeled person internalizes the deviant label, adopts a deviant self-concept, and acts accordingly. For example, the teenager who is caught selling drugs at school may be expelled and thus denied opportunities to participate in nondeviant school activities (e.g., sports, clubs) and to associate with nondeviant peer groups. The labeled and stigmatized teenager may also adopt the self-concept of a "druggie" or "pusher" and continue to pursue drug-related activities and membership in the drug culture.

The assignment of meaning and definitions learned from others is also central to the second symbolic interactionist theory of crime, **differential association**. Edwin Sutherland (1939) proposed that through interaction with others, individuals learn the values and attitudes associated with crime as well as the techniques and motivations for criminal behavior. Individuals who are exposed to more definitions favorable to law violation (e.g., "crime pays") than to unfavorable ones (e.g., "do the crime, you'll do the time") are more likely to engage in criminal behavior. Thus children who see their parents benefit from crime or who live in high-crime neighborhoods where success is associated with illegal behavior are more likely to engage in criminal behavior.

Unfavorable definitions come from a variety of sources. Of particular concern of late is the role of video games in promoting criminal or violent behavior. One particular game, *Grand Theft Auto III,* has players "head bashing, looting, drug-dealing, drive-by shooting, and running over innocent bystanders with a taxi" (Richtel 2003). In response to this and other violent video games, many states now require a video rating system that differentiates between cartoon violence, fantasy violence, intense violence, and sexual violence (Associated Press 2003).

Types of Crime

The FBI identifies eight index offenses as the most serious crimes in the United States. The **index offenses**, or street crimes as they are often called, can be against a person (called violent or personal crimes) or against property (see Table 4.2). Other types of crime include vice crime (such as drug use, gambling, and prostitution),

Table 4.2

Index Crime Rates, Percentage Change, and Clearance Rates, 2003

	Rate per 100,000, 2003	Percentage Change in Rate (2002–2003)	Percentage Cleared, 2003
Violent crime			
Murder	5.7	+ .1	62.4
Forcible rape	32.1	−2.8	44.0
Robbery	142.2	−2.7	26.3
Aggravated assault	205.0	−4.7	55.9
Total	475.0	−3.9	46.5
Property crime			
Burglary	740.5	−.9	13.1
Larceny/theft	2414.5	−1.5	18.0
Motor vehicle theft	433.4	+.1	13.1
Arson	37.1[a]	+2.2	16.7
Total[b]	3588.4	−1.2	16.4

[a] Arson rates per 100,000 are calculated independently because population coverage for arson is lower than for the other index offenses—1999 rate.
[b] Property crime totals do not include arson.

Source: FBI (2004)

organized crime, white-collar crime, computer crime, and juvenile delinquency. Hate crimes are discussed in Chapters 9 and 10.

Street Crime: Violent Offenses

Data from the FBI's Uniform Crime Reports indicate that the 2003 violent crime rate decreased from the previous year by 3.9%. Remember, however, that crime statistics represent only those crimes *reported* to the police: 1.38 million violent crimes in 2003. Victim surveys indicate that a little over half of all violent crimes are actually reported to the police (FBI 2004; U.S. Bureau of the Census 2004).

Violent crime includes homicide, assault, rape, and robbery. Homicide refers to the willful or nonnegligent killing of one human being by another individual or group of individuals. Although homicide is the most serious of the violent crimes, it is also the least common, accounting for less than 1% of all index crimes (FBI 2004). A typical homicide scenario includes a male killing a male with a handgun after a heated argument. The victim and offender are disproportionately young and of minority status. When a woman is murdered and the victim-offender relationship is known, she is most likely to have been killed by her husband or boyfriend (FBI 2004).

Serial killers are a unique category of murderers. Unlike mass murderers, who have more than one victim in a killing event, serial killers kill consecutively over a long period of time. The most well known serial killers, who were responsible for some of the most horrific episodes of homicide, are Ted Bundy, Kenneth Bianchi, and Jeffery Dahmer. Recently, Dennis Rader, the self-proclaimed "BTK" (bind, torture, kill) killer was captured. Accused of killing 10 people (two men and eight

Should Sex Offenders Be Required to Register?

Megan Kanka, an 8-year-old girl, was sexually molested and killed after she was lured into a neighbor's house with the promise of a puppy. Megan's Law requires that "sexually violent persons, persons convicted of sexually violent offenses, and those convicted of offenses against minors" register in the community where they live (Reid 2003, p. 521). Opponents argue that such a law leads to vigilante-style violence as community members take the law into their own hands. In 2003 the U.S. Supreme Court held that (1) states may require registration of sex offenders whose crimes took place before Megan's Law was enacted and that (2) states may publish the names of offenders even if no assessment of risk has taken place ("Major Rulings," 2003).

It Is Your Turn Now to Take a Stand!

What do you think? Should sex offenders be required to register? Why or why not?

women) between 1974 and 1991, Rader was convicted of murder and received 10 consecutive life sentences with no chance of parole for 175 years. (Romano 2005; Coates 2005). Another form of violent crime, aggravated assault, involves the attacking of a person with the intent to cause serious bodily injury. Like homicide, aggravated assault occurs most often between members of the same race and, as with violent crime in general, is more likely to occur in warm weather months. In 2003 the assault rate was more than 50 times the murder rate, with assaults making up an estimated 60% of all violent crimes (FBI 2004).

Rape is also classified as a violent crime and is also intraracial, that is, the victim and offender are from the same racial group. The FBI definition of rape contains three elements: sexual penetration, force or the threat of force, and nonconsent of the victim. In 2003 more than 93,000 forcible rapes were reported in the United States, a slight decrease from the previous year (FBI 2004). Rapes are more likely to occur in warm months, in part because of the greater ease of victimization. People are outside more and later, doors are open, windows are unlocked, and so forth.

Perhaps as much as 80% of all rapes are **acquaintance rapes**—rapes committed by someone the victim knows. Although acquaintance rapes are the most likely to occur, they are the least likely to be reported and the most difficult to prosecute. Unless the rape is what Williams (1984) calls a **classic rape**—that is, the rapist was a stranger who used a weapon and the attack resulted in serious bodily injury—women hesitate to report the crime out of fear of not being believed. The increased use of "rape drugs," such as Rohypnol, may lower reporting levels even further (see Chapter 3).

Robbery, unlike simple theft, also involves force or the threat of force or putting a victim in fear and is thus considered a violent crime. Officially, in 2003 more than 400,000 robberies took place in the United States. Robberies are most often committed by young adults with the use of a gun (FBI 2004). Robbers and thus robberies vary dramatically in type, from opportunistic robberies whose victims are easily accessible and that yield only a small amount of money to professional robberies of commercial establishments, such as banks, jewelry stores, and convenience stores.

ThomsonNOW™

Learn more about **Violent Crimes** by going through the Violent Crimes Map Exercise.

Street Crime: Property Offenses

Property crimes are those in which someone's property is damaged, destroyed, or stolen; they include larceny, motor vehicle theft, burglary, and arson. Property crimes have gone down since 1994, with a 14.0% decrease in the last decade. Larceny, or simple theft, is the most common property crime, accounting for more than half of all property arrests. The average dollar value lost per larceny incident is less than $1,000. Examples of larcenies include purse snatching, theft of a bicycle, pickpocketing, theft from a coin-operated machine, and shoplifting. In 2003 there were more than 7 million larcenies reported in the United States (FBI 2004).

Larcenies involving automobiles and auto accessories are the largest category of thefts. However, because of the cost involved, motor vehicle theft is considered a separate index offense. Numbering more than 1.2 million in 2003, the motor vehicle theft rate has decreased 26.7% since 1994. Because of insurance requirements, vehicle theft is one of the most highly reported index crimes, and, consequently, estimates between the FBI's Uniform Crime Reports and the National Crime Victimization Survey are fairly compatible. Less than 14% of motor vehicle thefts are cleared.

Burglary, which is the second most common index offense, entails entering a structure, usually a house, with the intent to commit a crime while inside. Official statistics indicate that in 2004 more than 2.1 million burglaries occurred, a rate of 740 per 100,000 population. Most burglaries are residential rather than commercial and take place during the day when houses are unoccupied. The most common type of burglary is forcible entry, followed by unlawful entry.

Arson involves the malicious burning of the property of another. Estimating the frequency and nature of arson is difficult given the legal requirement of "maliciousness." Of the reported cases of arson, 42.1% involved structures (most of which were residential) and 33.3% involved movable property (e.g., boat, car), with the remainder being miscellaneous property (e.g., crops, timber). In 2003 the average dollar amount of damage as a result of arson was $11,942 (FBI 2004).

Vice Crimes

Vice crimes are illegal activities that have no complaining party and therefore are often called **victimless crimes**. Vice crimes include using illegal drugs, engaging in or soliciting prostitution (except for legalized prostitution, which exists in Nevada), illegal gambling, and pornography.

Most Americans view drug use as socially disruptive (see Chapter 3). There is less consensus, nationally or internationally, that gambling and prostitution are problematic. For example, in the Netherlands prostitution is legal. Although a country of only 16 million, the Netherlands has an estimated 25,000 to 50,000 sex workers. Like other workers, sex workers in the Netherlands have access to the social security system and pay income tax. It is, however, illegal for anyone under the age of 16 to visit a prostitute, and brothels, although openly advertised, are also illegal (The Situation in the Netherlands 2003).

In the United States many states have legalized gambling, including casinos in Nevada, New Jersey, Connecticut, and other states, as well as state lotteries, bingo parlors, horse and dog racing, and jai alai. In addition, some have argued that there is little difference, other than societal definitions of acceptable and unacceptable behavior, between gambling and other risky ventures such as investing in the stock market. Conflict theorists are quick to note that the difference is who is making the wager.

Organized crime refers to criminal activity conducted by members of a hierarchically arranged structure devoted primarily to making money through illegal means. Although often discussed under victimless crimes because of its association with prostitution, drugs, and gambling, organized crime often uses coercive techniques. For example, organized crime groups may force legitimate businesses to pay "protection money" by threatening vandalism or violence.

The traditional notion of organized crime is the Mafia, a national band of interlocked Italian families, but members of many ethnic groups engage in organized crime.

> Chinese, Vietnamese, Korean, and Japanese gangs have been found on the East and West coasts, active in smuggling drugs and extorting money from businesses. . . . Scores of other groups can be found in various cities: Cubans running illegal gambling operations in Miami, Canadians engaging in gun smuggling and money laundering . . . and Russians carrying out extortion and contract murders in New York. (Thio 2004, p. 395)

Organized crime also occurs at the international level. One of the largest organized crime groups in the world is the Yakuza of Japan. It is made up of 10 crime families, the largest one composed of over 750 separate gangs, each with 25 to 30 members. The young men who join the Yakuza tend to be from the lower class and must undergo a training period of five years. It is during this apprenticeship that members learn absolute loyalty to their superiors as well as the other norms and values of the group. The Yakuza are involved in drugs, gambling, and prostitution as well as several legitimate businesses. Interestingly, the Yakuza are legal in Japan and proudly display their name at their office headquarters, and recruits wear lapel pins identifying themselves as members (Thio 2004).

White-Collar Crime

White-collar crime includes both *occupational crime,* in which individuals commit crimes in the course of their employment, and *corporate crime,* in which corporations violate the law in the interest of maximizing profit. Occupational crime is motivated by individual gain. Employee theft of merchandise, or pilferage, is one of the most common types of occupational crime. Other examples include embezzlement, forgery and counterfeiting, and insurance fraud. Price fixing, antitrust violations, and security fraud are all examples of corporate crime, that is, crime that benefits the organization. In recent years several officers of major corporations, including Enron, WorldCom, Adelphia, and Imclone, have been charged with securities fraud, tax evasion, and insider trading (Eisenberg 2002). WorldCom engaged in what has been called the "largest accounting fraud in history," exaggerating its worth by $9 billion. Shareholders lost more than $3 billion, and 17,000 employees are in danger of losing their jobs (Ripley 2003). In part as a response to the widespread scandals of late, the U.S. Sentencing Commission has approved stiffer penalties for white-collar criminals, including a new sentencing formula (Lichtblau 2003). Moreover, the financial crimes section of the FBI has become more vigilant in its pursuit of white-collar criminals. For example, insurance fraud probes have increased and are now one of the FBI's top investigative priorities. (FBI 2005).

Corporate violence, another form of corporate crime, refers to the production of unsafe products and the failure of corporations to provide a safe working environment for their employees. Corporate violence is the result of negligence, the pursuit of profit at any cost, and intentional violations of health, safety, and environmental regulations. For example, after more than a year of recalls in 16 countries,

Residents of Love Canal were victimized by corporate violence when toxic waste, dumped by Hooker Chemical Company, began to seep into the basements of homes and schools. Many residents of Love Canal moved out of the area; others complained of high rates of miscarriages, birth defects, and cancer. Although Love Canal was allegedly cleaned up, residents protested against resettling the area. After 20 years of litigation the last lawsuit was settled in 1998, with the City of Love Canal receiving $250,000 and the construction of a park near the spill.

Bridgestone/Firestone began a U.S. recall of more than 6.5 million tires. The tires, many of which were standard equipment on the popular Ford Explorer, had a 10-year history of tread separation. It was only after 88 U.S. traffic deaths were linked to the defective tires, prompting a congressional investigation, that Ford and Bridgestone/Firestone acknowledged the overseas recalls and the tires' questionable safety history (Pickler 2000). Subsequently, Ford Motor Company was asked by a federal judge to turn over data on their 15-passenger van. Several deaths occurred as a result of the van rolling over, and Ford was accused of hiding evidence of the problem. According to the National Highway Traffic Safety Administration, more than 400 people have died in passenger van accidents since 1990 (Chicago Tribune 2003). Table 4.3 summarizes some of the major categories of white-collar crime.

One of the problems in pursuing "corporate criminals" is the difficulty in assigning legal culpability to the offender. As Friedrichs (2004) notes:

> The absence of the direct intent to do harm, the difficulty in pinpointing the specific cause of the harm, the diffusion of responsibility for harm producing corporate decisions, and the economic and political clout of the corporations has combined to shield corporate employers from full fledged liability. (p. 72)

Some recent evidence, however, suggests that the white-collar crime culture of "boys will be boys" is being replaced, according to Marjorie Kelly, editor of *Business Ethics,* with a new and less tolerant "Puritanism" (Thomas 2005).

Computer Crime

Computer crime refers to any violation of the law in which a computer is the target or means of criminal activity. Sometimes called cybercrime, computer crime is one of the fastest growing crimes in the United States. Hacking, or unauthorized computer intrusion, is one type of computer crime. In just one month hackers

Table 4.3

Types of White-Collar Crime

Crimes against consumers	Crimes against employees
Deceptive advertising	Health and safety violations
Antitrust violations	Wage and hour violations
Dangerous products	Discriminatory hiring practices
Manufacturer kickbacks	Illegal labor practices
Physician insurance fraud	Unlawful surveillance practices
Crimes against the public	**Crimes against employers**
Toxic waste disposal	Embezzlement
Pollution violations	Pilferage
Tax fraud	Misappropriation of government funds
Security violations	Counterfeit production of goods
Police brutality	Business credit fraud

successfully attacked the computer systems of Walt Disney World, Yahoo, eBay, and Amazon.com through "denial of service" invasions (Kong & Swartz 2000). The cost of computer crimes, although difficult to set, is estimated to be in the billions of dollars (Jacobson & Green 2002).

The increase in computer break-ins has also led to an increase in **identity theft**, the use of someone else's identification (e.g., social security number, birth date) to obtain credit or other rewards. In 2002 the number of identity thefts doubled from the previous year, becoming the most frequent complaint to the Federal Trade Commission (Lee 2003). Although mail theft is one of the most common modes of obtaining the needed information, new technologies have contributed to the increased rate of identity theft. For example, weekly, thousands of stolen credit card numbers are sold online in "membership only cyberbazaars, operated largely by residents of the former Soviet Union who have become central players in credit card and identity theft" (Richtel 2002, p. 1). Recent accounts of unauthorized use of personal data from such companies as Lexis-Nexis, Bank of America, eBay, and ChoicePoint have led Congress to call for stricter treatment of "data brokers." In 2005 the personal information, including social security numbers, of nearly 100,000 University of Berkeley students and alumni was stolen (Associated Press 2005a). Given the more than 10 million victims a year (Harrow 2005), it is not surprising that a survey of Americans indicates a rise in concern about identity theft and privacy issues in general (Cohen 2005). In response to such public opinion, in 2004 President Bush signed the Identity Theft Penalty Enhancement Act (White House 2005).

Identity theft is just one category of computer crime. Another category is Internet fraud. According to a 2005 report by the Internet Crime Complaint Center (ICCC), the most common category of Internet fraud is Internet auction fraud, which represents 71.2% of all fraud complaints referred to law enforcement agencies. Internet auction fraud entails the misrepresentation of items for sale online and/or the nondelivery of products bought online. Other Internet fraud categories include pyramid schemes, investment fraud, credit card fraud, and counterfeit check schemes (ICCC 2005). In 2004 the ICCC received more than 200,000 complaints.

Conklin (1998), Reid (2003), Schiesel (2005), and Siegel (2006) identify other types of computer crime:

- Two individuals were charged with the theft of 80,000 cellular phone numbers. Using a device purchased from a catalogue, the thieves picked up radio waves from passing cars, determined private cellular codes, reprogrammed computer chips with the stolen codes, and then, by inserting the new chips into their own cellular phones, charged calls to the original owners.
- A programmer made $300 a week by programming a computer to round off each employee's paycheck down to the nearest dime and then to deposit the extra few pennies in the offender's account.
- A computer hacker broke into a telephone system and rigged the outcome of a radio station contest. Three hackers won a trip to Hawaii, a Porsche, and a cash prize.
- An oil company illegally tapped into another oil company's computer to get information that allowed the offending company to underbid the other company for leasing rights.
- After building a good credit record, a California man sold over $800,000 worth of merchandise on eBay and never delivered the goods.
- While sitting in his car, a suspect was caught viewing child pornography he had just downloaded onto his laptop from a neighbor's wireless system.

Juvenile Delinquency

In general, children under the age of 18 are handled by the juvenile courts, either as status offenders or as delinquent offenders. A status offense is a violation that can be committed only by a juvenile, such as running away from home, truancy, and underage drinking. A delinquent offense is an offense that would be a crime if committed by an adult, such as the eight index offenses. The most common status offenses handled in juvenile court are underage drinking, truancy, and running away. In 2003, 16.3% of all arrests (excluding traffic violations) were of offenders under the age of 18. The number of arrests of minors for violent crime remained fairly stable

© A. Ramey/PhotoEdit

Females who join gangs often do so to win approval from their boyfriends who are gang members. Increasingly, though, females are forming independent "girl gangs." The most common type of female gang member remains, however, a female auxiliary to a male gang.

between 2002 and 2003, although arrests for drug offenses increased nearly 4% between the two years (FBI 2004).

As is the case with adults, juveniles commit more property than violent offenses and the number of violent offenses has dropped in recent years. Nonetheless, Americans are concerned about the high rate of juvenile violence, including violence in schools (see Chapter 12) and gang-related violence. Gang-related crime is, in part, a function of two interrelated social forces: the increased availability of guns in the 1980s and the lucrative and expanding drug trade. It is estimated that the United States has 24,500 street gangs with 772,500 members, the highest proportion of which are racial and ethnic minorities (Shelden et al. 2004). The most dangerous gang in America is Mara Salvatrucha (MS-13), and it has members in more than 30 states in the United States (Campo-Flores 2005). Made up primarily of street-savvy Salvadorans and boasting a membership of more than 10,000, MS-13 has become one of the highest priorities of the FBI's criminal enterprise division.

ThomsonNOW™

Learn more about **Demographic Patterns of Crime** by going through the Measuring Crime Animation.

Demographic Patterns of Crime

Although virtually everyone violates a law at some time, individuals with certain demographic characteristics are disproportionately represented in the crime statistics. Victims, for example, are disproportionately young lower-class minority males from urban areas. Similarly, the probability of being an offender varies by gender, age, race, social class, and region (see Figure 4.2).

Gender and Crime

Both official statistics and self-report data indicate that females commit fewer violent crimes than males (see Figure 4.2). Why are females less likely to commit violent crimes? Some would argue that "girls are less violent than boys because they are controlled through subtle mechanisms, which include[s] learning that violence is incompatible with the meaning of their gender" (Heimer & DeCoster 1999, pp. 305–306).

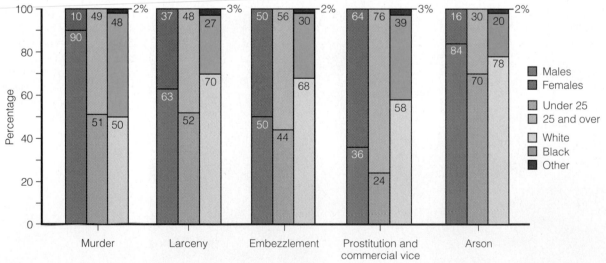

Figure 4.2
Percentage of arrests by sex, age, and race, 2003.
Source: U.S. Department of Justice (2004).

In 2003 males accounted for 77% of all arrests, 82.2% of all arrests for violent crime, and 69.2% of all arrests for property crimes (FBI 2004). Not only are females less likely than males to commit serious crimes, but the monetary value of female involvement in theft, property damage, and illegal drugs is typically less than that for similar offenses committed by males. Nevertheless, a growing number of women have become involved in characteristically male criminal activities, such as gang-related crime and drug use.

The recent increase in crimes committed by females has led to the development of a feminist criminology. Feminist criminology focuses on how the subordinate position of women in society affects their criminal behavior. For example, Chesney-Lind and Shelden (2004) report that arrest rates for runaway juvenile females are higher than those for males not only because girls are more likely to run away as a consequence of sexual abuse in the home but also because police with paternalistic attitudes are more likely to arrest female runaways than male runaways. Feminist criminology thus adds insights into understanding crime and violence that are often neglected by traditional theories by concentrating on gender inequality in society.

The subordinate position of women in the United States also affects their victimization rates. A report from the Harvard School of Public Health revealed that 70% of all female homicide victims in industrial countries are American (Harvard School of Public Health 2002). A female in the United States is five times more likely to be murdered than a female in Germany, eight times more likely to be murdered than a female in England, and three times more likely to be murdered than a female in Canada. When she is murdered, it is most likely by an ex-boyfriend, husband, or other intimate.

Age and Crime

In general, criminal activity is more prevalent among younger people than older people. The highest arrest rates are for individuals under the age of 25. Crimes committed by people in their teens or early 20s tend to be property crimes, such as burglary, larceny, arson, and vandalism. In 2003, 48% of all arrests in the United States were of people under the age of 25 (FBI 2004). The median age of people who commit more serious crimes, such as aggravated assault and homicide, is the late 20s.

Why is criminal activity more prevalent among individuals in their teens and early 20s? One reason is that juveniles are insulated from many of the legal penalties for criminal behavior. Younger individuals are also more likely to be unemployed or employed in low-wage jobs. Thus, as strain theorists argue, they have less access to legitimate means for acquiring material goods.

Some research suggests, however, that high school students who have jobs become more, rather than less, involved in crime (Felson 2002). In earlier generations teenagers who worked did so to support themselves and/or their families. Today, teenagers who work typically spend their earnings on recreation and "extras," including car payments and gasoline. The increased mobility associated with having a vehicle also increases opportunity for criminal behavior and reduces parental control.

Race, Social Class, and Crime

Race is a factor in who gets arrested. Minorities are disproportionately represented in official statistics. For example, although African Americans represent about 13% of the population, they account for more than 37% of all violent index offenses and

Race and Ethnicity in Sentencing Outcomes

Certainly one of the more pressing issues of recent times is the question of whether legally irrelevant information (e.g., race, sex of offender) affects criminal justice outcomes. Given recent accounts of alleged police brutality, accusations of racial profiling, and concern over race, class, and ethnic bias in death penalty cases, Steffensmeier and Demuth's (2000) investigation of the impact of extralegal variables in federal court sentencing is particularly timely.

Sample and Methods

Steffensmeier and Demuth (2000) investigated whether the criminal justice system discriminates on the basis of race and ethnicity. Although considerable research supports the contention that minorities are more likely to receive harsher sentences than other members of society (because minorities have fewer resources, are perceived as "dangerous" or as threatening to the status quo, etc.), Steffensmeier and Demuth's (2000) study examined both inter- and intragroup variations. The researchers argue that court decisions take place within a powerful cultural context, a "perceptual shorthand" of sorts, where the definitions of an offender's blameworthiness, the need to protect society, and the practical limitations of the sentence imposed affect the decision-making process. Given that federal sentencing guidelines allow for some discretion and that "Hispanic defendants may seem even more culturally dissimilar and be even more disadvantaged than their black counterparts" (Steffensmeier & Demuth 2000, p. 709), the researchers hypothesized (1) that Hispanics will receive more severe sentences than whites or blacks, (2) that in drug cases Hispanics specifically but minorities in general will receive harsher sentences than whites, and (3) that Hispanics identified as black will receive harsher sentences than Hispanics identified as white. The dependent variable, severity of sentence, was measured by whether the court imposed a prison sentence and, if so, the length of the sentence. The independent variables were race and ethnicity. The data included all convictions of U.S. citizens in federal courts between 1993 and 1996 ($N = 89,637$).

Findings and Conclusions

Steffensmeier and Demuth (2000) note the "considerable consistency" of sentencing of federal criminal defendants across racial and ethnic categories for *similar cases*.

> Judges, on balance, prescribe similar sentences for similar defendants convicted of the same offense. Whether they were white, black, or Hispanic, defendants who committed more serious crimes, had more extensive criminal histories, or were convicted at trial (as opposed to a guilty plea), were much more likely to be incarcerated and receive longer prison sentences. (p. 724)

The investigators also note small to moderate effects of race and ethnicity in terms of the *overall* sentencing, with Hispanics receiving harsher sentences than whites and blacks. These differences are likely the result of what is called "substantial assistance departures." When a defendant, for example, "substantially assists" the state in the prosecution or investigation of another offender, the judge may "depart" from the sentencing guidelines. Because Hispanics may be more likely to be involved in drug-trafficking circles that seek retribution against those who cooperate with authorities, they may be less likely to provide "substantial assistance" and thus are less likely to have their sentences reduced.

Finally, the researchers conclude that their results "provide strong evidence for the continuing significance of race and ethnicity in the larger society in general and in organizational decision-making processes in particular" (Steffensmeier and Demuth 2000, p. 726). Although acknowledging that the results are less an indication of discrimination than the unintended consequence of seemingly neutral policies, the investigators call for more research in the area of race and ethnic relations.

29% of all property index offenses (FBI 2004). In addition, black males between the ages of 20 and 39 make up about one-third of the prison population (BJS 2003).

Nevertheless, it is inaccurate to conclude that race and crime are causally related. First, official statistics reflect the behaviors and policies of criminal justice actors. Thus the high rate of arrests, conviction, and incarceration of minorities may be a consequence of individual and institutional bias against minorities (see this chapter's *Social Problems Research Up Close* feature). For example, blacks and Hispanics have higher perceptions of police misconduct than whites do (Weitzer &

Tuch 2004). Furthermore, blacks are sent to prison for drug offenses at a rate 8.2 times higher than the rate for whites. If present trends continue, by 2020 two out of every three black men between the ages of 18 and 34 will be in prison (Fletcher 2000; Dickerson 2000). These disturbing statistics have led to concerns over **racial profiling**—the practice of targeting suspects on the basis of race. Proponents of the practice argue that because race, like gender, is a significant predictor of who commits crime, the practice should be allowed. Opponents hold that racial profiling is little more than discrimination and should therefore be abolished.

Second, race and social class are closely related in that nonwhites are overrepresented in the lower classes. Because lower-class members lack legitimate means to acquire material goods, they may turn to instrumental, or economically motivated, crimes. In addition, although the "haves" typically earn social respect through their socioeconomic status, educational achievement, and occupational role, the "have-nots" more often live in communities where respect is based on physical strength and violence, as subcultural theorists argue. For example, Kubrin (2005) examined the "street code" of inner-city black neighborhoods by analyzing rap music lyrics. Her results indicate that song "lyrics instruct listeners that toughness and the willingness to use violence are central to establishing viable masculine identity, gaining respect, and building a reputation" (p. 375).

Thus the apparent relationship between race and crime may be, in part, a consequence of the relationship between these variables and social class. Philips (2002), in her investigation of white, black, and Latino homicide rates, concludes that

> it nonetheless remains clear from this study that a significant portion of the racial homicide differential could be reduced by improving socioeconomic conditions for minority populations. This conclusion provides some promising policy options. For example, improving levels of education, lowering levels of poverty, and reducing the extent of male unemployment among minority populations might well have an impact on levels of violence and reduce the striking racial homicide differential that currently exists in the United States. (p. 367)

A third hypothesis is that criminal justice system contact, which is higher for nonwhites, may actually act as the independent variable; that is, it may lead to a lower position in the stratification system. Kerley and colleagues (2004) found that "contact with the criminal justice system, especially when it occurs early in life, is a major life event that has a deleterious effect on individuals' subsequent income level" (p. 549).

Some research indicates, however, that even when social class backgrounds of blacks and whites are comparable, blacks have higher rates of criminality (D'Alessio and Stolzenberg 2003). In addition, to avoid the bias inherent in official statistics, researchers have compared race, class, and criminality by examining self-report data and victim studies. Barkan's (2006) findings indicate that although racial and class differences in criminal offenses exist, the differences are not as great as official data would indicate.

Region and Crime

In general, crime rates, and particularly violent crime rates, are higher in metropolitan areas than in nonmetropolitan areas (Moore 2005). In 2003, 90.1% of the estimated total of violent crime took place in a metropolitan statistical area (FBI 2004). Higher crime rates in urban areas result from several factors. First, social control is a function of small intimate groups that socialize their members to engage in law-abiding behavior, expressing approval for their doing so and disapproval for their noncompliance. In large urban areas people are less likely to know each other and thus are not influenced by the approval or disapproval of strangers. Demographic factors also explain why crime rates are higher in urban areas: Cities have large concentrations of poor, unemployed, and minority individuals.

Crime rates also vary by region of the country (see Figure 4.3). In 2003 both violent and property crimes were highest in southern states followed by western, midwestern, and northeastern states. The murder rate is particularly high in the South, with 44% of all murders recorded in southern states (FBI 2004). The high rate of

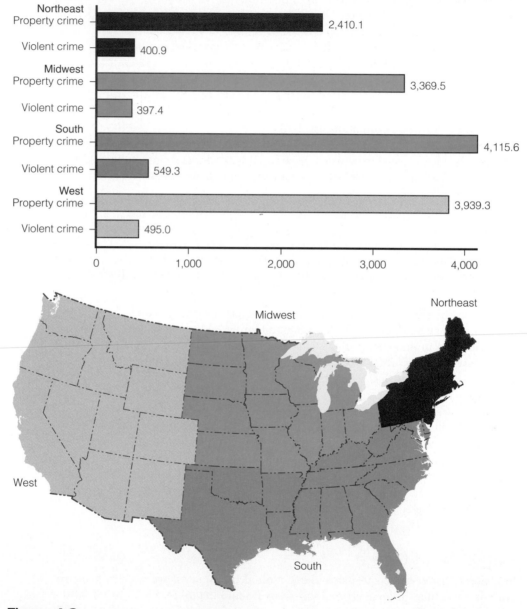

Figure 4.3
Regional crime rates, 2003: Violent and property crime per 100,000 inhabitants.
Source: FBI (2004).

southern lethal violence has been linked to high rates of poverty and minority populations in the South, a southern "subculture of violence," higher rates of gun ownership, and a warmer climate that facilitates victimization by increasing the frequency of social interaction.

ThomsonNOW™

Learn more about **Crime Rates** by going through the Measuring Crime Rates Data Experiment.

The Costs of Crime and Social Control

Crime often results in physical injury and loss of life. For example, in 2003 there were more than 16,500 victims of a homicide (FBI 2004). That number is dwarfed, however, by the deaths that take place as a consequence of white-collar crime. For example, criminologist Steven Barkin (2006), who collected data from a variety of sources, reports that annually there are

(1) 56,425 workplace-related deaths from illness or injury; (2) 9,600 deaths from unsafe products; (3) 35,000 deaths from environmental pollution; and (4) 12,000 deaths from unnecessary surgery. Adding these figures together, 113,025 people a year die from corporate and professional crime and misconduct. (p. 388)

Moreover, the U.S. Public Health Service now defines violence as one of the top health concerns facing Americans. In addition to death and physical injury, crime also has economic, social, and psychological costs.

Economic Costs

Conklin (1998, pp. 71–72) suggests that the financial costs of crime can be classified into at least six categories. First are direct losses from crime, such as the destruction of buildings through arson, of private property through vandalism, and of the environment by polluters. In 2003 the average dollar loss of destroyed or damaged property as a result of arson was $11,942 (FBI 2004). Second are costs associated with the transferring of property. Bank robbers, car thieves, and embezzlers have all taken property from its rightful owner at tremendous expense to the victim and society. For example, it is estimated that in 2003, $8.6 billion was lost as a result of motor vehicle theft; the average value per vehicle at the time of the theft was $6,797 (FBI 2004).

A third major cost of crime is that associated with criminal violence, for example, the medical cost of treating crime victims ($5 billion annually) or the loss of productivity of injured workers (Surgeon General 2002). Fourth are the costs associated with the production and sale of illegal goods and services, that is, illegal expenditures. The expenditure of money on drugs, gambling, and prostitution diverts funds away from the legitimate economy and enterprises and lowers property values in high-crime neighborhoods. Fifth is the cost of prevention and protection—the billions of dollars spent on locks and safes, surveillance cameras, guard dogs, and the like. It is estimated that Americans spend $65 billion annually on self-protection items (Surgeon General 2002).

Finally, there is the cost of social control—the criminal justice system, law enforcement, litigative and judicial activities, corrections, and victims' assistance. The cost of the criminal justice system is estimated to be $90 billion annually and growing (Surgeon General 2002). Reasons for such growth include increases in the rates of arrest and conviction, changes in the sentencing structure, public attitudes toward criminals, the growing number of young males, and the war on drugs. Regardless of the cause, however, the staggering cost of public institutions has led to the "privatization" of prisons, whereby the private sector increasingly supplies needed prison services.

What is the total economic cost of crime? One estimate suggests that the total cost of crime in the United States is more than $1.7 trillion a year (Anderson 1999). Although costs from "street crimes" are staggering, the costs from "crimes in the suites," such as tax evasion, fraud, false advertising, and antitrust violations, are greater than the cost of the FBI index crimes combined (Reiman 2004). For example, Barkan (2006), using an FBI estimate, reports that the total cost of property crime and robbery is $17.1 billion annually. This is less than the $44 billion price tag for employee theft alone.

Social and Psychological Costs

Crime entails social and psychological costs as well as economic costs. Despite falling crime rates, when a random sample of Americans were asked whether there was more or less crime today than there was a year ago, 53% of respondents reported that there was more crime, 28% responded that there was less crime, and 14% responded that the amount of crime was about the same. In the same survey, when asked how serious a problem U.S. crime is, 42% responded that it is an extremely or very serious problem. Interestingly, when asked about the "crime problem" "in the area where you live," less than 10% responded that the problem was extremely or very serious (Gallup Poll 2004).

Fear of crime may be fueled by media presentations that may not accurately reflect the crime picture. For example, in a content analysis of local television portrayals of crime victims and offenders, whites were overrepresented as victims and blacks were overrepresented as offenders compared with official statistics (Dixon & Linz 2000).

> Fear of crime and violence also affects community life. If frightened citizens remain locked in their homes instead of enjoying public spaces, there is a loss of public and community life, as well as a loss of "social capital," the family and neighborhood channels that transmit positive social values from one generation to the next. (National Research Council 1994, pp. 5–6)

This is particularly true of women and the elderly who restrict their activities, living "limited lives," as a consequence of their fear of victimization (Madriz 2000; Merkle 2004).

White-collar crimes also take a social and psychological toll at both the individual and the societal level. Rosoff and colleagues (2002) state that white-collar crime can produce "feelings of cynicism among the public, remove an essential element of trust from everyday interaction, de-legitimatize political institutions, and weaken respect for the law" (p. 346). In addition, the authors argue that white-collar crime "encourages and facilitates" other types of crime; that is, "there is a connection, both direct and indirect, between 'crime in the suites' and 'crime in the streets'" (p. 346).

Strategies for Action: Crime and Social Control

Clearly, one way to combat crime is to attack the social problems that contribute to its existence. Moreover, when a random sample of Americans were asked which of two views came closer to their own in dealing with the crime problem, increasing law enforcement or resolving social problems, the majority of respondents, 69%,

"The possibility of being a victim of a crime is ever present on my mind. . . . Thinking about it is as natural as . . . breathing."

40-year-old woman in New York City

"The eight blunders that lead to violence in society:

- Wealth without work
- Pleasure without conscience
- Knowledge without character
- Commerce without morality
- Science without humanity
- Worship without sacrifice
- Politics without principle
- Rights without responsibilities"

Mohandas K. Gandhi Indian nationalist leader and peace activist

selected resolving social problems (Gallup Poll 2004). In addition to policies that address social problems, numerous social programs have been initiated to alleviate the crime problem. These policies and programs include local initiatives, criminal justice policies, and legislative action.

Local Initiatives

Youth Programs. Early intervention programs acknowledge that it is better to prevent crime than to "cure" it once it has occurred. Preschool enrichment programs, such as the Perry Preschool Project, have been successful in reducing rates of aggression in young children. After a group of children were randomly assigned to either a control group or an experimental group, the experimental-group members received academically oriented interventions for one to two years, frequent home visits, and weekly parent-teacher conferences. When the control and experimental groups were compared, the experimental group had better grades, higher rates of high school graduation, lower rates of unemployment, and fewer arrests (Murray et al. 1997).

In recognition of the link between juvenile delinquency and adult criminality, many anticrime programs are directed toward at-risk youths. These prevention strategies, including youth programs such as Boys and Girls Clubs, are designed to keep young people "off the streets," provide a safe and supportive environment, and offer activities that promote skill development and self-esteem. According to Gest and Friedman (1994), housing projects with such clubs report 13% fewer juvenile crimes and a 25% decrease in the use of crack.

Finally, many youth programs are designed to engage juveniles in noncriminal activities and integrate them into the community. In Weed and Seed, a program under the Department of Justice, "law enforcement agencies and prosecutors cooperate in 'weeding out' criminals who participate in violent crime and drug abuse . . . and 'seeding' human services to the area, encompassing prevention, intervention, treatment and neighborhood revitalization" (Weed and Seed 2005, p. 1). As part of the program, "safe havens" are established in, for example, schools, where multi-agency services are provided for youth.

Community Programs. Neighborhood watch programs involve local residents in crime prevention strategies. For example, MAD DADS (Men Against Destruction Defending Against Drugs and Social-Disorder) patrol the streets in high-crime areas of the city on weekend nights, providing positive adult role models and fun community activities for troubled children. Members also report crime and drug sales to police, paint over gang graffiti, organize gun buy-back programs, and counsel incarcerated fathers. At present, there are 65,000 MAD DADS in 60 chapters located in 16 states (MAD DADS 2005). In 2004, 10,000 communities in 50 states—34 million people—participated in "National Night Out," a crime prevention event in which citizens, businesses, neighborhood organizations, and local officials joined together in outdoor activities to heighten awareness of neighborhood problems, promote anticrime messages, and strengthen community ties (National Night Out 2005).

Mediation and victim-offender dispute-resolution programs are also increasing, with more than 3,000 such programs in the United States and Canada. The growth of these programs is a reflection of their success rate: Two-thirds of cases referred result in face-to-face meetings, 95% percent of these cases result in a written restitution agreement, and 90% of the written restitution agreements are completed within one year (Victim-Offender Reconciliation Program 2005).

ThomsonNOW™

Learn more about **Criminal Justice Policy** by going through the Criminal Justice System Learning Module.

Finally, the Internet has provided several ways to fight crime locally. Community-sponsored Internet sites routinely post names, descriptions, and photographs of wanted criminals. Other community efforts have made the transition from traditional media (e.g., television) to cybercrime fighting. One such case is McGruff and the "Take a bite out of crime" campaign, which now hosts on its home page, among other things, the National Citizens' Crime Prevention Campaign (NCCPC), the first public education initiative on community crime prevention in the country. This campaign is "designed to stimulate community involvement, generate confidence in comprehensive crime prevention activities, and provide a national focus and resource for crime prevention programs across the country" (NCCPC 2005).

Criminal Justice Policy

The criminal justice system is based on the principle of **deterrence**, that is, the use of harm or the threat of harm to prevent unwanted behaviors. The criminal justice system assumes that people rationally choose to commit crime, weighing the rewards and consequences of their actions. Thus the recent emphasis on "get tough" measures holds that maximizing punishment will increase deterrence and cause crime rates to decrease. Research indicates, however, that the effectiveness of deterrence is a function of not only the severity of the punishment but also the certainty and swiftness of the punishment. Furthermore, get-tough policies create other criminal justice problems, including overcrowded prisons and, consequently, the need for plea bargaining and early-release programs.

Law Enforcement Agencies. In 2003 the United States had 663,796 full-time law enforcement officers and 285,146 civilian support staff, yielding 3.5 law enforcement employees per 1,000 inhabitants (FBI 2004). Ironically, despite recent increases in the number of law enforcement personnel, public opinion that "more police on the streets" will reduce violent crime has decreased (Pew Research Center 2000). In addition, accusations of racial profiling, police brutality, and discriminatory arrest practices have shaken public confidence in the police.

In response to such trends, the Crime Control Act of 1994 established the Office of Community Oriented Policing Services (COPS). Called the "most important development in policing in the past quarter century" (Skogan & Roth 2004, p. xvii), community-oriented policing involves collaborative efforts among the police, the citizens of a community, and local leaders. As part of community policing efforts, officers speak to citizen groups, consult with social agencies, and enlist the aid of corporate and political leaders in the fight against neighborhood crime (COPS 2005).

Officers using community policing techniques often employ "practical approaches" to crime intervention. Such solutions may include what Felson (2002) calls "situational crime prevention." Felson argues that much of crime could be prevented simply by minimizing the opportunity for its occurrence. For example, cars could be outfitted with unbreakable glass, flush-sill lock buttons, an audible reminder to remove keys, and a high-security lock for steering columns (Felson 2002). These techniques, and community-oriented policing in general, have been fairly successful. A recent assessment indicates that visible community policing is positively associated with quality-of-life measures at the both the individual level (e.g., perception of personal safety) and the neighborhood level (e.g., perception of gang activity) (Reisig and Parks 2004).

Rehabilitation Versus Incapacitation. An important debate concerns the primary purpose of the criminal justice system: Is it to rehabilitate offenders or to incapacitate them through incarceration? Both **rehabilitation** and **incapacitation** are concerned with recidivism rates, or the extent to which criminals commit another crime. Advocates of rehabilitation believe that recidivism can be reduced by changing the criminal, whereas proponents of incapacitation think that recidivism can best be reduced by placing the offender in prison so that he or she is unable to commit further crimes against the general public.

Societal fear of crime has led to a public emphasis on incapacitation and a demand for tougher mandatory sentences, a reduction in the use of probation and parole, support of a "three strikes and you're out" policy, and truth-in-sentencing laws. However, these tough measures have recently come under attack for two reasons. First, research indicates that incarceration may not deter crime. For example, a study by the U.S. Department of Justice reported that 67% of inmates re-offend within three years of being released (Butterfield 2002b). Second is the accusation that get-tough measures, such as California's "three strikes and you're out" policy, are patently unfair (Irwin 2005). Leandro Andrade stole $153 worth of children's videos from K-Mart and was sentenced to 50 years to life in prison. Is that equitable? As unjust as it sounds, those in favor of the policy would be quick to note that the man had prior convictions for burglary and shoplifting. Andrade appealed his sentence to the U.S. Supreme Court, which held that California's three-strikes law did not violate the constitutional ban on grossly disproportionate sentences ("Major Rulings," 2003). In 2004 Proposition 66, which would have, in effect, repealed California's three-strikes policy, was defeated by a slim margin.

Although incapacitation is clearly enhanced by longer prison sentences, deterrence, as discussed previously, and rehabilitation may not be (Irwin 2005). Rehabilitation assumes that criminal behavior is caused by sociological, psychological, and/or biological forces rather than being solely a product of free will. If such forces can be identified, the necessary change can be instituted. Rehabilitation programs include education and job training, individual and group therapy, substance abuse counseling, and behavior modification. As Siegel (2006) notes, from this perspective "even the most hardened criminal may be helped by effective institutional treatment plans and services" (p. 506) One example is the IMPACT program (inmates providing animals with care and training), in which prisoners of Georgia's Metro State Prison, over a 16-month period, raise and train guide dogs.

Corrections. While the debate between rehabilitation and incarceration continues, the number of inmates continues to grow. In June 2004 just over 2 million inmates were housed in the nation's federal, state, and local institutions (BJS 2005c). Jail officials were also supervising 70,000 women and men who were "in the community in work release programs, weekend reporting, electronic monitoring and other alternative programs" (BJS 2005c, p. 1). The United States has the highest incarceration rate in the world, and it comes with a $60 billion price tag, up from $9 billion just two decades ago ("Death Behind Bars," 2005) (see Figure 4.4).

Between 1971 and 2002 the prison population increased 600% (Clear 2006). Such increases have usually been met with building more prisons, but concerns with budget deficits have caused many states to "close prisons, lay off guards, and consider shortening sentences" (Butterfield 2002b). **Probation** entails the conditional release of an offender who, for a specific time period and subject to certain conditions, remains under court supervision in the community (Siegel 2006).

Figure 4.4

Prison population rates per 100,000 and rank in world (out of 211 countries) as of May 10, 2005.

Source: International Centre for Prison Studies, 2005. Available at http://www.kcl.ac.uk/depsta/rel/icps/worldbrief/highest_to_lowest_rates.php

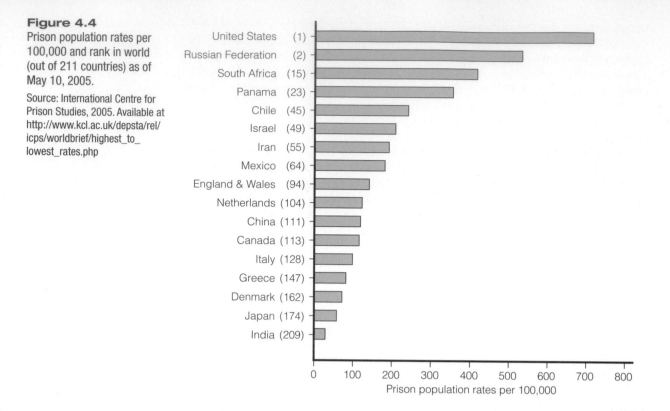

According to the U.S. Bureau of Justice Statistics, there were 2,131,180 prisoners held in federal or state prisons or local jails as of June 30, 2004. Typically, prisons confine individuals convicted of felonies, whereas jails confine people who have committed misdemeanors and have sentences of one year or less.

AP/Wide World Photos

In 2003, 4 million adult women and men were on probation at the federal, state, or local level, about half of whom had been convicted of a felony (the other half had been convicted of a misdemeanor). An estimated 56% were white, 30% black, and 12% Hispanic (BJS 2004).

Witness to an Execution

At 10:58 the prisoner entered the death chamber. He was, I knew from my research, a man with a checkered, tragic past. He had been grossly abused as a child, and went on to become grossly abusive of others. I was told he could not describe his life, from childhood on, without talking about confrontations in defense of a precarious sense of self, at home, in school, on the streets, in the prison yard. Belittled by life and choking with rage, he was hungry to be noticed. Paradoxically, he had found his moment in the spotlight. . . .

En route to the chair, the prisoner stumbled slightly, as if the momentum of the event had overtaken him. Were he not held securely by two officers, one at each elbow, he might have fallen. . . . Once the prisoner was seated, again with help, the officers strapped him into the chair.

Arms, legs, stomach, chest, and head were secured in a matter of seconds. Electrodes were attached to the cap holding his head and to the strap holding his exposed right leg. A leather mask was placed over his face. The last officer mopped the prisoner's brow, then touched his hand in a gesture of farewell. . . .

The strapped and masked figure sat before us, utterly alone, waiting to be killed . . . waiting for a blast of electricity that would extinguish his life. Endless seconds passed. His last act was to swallow, nervously, pathetically, with his Adam's apple bobbing. I was struck by the simple movement then, and can't forget it even now. It told me, as nothing else did, that in the prisoner's restrained body, behind that mask, lurked a fellow human being who, at some level, however primitive, knew or sensed himself to be moments from death.

. . . Finally, the electricity hit him. His body stiffened spasmodically, though only briefly. A thin swirl of smoke trailed away from his head and then dissipated quickly. The body remained taut, with the right foot raised slightly at the heel, seemingly frozen there. A brief pause, then another minute of shock. When it was over, the body was flaccid and inert.

Three minutes passed while the officials let the body cool. (Immediately after the execution, I'm told, the body would be too hot to touch and would blister anyone who did.) All eyes were riveted to the chair; I felt trapped in my witness seat, at once transfixed and yet eager for release. I can't recall any clear thoughts from that moment. One of the death watch officers later volunteered that he shared this experience of staring blankly at the execution scene. Had the prisoner's mind been mercifully blank before the end? I hope so.

The physician listened for a heartbeat. Hearing none, he turned to the warden and said, "This man has expired." The warden, speaking to the Director, solemnly intoned: "Mr. Director, the court order has been fulfilled." . . .

Source: Johnson, Robert, 1989. "This Man Has Expired: Witness to an Execution." *Commonwealth,* January 13, 1989, pp. 9–15. Copyright by the Commonwealth Foundation. Reprinted with permission.

Capital Punishment. With capital punishment the state (the federal government or a state) takes the life of a person as punishment for a crime. Thirty-eight states allow capital punishment. In 2004, 59 executions took place in 11 states, with over 3,374 inmates on death row (BJS 2005a). In the same year, 3,797 people were executed in 25 countries despite the global trend toward abolition of the death penalty (Amnesty International 2005). For example, more than 120 countries have banned state executions, including Australia, Canada, Italy, Denmark, and Ireland. The United States was the last country in the world to reject the execution of offenders who committed their crimes before the age of 18 (Wallis 2005). In this chapter's *The Human Side* feature, Robert Johnson, a professor at American University, describes his reaction to being witness to an execution.

Proponents of capital punishment argue that executions of convicted murderers are necessary to convey public disapproval and intolerance for such heinous crimes. Those against capital punishment believe that no one, including the state, has the right to take another person's life and that putting convicted murderers behind bars for life is a "social death" that conveys the necessary societal disapproval.

> "I feel morally and intellectually obligated to concede that the death penalty experiment has failed."
>
> **Justice Harry Blackmun**
> **U.S. Supreme Court Justice**

Proponents of capital punishment also argue that it deters individuals from committing murder. Critics of capital punishment hold, however, that because most homicides are situational and are not planned, offenders do not consider the consequences of their actions before they commit the offense. Critics also point out that the United States has a much higher murder rate than Western European nations that do not practice capital punishment and that death sentences are racially discriminatory. For example, a study of Maryland's death penalty practices found that "blacks who kill whites are significantly more likely to face the death penalty . . . than are blacks who kill blacks, or white killers" (Liptak 2003, p. 1).

Capital punishment advocates suggest that executing a convicted murderer relieves the taxpayer of the costs involved in housing, feeding, guarding, and providing medical care for inmates. Opponents of capital punishment argue that the principles that decide life and death issues should not be determined by financial considerations. In addition, taking care of convicted murderers for life may actually be less costly than sentencing them to death, because of the lengthy and costly appeals process for capital punishment cases (Garey 1985; Myths and Facts About the Death Penalty 1998).

Nevertheless, those in favor of capital punishment argue that it protects society by preventing convicted individuals from committing another crime, including the murder of another inmate or prison official. One study of the deterrent effect of capital punishment concluded that each execution is associated with at least eight fewer homicides (Rubin 2002). Opponents contend, however, that capital punishment may result in innocent people being sentenced to death. Since 1973, 119 death row inmates have been exonerated (Amnesty International 2005). In addition, a report by the Justice Project titled *A Broken System,* found reversible errors in two of every three death penalty cases reviewed over the study period (Herbert 2002). Stating that the system was riddled with error, in 2002 the governor of Illinois commuted the sentence of 167 inmates from the death penalty to life in prison or less (Wilgoren 2003; Napolitano 2004). Furthermore, some hypothesize that the recent reduction in U.S. executions is a result of exonerations based on new DNA evidence (see this chapter's *Focus on Technology* feature). Others maintain, however, that the reduction is merely a result of the recent decrease in the homicide rate (Lane 2004).

ThomsonNOW™

Learn more about **Capital Punishment** by going through the Death Penalty Map Exercise.

Legislative Action

Gun Control. Fueled by the Columbine school shooting and other recent images of children as both gun victims and offenders, in May 2000 tens of thousands of women descended on Washington, D.C., demanding that something be done about gun violence. Although the impact of the Million Mom March is still unknown, most Americans, and particularly women, support some restriction on handguns. When a national sample of U.S. adults were asked, "Would you like to see gun laws in this country made more strict, less strict, or remain as they are?" 52% responded more strict, 12% less strict, and 35% remain as they are (Gallup Poll 2005b).

Those against gun control argue that not only do citizens have a constitutional right to own guns but also that more guns may actually lead to less crime as would-be offenders retreat in self-defense (Lott 2003). Advocates of gun control, however, insist that the 200 million privately owned guns in this country, 70 million of which are handguns (Albanese 2000), significantly contribute to the violent crime rate in the United States and distinguish the country from other industrialized nations.

After a seven-year battle with the National Rifle Association (NRA), gun control advocates achieved a small victory in 1993 when Congress passed the

DNA Evidence

In 1954 Sam Sheppard, a prominent Cleveland physician, was accused of killing his wife, Marilyn Sheppard. More than 40 years later this case was the basis for the movie *The Fugitive* and the television series of the same name. The real-life drama carries on as Sam Sheppard Jr. continues to try to clear his father's name. This time, however, he is armed with DNA evidence, evidence that suggests that someone other than his father may have killed Marilyn Sheppard.

Increasingly, law enforcement officers in both the United States and Europe are using DNA evidence in the identification of criminal suspects. DNA stands for deoxyribonucleic acid; it is found in the nucleus of every cell and contains an individual's complete and unique genetic makeup. DNA fingerprinting is a general term used to describe the process, developed in the mid 1980s, of analyzing and comparing DNA from different sources, including evidence found at a crime scene (e.g., blood, semen, hair, saliva, fibers, skin tissue, and the DNA of a suspect). Since 1989, according to the Innocence Project, more than 150 people in the United States have been released as a result of post-conviction DNA evidence (Innocence Project 2005). Thirty-eight states provide convicted offenders access to DNA testing.

Concern over the number of post-conviction exonerations has led to the passage of the Innocence Protection Act of 2004. The act, which is part of a larger crime bill aimed at reducing the likelihood of innocent persons being executed, ensures that convicted offenders are given the opportunity to prove their innocence through DNA testing. The act also provides funds to help states defray the cost of post-conviction DNA testing (Death Penalty Information Center 2005). A recent report by a Washington-based law firm estimates that there are "over 500,000 DNA samples from crime scenes and from convicted criminals across the country awaiting processing, in which genetic material is extracted and a DNA profile created" (Associated Press 2005c, p. 1). A lack of staffing and money are the two primary reasons for the backlog.

Some concern exists, however, about the use of DNA evidence, specifically, how donors will be selected, what methods of data collection will be used, and potential abuses of analysis results. In California several hundred inmates brought suit against the state, refusing to give up their DNA. The appellate court refused to hear the case, but the governor signed a bill permitting the use of force in obtaining a sample (Kluger 2002). In addition, some law enforcement agencies been involved in what is being called "DNA dragnets" (Ripley 2005). For example, police in Truro, Massachusetts, have set out to collect DNA from every one of the town's nearly 800 males in hopes of finding the killer of a local resident. Although most of the citizens have been supportive, others have resisted but not without fear of being placed on a suspect list.

That is why, until recently, England had the only nationwide DNA data bank in the world. However, in 1994 the FBI initiated the Combined DNA Index System (CODIS), and in 1998 it became operational. With every state contributing to it, CODIS is a national database that permits various agencies access for investigative purposes. To date, 2.3 million profiles have been registered (CODIS 2005). The extent of the collection effort has led some to ask what will happen to the samples that, among other things, reveal a donor's genetic disposition for certain diseases.

It is likely that use of DNA will face many legal battles, but nonetheless the future of DNA evidence is bright. Scientists are already testing a tool that will allow eight DNA samples to be measured simultaneously, followed by a determination of matches. In addition, even if it does not survive the legal scrutiny that is likely, DNA fingerprinting remains a valuable identification technique used in biology, archeology, medical diagnosis, paleontology, and forensics.

Brady Bill. This law requires a five-day waiting period on handgun purchases so that sellers can screen buyers for criminal records or mental instability. Today, the law requires background checks of not just handgun users but also those who purchase rifles and shotguns (Reid 2003). At present, there is a bill before Congress that would grant gun dealers legal immunity from civil suits brought by gun victims (Brady Campaign 2005).

Other Legislation. Major legislative initiatives have been passed in recent years, including the 1994 Crime Control Act, which created community policing, "three strikes and you're out," and truth-in-sentencing laws. Significant crime-related legislation presently before Congress include the following bills (Orator 2005):

- *Rights of Abducted Children Act.* This act is intended "to provide protection and victim services to children abducted by family members."
- *Ex-Offenders Voting Rights Act.* This act would ensure that certain qualified ex-offenders who have served their sentences would be able secure their federal voting rights.
- *Gang Deterrence and Community Protection Act.* This legislation aims "to reduce gang violence and protect law-abiding citizens and communities from violent criminals."
- *Dru Sjodin National Sex Offender Public Database Act.* This act, also known as Dru's law, would, among other things, establish a national sex offender registration database.
- *Witness Security and Protection Act.* This bill is intended to provide short-term protection by U.S. marshals for witnesses in state and local districts in cases involving homicide and other serious crimes.

Understanding Crime and Social Control

ThomsonNOW™

Learn more about **Understanding Crime and Violence** by going through the Difference between Deviance and Crime Animation.

"To do justice, to break the cycle of violence, to make Americans safer, prisons need to offer inmates a chance to heal like a human, not merely heel like a dog."

Richard Stratton
Former inmate

What can we conclude from the information presented in this chapter? Research on crime and violence supports the contentions of both structural functionalists and conflict theorists. Inequality in society, along with the emphasis on material well-being and corporate profit, produces societal strains and individual frustrations. Poverty, unemployment, urban decay, and substandard schools—the symptoms of social inequality—in turn lead to the development of criminal subcultures and conditions favorable to law violation. Furthermore, criminal behavior is encouraged by the continued weakening of social bonds among members of society and between individuals and society as a whole, the labeling of some acts and actors as "deviant," and the differential treatment of minority groups by the criminal justice system.

There has been a general decline in crime over the last decade, making it tempting to conclude that get-tough criminal justice policies are responsible for the reductions. Other valid explanations exist and are likely to have contributed to the falling rates: changing demographics, community policing, stricter gun control, and a reduction in the use of crack cocaine. Nonetheless, "nail 'em and jail 'em" policies have been embraced by citizens and politicians alike. Get-tough measures include building more prisons and imposing lengthier mandatory prison sentences on criminal offenders. Advocates of harsher prison sentences argue that "getting tough on crime" makes society safer by keeping criminals off the streets and by deterring potential criminals from committing crime.

Prison sentences, however, may not always be effective in preventing crime. In fact, they may promote it by creating an environment in which prisoners learn criminal behavior, values, and attitudes from each other. Two-thirds of all parolees are rearrested within three years of release (Butterfield 2002a). With 90% of inmates being discharged into the community—600,000 annually, 1,700 a day—punitive policies may be a shortsighted temporary fix (Travis & Waul 2002).

Rather than getting tough on crime after the fact, some advocate getting serious about prevention. Reemphasizing the values of honesty and, most important, taking responsibility for one's actions is a basic line of prevention with which most agree. The movement toward **restorative justice**, a philosophy primarily concerned with repairing the victim-offender-community relation, is in direct response to the concerns of an adversarial criminal justice system that encourages offenders to deny, justify, or otherwise avoid taking responsibility for their actions.

Restorative justice holds that the justice system, rather than relying on "punishment, stigma and disgrace" (Siegel 2006, p. 275), should "repair the harm" (Sherman 2003, p. 10). Key components of restorative justice include restitution to the victim, remedying the harm to the community, and mediation. Restorative justice programs have been instituted in several states, including Vermont and New York. In 2002 the United Nations Economic and Social Council endorsed the "Basic Principles on the Use of Restorative Justice Programmes in Criminal Matters" around the world (United Nations 2003).

Chapter Review

ThomsonNOW™

Reviewing is as easy as ① ② ③

1. Before you do your final review, take the ThomsonNOW diagnostic quiz to help you identify the areas on which you should concentrate. You will find information on ThomsonNOW and instructions on how to access all of its great resources on the foldout at the beginning of the text.

2. As you review, take advantage of ThomsonNOW's study videos and interactive Map the Stats exercises to help you master the chapter topics.

3. When you are finished with your review, take ThomsonNOW's posttest to confirm you are ready to move on to the next chapter.

• **Are there any similarities between crime in the United States and crime in other countries?**

All societies have crime and have a process by which they deal with crime and criminals; that is, they have police, courts, and correctional facilities. Worldwide, most offenders are young males and the most common offense is theft; the least common offense is murder.

• **How can we measure crime?**

There are three primary sources of crime statistics. First are official statistics, for example, the FBI's Uniform Crime Reports, which are published annually. Second are victimization surveys designed to get at the "dark figure" of crime, crime that is missed by official statistics. Finally, there are self-report studies that have all the problems of any survey research. Investigators must be cautious about whom they survey and how they ask the questions.

• **What sociological theory of criminal behavior blames the schism between the culture and structure of society for crime?**

Strain theory was developed by Robert Merton (1957) and uses Durkheim's concept of *anomie,* or normlessness. Merton argued that when legitimate means (e.g., a job) of acquiring culturally defined goals (e.g., money) are limited by the structure of society, the resulting strain may lead to crime. Individuals, then, must adapt to the inconsistency between means and goals in a society that socializes everyone into wanting the same thing but provides opportunities for only some.

• **What are index offenses?**

Index offenses, as defined by the FBI, include two categories of crime: violent crime and property crime. Violent crimes include murder, robbery, assault, and rape; property crimes include larceny, car theft, burglary, and arson. Property crimes, although less serious than violent crimes, are the most numerous.

• **What is meant by white-collar crime?**

White-collar crime includes two categories: occupational crime, that is, crime committed in the course of one's occupation; and corporate crime, in which corporations violate the law in the interest of maximizing profits. In occupational crime the motivation is individual gain.

• **How do social class and race affect the likelihood of criminal behavior?**

Official statistics indicate that minorities are disproportionately represented in the offender population. Nevertheless, it is inaccurate to conclude that race and crime are causally related. First, official statistics reflect the behaviors and policies of criminal justice actors. Thus the high rate of arrests, conviction, and incarceration of minorities may be a consequence of individual and institutional bias against minorities. Second, race and social class are closely related in that nonwhites are overrepresented in the lower classes. Because lower-class members lack legitimate means to acquire material goods, they may turn to instrumental, or economically motivated, crimes. Thus the apparent relationship between race and crime may, in part, be a consequence of the relationship between these variables and social class.

- **What are some of the economic costs of crime?**

  First are direct losses from crime, such as the destruction of buildings through arson or of the environment by polluters. Second are costs associated with the transferring of property (e.g., embezzlement). A third major cost of crime is that associated with criminal violence, for example, the medical cost of treating crime victims. Fourth are the costs associated with the production and sale of illegal goods and services. Fifth is the cost of prevention and protection. Finally, there is the cost of the criminal justice system, law enforcement, litigative and judicial activities, corrections, and victims' assistance.

- **What are some of the most recent criminal justice policies in the fight against crime?**

  First, the number of police on the streets has been increased in the hope that their presence will increase deterrence and therefore decrease crime. Second, societal fear of crime has led to a public emphasis on incapacitation and a demand for tougher mandatory sentences, support of a "three strikes and you're out" policy, and truth-in-sentencing laws. Finally, there has been a slight increase in the use of probation and parole as get-tough policies contribute to prison overcrowding.

Critical Thinking

1. Crime statistics are sensitive to demographic changes. Explain why crime rates in the United States began to rise in the 1960s as baby-boom teenagers entered high school, and why they may increase again as we move further into the 21st century.
2. Some countries have high gun ownership and low crime rates. Others have low gun ownership and low crime rates. What do you think accounts for the differences between these countries?
3. The use of technology in crime-related matters is likely to increase dramatically over the next several decades. DNA testing and the use of heat sensors, blood-detecting chemicals, and computer surveillance are just some of the ways science will help fight crime. As with all technological innovations, however, there is the question, Who benefits? Do these new technologies have gender, race, and/or class implications?

Key Terms

acquaintance rape	labeling theory
Brady Bill	organized crime
classic rape	primary deviance
clearance rate	probation
computer crime	racial profiling
control theory	rehabilitation
corporate violence	restorative justice
crime	secondary deviance
crime rate	strain theory
deterrence	subcultural theory
differential association	transnational crime
identity theft	victimless crimes
incapacitation	white-collar crime
index offenses	

Media Resources

 The Companion Website for
Understanding Social Problems,
Fifth Edition

http://sociology.wadsworth.com/mooney_knox_schacht5e

Supplement your review of this chapter by going to the companion website to take one of the Tutorial Quizzes, use the flash cards to master key terms, and check out the many other study aids you'll find there. You'll also find special features such as *Wadsworth's Sociology Online Resources and Writing Companion,* GSS Data, and Census 2000 information, data, and resources at your fingertips to help you complete that special project or do some research on your own.

"We must recognize that there are healthy as well as unhealthy ways to be single or to be divorced, just as there are healthy and unhealthy ways to be married." *Stephanie Coontz, family historian*

Family Problems

helly, a student in one of our sociology classes, had been married to Clay for 23 years. During a class discussion on domestic violence, she shared her experience with the class, reporting that within six months of their marriage, "he turned into the devil." "He would beat me for anything—not ironing his shirt right, not having his food ready on time, and not keeping the children quiet." She talked of living in fear and stayed because she had no job and no way to support herself. To escape, she prepared a meatloaf "loaded" with rat poison ("enough to kill three pigs"). But, unexpectedly, her son came home early and sat down for dinner. So she "accidentally" spilled her drink on the food and threw it in the trash. "Of course, I got a beating for that," she said. Her eventual exit came "when he beat me so bad I went to the hospital. When I came out, my mamma took me to her house." Shelly (and her two children) stayed with her mother through her divorce. Two years later she remarried and reports, "My new husband is completely different, but I'm still messed up from my years of abuse. . . . I have had to learn to trust all over."

ThomsonNOW™

Reviewing is as easy as ❶ ❷ ❸

Use ThomsonNOW to help you make the grade on your next exam. When you are finished reading this chapter, go to the chapter review for instructions on how to make ThomsonNOW work for you.

Shelly's situation involves some of the issues discussed in this chapter: violence and abuse in relationships, divorce, and single parenting. We begin by defining families and reviewing family diversity worldwide.

The Global Context: Families of the World

The U.S. census defines *family* as a group of two or more persons related by blood, marriage, or adoption. This definition has been challenged because it does not include foster families or unmarried couples (heterosexual or homosexual) who live together. Sociologically, a **family** is a kinship system of all relatives living together or recognized as a social unit, including adopted individuals. Families are found in all societies. But, as we describe in the following section, family forms and patterns vary worldwide.

Monogamy, Polygamy, and Arranged Marriages

In many countries, including the United States, the only legal form of marriage is **monogamy**—a marriage between two partners. A common variation of monogamy is **serial monogamy**—a succession of marriages in which a person has more than one spouse over a lifetime but is legally married to only one person at a time. Some societies practice **polygamy**—a form of marriage in which one person may have two or more spouses. In the United States polygamy is illegal and is often referred to as **bigamy**—the criminal offense of marrying one person while still legally married to another. However, some Latter Day Saints in the United States practice what they call **plural marriage**, otherwise known as **polygyny**—the most common form of polygamy in which one husband is married to two or more wives. Polygyny tends to be practiced by the rich and the powerful and/or by men who are taking on the wife of a deceased brother as a familial duty (a practice mentioned in the Old Testament) (R. Bunger, personal communication, 2005). Although polygyny is diminishing worldwide, it is still practiced in some Asian and African societies. In societies in which polygyny is normative, monogamy has always been the predominant marriage form (Adams 2004). The second form of polygamy is **polyandry**—the concurrent marriage of one woman to two or more men. Polyandry is rare but does occur in tribal India (e.g., among the Todas), the Himalayan regions of Asia, and Polynesia.

ThomsonNOW™

Learn more about the **Global Context: Families of the World** by going through the What is Family? What is Marriage? Learning Module.

In Utah's first trial involving bigamy charges in 50 years, Thomas A. Green, shown here with his five wives, was sentenced to five years in prison after being convicted of four counts of bigamy and one count of criminal nonsupport. Mr. Green's wives had pleaded with Judge Guy R. Burningham to keep him out of prison and were relieved that Green was not sentenced to the 25 years sought by the prosecutor in the case. Although as many as 30,000 Utahans live in polygamous marriages, Green was singled out for prosecution because of his frequent media interviews about his illegal lifestyle.

© AFP/Getty, Images

These two Masai wives, shown with their children, share the same husband and live in two separate huts in Tanzania. Another Masai man may have sex with these women so long as he places his spear outside the hut to announce his presence inside.

© Caroline Schacht

In the United States the selection of a marital partner is considered a personal choice. However, arranged marriages, whereby the choice of a marital partner is a family decision rather than a personal choice, occur in other societies (e.g., India), although the practice is slowly diminishing, especially in urban areas. Societies that traditionally have embraced arranged marriages are likely to experience intergenerational conflict as the younger generation asserts its independence from this tradition (Adams 2004).

Role of Women in the Family

The roles of women also vary across societies. In some societies wives are commonly expected to be subservient to their husbands (e.g., rural Pakistan and India). In Northern European countries, specifically Norway and Sweden, women and men tend to have **egalitarian relationships,** in which partners share decision making and assign family roles based on choice rather than on traditional beliefs about gender (Lindsey 2005).

In most of the world the work of the home is still primarily the responsibility of the wife. In a study of husbands and wives in 13 nations, in all but 1 nation (Russia), respondents reported that women performed most of the household labor (Davis & Greenstein 2004). The percentage of wives reporting that they "always" or "usually" did the housework ranged from 98% in Japan to 38% in Russia; the percentage of husbands reporting that wives always or usually did the housework ranged from 93% in Japan to 31% in Russia. In the United States 60% of husbands and 67% of wives reported that the wives always or usually did the housework. When husbands help with domestic tasks, they are often viewed as "helping" the woman and not as performing a male role.

Same-Sex Relationships

Norms and policies concerning same-sex intimate relationships also vary around the world. In some countries homosexuality is punishable by imprisonment or even death (see Chapter 11).

In a number of countries (Denmark, Finland, Germany, Iceland, Greenland, the Netherlands, Norway, and Sweden), same-sex couples can sign up as "registered partners" to claim a status and benefits similar to marriage. As of 1999, couples in France (both same-sex and opposite-sex couples) can enter a type of marital arrangement called *Pacte civil de solidarite* (civil solidarity pact), or PAC. PACs offer some of the tax, welfare, and inheritance rights that married couples enjoy. Registered couples are able to share such things as joint auto insurance and can extend their social security coverage to each other, file joint tax returns, and leave each other property in their wills on favorable tax terms. Most couples (60%) opting for PACs in France are heterosexual couples (Demian 2004).

In 2001 the Netherlands became the first country in the world to offer legal marriage to same-sex couples, and in 2003 Belgium became the second. In 2005 Canada and Spain legalized same-sex marriage. In 2004 Massachusetts became the first U.S. state to offer same-sex marriage.

In the United States the 1996 federal Defense of Marriage Act defines marriage as the union between one man and one woman and denies federal recognition of same-sex marriages. As of spring 2005, 40 states have laws or state constitutional amendments that ban marriage between same-sex couples. California, the District of Columbia, Hawaii, Maine, and New Jersey have domestic partnership laws that provide some limited rights to same-sex couples. Vermont and Connecticut offer "civil unions" to same-sex couples; these unions provide all the state-level rights and responsibilities of marriage but none of the more than 1,100 federal protections. In 2005 a California superior court judge ruled that limiting marriage to a man and a woman violates the state's constitution and that same-sex marriages should be allowed. If upheld on appeal, California could be the second state to legalize same-sex marriages (Jones 2005).

It is clear from the previous discussion that families are shaped by the social and cultural context in which they exist. As we discuss the family issues addressed in this chapter—violence and abuse, divorce, and nonmarital and teenage childbearing—we refer to social and cultural forces that shape these events and the attitudes surrounding them. Next, we look at changing patterns and structures of U.S. families and households.

Changing Patterns and Structures in U.S. Families and Households

Over the last century dramatic changes have occurred in the patterns and structures of U.S. families and households, some of which are reflected in Figure 5.1. A **family household,** as defined by the U.S. Census Bureau, consists of two or more individuals related by birth, marriage, or adoption who reside together. **Nonfamily households** can consist of one person who lives alone, two or more people as roommates, or cohabiting heterosexual or homosexual couples involved in a committed relationship. The Census Bureau considers households composed of heterosexual cohabiting couples and same-sex couples as "nonfamily" households, even though such couples function economically and emotionally as a family. Some of the significant changes in U.S. families and households that have occurred over the last several decades include the following:

- *Increased singlehood and older age at first marriage.* U.S. women and men are staying single longer. From 1970 to 2003 the median age at first marriage increased from about 21 to 25 for women and from about 23 to 27 for men (Fields 2004). Today, 12.2% of women and 17.4% of men age 40 to 44 have

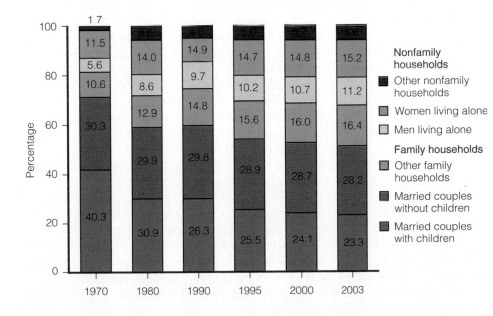

Figure 5.1
The many types of U.S. households, 1970 to 2003 (percentage distribution).

Source: U.S. Census Bureau

Figure 5.2

Number of cohabiting, unmarried couples of the opposite sex by year, United States.

Source: U.S. Bureau of the Census (2000) and earlier reports in the same series.

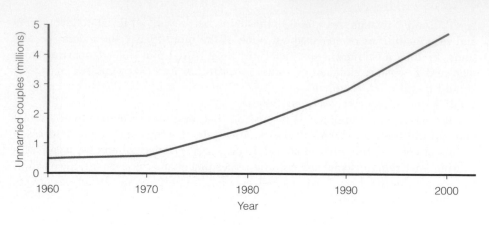

never been married—the highest figures in this nation's history (U.S. Bureau of the Census 2005). In effect, individuals are delaying marriage.

- *Delayed childbearing.* Women are also delaying having children. The 2003 birthrate for women ages 40 to 44 was 8.7 births per 1,000 women—more than double what it was in 1981 (Hamilton et al. 2004). For the first time, in 2003 births to women over age 40 topped 100,000 in a single year. First-birth rates for women ages 30 to 34 years, 35 to 39 years, and 40 to 44 years increased considerably in just one year (from 2002 to 2003) by 7%, 12%, and 11%, respectively (Hamilton et al. 2004).

- *Increased heterosexual and same-sex cohabitation.* Although women and men are staying single longer, they are not forgoing intimate relationships. From 1960 to 2000 the number of cohabiting unmarried couples skyrocketed (see Figure 5.2). Nationally, 9% of coupled households are unmarried partner households (Simmons & O'Connell 2003). Couples live together to assess their relationship, to reduce or share expenses, or to avoid losing pensions from previously deceased partners. For most unmarried couples who live together, cohabitation is a part of the modern courtship process; for some, it is an alternative to marriage. Most of the 5.5 million cohabiting couples in 2000 were heterosexual couples, but about 1 in 9 had partners of the same sex (Simmons & O'Connell 2003) (see Figure 5.3).

Increased cohabitation among adults means that children are increasingly living in families that may function as two-parent families but do not have the social or legal recognition that married-couple families have. About 4 in 10 (43%) opposite-sex unmarried partner households, one-fifth (22.3%) of gay male couples, and one-third (34.3%) of lesbian couples have children present in the home (Simmons & O'Connell 2003). When children are denied a legal relationship to both parents because of the parents' unmarried status and/or sexual orientation, they may be denied Social Security survivor benefits, health care insurance, or the ability to have either parent authorize medical treatment in an emergency, among other protections.

Some states, cities, counties, and employers allow unmarried partners (same-sex and/or heterosexual partners) to apply for a **domestic partnership** designation, which grants them some legal entitlements, such as health insurance benefits and inheritance rights, that have traditionally been reserved for married couples. Eight states, the District of Columbia, and some jurisdictions in 15 other states allow same-sex second-parent adoptions that allow a same-sex parent to adopt his or her partner's biological or adopted child (National Gay and Lesbian Task Force 2005).

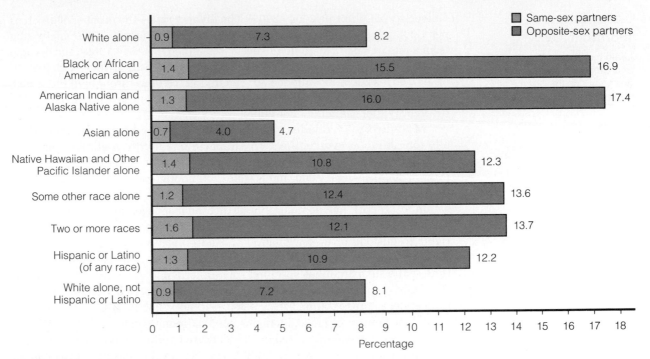

Figure 5.3
Unmarried same-sex and opposite-sex couple households, as percentages of all coupled households, by race and Hispanic origin: United States, 2000.

Source: Simmons & O'Connell (2003).

- *Increased births to unmarried women.* The percentage of births to unmarried women increased from 18.4% of total births in 1980 to 30.1% in 1991, to 34.6% in 2003 (Hamilton et al. 2004; Ventura 1995). Coontz (1997) emphasizes that we must be careful not to overdramatize the increase because "much illegitimacy was covered up in the past and reporting methods have become much more sophisticated" (p. 29). Nevertheless, today more than one-third of all U.S. births are to unmarried women. However, not all unwed mothers are single parents. Half of new unwed mothers are cohabiting with the fathers when their children are born (Sigle-Rushton & McLanahan 2002).
- *Increased single-parent families.* The rise in both divorce and nonmarital births has resulted in an increase in single-parent families. From 1970 to 2003 the proportion of single-mother families grew from 12% to 26% and single-father families grew from 1% to 6% (Fields 2004). As noted earlier, many children in what are considered to be "single-parent families" have two parental adults in the home who are living in an unmarried cohabiting relationship: 16% of children living with single fathers and 9% of children living with single mothers also live with their parents' partners (Forum on Child and Family Statistics 2000).
- *Fewer children living in married-couple families.* The percentage of children living in married-couple families decreased from 77% in 1980 to 68% in 2003. This percentage has remained stable since 1995, reversing a long-standing downward trend (Federal Interagency Forum on Child and Family Statistics 2004). White children are more than twice as likely as black children to live with two unmarried parents (77% versus 36% in 2003).
- *Increased divorce and blended families.* About one-fourth of U.S. first-year college students have parents who are divorced (American Council on Education &

ThomsonNOW™

Learn more about **Divorce and Blended Families** by going through the Divorce and Remarriage Learning Module.

University of California 2004–2005). The **divorce rate**—the number of divorces per 1,000 population—doubled from 1950 to its peak around 1980, increasing from a rate of 2.6 to 5.3. In nearly every year since the early 1980s, the divorce rate has decreased, and in 2003 it was 3.8 (Munson & Sutton 2004; U.S. Bureau of the Census 2005). Despite the decline in divorce rates throughout the 1990s, at today's rate an estimated half of marriages started today are expected to end in divorce (Whitehead & Popenoe 2005).

Most divorced individuals remarry and create blended families, traditionally referred to as stepfamilies. An estimated one-fourth of all children born in the United States will live with a stepparent before they reach adulthood (Mason 2003). Federal, state, and private sector policies have not kept pace with the concerns of modern stepfamilies. In most states stepparents have no obligation to support their stepchildren during the marriage, nor do they have any right of custody or control. In the event of divorce stepparents usually have no rights to custody or even visitation and no obligation to pay child support. Stepchildren have no right of inheritance in the event of the stepparent's death (unless the stepparent has specified such inheritance in a will) (Mason 2003).

- *Increased employment of mothers.* Employment of married women with children under age 18 rose from 24% in 1950 to 40% in 1970 to 70% in 2004 (Gilbert 2003; U.S. Bureau of Labor Statistics 2005). Among single-parent families in 2004, the mother was employed in 72% of those families maintained by women (U.S. Bureau of Labor Statistics 2005). The idea of the traditional two-parent family in which the husband is a breadwinner and the wife is a homemaker does not reflect the reality of American families. In 2004 two-thirds (61%) of U.S. married-couple families with children under age 18 were dual-earner marriages; less than one-third (31%) of these families involved an employed father and an unemployed mother (U.S. Bureau of Labor Statistics 2005). Yet work, school, and medical care in the United States tend to be organized around the expectation that every household has a full-time mother at home who is available to transport children to medical appointments, pick up children from school on early dismissal days, and stay home when a child is sick (Coontz 1992).

The Marital Decline and Marital Resiliency Perspectives on the American Family

Do the recent transformations in American families signify a collapse of marriage and family in the United States? Does the trend toward diversification of family forms mean that marriage and family are disintegrating, falling apart, or even disappearing? Or, as in other aspects of social life, has family simply undergone transformations in response to changes in socioeconomic conditions, gender roles, and cultural values?

The answers to these questions depend on whether we adopt the marital decline perspective or the marital resilience perspective. According to the **marital decline perspective,** (1) personal happiness has become more important than marital commitment and family obligations and (2) the decline in lifelong marriage and the increase in single-parent families have contributed to a variety of social problems, such as poverty, delinquency, substance abuse, violence, and the erosion of neighborhoods and communities (Amato 2004). According to the **marital resiliency perspective,** "poverty, unemployment, poorly funded schools, discrimination, and the lack of basic services (such as health insurance and child care) represent more serious threats to the well-being of children and adults than does the decline in married two-parent families" (Amato 2004, p. 960). According to this perspective, many marriages in the past were troubled, but because divorce was not socially acceptable, these problematic marriages remained intact. Rather than view divorce as a sign of the decline of marriage, divorce provides a second chance at happiness for adults and an escape from dysfunctional and aversive home environments for many children.

Family scholar Stephanie Coontz (1997) observed that "marriage is certainly a transformed institution, and it plays a smaller role than ever before in organizing social and personal life" (p. 31). Yet a national Gallup survey shows that Americans rank their family as the most important aspect of life (Moore 2003). And in a national survey of first-year U.S. college students, "raising a family" was ranked as the most important value (the second most important value was "being well-off financially") (American Council on Education & University of California 2004–2005). Marriage continues to be a normative part of adult life, with about 95% of women and men in their early 60s having been married at least once (Hacker 2003). Although women and men are marrying later than they did in the past and although their marriages may not last as long as they once did, most people eventually do get married. And despite high rates of cohabitation, living together tends to be a precursor to marriage rather than a permanent substitute (Goldstein & Kenney 2001).

Although the high rate of marital dissolution seems to suggest a weakening of marriage, divorce may also be viewed as resulting from placing a high value on marriage, such that a less than satisfactory marriage is unacceptable. In effect, people who divorce may be viewed not as incapable of commitment but as those who would not settle for a bad marriage. Indeed, the expectations that young women and men have of marriage have changed. Whereas once the main purpose of marriage was to have and raise children, today women and men want marriage to provide adult intimacy and companionship (Coontz 2000).

The high rate of out-of-wedlock childbirth and single parenting is also not necessarily indicative of a decline in the value of marriage. Edin (2000) found that in a sample of low-income single women with children, most said they would like to be married but just have not found "Mr. Right."

Is the well-being of a family measured by the degree to which that family conforms to the idealized married, two-parent, stay-at home mom model of the 1950s? Or is family well-being measured by function rather than form? As suggested by family scholars Mason, Skolnick, and Sugarman (2003), "the important question to ask about American families . . . is not how much they conform to a particular image of the family, but rather how well do they function—what kind of love, care, and nurturance do they provide?" (p. 2).

Finally, it is important to have a perspective that takes into account the historical realities of families. Family historian Stephanie Coontz (2004) explains:

> I have spent much of my career as a historian explaining to people that many things that seem new in family life are actually quite traditional. Two-provider families, for example, were the norm through most of history. Stepfamilies were more numerous in much of history than they are today. There have been several times and places when cohabitation, out-of-wedlock births, or nonmarital sex were more widespread than they are today. Divorce was higher in Malaysia during the 1940s and 1950s than it is today in the United States. Even same-sex marriage, though comparatively rare, has been accepted in some cultures under certain conditions. (p. 974)

After a brief discussion of sociological theories of family problems, we look at what happens when children and adults experience violence and abuse in their intimate and family relationships.

Sociological Theories of Family Problems

Three major sociological theories—structural functionalism, conflict theory, and symbolic interactionism—help to explain different aspects of the family institution and the problems in families today.

Structural-Functionalist Perspective

The structural-functionalist perspective views the family as a social institution that performs important functions for society, including producing new members, regulating sexual activity and procreation, socializing the young, and providing physical and emotional care for family members. According to the structural-functionalist perspective, traditional gender roles contribute to family functioning: Women perform the "expressive" role of managing household tasks and providing emotional care and nurturing to family members, and men perform the "instrumental" role of earning income and making major family decisions.

According to the structural-functionalist perspective, the high rate of divorce and the rising number of single-parent households constitute a "breakdown" of the family institution that has resulted from rapid social change and social disorganization. The structural-functionalist perspective views the breakdown of the family as a primary social problem that leads to such secondary social problems as crime, poverty, and substance abuse.

Structural-functionalist explanations of family problems examine how changes in other social institutions contribute to family problems. For example, a structural-functionalist view of divorce examines how changes in the economy (such as more dual-earner marriages) and in the legal system (such as the adoption of "no-fault" divorce) contribute to high rates of divorce. Changes in the economic institution, specifically falling wages among unskilled and semiskilled men, also contribute to both intimate partner abuse and the rise in female-headed single-parent households (Edin 2000).

The structural-functionalist perspective is also concerned with latent dysfunctions—unintended and unrecognized negative consequences. For example, one of the latent dysfunctions of marriage-promotion programs is that they may encourage battered women to stay in their abusive relationships.

> By stigmatizing single parents, stigmatizing divorce, or encouraging women to believe that they are harming their children if they leave their partners, [marriage-promotion] programs make it more difficult for some women to leave violent relationships or encourage them, intentionally or not, to remain with abusive partners. (Family Violence Prevention Fund 2005, p. 5)

Conflict and Feminist Perspectives

Conflict theory focuses on how capitalism, social class, and power influence marriages and families. Feminist theory is concerned with how gender inequalities influence and are influenced by marriages and families. Feminists are critical of the traditional male domination of families—a system known as **patriarchy**—that is reflected in the tradition of wives taking their husband's last name and children taking their father's name. Patriarchy is also reflected in the view that wives and children are the property of fathers and husbands.

The overlap between conflict and feminist perspectives is evident in views on how industrialism and capitalism have contributed to gender inequality. With the onset of factory production during industrialization, workers—mainly men—left the home to earn incomes and women stayed home to do unpaid child care and domestic work. This arrangement resulted in families founded on what Engels calls "domestic slavery of the wife" (quoted by Carrington 2002, p. 32). Modern society, according to Engels, rests on gender-based slavery, with women doing household labor for which they receive neither income nor status while men leave the home to earn an income. Times have certainly changed since Engels made his observations, with most wives today leaving the home to earn incomes. However, wives employed full-time still do the bulk of unpaid domestic labor, and women are more likely than men to compromise their occupational achievement to take on child care and other domestic responsibilities. The continuing unequal distribution of wealth that favors men contributes to inequities in power and fosters economic dependence of wives on husbands. When wives do earn more money than their husbands (which is the case in 30% of marriages), the divorce rate is higher—the women can afford to leave abusive or inequitable relationships (Jalovaara 2003).

Economic factors have also influenced norms concerning monogamy. In societies in which women and men are expected to be monogamous within marriage, there is a double standard that grants men considerably more tolerance for being nonmonogamous. Engels explains that monogamy arose from the concentration of wealth in the hands of a single individual—a man—and from the need to bequeath this wealth to children of his own, which requires that his wife be monogamous. The "sole exclusive aims of monogamous marriage were to make the man supreme in the family and to propagate, as the future heirs to his wealth, children indisputably his own" (quoted by Carrington 2002, p. 32).

Feminist and conflict perspectives on domestic violence suggest that the unequal distribution of power among women and men and the historical view of women as the property of men contribute to wife battering. When wives violate or challenge the male head-of-household's authority, the male may react by "disciplining" his wife or using anger and violence to reassert his position of power in the family.

Although modern gender relations within families and within society at large are more egalitarian than in the past, male domination persists, even if less obvious. Lloyd and Emery (2000) note that "one of the primary ways that power disguises itself in courtship and marriage is through the 'myth of equality between the sexes.' . . . The widespread discourse on 'marriage between equals' serves as a cover for the presence of male domination in intimate relationships . . . and allows couples to create an illusion of equality that masks the inequities in their relationships" (pp. 25–26).

Conflict theorists emphasize that social programs and policies that affect families are largely shaped by powerful and wealthy segments of society. The interests of corporations and businesses are often in conflict with the needs of families. Corporations and businesses strenuously fought the passage of the 1993 Family and Medical Leave Act, which gives people employed full-time for at least 12 months in companies with at least 50 employees up to 12 weeks of unpaid time off for parenting leave, illness or death of a family member, and elder care. Government, which is largely influenced by corporate interests through lobbying and political financial contributions, enacts policies and laws that serve the interests of for-profit corporations rather than families.

Symbolic Interactionist Perspective

Symbolic interactionism emphasizes that human behavior is influenced by meanings and definitions that emerge from small-group interaction. Divorce, for example, was once highly stigmatized and informally sanctioned through the criticism and rejection of divorced friends and relatives. As societal definitions of divorce became less negative, however, the divorce rate increased. The social meanings surrounding single parenthood, cohabitation, and delayed childbearing and marriage have changed in similar ways. As the definitions of each of these family variations became less negative, the behaviors became more common.

The symbolic interactionist perspective is concerned with how labels affect meaning and behavior. For example, when a noncustodial divorced parent (usually a father) is awarded "visitation" rights, he may view himself as a visitor in his children's lives. The meaning attached to the visitor status can be an obstacle to the father's involvement because the label "visitor" minimizes the importance of the noncustodial parent's role and results in conflict and emotional turmoil for fathers (Pasley & Minton 2001). Fathers' rights advocates suggest replacing the term *visitation* with such terms as *parenting plan* or *time-sharing arrangement,* because these terms do not minimize either parent's role.

Symbolic interactionists also point to the effects of interaction on one's self-concept, especially the self-concept of children. In a process called the "looking glass self," individuals form a self-concept based on how others interact with them. Family members, such as parents, grandparents, siblings, and spouses, are significant others who have a powerful effect on our self-concepts. For example, negative self-concepts may result from verbal abuse in the family, whereas positive self-concepts may develop in families where interactions are supportive and loving. The importance of social interaction in children's developing self-concept suggests a compelling reason for society to accept rather than stigmatize nontraditional family forms. Imagine the effect on children who are called "illegitimate" or who are teased for having two moms or dads.

The symbolic interactionist perspective is useful in understanding the dynamics of domestic violence and abuse. For example, some abusers and their victims learn to define intimate partner violence as an expression of love (Lloyd 2000). Emotional abuse often involves using negative labels (e.g., "stupid," "whore," "bad") to define a partner or family member. Such labels negatively affect the self-concept of abuse victims, often convincing them that they deserve the abuse. In the next section we discuss violence and abuse in intimate and family relationships, noting the scope, causes, and consequences of this troubling social problem.

Violence and Abuse in Intimate and Family Relationships

Although intimate and family relationships provide many individuals with a sense of intimacy and well-being, for others these relationships involve physical violence, verbal and emotional abuse, sexual abuse, and/or neglect. Indeed, in U.S. society people are more likely to be physically assaulted, abused and neglected, sexually assaulted and molested, or killed in their own homes rather than anywhere else and by other family members rather than by anyone else (Gelles 2000). Before reading further, you may want to take the Abusive Behavior Inventory in this chapter's *Self and Society* feature.

Abusive Behavior Inventory

Circle the number that best represents your closest estimate of how often each of the behaviors happened in your relationship with your partner or former partner during the previous six months (1, never; 2, rarely; 3, occasionally; 4, frequently; 5, very frequently).

1. Called you a name and/or criticized you.   1 2 3 **4** 5
2. Tried to keep you from doing something you wanted to do (e.g., going out with friends, going to meetings).   **1** 2 3 4 5
3. Gave you angry stares or looks.   1 2 **3** 4 5
4. Prevented you from having money for your own use.   1 2 **3** 4 5
5. Ended a discussion with you and made the decision himself/herself.   1 2 **3** 4 5
6. Threatened to hit or throw something at you.   **1** 2 3 4 5
7. Pushed, grabbed, or shoved you.   **1** 2 3 4 5
8. Put down your family and friends.   1 **2** 3 4 5
9. Accused you of paying too much attention to someone or something else.   1 2 3 **4** 5
10. Put you on an allowance.   1 2 **3** 4 5
11. Used your children to threaten you (e.g., told you that you would lose custody, said he/she would leave town with the children).   **1** 2 3 4 5
12. Became very upset with you because dinner, housework, or laundry was not done when he/she wanted it done or done the way he/she thought it should be.   1 2 **3** 4 5
13. Said things to scare you (e.g., told you something "bad" would happen, threatened to commit suicide).   1 2 **3** 4 5
14. Slapped, hit, or punched you.   **1** 2 3 4 5
15. Made you do something humiliating or degrading (e.g., begging for forgiveness, having to ask his/her permission to use the car or to do something).   **1** 2 3 4 5
16. Checked up on you (e.g., listened to your phone calls, checked the mileage on your car, called you repeatedly at work).   1 **2** 3 4 5
17. Drove recklessly when you were in the car.   **1** 2 3 4 5
18. Pressured you to have sex in a way you didn't like or want.   **1** 2 3 4 5
19. Refused to do housework or child care.   1 **2** 3 4 5
20. Threatened you with a knife, gun, or other weapon.   **1** 2 3 4 5
21. Spanked you.   **1** 2 3 4 5
22. Told you that you were a bad parent.   1 2 **3** 4 5
23. Stopped you or tried to stop you from going to work or school.   1 **2** 3 4 5
24. Threw, hit, kicked, or smashed something.   **1** 2 3 4 5
25. Kicked you.   **1** 2 3 4 5
26. Physically forced you to have sex.   **1** 2 3 4 5
27. Threw you around   **1** 2 3 4 5
28. Physically attacked the sexual parts of your body.   **1** 2 3 4 5
29. Choked or strangled you.   **1** 2 3 4 5
30. Used a knife, gun, or other weapon against you.   **1** 2 3 4 5

Scoring:

Add the numbers you circled and divide the total by 30 to find your score. The higher your score, the more abusive your relationship.

The inventory was given to 100 men and 78 women equally divided into groups of abusers/abused and nonabusers/nonabused. The men were members of a chemical dependency treatment program in a veterans' hospital and the women were partners of these men. Abusing or abused men earned an average score of 1.8; abusing or abused women earned an average score of 2.3. Nonabusing or nonabused men and women earned scores of 1.3 and 1.6, respectively.

Source: Shepard, Melanie F., and James A. Campbell, 1992. "The Abusive Behavior Inventory: A Measure of Psychological and Physical Abuse." *Journal of Interpersonal Violence,* September 1992, 7(3): 291–305. Inventory is on pages 303–304. Used by permission of Sage Publications, 2455 Teller Road, Newbury Park, CA 91320.

Intimate Partner Violence and Abuse

Abuse in relationships can take many forms, including emotional and psychological abuse, physical violence, and sexual abuse. **Intimate partner violence** refers to actual or threatened violent crimes committed against individuals by their current or former spouses, cohabiting partners, boyfriends, or girlfriends.

Prevalence and Patterns of Intimate Partner Violence. Globally, one woman in every three has been subjected to violence in an intimate relationship (United Nations Development Programme 2000). Most acts of intimate partner violence (85%) are committed against women and slightly more than one in five U.S. women (22%) has been assaulted by an intimate partner during her lifetime (Mercy et al. 2003; Rennison 2003). Although women also assault their male partners, these assaults tend to be acts of retaliation or self-defense (Johnson 2001). Most research on intimate partner abuse has been conducted on heterosexuals, but more than a third of gay women and gay men in one study reported physical violence in their relationships in the past year (McKenry et al. 2004).

Factors associated with higher rates of intimate partner victimization against women include being young (ages 16 to 24), black, divorced or separated, and earning lower incomes. The rate and severity of physical violence tend to be higher among cohabiting couples than among dating and marital partners (Johnson & Ferraro 2003; Magdol et al. 1998). Characteristics of abusive partners include alcohol or substance abuse, a history of trauma, limited support systems, emotional dependency and jealousy, male unemployment (which creates feelings of inadequacy), and having a traditional role-relationship ideology (Loy et al. 2005; Umberson et al. 2003).

Four patterns of partner violence have been identified: common couple violence, intimate terrorism, violent resistance, and mutual violent control (Johnson & Ferraro 2003).

1. **Common couple violence** refers to occasional acts of violence arising from arguments that get "out of hand." Common couple violence usually does not escalate into serious or life-threatening violence.
2. **Intimate terrorism** is violence that is motivated by a wish to control one's partner and involves the systematic use of not only violence but also economic subordination, threats, isolation, verbal and emotional abuse, and other control tactics. Intimate terrorism is almost entirely perpetrated by men and is more likely to escalate over time and to involve serious injury.
3. **Violent resistance** refers to acts of violence that are committed in self-defense. Violent resistance is almost exclusively perpetrated by women against a male partner.
4. **Mutual violent control** is a rare pattern of abuse "that could be viewed as two intimate terrorists battling for control" (Johnson & Ferraro 2003, p. 169).

Intimate partner abuse also takes the form of sexual aggression, which refers to sexual interaction that occurs against one's will through use of physical force, threat of force, pressure, use of alcohol or drugs, or use of position of authority. An estimated 7–14% of married women have been raped by their husbands (Monson et al. 1996). A national study found that about 3% of college women experienced a completed or attempted rape during a college year (Fisher et al. 2000). Nearly 90% of the sexually assaulted college women knew the person who assaulted them. Based on data from the National Violence Against Women survey, half of the women raped by an intimate partner and two-thirds of the women physically assaulted by an intimate partner had been victimized multiple times (Rand 2003).

Effects of Intimate Partner Violence and Abuse. Battering results in physical, psychological, economic, and marital consequences. Each year, intimate partner violence results in nearly 2 million injuries and more than 1,000 deaths (National Center for Injury Prevention and Control 2004). In 2003 about one-third of U.S. female murder victims were killed by husbands or boyfriends; only 2.5% of male murder victims were killed by wives or girlfriends (FBI 2004). Many battered women are abused during pregnancy, resulting in a high rate of miscarriage and birth defects. Psychological consequences for victims of intimate partner violence can include depression, anxiety, suicidal thoughts and attempts, lowered self-esteem, and substance abuse (National Center for Injury Prevention and Control 2004).

Battering also interferes with women's employment. Some abusers prohibit their partners from working. Other abusers "deliberately undermine women's employment by depriving them of transportation, harassing them at work, turning off alarm clocks, beating them before job interviews, and disappearing when they promise to provide child care" (Johnson & Ferraro 2003, p. 508). Battering also undermines employment by causing repeated absences, impairing women's ability to concentrate, and lowering their self-esteem and aspirations.

Women who have experienced physical or sexual abuse as adults or children are also less likely to be married or in a stable, long-term relationship. In a study of more than 2,000 women living in low-income neighborhoods, 42% of women who did not report abuse were married, compared to only 22% of the women who reported past abuse (Cherlin et al. 2005). Abuse, whether physical or emotional, is also a factor in many divorces, which often results in a loss of economic resources. Women who flee an abusive home and who have no economic resources may find themselves homeless. In a survey of U.S. mayors domestic violence was identified as a primary cause of homelessness in 12 out of 27 cities (U.S. Conference of Mayors 2004).

Many children who witness domestic violence get involved by yelling, calling for help, or intervening to try to stop the abuse (Edleson et al. 2003). Children who witness domestic violence are at risk for emotional, behavioral, and academic problems as well as future violence in their own adult relationships (Kitzmann et al. 2003; Parker et al. 2000). Children may also commit violent acts against a parent's abusing partner.

Why Do Some Adults Stay in Abusive Relationships? Adult victims of abuse are commonly blamed for tolerating abusive relationships and for not leaving the relationship as soon as the abuse begins. But from the point of view of the victims, there are compelling reasons to stay. These reasons include love, emotional dependency, commitment to the relationship, hope that things will get better, the view that violence is legitimate because they "deserve" it, guilt, fear, economic dependency, and feeling stuck.

Few and Rosen (2005) interviewed 28 women (7 blacks, 21 whites) who were victims of chronic abuse from a male dating partner and found that 24 of them had played a caretaker role in their families of origin that made them more vulnerable to being seduced by abusive boyfriends and to becoming trapped by their commitment to rescue them. One woman in the study reported that she tended to be attracted to needy men and tried to make them feel good about themselves. She explained, "I always was a rescuer in my family. I felt that I was rescuing him

ThomsonNOW™

Learn more about **Effects of Intimate Partner Violence and Abuse** by going through the Rape in Intimate Relationships Learning Module.

[her boyfriend] and taking care of him. He never knew what it was like to have a good, positive home environment, so I was working hard to create that for him" (Few & Rosen 2005, p. 272). Women in this study also reported that witnessing abuse in their parents' adult relationships taught them the notion that oppression and abuse of power were normal within intimate relationships.

Most of the women in Few and Rosen's (2005) study reported feeling pressure to be in a serious relationship or to have a husband and feared that if they ended the abusive relationship, they might not find someone else and would end up alone. In another study of why abused women stay in abusive relationships, the most important reason reported by women in these relationships was fear of loneliness (Hendy et al. 2003).

Six out of the seven black women in Few and Rosen's (2005) study talked about the scarcity of eligible black men as the reason some black women settle for abusive men. Another reason black women remained in their abusive relationships was to prove that black relationships were not inherently dysfunctional. One black woman explained, "There's so much negativity out there about Black relationships. Our relationships are always portrayed as being so adversarial. . . . We have so few positive images of Black healthy relationships. . . . So you got to settle . . . as much as you can" (Few & Rosen 2005, p. 273). In addition, abused black women who hesitated to seek help from police cited a distrust of police and talked explicitly about the historical mistreatment of black men by police in the South. Christina, a black woman, who reluctantly called the police as a last resort, explained,

> He was going to lose everything if I made this report. One, because he was Black. Two, because he was a Black man in a very White town. I was afraid he would be mistreated by these hick officers . . . that he would lose his job. . . . I decided to call the police and found when I did, none of my girlfriends would support me. (Few & Rosen 2005, p. 273)

Some of Christina's black girlfriends told that her she was a "traitor" for calling the police. None of her girlfriends would testify that her boyfriend "terrorized" her at parties.

Some victims of intimate partner abuse stay because they fear retribution from their abusive partner if they leave. Approximately one in four victims delays leaving a violent home because she fears the abuser will abuse or neglect a family pet (Fogle 2003).

Victims also stay because abuse in relationships is usually not ongoing and constant but rather occurs in cycles. The **cycle of abuse** involves a violent or abusive episode followed by a makeup period when the abuser expresses sorrow and asks for forgiveness and "one more chance." The honeymoon period may last for days, weeks, or even months before the next outburst of violence occurs.

Child Abuse

Child abuse refers to the physical or mental injury, sexual abuse, negligent treatment, or maltreatment of a child under the age of 18 by a person who is responsible for the child's welfare. The most common form of child maltreatment is **neglect**—the caregiver's failure to provide adequate attention and supervision, food and nutrition, hygiene, medical care, and a safe and clean living environment (see Figure 5.4). Slightly more than half (52%) of child abuse and neglect victims in 2003 were girls (U.S. Department of Health and Human Services, Administration on Children, Youth, and Families 2005). Children at highest risk for victimization include children

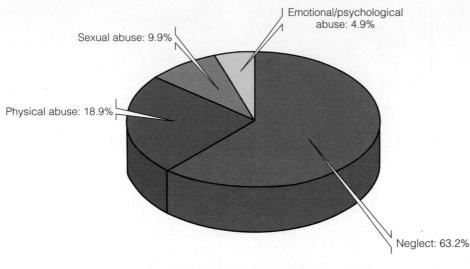

Sexual abuse: 9.9%

Emotional/psychological abuse: 4.9%

Physical abuse: 18.9%

Neglect: 63.2%

Total number of child victims of abuse/neglect in 2003: 906,000

Figure 5.4
Types of child maltreatment, 2003.

Source: U.S. Department of Health and Human Services, Administration on Children, Youth, and Families (2005).

Table 5.1
Rates of Child Abuse and Neglect by Race and Ethnicity: United States, 2003

Race or Ethnic Group	Rate per 1,000 Children
Pacific Islander	21.4
American Indian/Alaska Native	21.3
African American	20.4
White	11.0
Hispanic	9.9
Asian	2.7

Source: U.S. Department of Health and Human Services, Administration on Children, Youth, and Families (2005).

in the age group birth to 3 years, children with disabilities, and some minorities. Although more than half of child abuse and neglect victims in 2003 were white (54%), *rates* of victimization are higher among Pacific Islander children, American Indian or Alaska Native children, and African American children (see Table 5.1).

Perpetrators of child abuse are most often parents of the victim (see Figure 5.5). Parents who are at the greatest risk of child maltreatment are those who are socially isolated, poor or unemployed, young and single, suffering from mental illness, lack an understanding of children's needs and child development, and have a history of domestic abuse (National Center for Injury Prevention and Control 2005).

Effects of Child Abuse. The effects of child abuse and neglect vary according to the frequency and intensity of the abuse or neglect. Physical injuries sustained by child abuse cause pain, disfigurement, scarring, physical disability, and death. In

Figure 5.5

Child abuse or neglect victims by parental status of perpetrator: United States, 2003.

Source: U.S. Department of Health and Human Services, Administration on Children, Youth, and Families (2005).

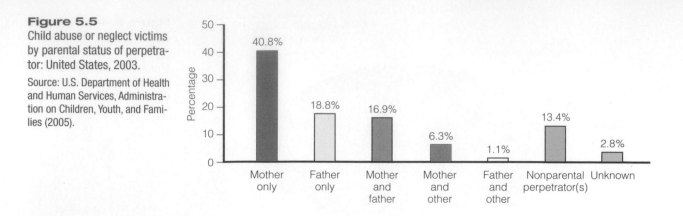

In his memoir, *A Private Family Matter,* Victor Rivas Rivers describes the horrific physical abuse he endured at the hands of his violent and controlling father and the pain of witnessing his father abuse his mother and siblings. Rivers overcame the rage and self-loathing that developed in his childhood and became a successful football player, actor, and national spokesperson for the National Network to End Domestic Violence.

2003 an estimated 1,500 U.S. children died of abuse or neglect (U.S. Department of Health and Human Services, Administration on Children, Youth, and Families 2005). Most of these children (79%) were under age 4. More than half of U.S. murder victims under age 12 are killed by a parent (FBI 2004).

Head injury is the leading cause of death in abused children (Rubin et al. 2003). **Shaken baby syndrome,** whereby the caretaker, most often the father, shakes the baby to the point of causing the child to experience brain or retinal hemorrhage, most often occurs in response to a baby, who typically is younger than 6 months, who won't stop crying (Ricci et al. 2003; Smith 2003). Battered or shaken babies are often permanently handicapped.

In addition to risk of immediate injury and death, abuse during childhood is associated with risk factors and risk-taking behaviors later in life, including depression, low academic achievement, smoking, obesity, teen pregnancy, alcohol and drug use, sexually transmitted diseases, low self-esteem, aggressive behavior, juvenile delinquency, adult criminality, and suicide (Administration for Children and Families 2003; Mercy et al. 2003). Among females early forced sex is associated with decreased self-esteem, increased levels of depression, running away from home, alcohol and drug use, and multiple sexual partners (Jasinski et al. 2000; Whiffen et al. 2000). Compared to nonabused peers, sexually abused girls are also more likely to experience teenage pregnancy, to have higher numbers of sexual partners in adulthood, and to acquire sexually transmitted infections and experience forced sex (Browning & Laumann 1997; Stock et al. 1997). Women who were sexually abused as children also report a higher frequency of post-traumatic stress disorder (Spiegel 2000) and suicide ideation (Thakkar et al. 2000). A review of the research suggests that sexual abuse of boys produces many of the same reactions that sexually abused girls experience, including depression, sexual dysfunction, anger, self-blame, suicidal feelings, guilt, and flashbacks (Daniel 2005). Married adults who were physically and sexually abused as children report lower marital satisfaction, higher stress, and lower family cohesion than married adults with no abuse history (Nelson & Wampler 2000).

Effects of child sexual abuse are likely to be severe when the sexual abuse is forceful, is prolonged, and involves intercourse and when the abuse is perpetrated by a father or stepfather (Beitchman et al. 1992). Not only has the child been violated physically, but she or he also has lost an important social support. One woman who had been sexually abused by her father described feeling that she had lost her father; he was no longer a person to love and protect her (Spiegel 2000).

Elder Abuse, Parent Abuse, and Sibling Abuse

Domestic violence and abuse may involve adults abusing their elderly parents or grandparents, children abusing their parents, and siblings abusing each other.

Elder Abuse. Elder abuse in domestic settings affects hundreds of thousands of elderly Americans each year. **Elder abuse** includes physical abuse, sexual abuse, psychological abuse, financial abuse (such as improper use of the elder's financial resources), and neglect. The most common form of elder abuse is neglect—failure to provide basic health and hygiene needs, such as clean clothes, doctor visits, medication, and adequate nutrition. Neglect also involves unreasonable confinement, isolation of elderly family members, lack of supervision, and abandonment.

Adult children are the most frequent abusers of the elderly, followed by other family members and spouses. A number of risk factors are associated with elder abuse, including social isolation, a history of a poor-quality relationship between the abused and abuser, a pattern of family violence, and a history of mental health or substance abuse problems in the abuser (Newton 2005).

Parent Abuse. Some parents are victimized by their children's violence, ranging from hitting, kicking, and biting to pushing a parent down the stairs and using a weapon to inflict serious injury to or even kill a parent. More violence is directed against mothers than against fathers, and sons tend to be more violent toward parents than are daughters (Ulman 2003). In most cases of children being violent toward their parents, the parents had been violent toward the children.

Sibling Abuse. The most prevalent form of abuse in families is sibling abuse. Ninety-eight percent of the females and 89% of the males in one study reported having been emotionally abused by a sibling, and 88% of the females and 71% of the males reported having been physically abused by a sibling (Simonelli et al. 2002). Sexual abuse also occurs in sibling relationships.

Factors Contributing to Intimate Partner and Family Violence and Abuse

Research suggests that cultural, community, and individual and family factors contribute to domestic violence and abuse.

Cultural Factors. Violence in the family stems from our society's acceptance of violence as a legitimate means of enforcing compliance and solving conflicts at personal, national, and international levels (Viano 1992). Violence and abuse in the family may be linked to cultural factors, such as violence in the media (see Chapter 4), gender inequality and gender socialization, and acceptance of corporal punishment.

Gender Inequality and Gender Socialization. In patriarchal societies "the beating of women and children has been taken for granted, with both men and women accepting it as normal" (Adams 2004, p. 1079). In sub-Saharan Africa, for example, widespread spousal abuse is linked to cultural views of women. One Nigerian woman who was beaten unconscious by her husband explained, "He doesn't believe I have any rights of my own. . . . If I say no, he beats me" (quoted by LaFraniere 2005, p. A1). According to Nigeria's minister for women's affairs, "It is like it is a normal thing for women to be treated by their husbands as punching bags. . . . The Nigerian man thinks that a woman is his inferior. . . . Even when they marry out of love, they still think the woman is below them and they do whatever they want" (quoted by LaFraniere 2005, p. A1).

In the United States before the late 19th century a married woman was considered the property of her husband. A husband had a legal right and marital obligation to discipline and control his wife through the use of physical force. This traditional view of women as property may contribute to men's doing with their "property" as they wish. A recent study of men in battering intervention programs found that about half of the men viewed battering as acceptable in certain situations (Jackson et al. 2003).

The view of women and children as property also explains marital rape and father-daughter incest. Historically, the penalties for rape were based on property rights laws designed to protect a man's property—his wife or daughter—from rape by other men; a husband or father "taking" his own property was not considered rape (Russell 1990). In the past a married woman who was raped by her husband could not have her husband arrested because marital rape was not considered a crime. Today, marital rape is considered a crime in all 50 states.

Traditional male gender roles have taught men to be aggressive and to be dominant in male-female relationships. Male abusers are likely to hold traditional attitudes toward women and male-female roles (Lloyd & Emery 2000). Traditional male gender socialization also discourages men from verbally expressing their feelings, which increases the potential for violence and abusive behavior (Umberson et al. 2003). Anderson (1997) found that men who earn less money than their partners are more likely to be violent toward them: "Disenfranchised men then must rely on other social practices to construct a masculine image. Because it is so clearly associated with masculinity in American culture, violence is a social practice that enables men to express a masculine identity" (p. 667). Traditional female gender roles have also taught women to be submissive to their male partner's control.

Acceptance of Corporal Punishment. **Corporal punishment** is the intentional infliction of pain for a perceived misbehavior (Block 2003). Many mental health professionals and child development specialists argue that it is ineffective and damaging to children. Children who experience corporal punishment display more antisocial behavior, are more violent, and have an increased incidence of depression as adults (Straus 2000). Yet many parents accept the cultural tradition of spanking as an appropriate form of child discipline.

More than 90% of parents of toddlers reported using corporal punishment (Straus 2000). Eighty-three percent of more than 11,000 undergraduate students at the University of Iowa reported that they had experienced some form of physical punishment during their childhood (Knutson & Selner 1994). Although not everyone agrees that all instances of corporal punishment constitute abuse, undoubtedly, some episodes of parental "discipline" are abusive.

Community Factors. Community factors that contribute to violence and abuse in the family include social isolation and inaccessible or unaffordable community services, such as health-care, day-care, elder-care, and respite-care facilities.

Living in social isolation from extended family and community members increases a family's risk for abuse. Isolated families are removed from material benefits, care-giving assistance, and emotional support from extended family and community members.

Failure to provide medical care to children and elderly family members (a form of neglect) is sometimes a result of the lack of accessible or affordable health care services in the community. Failure to provide supervision for children and adults may result from inaccessible day care and elder care services. Without elder care and respite care facilities, socially isolated families may not have any help with the stresses of caring for elderly family members and children with special needs.

Individual and Family Factors. Individual and family factors associated with intimate partner and family violence and abuse include a history of family violence, drug and alcohol abuse, and poverty.

Men who witnessed their fathers abusing their mothers and women who witnessed their mothers abusing their fathers are more likely to become abusive partners themselves (Babcock et al. 2003; Heyman & Slep 2002). Individuals who were abused as children are more likely to report being abused in an adult domestic relationship (Heyman & Slep 2002). Mothers who have been sexually abused as children are more likely to physically abuse their own children (DiLillo et al. 2000). Although a history of abuse is associated with an increased likelihood of being abusive as an adult, most adults who were abused as children do not continue the pattern of abuse in their own relationships (Gelles 2000).

Alcohol use is reported as a factor in 50–75% of incidents of physical and sexual aggression in intimate relationships (Lloyd & Emery 2000). Alcohol and other drugs increase aggression in some individuals and enable the offender to avoid responsibility by blaming his or her violent behavior on drugs/alcohol.

Although abuse in adult relationships occurs among all socioeconomic groups, it is more prevalent among the poor. Studies show that at least 50–60% of women receiving welfare have experienced physical abuse by an intimate partner, compared to 22% of the general population (Family Violence Prevention Fund 2005). However, Kaufman and Zigler (1992) note that "although most poor people do not maltreat their children, and poverty, per se, does not cause abuse and neglect, the correlates of poverty, including stress, drug abuse, and inadequate resources for food and medical care, increase the likelihood of maltreatment" (p. 284)

Strategies for Action: Preventing and Responding to Violence and Abuse in Intimate and Family Relationships

Strategies to prevent family violence and abuse include **primary prevention** strategies, which target the general population; **secondary prevention** strategies, which target groups at high risk for family violence and abuse; and **tertiary prevention** strategies, which target families who have experienced abuse (Gelles 1993; Harrington & Dubowitz 1993).

Primary Prevention Strategies

Preventing violence and abuse may require broad, sweeping social changes, such as eliminating the norms that legitimize and glorify violence in society and changing the sexist character of society (Gelles 2000). Specific abuse-prevention strategies

Table 5.2

Effective Discipline Techniques for Parents: Alternatives to Spanking

Punishment is a "penalty" for misbehavior, but discipline is a method of teaching a child right from wrong. Alternatives to physical discipline include the following:

1. Be a positive role model.

Children learn behaviors by observing their parents' actions, so parents must model the ways in which they want their children to behave. If a parent yells or hits, the child is likely to do the same.

2. Set rules and consequences.

Make rules that are fair, realistic, and appropriate to a child's level of development. Explain the rules to children along with the consequences of not following them. If children are old enough, they can be included in establishing the rules and consequences for breaking them.

3. Encourage and reward good behavior.

When children are behaving appropriately, give them verbal praise and occasionally reward them with tangible objects, privileges, or increased responsibility.

4. Create charts.

Charts to monitor and reward behavior can help children learn appropriate behavior. Charts should be simple and should focus on one behavior at a time, for a certain length of time.

5. Give time-outs.

A "time-out" involves removing a child from a situation following a negative behavior. This can help the child calm down, end the inappropriate behavior, and reenter the situation in a positive way. Explain what the inappropriate behavior is, why the time-out is needed, when the time-out will begin, and how long it will last. Set an appropriate length of time for the time-out based on age and level of development, usually just a few minutes.

Source: Based on National Mental Health Association, 2003. "Effective Discipline Techniques for Parents: Alternatives to Spanking." Strengthening Families Fact Sheet. www.nmha.org. Reprinted with permission.

include public education and media campaigns, which may help to reduce domestic violence by conveying the criminal nature of domestic assault and offering ways to prevent abuse.

Other abuse-prevention efforts focus on parent education to teach parents realistic expectations about child behavior and methods of child discipline that do not involve corporal punishment. For example, the National Mental Health Association (2003) distributes a fact sheet on alternatives to spanking (see Table 5.2). In 1979 Sweden became the first country in the world to ban corporal punishment. Fifteen countries (Austria, Bulgaria, Croatia, Cyprus, Denmark, Finland, Germany, Hungary, Iceland, Israel, Italy, Latvia, Norway, Romania, and Ukraine) have followed Sweden's lead and have banned corporal punishment in all settings, including the home (Global Initiative to End All Corporal Punishment of Children 2005). In the United States it is legal in all 50 states for a parent to spank, hit, belt, paddle, whip, or otherwise inflict punitive pain on a child, so long as the corporal punishment does not meet the individual state's definition of child abuse. Corporal punishment is also permitted in public schools in 27 states and in some large-city school districts.

Another abuse-prevention strategy involves reducing violence-provoking stress by reducing poverty and unemployment and providing adequate housing, child-care programs and facilities, nutrition, medical care, and educational opportunities. However, rather than strengthening the supports for poor families with children, welfare reform legislation enacted in 1996 limits cash assistance to poor single parents to two consecutive years with a five-year lifetime limit (some exceptions are granted) and forces women into the labor force. Many women going from welfare to work will experience greater hardships as a result of a loss of food stamp benefits, increases in federal housing rent, loss of Medicaid benefits, cost of transportation to work, and child-care costs (Edin & Lein 1997). Some women forced to go to work and unable to afford child care will leave their children unattended, increasing child neglect. The cumulative stresses and hardships that the welfare-to-work legislation will have on single parents may very well contribute to increases in child neglect and abuse.

Secondary Prevention Strategies

Families at risk of experiencing violence and abuse include low-income families, parents with a history of depression or psychiatric care, single parents, teenage mothers, parents with few social and family contacts, individuals who experienced abuse in their own childhood, and parents or spouses who abuse drugs or alcohol. Secondary prevention strategies, designed to prevent abuse from occurring in high-risk families, include parent education programs, parent support groups, individual counseling, substance abuse treatment, and home visiting programs.

Tertiary Prevention Strategies

The National Domestic Violence Hotline (1-800-799-SAFE) is a 24-hour, toll-free service that provides crisis assistance and local domestic violence shelter and safe-house referrals for callers across the country. Shelters provide abused women and their children with housing, food, and counseling services. Safe houses are private homes of individuals who volunteer to provide temporary housing to abused persons who decide to leave their violent homes. Some communities have abuse shelters for victims of elder abuse. Some programs offer a safe shelter for pets of domestic violence victims. Because one in four victims reports a delay in leaving dangerous domestic situations because of concerns over the safety of a pet, some domestic abuse agencies have paired with veterinary schools, humane societies, and community organizations to help victims and their pets escape violent homes (Fogle 2003).

Children who are abused in the family may be removed from their homes and placed in foster care or in the care of another family member, such as a grandparent. However, every state has various types of programs to prevent family breakup when desirable and possible without jeopardizing the welfare of children in the home. **Family preservation programs** are in-home interventions for families who are at risk of having a child removed from the home because of abuse or neglect.

Alternatively, a court may order an abusing spouse or parent to leave the home. Abused spouses or cohabiting partners may obtain a restraining order that prohibits the perpetrator from going near the abused partner. About half of the states and Washington, D.C., now have mandatory arrest policies that require police to arrest abusers, even if the victim does not want to press charges. However, many victims of intimate partner violence do not report the violence to law enforcement authorities. Reasons that victims do not report intimate partner violence to the police include (1) the belief that such violence is a private or personal matter, (2) fear of retaliation, (3) viewing the violence as a "minor" crime, (4) protecting the offender, and (5) the belief that the police will not help or will be ineffective.

Treatment for abusers—which may be voluntary or mandated by the court—typically involves group and/or individual counseling, substance abuse counseling, and/or training in communication, conflict resolution, and anger management. Treatment for men who sexually abuse children typically involves cognitive behavior therapy (changing the thoughts that lead to sex abuse) and medication to reduce the sex drive (Stone 2004).

Men who stop abusing their partners learn to take responsibility for their abusive behavior, develop empathy for their partner's victimization, reduce their dependency on their partners, and improve their communication skills (Scott & Wolfe 2000). However, recent evaluations of batterer intervention programs found no significant differences between treatment and control groups on re-offense rates or men's attitudes toward domestic violence (Jackson et al. 2003).

Problems Associated with Divorce

The United States has the highest divorce rate among Western nations. Despite the decline in divorce rates in recent years, 40% of first marriages end in divorce and 60% of those marriages involve children (Kimmel 2004). Divorce is considered problematic not only because of the negative effects it has on children but also because of the difficulties it causes for adults. However, in some societies legal and social barriers to divorce are considered problematic because such barriers limit the options of spouses in unhappy and abusive marriages. Ireland did not allow divorce under any condition until 1995, and Chile did not allow divorce until 2004. Even when divorce is a legal option, social barriers often prevent spouses from divorcing. Hindu women, for example, experience great difficulty leaving a marriage, even when the husband is abusive. Loss of status, possible loss of custody of her children, homelessness, poverty, and being labeled a "loose" woman are enough to keep women locked in the "confines of a tyrannous family as silent sufferers" (Laungani 2005, p. 88).

ThomsonNOW™

Learn more about **Problems Associated with Divorce** by going through the Divorce Process Learning Module.

Social Causes of Divorce

When we think of why a particular couple gets divorced, we typically think of a number of individual and relationship factors that might have contributed to the marital breakup: incompatibility in values or goals, poor communication, lack of conflict resolution skills, sexual incompatibility, extramarital relationships, substance abuse, emotional or physical abuse or neglect, boredom, jealousy, and difficulty coping with change or stress related to parenting, employment, finances, in-laws, and illness. But understanding the high rate of divorce in U.S. society requires awareness of how the following social and cultural factors contribute to marital breakup:

1. *Changing family functions.* Before the Industrial Revolution the family constituted a unit of economic production and consumption, provided care and protection to its members, and was responsible for socializing and educating

children. During industrialization, other institutions took over these functions. For example, the educational institution has virtually taken over the systematic teaching and socialization of children. Today, the primary function of marriage and the family is the provision of emotional support, intimacy, affection, and love. When marital partners no longer derive these emotional benefits from their marriage, they may consider divorce with the hope of finding a new marriage partner to fulfill these affectional needs.

2. *Increased economic autonomy of women.* As noted earlier, before 1940 most wives were not employed outside the home and depended on their husband's income. Today, about two-thirds of married women are in the labor force (U.S. Bureau of Labor Statistics 2005). A wife who is unhappy in her marriage is more likely to leave the marriage if she has the economic means to support herself (Jalovaara 2003). An unhappy husband may also be more likely to leave a marriage if his wife is self-sufficient and can contribute to the support of the children.

3. *Increased work demands and economic stress.* Another factor influencing divorce is increased work demands and the stresses of balancing work and family roles. Workers are putting in longer hours, often working overtime or taking second jobs. And as discussed in Chapters 6 and 7, many families struggle to earn enough money to pay for rising housing, health care, and child-care costs. Financial stress can cause marital problems. Whitehead and Popenoe (2005) report that couples with an annual income under $25,000 are 30% more likely to divorce than couples with incomes over $50,000.

4. *Dissatisfaction with marital division of labor.* Many employed parents, particularly mothers, come home to work a **second shift**—the work involved in caring for children and household chores (Hochschild 1989). Wives are more likely than husbands to perceive the marital division of labor—household chores and child care—as unfair (Nock 1995). This perception of unfairness can lead to marital tension and resentment, as reflected in the following excerpt:

> My husband's a great help watching our baby. But as far as doing housework or even taking the baby when I'm at home, no. He figures he works five days a week; he's not going to come home and clean. But he doesn't stop to think that I work seven days a week. Why should I have to come home and do the housework without help from anybody else? My husband and I have been through this over and over again. Even if he would just pick up from the kitchen table and stack the dishes for me, that would make a big difference. He does nothing. On his weekends off, I have to provide a sitter for the baby so he can go fishing. When I have a day off, I have the baby all day long without a break. He'll help out if I'm not here, but the minute I am, all the work at home is mine. (quoted by Hochschild 1997, pp. 37–38)

Women are increasingly looking for egalitarianism in relationships. Women want to be equal partners in their marriages, not just in earning income but in sharing the work of household chores, child rearing, and marital communication and in making decisions for the family. They are also looking for partners who are considerate and dependable (O'Reilly et al. 2005). Frustrated by men's lack of participation in marital work, women who desire relationship egalitarianism may see divorce as the lesser of two evils (Hackstaff 2003).

5. *Liberalized divorce laws.* Before 1970 the law required a couple who wanted a divorce to prove that one of the spouses was at fault and had committed an act defined by the state as grounds for divorce—adultery, cruelty, or desertion. In 1969 California became the first state to initiate **no-fault divorce,** which

"It is now widely accepted that men and women have the right to expect a happy marriage, and that if a marriage does not work out, no one has to stay trapped.**"**

Sylvia Ann Hewlett
Family advocate

ThomsonNOW™

Learn more about **Dissatisfaction with Marital Division of Labor** by going through the Familial Division of Labor Learning Module.

permitted a divorce based on the claim that there were "irreconcilable differences" in the marriage. No-fault divorce law has contributed to the U.S. divorce rate by making divorce easier to obtain. Although U.S. divorce rates started climbing before California instituted the first no-fault divorce law, the widespread adoption of such laws has probably contributed to their continued escalation. Today, all 50 states recognize some form of no-fault divorce.

6. *Increased individualism.* U.S. society is characterized by **individualism**—the tendency to focus on one's individual self-interests and personal happiness rather than on the interests of one's family and community. "Because people no longer wish to be hampered with obligations to others, commitment to . . . marriage has eroded. . . . Marital commitment lasts only as long as people are happy and feel that their own needs are being met" (Amato 2004, p. 960). **Familism**—the view that the family unit is more important than individual interests—is still prevalent among Asian Americans and Mexican Americans, which helps to explain why the divorce rate is lower among these groups than among whites and African Americans (Mindel et al. 1998).

7. *Increased life expectancy.* Finally, more marriages today end in divorce, in part, because people live longer than they did in previous generations. Because people live longer today than in previous generations, "till death do us part" involves a longer commitment than it once did. Indeed, one can argue that "marriage once was as unstable as it is today, but it was cut short by death not divorce" (Emery 1999, p. 7).

Consequences of Divorce

Divorce is considered a social problem because of the distress and difficulties associated with it. When parents have bitter and unresolved conflict and/or if one parent is abusing the child or the other parent, divorce may offer a solution to family problems. But divorce often has negative effects for ex-spouses and their children and also contributes to problems that affect society as a whole.

Physical and Mental Health Consequences. In a review of research on the consequences of divorce for adults, Amato (2003) cites numerous studies that found that divorced individuals have more health problems and a higher risk of mortality than married individuals; and divorced individuals also experience lower levels of psychological well-being, including more unhappiness, depression, anxiety, and poorer self-concepts. Both divorced and never-married individuals are, on average, more distressed than married people because unmarried people are more likely than married people to have low social attachment, low emotional support, and increased economic hardship (Walker 2001).

However, Amato's (2003) review also cites studies in which divorced individuals report higher levels of autonomy and personal growth than married individuals do. For example, many divorced mothers report improvements in career opportunities, social lives, and happiness following divorce; some divorced women report more self-confidence, and some men report more interpersonal skills and a greater willingness to self-disclose.

Economic Consequences. Compared to married individuals, divorced individuals have a lower standard of living, have less wealth, and experience greater economic hardship, although this difference is considerably greater for women than for men (Amato 2003). In general, the economic costs of divorce are greater for women

ThomsonNOW

Learn more about **Physical and Mental Health Consequences** by going through the Post Divorce Identity Animation.

and children because women tend to earn less than men (see Chapter 10) and because mothers devote substantially more time to household and child-care tasks than fathers do. The time women invest in this unpaid labor restricts their educational and job opportunities as well as their income. Men are less likely than women to be economically disadvantaged after divorce, because they continue to profit from earlier investments in education and career.

Following divorce, both parents are responsible for providing economic resources to their children. However, some nonresident parents fail to provide child support. In some cases failure to pay child support is not due to fathers being "deadbeats" but rather to the fact that many fathers are "dead broke." About one-third of nonresident fathers in 1999 lived in households with incomes below the poverty line, or their personal income was below the poverty level for a single person (Sorensen & Oliver 2002). Not surprisingly, poor fathers are less likely to pay child support. In 1999, 70% of poor fathers and 28% of nonpoor fathers failed to pay child support (Sorensen & Oliver 2002). More than one-fourth of poor fathers who paid child support in 1999 spent half or more of their personal income on child support.

In some cases a man's economic support of children following divorce may end if the man can prove that he is not the biological father of the child(ren). This chapter's *Focus on Technology* feature examines issues surrounding DNA testing and child support.

> "Divorce is definitely not a single event but a long-lasting process of radically changing family relationships that begins in the failing marriage, continues through the often chaotic period of the marital rupture and its immediate aftermath, and extends even further, often over many years of disequilibrium."
>
> **Judith S. Wallerstein**
> **Divorce researcher and scholar**

Effects on Children and Young Adults. Parental divorce is a stressful event for children, and is often accompanied by a variety of stressors, such as continuing conflict between parents, a decline in the standard of living, moving and perhaps changing schools, separation from the noncustodial parent (usually the father), and parental remarriage. These stressors place children of divorce at higher risk for a variety of emotional and behavioral problems. Reviews of research on the consequences of divorce for children have found that children with divorced parents score lower on measures of academic success, psychological adjustment, self-concept, social competence, and long-term health; they also have higher levels of aggressive behavior and depression (Amato 2003; Wallerstein 2003).

Many of the negative effects of divorce on children are related to the economic hardship associated with divorce. Economic hardship is associated with less effective and less supportive parenting, inconsistent and harsh discipline, and emotional distress in children (Demo et al. 2000).

One study found that divorce in one generation has adverse effects not only on that generation's children but also on future grandchildren who are not yet born (Amato & Cheadle 2005). Divorce in the first generation was associated with lower education, more marital discord, more divorce, and greater parent-child tensions in the second generation, which contributed to lower education, more marital discord, and weaker ties with parents in the third generation.

Despite the adverse effects of divorce on children, current research findings suggest that "most children from divorced families are resilient, that is, they do not suffer from serious psychological problems" (Emery et al. 2005, p. 24). Other researchers who study the effects of divorce on children conclude that "most offspring with divorced parents develop into well-adjusted adults," despite the pain they feel associated with the divorce (Amato & Cheadle 2005, p. 191).

Divorce can also have positive consequences for children and young adults. In highly conflictual marriages divorce may actually improve the emotional well-being of children relative to staying in a conflicted home environment (Jekielek 1998).

DNA Paternity Testing and Child Support

DNA paternity testing involves using samples of blood, body tissue, fluid, sperm, bone, or most commonly, buccal swabs (from inside the cheek) to determine whether the genetic makeup of a particular child matches that of the man tested. DNA (deoxyribonucleic acid) is the chemical found in almost every cell in our bodies. Genetic information coded in the DNA determines our physical characteristics and is passed from one generation to the next. Except for identical twins, each person's DNA is unique. Half of our DNA is inherited from our father, and the other half comes from our mother.

With more than 99.9% accuracy, DNA testing is the most advanced and accurate test available to determine the identity of a child's father. Most laboratories can have results in 10 days and charge about $290 for a basic paternity verification test (McFraser 2005).

The American Association of Blood Banks (2002) reports that the 310,490 paternity tests conducted on men in 2001 ruled out 29% as the biological father. Courts and legislatures have had to face the question, Should a man have to continue paying child support when DNA test results prove that he is not the biological parent?

When men voluntarily acknowledge paternity, federal law gives them 60 days to rescind it. After 60 days states set their own limits. For example, Florida allows a year after a child support order, and California allows two years after a birth. Four states—Maryland, Alabama, Ohio, and Georgia—have laws that release men from child support obligations for children proven by DNA testing not to be theirs biologically, with no time constraint on when relief can be obtained.

Many unwed fathers who pay child support never admitted paternity. A 1996 federal welfare law requires an unwed mother to identify a father when she applies for public assistance. A court summons can be mailed to his last known address, but many men do not receive the notice. Indeed, they may know nothing of a paternity claim until they discover that their paycheck is being docked under "default" judgments of paternity that cannot be contested after six months.

If a man believes that he has been falsely accused of fathering a child, he should act promptly to protect himself. The federal government requires states to provide genetic DNA testing in all contested paternity proceedings, at no cost to the man if he is indigent. However, the courts may or may not grant relief from child support based on DNA paternity testing, and some courts will not even consider DNA testing in child support hearings. California judges, for example, would not consider tests that showed that Air Force Master Sgt. Raymond Jackson's three children from his former marriage were fathered by other men. Jackson's resentment is reflected in comments he made in an interview with *USA Today*: "It's like they are saying, 'Let your wife cheat on you, have children by other men, divorce you, and now you have to pay for it all'" (Kasindorf 2002, p. A3). After DNA testing revealed that a Texas man was not the biological father of a child he was financially supporting, the man went to court to seek an end to his financial obligation. The court did not relieve him of the economic obligation to the child, but took away all visitation rights based on the DNA evidence. In contrast, an Oregon judge excused an ex-husband from child support payments after DNA tests showed he was not the father of his former wife's son. The judge nevertheless granted him visitation rights.

Various men's groups are lobbying state legislatures to change paternity laws as a result of the new DNA technology. Bert Riddick, a teacher in California who is paying $1,400 a month to support a teenage girl born out of wedlock and whom he has never met, says, "Think of it. I can get out of jail for murder based on DNA evidence, but I can't get out of child support payments" (quoted by Kasindorf 2002, p. A3). Groups such as U.S. Citizens Against Paternity Fraud (http://www.paternity-fraud.com) and New Jersey Citizens Against Paternity Fraud applaud the states (Alabama, Arkansas, Georgia, Iowa, Ohio, and Virginia) that have reshaped paternity laws to permit ex-husbands and out-of-wedlock alleged fathers to end child support on the basis of DNA evidence. Maryland has made the same change through court decisions. But some women's groups and child advocates have concerns about how such laws affect children. Valerie Ackerman of the National Center for Youth Law says, "Families are more complicated than who's biologically related to whom. . . . If there has been a relationship between a father and child, the man can't just abdicate the responsibility that he's taken on" (quoted by Kasindorf 2002, p. A3). Family law professor Carol Sanger (Columbia University) suggests that the law should be more lenient to men who may not even know a child than to dads who have been living with the children they did not father. Finally, Geraldine Jensen, president of the Association for Children for Enforcement of Support, suggests that all children should be DNA-tested at birth or at the time of divorce and that maternity wards should distribute pamphlets telling men, "Get tested now if you have any questions, because doing it later will disrupt this child's life" (quoted by Kasindorf 2002, p. A3).

In interviews with 173 grown children whose parents divorced years earlier, Ahrons (2004) found that most of the young adults reported positive outcomes for their parents as well as for themselves. More than half the young adults in this study reported that their relationships with their fathers actually improved after the divorce. In this chapter's *Human Side* feature one of the interviewees in Ahrons study describes her experience of her parents' divorce.

Most divorced individuals remarry, which necessitates (for children) adaptation to new parents and step-siblings. Research confirms that of children growing up in stepfamilies, "80 percent . . . are doing well. Children in stable stepfamilies look very much like those raised in stable first families" (Pasley 2000, p. 6).

Effects on Father-Child Relationships. Children who live with their mothers may suffer from a damaged relationship with their nonresidential father, especially if he becomes disengaged from their lives. Although some research has found that young adults whose parents divorced are less likely to report having a close relationship with their father compared to children whose parents are together (DeCuzzi et al. 2004), in a study of 173 adult children of divorce, more than half felt that their relationships with their fathers improved after the divorce (Ahrons 2004). Children may benefit from having more quality time with their fathers after parental divorce. Some fathers report that they became more active in the role of father after divorce. One father commented:

> Since my divorce I have been able to take my children on camping trips alone. We have spent weeks in the wilderness talking, cooperating, sharing in ways that would never have been possible on a "family" trip. I am sad for the divorce but the bonding with my children has been an unforeseen advantage. (Author's files)

The mother's attitude toward the father's continued contact with the child can have a dramatic effect on the father-child relationship. "If the mother approves of the close contact between her child and his/her father, the child will benefit both from the continued affection of the father and from the parental harmony. If the mother disapproves of the father's influence, the child, feeling torn by conflicting loyalties, may fail to benefit" (Wallerstein 2003, pp. 76–77).

Some divorced mothers not only fail to encourage their children's relationships with their fathers but also actively attempt to alienate the children from their father. (We note that some divorced fathers do likewise.) Thus some children of divorce suffer from **parental alienation syndrome (PAS)**, defined as an emotional and psychological disturbance in which children engage in exaggerated and unjustified denigration and criticism of a parent (Gardner 1998). "Children of PAS show negative parental reactions and perceptions which can be grossly exaggerated. . . . Put simply, they profess rejection and hatred of a previously loved parent, most often in the context of divorce and child custody conflicts" (Family Court Reform Council of America 2000). Parental alienation syndrome has been described as a form of "psychological kidnapping," whereby one parent manipulates children's psyches to make them hate and reject the other parent. Children who suffer from PAS are victims of a form of child abuse in which one parent essentially brainwashes the child to hate the other parent. A parent may alienate his or her child from the other parent by engaging in the following behaviors (Schacht 2000):

- Minimizing the importance of contact and relationship with the other parent.
- Being rude to the other parent; refusing to speak to or tolerate the presence of the other parent, even at events important to the child; refusing to allow the other parent near the home for drop-off or pick-up visitations.

My Parents' Divorce

Dr. Constance Ahrons conducted interviews with 173 grown children whose parents divorced years earlier. These interviews became the basis for her book We're Still Family: What Grown Children Have to Say About Their Parents' Divorce *(2004). This* Human Side *feature contains an excerpt from this book in which one of the interviewees talks about her parents' divorce.*

Sure, I would have liked to have had that perfect family that's on the cover of every magazine at Christmas. None of my friends had this perfect family but it's the one that every kid imagines the most popular kid at school has. I was only seven when my parents separated and I don't really remember much about what it was like when we all lived together, but I remember feeling sad and confused when they told me.

My parents were really young when they got married . . . it's hard for me to even imagine them together. I think the divorce

was a good decision, a necessary one, and I think we're all better off because of it today. I'm pretty lucky because my mom and dad told me that no matter what happened between them they both still loved me. . . . I always knew I was very valued. In some ways I think the divorce made both my parents really emphasize how much they cared about me. Some friends of mine with married parents didn't know where they stood in terms of their parents' affection or felt neglected or had pretty bad living situations.

I'm not saying it was always easy. I remember times when my parents disagreed about some decision that involved me and I felt caught in the middle. Sometimes I felt angry about the scheduling and going back and forth. I remember feeling really jealous when my mom told me her boyfriend Dan was moving in. I was surprised when my dad told me he was getting remarried and I really resented my stepmom and her kids.

Now I'm really close with my stepmom and I think she makes a much better mate for my dad than my mom did. I'm also close with my "stepdad," even though he and my mom never married and he's now married to someone else. It's confusing to explain all the relationships, and I used to be embarrassed about it, but now I feel lucky to have four parents. They were all there at my college graduation and I think it's widened my view of what I think a family is . . . it's helped me to communicate better and more freely with people who are important to me.

Source: Ahrons, Constance, 2004. *We're Still Family: What Grown Children Have to Say about Their Parents' Divorce*, pp. 23–34. New York: HarperCollins.

- Failing to express concern for missed visits with the other parent.
- Failing to display any positive interest in the child's activities or experiences during visits.
- Expressing disapproval or dislike of the child's spending time with the other parent and refusing to discuss anything about the other parent ("I don't want to hear about . . .") or selective willingness to discuss only negative matters.
- Making innuendoes and accusations against the other parent, including statements that are false.
- Demanding that the child keep secrets from the other parent.
- Destruction of gifts or memorabilia from the other parent.
- Promoting loyalty conflicts (e.g., offering an opportunity for a desired activity that conflicts with scheduled visitation).

Long-term effects of PAS on children are extremely serious and can include long-term depression, inability to function, guilt, hostility, alcoholism and other drug abuse, and other symptoms of internal distress (Family Court Reform Council of America 2000). Indeed, the effects on the rejected parent are equally devastating.

Some noncustodial divorced fathers discontinue contact with their children as a coping strategy for managing emotional pain (Pasley & Minton 2001). Many divorced fathers are overwhelmed with feelings of failure, guilt, anger, and sadness over the separation from their children (Knox 1998). Hewlett and West (1998) explain that "visiting their children only serves to remind these men of their painful loss, and they respond to this feeling by withdrawing completely" (p. 69). Divorced fathers commonly experience the legal system as favoring the mother in child-related matters. One divorced father commented:

> I believe that the system [judges, attorneys, etc.] have [sic] little or no consideration for the father. At some point the system creates an environment where the father loses any natural desire to see his children because it becomes so difficult, both financially and emotionally. At that point, he convinces himself that the best thing to do is wait until they are older. (quoted by Pasley & Minton 2001, p. 242)

As we have seen, the effects of divorce on adults and children are mixed and variable. In a review of research on the consequences of divorce for children and adults, Amato (2003) concludes that "divorce benefits some individuals, leads others to experience temporary decrements in well-being that improve over time, and forces others on a downward cycle from which they might never fully recover" (p. 206).

Strategies for Action: Strengthening Marriage and Alleviating Problems of Divorce

Two general strategies for responding to the problems of divorce are those that prevent divorce by strengthening marriages and those that strengthen post-divorce families.

Strategies to Strengthen Marriage and Prevent Divorce

A growing "marriage movement" involves efforts by religion and government to strengthen marriage and prevent divorce through a number of strategies, including the promotion of marriage education. Promoting gender equality and providing workplace and economic supports are also important in strengthening marriages and preventing divorce.

Marriage Education. Marriage education, also known as family life education, includes various types of workshops, classes, and encounter groups that (1) teach relationships skills, communication, and problem solving; (2) convey that sustaining healthy marriages requires effort; and (3) convey the importance of having realistic expectations of marriage, commitment, and a willingness to make personal sacrifices (Hawkins et al. 2004). A valuable alternative or supplement to face-to-face family life education is Web-based education. In the first month of its official launch, about 21,900 individuals visited the Forever Families website, a faith-based family education website (Steimle & Duncan 2004).

Marriage education is promoted by government and faith-based organizations. For example, the Community Marriage Policy strategy asks religious officials (who perform 75% of all U.S. weddings) to follow the Common Marriage Policy of the American Roman Catholic Church, which includes the following five components

(Browning 2003): (1) a six-month minimum marriage preparation period; (2) the use of a premarital questionnaire to identify problems or potential problems the couple may have; (3) the practice of mentoring engaged and newlywed couples; (4) the use of marriage education for engaged and married couples for the purpose of exploring the relationship, identifying problems, and learning effective communication and conflict resolutions techniques; and (5) engagement ceremonies held before the entire congregation (Browning 2003).

The Community Marriage Policy seeks to establish a common marriage policy across religious denominations to provide a united front on standards of marriage preparation. Ministers from more than 147 cities have adopted the Community Marriage Policy (Browning 2003). Of course, couples who are not required by their minister, priest, or rabbi to participate in premarital or marital education can voluntarily choose to participate in such programs offered by their religious organization, local mental health organization, or private marriage and family counselors.

States also promote marriage education. For example, the Oklahoma Marriage Initiative trained hundreds of individuals to provide free premarital education. A 1998 Florida law requires all ninth- or tenth-grade public school students to take a course in marriage and relationship education (Browning 2003). Florida also cuts the cost of a marriage license by half if the couple can show that they have taken a four-hour marriage education course. Since Florida passed its marriage education law, similar bills have been introduced in Minnesota, Maryland, Arizona, Connecticut, Kansas, and Michigan.

How effective are marriage education programs? Researchers have found that couples who participated in a widely used couples' education program called PREP (Prevention and Relationship Enhancement Program) had a lower divorce and separation rate five years after completing the program than couples who did not participate (Stanley et al. 1995). Marriage education programs for newlyweds have also been found to be effective in improving communication and conflict resolution and overall marital satisfaction (Cole et al. 2000).

Covenant Marriage and Divorce Law Reform. With the passing of the 1996 Covenant Marriage Act, Louisiana became the first state to offer two types of marriage contracts. Under the new law couples can voluntarily choose the standard marriage contract that allows a no-fault divorce (after a six-month separation) or a **covenant marriage,** which permits divorce only under condition of fault (e.g., abuse, adultery, felony conviction) or after a two-year separation. Couples who choose a covenant marriage must also get premarital counseling. Variations of the covenant marriage have also been adopted in Arizona and Arkansas. Only 3% of couples in states with covenant marriage laws have chosen the covenant marriage option (Coontz 2005).

Advocates of the covenant marriage believe that such marriages will strengthen marriages and decrease divorce. However, critics argue that covenant marriage may increase family problems by making it more emotionally and financially difficult to terminate a problematic marriage and by prolonging the exposure of children to parental conflict (Applewhite 2003).

Several states are considering divorce reform legislation that is intended to decrease the number of divorces by extending the waiting period required before a divorce is granted or requiring proof of fault (e.g., adultery, abuse). Opponents argue that divorce law reform measures would increase acrimony between divorcing spouses (which harms the children as well as the adults involved), increase the legal costs of getting a divorce (which leaves less money to support any children), and

Table 5.3

Factors That Decrease Women's Risk of Separation or Divorce During the First 10 Years of Marriage

Factor	Percentage Decrease in Risk of Divorce or Separation
Annual income over $50,000 (vs. under $25,000)	30
Having a baby 7 months or more after marriage (vs. before marriage)	24
Marrying over 25 years of age (vs. under 18)	24
Having an intact family of origin (vs. having divorced parents)	14
Religious affiliation (vs. none)	14
Some college (vs. high school dropout)	13

Source: Whitehead & Popenoe (2005).

delay court decisions on child support and custody and distribution of assets. Efforts in many state legislatures to repeal no-fault divorce laws have largely failed.

Promotion of Gender Equality. One study found that decision-making equality between spouses was one of the strongest correlates of husbands' and wives' marital happiness and perceived marital stability (Amato et al. 2003). Thus strategies that promote gender equality and egalitarianism in marriage relationships may contribute to marital happiness and stability

Workplace and Economic Supports. In a nationwide poll of parents' political priorities (National Parenting Association 1996), parents indicated that they want economic help in raising their children and work policies that help them balance their work and family roles. Indeed, one of the most important divorce prevention measures may be those that maximize employment and earnings. Of the factors associated with a decreased risk of divorce or separation listed in Table 5.3, the most important factor is having an annual income of more than $50,000 (versus less than $25,000) (Whitehead & Popenoe 2005).

Strengthening marriages may be achieved through policies and services that provide greater economic resources to couples whose marriages are at risk. These supports include flexible workplace policies that decrease work-family conflict (see Chapter 7), affordable child care, and economic support, such as the earned income tax credit (see Chapter 6).

Strategies to Strengthen Families During and After Divorce

When one or both marriage partners decide to divorce, what can the couple do to achieve a "friendly divorce" and minimize the negative consequences of divorce for their children? According to Ahrons (2004), it is the post-divorce conflict between

ThomsonNOW™

Learn more about **Strategies to Strengthen Families During and After Divorce** by going through the Characteristics of Strong Families Learning Module.

parents and not the divorce itself that is traumatic for children. A review of the literature on the effects of parental conflict on children suggests that children who are exposed to high levels of parental conflict are at risk for anxiety, depression, and disruptive behavior; they are more likely to be abusive toward romantic partners in adolescence and adulthood and are likely to have higher rates of divorce and maladjustment in adulthood (Grych 2005). Two strategies that help to reduce parental conflict and promote cooperative parenting following divorce are divorce mediation and divorce education programs.

Divorce Mediation. In **divorce mediation** divorcing couples meet with a neutral third party, a mediator, who helps them resolve issues of property division, child support, child custody, and spousal support (i.e., alimony) in a way that minimizes conflict and encourages cooperation. In a longitudinal study researchers compared two groups of divorcing parents who were petitioning for a court custody hearing: parents who were randomly assigned to try mediation and those who were randomly assigned to continue the adversarial court process (Emery et al. 2005). If mediation did not work, the parents in the mediation group could still go to court to resolve their case. The study found that the parents who mediated were much more likely than the parents who did not mediate to settle their custody dispute outside court. The researchers also found that mediation can speed settlement, save money, and increase compliance. Most important, mediation led to improved relationships between nonresidential parents and children as well as between divorced parents 12 years after the dispute settlement. An increasing number of jurisdictions and states have mandatory child custody mediation programs, whereby parents in a custody or visitation dispute must attempt to resolve their dispute through mediation before a court will hear the case.

Divorce Education Programs. Another trend aimed at strengthening post-divorce families is the establishment of divorce education programs that emphasize the importance of cooperative co-parenting for the well-being of children. Parents are taught about children's reactions to divorce, learn nonconflictual co-parenting skills, and learn how to avoid negative behavior toward their ex-spouse. In some programs children are taught that they are not the cause of the divorce, learn how to deal with grief reactions to divorce, and learn techniques for talking to parents about their concerns. Grych (2005) reports that court-connected programs for divorcing couples are available in nearly half the counties in the United States. Many counties and some states (e.g., Arizona and Hawaii) require divorcing spouses to attend a divorce education program, whereas it is optional in other jurisdictions.

Nonmarital and Teenage Childbearing

Social norms concerning having children out of wedlock vary widely throughout the world. In Oslo, Norway, where out-of-wedlock childbirth is socially accepted, half of all births are to unmarried women (Roopnarine & Gielen 2005), whereas in India it is almost unheard of for a Hindu woman to have a child outside marriage and doing so would bring great shame to the woman and her family (84% of the population of India is Hindu) (Laungani 2005).

As noted earlier in this chapter, more than one-third of all U.S. births are to unmarried women. Compared to some countries, the United States has a low proportion of nonmarital births. Two-thirds of births in Iceland and at least half of births in Norway and Sweden are out of wedlock. Other countries with higher proportions of nonmarital births than the United States include Denmark, France, the United Kingdom, and Finland. In other countries, such as Germany, Italy, Greece, and Japan, less than 15% of births occur out of wedlock (Ventura & Bachrach 2000).

Over the last several decades the rates of nonmarital childbearing have increased substantially, and although most nonmarital births are among whites (64% in 2000), blacks have a much higher rate of nonmarital childbearing (see Figure 5.6). However, between 1995 and 2001 the rate of nonmarital births decreased slightly among blacks while continuing an upward trend among whites.

In recent years the U.S. teenage birthrate (per 1,000) steadily dropped from 61.8 in 1991 to a record low of 41.7 in 2003 (National Campaign to Prevent Teen Pregnancy 2004). Although the largest number of teenage births are to non-Hispanic whites (42.6%), the birthrate per 1,000 for non-Hispanic whites ages 15 to 19 is less than half that of Hispanic, black, and American Indian/Alaska Native groups. The lowest teenage birthrates are among Asians and Pacific Islanders (see Figure 5.7).

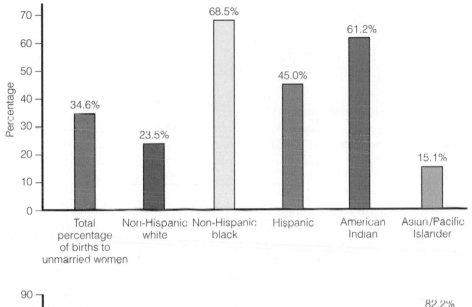

Figure 5.6
Percentage of all births to unmarried women by race and Hispanic origin: United States, 2003.

Source: Hamilton et al. (2004, Table A).

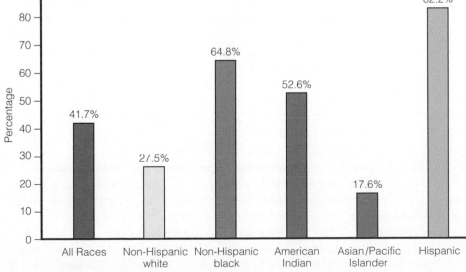

Figure 5.7
Birthrates (per 1,000) of U.S. teenage females, ages 15–19, by race and Hispanic origin, 2003.

Source: Hamilton et al. (2004, Table 1).

In the 1950s, when most teen mothers were married and were expected to be stay-at-home wives, teenage childbearing was not a public concern. Today, most teenage births (80%) occur outside wedlock (Mauldon 2003). Teenage births today are considered problematic because early parenthood interferes with the acquisition of education and job-related skills, and "it seems to guarantee a lifetime of poverty and hardship for a teenage mother and her baby" (Mauldon 2003, p. 41).

Nonmarital births to teens have decreased over the last decade. This decline has been attributed to teens delaying first intercourse, increased use of contraception, and education about HIV and pregnancy prevention (As-Sanie et al. 2004). Nevertheless, 1 million U.S. females between the ages of 15 and 19 become pregnant annually. About half these teens carry their babies to term, 35% have an abortion, and the remainder miscarry. Most teens today who carry their babies to term (95%) keep the baby rather than place it for adoption (Jorgensen 2000).

Teen pregnancy, birth, and abortion rates in the United States are higher than in any other industrialized country. In much of Western Europe low teen pregnancy and birth rates are attributed to the widespread availability and use of effective contraception among sexually active teens.

Social Factors Related to Nonmarital and Teenage Childbearing

High rates of childbearing among teenage and unmarried women are related to several factors, including increased social acceptance of unwed childbearing; increased singlehood, cohabitation, and same-sex relationships; and poverty. This chapter's *Social Problems Research Up Close* feature examines attitudes of low-income single mothers toward marriage that help to explain why they are not married.

Increased Social Acceptance of Unwed Childbearing. Having a baby outside marriage has become more socially acceptable and does not carry the stigma it once did. In the 1950s and 1960s more than half of U.S. women who gave birth to a baby conceived out of wedlock married before the birth of the baby. In the 1990s less than one in four (23%) of such women married before the birth of the baby (Ventura & Bachrach 2000). In a 2001 Gallup survey of women and men in their 20s, less than half (44%) of respondents agreed that "it is wrong to have a child outside of marriage" (Whitehead & Popenoe 2005). In the same survey 40% of single women agreed with the statement that "although it might not be the ideal option, you would consider having a child on your own if you reached your mid-thirties and had not found the right man to marry."

College students are also accepting of single motherhood. In a study of 248 university students 79% agreed that "it is perfectly OK for a woman to decide to have and to raise a child without a husband," with significantly more women (85%) than men (73%) in agreement (Knox et al. 2000).

Increased Singlehood, Cohabitation, and Same-Sex Relationships. The dictionary once defined a spinster as an unmarried woman older than age 30. In previous generations being a spinster meant the fear of isolation, living alone, and being something of a social outcast. Today, women are "more confident, more self-sufficient, and more choosy than ever [and] no longer see marriage as a matter of survival and acceptance" (Edwards 2000, p. 48). The women's movement created both new opportunities for women and expectations of egalitarian relationships with

ThomsonNOW™

Learn more about **Nonmarital and Teenage Childbearing** by going through the Births to Teenage Mothers Map Exercise.

ThomsonNOW™

Learn more about **Social Factors Related to Nonmarital and Teenage Childbearing** by going through the Conducting Research on Families Learning Module.

men. A Time/CNN poll of single women found that 61% felt that men were "too controlling" (Edwards 2000, p. 50). Increasingly, college-educated women in their 30s and 40s are making "the conscious decision to have a child on their own because they haven't found Mr. Adequate, let alone Mr. Right" (Drummond 2000, p. 54).

Increased acceptance of cohabitation and same-sex relationships has also contributed to the high rate of nonmarital births. In the 1990s about 40% of out-of-wedlock births were to women who were cohabiting, presumably with the child's father (Bachu 1999), but in some cases with the mother's same-sex partner.

Poverty. Teenage pregnancy has been related to a variety of factors, including low self-esteem and hopelessness, low parental supervision, and perceived lack of future occupational opportunities (Aassve 2003; Jorgensen 2000; Luker 1996). Although lack of information about and access to contraceptives contributes to unintended teenage pregnancy, 30–40% of adolescent pregnancies are intended (Jorgensen 2000). Teenage females who do poorly in school may have little hope of success and achievement in pursuing educational and occupational goals. They may think that their only remaining option for a meaningful role in life is to become a parent. In addition, some teenagers feel lonely and unloved and have a baby to create a sense of feeling needed and wanted. Teens less likely to get pregnant see a bright future for themselves, feel connected to parents and school, and take fewer risks of any kind (Realini 2004).

Social Problems Related to Nonmarital and Teenage Childbearing

Teenage and unmarried childbirth are considered social problems because of their adverse consequences for women and children. These consequences include poverty, poor health outcomes, low academic achievement, and the absence of fathers in the lives of children.

> "For many disadvantaged teenagers, childbearing reflects— rather than causes— the limitations of their lives."
>
> **Ellen W. Freeman and Karl Rickels**
> **Researchers**

Poverty for Single Mothers and Children. Many unmarried mothers, especially teenagers, have no means of economic support or have limited earning capacity. In 2003 the poverty rate for female-headed households was 28%, compared to 13.5% for male-headed households and only 5.4% for married couples (Fields 2004). Single mothers and their children often live in substandard housing and have inadequate nutrition and medical care. Of the 17.2% of U.S. children living in poverty in 2003, more than half (57%) lived in single-mother homes (U.S. Census Bureau 2004).

Poor Health Outcomes. Compared with older pregnant women, pregnant teenagers are less likely to receive timely prenatal care and to gain adequate weight and are more likely to smoke and use alcohol and drugs during pregnancy (Jorgensen 2000; Ventura et al. 2000). As a consequence of these and other factors, infants born to teenagers are at higher risk of low birth weight, of premature birth, and of dying in the first year of life.

Low Academic Achievement. Low academic achievement is both a contributing factor and a potential outcome of teenage parenthood. Teens whose parents have not graduated from high school are at higher risk for becoming pregnant (Hogan

Perceptions of Marriage Among Low-Income Single Mothers

Single mothers with low incomes face both economic hardships and parenting challenges. These hardships and challenges could potentially be alleviated by marriage to a partner who contributes income to the family and who shares the responsibilities of housework and child care and supervision. Yet low-income single women have low rates of marriage and remarriage. Sociologist Kathryn Edin (2000) conducted research on the perceptions of marriage among low-income single mothers that helps to explain why these mothers are not married. In this feature we summarize her research and its findings.

Sample and Methods

Edin's (2000) study focused on transcripts and field notes collected from qualitative interviews with 292 low-income single mothers. Initially, interviewers used the resources of community groups and local institutions to recruit eligible mothers. They then relied on references from the mothers themselves to gain access to others who might also participate in the interviews. The sample was evenly divided between African Americans and whites in three U.S. cities. Approximately half the participants in each city depended on welfare, and about half worked at low-wage jobs earning less than $7.50 an hour.

Researchers scheduled a minimum of two interviews with each mother to maximize rapport and to encourage them to expand and clarify responses that may have been initially unclear. Interviewers asked respondents to describe the circumstances surrounding the births of their children, their current family situations and relationship with each child's father, their views about how their family situations might change over time, and their views of marriage in general. Interviewers focused on two issues in their conversations with mothers: why women with few economic resources

choose to have children and why these same women fail to marry or remarry.

Findings and Conclusions

Edin (2000) found four primary reasons that low-income single mothers do not marry. These were the low economic status of available men; the desire for parental, economic, and relationship control; mistrust; and domestic violence.

LOW ECONOMIC STATUS OF MEN

Edin (2000) found that men's income is a significant factor in poor single mothers' willingness to marry, because "they simply could not afford to keep an unproductive man around the house" (p. 118). Nearly every mother in Edin's sample said that it was important for any man they would marry to have a "good job." Although total earnings is the most important aspect for single mothers, also important are the regularity of those earnings, the effort men make to find and keep employment, and the source of the income. "Women whose male partners couldn't or wouldn't find work, often lost respect for them and 'just couldn't stand' to keep them around" (p. 119). Mothers did not consider earnings from crime (e.g., drug dealing) as legitimate earnings and reported that they would not marry such a man no matter how much money he made from crime.

Respectability was also an issue for poor single mothers:

Mothers said that they could not achieve respectability by marrying someone who was frequently out of work, otherwise underemployed, supplemented his income through criminal activity, and had little chance of improving his situation over time. Mothers believed that marriage to such a man would diminish their respectability, rather than enhance it. (Edin 2000, p. 120)

Mothers also believed that marriage to a lower-class man would be unlikely to last

because the economic pressures on the relationship would be too great. Mothers talked about the "sacred" nature of marriage and believed that no "respectable" woman would marry a man whose economic situation would likely lead to a future divorce.

Edin (2000) notes that:

In interview after interview, mothers stressed the seriousness of the marriage commitment and their belief that "it should last forever." Thus, it is not that mothers held marriage in low esteem, but rather the fact that they held it in such high esteem that convinced them to forego marriage, at least until their prospective marriage partner could prove himself worthy economically or they could find another partner who could. (pp. 120–121)

DESIRE FOR PARENTAL, ECONOMIC, AND RELATIONSHIP CONTROL

When asked what they liked best about being a single parent, the most common response was "I am in charge" or "I am in control." Most mothers felt that the presence of fathers often interfered with their parental control. Single mothers in Edin's (2000) study felt that a husband might be "too demanding" of them and thus impede their efforts to spend time with their children.

Mothers were also concerned about losing control of the family's economic situation. One mother said she would not marry because "the men take over the money. I'm too afraid to lose control of my money again" (Edin 2000, p. 121).

Regardless of whether the prospective wife worked, mothers feared that prospective husbands would expect to be the "head of the house," who, being "in charge," would want to make the "final" decisions about child rearing, finances, and other matters. Mothers also expressed the view that if they married, their husbands would expect them to do all the household chores. Some women described their relationships with their ex-partners as "like having one more kid to take care of" (Edin 2000, p. 122).

Most single mothers in Edin's sample wanted to marry eventually and thought that the best way to maximize their control in a marriage was by working and earning an income. One African American mother explained:

> One thing my mom did teach me is that you must work some and bring some money into the household so you can have a say in what happens. If you completely live off a man, you are helpless. That is why I don't want to get married until I get my own [career] and get off of welfare. (Edin 2000, p. 122)

Mothers also wanted to develop their labor market skills before marriage to ensure against destitution in the event of a divorce and to increase their bargaining power in their relationships. Mothers believed that women who were economically dependent on men had to "put up with all kinds of behavior" because they could not legitimately threaten to leave.

> Mothers felt that if they became more economically independent . . . they could legitimately threaten to leave their husbands if certain conditions (i.e., sexual fidelity) weren't met. These threats would, in turn they believed, keep a husband on his best behavior. (Edin 2000, p. 122)

Many of the single mothers in Edin's study had held traditional gender role attitudes when they were younger and still in a relationship with their children's fathers. When the men for whom they sacrificed so much gave them nothing but pain and anguish, they felt they had been "duped" and were no longer willing to be dependent or subservient to men.

MISTRUST

Many of the single mothers in Edin's sample did not marry because they did not trust

their partners or men in general and/or because their partners did not trust them. Some fathers claimed that the child was not theirs because the mother was "a whore." One partner of a pregnant woman in the sample told the interviewer, "How do I know the baby's mine? Who knows if she hasn't been stepping out on me with some other man and now she wants me to support another man's child!" (Edin 2000, p. 124).

Mothers tended to mistrust men because their previous boyfriends and partners had been unfaithful. This experience was so common among respondents that many simply did not believe men could be faithful to only one woman. Most said they would rather never marry than to "let him make a fool out of me" (Edin 2000, p. 124). "Nonetheless, many of these same women often held out hope of finding a man who was 'different,' one who could be trusted" (p. 125).

DOMESTIC VIOLENCE

For some mothers in Edin's sample, domestic violence in either their childhood or adult lives played a role in their negative attitudes toward marriage. Many mothers reported physical abuse during pregnancy, and several mothers had miscarriages because of such abuse. One woman who had been in the hospital three times because of abuse said, "I was terrified to leave because I knew it would mean going on welfare. . . . But that is okay. I can handle that. The thing I couldn't deal with is being beat up" (Edin 2000, p. 126). Edin commented,

> The fact that women tended to experience repeated abuse from their children's fathers before they decided to leave attests to their strong desire to make things work with their children's fathers. Many women finally left when they saw the abuse beginning to affect their children's well-being. (p. 126)

In conclusion, low-income single mothers in Edin's (2000) study were reluctant to marry the father of their children because these men had low economic status, traditional notions of male domination in household and parental decisions, and patterns of untrustworthy and even violent behavior. Given the low level of trust these mothers have of men and given their view that husbands want more control than the women are willing to give them, women realize that a marriage that is also economically strained is likely to be conflictual and short-lived.

"Interestingly, mothers say they reject entering into economically risky marital unions out of respect for the institution of marriage, rather than because of a rejection of the marriage norm" (Edin 2000, p. 130). Low-income single mothers in Edin's study "say they are willing, even eager, to marry if the marriage represents an increase in their class standing and if . . . their prospective husband's behavior indicates he won't beat them, abuse their children, refuse to share household tasks, insist on making all decisions, be sexually unfaithful, or abuse alcohol or drugs" (p. 113).

> In short, the mothers interviewed here believe that marriage will probably make their lives more difficult than they are currently. . . . If they are to marry, they want to get something out of it. If they cannot enjoy economic stability and gain upward mobility from marriage, they see little reason to risk the loss of control and other costs they fear marriage might exact from them. Unless low-skilled men's economic situations improve and they begin to change their behaviors toward women, it is quite likely that most low-income women will continue to resist marriage. (p. 130)

Not all nonmarital births imply the absence of a father. Actress Goldie Hawn and actor Kurt Russell have been in a committed cohabiting relationship for more than 20 years. Although their child Wyatt was born "out of wedlock," he has been raised in a stable, loving family with his mother and his father. Kurt Russell also helped Goldie raise her two children (Kate and Oliver) from a previous marriage.

et al. 2000). Three-fifths of teenage mothers drop out of school and, as a consequence, have a much higher probability of remaining poor throughout their life. Because poverty is linked to unmarried parenthood, a cycle of successive generations of teenage pregnancy may develop.

A study based on a large nationally representative sample of U.S. children found that children born to mothers age 17 and younger began kindergarten with lower levels of school readiness—including lower math and reading scores, language and communication skills, social skills, and physical and emotional well-being—compared to children born to older mothers (Terry-Humen et al. 2005). When the researchers controlled for the mother's marital status and socioeconomic status, these effects were diminished but still important.

Children Without Fathers. About one-third of U.S. children who live in households without their fathers are the products of unmarried childbirth (Hewlett & West 1998). Shapiro and Schrof (1995) report that children who grow up without fathers are more likely to drop out of school, be unemployed, abuse drugs, experience mental illness, and be a target of child sexual abuse. They also note that "a missing father is a better predictor of criminal activity than race or poverty" (p. 39). Popenoe (1996) believes that fatherlessness is a major cause of the degenerating conditions of our young. However, others argue that the absence of a father is not, in and of itself, damaging to children. Rather, other conditions associated with female-headed single-parent families, such as low educational attainment of the mother, poverty, and lack of child supervision, contribute to negative outcomes for children. Family historian Stephanie Coontz put a different twist on the issue of single parenthood when she argued that "the debate over whether one parent can raise a child alone . . . diverts attention from the fact that good child rearing has always required more than two parents" (1992, p. 230).

Strategies for Action: Interventions in Teenage and Nonmarital Childbearing

Interventions in teenage and nonmarital childbearing include efforts to prevent such births and strategies to minimize the negative effects of such births by providing various types of support to single and teenage mothers and their children.

Teenage Pregnancy Prevention

Efforts to prevent teenage pregnancy include providing sexuality education and access to contraceptives.

Sexuality Education. Sex education is provided through schools, faith-based groups, and community organizations throughout the United States. However, under the Bush administration federal support for "abstinence-only" education programs, which promote abstinence from sexual activity without teaching basic facts about contraception or providing access to contraception, has expanded rapidly. Supporters of abstinence-only programs believe that promoting condoms sends the "wrong message" that sex outside of marriage is OK. Abstinence-only sex education programs have not been shown to be effective in preventing teenage pregnancy (Realini 2004), and they are criticized for failing to provide youth with potentially life-saving information. As we noted in Chapter 2, a report released by Representative Henry Waxman found that more than 80% of the abstinence-only curricula contain false, misleading, or distorted information about reproductive health (Waxman 2004). For example, the report found that "many of the curricula misrepresent the effectiveness of condoms in preventing sexually transmitted diseases and pregnancy" (Waxman 2004, p. i).

Although the benefits of abstinence emphasized in the abstinence-only programs do influence some youth to delay onset of sexual activity, the many teens who are sexually active need to know how to prevent pregnancy. Nearly one-fifth (17%) of sexually active females age 15 to 19 and 9% of males in the same age group said that they used no method of contraception the last time they had sex (Kaiser Family Foundation 2005). In 2003 nearly half (46.7%) of high school students had ever had sexual intercourse (Centers for Disease Control and Prevention 2004). According to the Sex Information and Education Council of the United States, most parents want schools to provide **comprehensive sexuality education** that includes such topics as abstinence, contraception, sexually transmitted diseases, HIV/AIDS, and disease-prevention methods (SIECUS 2004).

Providing Access to Contraception. The European approach to teenage sexual activity is to provide widespread confidential and accessible contraceptive services to adolescents. The provision of contraceptive services to European teens is believed to be a central factor in explaining the more rapid declines in teenage childbearing in Northern and Western European countries, in contrast to slower declines in the United States (Singh & Darroch 2000).

In the United States there are significant barriers to obtaining contraception. As we have already discussed, most high schools do not provide students with contraceptive health services. Although 21 states and the District of Columbia allow minors to consent to contraceptive services, 14 states allow only certain categories of minors access to such services (e.g., those who have had a previous birth), and in other states the decision of whether to allow minors access to contraceptive services is at the discretion of the health care provider (Jones et al. 2005). Proposed legislation that would require parental notification for minors obtaining prescription contraception from federally funded family planning clinics further threatens minors' access to contraception. In a survey of teenage females seeking reproductive health services at 79 family planning clinics across the country, one in five teens said that they would use no contraception or rely on withdrawal if parental notification were required (Jones et al. 2005).

Another barrier to access to contraception for all U.S. women is the refusal of some pharmacists to fill prescriptions for birth control. See this chapter's *Taking a Stand* feature.

ThomsonNOW

Learn more about **Providing Access to Contraception** by going through the Contraceptives Map Exercise.

Should Pharmacists Be Allowed to Refuse to Fill Prescriptions for Birth Control Products?

In 2005 the media reported that pharmacists across the country were refusing to fill prescriptions for birth control and morning-after "emergency birth control" pills because dispensing the medications violated their personal moral or religious beliefs. Some pharmacists will give birth control pills to a woman only if she is married; others believe that contraception is a form of abortion and refuse to dispense it to anyone. As of spring 2005 four states had laws that allow pharmacists to refuse to fill prescriptions that violate their beliefs, eleven states were considering "conscience clause" laws that would allow pharmacists to refuse to fill prescriptions, and four states were considering laws that would require pharmacists to fill all prescriptions (Stein 2005).

It Is Your Turn Now to Take a Stand!

Do you think that pharmacists should be required to fill all prescriptions? Or should they be allowed to refuse to fill prescriptions that violate their beliefs?

Increase Life Options of Teenagers and Unmarried Mothers

Other programs aim at both preventing teenage and unmarried childbearing and minimizing its negative effects by increasing the life options of teenagers and unmarried mothers. Luker (1996) argued that the most effective approach is to target social policies that address community-wide malaise that leads to "discouraged and disadvantaged" youth. Such programs include educational programs, job training, and skill-building programs. Other programs designed to help teenage and unwed mothers and their children include prenatal programs to help ensure the health of the mother and baby and public welfare, such as WIC (the Special Supplemental Food Program for Women, Infants, and Children) and TANF (Temporary Assistance to Needy Families). However, public assistance to low-income single mothers has been blamed for contributing to the problem by discouraging the poor from marrying. Up until 1996 single mothers with no or little income could receive welfare so long as they had a dependent child living with them. If single mothers on welfare married, they could lose welfare benefits. In 1996 the Personal Responsibility and Work Opportunity Reconciliation Act (PRWORA) reformed the welfare program (see Chapter 6) by limiting cash benefits to adults. For example, adults can receive welfare benefits for no more than two consecutive years (unless they work at least 20 hours per week) and are limited to five years of welfare cash benefits over their lifetime. One goal of welfare reform is to encourage marriage by decreasing the benefits that an unmarried mother can claim. Despite the new time limitations on welfare benefits, marriage rates among low-income single mothers continue to be low.

Increase Men's Involvement with Children

Strategies to increase and support fathers' involvement with their children are relevant to both children of unwed mothers and children of divorce. At the federal and state levels fatherhood initiative programs encourage fathers' involvement with children through a variety of means (U.S. Department of Health and Human Services 2000). These include promoting responsible fatherhood by improving work opportunities for low-income fathers, increasing child support collections, providing parent education training for men, supporting access and visitation by noncustodial parents, and involving boys and young men in teenage pregnancy prevention and early parenting programs. Because teenage parents are less likely than older parents to use positive and effective child-rearing techniques, parent education programs for teen mothers and fathers are an important component of improving the lives of young parents and their children.

Pregnancy: a lifetime commitment
Are you ready?

El Embarzo: un compromiso para toda la vida
¿Ya estás listo tú?

LATINO MEN'S HEALTH SERIES
SERIE SOBRE LA SALUD DEL HOMBRE LATINO

Courtesy of The East Los Angeles Men's Health Center, a project of Bienvenidos Children's Center, Inc. www.bienvenidos.org

This brochure, distributed by the National Latino Fatherhood and Family Initiative, is a teen pregnancy prevention effort targeting young Latino men.

Understanding Family Problems

Family problems can best be understood within the context of the society and culture in which they occur. Although domestic violence, divorce, and teenage pregnancy and unmarried parenthood may appear to result from individual decisions, these decisions are influenced by myriad social and cultural forces.

The impact of family problems, including divorce, abuse, and nonmarital childbearing, is felt not only by family members but also by society at large. Family members experience such life difficulties as poverty, school failure, low self-esteem, and mental and physical health problems. Each of these difficulties contributes to a cycle of family problems in the next generation. The impact on society includes public expenditures to assist single-parent families and victims of domestic violence and neglect, increased rates of juvenile delinquency, and lower worker productivity.

For some the solution to family problems implies encouraging marriage and discouraging other family forms, such as single parenting, cohabitation, and same-sex unions. But many family scholars argue that the fundamental issue is making sure that children are well cared for, regardless of their parents' marital status or sexual orientation. Some even suggest that marriage is part of the problem, not part of the solution. Martha Fineman of Cornell Law School said, "This obsession with marriage prevents us from looking at our social problems and addressing them. . . . Marriage is nothing more than a piece of paper, and yet we rely on marriage to do a lot of work in this society: It becomes our family policy, our police in regard to welfare and children, the cure for poverty" (quoted by Lewin 2000, p. 2).

Strengthening marriage is a worthy goal because strong marriages offer many benefits to individuals and their children. However, "strengthening marriage does not have to mean a return to the patriarchal family of an earlier era. . . . Indeed, greater marital stability will only come about when men are willing to share power, as well as housework and child care, equally with women" (Amato 1999, p. 184). And strengthening marriage does not mean that other family forms should not also be supported. In their book *Joined at the Heart,* Al and Tipper Gore (2002) suggest that the first and most important step to helping families is to change our way of

ThomsonNOW™

Learn more about **Diversity of Families** by going through the Diversity among Families Animation.

thinking about families so that our view of family encompasses those who are connected emotionally and committed to one another as family—those who are "joined at the heart" (p. 327). The reality is that the postmodern family comes in many forms, each with its strengths, needs, and challenges. Given the diversity of families today, social historian Stephanie Coontz (2004) suggests that "the appropriate question . . . is not what single family form or marriage arrangement we would prefer in the abstract, but how we can help people in a wide array of different committed relationships minimize their shortcomings and maximize their solidarities" (p. 979).

Finally, the health of marriages and families depends on a variety of structural and cultural factors, including the health of the economy, continued progress toward gender equality, and resolution of work-family dilemmas facing many parents (Amato et al. 2003). These social issues, as well as others that affect families, are discussed in subsequent chapters in this book.

Chapter Review

ThomsonNOW™

Reviewing is as easy as ❶ ❷ ❸

1. Before you do your final review, take the ThomsonNOW diagnostic quiz to help you identify the areas on which you should concentrate. You will find information on ThomsonNOW and instructions on how to access all of its great resources on the foldout at the beginning of the text.

2. As you review, take advantage of ThomsonNOW's study videos and interactive Map the Stats exercises to help you master the chapter topics.

3. When you are finished with your review, take ThomsonNOW's posttest to confirm you are ready to move on to the next chapter.

- **What are some examples of diversity in families around the world?**

Some societies recognize monogamy as the only legal form of marriage, whereas other societies permit polygamy. Societies also vary in their policies regarding same-sex couples and their norms regarding the roles of women, men, and children in the family.

- **What are some of the major changes in U.S. households and families that have occurred in the last several decades?**

Some of the major changes in U.S. households and families that have occurred in recent decades include increased singlehood and older age at first marriage, increased heterosexual and same-sex cohabitation, increased births to unmarried women, increased single-parent families, fewer children living in married-couple families, increased divorce and blended families, and increased employment of married mothers. Although some view these changes as signaling a breakdown of marriage and family, others emphasize that the forms and patterns of families are less important than the quality of love, care, and nurturance that families provide for their members.

- **Feminist theories of family are most similar to which of the three main sociological theories: structural functionalism, conflict theory, or symbolic interactionism?**

Feminist theories of family are most aligned with conflict theory. Both feminist and conflict theories are concerned with how gender inequality influences and results from family patterns.

- **What are the four patterns of partner violence identified by Johnson and Ferraro (2003)?**

The four patterns of partner violence are (1) common couple violence (occasional acts of violence arising from arguments that get "out of hand"); (2) intimate terrorism (violence that is motivated by a wish to control one's partner); (3) violent resistance (acts of violence that are committed in self-defense); and (4) mutual violent control (both partners battling for control).

- **What are the differences between primary prevention, secondary prevention, and tertiary prevention strategies with regard to preventing and responding to domestic violence and abuse?**

Primary prevention strategies target the general population, secondary prevention strategies target groups at high risk for family violence and abuse, and tertiary prevention strategies target families who have experienced abuse.

- **Why do many abused adults stay in abusive relationships?**

Adult victims of abuse are commonly blamed for choosing to stay in their abusive relationships. From the point of view of the victim, reasons to stay in the relationship include love, emotional dependency, commitment to the relationship, hope that things will get better, the view that violence is legitimate because they "deserve" it, guilt, fear, economic dependency, feeling stuck, and fear of loneliness. Some victims stay because they fear the abuser will abuse or neglect a family pet. Victims also stay because abuse in relationships is usually not ongoing and constant but rather occurs in cycles in which a violent or abusive episode is followed by a makeup period in which the abuser expresses sorrow and asks for forgiveness and "one more chance."

- **What are some of the effects of divorce on children?**

Reviews of recent research on the consequences of divorce for children find that children with divorced parents score lower on measures of academic success, psychological adjustment, self-concept, social competence, and long-term health and that they have higher levels of aggressive behavior and depression. Such effects are related to the economic hardship associated with divorce, the reduced parental supervision resulting from divorce, and parental conflict during and following divorce. In highly conflictual marriages divorce may actually improve the emotional well-being of children relative to staying in a conflicted home environment.

- **What is divorce mediation?**

In divorce mediation divorcing couples meet with a neutral third party, a mediator, who helps them resolve issues of property division, child custody, child support, and spousal support in a way that minimizes conflict and encourages cooperation. In some states, counties, and jurisdictions, divorcing couples who are disputing child custody issues are required to participate in divorce mediation before their case can be heard in court.

- **What are some of the social problems related to nonmarital and teenage childbearing?**

Teenage and unmarried childbirth are considered social problems because of the adverse consequences for women and children that are associated with such births, including (1) increased risk of poverty for single mothers and their children, (2) risk of poor health outcomes for babies born to teenage women, (3) risk of dropping out of school for teenage mothers and for low academic achievement of their children, and (4) increased risks for children who grow up without fathers.

- **How does the European approach to teenage sexuality compare with the U.S. approach?**

The European approach to teenage sexual activity involves providing widespread confidential and accessible contraceptive services to adolescents. Although sex education is provided in schools throughout the United States, most programs emphasize abstinence and do not provide students with access to contraception. Research suggests that comprehensive sexuality education that includes such topics as abstinence, sexually transmitted diseases, HIV/AIDS, contraception, and disease-prevention methods is more effective at preventing pregnancy, as well as disease.

Critical Thinking

1. Some scholars and politicians argue that "stable families are the bedrock of stable communities." Others argue that "stable communities and economies are the bedrock of stable families." Which of these two positions would you take and why?
2. What role should U.S. colleges play in preparing adults for family life? In what ways could high schools and colleges help reduce and alleviate problems of domestic violence and abuse, divorce, and teenage and nonmarital pregnancy and parenting?
3. In the United States women are more likely than men to initiate divorce. Why do you think this is so?

Key Terms

bigamy
child abuse
common couple violence
comprehensive sexuality
  education
corporal punishment
covenant marriage
cycle of abuse
divorce mediation
divorce rate
domestic partnership
egalitarian relationship
elder abuse
familism
family
family household
family preservation program
individualism
intimate partner violence
intimate terrorism
marital decline perspective

marital resiliency
  perspective
monogamy
mutual violent control
neglect
no-fault divorce
nonfamily household
parental alienation
  syndrome (PAS)
patriarchy
plural marriage
polyandry
polygamy
polygyny
primary prevention
second shift
secondary prevention
serial monogamy
shaken baby syndrome
tertiary prevention
violent resistance

Media Resources

The Companion Website for
Understanding Social Problems,
Fifth Edition

http://sociology.wadsworth.com/mooney_knox_schacht5e

Supplement your review of this chapter by going to the companion website to take one of the Tutorial Quizzes, use the flash cards to master key terms, and check out the many other study aids you'll find there. You'll also find special features such as *Wadsworth's Sociology Online Resources and Writing Companion,* GSS Data, and Census 2000 information, data, and resources at your fingertips to help you complete that special project or do some research on your own.

> " We are the first generation that can look extreme poverty in the eye, and say this and mean it— we have the cash, we have the drugs, we have the science. Do we have the will to make poverty history?"

Bono, U2 (rock music group)

Poverty and Economic Inequality

eric Dunbar, a 54-year-old resident of New Orleans, was one of millions of victims of the 2005 Hurricane Katrina—the worst natural disaster to hit the United States. Like many other residents of New Orleans, where one-third of the population lived below the poverty line, Eric Dunbar and his wife did not have the resources to evacuate the city before the hurricane hit. "I don't own a car," Dunbar explained. "Me and my wife, we travel by bus, public transportation. The most money I ever have on me is $400. And that goes to pay the rent" (quoted by Haygood 2005). Another New Orleans hurricane victim, 47-year-old grandmother Carmita Stephens, remarked, "These people look at us and wonder why we stayed behind. . . . Well, would they leave their grandparents and children behind? Look around and say, 'See you later'? . . . We had one vehicle. A truck. I wanted my family to be together. They all couldn't fit in the truck. We had to decide on leaving family members—or staying" (quoted by Haygood 2005).

Although natural disasters such as hurricanes, tsunamis, floods, and earthquakes strike indiscriminately—rich and poor alike—the poor are more vulnerable to devastation from such disasters. After a tsunami devastated a large part of South and Southeast Asia in December 2004—killing thousands and tearing apart families, villages, and communities—Oxfam director Barbara Stocking noted that "it is not mere chance that most of those who died or have been left homeless and destitute were already among the world's poorest. Poor families are always much more severely affected by natural disasters. . . . They live in flimsier homes" (Stocking 2005, n.p.).

The poor have few to no resources to help them avoid or cope with natural disasters. As Eric Dunbar exemplifies in this chapter's opening vignette, many poor people in the path of Hurricane Katrina lacked a means of transportation to evacuate. And when the poor lose their homes and their livelihoods in a natural disaster, they do not have the resources to rebuild. For example, when poor fishing communities lost their boats and nets—their very means of survival—in the Asian tsunami, they had no bank accounts or insurance policies to replace their losses.

Poverty also affects natural disaster relief efforts. Because the poorest of the tsunami victims lived in areas with weak or nonexistent infrastructure, hundreds of thousands of tsunami victims were cut off from aid because they did not live near a road or airport. In addition, just as upper-class passengers on the *Titanic* were given priority access to lifeboats, some reports claimed that the tsunami-affected areas that catered to well-off tourists received more assistance than the thousands of poor people who lived in the villages (Roberts 2005). Similarly, in the wake of Hurricane Katrina, some of the local poor residents waited at the back of an evacuation line while 700 guests and employees of the Hyatt Hotel were bused out first (Dowd 2005). Other victims of Hurricane Katrina were stranded for days without food, water, and medical supplies. Five days after the hurricane hit, 20,000 hungry, dehydrated, desperate people stranded in the rain-soaked and sweltering hot Louisiana Superdome in New Orleans, where overflowed toilets forced people to relieve themselves in hallways and stairwells, waited to be evacuated. For days, in some cases a week or more, some hurricane victims waited to be rescued from rooftops or from neck-high floodwater in their attics. For many, rescue came too late. Scores of media interviews and editorials suggested that the government's slow response was at least partly due to the fact that Katrina's victims were predominantly poor. A church pastor lamented, "I think a lot of it has to do with race and class. The people affected were largely poor people. Poor, black people" (quoted by Gonzales 2005). A physician suggested, "Maybe the response would have been quicker if [Katrina] had

ThomsonNOW™

Reviewing is as easy as
① ② ③
Use ThomsonNOW to help you make the grade on your next exam. When you are finished reading this chapter, go to the chapter review for instructions on how to make ThomsonNOW work for you.

Natural disasters, such as the December 2004 tsunami, are more devastating to the poor, who live in flimsier housing, have little to no infrastructure, and lack resources to cope with and recover from devastation.

REUTERS/Amit Dave/Landov

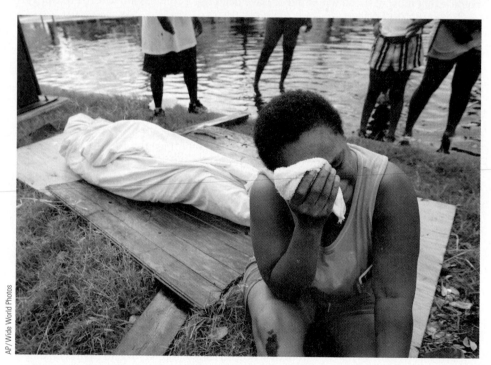

Many of the more than 1,300 people who died in the wake of Hurrican Katrina were poor.

AP/Wide World Photos

occurred in some other area of the country, for example in New York or California where there's more money, more people who are going to object, raise their voices" (quoted by Purdum 2005).

In this chapter we examine the extent of poverty globally and in the United States, focusing on the consequences of poverty for individuals, families, and societies. We present theories of poverty and economic inequality and consider strategies for rectifying economic inequality and poverty.

The Global Context: Poverty and Economic Inequality Around the World

Who are the poor? Are rates of world poverty increasing, decreasing, or remaining stable? The answers depend on how we define and measure poverty.

Defining and Measuring Poverty

Poverty has traditionally been defined as the lack of resources necessary for material well-being—most important, food and water, but also housing, land, and health care. This lack of resources that leads to hunger and physical deprivation is known as **absolute poverty.** In contrast, **relative poverty** refers to a deficiency in material and economic resources compared with some other population. Although many lower-income Americans, for example, have resources and a level of material well-being that millions of people living in absolute poverty can only dream of, they are relatively poor compared with the American middle and upper classes.

Various measures of poverty are used by governments, researchers, and organizations. Next, we describe international and U.S. measures of poverty.

International Measures of Poverty. The World Bank sets a "poverty threshold" of $1 per day to compare poverty in most of the developing world, $2 per day in Latin America, $4 per day in Eastern Europe and the Commonwealth of Independent States (CIS), and $14.40 per day in industrial countries (which corresponds to the income poverty line in the United States). Another poverty measure is based on whether individuals are experiencing hunger, which is defined as consuming less than 1,960 calories a day.

In industrial countries national poverty lines are sometimes based on the median household income of a country's population. According to this relative poverty measure, members of a household are considered poor if their household income is less than 50% of the median household income in that country.

Recent poverty research has concluded that poverty is multidimensional and includes such dimensions as food insecurity, poor housing, unemployment, psychological distress, powerlessness, hopelessness, vulnerability, and lack of access to health care, education, and transportation (Narayan 2000). To capture the multidimensional nature of poverty, the United Nations Development Programme developed a composite measure of poverty: the **human poverty index (HPI)** (UNDP 1997). Rather than measure poverty by income, three measures of deprivation are combined to yield the HPI: (1) deprivation of a long, healthy life, (2) deprivation of knowledge, and (3) deprivation in decent living standards. As shown in Table 6.1, the HPI for developing countries **(HPI-1)** is measured differently from the HPI for industrialized countries **(HPI-2).** Among the 17 industrialized countries for which the HPI-2 was calculated, Sweden has the lowest level of human poverty (6.5%), followed by Norway (7.2%) and the Netherlands (8.4%) (UNDP 2003). The highest rate of human poverty is in the United States (15.8%), followed by Ireland (15.3%) and the United Kingdom (14.8%). The HPI is a useful complement to income measures of poverty and "will serve as a strong reminder that eradicating poverty will always require more than increasing the income of the poorest" (UNDP 1997, p. 19).

> **"**Human poverty is more than income poverty—it is the denial of choices and opportunities for living a tolerable life.**"**
>
> **UNDP, *Human Development Report 1997***

U.S. Measures of Poverty. In 1964 the Social Security Administration devised a poverty index based on a 1955 Agriculture Department survey that estimated the cost of an economy food plan for a family of four. Because families with three

Table 6.1

Measures of Human Poverty in Developing and Industrialized Countries

	Longevity	Knowledge	Decent Standard of Living
For developing countries	Probability at birth of not surviving to age 40	Adult illiteracy	A composite measure based on 1. Percentage of people without access to safe water 2. Percentage of people without access to health services 3. Percentage of children under five who are underweight
For industrialized countries	Probability at birth of not surviving to age 60	Adult functional illiteracy rate	Percentage of people living below the income poverty line, which is set at 50% of median disposable income

Source: Adapted from UNDP (2000).

Table 6.2

Poverty Thresholds: 2004 (Householder Younger Than 65 Years Old)

Household Makeup	Poverty Threshold
One adult	$9,827
Two adults	$12,649
One adult, one child	$13,020
Two adults, one child	$15,205
Two adults, two children	$19,157

Source: U.S. Census Bureau (2005b).

or more members spent one-third of their income on food at the time, the poverty line was set at three times the minimum cost of an adequate diet. Poverty thresholds differ by the number of adults and children in a family and by the age of the family head of household (see Table 6.2). Poverty thresholds are adjusted each year for inflation. Anyone living in a household with pretax income below the official poverty line is considered "poor." Individuals living in households that are above the poverty line, but not very much above it, are classified as "near poor," and those living below 50% of the poverty line live in "deep poverty." A common working definition of "low-income" households are those with incomes that are between 100% and 200% of the federal poverty line, or up to twice the poverty level.

The U.S. poverty line has been criticized as underestimating the extent of material hardship in the United States (Mishel et al. 2005). The poverty line is based on the assumption that low-income families spend one-third of their household income on food. That was true in 1955, but because other living costs (e.g., housing, medical care, and child care) have risen more rapidly than food costs, low-income families today spend less than one-fifth (rather than one-third) of their income on food (Schiller 2004). Thus current poverty lines should be based on multiplying food costs by five rather than three. This would raise the official poverty line by two-thirds, making the poverty level consistent with public opinion regarding what a family needs to escape poverty.

Another shortcoming of the official poverty line is that it is based solely on money income and does not take into consideration noncash benefits received by many low-income individuals, such as food stamps, Medicaid, and public housing. Family assets, such as savings and property, are also excluded in official poverty calculations. In addition, the poverty index fails to account for tax burdens that affect the amount of disposable income available to meet basic needs. The U.S. poverty line also disregards regional differences in the cost of living, and because poverty rates are based on surveys of households, the homeless—the most destitute of the poor—are not counted among the poor.

ThomsonNOW

Learn more about **U.S. Measures of Poverty** by going through the Poverty Level Map Exercise.

The Extent of Global Poverty and Economic Inequality

Globally, 2.8 billion people—nearly half the world's population—survive on less than $2 a day (United Nations Population Fund 2004). More than 1.2 billion people—1 in 5 people on this planet—survive on less than $1 a day. In sub-Saharan Africa nearly half the population lives on less than $1 a day (World Bank 2005). South Asia has the greatest number of people affected by poverty, and sub-Saharan Africa has the highest proportion of people in poverty. Every day, nearly 1 in 5 (18%) of the world's population goes hungry. In South Asia 1 in 4 goes hungry, and in sub-Saharan Africa as many as 1 in 3 goes hungry (UNDP 2003).

Global economic inequality has reached unprecedented levels. In 2000 the average income of the richest 20 countries was 37 times that of the poorest 20 countries—a gap that doubled in the past 40 years (World Bank 2001). From 1960 to 2002 income per person in the world's poorest countries rose only slightly, from $212 to $267, whereas income in the richest 20 nations tripled from $11,417 to $32,339 (Schifferes 2004). Although inequality between nations accounts for most of the inequality in global distribution of income, within-nation income differences are growing (Goesling 2001).

As we discuss later, the United States has the greatest gap between the rich and the poor among the developed countries of the world. But income inequality is much greater in poor countries than it is in the United States. The income share of the top 10% of Americans is 30% of national income—the highest share among developed nations. But in several undeveloped countries in Asia, Latin America, and Africa, the top 10% of income earners receive more than 40% of national income (Sanderson & Alderson 2005).

ThomsonNOW

Learn more about the **Extent of Global Poverty and Economic Inequality** by going through the Theories of Global Stratification Learning Module.

Sociological Theories of Poverty and Economic Inequality

The three main theoretical perspectives in sociology—structural functionalism, conflict theory, and symbolic interactionism—offer insights into the nature, causes, and consequences of poverty and economic inequality.

Structural-Functionalist Perspective

According to the structural-functionalist perspective, poverty results from institutional breakdown: economic institutions that fail to provide sufficient jobs and pay, educational institutions that fail to provide adequate education in low-income school districts, family institutions that do not provide two parents, and government institutions that do not provide sufficient public support. These institutional breakdowns create a "culture of poverty" whereby, over time, the poor develop norms, values, beliefs, and self-concepts that contribute to their own plight. According to Lewis, the **culture of poverty** is characterized by female-centered households, an emphasis on gratification in the present rather than in the future, and a relative lack of participation in society's major institutions (Lewis 1966). "The people in the culture of poverty have a strong feeling of marginality, of helplessness, of dependency, of not belonging. . . . Along with this feeling of powerlessness is a widespread feeling of inferiority, of personal unworthiness" (Lewis 1998, p. 7). Early sexual activity, unmarried parenthood, joblessness, reliance on public assistance, illegitimate income-producing activities (e.g., selling drugs), and substance abuse are common among the **underclass**—people living in persistent poverty. The culture of poverty view emphasizes that the behaviors, values, and attitudes exhibited by the chronically poor are transmitted from one generation to the next, perpetuating the cycle of poverty. Behaviors, values, and attitudes of the underclass emerge from the constraints and blocked opportunities that have resulted largely from the failure of the economic institution to provide employment, as jobs moved out of inner-city areas to the suburbs (Jargowsky 1997; Van Kempen 1997; Wilson 1996).

> "Many of today's problems in the inner-city ghetto neighborhoods— crime, family dissolution, welfare, low levels of social organization, and so on— are fundamentally a consequence of the disappearance of work."
>
> **William Julius Wilson**
> **Sociologist**

Where jobs are scarce . . . and where there is a disruptive or degraded school life purporting to prepare youngsters for eventual participation in the workforce, many people eventually lose their feeling of connectedness to work in the formal economy; they no longer expect work to be a regular, and regulating, force in their lives. . . . These circumstances also increase the likelihood that the residents will rely on illegitimate sources of income, thereby further weakening their attachment to the legitimate labor market. (Wilson 1996, pp. 52–53)

From the late 1960s through the 1980s, poverty became more and more concentrated in inner-city neighborhoods, and conditions in those neighborhoods steadily deteriorated. But during the economic boom of the late 1990s, when unemployment was low, poverty became less concentrated (Kingsley & Pettit 2003). This decrease in the concentration of poverty during strong economic times supports the view that economic conditions underlie the culture of poverty.

From a structural-functionalist perspective economic inequality within a society can be beneficial for society. A system of unequal pay, argued Davis and Moore (1945), motivates people to achieve higher levels of training and education and to take on jobs that are more important and difficult by offering higher rewards for higher achievements. If physicians were not offered high salaries, for example, who would want to endure the arduous years of medical training and long, stressful hours at the hospital? However, this argument is criticized on the grounds that many important occupational roles, such as child-care workers, are poorly paid (the average salary of a child-care worker is less than $18,000 per year) (Bureau of Labor Statistics 2005), whereas many individuals in nonessential roles (e.g., professional sports stars and entertainers) earn astronomical sums of money. The structural-functionalist argument that CEO pay is high because of the risks and responsibilities of the job falls apart when one considers that the average CEO pay is 56 times the pay of a U.S. army general with 20 years of experience (Anderson et al. 2004). If pay is based on risk and responsibility, does it make sense that the annual pay of 919 U.S. soldiers killed in Iraq (as of early August 2004) was about equal to the combined pay of just five average U.S. CEOs?

Conflict Perspective

Karl Marx (1818–1883) proposed that economic inequality results from the domination of the **bourgeoisie** (owners of the means of production) over the **proletariat** (workers). The bourgeoisie accumulate wealth as they profit from the labor of the proletariat, who earn wages far below the earnings of the bourgeoisie. Modern conflict theorists recognize that the power to influence economic outcomes comes not only from ownership of the means of production but also from management position, interlocking board memberships, control of media, and financial contributions to politicians.

For example, wealthy corporations use financial political contributions to influence politicians to enact policies that benefit the wealthy. Laws and policies that favor the rich—sometimes referred to as **wealthfare** or **corporate welfare**—include low-interest government loans to failing businesses, special subsidies and tax breaks to corporations, and other laws and policies that benefit corporations and the wealthy. In 2003 corporate tax breaks cost American taxpayers more than $170 billion (Citizens for Tax Justice 2002). A study of 252 of America's largest and most profitable corporations found that from 2001 to 2003 those companies avoided paying state income taxes on nearly two-thirds of their U.S. profits—at a cost to state governments of $42 billion (McIntyre 2005). In addition, 275 large U.S. corporations paid, on average, only 17.3% of their U.S. profits in federal income taxes in 2002—less than half the 35% rate that the tax code requires. From 2001 to 2003, 82 of these 275 corporations enjoyed at least one year in which they paid no federal income tax, despite pretax profits in those no-tax years of $102 billion (McIntyre 2005). Corporate income taxes in the United States have fallen so much in the last few decades that they are nearly the lowest among the world's developed countries. U.S. corporate taxes were 4% of gross domestic product (GDP) in 1965. This figure fell to 1.5% of GDP in 2003, whereas in other wealthy industrialized countries that belong to the Organization for Economic Cooperation and Development (OECD), corporate taxes were 3% of GDP (Citizens for Tax Justice 2005). In 2003 corporate tax breaks cost American taxpayers more than $170 billion (Citizens for Tax Justice 2002).

Recent federal tax cuts enacted by the Bush administration also benefit the wealthy. Households in the top 1% of the income scale saved about $67,000 as a result of tax cuts made between 2001 and 2003, whereas middle-income families saved just under $600 and the lowest fifth of households saved $61.00 (Mishel et al. 2005).

Conflict theorists also note that throughout the world "free-market" economic reform policies have been hailed as a solution to poverty. Yet, although such economic reform has benefited many wealthy corporations and investors, it has also resulted in increasing levels of global poverty. As companies relocate to countries with abundant supplies of cheap labor, wages decline. Lower wages lead to decreased consumer spending, which leads to more industries closing plants, going bankrupt, and/or laying off workers (downsizing). This results in higher unemployment rates and a surplus of workers, enabling employers to lower wages even more.

Symbolic Interactionist Perspective

Symbolic interactionism focuses on how meanings, labels, and definitions affect and are affected by social life. This view calls attention to ways in which wealth and poverty are defined and the consequences of being labeled "poor." Individuals who are viewed as poor—especially those receiving public assistance (i.e., welfare)—are often stigmatized as lazy, irresponsible, and lacking in abilities, motivation, and moral values. Wealthy individuals, on the other hand, tend to be viewed as capable, motivated, hardworking, and deserving of their wealth.

The symbolic interactionist perspective also focuses on the meanings of being poor. A qualitative study of more than 40,000 poor women and men in 50 countries around the world explored the meanings of poverty from the perspective of those who live in poverty (Narayan 2000). Among the study's findings is that the experience of poverty involves psychological dimensions such as powerlessness, voicelessness, dependency, shame, and humiliation.

Meanings and definitions of wealth and poverty vary across societies and across time. Although many Americans think of poverty in terms of income level, for millions of people poverty is not primarily a function of income but of their alienation from sustainable patterns of consumption and production. For indigenous women living in the least developed areas of the world poverty and wealth are determined primarily by access to and control of their natural resources (such as land and water) and traditional knowledge, which are the sources of their livelihoods (Susskind 2005).

By global standards the Dinka, the largest ethnic group in the sub-Saharan African country of Sudan, are among the poorest of the poor, being among the least modernized peoples of the world. In Dinka culture wealth is measured in large part by how many cattle a person owns. To the Dinka cattle have a social, moral, and spiritual value as well as an economic value. In Dinka culture a man pays an average "bridewealth" of 50 cows to the family of his bride. Thus men use cattle to obtain a wife to beget children, especially sons, to ensure continuity of their ancestral lineage, and, according to Dinka religious beliefs, to strengthen their linkage with God. Although modernized populations might label the Dinka as poor, the Dinka view themselves as wealthy. As one Dinka elder explained, "It is for cattle that we are admired, we, the Dinka. . . . All over the world, people look to us because of cattle . . . because of our great wealth; and our wealth is cattle" (Deng 1998, p. 107).

Definitions of poverty also vary within societies. For example, in Ghana men associate poverty with a lack of material assets, whereas for women poverty is defined as food insecurity (Narayan 2000).

Economic Inequality and Poverty in the United States

The United States is a nation of tremendous economic variation, ranging from the very rich to the very poor. Signs of this disparity are visible everywhere, from opulent mansions perched high above the ocean in California to shantytowns in the rural South where people live with no running water or electricity.

Economic Inequality in the United States

The 1990s was a decade of U.S. economic growth: Interest rates were down, unemployment was low, and stock market averages reached record levels before declining at the end of 1999. At the close of the 20th century the United States had experienced the longest period of peacetime economic expansion in history. But contrary to the adage that "a rising tide lifts all boats," economic prosperity has not been equally distributed in the United States.

For example, between 1979 and 2002 average after-tax income of the top 1% of the population more than doubled, rising from $298,000 to $631,700—an increase of 111%. By contrast, the after-tax income of the poorest fifth of the population rose just 5%, or $600 over the same period (Shapiro 2005). In 1979 the top 1% of the population received 7.5% of national after-tax income, increasing to 11.4% in 2002. In the same period the income of the bottom fifth fell from 6.8% to 5.1% (Shapiro 2005).

Income inequality in the United States is reflected in comparisons of the average income of CEOs with that of production workers. In 1982 the average pay of a CEO was 42 times that of the average production worker. In 1999 the average CEO compensation—$12.4 million—was 475 times the pay of the average production worker (Anderson et al. 2000). By 2003 average CEO pay had declined to $8.1 million while worker pay rose slightly, lowering the CEO-worker pay ratio to 301:1 (Anderson et al. 2004). In 2003, although the average CEO earned $155,769 per week, the average production worker took home $517 in his or her weekly paycheck. If the federal minimum wage had grown at the same rate as CEO pay since 1990, it would have been $15.76 in 2003, instead of $5.15. Likewise, if the average annual salary of production workers had increased at the same rate as CEO pay since 1990, it would have been $75,388 per year in 2003 rather than $26,899 (Anderson et al. 2004).

The distribution of wealth is much more unequal than the distribution of wages or income. **Wealth** refers to the total assets of an individual or household minus liabilities (mortgages, loans, and debts). Wealth includes the value of a home, investment real estate, the value of cars, unincorporated business, life insurance (cash value), stocks, bonds, mutual funds, trusts, checking and savings accounts, individual retirement accounts (IRAs), and valuable collectibles. In 2001 the wealthiest 1% of households owned more than 33% of all national wealth, whereas the bottom 80% of households owned only 16% (Mishel et al. 2005). For another illustration of inequality in wealth, consider that in 2001 the net worth of families in the top 10% of incomes was $833,600, compared to $7,900 net worth of families in the lowest fifth of income earners (Andrews 2003).

Patterns of Poverty in the United States

Poverty is not as widespread or severe in the United States as it is in many less developed countries. Nevertheless, poverty is a significant social problem in the United States. In 2004 the official U.S. poverty rate was 12.7%—the fourth consecutive annual increase in the poverty rate since its most recent low of 11.3% in 2000 (DeNavas-Walt et al. 2005). This translates into 37 million Americans living in poverty in 2004. Poverty rates (three-year average, 2002–2004) vary considerably among the states, from 5.7% in New Hampshire to more than 17% in Arkansas, Mississippi, and New Mexico. Poverty rates also vary according to age, education, sex, family structure, race and ethnicity, and labor force participation.

"The social class you belong to really matters— it determines your health, how long you live, where you live, your exposure to crime, your success in school, and the likely success of your children."

Lawrence Mishel
Economic Policy Institute

ThomsonNOW™

Learn more about **Economic Inequality in the United States** by going through the American Class Structure Learning Module.

ThomsonNOW™

Learn more about **Economic Inequality in the United States** by going through the Social Stratification in the United States Data Experiment.

Table 6.3
U.S. Poverty Rates by Age, 2004

Age (Years)	Poverty Rate
Under 18	17.8
18 to 64	11.3
65 and older	9.8
All ages	12.7

Source: DeNavas-Walt et al. (2005).

Children are more likely than adults to live in poverty.

Age and Poverty. Children are more likely than adults to live in poverty (see Table 6.3). More than one-third (35.2%) of the U.S. poor population are children (DeNavas-Walt et al. 2005). The United States has the highest child poverty rate (and highest overall poverty rate) of the 20 wealthy, industrialized countries that belong to the OECD (Mishel et al. 2005).

Since the late 1950s the poverty rate among the elderly has experienced a downward trend, largely as a result of more Social Security benefits and the growth of private pensions (see also Chapter 9). In 1959 the poverty rate among U.S. elderly was 35.2%; this rate fell to 25.3% in 1969, 15.2% in 1979, and 11.4% in 1989 and reached a record low of 9.7% in 1999. The elderly poverty rate in 2004 was 9.8% (DeNavas-Walt et al. 2005).

Education and Poverty. Education is one of the best insurance policies for protecting an individual against living in poverty. In general, the higher a person's level of educational attainment, the less likely that person is to be poor (see also Chapter 8). Adults without a high school diploma are the most vulnerable to poverty, followed by those with a high school diploma but no college degree. Nearly two-thirds (62%) of children in low-income families (100–200% of the poverty level) have parents with no college education (National Center for Children in Poverty 2004).

Sex and Poverty. Women are more likely than men to live below the poverty line—a phenomenon referred to as the **feminization of poverty.** The 2004 poverty rate for U.S. females was 13.9%, compared to 11.5% for males (U.S. Census Bureau 2005a). As discussed in Chapter 10, women are less likely than men to pursue advanced educational degrees and tend to be concentrated in low-paying jobs, such as service and clerical jobs. However, even with the same level of education and the same occupational role, women still earn significantly less than men. Women who are minorities and/or who are single mothers are at increased risk of being poor.

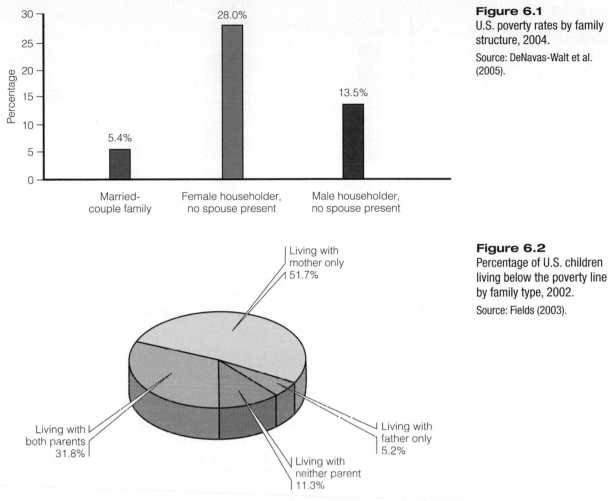

Figure 6.1
U.S. poverty rates by family structure, 2004.
Source: DeNavas-Walt et al. (2005).

Figure 6.2
Percentage of U.S. children living below the poverty line by family type, 2002.
Source: Fields (2003).

Family Structure and Poverty. Poverty is much more prevalent among female-headed single-parent households than among other types of family structures (see Figure 6.1). The relationship between family structure and poverty helps to explain why women and children have higher poverty rates than men (see also Chapter 5). As shown in Figure 6.2, most children living below the poverty line live with their single mothers.

In other industrialized countries poverty rates of female-headed families are lower than those in the United States. Unlike the United States, other developed countries offer a variety of supports for single mothers, such as income supplements, tax breaks, universal child care, national health care, and higher wages for female-dominated occupations.

Race or Ethnicity and Poverty. Nearly half (45.6%) of the poor in the United States are non-Hispanic whites (DeNavas-Walt et al. 2005). However, as displayed in Figure 6.3, poverty rates are higher among blacks, Hispanics, and Native American/Alaska Natives than among non-Hispanic whites. As discussed in Chapter 9, past and present discrimination has contributed to the persistence of poverty among minorities. Other contributing factors include the loss of manufacturing jobs from the inner city, the movement of whites and middle-class blacks out of the inner city,

Figure 6.3
U.S. poverty rates by race and
Hispanic origin, 2004.

*Two-year average (2003–2004).

Source: DeNavas-Walt et al.
(2005).

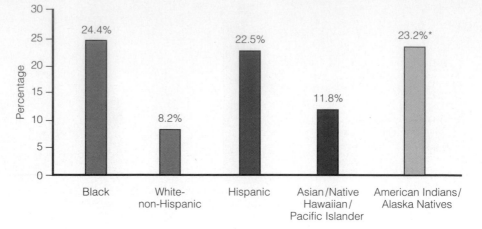

*3 year average (2001–2003)

and the resulting concentration of poverty in predominantly minority inner-city neighborhoods (Massey 1991; Wilson 1987, 1996). Finally, blacks and Hispanics are more likely to live in female-headed households with no spouse present—a family structure that is associated with high rates of poverty.

Labor Force Participation and Poverty. A common image of the poor is that they are jobless and unable or unwilling to work. Although the poor in the United States are primarily children and adults who are not in the labor force, many U.S. poor are classified as working poor. The **working poor** are individuals who spend at least 27 weeks per year in the labor force (working or looking for work) but whose income falls below the official poverty level. In 2003, 36.3% of all U.S. poor (ages 16 and over) worked; 10.9% worked year-round full-time (U.S. Census Bureau 2004).

Consequences of Poverty and Economic Inequality

Poverty is associated with health problems, problems in education, problems in families and parenting, and housing problems. These various problems are interrelated and contribute to the perpetuation of poverty across generations, feeding a cycle of intergenerational poverty. In addition, poverty and economic inequality breed social conflict and war.

Health Problems, Hunger, and Poverty

In Chapter 2 we noted that poverty has been identified as the world's leading health problem. In developing countries absolute poverty is associated with unsafe water and sanitation, indoor air pollution from heating and cooking fumes, high rates of maternal and infant deaths, and hunger and malnutrition. (World Health Organization 2002). More than 850 million people go hungry each day, equivalent to the combined populations of North America, Japan, and Europe. Hunger kills more than 5 million children each year—about one child every five seconds (Halweil 2005).

“If all of the afflictions of the world were assembled on one side of the scale and poverty on the other, poverty would outweigh them all.”

Exodus Rabbah, Mishpatim **31:14**

Students at universities across the country participate in Hunger Banquets—an event created by Oxfam, an international nongovernmental organization dedicated to eliminating hunger and poverty. The intent of the banquet is to increase people's awareness of the economic and nutritional inequalities through an experiential exercise. Based on worldwide statistics, 55% of the attendees, representing the poorest people of the world, are randomly chosen to receive rice and water; 30% (representing the world's middle-income population) receive beans and rice; and the wealthiest 15% receive a full-course meal. Hunger Banquets are hosted at high schools, colleges, community centers, and faith-based organizations.

In the United States low socioeconomic status is associated with higher incidence and prevalence of health problems, disease, and death (Malatu & Schooler 2002). Despite the increasing rates of obesity among Americans, many U.S. poor do not have enough to eat. A Department of Agriculture report found that 11.2% of U.S. households experienced food insecurity at some time during 2003 (which means that the head of household was uncertain of having or unable to acquire enough food for all household members) and 3.5% of U.S. households families were hungry to the point that someone in the household skipped meals because the family could not afford them (Nord 2005). Assess your own degree of food security in this chapter's *Self and Society* feature.

Poor children and adults also receive inadequate and inferior health care, which exacerbates their health problems. Finally, poverty is linked to higher levels of mental health problems, including stress, depression, and anxiety (Leventhal & Brooks-Gunn 2003).

Economic inequality also affects psychological and physical health. Streeten (1998) cited research suggesting that "perceptions of inequality translate into psychological feelings of lack of security, lower self-esteem, envy, and unhappiness, which, either directly or through their effects on life-styles, cause illness" (p. 5). Poor and middle-income adults who live in states with the greatest gap between the rich and the poor are much more likely to rate their own health as poor or fair than people who live in states where income is more equitably distributed (Kennedy et al. 1998).

Food Security Scale

The United States Department of Agriculture conducts national surveys to assess the degree to which U.S. households experience food security, food insecurity, and food insecurity with hunger. To assess your own level of food security, respond to the following items and use the scoring key to interpret your results.

1. In the last 12 months, the food that (I/we) bought just didn't last, and (I/we) didn't have money to get more.
   (a) **Often true**
   (b) **Sometimes true**
   (c) Never true

2. In the last 12 months, (I/we) couldn't afford to eat balanced meals.
   (a) **Often true**
   (b) **Sometimes true**
   (c) Never true

3. In the last 12 months, did you ever cut the size of your meals or skip meals because there wasn't enough money for food?
   (a) **Yes**
   (b) No (skip Question 4)

4. If you answered yes to Question 3, how often did this happen in the last 12 months?
   (a) **Almost every month**
   (b) **Some months but not every month**
   (c) Only 1 or 2 months

5. In the last 12 months, did you ever eat less than you felt you should because there wasn't enough money to buy food?
   (a) **Yes**
   (b) No

6. In the last 12 months, were you ever hungry but didn't eat because you couldn't afford enough food?
   (a) **Yes**
   (b) No

Scoring and Interpretation: The answer responses in boldface type indicate affirmative responses. Count the number of affirmative responses you gave to the items, and use the following scoring key to interpret your results.

Number of Affirmative Responses and Interpretation

0 or 1 item: *Food secure* (In the last year you have had access to enough food for an active, healthy life.)

2, 3, or 4 items: *Food insecure* (In the last year you have had limited or uncertain availability of food and have been worried or unsure you would get enough to eat.)

5 or 6 items: *Food insecure with hunger evident* (In the last year you have experienced more than isolated occasions of involuntary hunger as a result of not being able to afford enough food.)

If you scored as food insecure (with or without hunger), you might consider exploring whether you are eligible for public food assistance (e.g., food stamps) or whether there is a local food assistance program (e.g., food pantry or soup kitchen) that you could use.

Source: Based on the short form of the 12-month Food Security Scale found in Bickel et al. (2000).

Educational Problems and Poverty

In recent years access to education has risen in developing countries. However, in many countries most children from the poorest households have no schooling. A study of 35 countries in West and Central Africa and in South Asia found that in 10 countries half or more of 15- to 19-year-olds from poor households never completed first grade (United Nations Population Fund 2002). "Children aged 6–14 from the wealthiest 20 percent of households are substantially more likely to be enrolled in school than children from the poorest 40 percent of households in almost all [developing] countries" (p. 47).

In the United States children living in poverty are more likely to suffer academically than are children who are not poor. "Overall, poor children receive lower grades, receive lower scores on standardized tests, are less likely to finish high school, and are less likely to attend or graduate from college than are nonpoor youth" (Seccombe 2001, p. 323). Health problems associated with childhood poverty, including poorer vision, lead poisoning, asthma, and inadequate nutrition, contribute to poor academic performance (Rothstein 2004). The poor often attend schools that are characterized by lower quality facilities, overcrowded classrooms, and a higher teacher turnover rate (see also Chapter 8). Children living in poor inner-city ghettos "have to contend with public schools plagued by unimaginative curricula, overcrowded classrooms, inadequate plant and facilities, and only a small proportion of teachers who have confidence in their students and expect them to learn" (Wilson 1996, p. xv). Because poor parents have less schooling on average than do nonpoor parents, they may be less able to encourage and help their children succeed in school. However, research suggests that family income is a stronger predictor of ability and achievement outcomes than are measures of parental schooling or family structure (Duncan & Brooks-Gunn 1997). Poor parents have fewer resources to provide their children with books, computers, travel, and other goods and experiences that promote cognitive development and educational achievement (Sobolewski & Amato 2005). And with the skyrocketing costs of tuition and other fees, many poor parents cannot afford to send their children to college. The cost for low-income families of sending a child to a four-year public college or university rose from 13% of family income in 1980 to 25% in 2000 (Washburn 2004).

Poverty also presents obstacles to educational advancement among poor adults. Women and men who want to further their education to escape poverty may have to work while attending school or may be unable to attend school because of unaffordable child care, transportation, and/or tuition, fees, and books.

Family Stress and Parenting Problems Associated with Poverty

Poverty is both a cause and a consequence of family problems. For example, domestic violence causes some women to flee from their homes and live in poverty without the economic support of their husbands. The stresses associated with low income contribute to substance abuse, domestic violence, child abuse and neglect, divorce, and questionable parenting practices. For example, economic stress is associated with greater marital discord (Sobolewski & Amato 2005), and couples with incomes less than $25,000 are 30% more likely to divorce than couples with incomes greater than $50,000 (Whitehead & Popenoe 2004). Poor parents unable to afford child care or medical expenses are more likely to commit child neglect by leaving children at home without adult supervision or by failing to provide needed medical care. Poor parents are more likely than other parents to use harsh physical disciplinary techniques, and they are less likely to be nurturing and supportive of their children (Mayer 1997; Seccombe 2001).

Another family problem associated with poverty is teenage pregnancy. Poor adolescent girls are more likely to have babies as teenagers or to become young single mothers. Early childbearing is associated with numerous problems, such as increased risk of premature or low-birth-weight babies, dropping out of school, and lower future earning potential as a result of lack of academic achievement. Luker (1996) notes that "the high rate of early childbearing is a measure of how bleak life is for young people who are living in poor communities and who have no obvious arenas for success" (p. 189). For poor teenage women who have been excluded from the American dream and disillusioned with education, "childbearing . . . is one of the few ways . . . such women feel they can make a change in their lives" (p. 182).

> Having a baby is a lottery ticket for many teenagers: it brings with it at least the dream of something better, and if the dream fails, not much is lost. . . . In a few cases it leads to marriage or a stable relationship; in many others it motivates a woman to push herself for her baby's sake; and in still other cases it enhances the woman's self-esteem, since it enables her to do something productive, something nurturing and socially responsible. . . . To the extent that babies can be ill or impaired, mothers can be unhelpful or unavailable, and boyfriends can be unreliable or punitive, childbearing can be just another risk gone wrong in a life that is filled with failures and losses. (Luker 1996, p. 182)

Housing Problems and Homelessness

The following description of housing in a low-income inner-city neighborhood is not atypical of U.S. housing conditions for the poor:

> From the outside, Jamal's building looks like an ordinary house that has seen better days. . . . But once you walk through the front door, all resemblance to a real home disappears. . . . The building has been broken up into separate living quarters, a rooming house with whole families squeezed into spaces that would not even qualify as bedrooms in most homes. . . . Six families take turns cooking their meals in the only kitchen. . . . The plumbing breaks down without warning. . . . Windows . . . are cracked and broken, pieced together by duct tape that barely blocks the steady, freezing draft blowing through on a winter evening. Jamal is of the opinion that for the princely sum of $300 per month, he ought to be able to get more heat. (Newman 1999, pp. 3–9)

Many poor families and individuals live in housing units that lack central heating and air conditioning, sewer or septic systems, and electric outlets in one or more

More than 70,000 people in the United States alone are homeless each night.

© Elena Rooraid/PhotoEdit

Table 6.4

Characteristics of the Homeless in U.S. Cities

- 41% are single men
- 40% are families with children
- 23% are mentally ill
- 17% are employed
- 10% are veterans
- 14% are single women
- 5% are unaccompanied minors
- 30% are substance abusers

Source: U.S. Conference of Mayors (2004).

rooms; many have no telephones. Housing units of the poor are also more likely to have holes in the floor, leaky roofs, and open cracks in the walls or ceiling. In addition, poor individuals are more likely than the nonpoor to live in high-crime neighborhoods.

Even substandard housing would be a blessing to the hundreds of thousands of men, women, and children in the United States who are homeless on a given day. Homelessness, a growing problem in the United States, affects men, women, and children (see Table 6.4). In an annual survey of 27 U.S. cities, city officials cited lack of affordable housing as the major cause of homelessness (U.S. Conference of Mayors 2004). Other causes of homelessness include mental illness and the lack of mental health services, substance abuse and the lack of substance abuse services, low-paying jobs, unemployment, domestic violence, poverty, and prison release. This chapter's *Human Side* feature presents glimpses of what it is like to be homeless.

Intergenerational Poverty

As we have seen, problems associated with poverty, such as health and educational problems, create a cycle of poverty from one generation to the next. Poverty that is transmitted from one generation to the next is called **intergenerational poverty**. In a study of intergenerational poverty using a national longitudinal survey of families, Corcoran & Adams (1997) found considerable mobility out of childhood poverty: three-quarters of white poor children and more than half of black poor children escaped poverty in early adulthood. However, both white and black children in poor families were still much more likely to be poor in early adulthood than were children raised in nonpoor families.

Intergenerational poverty creates a persistently poor and socially disadvantaged population, sometimes referred to as the underclass. The term **underclass** usually refers to impoverished individuals, often those who live in economically distressed neighborhoods (ghettos, slums, or barrios) with low educational attainment, chronic unemployment or underemployment, criminal involvement, unstable family structures, and welfare dependency. Although the underclass is stereotyped as being composed of minorities living in inner-city or ghetto communities, the underclass is a heterogeneous population that includes poor whites living in urban and nonurban communities (Alex-Assensoh 1995).

Life on the Streets: New York's Homeless

While a sociology graduate student at Columbia University, Gwendolyn Dordick undertook a study of homeless people living in New York City. She spent 15 months with four groups of homeless people: inhabitants of a large bus terminal, residents of a shantytown, occupants of a large public shelter, and clients of a small church-run private shelter. Dordick published the results of her study in her book Something Left to Lose: Personal Relations and Survival Among New York's Homeless *(1997). In this* Human Side *feature we present some of Dordick's observations, as well as excerpts from her conversations with homeless individuals she encountered at the Station and the Shanty.*

The Station

The Station, located on Manhattan's West Side, encompasses bus terminals and depots, ticketing windows, shops, and fast food restaurants. Scattered among the commuters and visitors are homeless women and men and young adolescent boys and girls who have either run away or were kicked out by their guardians. One homeless man described living on the edge:

> Living on the streets makes you do a lot of things that you wouldn't normally do. . . .

Comin' into this environment I've done a lot of things I said I wouldn't do. . . . There was some people that came along in a van and just threw sandwiches on the street and I picked them up and ate them. . . . The guilt almost killed me . . . but my stomach said, "Hey, listen, you better eat this food." (Dordick 1997, pp. 5–6)

Homeless individuals often rely on one another, offering each other companionship, friendship, and protection. Although they have little to share, they often share what they have with their fellow homeless friends.

> These people, when I got down here, these people reached out to me because they knew, they already knew what it was like. They're not afraid to help their fellow man. As soon as I got down here I met Ron and the fellows and they didn't push me away. I mean I didn't know where to go, I didn't know where to eat, I didn't know where to sleep. They just invited me right in. And ever since then at least I've been healthy, and I've been clean since I've met them. (Dordick 1997, p. 13)

Homeless individuals are often treated harshly by police. One homeless man lamented: "You may have an invalid laying down here. He's got problems and the [Station] cops will come up and kick him. Like he's an animal with no rights" (Dordick 1997, p. 11).

Pregnancy is common among homeless women who do not have access to or cannot afford contraception. Pregnancy can have disastrous consequences for these women, who fear having a baby and having to care for a baby. According to one man in the Station:

> It's one thing being homeless, but pregnant and homeless? Some women have their babies right out here; others get rid of them. . . . Some of them abort theirself by sticking hangers up their vaginas. I've seen that myself. This young girl didn't want a baby and she stuck a hanger up her vagina. She had to go to the hospital. (Dordick 1997, p. 25)

The Shanty

A barricaded makeshift community of 20 or so residents sits on a formerly vacant lot visible from the nearby streets and a bridge that crosses the East River. The Shanty consists of 15 makeshift dwellings, or "huts" as they are called by the residents, which are made of a variety of discarded materials such as pieces of wood and boards,

---

William Julius Wilson attributes intergenerational poverty and the underclass to a variety of social factors, including the decline in well-paid jobs and their movement out of urban areas, the resultant decline in the availability of marriageable males able to support a family, declining marriage rates and an increase in out-of-wedlock births, the migration of the middle class to the suburbs, and the effect of deteriorating neighborhoods on children and youth (Wilson 1987, 1996).

War and Social Conflict

Poverty is often the root cause of conflict and war within and between nations. In the developing world most of the people recruited for armed conflict are unemployed. "They don't have education opportunities and they don't really see what the

cardboard, mattresses, fabric, and plastic tarps.

The materials are fastened with nails, twine, or fabric. One of the residents has tapped into a source of electricity by running a wire from a lamppost into several huts, providing electricity for light and heat. Dordick noted that, as in the case of the residents of the Station, welfare plays a minimal role in the lives of residents of the Shanty. "So difficult is negotiating the system that most forgo their entitlements" (Dordick 1997, p. 58). One resident explained:

> I don't get welfare. I just can't . . . do it. I hate those people in there. They make you . . . sit and ask you questions that don't make any sense. . . . You're homeless but you have to have an address. What kind of shit is that? Give me a break. They want you to get so . . . upset that you do get up and walk out. (Dordick 1997, p. 58)

Although drug use is common among residents of the Shanty, using drugs in public is a violation of norms of "etiquette." One resident explained that using drugs in the presence of a nonuser is disrespectful: "For me to just take my works out and shoot, I would feel uncomfortable in front of you. It's not right . . . very disrespectful. God forbid, I could be an influence. I could cause you to do it" (Dordick 1997, pp. 71–72).

Although survival among the homeless requires hustling, buying, trading, and selling, not everything is for sale. Some belongings have sentimental value that outweighs their economic value. One resident of the Shanty treasures a small gold key:

> There's a golden rule about gifts. You treasure them. You don't give them away, you don't sell them. I have right here a little key, a skeleton key. A little kid handed it to me four years ago. And every time somebody see that and say, "what is that?" I say it's a key to the world. I wouldn't give it to anyone if it was given to me. I treasure it. (Dordick 1997, p. 78)

Friendships and love relationships among the homeless suffer from the stresses of drug addictions and impoverished conditions. Nevertheless, Dordick explains that the homeless "survive through their personal relationships" (Dordick 1997, p. 193). Relationships are critical to securing the material resources needed to survive and to creating—to the extent that it is possible—a safe and secure environment. One resident of the Shanty, Richie, conveys the importance of a love relationship:

> Regardless of what people might think and say, most of us that might have a woman it's all we want. We don't look for anyone else. We really don't. . . . I'm happy just to take care of my woman. . . . I happen to love my girl. . . . Really, I know it sounds corny, but that's the truth. . . . As a matter of fact, we're gonna be married soon.

A Final Note

Virtually all the homeless individuals that Dordick encountered expressed the desire to escape homelessness and be self-reliant —and they want to be understood. In the words of one homeless man at the Station:

> You never see me sleep in the street. I worked 32 years of my life. Went to prison in '85. . . . I was brought up with a certain degree of independence. . . . And now all I need is two dollars to go and sit in a movie all night long. My pride is too good to beg. I don't want you to help me, Miss. I want you to understand me. (Dordick 1997, p. 5)

Source: Excerpted and reprinted by permission from Dordick, Gwendolyn, 1997. *Something Left to Lose: Personal Relations and Survival among New York's Homeless.* Philadelphia: Temple University Press.

future holds for them other than war and misery" (World Population News Service 2003, p. 4). Tanzania president Benjamin Mkapa said that "countries with impoverished, disadvantaged and desperate populations are breeding grounds for present and future terrorists" (quoted by Schifferes 2004, n.p.).

Not only does poverty breed conflict and war, but war also contributes to poverty. War devastates infrastructures, homes, and businesses and leaves widows and orphans to fend for themselves. Military spending associated with war diverts resources away from economic development and social spending on health and education. Among the 21 countries with extreme food emergencies in 2002, these emergencies were sparked by war, civil unrest, and the lingering effects of past conflicts in 15 of them (UNDP 2003).

In the United States the widening gap between the rich and the poor may lead to class warfare (hooks 2000). Briggs (1998) asked how long the United States can maintain social order "when increasing numbers of persons are left out of the banquet while a few are allowed to gorge?" (p. 474). Although Karl Marx predicted that the have-nots would revolt against the haves, Briggs did not foresee a revival of Marxism. "The means of surveillance and the methods of suppression by the governments of industrialized states are far too great to offer any prospect of success for such endeavors" (p. 476). Instead, Briggs predicted that American capitalism and its resulting economic inequalities would lead to social anarchy—a state of political disorder and weakening of political authority.

ThomsonNOW™

Learn more about **Class Conflict** by going through the American Class Structure Learning Module.

Strategies for Action: Antipoverty Programs, Policies, and Proposals

In this section we provide an overview of government programs designed to alleviate poverty, including various types of welfare and public assistance, the earned income tax credit, and policies and proposals that involve increasing wages. We also look at faith-based initiatives and international responses to poverty.

Government Public Assistance and Welfare Programs in the United States

Many public assistance programs stipulate that households are not eligible for benefits unless their income and/or assets fall below a specified guideline. Programs that have eligibility requirements based on income are called **means-tested programs.** Government public assistance programs designed to help the poor include Supplemental Security Income, Temporary Assistance to Needy Families, food programs, housing assistance, medical care, educational assistance, child care, child support enforcement, and the earned income tax credit (EITC).

Supplemental Security Income. Supplemental Security Income, federal SSI, administered by the Social Security Administration, provides a minimum income to poor people who are age 65 or older, blind, or disabled. Under the 1996 welfare reforms the definition of disability has been sharply restricted, and the eligibility standards have been tightened.

Temporary Assistance to Needy Families. Before 1996 a cash assistance program called **Aid to Families with Dependent Children (AFDC)** provided single parents (primarily women) and their children with a minimum monthly income. In 1996 Congress passed the **Personal Responsibility and Work Opportunity Reconciliation Act (PRWORA),** commonly referred to as "welfare reform," which ended AFDC and replaced it with a program called **Temporary Assistance to Needy Families (TANF).** In 2002 TANF families received an average monthly amount (of cash and cash equivalent assistance) of $355 (Office of Family Assistance 2004). Within two years of receiving benefits adult TANF recipients must be either employed or involved in work-related activities, such as on-the-job training, job search, and vocational education. A lifetime limit of five years is set for families receiving benefits, and able-bodied recipients ages 18 to 50 without dependents have a two-year lifetime limit. Some exceptions to these rules are made for individuals with disabilities, victims of domestic violence, residents of high unemployment areas, and those caring for young children. To qualify for TANF benefits, unwed mothers under the age of 18 are required to live in an adult-supervised environment (e.g., with their parents) and to receive education and job training. Legal immigrants who entered the United States before August 22, 1996, can receive TANF, but those who entered after this date can receive services only after they have been in the country for five years.

Food Assistance. The largest food assistance program in the United States is food stamps, followed by school meals and the Special Supplemental Food Program for Women, Infants, and Children (WIC). The food stamp program issues monthly benefits through coupons or electronic benefits transfer (EBT), using a plastic card similar to a credit card and a personal identification number (PIN). In 2003 the typical household receiving food stamps had a gross income of $640 per month, received a monthly food stamp benefit of $185, and did not receive cash welfare benefits (USDA 2004). The Farm Bill of 2002 restored food stamp benefits for all immigrant children, no matter what their date of entry to the United States.

To supplement food stamps, school meals, and WIC, many communities have food pantries (which distribute food to poor households), "soup kitchens" (which provide cooked meals on-site), and food assistance programs for the elderly (such as Meals on Wheels). Despite the various forms of food assistance, a significant share of poor U.S. children (18% of young children and 12% of school-age children) receives no food assistance (Zedlewski & Rader 2005).

Housing Assistance. Housing costs are a major burden for the poor. In 27 U.S. cities low-income households spend, on average, nearly half (45%) of their income on housing (U.S. Conference of Mayors 2004). Federal housing programs include public housing, Section 8 housing, and other private project-based housing.

The **public housing** program, initiated in 1937, provides federally subsidized housing that is owned and operated by local public housing authorities (PHAs). To save costs and avoid public opposition, high-rise public housing units were built in

A SMALL REASON TO FIND OUT IF YOU QUALIFY FOR FOOD STAMPS.

Call **1-800-221-5689**

United States Department of Agriculture
Food and Nutrition Service
USDA is an equal opportunity provider and employer.

USDA

USDA Food Stamp Program

This poster is a public service message designed to encourage economically distressed families to apply for food stamps.

inner-city projects that have been plagued by poor construction, managerial neglect, inadequate maintenance, and rampant vandalism.

> Distressed public housing subjects families and children to dangerous and damaging living environments that raise the risks of ill health, school failure, teen parenting, delinquency, and crime—all of which generate long-term costs that taxpayers ultimately bear. . . . These severely distressed developments are not just old, outmoded, or run down. Rather, many have become virtually uninhabitable for all but the most vulnerable and desperate families. (Turner et al. 2005, pp. 1–2)

The HOPE VI Urban Demonstration Program was established in 1992 to transform the nation's most distressed public housing projects by rebuilding the physical structure of public housing developments, expanding the opportunities of its residents, and building a sense of community among residents. But improving the conditions in 47,000 to 82,000 severely distressed public housing units is jeopardized by a Department of Housing and Urban Development proposal that would cut back dramatically or even eliminate the HOPE VI program (Turner et al. 2005).

Rather than building new housing units for low-income families, **Section 8 housing** relies on existing housing. With Section 8 housing federal rent subsidies are provided either to tenants (in the form of certificates and vouchers) or to private landlords. Other private project-based housing includes privately owned housing units that do not receive rent subsidies but receive other federal subsidies, such as interest rate reductions. Unlike public housing that confines low-income families to high-poverty neighborhoods, Section 8 housing and other private project-based housing attempt to disperse low-income families throughout the community. However, because of opposition by residents in middle-class neighborhoods, most Section 8 housing units remain in low-income areas.

The level of housing assistance available is sorely inadequate to meet the housing needs of low-income Americans. In 27 U.S. cities only one-third (32%) of eligible low-income households received housing assistance. Applicants for housing assistance must wait an average of 20 months from the time of application to the time they receive assistance, and more than half of the survey cities had stopped accepting applications for at least one assisted housing program because of the excessive length of the waiting list (U.S. Conference of Mayors 2004). Emergency housing for homeless individuals is also inadequate, leaving many homeless individuals to survive on the streets.

Medicaid. The largest U.S. public medical care assistance program is Medicaid, which provides medical services and hospital care for the poor through reimbursements to physicians and hospitals. However, many low-income individuals and families do not qualify for Medicaid and either cannot afford health insurance or cannot pay the deductible and copayments under their insurance plan. Out-of-pocket medical expenses for poor adult Medicaid beneficiaries have grown twice as fast as their incomes in recent years, causing low-income beneficiaries to cut back on essential medical care (Center on Budget and Policy Priorities 2005).

In the earlier AFDC welfare program all recipients were automatically entitled to Medicaid. Under the TANF program states decide who is eligible for Medicaid; eligibility for cash assistance does not automatically convey eligibility for Medicaid. A provision of the 1996 welfare reform legislation guarantees welfare recipients at least one year of transitional Medicaid when leaving welfare for work (see also Chapter 2).

Do Laws That Prohibit Homeless People from Begging and Sleeping in Public Unfairly Punish the Homeless?

In the United States there are at least 840,000 homeless people—on the street or in temporary housing—on any given day (National Law Center on Homelessness and Poverty in America 2004). The number of spaces in homeless shelters is grossly inadequate to accommodate the numbers of homeless individuals, which means that hundreds of thousands of homeless people have no place to be, except in public. Many cities have passed laws that prohibit homeless people from begging as well as sleeping and even sitting in public.

It Is Your Turn Now to Take a Stand!

Do you think that such laws unfairly punish the homeless? Or are these laws necessary to protect the public?

Educational Assistance. Educational assistance includes Head Start and Early Head Start programs and college assistance programs (see also Chapter 8). Head Start and Early Head Start programs provide educational services for disadvantaged infants, toddlers, and preschool-age children and their parents. Evaluations of Head Start and Early Head Start programs indicate that they improve children's cognitive, language, and social-emotional development and strengthen parenting skills (Administration for Children and Families 2002). According to the Children's Defense Fund (2003), every $1 invested in high-quality early childhood care and education saves as much as $7 by increasing the likelihood that children will be literate, go to college, and be employed and by decreasing the likelihood that they will drop out of schools, be dependent on welfare, or be arrested for criminal activity.

To alleviate economic barriers for low-income individuals wanting to attend college, the federal government offers grants, loans, and work opportunities. The Pell grant program aids students from low-income families. The guaranteed student loan program enables college students and their families to obtain low-interest loans with deferred interest payments. The federal college work-study program provides jobs for students with "demonstrated need."

Child Care Assistance. In the United States lack of affordable, good child care is a major obstacle to employment for single parents and a tremendous burden on dual-income families and employed single parents. The cost of full day care in a day-care center ranges from $4,000 to $10,000 per year (Children's Defense Fund 2002). In many cases low-income families have placed their children in low-cost, often lower-quality, and unstimulating care, and nearly 7 million children are left home alone each week.

Some public- and private-sector programs and policies provide limited assistance with child care. The Dependent Care Assistance Plan provisions of the 1981 Economic Recovery Tax Act permits individuals to exclude the value of employer-provided child-care services from their gross income. However, few employers provide on-site child care or subsidies for child care. At the same time Congress increased the amount of the child-care tax credit and modified the federal tax code to allow taxpayers to shelter pretax dollars for child care in "flexible spending plans." The Family Support Act of 1988 offered additional funding for child-care services for the poor (in conjunction with mandatory work requirements). The Child Care and Development Block Grant, which became law in 1990, targeted child-care funds to low-income groups, and the Personal Responsibility and Work Opportunity Reconciliation Act of 1996 appropriated funds for child care. But child-care assistance is inadequate; only 14% of the 16 million children under age 13 who are eligible for child-care assistance receive any help (Children's Defense Fund 2002). With recent budget shortfalls states have made drastic cuts in child-care services for low-income employed parents. According to Sonya Michel (1998), "the reluctance to make adequate provision for childcare is . . . symptomatic of a deeper aversion on the part of many legislators and public officials to helping poor and low-income women become truly economically independent, a status which is, in turn, essential to their ability to form autonomous households" (pp. 47–48).

Child Support Enforcement. To encourage child support from absent parents, the Personal Responsibility and Work Opportunity Reconciliation Act of 1996 requires states to set up child support enforcement programs, and single parents who receive TANF are required to cooperate with child support enforcement efforts. The welfare reform law established a Federal Case Registry and National Directory of New Hires to track delinquent parents across state lines, increased the use of wage withholding to collect child support, and allowed states to seize assets and to revoke driving licenses, professional licenses, and recreational licenses of parents who fall behind in their child support. These efforts to improve child support compliance have been modestly successful: The percentage of poor children in single-mother households receiving child support increased from 31% in 1996 to 36% in 2001 (Sorensen 2003). Among poor families receiving some support, child support as a share of a poor family's income was 30% in 2001 (Sorensen 2003).

Although gains in child support are good news for poor children and their single parents, more than 60% of poor children in single-mother households do not receive child support. One reason is that the fathers of these children tend to be unemployed or have low incomes themselves, limiting their ability to pay child support. Another reason that child support receipt is low among poor families is that most of the support paid to children receiving public assistance goes to the government rather than to the children, to recoup part of the cost of assistance already paid to the family; this reduces the incentive for fathers to pay child support.

Earned Income Tax Credit. The federal **earned income tax credit (EITC),** created in 1975, is a refundable tax credit based on a working family's income and number of children. The EITC is designed to offset Social Security and Medicare payroll taxes on working poor families and to strengthen work incentives. In 2004 an eligible family of four with two children could receive a credit of up to $4,300. Almost one out of every six families who file federal income tax returns claim the federal EITC, which lifts more children out of poverty than any other program (Llobrera &

Zahradnik 2004). In 2005, 17 states and the District of Columbia offered state EITCs (The Hatcher Group 2005).

Welfare in the United States: Myths and Realities

American attitudes toward welfare assistance and welfare recipients are generally negative. Hostile attitudes toward welfare are common among the poor and the wealthy, blacks and whites, men and women, and the variety of ethnic minorities (Epstein 2004). What are some of the common myths about welfare that perpetuate negative images of welfare and welfare recipients?

Myth 1. People who receive welfare are lazy, have no work ethic, and prefer to have a "free ride" on welfare rather than work.

Reality. Most recipients of TANF cash benefits and food stamps are children and therefore are not expected to work. Unemployed adult welfare recipients experience a number of barriers that prevent them from working, including poor health, job scarcity, lack of transportation, lack of education, and/or the desire to stay home and care for their children (which often stems from the inability to pay for child care or the lack of trust in child-care providers) (Zedlewski 2003). Welfare recipients who stay home to care for children *are* doing very important work: parenting. "Raising children is work. It requires time, skills, and commitment. While we as a society don't place a monetary value on it, it is work that is invaluable—and indeed, essential to the survival of our society" (Albelda & Tilly 1997, p. 111).

It is also important to note that many adults receiving public assistance are either employed or in the labor force looking for work. In 2002 one-fourth of adult TANF recipients were employed, earning an average monthly income of $678 (Office of Family Assistance 2004). More than one-fourth of food stamp recipients in 2003 had earnings, typically $640 per month (USDA 2004).

Finally, most adult welfare recipients would rather be able to support themselves and their families than rely on public assistance. The image of a welfare "freeloader" lounging around enjoying life is far from the reality of the day-to-day struggles and challenges of supporting a household on a monthly TANF check of $355 (which was the average monthly cash and cash-equivalent assistance to TANF families in 2002) (Office of Family Assistance 2004).

Myth 2. Most welfare mothers have large families with many children.

Reality. Mothers receiving welfare have no more children, on average, than mothers in the general population. In fiscal year 2003 the average number of individuals in TANF families was 2.5, including an average of 1.9 children (Office of Family Assistance 2004). The average size of households receiving food stamps in 2003 was 2.3 (USDA 2004).

Myth 3. Welfare benefits are granted to many people who are not really poor or eligible to receive them.

Reality. Although some people obtain welfare benefits through fraudulent means, it is much more common for people who are eligible to receive welfare not to receive benefits. Only about half of families poor enough to qualify for TANF receive monthly cash assistance, and 6 out of 10 of those eligible for the food stamp program receive benefits (Food Research and Action Center 2004; Fremstad 2004). As noted earlier, in 27 U.S. cities only 32% (on average) of eligible low-income

"People on welfare are just like you and me. They have the same basic hopes and fears. They want a job that brings self-worth and validation. They want to support their families and contribute to their communities. They want pride and dignity, just like those of us who have been lucky enough never to need public assistance."

Alexis M. Herman
Secretary, U.S. Department of Labor

Table 6.5

Percentage of Individuals Living Below the Poverty Level in Households That Receive Means-Tested Assistance: 2001

Type of Assistance	Percentage
Total	67.5
Receiving cash assistance	22.8
Receiving food stamps	33.6
One or more individuals in household covered by Medicaid	52.5
Live in public or subsidized housing	17.9

Source: U.S. Census Bureau (2002).

households were receiving public housing assistance (U.S. Conference of Mayors 2004). As surmised from the data in Table 6.5, one-third of individuals living below the poverty level in 2001 lived in households that did not receive any form of means-tested assistance.

A main reason for not receiving benefits is lack of information; people do not know they are eligible. Many people who are eligible for public assistance do not apply for it because they do not want to be stigmatized as lazy people who just want a "free ride" at the taxpayers' expense—their sense of personal pride prevents them from receiving public assistance. Others want to avoid the administrative and transportation hassles involved in obtaining assistance (Zedlewski et al. 2003). Finally, some individuals who are eligible for public assistance do not receive it because it is not available. As noted earlier, in 27 U.S. cities applicants for housing assistance must wait an average of 20 months before housing becomes available, and more than half of the survey cities had stopped accepting applications for at least one assisted housing program because of the excessive length of the waiting list (U.S. Conference of Mayors 2004).

Minimum Wage Increase and "Living Wage" Laws

In 2004, 2 million workers (2.7% of all hourly paid workers) earned wages at or below the minimum wage of $5.15. About half of minimum wage workers were under age 25, and twice as many women as men reported earning $5.15 or less (4% versus 2%) (Bureau of Labor Statistics 2005). A full-time worker earning the $5.15 an hour would earn $10,712 per year, well below the 2004 federal poverty line of $15,205 for a family of three. Furthermore, the purchasing power of the minimum wage has declined because increases in the minimum wage have not kept up with inflation. The result is that the minimum wage, when adjusted for inflation, is worth less today than it was in 1979 (Mishel et al. 2005).

A nationwide poll by the Pew Research Center found that 77% of Americans agreed that increasing the minimum wage is an important priority (Chasanov & Chapman 2005). Some states have established a minimum wage that is higher than the federal minimum wage. Between 1997 and spring 2005, 15 states and the District of Columbia raised their minimum wage while the federal minimum wage remained stagnant at $5.15 an hour (Chasanov & Chapman 2005). Those opposed to increasing the minimum wage argue that such an increase would result in higher unemployment, because businesses would reduce wage costs by hiring fewer employees. However, research has failed to find any systematic, significant job loss associated with minimum wage increases (Economic Policy Institute 2000).

As of August 2005, 130 cities and counties had living wage laws (Living Wage Resource Center 2005). **Living wage laws** require state or municipal contractors, recipients of public subsidies or tax breaks, or, in some cases, all businesses to pay employees wages that are significantly above the federal minimum, enabling families to live above the poverty line. Living wage laws are good not only for individuals and families but also for business. Research findings show that businesses that pay their employees a living wage have lower worker turnover and absenteeism, reduced training costs, higher morale and higher productivity, and a stronger consumer market (Kraut et al. 2000).

Currently, a campaign is under way to establish a **universal living wage**, based on a single national formula that indexes the minimum wage to the local cost of housing as set each year by the U.S. Department of Housing and Urban Development. The proposed universal living wage is based on the premise that a person who works 40 hours a week should be able to afford basic housing, spending no more than 30% of income on housing (Universal Living Wage 2005).

As more individuals receiving welfare (TANF) reach their time limits and are forced to enter the job market, it is increasingly important to provide jobs that pay a living wage. As shown in this chapter's *Social Problems Research Up Close* feature, single mothers who work in low-wage jobs often have more hardships than those who are dependent on welfare.

Faith-Based Services for the Poor

President G. W. Bush's **Faith-Based and Community Initiative** views faith-based (i.e., religious) organizations as effective social service providers and allows such organizations to compete for federal funding for programs that serve the needy. In fiscal year 2004, $2 billion in federal competitive grants were awarded to faith-based organizations—nearly double the amount awarded in 2003 (Fletcher 2005). Some of these funds help religious organizations provide services to the poor, such as homeless services and food aid.

Critics point out that the faith-based initiative is not a serious effort to help the poor but rather a political tool to engender support among Bush's conservative base. "Bush made it clear from the beginning that no new money would be allocated for social programs aimed at the needy. Instead, the existing pie would simply be carved up in different ways, with religious groups getting bigger slices" (Boston 2005, n.p.). Indeed, although Bush has pushed for increased funding of religious-based groups, he has also proposed deep cuts for many traditional antipoverty programs, such as public housing subsidies and food stamps.

Critics are also concerned about the degree to which the faith-based initiative violates the separation of church and state and affects the rights of clients seeking services. "How are the lives of the jobless improved when they are told they won't get work until they first get right with God?" (Boston 2005, n.p.). Although religious groups are prohibited from using government funding to promote religion, this policy is difficult to police. Furthermore, Bush seems to endorse proselytizing in faith-based social services by comments such as those made in a January 2004 speech in New Orleans when he said, "We want to fund programs that save Americans one soul at a time." Faith-based groups that impose religion on social service clients risk losing their funding, however. In 2005 a federal judge blocked the Bush

Making Ends Meet: Survival Strategies Among Low-Income and Welfare Single Mothers

As welfare recipients reach the time limit established by welfare legislation of 1996 for receiving welfare benefits, they are forced into the workforce. But as individuals leave welfare for work, they often find themselves in low-paying jobs, often with no or few benefits. How do individuals in low-income jobs compare with those dependent on welfare in terms of their well-being? And how do both low-wage earners and welfare recipients survive on income that does not meet their basic needs? Researchers Kathryn Edin and Laura Lein (1997) conducted research to answer these questions.

Sample and Methods

The sample consisted of 379 African American, white, and Mexican American single mothers from four cities (Chicago, San Antonio, Boston, and Charleston, South Carolina). The mothers either received welfare cash assistance (Aid to Families with Dependent Children, or AFDC) ($N = 214$) or were non-recipients who held low-wage jobs earning $5 to $7 an hour between 1988 and 1992 ($N = 165$). Edin and Lein (1997) used a "snowball sampling" technique in which each mother who was interviewed was asked to refer researchers to one or two friends who might also participate in interviews. Nearly 90% of the mothers contacted agreed to be interviewed.

Edin and Lein collected data by conducting multiple semistructured in-depth interviews with women in the sample. Interview topics included the mothers' income and job experience, types and amount of welfare benefits they received, spending behavior, housing situation, use of medical care and child care, and hardships the women and their children experienced because of lack of financial resources.

Interviewing the mothers more than once was an important research strategy in gathering accurate information. Mothers who were unclear about their expenditures in the first interview could keep careful track of what they spent between interviews and give a more precise accounting of their spending in a later interview. Also, some mothers who insisted that they received no child support later revealed that the child's father "helped out" every week by providing cash. "Most mothers termed absent fathers' cash contributions as 'child support' only if it was collected by the state" (Edin & Lein 1997, p. 13).

Findings and Conclusions

Low-wage-earning single mothers had a higher monthly reported income than welfare-reliant mothers. However, the expenses of wage-earning mothers were also higher. This is because employed mothers usually have to pay for child care, transportation to work, and additional clothing to wear to work. If newly employed mothers have a federal housing subsidy, some of their new income is spent on the increase in the rent they must pay. And employed mothers are usually not eligible for Medicaid, which means that they have more out-of-pocket medical expenses and often go uninsured.

The monthly expenses of both groups of women exceeded their reported monthly income, forcing women to use various strategies to make ends meet. Cash welfare and food stamps covered only three-fifths of welfare-reliant mothers' expenses. The main job of low-wage-earning mothers covered only 63% of their expenses. Edin and Lein (1997) found that women relied on three basic strategies to make ends meet: work

in the formal, informal, or underground economy; cash assistance from absent fathers, boyfriends, relatives, and friends; and cash assistance or help from agencies, community groups, or charities in paying overdue bills. Welfare recipients had to keep their income-generating activities hidden from their welfare caseworkers and other government officials. Otherwise, their welfare checks would be reduced by nearly the same amount as their earnings. Many of the wage-earning mothers also concealed income generated "on the side" in order to maintain food stamps, housing subsidies, or other benefits that would have been reduced or eliminated if they had reported this additional income.

Most of the single mothers in the study described experiencing serious material hardship during the previous 12 months. Material hardships included not having enough food and clothes, not receiving needed medical care, not having health insurance, having the utilities or phone cut off, not having a phone, and being evicted and/or homeless. An important finding was that wage-reliant mothers experienced more hardship than welfare-reliant mothers. In addition to the increased financial pressures of child-care costs, transportation, health care, and work clothing, employed mothers worried about not providing adequate supervision of their children and struggled with balancing work and parenting responsibilities, especially when their children were sick. Nevertheless, almost all the mothers said they would rather work than rely on welfare. They believed that work provided important psychological benefits and increased self-esteem, avoided the stigma of welfare, and enabled them to be good role models for their children.

Source: Based on Edin and Lein (1997).

Table 6.6
The Millennium Development Goals

1. Eradicate extreme poverty and hunger.
2. Achieve universal primary education.
3. Promote gender equality and empower women.
4. Reduce child mortality.
5. Improve maternal health.
6. Combat HIV/AIDS, malaria, and other diseases.
7. Ensure environmental sustainability.
8. Develop a global partnership for development.

administration from providing future grants to a faith-based mentoring program that promoted religious worship and instruction in its program (Associated Press 2005). It is of some reassurance, then, that religious organizations that express their religiosity more publicly and place a higher value on evangelizing are less likely to apply for and receive government funding than religious organizations that have less religious policies and practices (Ebaugh et al. 2005).

International Responses to Poverty

In 2000 leaders from 191 United Nations member countries pledged to achieve eight **Millennium Development Goals**—an international agenda for reducing poverty and improving lives. One of the Millennium Development Goals (MDGs) is to reduce by half the proportion of people who live on $1 a day and who suffer from hunger. As can be seen in Table 6.6, several other MDGs involve alleviating problems related to poverty, such as disease, child and maternal mortality, and lack of access to education. The target date for achieving most of the MDGs is 2015, with 1990 as the benchmark. Although China and India have made significant progress toward achieving the MDG of reducing poverty, in sub-Saharan Africa poverty *rose* between 1990 and 2001 (World Bank 2005). Approaches for achieving poverty reduction throughout the world include promoting economic growth, investing in "human capital," providing financial aid and debt cancellation, and changing policies that increase poverty.

Promoting Economic Growth. In 2004 growth in the gross domestic product of developing countries averaged 6.7%—the highest level in three decades (World Bank 2005). Over the long term economic growth is believed to reduce poverty (United Nations 1997). An expanding economy creates new employment opportunities and increased goods and services. As employment prospects improve, individuals are able to buy more goods and services. The increased demand for goods and services, in turn, stimulates economic growth. As emphasized in Chapters 13 and 14, economic development requires controlling population growth and protecting the environment and natural resources, which are often destroyed and depleted in the process of economic growth.

However, economic growth does not always reduce poverty; in some cases it increases it. Policies that involve cutting government spending, privatizing basic services, liberalizing trade, and producing goods primarily for export may increase economic growth at the national level, but the wealth ends up in the hands of the political and corporate elite at the expense of the poor. Growth does not help poverty reduction when public spending is diverted away from meeting the needs of the poor and instead is used to pay international debt, finance military operations, and support corporations that do not pay workers fair wages and that are hostile to unionization. The World Bank lends $30 billion a year to developing nations to pay primarily for roads, bridges, and industrialized agriculture that mostly benefit corporations. "Relatively little attention or money has been given to developing basic social services, building schools and clinics, and building decent public sanitation and clean water systems in some of the world's poorest countries" (Mann 2000, p. 2). Thus "economic growth, though essential for poverty reduction, is not enough. Growth must be pro-poor, expanding the opportunities and life choices of poor people" (UNDP 1997, pp. 72–73). Because three-fourths of poor people in most developing countries depend on agriculture for their livelihoods, economic growth to reduce poverty must include raising the productivity of small-scale agriculture. Not only does improving the productivity of small-scale agriculture create employment, but it also reduces food prices. This chapter's *Focus on Technology* feature examines genetically modified food as a strategy for alleviating global hunger.

Investing in Human Capital. Promoting economic development in a society requires having a productive workforce. Yet in many poor countries large segments of the population are illiterate and without job skills and/or are malnourished and in poor health. Thus a key feature of poverty reduction strategies involves investing in human capital. The term **human capital** refers to the skills, knowledge, and capabilities of the individual. Investments in human capital involve programs and policies that provide adequate nutrition, sanitation, housing, health care (including reproductive health care and family planning), and educational and job training.

Poor health is both a consequence and a cause of poverty; improving the health status of a population is a significant step toward breaking the cycle of poverty. A cross-country comparison of children living in households that survive on $1 a day found that children living in countries with higher levels of per capita public spending on health had significantly lower levels of mortality and malnutrition (Wagstaff 2003).

Investments in education are also critical for poverty reduction. Increasing the educational levels of a population better prepares individuals for paid employment and for participation in political affairs that affect poverty and other economic and political issues. Improving the educational level and overall status of women in developing countries is also associated with lower birthrates, which in turn fosters economic development.

Providing Financial Aid and Debt Cancellation. To pay for investments in human capital, poor countries depend on financial aid from wealthier countries. To meet the MDG of halving poverty by 2015, the United Nations recommends that wealthier countries allocate just 0.7% of national income to aid. But only 5 of 22 major donors (none of them from the most powerful G7 nations) are meeting that target. Of 22 OECD countries the United States is the *least* generous donor in terms of aid as a proportion of its national income (Oxfam 2005).

Another way to help poor countries invest in human capital and reduce poverty is to provide debt relief. In poor countries with large debts money needed for health and education is instead spent on debt repayment. Poor countries are often caught in a vicious cycle, paying out more in debt repayments than they receive in aid. In 2003 low-income nations spent $39 billion on debt repayment and received

$29 billion in aid (Oxfam 2005). Canceling the debts of 32 of the poorest countries would cost citizens of the richest countries of the world just $2.10 a year for each citizen for 10 years (Oxfam 2005).

Understanding Poverty and Economic Inequality

After George W. Bush was reelected in the 2004 presidential election, numerous commentators remarked that Bush's victory was largely due to his focus on "values" and the appeal that this focus had for the American public, especially those with strong religious ties. In his campaign Bush emphasized such values as freedom and democracy, of marriage being defined as one man and one woman, and of the sanctity of life (implying disapproval of abortion and stem-cell research). However, Economic Policy Institute president Lawrence Mishel reminds us that "economic inequality and how it is addressed is as much a 'values' issue as any of those that are more frequently discussed" (2005, p. 1).

As we have seen in this chapter, economic prosperity has not been evenly distributed; the rich have become richer while the poor have become poorer. Meanwhile, the United States has implemented welfare reform measures that essentially weaken the safety net for the impoverished segment of the population—largely children. Welfare reform legislation of 1996 has achieved its goal of reducing welfare rolls across the country. From 1996 to 2003 the number of families receiving TANF dropped by 54% (U.S. Department of Health and Human Services 2004). Advocates of welfare reform argue that transitions from welfare to work benefit children by creating positive role models in their working mothers, promoting maternal self-esteem, and fostering career advancement and higher family earnings. Critics of welfare reform argue that reforms increase stress on parents, force young children into inadequate child care, reduce parents' abilities to monitor the behavior of their adolescents, and deepen the poverty of many families. Of those who leave TANF, only 60% are able to find employment, and of those only a fraction are able to earn a living wage (Parisi et al. 2003). Although the long-term effects of welfare reform are not yet known, one study of the impact of welfare reform on children concluded that reforms will help some children and hurt others (Duncan & Chase-Lansdale 2001). As we discuss in the next chapter ("Work and Unemployment"), many of the jobs available to those leaving welfare for work are low paying, have little security, and offer few or no benefits. Without decent wages and without adequate assistance in child care, housing, health care, and transportation, many families who leave welfare for work will find that their situation becomes worse, not better.

A common belief among U.S. adults is that the rich are deserving and the poor are failures. Blaming poverty on the individual rather than on structural and cultural factors implies not only that poor individuals are responsible for their plight but also that they are responsible for improving their condition. If we hold individuals accountable for their poverty, we fail to make society accountable for making investments in human capital that are necessary to alleviate poverty. Such human capital investments include providing health care, adequate food and housing, education, child care, and job training. Economist Lewis Hill (1998) believes that "the fundamental cause of perpetual poverty is the failure of the American people to invest adequately in the human capital represented by impoverished children" (p. 299). Blaming the poor for their plight also fails to recognize that there are not enough jobs for those who want to work and that many jobs fail to pay wages

"All too many of those who live in affluent America ignore those who exist in poor America; in doing so, the affluent American will eventually have to face themselves with the question . . . : How responsible am I for the well-being of my fellows? To ignore evil is to become an accomplice to it."

Martin Luther King Jr.
Civil rights activist

Is Genetically Modified Food the Solution for Global Hunger?

Genetically modified food, also known as genetically engineered (GE) food and genetically modified organisms (GMOs), involves the process of DNA recombination, in which scientists transfer genes from one plant into the genetic code of another plant. The most common genetically modified (GM) trait is herbicide tolerance (which produces crops that tolerate weed-killing chemicals), followed by insect resistance.

Between 1996 and 2004 the global area of GM crops increased by 4.3 million, to more than 200 million acres. In 2004, 17 countries grew GM crops, with most grown in the United States (59%), followed by Argentina (20%), Canada and Brazil (each 6%), and China (4%). The most common GM crop is soybean, followed by corn, cotton, and canola (Global Knowledge Center on Crop Biotechnology 2005).

An estimated 60–70% of processed foods in U.S. markets contain some form of GM ingredient, most often corn or soy, followed by canola and cotton (in cottonseed oil). Yet a national survey of U.S. adults found that less than half (48%) were aware that foods containing GM ingredients are currently sold in stores, and although most Americans are likely to consume foods with GM ingredients every day, less than one-third (31%) of the survey sample said they had consumed food containing GM ingredients (Hallman et al. 2004).

Scientists, academics, environmentalists, public health officials, policy makers, corporations, farmers, and citizens throughout the world are deeply divided over GM crops and food. Not surprisingly, supporters of GM food technology emphasize its potential benefits, whereas critics focus on the potential risks.

Can Genetically Modified Foods Alleviate Hunger and Malnutrition?

Supporters of GM food commonly cite the alleviation of hunger and malnutrition as a main benefit, claiming that this technology can enable farmers to produce crops with higher yields. However, research findings suggest that GM crop yields are not significantly higher than conventional crop yields and in some cases are *lower* than conventional crop yields (Mendelson 2002).

Biotechnology companies promote the use of genetic engineering to enhance the nutritional value of foods as a strategy for alleviating nutritional deficiencies in the diets of poor populations. "Golden rice," genetically modified to contain vitamin A, has been touted as a remedy for vitamin A deficiencies among poor children in developing countries. However, golden rice has never entered the market because, to get an adequate level of vitamin A from this rice, a 4-year-old child would have to eat 27 cups of rice per day (Mendelson 2002).

Critics of GM foods argue that the world already produces enough food for all people to have a healthy diet. According to the United Nations Development Programme, if all the food produced worldwide were distributed equally, every person would be able to consume 2,760 calories a day (UNDP 2003). Biotechnology, critics argue, will not alter the fundamental causes of hunger, which are poverty and lack of access to food and to land on which to grow food.

Critics argue that biotechnology companies use the issue of poverty and hunger in the developing world to justify GM crops. Such claims are promoted by a consortium created by the biotechnology industry, the Council for Biotechnology Information, which has a multimillion-dollar public relations budget to tout the benefits of GE foods (Altieri 2003). In fact, most GM food products and research dollars for the development of GM foods target the more affluent nations' agriculture and consumers.

Skeptics suggest that transgenic crop technology can actually *increase* hunger and poverty. The corporate control of GM seeds, protected by patents and intellectual property rights, threatens the age-old farming practice of saving seeds from one crop to use for the next season. Instead, farmers who use GM seeds must purchase new seeds each season—an expense that many subsistence farmers cannot afford.

Are Genetically Modified Foods Safe for Human and Environmental Health?

Biotechnology companies claim that GM foods that have been approved by the Food and Drug Administration are safe for human consumption, and they even cite potential health benefits, such as the use of genetic modification to remove allergens that naturally occur in foods such as nuts, making these foods safer to eat for individuals who have allergies to these foods (Bailey 2004). But critics claim that research on the effects of GM crops and foods on human health is inadequate, especially concerning long-term effects. Human health concerns include possible toxicity, carcinogenicity, food intolerance, antibiotic resistance buildup, decreased nutritional value, and food allergens in GM foods.

Biotechnology skeptics are also concerned about the environmental effects of GM crops. Biotechnology companies claim that crops that are genetically designed to repel insects negate the need for chemical (pesticide) control and thus reduce pesticide poisoning of land, water, animals, foods, and farmworkers. However, critics are concerned that insect populations can build up resistance to GM plants with insect-repelling traits, which would necessitate increased rather than decreased use of pesticides. Indeed, a recent study found that although GE crops substantially reduce pesticide use in the first few years of planting, GE crops have increased the overall volume of pesticides applied to corn, soybeans, and cotton over a nine-year period (Benbrook 2004).

Another health and environmental risk is the spread of traits from GM plants to non-GM plants, the effects of which are unknown. In 2003 an analysis of corn grown in nine Mexican states found that 24% of the samples tested positive for contamination by several varieties of GM corn, including Starlink, produced as cattle feed and deemed unfit for human consumption because of the presence of a bacterial protein that is not broken down by the human digestive system and is therefore a potential allergen (ETC Group 2003b; Ruiz-Marrero 2002). Mexican farmers and community members view the contamination of Mexican corn as an attack on Mexican culture. In the words of one Mexican citizen, "Our seeds, our corn, is the basis of the food sovereignty of our communities. It's much more than a food, it's part of what we consider sacred, of our history, our present and future" (quoted by ETC Group 2003b, p. 2).

GM seed contamination is of particular concern with regard to seed sterility technology. To maintain control over their products, biotechnology companies have developed "terminator" seeds, which cause the plant to produce sterile seeds. Because of public opposition to terminator seeds, in 1999 Monsanto agreed not to market its terminator technology. However, Monsanto later adopted a positive stance on genetic seed sterilization, suggesting that the commercialization of terminator technology may occur in the future (ETC Group 2003a). Could the seed sterility trait in terminator crops inadvertently contaminate both traditional crops and wild plant life? The possible ramifications of widespread plant sterility could be devastating to life on earth.

Biotechnology critics also raise concerns about insufficient safeguards and regulatory mechanisms. In 2000 Taco Bell taco shells, made by Kraft Foods, were recalled after traces of Starlink corn were found in the taco shells. No one—from farmers to grain dealers to Kraft—could explain how the Starlink corn got mixed into corn meant for taco shells. The traces of unapproved corn were not found by the U.S. Department of Agriculture's Food Safety and Inspection Service or by the Department of Health and Human Service's Food and Drug Administration. Rather, the traces were discovered by Genetically Engineered Food Alert—a coalition of biotechnology skeptics. In addition to taco shells, Starlink contamination caused a recall of more than 300 corn-based foods. Another example of contamination occurred in 2002, when traces of corn genetically engineered to produce an "edible vaccine" to prevent piglets from getting diarrhea were found mixed in with soybeans that would be processed into dozens of food items (Hickey & Mittal 2003). These incidents of contamination raise disturbing questions about the regulatory oversight of GM foods that is meant to govern food safety, assess risks, monitor compliance, and enforce regulations. But such safeguards are nonexistent in some countries and, as the aforementioned contamination incidents suggest, even when regulatory systems are in place, they are not foolproof.

The Labeling of Foods with Genetically Modified Ingredients

A 2004 national survey of U.S. adults and a compilation of 18 public opinion polls on GM foods revealed that the overwhelming majority of U.S. adults say that GE foods should be labeled as such (Center for Food Safety and International Forum on Globalization 2005; Hallman et al. 2004).Without such labeling consumers cannot exercise their right to make their own choices regarding the purchase and consumption of GM foods. More than 30 countries require labeling of products containing GE ingredients and of GE whole foods. In 2005 Alaska passed the first U.S. law requiring GE fish to be labeled as such. There is, however, no U.S. federal requirement for labeling GM foods (although many food products that do not contain GM foods label their products as "GMO-free"). The Genetically Engineered Food Right to Know Act of 2005 calls for food companies to label all foods containing GM ingredients.

Worldwide Government Reactions to Genetically Modified Crops and Food

In 2000 worldwide concern about the safety of GM crops resulted in 130 nations signing the landmark Biosafety Protocol, which requires producers of a GM food to demonstrate that it is safe before it is widely used. The Biosafety Protocol also allows countries to ban the import of GM crops based on suspected health, ecological, or social risks. As of 2005, several countries had banned the commercial planting of GM crops, and a handful of countries had banned the importation of GM crops (Center for Food Safety and International Forum on Globalization

2005). In 2004 the European Commission ended a 6-year moratorium on the commercial planting of GM crops and also voted to allow imports of GE maize, ending a 6-year moratorium on GM imports.

Zambia, a country facing widespread famine, refused GM food aid from the United States after its scientists concluded that there was insufficient evidence to demonstrate its safety. Other African countries, although plagued by hunger and malnutrition, also reject GM foods. In 1998 all African delegates (except those from South Africa) to the UN Food and Agriculture Organization released the following statement against GM foods:

> European citizens have been exposed to an aggressive publicity campaign . . . to convince [us] that the world needs genetic engineering to feed the hungry. Organized and financed by Monsanto, one of the world's biggest chemical companies . . . this campaign gives a totally distorted and misleading picture of the potential of genetic engineering to feed developing countries. We, the undersigned delegates of African countries . . . strongly object that the image of the poor and hungry from our countries is being used by giant multinational corporations to push a technology that is neither safe, environmentally friendly, nor economically beneficial to us. . . . We think it will destroy the diversity, the local knowledge and the sustainable agricultural systems that our farmers have developed for millennia and that it will thus undermine our capacity to feed ourselves. (Hickey & Mittal 2003, p. 4)

Concluding Remarks

Many citizens have clearly taken a stand either for or against GM food. However, many more are uncertain and struggle to make sense out of the competing claims of the benefits and safety versus the potential hazards of GM food and the complex ethical and sociopolitical implications related to this technology.

Lester Brown (2001) of the World Watch Institute suggests that "perhaps the largest question hanging over the future of biotechnology is the lack of knowledge about the possible environmental and human health effects of using genetically modified crops on a large scale over the long term" (p. 52). This lack of knowledge calls for more research to answer questions about the potential risks of GM crops. In the meantime, providing food for hungry populations can be achieved through promoting sustainable agricultural practices that have already been shown to be effective in increasing food crops for small farmers in developing countries (Altieri 2003).

that enable families to escape poverty. And last, blaming the poor for their condition diverts attention away from the recognition that the wealthy—individuals and corporations—receive far more benefits in the form of wealthfare or corporate welfare, without the stigma of welfare.

Ending or reducing poverty begins with the recognition that doing so is a worthy ideal and an attainable goal. Imagine a world where everyone had comfortable shelter, plentiful food, adequate medical care, and education. If this imaginary world were achieved and if absolute poverty were effectively eliminated, what would the effects be on such social problems as crime, drug abuse, family problems (e.g., domestic violence, child abuse, divorce, and unwed parenthood), health problems, prejudice and racism, and international conflict? In the current global climate of conflict and terrorism, we might consider that "reducing poverty and the hopelessness that comes with human deprivation is perhaps the most effective way of promoting long-term peace and security" (World Bank 2005).

According to one source, the cost of eradicating poverty worldwide would be only about 1% of global income—and no more than 2–3% of national income in all but the poorest countries (UNDP 1997). Certainly the costs of allowing poverty to continue are much greater than that.

Chapter Review

ThomsonNOW™

Reviewing is as easy as ① ② ③

1. Before you do your final review, take the ThomsonNOW diagnostic quiz to help you identify the areas on which you should concentrate. You will find information on ThomsonNOW and instructions on how to access all of its great resources on the foldout at the beginning of the text.

2. As you review, take advantage of ThomsonNOW's study videos and interactive Map the Stats exercises to help you master the chapter topics.

3. When you are finished with your review, take ThomsonNOW's posttest to confirm you are ready to move on to the next chapter.

- **What is the difference between absolute poverty and relative poverty?**

  Absolute poverty refers to a lack of basic necessities for life, such as food, clean water, shelter, and medical care. In contrast, relative poverty refers to a deficiency in material and economic resources compared with some other population.

- **What share of the world's population lives on less than $2 per day?**

  According to the World Health Organization, nearly half the world's population lives on less than $2 per day.

- **Which sociological perspective criticizes wealthy corporations for using financial political contributions to influence politicians to enact policies that benefit corporations and the wealthy?**

  The conflict perspective is critical of wealthy corporations that use financial political contributions to influence laws and policies that favor corporations and the rich. Such laws and policies, sometimes referred to as wealthfare or corporate welfare, include low-interest government loans to failing businesses and special subsidies and tax breaks to corporations.

- **In the United States what age group has the highest rate of poverty?**

  U.S. children are more likely than adults to live in poverty. More than one-third of the U.S. poor population are children. Child poverty rates are much higher in the United States than in Canada or any other Western European industrialized country.

- **According to officials in 27 U.S. cities, what is the main cause of homelessness?**

  City officials in 27 U.S. cities cited lack of affordable housing as the major cause of homelessness. Other causes of homelessness include mental illness and the lack of mental health services, substance abuse and the lack of substance abuse services, low-paying jobs, unemployment, domestic violence, poverty, and prison release.

- **What are three common myths about welfare and welfare recipients?**

  Common myths about welfare and welfare recipients are (1) that welfare recipients are lazy, have no work ethic, and prefer to have a "free ride" on welfare rather than work; (2) that most welfare mothers have large families with many children; and (3) that welfare benefits are granted to many people who are not really poor or eligible to receive them.

- **Which federal program lifts more children out of poverty than any other program?**

  The federal earned income tax credit (EITC), created in 1975, is a refundable tax credit based on a working family's income and number of children. The EITC lifts more children out of poverty than any other program.

Critical Thinking

1. Should someone receiving welfare benefits be entitled to spend some of his or her money on "nonessentials," such as cosmetics, eating out, lottery tickets, and cable TV? Why or why not?

2. In our sociology classes we introduce the topic of U.S. poverty by asking students to think of an image of a person who represents poverty in America and to draw that imaginary person. Students are asked to give the person a name (to indicate their sex), and to write down the age of the person. In every semester and in every class most students draw a picture of a middle-aged man. Yet U.S. poverty statistics reveal that the higher poverty rates are among women, not men, and among youth, not middle-aged adults. Why do you think the most common image of a U.S. poor person is a middle-aged man?

3. The poor in the United States have low rates of voting and thus have minimal influence on elected government officials and the policies they advocate. What strategies might be effective in increasing voter participation among the poor?

Key Terms

absolute poverty
Aid to Families with Dependent Children (AFDC)
bourgeoisie
corporate welfare
culture of poverty
earned income tax credit (EITC)
Faith-Based and Community Initiative
feminization of poverty

genetically modified
  food
HPI-1
HPI-2
human capital
human poverty index
  (HPI)
intergenerational poverty
living wage laws
means-tested programs
Millennium Development
  Goals
Personal Responsibility
  and Work Opportunity

Reconciliation Act
  (PRWORA)
poverty
proletariat
public housing
relative poverty
Section 8 housing
Temporary Assistance to
  Needy Families (TANF)
underclass
universal living wage
wealth
wealthfare
working poor

Media Resources

The Companion Website for
Understanding Social Problems, Fifth
Edition

http://sociology.wadsworth.com/mooney_knox_schacht5e

Supplement your review of this chapter by going to the companion website to take one of the Tutorial Quizzes, use the flash cards to master key terms, and check out the many other study aids you'll find there. You'll also find special features such as *Wadsworth's Sociology Online Resources and Writing Companion,* GSS Data, and Census 2000 information, data, and resources at your fingertips to help you complete that special project or do some research on your own.

> " When a man tells you that he got rich through hard work, ask him whose. " *Don Marquis, Journalist*

Work and Unemployment

i n 1998 a fifth-grade class in Denver was in the middle of a history unit on American slavery when they learned that black people were still slaves in Sudan. The students were shocked and decided to take action. They began saving their lunch money to purchase the freedom of slaves, and some students contributed cash given to them as birthday presents. In the end they raised enough money to free two Sudanese slaves. With the help of their teacher, Barbara Vogel, this fifth-grade class founded STOP—a student group dedicated to ending slavery, which has grown into a nationwide campaign to free slaves.

The persistence of modern-day slavery is just one of many concerns regarding the well-being of workers throughout the world. In this chapter we examine problems of work and unemployment, including forced labor, sweatshop labor, health and safety hazards in the workplace, job dissatisfaction and alienation, work-family concerns, and declining labor strength and representation. We begin by looking at the global economy.

The Global Context: The Economy in the 21st Century

In 2004 the European Union (EU) accepted 10 new member nations, forming the largest single trading bloc in the world and representing one-fourth of the world's wealth. Residents of EU countries can buy and sell goods and services in any of the 25 member countries without tariff barriers, and most of the EU countries share a common currency, the euro. The EU reflects the increasing globalization of economic institutions. The **economic institution** refers to the structure and means by which a society produces, distributes, and consumes goods and services.

In recent decades innovations in communication and information technology have spawned the emergence of a **global economy**—an interconnected network of economic activity that transcends national borders and spans the world. The globalization of economic activity means that increasingly our jobs, the products and services we buy, and our nation's political policies and agendas influence and are influenced by economic activities occurring around the world. After summarizing the two main economic systems in the world—capitalism and socialism—we describe how industrialization and post-industrialization have changed the nature of work, and we look at the emergence of free trade agreements and transnational corporations.

Socialism and Capitalism

Socialism is an economic system in which the means of producing goods and services are collectively owned. In a socialist economy the government controls income-producing property. Theoretically, goods and services are equitably distributed according to the needs of the citizens. Socialist economic systems emphasize collective well-being rather than individualistic pursuit of profit.

Under **capitalism** private individuals or groups invest capital (money, technology, machines) to produce goods and services to sell for a profit in a competitive market. Whereas socialism emphasizes social equality, capitalism emphasizes individual freedom. Capitalism is characterized by economic motivation through profit, the determination of prices and wages primarily through supply and demand, and the absence of government intervention in the economy. More people are working in a capitalist economy today than ever before in history (Went 2000). Critics of capitalism argue that it creates too many social evils, including alienated workers, poor working conditions, near-poverty wages, unemployment, a polluted and depleted environment, and world conflict over resources.

In reality, there are no pure socialist or capitalistic economies. Rather, most countries are hybrids of some kind, incorporating elements of both capitalism and socialism. For example, Western European countries have a strong social welfare system, but they still have elements of a capitalistic free market. The U.S. economy is dominated by capitalism, but there are elements of socialism in our welfare system and in government subsidies to industry.

ThomsonNOW™

Learn more about **Socialism and Capitalism** by going through the Capitalism versus Socialism Learning Module.

Industrialization, Post-Industrialization, and the Changing Nature of Work

The nature of work has been shaped by the Industrial Revolution, the period between the mid-18th century and the early 19th century when the factory system was introduced in England. **Industrialization** dramatically altered the nature of work: Machines replaced hand tools; and steam, gasoline, and electric power replaced human or animal power. Industrialization also led to the development of the assembly line and an increased division of labor as goods began to be mass produced. The development of factories contributed to the emergence of large cities where the earlier informal social interactions dominated by primary relationships were replaced by formal interactions centered on secondary groups. Instead of the family-centered economy characteristic of an agricultural society, people began to work outside the home for wages.

Post-industrialization refers to the shift from an industrial economy dominated by manufacturing jobs to an economy dominated by service-oriented, information-intensive occupations. Post-industrialization is characterized by a highly educated workforce, automated and computerized production methods, increased government involvement in economic issues, and a higher standard of living.

Primary, Secondary, and Tertiary Work Sectors. The three fundamental work sectors (primary, secondary, and tertiary) reflect the major economic transformations in society—the Industrial Revolution and the Post-Industrial Revolution. The *primary work sector* involves the production of raw materials and food goods. In developing countries agriculture employs 40% of workers; in the United States less than 1% of employees work in farming, fishing, and forestry (International Labour Organization 2005c; U.S. Bureau of the Census 2004). The *secondary work sector* involves the production of manufactured goods from raw materials (e.g., paper from wood). The *tertiary work sector* includes professional, managerial, technical support, and service jobs. The transition to a post-industrialized society is marked by a decrease in manufacturing jobs and an increase in service and information technology jobs in the tertiary work sector. Between 1994 and 2000 there were 17 million U.S. manufacturing jobs, but this number declined sharply to 14.3 million in 2004 (Bureau of Labor Statistics 2005c).

A large number of U.S. workers are not educated and skilled enough for many of these tertiary-level positions, however (Koch 1998). In developing countries many individuals with the highest level of skill and education leave the country in search of work abroad, leading to the phenomenon known as the **brain drain.** Although U.S. employers benefit because they pay lower wages to foreign workers, U.S. workers are displaced and developing countries lose valuable labor. In recent years highly skilled foreign workers have been finding employment with U.S. companies in their own countries, because work sent abroad includes not only manufacturing jobs but also highly skilled jobs, for example, aeronautical engineer, software designer, and stock analyst (Uchitelle 2003).

McDonaldization of the Workplace. Sociologist George Ritzer (1995) coined the term **McDonaldization** to refer to the process by which the principles of the fast food industry are being applied to more and more sectors of society, particularly the workplace. McDonaldization involves four principles:

1. *Efficiency.* Tasks are completed in the most efficient way possible by following prescribed steps in a process overseen by managers.
2. *Calculability.* Quantitative aspects of products and services (such as portion size, cost, and the time it takes to serve the product) are emphasized over quality.
3. *Predictability.* Products and services are uniform and standardized. A Big Mac in Albany is the same as a Big Mac in Tucson. Workers behave in predictable ways. For example, servers at McDonald's learn to follow a script when interacting with customers.
4. *Control through technology.* Automation and mechanization are used in the workplace to replace human labor.

The principles of McDonaldization are not new. Henry Ford's assembly line was designed to produce automobiles efficiently and predictably by using technology to replace human labor. But the degree to which these rational principles characterize the workplace today is unprecedented.

What are the effects of McDonaldization on workers? In a McDonaldized workplace employees are not permitted to use their full capabilities, be creative, or engage in genuine human interaction. Workers are not paid to think, just to follow a predetermined set of procedures. Because human interactions are unpredictable and inefficient (they waste time), "we're left with either no interaction at all, such as at ATMs, or a 'false fraternization.' Rule number 17 for Burger King workers is to smile at all times" (Ritzer, quoted by Jensen 2002, p. 41). Workers also may feel that they are merely extensions of the machines they operate. The alienation that workers feel—the powerlessness and meaninglessness that characterizes a "McJob"—may lead to dissatisfaction with one's job, and more generally, with one's life. Worker dissatisfaction and alienation are discussed later in this chapter.

The Globalization of Trade and Free Trade Agreements

Just as industrialization and post-industrialization changed the nature of economic life, so has the globalization of trade—the expansion of trade of raw materials, manufactured goods, and agricultural products across national and hemispheric borders. The first set of global trade rules were adopted through the General Agreement on Tariffs and Trade (GATT) in 1947. GATT members met periodically to revise trade agreements in negotiations called "rounds." In 1995 the World Trade Organization (WTO) replaced GATT as the organization overseeing the multilateral trading system.

In the 1980s and early 1990s U.S. officials began negotiating regional free trade agreements that would open doors to U.S. goods in neighboring countries and reduce the massive U.S. trade deficit, which had grown from $25.3 billion in 1980 to $122 billion in 1985 (Schaeffer 2003). **Free trade agreements** are pacts between two countries or among a group of countries that make it easier to trade goods across national boundaries. Free trade agreements reduce or eliminate foreign restrictions on exports, reduce or eliminate tariffs (or taxes) on imported goods, and prevent U.S. technology from being copied and used by competitors through protection of "intellectual property rights." Treaties such as the Canada–U.S. Free Trade Agreement, the North American Free Trade Agreement (NAFTA), the Free Trade Area of the Americas (FTAA), and the Central American Free Trade Agreement (CAFTA) (which was approved by Congress in July 2005) are designed to accomplish these trade goals.

U.S. officials have also used Section 301 of the Trade Acts of 1984 and 1988 to force trade negotiations with individual countries. If U.S. trade officials determine that other countries have denied U.S. corporations "reasonable" access to domestic markets, sold their goods in the United States at below-market prices, or failed to protect the patents and copyrights of U.S. companies, Section 301 allows the United States to impose retaliatory sanctions and tariffs on goods from these countries (Schaeffer 2003).

Through GATT and the WTO, free trade agreements, and Section 301, U.S. trade officials have expanded trading opportunities, benefiting large export manufacturing and service industries in the global north, specifically aircraft, auto, computer, pharmaceutical, and entertainment industries in Western Europe, the United States, and Japan. But trade globalization also hurt the U.S. steel and textile-apparel industries and the workers employed in them, small businesses that cannot compete with large retail chain stores, supermarkets, and franchises, and small farmers (Schaeffer 2003). Since NAFTA was signed in 1993, the growth of U.S. exports supported 1 million U.S. jobs, but the growth of imports from Mexico and Canada displaced production that would have supported 2 million jobs. Thus NAFTA has resulted in the loss of 1 million U.S. jobs, two-thirds of which are in the manufacturing industries (Scott & Ratner 2005).

Foreign workers have also been hurt by trade agreements. Since NAFTA took effect, real wages of Mexican manufacturing workers have fallen, and more than 1 million agricultural jobs in Mexico have been lost (Scott & Ratner 2005).

As discussed in Chapters 2 and 14, free trade agreements also undermine the ability of national, state, and local governments to implement environmental and food or product safety policies.

Transnational Corporations

Although free trade agreements have increased business competition around the world, resulting in lower prices for consumers for some goods, they have also opened markets to monopolies (and higher prices) because they have facilitated the development of large-scale transnational corporations. **Transnational corporations**, also known as *multinational corporations,* are corporations that have their home base in one country and branches, or affiliates, in other countries. In less than 20 years the number of transnational corporations has increased from 7 to more than 45,000. The top 100 economies around the world are transnational corporations rather than nations, and the combined yearly revenues of the largest corporations are greater than those of 182 nations, which are home to more than four-fifths of the world's population (Clarke 2002). Three to six transnational corporations control 85–90% of world wheat, corn, coffee, cotton, and tobacco exports, 90% of forest product exports, and about 90% of iron ore exports (Schaeffer 2003).

Transnational corporations provide jobs for U.S. managers, secure profits for U.S. investors, and help the United States compete in the global economy. Transnational corporations benefit from increased access to raw materials, cheap foreign labor, and the avoidance of government regulations.

> By moving production plants abroad, business managers may be able to work foreign employees for long hours under dangerous conditions at low pay, pollute the environment with impunity, and pretty much have their way with local communities. Then the business may be able to ship its goods back to its home country at lower costs and bigger profits. (Caston 1998, pp. 274–275)

Transnational corporations contribute to the trade deficit in that more goods are produced and exported from outside the United States than from within. Transnational corporations also contribute to the budget deficit, because the United States does not get tax income from U.S. corporations abroad, yet transnational corporations pressure the government to protect their foreign interests; as a result, military spending increases. Third, transnational corporations contribute to U.S. unemployment by letting workers in other countries perform labor that could be performed by U.S. employees. Finally, transnational corporations are implicated in an array of other social problems, such as poverty resulting from fewer jobs, urban decline resulting from factories moving away, and racial and ethnic tensions resulting from competition for jobs.

Sociological Theories of Work and the Economy

Numerous theories in economics, political science, and history address the nature of work and the economy. In sociology structural functionalism, conflict theory, and symbolic interactionism serve as theoretical lenses through which we may better understand work and economic issues and activities.

Structural-Functionalist Perspective

According to the structural-functionalist perspective, the economic institution is one of the most important of all social institutions. It provides the basic necessities common to all human societies, including food, clothing, and shelter. By providing for the basic survival needs of members of society, the economic institution contributes to social stability. After the basic survival needs of a society are met, surplus materials and wealth may be allocated to other social uses, such as maintaining military protection from enemies, supporting political and religious leaders, providing formal education, supporting an expanding population, and providing entertainment and recreational activities. Societal development is dependent on an economic surplus in a society (Lenski & Lenski 1987).

Although the economic institution is functional for society, elements of it may be dysfunctional. For example, before industrialization agrarian societies had a low degree of division of labor, in which few work roles were available to members of society. Limited work roles meant that society's members shared similar roles and thus developed similar norms and values (Durkheim 1966 [1893]). In contrast, industrial societies are characterized by many work roles, or a high degree of division of labor, and cohesion is based not on the similarity of people and their roles but on their interdependence. People in industrial societies need the skills and services that others provide. The lack of common norms and values in industrialized societies may result in **anomie**—a state of normlessness—which is linked to a variety of social problems, including crime, drug addiction, and violence (see Chapters 3 and 4).

The structural-functionalist perspective is also concerned with how changes in one aspect of society affect other aspects. How, for example, does the level of economic development of a country affect the subjective life satisfaction and core values of its population? Data from the World Value Survey indicate that as one moves from subsistence-level economies in developing countries to advanced industrialized societies, there is a large increase in the percentage of the population who consider themselves happy or satisfied with their lives as a whole. Above a certain economic point, however, the curve levels off. In other words, "moving from a starvation level to a reasonably comfortable existence makes a big difference," but once this level is reached, further economic development does not increase subjective well-being (Inglehart 2000, p. 219). Economic development level also affects core values, because economic insecurity breeds xenophobia and deference to authority, whereas a sense of basic security fosters values such as self-expression (rather than deference to authority) and not only a tolerance of cultural diversity but also a sense that cultural differences are stimulating and exotic (Inglehart 2000).

Conflict Perspective

According to Karl Marx, the ruling class controls the economic system for its own benefit and exploits and oppresses the working masses. Whereas structural functionalism views the economic institution as benefiting society as a whole, conflict theory holds that capitalism benefits an elite class that controls not only the economy but other aspects of society as well—the media, politics and law, education, and religion.

The conflict perspective is critical of ways that the government caters to the interests of big business at the expense of workers, consumers, and the public interest. This system of government that serves the interests of corporations—known as **corporatocracy**—involves ties between government and business. For example, George W. Bush, the first president with an MBA (master's degree in business administration), is a former Texas oilman, and Dick Cheney was the CEO of Halliburton, the world's largest oil field services company, until he joined the Bush ticket in 2000. The majority of Bush's cabinet and advisers have ties to a number of corporations, including General Motors, Bank of America, Chevron, Sears, Goldman Sachs, and Boeing (Center for Responsive Politics 2005a).

Corporate interests also find their way into politics through large political contributions and **soft money**, which is money that flows through a loophole to provide political parties, candidates, and contributors a means to evade federal limits on political contributions. Critics of this system of campaign financing argue that corporations and interest groups purchase political influence through financial contributions. Because of the high cost of political campaigns, many candidates rely on funds from special interest groups, who then expect (and often get) special treatment. The special treatment can be in the form of lower business taxes, environmental loopholes, subsidies, or lower standards of consumer and worker protection. For example, the Bankruptcy Abuse Prevention and Consumer Protection Act, which was passed by the House of Representatives in 2005, makes it more difficult for individuals to escape from debt by filing for bankruptcy protection. Finance and credit companies, which benefit from the new bankruptcy law, contributed more than $8 million (64% to Republicans) in the 2004 election cycle, and spent millions more on federal lobbying (Center for Responsive Politics 2005b).

A survey of business leaders' views on political fund-raising found that the main reasons U.S. corporations make political contributions are fear of retribution and to buy access to lawmakers ("Big Business for Reform," 2000). Although 75% of the surveyed business leaders said that political donations give them an advantage in shaping legislation, nearly three-quarters (74%) said that business leaders are pressured to make large political donations. Half of the executives said that their colleagues "fear adverse consequences for themselves or their industry if they turn down requests" for contributions.

Penalties for violating health and safety laws in the workplace provide an example of legal policy that favors corporate interests. Suppose that a corporation is guilty of a serious violation of health and safety laws, in which "serious violation" is defined as one that poses a substantial probability of death or serious physical harm to workers. What penalty do you think that corporation should pay for such a violation? According to a report by the AFL-CIO (2005), serious violations of workplace health and safety laws carry an average penalty of only $873. As discussed in Chapter 4, penalties for corporate crimes tend to be much less severe than those applied to individuals who violate the law. For example, under federal law, causing the death of a worker by willfully violating safety rules—a misdemeanor with a six-month maximum prison term—is a less serious crime than harassing a wild burro on federal lands, which is punishable by one year in prison (Barstow & Bergman 2003).

Some social critics believe that the influence of corporations in U.S. government is so strong that "the corporate leaders run the system more than the president does. The Republicans tend to be more closely tied to the interests of certain corporations, but the leaders of the corporatocracy will find some way to render ineffective any president who fails to advocate for what they want or who tries to stand in their way" (Perkins, quoted by MacEnulty 2005, p. 10). The pervasive influence of corporate power in government exists not only in the United States but also throughout the world. On a global scale the policies of the International Monetary Fund (IMF) and the World Bank pressure developing countries to open their economies to foreign corporations, promoting export production at the expense of local consumption, encouraging the exploitation of labor as a means of attracting foreign investment, and hastening the degradation of natural resources as countries sell their forests and minerals to earn money to pay back loans.

In his book *Confessions of an Economic Hit Man,* John Perkins (2004) describes his prior job as an "economic hit man"—a highly paid professional who would convince leaders of poor countries to accept huge loans (primarily from the World Bank) that were much bigger than the country could possibly repay. The loans would be used to help develop the country by paying for needed infrastructure, such as roads, electrical plants, airports, shipping ports, and industrial plants. One of the conditions of the loan was that the borrowing country had to give 90% of the loan back to U.S. companies (such as Halliburton or Bechtel) to build the infrastructure. The result: The wealthiest families in the country benefit from additional infrastructure, and the poor masses are stuck with a debt they cannot repay. The United States uses the debt as leverage to ask for "favors," such as land for a military base or access to natural resources such as oil. According to Perkins, large corporations want "control over the entire world and its resources, along with a military that enforces that control" (quoted by MacEnulty 2005, p. 10).

How Do You Define the American Dream?

From the following list, choose two items that you personally believe best represent or define the American Dream. After you make your choices, compare your responses to those of a national sample of U.S. adults.

Owning a home
Having a family
Obtaining a quality education for your children
Being financially secure
Living in freedom
Having a secure retirement
Enjoying good health
Living in a good community
Having a good job

Comparison: In a national survey of U.S. adults, respondents were asked to select two items from the given list that best represent their definition of the American Dream (National League of Cities 2004). As shown in the following table, the most common choice was "living in freedom" (33%), followed by "being financially secure" (26%).

Item	Percentage of U.S. Adults Who Chose Item
Living in freedom	33
Being financially secure	26
Obtaining a quality education for your children	17
Having a family	17
Enjoying good health	16
Having a good job	9
Owning a home	9
Living in a good community	6
Having a secure retirement	6

Symbolic Interactionist Perspective

According to symbolic interactionism, the work role is a central part of a person's identity. When making a new social acquaintance, one of the first questions we usually ask is "What do you do?" The answer largely defines for us who that person is. For example, identifying a person as a truck driver provides a different social meaning than identifying someone as a physician. In addition, the title of one's work status—maintenance supervisor or university professor—also gives meaning and self-worth to the individual. An individual's job is one of his or her most important statuses; for many it represents a "master status," that is, the most significant status in a person's social identity.

Symbolic interactionism emphasizes that attitudes and behavior are influenced by interaction with others. The applications of symbolic interactionism in the workplace are numerous: Employers and managers are concerned with using interpersonal interaction techniques that elicit the attitudes and behaviors they want from their employees; union organizers are concerned with using interpersonal interaction techniques that persuade workers to unionize; and job-training programs are concerned with using interpersonal interaction techniques that are effective in motivating participants.

From a symbolic interactionist perspective we might look at the meanings of work that members of a society learn. In the United States work and the rewards of

> "No race can prosper 'til it learns there is as much dignity in tilling a field as in writing a poem."
>
> **Booker T. Washington**
> **Address to the Atlanta Exposition, September 18, 1895**

work are viewed as part of the American Dream. Does work play a major part in your view of the American Dream? Assess your views by reading this chapter's *Self and Society* feature.

Problems of Work and Unemployment

In this section we examine unemployment and other problems associated with work. Problems of workplace discrimination based on gender, race and ethnicity, and sexual orientation are addressed in Chapters 9, 10, and 11. Minimum wage and living wage issues are discussed in Chapter 6. Here, we discuss problems concerning forced labor, sweatshop labor, health and safety hazards in the workplace, job dissatisfaction and alienation, work-family concerns, unemployment and underemployment, and labor unions and the struggle for workers' rights.

Forced Labor and Slavery

> "Everything that is really great and inspiring is created by the individual who can labor in freedom."
>
> **Albert Einstein**

Worldwide at least 12.3 million people are victims of forced labor (International Labour Organization 2005a). **Forced labor,** also known as *slavery,* refers to any work that is performed under the threat of punishment and is undertaken involuntarily. Slavery expert Kevin Bales (1999) explained that the resurgence of slavery around the world is linked to three main factors: (1) rapid growth in population, especially in the developing world; (2) social and economic changes that have displaced many rural dwellers to urban centers and their outskirts, where people are powerless and jobless and are vulnerable to exploitation and slavery; and (3) government corruption that allows slavery to go unpunished, even though it is illegal in every country.

Forced prison labor is a type of forced labor that is controlled by the state. Forced prison labor is particularly widespread in China.

© Mark Peterson /Corbis

Forced labor exists all over the world but is most prevalent in India, Pakistan, Bangladesh, and Nepal. Most forced laborers work in agriculture, mining, prostitution, and factories. Forced labor produces goods we use every day, including sugar from the Dominican Republic, chocolate from Ivory Coast, paper clips from China, carpets from Nepal, and cigarettes from India. Between 40% and 50% of all victims of forced labor are children (International Labour Organization 2005a). Child labor is discussed in Chapter 12.

Forms of Slavery and Forced Labor. Modern slavery is different from the old form most people know, which is **chattel slavery.** In chattel slavery slaves were considered property that could be bought and sold. In the past the high cost of purchasing a slave (about $40,000 in today's money) gave the master incentive to provide a minimum standard of care to ensure that the slave would be healthy enough to work and generate profit for the long term. Today, slaves are cheap, costing an average of $100 (Cernasky 2002). Because they are so cheap and abundant, slaves are no longer a major investment worth maintaining. If slaves become ill or injured, too old to work, or troublesome to the slaveholder, they are dumped or killed and replaced with another slave (Cernasky 2002).

Although chattel slavery still exists in some areas, most forced laborers today are not "owned" but are rather controlled by violence or the threat of violence. The most common form of forced labor today is called *bonded labor*. Bonded laborers are usually illiterate landless rural poor who take out a loan simply to survive or to pay for a wedding, funeral, medicines, fertilizer, or other necessities. Debtors must work for the creditor to pay back the loan, but often they are unable to repay it. Creditors can keep debtors in bondage indefinitely in two main ways. First, they can charge the debtors illegal fines (for workplace "violations" or for poorly performed work) or charge laborers for food, tools, and transportation to the work site while keeping wages too low for the debt to ever be repaid. Alternatively, creditors can claim that all the labor performed by the debtor is collateral for the debt and cannot be used to reduce it (Miers 2003).

Another common form of forced labor involves luring individuals with the promise of a good job and instead holding them captive and forcing them to work. Migrant workers are particularly vulnerable because if they try to escape and report their abuse, they risk deportation. Organized crime rings are sometimes involved in the international trafficking of human beings, which often flows from developing nations to the West. A form of forced labor most common in South Asia is sex slavery, in which girls are forced into prostitution by their own husbands, fathers, and brothers to earn money to pay family debts. Other girls are lured by offers of good jobs and then forced to work in brothels under the threat of violence.

Another type of forced labor is conducted by the state or military. Forced military service, which has been reported in parts of Africa, and forced prison labor (common in China) are examples of state and military forced labor.

Slavery and Forced Labor in the United States. Kevin Bales estimates that there are 100,000 to 150,000 slaves in the United States today, mostly because of human trafficking for domestic work, migrant farm labor, or work in the sex industry (Cockburn 2003). Migrant workers are tricked into working for little or no pay as a means of repaying debts from their transport across the U.S. border, similar to debt bondage in South Asia. Traffickers posing as employment agents lure women into the United States with the promise of good jobs and education but then place them in "jobs" where they are forced to do domestic or sex work.

Sweatshop Labor

A U.S. Department of Labor investigation of the Daewoosa Samoa garment factory in American Samoa—a factory that produces men's sportswear for J. C. Penney—found that garment factory workers lived and worked under conditions of poor sanitation, malnutrition, electrical hazards, fire hazards, machinery hazards, illegally low wages, sexual harassment and invasion of privacy, workplace violence and corporal punishment, and overcrowded barracks in which two workers were forced to share each bed (National Labor Committee 2001). Female workers reported that the company owner routinely entered their barracks to watch them shower and dress. Workers reported incidents in which security guards slapped and kicked workers. The food provided to the workers at the Daewoosa Samoa garment factory consisted of a watery broth of rice and cabbage.

The workers at the Daewoosa Samoa garment factory are among the millions of people worldwide who work in **sweatshops**—work environments that are characterized by less-than-minimum-wage pay, excessively long hours of work (often without overtime pay), unsafe or inhumane working conditions, abusive treatment of workers by employers, and/or the lack of worker organizations aimed at negotiating better working conditions. Sweatshop labor conditions occur in a wide variety of industries, including garment production, manufacturing, mining, and agriculture. The dangerous conditions in sweatshops result in high rates of illness, injury, and death. In one tragic example of death resulting from sweatshop conditions, at least 53 workers, including 10 children, were burned to death in a fire at a Sagar Chowdury garment factory in Bangladesh (Hargis 2001). The fire, caused by an electrical short circuit, engulfed the entire factory; the factory's 900 workers were *locked inside.* Local residents and firefighters broke open the locked gates of the building and rescued survivors.

Sweatshop labor commonly occurs in the garment industry.

© Guenter Stand/VISUM/The Image Works

Many products in the U.S. consumer market are made under sweatshop conditions. An investigative report on working conditions in five Chinese factories that produce products for Disney, Wal-Mart, K-Mart, Mattel, and McDonalds revealed sweatshop conditions that violate Chinese labor laws (Students and Scholars Against Corporate Misbehavior 2005). Workers are forced to work grueling 12- to 15-hour days, earning just 33 to 41 cents an hour. In some factories women are denied their legal maternity rights. Workers are housed in overcrowded dorm rooms and fed horrible food at the factory canteen. Workers are charged for the housing and food provided at the factory (even if they live and eat elsewhere), often costing them one-fifth to one-third of their monthly wages. Some factories have no fans and become oppressively hot. Workers often faint from exhaustion and the unbearably stifling heat. Some workers are exposed to strong-smelling gases from working with glue, with no protective masks or ventilation system. Crushed fingers and other injuries are common in some factory departments. Workers have no health insurance, no pension, and no right to freedom of association or to organize.

Sweatshop Labor in the United States. Sweatshop conditions in overseas industries have been widely publicized. However, many Americans do not realize the extent to which sweatshops exist in the United States. The Department of Labor estimates that more than half of the country's 22,000 sewing shops violate minimum wage and overtime laws and that 75% violate safety and health laws ("The Garment Industry," 2001). Most garment workers in the United States are immigrant women who typically work 60 to 80 hours a week, often earning less than minimum wage, with no overtime, and many face verbal and physical abuse.

Migrant farmworkers, who process 85% of the fruits and vegetables grown in the United States, also work under sweatshop conditions. Many live in substandard and crowded housing provided by their employer and lack access to safe drinking water as well as bathing and sanitary toilet facilities. Farmworkers commonly suffer from heat exhaustion, back and muscle strains, injuries resulting from the use of sharp and heavy farm equipment, and illness resulting from pesticide exposure (Austin 2002). Working 12-hour days under hazardous conditions, farmworkers have the lowest annual family incomes of any U.S. wage and salary workers, and more than 60% of them live in poverty (Thompson 2002).

Health and Safety Hazards in the U.S. Workplace

Many workplaces are safer today than in generations past. Nevertheless, fatal and disabling occupational injuries and illnesses still occur in troubling numbers. The incidence of illnesses resulting from hazardous working conditions is probably much higher than the reported statistics show, because long-term latent illnesses caused by, for example, exposure to carcinogens often are difficult to relate to the workplace and are not adequately recognized and reported.

Workplace Fatalities. The International Labour Organization estimates that 1.1 million workers worldwide die on the job or from occupational disease each year ("Editorial," 2000). In 2004, 5,703 U.S. workers—93% of whom were men—died of fatal work-related injuries (Bureau of Labor Statistics 2005d). The most common type of job-related fatality involves transportation accidents (see Figure 7.1). Industries with the highest rates (per 100,000 workers) of fatal injuries include agriculture/forestry/fishing and hunting, mining, transportation, and construction.

Figure 7.1
Causes of workplace fatalities, 2004.

Source: Bureau of Labor Statistics (2005d).

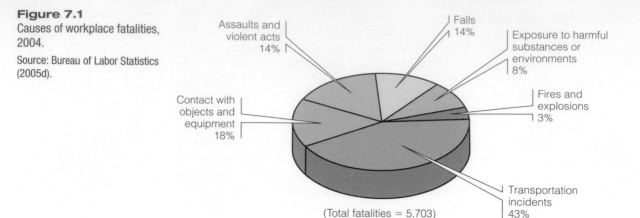

Assaults and violent acts 14%

Falls 14%

Exposure to harmful substances or environments 8%

Contact with objects and equipment 18%

Fires and explosions 3%

Transportation incidents 43%

(Total fatalities = 5,703)

Occupational Illnesses and Nonfatal Injuries. The Bureau of Labor Statistics (2004) reported 4.4 million nonfatal occupational injuries and illnesses in private industry in 2003—a rate of 5 cases per 100 full-time workers. Sprains and strains are the most common nonfatal occupational injury or illness involving days away from work. Laborers and material movers and truck drivers suffer the most injuries and illnesses involving days away from work, followed by nursing aides, orderlies, and attendants, who commonly experience back strains from lifting and moving patients (Bureau of Labor Statistics 2005b).

The most common types of workplace illness are disorders associated with repeated motion or trauma, such as carpal tunnel syndrome (a wrist disorder that can cause numbness, tingling, and severe pain), tendonitis (inflammation of the tendons), and noise-induced hearing loss. Such disorders—referred to by a number of terms, including **cumulative trauma disorders** and **repetitive motion disorders**—are muscle, tendon, vascular, and nerve injuries that result from repeated or sustained actions or exertions of different body parts. Jobs that are associated with high rates of upper-body cumulative trauma disorders include computer programming, manufacturing, meatpacking, poultry processing, and clerical and office work. Repetitive motion disorders are classified as illness, not injury, because they are not sudden, instantaneous traumatic events. Carpal tunnel syndrome results in more days absent from work than fractures or amputations.

Job Stress. Another work-related health problem is job stress. In a national sample of U.S. employees more than one-fourth (26%) felt "overworked" and 27% felt "overwhelmed" by how much work they had to do often or very often in the past month (Galinsky et al. 2005). A national Gallup poll found that slightly more than one-third said they are "somewhat (23%) or "completely" (11%) dissatisfied with the amount of on-the-job stress in their jobs (Kiefer 2003).

Prolonged job stress, also known as **job burnout,** can cause or contribute to physical and mental health problems, such as high blood pressure, ulcers, headaches, anxiety, and depression. Taking time off to heal and "recharge one's batteries" is not an option for many workers: One-half of the U.S. workforce has no paid sick leave and one-fourth has no paid vacation (Watkins 2002). And pressures in the workplace often mean that workers bring their jobs with them on vacation. One in five employees (21%) works sometimes, often, or very often while on vacation (Galinsky et al. 2005).

This man, vacationing on the Greek island of Santorini, is among the one in five U.S. workers who works while on vacation.

Job Dissatisfaction and Alienation

A 2005 Gallup poll found that 5% of non-Hispanic whites, 10% of Hispanics, and 16% of blacks are somewhat or very dissatisfied with their job or the work they do (Gallup Organization 2005). One source of dissatisfaction is declining wages. In 2003 nearly one-fourth (24.3%) of U.S. workers earned poverty-level hourly wages (Mishel et al. 2005). Workers in low-wage, low-status jobs with few or no benefits and little job security are vulnerable to feeling dissatisfaction not only with their jobs but also with their limited housing and lifestyle options. In this chapter's *Human Side* feature some of the dissatisfactions associated with low-wage work are expressed.

Some workers may be dissatisfied with the level of on-the-job monitoring and lack of privacy they experience at work. This chapter's *Taking a Stand* feature deals with the monitoring of employees' e-mails at work.

One form of job dissatisfaction is a feeling of **alienation.** Work in industrialized societies is characterized by a high degree of division of labor and specialization of work roles. As a result, workers' tasks are repetitive and monotonous and often involve little or no creativity. Limited to specific tasks by their work roles, workers are unable to express and utilize their full potential—intellectual, emotional, and physical. According to Marx, when workers are merely cogs in a machine, they become estranged from their work, the product they create, other human beings, and themselves. Marx called this estrangement "alienation." As we discussed earlier, the McDonaldization of the workplace also contributes to alienation.

Alienation usually has four components: powerlessness, meaninglessness, normlessness, and self-estrangement. Powerlessness results from working in an environment in which one has little or no control over the decisions that affect one's work. Meaninglessness results when workers do not find fulfillment in their work. Workers may experience normlessness if workplace norms are unclear or conflicting. For example, many companies that have family leave policies informally discourage workers from using them. Or workplaces that officially promote

"Clearly the most unfortunate people are those who must do the same thing over and over again, every minute, or perhaps twenty to the minute. They deserve the shortest hours and the highest pay.**"**

John Kenneth Galbraith
American economist

Excerpts from an Interview with Barbara Ehrenreich, Author of *Nickel and Dimed: On (Not) Getting By in America*

Barbara Ehrenreich's best-selling book *Nickel and Dimed* describes the day-to-day struggles of low-wage work and surviving on a low-wage income.

Barbara Ehrenreich, a successful journalist with a Ph.D., wondered how anyone could survive on low-income wages: $6 to $7 an hour. To find out, she lived for one year on wages she earned doing low-wage work, moving from Florida to Maine to Minnesota, taking jobs as a waitress, hotel maid, house cleaner, nursing home aide, and Wal-Mart

sales person, and living in the cheapest lodging she could find that offered an acceptable level of safety. Ehrenreich chronicled her experiences in Nickel and Dimed: On (Not) Getting By in America—*a book that became a* New York Times *bestseller. In this* Human Side *feature we present excerpts from an interview of Barbara Ehrenreich by Jamie Passaro in which Ehrenreich talks about her experiences as a low-wage earner and her viewpoints concerning low-wage work in America (Passaro 2003).*

Passaro: What surprised you most during your months of low-wage work?

Ehrenreich: It was a surprise to me how challenging these jobs were. I was expecting that I would be doing dull, repetitive work, that I would be bored out of my mind. Instead I was struggling all the time, physically and mentally, to master these jobs. At Wal-Mart I had to memorize the locations of hundreds of clothing items so I could put everything back in its exact place. In the nursing home I had about fifteen minutes to learn the names and dietary requirements of thirty patients. It took all the concentration I had. So I no longer use the word *unskilled* to describe any job.

Passaro: How do you think your experience would have been different if you were a man?

Ehrenreich: A lot of low-wage jobs are really for either sex now because, as heavy industry declines, the "masculine" jobs of the past are not there anymore. There are men working at Wal-Mart and in restaurants and in nursing homes. The only difference for me is that a man probably would not have been as fearful as I was about living in a creepy residential motel with no privacy or security. . . .

Passaro: When you went back to your middle-class life after working low-wage jobs, how were you different?

Ehrenreich: I was more impatient with affluent people who don't see these problems or who aren't particularly interested and brush them off. . . .

Passaro: Why do you think class inequality is such a taboo subject in the mainstream media?

Ehrenreich: It undercuts the American myth that anybody can become rich, that it's just a matter of personal ability and determination. . . . We like to tell ourselves that everybody is equal. To admit that large numbers of people are systematically held back is hard, because it means upward mobility

nondiscrimination in reality practice discrimination. Alienation also involves a feeling of self-estrangement, which stems from the workers' inability to realize their full human potential in their work roles and lack of connections to others.

> **"**Don't sacrifice your life to work and ideals. The most important things in life are human relations. I found that out too late.**"**
>
> **Katharinde Susannah Prichard**
> **Australian author**

Work-Family Concerns

Spouses, parents, and adult children caring for elderly parents increasingly struggle to balance their work and family responsibilities. When Hochschild (1997) asked a sample of employed parents, "Overall, how well do you feel you can balance the demands of your work and family?" only 9% said "very well" (pp. 199–200).

is not an option for everybody. But that's the way it is. There are just too many things pressing poor people down, keeping them where they are. . . . The poor have become "invisiblized" in our society. They're given very little mention in the news and entertainment media. You just don't hear about them. The media system is fed by corporate advertising, and advertisers want "good demographics"—that is, they want to reach mostly the upper middle-class. . . .

Passaro: Many editors claim the middle class isn't interested in reading about poverty or the working poor. So why did *Nickel and Dimed* grab the attention of the media and the middle-class people who are presumably reading it?

Ehrenreich: One reason is that *Nickel and Dimed* is very personal and subjective, not preachy. It's not about the poor in general. It's just about me trying to survive. So people who are completely unfamiliar with the world of low-wage work can see it through the eyes of someone who is somewhat like them . . .

Passaro: Do you think that we should boycott chain stores and restaurants that don't pay a living wage?

Ehrenreich: And then where are you going to shop or eat? At an upscale restaurant where the busboys and the dishwashers still earn little above minimum wage and the coffee beans have been picked by children in Central America? A lot of people come up to me and say, "I'll never go to Wal-Mart again." Well, terrific. So you go to a nice little boutique, which also pays its retail clerks seven dollars an hour and maybe gets its very expensive clothes from sweatshops, too. You could pay two hundred dollars for a dress that some poor seamstress made five dollars for sewing. These problems are so widespread, it's hard for me to see how boycotting a single business would help much.

That said, if a boycott were called on some particular business, and there were a focused campaign surrounding it, I would respect it. . . .

Passaro: I feel guilty wherever I shop.

Ehrenreich: There's no avoiding that guilt. What are you going to do? Weave your own cloth like Gandhi tried to do?

What we can do to help hardworking people trapped in poverty is fight for increasing social benefits, universal health insurance, and a universal child-care subsidy. We can demand that cities build affordable housing. . . . Another possibility would be to tell the courts to get serious about enforcing the law against firing people for union activity. That's the law, but it's not enforced. You could also join the living-wage movement, which is using whatever leverage it has to convince individual cities to raise wages. . . .

Passaro: Education is often seen as the best way to move people out of poverty, yet menial jobs are always going to exist. Does the nature of these jobs need to change?

Ehrenreich: I get a little annoyed when someone says, "What's wrong with these people? Why don't they get an education?" Well, great, but then who's going to take care of your elderly grandmother in the nursing home? Who's going to wait on you when you go to a restaurant or a discount store? These are important jobs, jobs that need to be done, jobs that take intelligence and concentration and sometimes a great deal of compassion. Why don't we just pay people decently for doing them?

Source: Passaro, Jamie, 2003 (January). "Fingers to the Bone: Barbara Ehrenreich on the Plight of the Working Poor," *The Sun*, pp. 4–10. Used by permission.

In nearly two-thirds (61%) of married couples with children under age 18 and in more than half (53%) of married couples with children under age 6, both parents are employed. In addition, 72% of women in female-headed single-parent households and 84% of men in male-headed single-parent households are employed (Bureau of Labor Statistics 2005a).

A major concern of employed parents is arranging and paying for child care. About 3.3 million children under age 13 (15% of 6- to 12-year-olds) are left without adult supervision for some period of time each week (Vandivere et al. 2003). In some two-parent households spouses or partners work different shifts so that one adult can be home with the children. One-third of U.S. working women work shifts different from their spouses or partners (AFL-CIO 2004). Working different shifts

Should Employers Have the Right to Monitor Their Employees' E-Mails from the Work Site?

In one survey three out of four workers admitted to sending personal e-mails from work (Werhane et al. 2004). The courts and legislatures have determined that employers have the right to monitor employees' e-mails that are sent or received on technology equipment (i.e., computers, servers) owned or managed by the employer. More than 50% of all companies monitor Internet use and e-mail of employees (Werhane et al. 2004).

It Is Your Turn Now to Take a Stand!

Do you think that employers should have the right to monitor employees' e-mails at work? Or is it an invasion of employees' privacy?

Figure 7.2

Annual hours worked per worker in 2002 in OECD countries.

Source: OECD (2004).

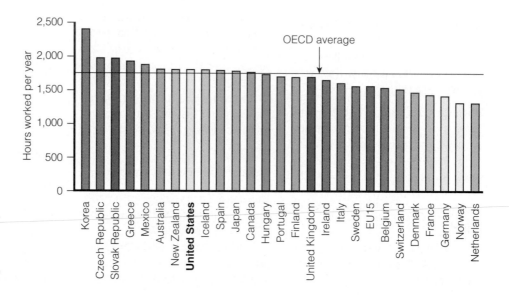

puts strain on marriage relationships, because the partners rarely have time off together. Another work-family concern involves caring for elderly family members. More than one-third (35%) of workers say that they have provided care for a relative or in-law age 65 or older in the past year (Bond et al. 2002).

As shown in Figure 7.2, Americans work more hours annually than workers in many other developed countries do. More time at work means less time to care for and be with one's family. One reason that the average U.S. worker puts in more hours on the job than the average European worker is that the typical European worker enjoys significantly more paid vacations each year (6 to 8 weeks) than the typical U.S. worker, who has an average of 16.6 paid vacation days (Galinsky et al. 2005). The typical European workweek also tends to be shorter (such as the 35-hour week in France). And there are possible cultural differences: U.S. workers tend to be more willing to work longer hours to achieve higher earnings, whereas European

Table 7.1

U.S. Unemployment Rates by Race and Hispanic Origin: 2004

Group	Unemployment Rate
Total	5.5
White	4.8
African American	10.4
Hispanic	7.0
Asian	4.0

Source: Mishel et al. (2005).

workers tend to be more willing to sacrifice potential earnings in order to have more leisure time (OECD 2004).

Unemployment and Underemployment

The International Labour Organization (2005b) reported that in 2004 an estimated 185 million people worldwide—6.1% of the labor force—were unemployed. Unemployment rates ranged from 3.3% in East Asia to 11.7% in the Middle East and North Africa. Nearly half of the jobless worldwide are young people ages 15–24 (United Nations 2005).

In 2000 the U.S. unemployment rate dipped to a 31-year low of 4% but rose following the events of September 11, 2001, and in 2004 the unemployment rate was 5.5% (Mishel et al. 2005). Rates of unemployment are higher among racial and ethnic minorities (see Table 7.1) and among those with lower levels of education (see Chapter 8).

The causes of unemployment are complex. Joblessness is linked to high population growth and depletion of natural resources (water, soil, forests, fish) as well as to policies that favor corporate profits over sustenance farming. Another cause of unemployment is increased global competition. In 2005 General Motors, the world's largest automaker, announced that it would cut 25,000 jobs, or about 23% of its workforce, largely because of increased competition in the automobile market. The corporate practice of laying off large numbers of employees is called **corporate downsizing.** Another cause of U.S. unemployment is **job exportation,** the relocation of jobs to other countries where products can be produced more cheaply. **Automation,** or the replacement of human labor with machinery and equipment, also contributes to unemployment.

Long-Term Unemployment. The **long-term unemployment rate** refers to the share of the unemployed who have been out of work for 27 weeks or more. In 2004, 20%, or 1 in 5, of the unemployed in the United States had been out of work for six months or more (Stettner & Allegretto 2005). Although less educated workers are most vulnerable to unemployment, from the 1990s to the 2000s the percentage of long-term jobless Americans who had completed at least a bachelor's degree increased from 11% to 17.6%, and the percentage of long-term jobless in white-collar jobs increased from 38.5% to 44.4%. This means that "higher levels of education

ThomsonNOW

Learn more about **Unemployment and Underemployment** by going through the Unemployment Map Exercise.

and white collar jobs are no longer providing insulation against severe joblessness" (Stettner & Allegretto 2005, p. 2). Women in today's economy are also more vulnerable to long-term joblessness. From the 1990s to the 2000s the percentage of long-term unemployed American women increased from 34.7% to 43%. "The increase in long-term unemployment among women has detrimental implications for children and families. . . . These problems are most dramatic for single mothers, who have neither spousal income nor welfare to fall back on" (Stettner & Allegretto 2005, p. 7).

Underemployment. Measures of **unemployment** in the United States consider an individual to be unemployed if he or she is currently without employment, is actively seeking employment, and is available for employment. Unemployment figures do not include "discouraged" workers, who have given up on finding a job and are no longer looking for employment. **Underemployment** is a broader term that includes unemployed workers as well as (1) those working part-time but who wish to work full-time ("involuntary" part-timers), (2) those who want to work but have been discouraged from searching by their lack of success ("discouraged" workers), and (3) others who are neither working nor seeking work but who indicate that they want and are available to work and have looked for employment in the last 12 months. The underemployment rate, which tends to be higher than the unemployment rate, was 9.6% in 2004 (Mishel et al. 2005).

Effects of Unemployment on Individuals, Families, and Societies. Unemployment causes serious difficulties for individuals and families. Workers who lose their jobs and receive unemployment insurance receive less than 40% of their prior wages (Stettner & Allegretto 2005). Typically, unemployment benefits are exhausted after six months. Long-term unemployment can have lasting effects, such as increased debt, diminished retirement and savings accounts (which are depleted to meet living expenses), and/or relocation from secure housing and communities to unfamiliar places in order to find a job. Unemployment is also a risk factor for substance abuse, poor health, and homelessness. In addition, unemployment, along with bleak job prospects, propel some individuals to turn to illegitimate, criminal sources of income, such as theft, drug dealing, and prostitution. In families, unemployment is a risk factor for child and spousal abuse and marital instability.

High numbers of unemployed adults create a drain on societies that provide support to those without jobs. The high numbers of young adults without jobs create a risk for crime, violence, and unrest (United Nations 2005). As discussed in Chapters 6 and 13, unemployed young adults are targeted for recruitment into terrorist groups.

Labor Unions and the Struggle for Workers' Rights

Labor unions originally developed to protect workers and represent them at negotiations between management and labor. Labor unions have played an important role in fighting for fair wages and benefits, healthy and safe work environments, and other forms of worker advocacy. In 2004 the median earnings of full-time wage and salary workers who were union members was $781, compared to a median of $612 for nonunion workers (Bureau of Labor Statistics 2005e). Unionized workers also received insurance and pension benefits worth more than those of nonunion employees, and union workers also get more paid time off than nonunion workers.

In the Justice for Janitors 2000 campaign, approximately 100,000 janitors who are members of the Service Employees International Union (SEIU) participated in strikes, rallies, and/or protests in cities across the country in an effort to improve employment conditions (Wright 2001). The janitors won increased wages, expanded health care benefits, and the restoration of full-time jobs. The wage raises will help janitors lift their families out of poverty, which was a major goal of the national campaign.

Labor unions are also influential in achieving better working conditions. For example, the United Food and Commercial Workers (UFCW), the country's largest union representing poultry-processing workers, was instrumental in the formation of an Occupational Safety and Health Administration (OSHA) rule that established a federal workplace "potty" policy governing when employees can use the bathroom while on the job. According to UFCW international president Doug H. Dority, "For years workers in food processing industries have had to suffer the indignity of being denied the right to go to the bathroom when needed, just to maintain ever-increasing assembly-line speeds" (*New OSHA Policy Relieves Employees,* 1998, p. 8). Dority claimed that "poultry processors often have no other choice than to relieve themselves where they stand on the assembly line because their floor boss will not let them leave their workstation" (p. 8). Now, OSHA mandates that employers must make toilet facilities available so that employees can use them when they need to.

The increasing number of women in labor unions, which nearly doubled from 20% to 39% between 1960 and 1998, has helped to strengthen labor unions' advocacy for women (*Labor's "Female Friendly" Agenda,* 1998). For example, a number of unions have been successful in bargaining for expanded family leave benefits, subsidized child care, elder care, and pay equity.

Despite the successes of labor unions' efforts to improve wages, benefits, and working conditions, the strength and membership of unions in the United States have declined over the last several decades. **Union density**—the percentage of workers who belong to unions—grew in the 1930s and peaked in the 1940s and 1950s, when 35% of U.S. workers were unionized. In the 1960s and 1970s U.S. corporations mounted an offensive attack on labor unions, "aiming to tame them or maim them" (Gordon 1996, p. 207). Corporations hired management consultants to help them develop and implement antiunion campaigns. They threatened unions with decertification, fired union leaders and organizers, and threatened to relocate their plants unless the unions and their members "behaved."

> One management consultant firm . . . was unusually blunt in broadcasting its methods. A late-1970s blurb promoting its manual promised: "We will show you how to screw your employees (before they screw you)—how to keep them smiling on low pay—how to maneuver them into low-pay jobs they are afraid to walk away from." (Gordon 1996, p. 208)

In 2004 the percentage of American workers belonging to unions had fallen to 12.5%, its lowest point in decades (Bureau of Labor Statistics 2005e). One reason for the decline in union representation is the loss of manufacturing jobs, which tend to have higher rates of unionization than other industries. Job growth has occurred in high technology and financial services, where unions have little presence. In addition, globalization has led to layoffs and plant closings at many unionized work sites as a result of companies moving to other countries to find cheaper labor.

Annual Survey of Violations
of Trade Union Rights

The International Confederation of Free
Trade Unions (ICFTU) represents more than
151 million workers in 233 affiliated organizations in 154 countries and territories.
Each year, the ICFTU publishes a survey that
describes severe abuses of workers' rights
in countries around the world.

Methods and Sample
The *2004 Annual Survey of Violations of
Trade Union Rights* (International Confederation of Free Trade Unions 2005) is based on
secondary data collected during 2003 from
134 countries. As explained in Chapter 1,
secondary data are data that have already
been collected by other researchers or government agencies or that exist in forms
such as police reports and other sources.

   Data for the annual survey are collected
through a number of sources. The principal
source of information for the survey is the
ICFTU's 233 affiliated organizations (national
trade union confederations) in 154 countries
and territories around the world. The ICFTU
also receives information from other trade
union sources, notably the Global Union

Federations (representing the different sectors and industries), trade union research
bodies, and the trade union press. Another
source of information is the International
Labour Organization, notably the reports of
its key supervisory bodies, as well as
friendly nongovernmental organizations
(NGOs) (e.g., human rights NGOs). In addition, the annual survey contains data from
international media press reports on labor
issues that are published by the newswires
(Reuters, etc.). The ICFTU also gathers data
using the Internet, particularly the Labour
Start website, which has countless links to
labor-related press reports. Finally, the *Annual Survey of Violations of Trade Union
Rights* includes data from government
sources, such as the U.S. State Department.
The ICFTU verifies the accuracy of their data
by sending their reports to their regional organizations and officers and to the various
trade unions in the countries concerned for
checking. Next, we present selected findings from the *2004 Annual Survey of Violations of Trade Union Rights* (International
Confederation of Free Trade Unions 2005).

Findings and Conclusions
In regions throughout the world workers
who form or join trade unions present collective demands or go on strike and thus
risk dismissal, arrest, physical intimidation,
death threats, and even assassination. In
2003, 129 trade unionists were murdered.
Colombia remains the most dangerous
place in the world to be a trade unionist,
with 90 people killed in 2003 for their trade
union activity. In some cases the families of
the unionist murder victims were also killed.
Most reported cases of trade unionist assassinations (95%) are not properly investigated, and the murderers are not caught or
punished.

   Trade union rights are violated by both
employers and governments. Governments
in numerous countries hamper trade union
activity or strike action and fail to enforce
existing national and international laws that
protect workers' rights. Eager to secure
financial benefits from participating in the
global market, governments see trade
unions as an obstacle to their economic
development. For example, in Uganda,

Although the 1935 National Labor Relations Act guarantees the right to unionize, bargain collectively, and to strike, this legislation does not cover agricultural or
domestic workers, supervisors, or independent contractors. As a result, 25 million
private civilian workers, as well as 40% of government employees, do not have the
right under U.S. law to negotiate their wages, hours, or employment terms (International Confederation of Free Trade Unions 2005). And workers who do have the
right to unionize and to strike risk their jobs by doing so. In the 1990s more than
20,000 U.S. workers each year were fired or discriminated against because of their
union-related activities (Human Rights Watch 2001). In the United States at least
1 in 10 union supporters campaigning to form a union is illegally fired. For every
30 people who vote for a union in elections in any one year, 1 will be illegally fired.
These workers can seek reinstatement and back pay by reporting their case to the
National Labor Relations Board (NLRB).

President Museveni publicly admitted to the mass dismissal of striking textile workers because their "action would scare off investors." Equally, many employers also resist union organizing and intimidate or fire workers who take collective action. In Venezuela 19,000 oil workers were fired for participating in a strike, serving as a warning to other Venezuelan workers.

In Africa documented violations of trade union rights include 25 murdered workers, 218 injured, beaten, or tortured workers, 903 arrested workers, and 6,556 workers dismissed from their jobs. The most common form of repression against workers in Asia and the Pacific is job dismissal. In 2003, 353,128 workers were laid off in the Asian and Pacific countries (mostly in India) for defending their trade union rights. China, Vietnam, and Burma (the country renamed Myanmar by its military junta) prohibit any independent association of workers. Three representatives of the underground Federation of Trade Unions–Burma (FTUB) were sentenced to death for their labor activities. Chinese authorities continue to suppress trade union activity, and send people to prison for their trade union activities.

In Europe the worst violations of trade union rights have occurred outside the European Union. In Belarus anyone who revolts or protests is liable to imprisonment, and in Bosnia and Herzegovina gangs of thugs have harassed and threatened trade union leaders.

In Canada public sector workers continue to be denied their trade union rights. Violations of labor rights also occur in the United States. For example, Wal-Mart has consistently stated that it will not bargain with any union and has repeatedly taken drastic steps to prevent workers from organizing in stores across North America. During organizing campaigns in U.S. workplaces, about 75% of employers hire consultants and security firms to run antiunion campaigns aimed at suppressing the freedom of workers to form unions and bargain collectively. More than 90% of employers, when faced with employees who want to join together in a union, force employees to attend closed-door meetings to hear antiunion propaganda. Sometimes, employers threaten to move the plant during antiunion campaigns, and employers often challenge the results of union elections, which can delay union representation and contract negotiations for months, if not years. Last, migrant workers in the Gulf states have no trade union rights whatsoever.

The annual survey of trade union violations is frightening documentation of the lack of workers' rights around the world and of the abuses of governments and employers in the continued suppression of workers' rights. Looking over the results of the survey in the last few years, we can conclude that between 100 and several hundred trade unionists are killed each year. Several thousands more are imprisoned, beaten in demonstrations, tortured by security forces or others, and often sentenced to long prison terms. And each year hundreds of thousands of workers lose their jobs merely for attempting to organize a trade union. The survey is a wake-up call to the global community to prioritize the protection of workers' rights.

One study found that in more than half of all union-organizing drives, employers threatened to close the plant (Mokhiber & Weissman 1998). Where union-organizing drives are successful, employers carry out the threat and close the plant, in whole or in part, 15% of the time. In today's climate of expanding trade agreements and skyrocketing levels of corporate migration, plant-closing threats continue to be one of the most powerful antiunion strategies.

Labor Union Struggles Around the World. International norms established by the United Nations and the International Labour Organization declare the rights of workers to organize, negotiate with management, and strike (Human Rights Watch 2000). In European countries labor unions are generally strong. However, in many less-developed countries and in countries undergoing economic transition, workers and labor unions struggle to have a voice in matters of wages and working conditions. A survey of 134 countries by the International Confederation of Free Trade

Unions (2005) found that both corporations and governments repressed union efforts. Findings of this survey are presented in this chapter's *Social Problems Research Up Close* feature.

Strategies for Action: Responses to Workers' Concerns

Government, private business, human rights organizations, labor organizations, college student activists, and consumers play important roles in responding to the concerns of workers. Next we look at responses to slavery and sweatshop labor, health and safety concerns, work-family policies and programs, workforce development programs, efforts to strengthen labor, and challenges to corporate power and economic globalization.

Efforts to End Slavery

More than 50 years ago the United Nations stated in Article 4 of its Universal Declaration of Human Rights that "no one shall be held in slavery or servitude; slavery and the slave trade shall be prohibited in all their forms." Yet slavery persists throughout the world. The international community has drafted treaties on slavery, but many countries have yet to ratify and implement the different treaties.

One strategy to fight slavery is punishment. Slave traffickers often avoid punishment because, as a former official of the U.S. Agency for International Development explained, "government officials in dozens of countries assist, overlook, or actively collude with traffickers" (quoted by Cockburn 2003, p. 16). In many countries the justice system is more likely to jail or expel sex slaves than to punish traffickers ("Sex Trade Enslaves Millions of Women, Youth," 2003). In 25 countries, however, slave trafficking is actively prosecuted and treated as a serious crime.

In the United States the Victims of Trafficking and Violence Protection Act, passed by Congress in 2000, protects slaves against deportation if they testify against their former owners. Convicted slave traffickers in the United States are subject to prison sentences, as shown in the following examples (Cockburn 2003):

- Louisa Satia and Kevin Waton Nanji each received nine years for luring a 14-year-old girl from Cameroon with promises of schooling and then isolating her in their Maryland home, raping her, and forcing her to work as their domestic servant for three years.
- Sardar and Nadira Gasanov were sentenced to five years each for recruiting women from Uzbekistan with promises of jobs, taking their passports, and forcing them to work in strip clubs and bars in Texas.
- Juan, Ramiro, and Jose Ramos each received 10 to 12 years for transporting Mexicans to Florida and forcing them to work as fruit pickers.

U.S. corporations are also being held accountable for enterprises that involve forced labor and other human rights and labor violations. In 2003 the Unocal oil company became the first corporation in history to stand trial in the United States for human rights violations abroad (George 2003). Unocal was accused of involvement in a pipeline project that used Myanmar (formerly Burma) military personnel to provide "security" for a natural gas pipeline project in the remote Yadana region near the Thai border. According to the Ninth U.S. Circuit Court of Appeals, the soldiers' true role was to force villagers in the pipeline region to work without pay—a modern form of slavery. The military also forced villagers living along the pipeline route to relocate without compensation, raped and assaulted villagers, and imprisoned and/or executed those who opposed them. In 2005 Unocal announced that it had reached a settlement with the parties who alleged that Unocal was complicit in the human rights violations committed by the Myanmar military.

Responses to Sweatshop Labor

The Fair Labor Association (FLA) is a coalition of companies, universities, and nongovernmental organizations that works to promote adherence to international labor standards and improve working conditions worldwide. In 2005, 16 apparel and footwear companies and more than 1,900 collegiate licensees (companies that manufacture logo-carrying goods for colleges and universities) voluntarily participated in FLA's monitoring system, which inspects their overseas factories and requires them to meet minimum labor standards, such as not requiring workers to work more than 60 hours a week. In its first few years of operation the FLA was criticized for allowing firms to select and directly pay their own monitors and to have a say in which factories were audited. In 2002 the FLA responded to these criticisms by taking much more control over external monitoring, with the FLA staff selecting factories for evaluation, choosing the monitoring organization, and requiring that inspections be unannounced (O'Rourke 2003). The FLA continues to be criticized, however, for having low standards in allowing below-poverty wages and excessive overtime and for requiring that only a small percentage of a manufacturer's supplier factories be inspected each year. Critics also suggest that companies use their participation in the FLA as a marketing tool. Once "certified" by the FLA, companies can sew a label into their products saying that the products were made under fair working conditions (Benjamin 1998).

Pressure from college students and other opponents of sweatshop labor and consumer boycotts of products made by sweatshop labor have resulted in some improvements in factories that make goods for companies such as Nike and Gap, which have cut back on child labor, use less dangerous chemicals, and require fewer employees to work 80-hour weeks (Greenhouse 2000). At many factories supervisors have stopped hitting employees, have improved ventilation, and have stopped requiring workers to obtain permission to use the toilet. But improvements are not widespread, and oppressive forms of labor continue throughout the world. According to the National Labor Committee, two areas where "progress seems to grind to a halt" are efforts to form unions and efforts to achieve wage increases (Greenhouse 2000).

Responses to Worker Health and Safety Concerns

Over the last few decades health and safety conditions in the U.S. workplace have improved as a result of media attention, demands by unions for change, more white-collar jobs, and regulations by OSHA. Through OSHA the government develops, monitors, and enforces health and safety regulations in the workplace. Since OSHA was created three decades ago, workplace fatalities have dropped by 75% ("Editorial," 2000). But much work remains to be done to improve worker safety and health. Inadequate funding leaves OSHA unable to do its job effectively, with only 2,138 federal and state inspectors responsible for monitoring and enforcing job safety laws at 8 million workplaces (AFL-CIO 2005). At current staffing levels it would take federal OSHA employees 108 years to inspect each workplace under its jurisdiction just once.

Because "the task of monitoring and enforcement simply cannot be effectively carried out by a government administrative agency," Kenworthy (1995) suggested that the United States follow the example of many other industrialized countries: Turn over the bulk of responsibility for health and safety monitoring to the workforce (p. 114). Worker health and safety committees are a standard feature of companies in many other industrialized countries and are mandatory in most of Europe. These committees are authorized to inspect workplaces and cite employers for violations of health and safety regulations.

In developing countries governments fear that strict enforcement of workplace regulations will discourage foreign investment ("Editorial,"2000). Investment in workplace safety in developing countries, whether by domestic firms or foreign multinationals, is far below that in the rich countries. Unless global standards of worker safety are implemented and enforced in *all* countries, millions of workers throughout the world will continue to suffer under hazardous work conditions. Low unionization rates and workers' fears of losing their jobs—or their lives—if they demand health and safety protections leave most workers powerless to improve their working conditions.

Business and industry often fight efforts to improve safety and health conditions in the workplace. For example, in 1999, after a 10-year struggle between labor and business, OSHA issued ergonomic standards requiring employers to implement ergonomic programs in jobs where musculoskeletal disorders occur. **Ergonomics** refers to the designing or redesigning of the workplace to prevent and reduce cumulative trauma disorders. According to OSHA, the new ergonomic standards would prevent 4.6 million workers over the next 10 years from experiencing painful, potentially debilitating work-related musculoskeletal disorders (U.S. Department of Labor 2001). However, business and industry representatives pressured Congress and President George W. Bush to repeal the ergonomic standard soon after Bush took office in 2000. Now, there are no regulations requiring employers to assess ergonomic hazards in the workplace (such as excessive repetition or poor workstation design) or to take steps to reduce these hazards.

Behavior-Based Safety Programs. A controversial health and safety strategy used by business management is behavior-based safety programs. Instead of examining how work processes and conditions compromise health and safety on the job, **behavior-based safety programs** direct attention to workers themselves as the problem. Behavior-based safety programs claim that 80–96% of job injuries and illnesses are caused by workers' own carelessness and unsafe acts (Frederick & Lessin 2000). These programs focus on teaching employees and managers to identify, "discipline," and change unsafe worker behaviors that cause accidents and encourage a work culture that recognizes and rewards safe behaviors.

Critics contend that behavior-based safety programs divert attention away from the employer's failure to provide safe working conditions. They also say that the real goal of behavior-based safety programs is to discourage workers from reporting illness and injuries. Workers whose employers have implemented behavior-based safety programs describe an atmosphere of fear in the workplace, such that workers are reluctant to report injuries and illnesses for fear of being labeled an "unsafe worker." At one factory that had implemented a behavior-based safety program,

Table 7.2

Percentage of U.S. Women Age 16 and Older in the Labor Force: 1970–2004

Year	Percentage of Women in Labor Force
1970	43.3
1980	51.5
1990	57.5
2004	59.2

Source: Bureau of Labor Statistics (2005f).

when a union representative asked workers during shift meetings to raise their hands if they were afraid to report injuries, about half of 150 workers raised their hands (Frederick & Lessin 2000). Worried that some workers feared even raising their hand in response to the question, the union representative asked a subsequent group to write yes on a piece of paper if they were afraid to report injuries. Seventy percent indicated they were afraid to report injuries. Asked why they would not report injuries, workers said, "We know that we will face an inquisition," "We would be humiliated," and "We might be blamed for the injury."

A new rule issued by OSHA protects workers by prohibiting discrimination against an employee for reporting a work-related fatality, injury, or illness ("Workers at Risk," 2003). This rule also prohibits discrimination against an employee for filing a safety and health complaint or asking for health and safety records.

Work-Family Policies and Programs

The increase in women in the workforce over the last several decades (see Table 7.2) has been accompanied by an increase in government and company policies designed to help women and men balance their work and family roles. Such policies are referred to by a number of terms, including "work-family," "work-life," and "family-friendly" policies.

Federal and State Family and Medical Leave Initiatives. In 1993 President Clinton signed into law the **Family and Medical Leave Act** (FMLA), which requires all companies with 50 or more employees to provide eligible workers (who work at least 25 hours a week and have been working for at least a year) with up to 12 weeks of job-protected, *unpaid* leave so that they can care for a seriously ill child, spouse, or parent; stay home to care for their newborn, newly adopted, or newly placed child; or take time off when they are seriously ill. Yet 40% of the workforce is not covered by the FMLA (Watkins 2002). Lower-wage earners are the least likely to have family and medical leave benefits and typically have few if any resources to fall back on in times of family illness or crisis.

In 2002 California became the first state in the country to adopt a comprehensive family leave policy that provides workers up to six weeks of time off with about 55% of their regular pay while caring for a newborn or newly adopted child, or when a family member is seriously ill (Watkins 2002). A number of other states have adopted state family leave policies.

In 2004 the Healthy Families Act was introduced into Congress. This act would require all employers with at least 15 employees to provide 7 days of paid sick leave

> "Our leaders talk as though they value families, but act as though families were a last priority."
>
> **Sylvia Hewlett and Cornel West**
> Family advocates

annually for full-time employees who work at least 30 hours a week (or 1,500 hours a year). One study found that, if passed, the Healthy Families Act would save employers money, largely because of reduced employee turnover (Lovell 2005).

The proposed federal Family Leave Insurance bill, introduced in Congress in 2004 and 2005, would provide partial wage replacement ($250 a week) and job protection for eligible workers caring for (1) a newborn, newly adopted, or newly placed foster child or (2) a seriously ill child, spouse, parent, or parent-in-law. If passed, this policy would be funded by a payroll deduction of 2 cents per hour worked.

Employer-Provided Work-Family Policies. Aside from government-mandated work-family policies, some corporations and employers have "family-friendly" work policies and programs, including unpaid or paid family and medical leave, child-care assistance (e.g., plans that allow employees to pay for child care with pretax dollars), assistance with elderly parent care, and flexible work options.

Offering employees more flexibility in their work hours helps parents balance their work and family demands. Flexible work arrangements, which benefit child-free workers as well as employed parents, include flextime, job sharing, a compressed workweek, and teleworking. **Flextime** allows the employee to begin and end the workday at different times so long as 40 hours per week are maintained. Although 8 in 10 working women say control over their work hours is an important benefit, most (64%) do not have control over their work hours (AFL-CIO 2004). A **compressed workweek** allows employees to condense their work into fewer days (e.g., four 10-hour days each week). With **job sharing** two workers share the responsibility of one job. **Telework** allows employees to work part- or full-time at home or at a satellite office (see this chapter's *Focus on Technology* feature). A study of U.S. companies found that the more women and minorities a company has in managerial positions, the more likely that company is to offer flexible work options (Galinsky & Bond 1998).

Barriers to Employees' Use of Work-Family Benefits. Even when work-family policies and programs are available, many employees do not take advantage of them. One barrier to using work-family benefits is lack of awareness: Many employees do not know what benefits are available to them. In a survey of employees covered by the FMLA, only 38% correctly reported that the FMLA applied to them and about one-half said they did not know whether it did (Cantor et al. 2001). In a study of employees in seven organizations, more than two-thirds of employees were unaware of or mistaken about at least one work-life policy or practice. These employees believed that a policy was in operation when in fact it was not or that a policy that did formally exist did not (Still & Strang 2003).

Other barriers that discourage employee use of work-family benefits are related to economic and job security. Many workers who are covered by the FMLA do not use it because they cannot afford to take leave without pay. One survey found that 88% of workers who needed time off but did not take it said they would have taken leave if they could have received pay during their absence; nearly one-third of workers who needed leave but did not take it cited worry about losing their job as a reason for not taking leave (Cantor et al. 2001). Professor Phyllis Moen (2003) explained, "Many are afraid to take advantage of existing [work-family] options because doing so might signal a less-than-full commitment to their jobs, exacting costs in future promotions, salary increases, and even job security" (p. 335).

Corporate initiatives in work-family benefits are a step in the right direction. However, companies that innovate in the area of work life by adopting many family-friendly programs do not show particularly high levels of benefit use among their employees (Still & Strang 2003). Although several magazines and professional organizations give awards and recognition to companies that offer family-friendly benefits, Moen (2003) suggested that "a better gauge of the worker friendliness of corporations might be the proportion of their workforces actually using them" (p. 335). In addition, "corporate leaders, along with those handing out rewards, need to recognize that job security is also a key component of friendliness or work-life quality" (Moen 2003, p. 335).

Workforce Development and Job-Creation Programs

The International Labour Organization (2003) estimated that at least 1 billion new jobs are needed in the coming decade to meet the United Nations goal of halving extreme poverty by 2015. Developing a workforce and creating jobs involves far-reaching efforts, including those designed to improve health and health care, alleviate poverty and malnutrition, develop infrastructures, and provide universal education.

In the United States workforce development programs have provided a variety of services, including assessment to evaluate skills and needs, career counseling, job search assistance, basic education, occupational training (classroom and on-the-job), public employment, job placement, and stipends or other support services for child care and transportation assistance (Levitan et al. 1998). Workforce development programs primarily assist youths, the handicapped, welfare recipients, displaced workers, the elderly, farmworkers, Native Americans, and veterans. Numerous studies have looked at the effectiveness of workforce development programs. In general, "evaluations indicate that employment and training programs enhance the earnings and employment of participants, although the effects vary by service population, are often modest because of brief training durations and the inherent difficulty of alleviating long-term deficiencies, and are not always cost effective" (Levitan et al. 1998, p. 199).

Efforts to prepare high school students for work include the establishment of technical and vocational high schools and high school programs and school-to-work programs. School-to-work programs involve partnerships between business, labor, government, education, and community organizations that help prepare high school students for jobs (Leonard 1996). Although school-to-work programs vary, in general, they allow high school students to explore different careers, and they provide job skill training and work-based learning experiences, often with pay (Bassi & Ludwig 2000).

In one strategy for creating jobs, local, state, and federal governments provide benefits to corporations in the form of subsidies, tax breaks, real estate, and low-interest loans with the hope that this "corporate welfare" will result in new jobs (see also Chapter 6). However, most recent job creation in the United States is with small and medium-size companies. Although Fortune 500 companies are the biggest beneficiaries of corporate welfare, they have eliminated more jobs than they have created in the past decade (Barlett & Steele 1998). In addition, many of the jobs that are created are part-time or temporary jobs.

With the new limits on welfare (see Chapter 6), more adults with low levels of education and job training are entering the workforce. The cuts in welfare benefits exacerbate the need to provide not only workforce development programs but also jobs that pay a living wage. As long as our economy allows people who work full-time to earn poverty-level wages, having a job is not necessarily the answer to

Telework: The New Workplace of the 21st Century

The ever-widening use of modern technology in the workplace, such as computers, the Internet, e-mail, fax machines, copiers, mobile phones, and personal digital assistants, makes it possible for many workers to perform their jobs at a variety of locations. The term **telework** (also known as **telecommuting**) refers to flexible and alternative work arrangements that involve use of information technology. There are four types of telework (Pratt 2000): (1) home-based telework; (2) satellite offices where all employees telework for one employer; (3) telework centers, which are occupied by employees from more than one organization; and (4) mobile workers. Most teleworkers (89%) are home based (Bowles 2000). Some people telework full-time, but a larger number telework one or two days a week. In 2003 nearly one-fifth (18.3%) of the U.S. workforce worked at home during business hours at least one day per month (International Telework Association and Council 2004).

Benefits of Telework for Employers

Telework provides several benefits for the employer.

- *Telework attracts and helps retain employees.* Companies regard telework and other flexible work arrangements as important in recruiting and maintaining good employees. Telework can also lower turnover and thus save companies expenses associated with hiring and training replacement employees. A 1997 AT&T survey of telecommuters showed that 36% of employees would quit or find another work-at-home job if their employer decided they could not work at home (cited by Lovelace 2000). However, Bowles (2000) reported that some companies express dissatisfaction with telework "because they believe that it causes resentment among office-bound colleagues and weakens corporate loyalty" (p. 2).

- *Telework reduces costs.* Companies can save money by eliminating offices that teleworkers do not need, consolidating other work spaces, and reducing related overhead costs.

- *Telework increases worker productivity.* Several studies of managers and employees at large companies concluded that telework increases worker productivity (Lovelace 2000).

Benefits of Telework for Employees

Telework is valued by employees for several reasons.

- *Telework increases job and life satisfaction.* Studies have shown that employee satisfaction among teleworkers is higher than for their nonteleworking counterparts (Lovelace 2000). Much of the job satisfaction among telecommuters is related to the job flexibility that enables them to balance work and family demands.

- *Telework helps balance work-family demands.* Telework can provide flexibility to working parents and adults caring for aging parents, thus reducing role conflict and strengthening family life. One father of three children described his being home when his children came back from school as being "the most significant impact" of his telecommuting (Riley et al. 2000, p. 5). He also took time during the day to take his children to school, to the doctor's office, and to run errands. However, one national study of children whose parents work at home found that older children (grades 7 to 12) were more likely to agree that "my father does not have the energy to do things with me because of his job" and "my father has not been in a good mood with me because of his job" than the children of fathers who work in an office (Galinsky & Kim 2000). The effects of telework on parent-child relationships seems to depend, then, on how each parent interacts with his or her children.

economic self-sufficiency. As David Shipler (2005) explained in his best-selling book *The Working Poor*:

> A job alone is not enough. Medical insurance alone is not enough. Good housing alone is not enough. Reliable transportation, careful family budgeting, effective parenting, effective schooling are not enough when each is achieved in isolation from the rest. There is no single variable that can be altered to help working people move away from the edge of poverty. Only where the full array of factors is attacked can America fulfill its promise. (p. 11)

- *Telework expands work opportunities for Americans outside the economic mainstream.* Telework may expand job opportunities for rural job seekers who lack local employment opportunities and for low-income urban job seekers who lack access to suburban jobs (Kukreja & Neely 2000). Telework can also bring work opportunities to individuals with disabilities. Some of the technologies that have been developed for individuals with serious disabilities include Eye Gaze (a communication system that allows people to operate a computer with their eyes), the Magic Wand Keyboard (a computer keyboard for people with limited or no hand movement), and Switched Adapted Mouse and Trackball (a computer mouse that allows clicking the mouse with parts of the body other than the hand) (Bowles 2000).

- *Telework avoids the commute.* For many teleworkers the primary motivation for working from home is to reduce or eliminate the long and stressful commute that so many Americans now endure. In one AT&T unit the average teleworker gained nearly five weeks per year by eliminating a 50-minute daily commute (Lovelace 2000).

Environmental Benefits of Telework

Telework can reduce pollution by reducing the need for transportation to the workplace, thus reducing the pollution associated with vehicle emissions. The National Environmental Policy Institute says that "telecommuting presents a non-coercive way for corporations to help the nation achieve environmental goals and improve quality of life" (quoted by Lovelace 2000, p. 3).

Concerns About Telework

With all the benefits of telework for both employers and employees, there are still some concerns.

- *Blurred boundaries between home and work.* People who work at home may find themselves on call around the clock, responding to e-mail, pagers, faxes, and voice mail. Without clear boundaries between home and work, teleworkers may feel that they are unable to escape the work environment and mind-set (Pratt 2000). Questions about overtime pay may arise when work spills over into personal time. Having a separate office within the home and a routine work schedule may help to create the psychological boundary between work and family and leisure. But for some teleworkers, learning to "log off" is a challenge.

- *Zoning regulations.* Teleworkers who work at home full-time must contend with zoning regulations that may prohibit residents from having an "office" in their home.

- *Losing benefits as a contract employee.* Some employers attempt to convert the teleworker into a contract worker. This type of worker lacks job protections and benefits (Bowles 2000).

- *Social isolation.* Does telework lead to social isolation for those who live and work at home? Evidence suggests that teleworkers are able to maintain personal relationships with co-workers and are included in office networks. However, for rural and disabled individuals telework may contribute to social isolation.

- *Exclusion of the disenfranchised.* Lower socioeconomic groups are less likely than more affluent populations to have access to and skills in the Internet and other modern forms of information technology. Bowles (2000) suggested that "the eventual success of telework programs in the future must . . . account for the masses of people left behind. . . . All must be included in the new economy; it is not a luxury, but a must" (p. 9).

Efforts to Strengthen Labor

Although efforts to strengthen labor are viewed as problematic to corporations, employers, and some governments, such efforts have the potential to remedy many of the problems facing workers. In an effort to strengthen their power, some labor unions have merged with one another. Labor union mergers result in higher membership numbers, thereby increasing the unions' financial resources, which are needed to recruit new members and to withstand long strikes. Because workers must fight for labor protections within a globalized economic system, their unions must

College student groups across the country have participated in boycotts against Coca-Cola in protest of the violence against union leaders at Colombian Coca-Cola plants.

cross national boundaries to build international cooperation and solidarity. Otherwise, employers can play working and poor people in different countries against each other. An example of international union cooperation occurred when leaders from 21 unions in 11 countries on 5 continents resolved to form a global union network at International Paper Company (IP), the largest paper company in the world. One union leader remarked, "IP crosses national borders in search of the highest profits, and the unions . . . have resolved to match that corporate globalization with a globalization of workers' solidarity" ("Unions Forge Global Network," 2002).

Strengthening labor unions requires combating the threats and violence against workers who attempt to organize or who join unions. One way to do this is to pressure governments to apprehend and punish the perpetrators of such violence. Although about 3,500 trade unionists were murdered between 1990 and 2005, only 600 cases were investigated, resulting in just 6 convictions (Moloney 2005). Another tactic is to stop doing business with countries where government-sponsored violations of free-trade union rights occur.

In the United States the NLRB and the courts play an important role in upholding workers' rights to unionize and sanctioning employers who violate these rights. The NLRB has the authority to issue job reinstatement and "back-pay" orders or other remedial orders to workers wrongfully fired or demoted for participating in union-related activities. The NLRB and the courts have held that employer threats to close the plant if the union succeeds in organizing can be unlawful under certain circumstances (Bronfenbrenner 2000). For example, *Guardian Industries Corp.* v. *NLRB* held that it was unlawful for a supervisor to say to an employee, "If we got a union in there, we'd be in the unemployment line." However, under the employer free-speech provisions of the Taft-Hartley Act, the courts have permitted the employer to predict a plant closing in situations where it is based on an objective assessment of the economic consequences of unionization.

Proposed legislation called the Employee Free Choice Act would allow employees to choose freely whether to form unions by signing cards authorizing union representation. The act would also provide mediation and arbitration for first-contract disputes and would establish stronger penalties for violation of employee rights when workers seek to form a union and during first-contract negotiations (International Confederation of Free Trade Unions 2005). At the time of this writing, this legislation is before Congress.

After numerous unsuccessful attempts to unionize workers at Wal-Mart, labor unions found a creative way to advocate for workers' rights at the chain—they helped form a new and unusual type of workers' association aimed at improving working conditions at Wal-Mart Stores Inc. The new group, called the Wal-Mart Workers Association, has urged the state of Florida to grant unemployment benefits to workers whose hours have been cut by Wal-Mart. The group is also pressing Wal-Mart to improve its wages and benefits (Greenhouse 2005).

Challenges to Corporate Power and Globalization

Challenges to corporate power and globalization include campaign finance reform and the antiglobalization movement.

Campaign Finance Reform. In the United States advocates for campaign finance reform have challenged the power that corporations have in influencing laws and policies. Campaign finance reform efforts were rewarded when Congress passed the McCain-Feingold bill known as the Bipartisan Campaign Reform Act of 2002. This law helps to remove the corrupting influence of soft money from federal elections so that corporations, labor unions, and wealthy donors will no longer be able to buy political influence and access. The law also prohibits corporations and unions from funding broadcast ads, run shortly before elections, that are designed to influence the election or defeat of candidates. Such ads may be funded by individual contributions, but broadcast stations must keep a public record of political ads and who paid for them.

The Antiglobalization Movement. Challenges to corporate globalization have also taken root in the United States and throughout the world. Antiglobalization activists have targeted the World Trade Organization, the International Monetary Fund, and the World Bank as forces that advance corporate-led globalization at the expense of social goals such as justice, community, national sovereignty, cultural diversity, ecological sustainability, and workers' rights.

In 1999, 50,000 street protesters and third world delegates demonstrated in Seattle in opposition to the policies of the World Trade Organization that promoted corporate-led globalization. The brutal assaults on largely peaceful demonstrators by Seattle police dressed in their Darth Vader—like uniforms in full view of television cameras has made the Seattle protest the "grand symbol of the crisis of globalization" (Bello 2001).

At the 2000 meeting of the International Monetary Fund (IMF) and the World Bank in Washington, D.C., 30,000 protesters descended on America's capital and found a large section of the northwest part of the city walled off by some 10,000 police. For four days the protesters tried, unsuccessfully, to break through the police barrier to reach the IMF-World Bank complex, resulting in hundreds of arrests.

In 2003 the fifth meeting of the World Trade Organization in Cancun, Mexico, collapsed when 21 developing nations walked out of the meeting after the United States and the European Union refused to concede on agricultural subsidies that hurt poorer nations. Two months later an estimated 20,000 union members, environmentalists, and religious and human rights activists from North, Central, and South America marched through the streets of Miami to protest the Free Trade Area of the Americas (FTAA). More recently, organized protests have occurred against

Opponents of free trade agreements such as CAFTA (U.S./Central America Free Trade Agreement) emphasize that "free trade" benefits corporations at the expense of workers and the environment.

UPI Photo/Roger L. Wollenberg/Landov

CAFTA—the Central American Free Trade Agreement. Media attention to these events contributes to the growing worldwide awareness of the forces of corporate globalization and its social, environmental, and economic effects.

Understanding Work and Unemployment

On December 10, 1948, the General Assembly of the United Nations adopted and proclaimed the Universal Declaration of Human Rights. Among the articles of that declaration are the following:

> "What the public wants is called 'politically unrealistic.' Translated into English, that means power and privilege are opposed to it."
>
> **Noam Chomsky**

Article 23. Everyone has the right to work, to free choice of employment, to just and favourable conditions of work and to protection against unemployment.

Everyone, without any discrimination, has the right to equal pay for equal work.

Everyone who works has the right to just and favourable remuneration ensuring for himself and his family an existence worthy of human dignity, and supplemented, if necessary, by other means of social protection.

Everyone has the right to form and to join trade unions for the protection of his interests.

Article 24. Everyone has the right to rest and leisure, including reasonable limitation of working hours and periodic holidays with pay.

More than half a century later, workers around the world are still fighting for these basic rights as proclaimed in the Universal Declaration of Human Rights.

To understand the social problems associated with work and unemployment, we must first recognize that corporatocracy—the ties between government and corporations—serves the interests of corporations over the needs of workers. We must also be aware of the roles that technological developments and post-industrialization play on what we produce, how we produce it, where we produce it, and who does the producing. With regard to what we produce, the United States is moving away from producing manufactured goods to producing services. With regard to production methods, the labor-intensive blue-collar assembly line is declining in importance, and information-intensive white-collar occupations are increasing. Because of increasing corporate multinationalization, U.S. jobs are being exported to foreign countries where labor and raw materials are cheap and regulations are lax. In developing countries investment in workplace safety is far below that in the rich nations.

Decisions made by U.S. corporations about what and where to invest influence the quantity and quality of jobs available in the United States. As conflict theorists argue, such investment decisions are motivated by profit, which is part of a capitalist system. Profit is also a driving factor in deciding how and when technological devices will be used to replace workers and increase productivity. If goods and services are produced too efficiently, however, workers are laid off and high unemployment results. When people have no money to buy products, sales slump, recession ensues, and social welfare programs are needed to support the unemployed. When the government increases spending to pay for its social programs, it expands the deficit and increases the national debt. Deficit spending and a large national debt make it difficult to recover from the recession, and the cycle continues.

What can be done to break the cycle? Those adhering to the classic view of capitalism argue for limited government intervention on the premise that business will regulate itself by means of an "invisible hand" or "market forces." For example, if corporations produce a desired product at a low price, people will buy it, which means workers will be hired to produce the product, and so on.

Ironically, those who support limited government intervention also sometimes advocate government intervention to bail out failed banks and to lend money to troubled businesses. Such government help benefits the powerful segments of our society. Yet when economic policies hurt less powerful groups, such as minorities, there has been a collective hesitance to support or provide social welfare programs. It is also ironic that such bail-out programs, which contradict the ideals of capitalism, are needed because of capitalism. For example, the profit motive leads to multinationalization, which leads to unemployment, which leads to the need for government programs. The answers are as complex as the problems.

Chapter Review

ThomsonNOW™

Reviewing is as easy as ① ② ③

1. Before you do your final review, take the ThomsonNOW diagnostic quiz to help you identify the areas on which you should concentrate. You will find information on ThomsonNOW and instructions on how to access all of its great resources on the foldout at the beginning of the text.

2. As you review, take advantage of ThomsonNOW's study videos and interactive Map the Stats exercises to help you master the chapter topics.

3. When you are finished with your review, take ThomsonNOW's posttest to confirm you are ready to move on to the next chapter.

• **The United States is described as a "post-industrialized" society. What does that mean?**

Post-industrialization refers to the shift from an industrial economy dominated by manufacturing jobs to an economy dominated by service-oriented, information-intensive occupations. The U.S. post-industrialized economy is characterized by a highly educated workforce, automated and computerized production methods, increased government involvement in economic issues, and a higher standard of living.

• **What are transnational corporations?**

Transnational corporations are corporations that have their home base in one country and branches, or affiliates, in other countries. Transnational corporations dominate the world economy today. In less than 20 years the number of transnational corporations has increased from 7 to more than 45,000, and the top 100 economies around the world are transnational corporations rather than nations.

• **What are the four principles of McDonaldization?**

The four principles of McDonaldization are (1) efficiency, (2) predictability, (3) calculability, and (4) control through technology.

- **According to data from the World Value Survey, how does the level of economic development of a country affect the subjective life satisfaction of its population?**

Data from the World Value Survey indicate that as one moves from subsistence-level economies in developing countries to advanced industrialized societies, there is a large increase in the percentage of the population who consider themselves happy or satisfied with their lives as a whole. But once a society moves from a starvation level to a reasonably comfortable existence, the increase in life satisfaction levels off.

- **Does slavery still exist today? If so, where?**

Forced labor, commonly known as slavery, exists today all over the world, including in the United States, but it is most prevalent in India, Pakistan, Bangladesh, and Nepal. Most forced laborers work in agriculture, mining, prostitution, and factories.

- **What is the most common cause of job-related fatality?**

The most common type of job-related fatality involves transportation accidents.

- **How does unionization benefit employees?**

Compared to nonunion workers, union workers have higher average wages, receive more insurance and pension benefits, and get more paid time off.

- **What is ergonomics?**

Ergonomics refers to the designing or redesigning of the workplace to prevent and reduce cumulative trauma disorders. After OSHA instituted ergonomic standards in the workplace to help prevent painful, potentially debilitating work-related musculoskeletal disorders, business and industry representatives pressured Congress and President George W. Bush to repeal the ergonomic standard soon after Bush took office in 2000.

- **What is the federal Family and Medical Leave Act?**

In 1993 President Clinton signed into law the Family and Medical Leave Act (FMLA), which requires all companies with 50 or more employees to provide eligible workers (who work at least 25 hours a week and have been working for at least a year) with up to 12 weeks of job-protected, *unpaid* leave so that they can care for a seriously ill child, spouse, or parent; stay home to care for their newborn, newly adopted, or newly placed child; or take time off when they are seriously ill.

Critical Thinking

1. Union membership is higher among black employees than among white employees. For example, in 2004, 15.1% of black workers belonged to a union, compared to 12.2% of white workers (Bureau of Labor Statistics

2005e). What might explain the higher rate of unionization among black workers?
2. Public approval of labor unions is higher among Democrats than among Republicans. A 2003 Gallup poll found that 74% of Democrats approved of unions, compared to only 41% of Republicans (Moore 2002). Why do you think this is so?
3. The economic health of a country is commonly measured by how much the country is producing (the total value of goods and services) and how much money consumers are spending on the purchase of goods and services. In what ways might high levels of production and consumption contribute to individual and social ills rather than to health and well-being?

Key Terms

alienation	job burnout
anomie	job exportation
automation	job sharing
behavior-based safety	labor unions
programs	long-term
brain drain	unemployment rate
capitalism	McDonaldization
chattel slavery	post-industrialization
compressed workweek	repetitive motion
corporate downsizing	disorders
corporatocracy	socialism
cumulative trauma disorders	soft money
economic institution	sweatshop
ergonomics	telecommuting
Family and Medical Leave Act	telework
flextime	transnational
forced labor	corporations
free trade agreements	underemployment
global economy	unemployment
industrialization	union density

Media Resources

The Companion Website for *Understanding Social Problems*, Fifth Edition

http://sociology.wadsworth.com/mooney_knox_schacht5e

Supplement your review of this chapter by going to the companion website to take one of the Tutorial Quizzes, use the flash cards to master key terms, and check out the many other study aids you'll find there. You'll also find special features such as *Wadsworth's Sociology Online Resources and Writing Companion,* GSS Data, and Census 2000 information, data, and resources at your fingertips to help you complete that special project or do some research on your own.

> "A memorable change must be made in the system of education, and knowledge must become so general as to raise the lower ranks of society nearer to the higher. The education of a nation, instead of being confined to a few schools and universities for the instruction of the few, must become the national care and expense for the formation of the many." *John Adams, second president of the United States*

Problems in Education

The Global Context: Cross-Cultural Variations in Education

Sociological Theories of Education

Who Succeeds? The Inequality of Educational Attainment

Problems in the American Educational System

Strategies for Action: Trends and Innovations in American Education

Understanding Problems in Education

Chapter Review

S ean, Dan, and Lance were eating lunch together in the school cafeteria when they decided to go outside. As they were walking up a hill, they saw two figures with guns in the distance. Must be Annihilation, Sean thought, a paintball game that seniors played. Odd. Those guns looked real—not like the plastic models he had seen before. "Pop, Pop, Pop." The guns were suddenly turned toward the school and before Sean knew it, he was the only one of his three friends standing. Sean turned to look for the paintball that had just grazed his neck. They must be frozen, he thought—he was bleeding—but as he turned he was shot three times in the abdomen. He began to run. Why am I running from paintballs, he thought, as he headed toward school. Then it hit—the bullet that really hurt—in the back, striking the spine, exiting through the hip.

Sean survived the attack at Columbine High School in Littleton, Colorado, as did Lance. But Dan Rohrbough did not, leaving others the grim task of trying to make some sense out of his death and the deaths of 12 others. Sean tries not to think about it, preferring to pretend it never happened, wanting to put it and his wheelchair behind him. But, embarrassingly, people keep staring at him. What are they looking at, he wonders. Then he thought, maybe that's how Dylan and Eric felt (Pollock 2000).

In March 2005, 16-year-old Jeff Weise walked into Red Lake High School in Red Lake, Minnesota, and, after killing nine people, committed suicide. He had been teased—no, "terrorized"—by his classmates for being "weird." A "loner" and a "Goth," like Dylan Klebold and Eric Harris, Jeff had worn a long black trench coat. He had been fascinated with death and violence and was filled with "intense anger" and "self-loathing" (Benson 2005). He fit, as one observer noted, "the profile of a Columbine shooter" (Haga et al. 2005).

Violence is just one of the many issues that must be addressed in today's schools (see this chapter's *Social Problems Research Up Close* feature). Students continue to graduate from high school unable to read, work simple math problems, or write grammatically correct sentences. Graduates discover that they are ill prepared for corporations that demand literate, articulate, informed employees. Teachers leave the profession because of uncontrollable discipline problems, inadequate pay, and overcrowded classrooms. Students and teachers alike are "dumbing down"—lowering their standards, expectations, and role performances to fit increasingly undemanding and unresponsive systems of learning.

And yet it is education that is often claimed as a panacea—the cure-all for poverty and prejudice, drugs and violence, war and hatred, and the like. Can one institution, riddled with problems, be a solution to other social problems? In this chapter we focus on this question and on what is being called an educational crisis. We begin with a look at education around the world.

The Global Context: Cross-Cultural Variations in Education

Looking only at the American educational system might lead one to conclude that most societies have developed some method of formal instruction for their members. After all, the United States has more than 98,000 schools, 4.2 million primary and secondary school teachers and college faculty, 4.8 million administrators and support staff, and nearly 75 million students (NCES 2004b; U.S. Census Bureau 2005). In reality, many societies have no formal mechanism for educating the masses.

Guns, Kids, and Schools

Reducing school violence is consistently listed as one of the top education priorities in public opinion polls and is one component of President Bush's educational reform package titled "No Child Left Behind." In an effort to curb violence, schools have established "zero-tolerance policies." For example, nationwide, more than 90% of schools have zero tolerance for students carrying firearms or other weapons to school. As a result, in part, the percentage of students carrying weapons to school has decreased from 12% to 6% over the last decade (NCES 2005b). Sociologist Pamela Roundtree (2000) addresses the issue of violence in schools by asking adolescents why they carry weapons.

Sample and Methods

The respondents in Roundtree's (2000) study were sixth- through twelfth-grade students who had participated in a state-sponsored research project, the Kentucky Youth Survey. Because weapon carrying is likely to vary by region, data from three distinct areas of the state were collected: (1) "Urban County" (from the wealthier, north-central part of Kentucky), (2) "Western County" (from the rural tobacco-growing area), and (3) "Eastern County" (from the poorer, high-unemployment mining region). Race and sex distributions varied between county samples, but whites and females were the majority in each of the three samples.

The dependent variable was possession of weapons at school, measured by whether a student reported carrying a weapon to school in the 30 days before the survey. In addition to the standard demographic variables of sex, race, age, and family socioeconomic background, variables thought to be predictive of carrying weapons were measured. The independent variables include (1) *fear-of-crime indicators* (prior victimization, fear of victimization), (2) *criminal-involvement indicators* (previous arrest, drug involvement), (3) *pro-weapon socialization indicators* (weapon ownership or use by respondent, weapon ownership by parent, weapon carrying by peers), and (4) *social-isolation indicators* (disattachment from school, church). Each category of indicators was predicted to be directly related to the likelihood of carrying a weapon to school; that is, as fear of crime, criminal involvement, pro-weapon socialization, and social isolation increase, the likelihood of carrying weapons increases.

Findings and Conclusion

Roundtree (2000) found that carrying a weapon to school was a relatively rare event, with 5% or less of students reporting that they had carried a weapon to school in the previous 30 days. Possession was slightly lower in Urban County than in Eastern or Western County, with the Urban County sample being the most industrialized and the wealthiest of the three.

In general, age, race, and sex were unrelated to the likelihood of a student taking a weapon to school. Only in Eastern County did sex significantly predict carrying a weapon, with males 700% more likely than females to possess a weapon in school. Surprisingly, prior victimization and fear of crime were unrelated to weapon carrying. However, drug dealing was predictive of weapon carrying in both Eastern and Western counties, and student drug use as an indicator of criminal involvement was significantly related to weapon possession in each of the three county samples.

Pro-weapon socialization had an even stronger effect than criminal involvement. Of the three measures of this variable—weapon ownership or use by respondent, ownership by parent, or carrying by peers—carrying by peers had the strongest relationship with carrying a weapon. With each "best friend" the respondent reported as having carried a gun to school, the likelihood of the respondent carrying a gun to school increased by 75–100%. Social isolation variables were not related to carrying a weapon in any consistent way.

Unlike many studies on adult weapon carrying, the results of Roundtree's (2000) study indicate that peer-based socialization has a much larger impact on the probability of carrying a weapon than fear-of-crime variables. However, consistent with research on adults, carrying a weapon to school was significantly related to criminal involvement and, specifically, to drug involvement.

There are 860 million illiterate adults around the world (see Figure 8.1), and 100 million children have little or no access to schools (UNESCO 2005).

A comparison of the United States to seven other developed nations (Canada, France, Germany, Italy, Japan, the Russian Commonwealth of Independent States, and the United Kingdom) reveals some interesting findings (NCES 2005a). First, in 2001 all "the countries, except the Russian Commonwealth, had close to universal

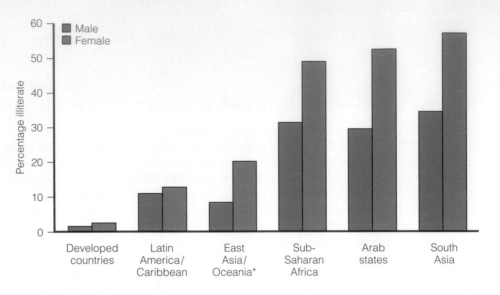

Because teaching is a highly respected vocation in Japan, students treat their teachers with respect and obedience. Teaching is
less well regarded in the United
States, and the consequences are
felt by our teachers every day.

© Karen Kasmauski / Woodfin Camp & Associates.

participation in formal education for youths 5 to 14" years of age (NCES 2005a,
p. 9). Second, the age of compulsory education ranged from 18 in Germany to 15 in
Italy and Japan. Third, in 2001, 64% of U.S. children ages 3 to 5 years were enrolled
in preprimary or primary education, a rate higher than all the other nations except
Canada. Fourth, in a comparison of reading literacy among fourth graders in the
eight countries, only England scored higher than the United States. Not surprisingly, fourth graders in all countries who reported 0–10 books in the home had
lower levels of reading proficiency that their counterparts who reported 11 or more
books in the home. Fifth, enrollment of 17-year-olds in secondary schools was high
in the United States—75%—although it was lower than that of Canada, Germany,
and Japan at 90%. Sixth, U.S. school teachers have a higher-than-average starting
salary compared to comparable teachers in the other countries, although they teach

Should Schools Be Privatized?

Given the less-than-stellar performance of many U.S. schools, some have argued that what is needed is privatization. **Privatization** entails states hiring corporations to operate local institutions. For example, Edison Schools—the largest of the for-profit corporations—has just renewed its contract with the Philadelphia school district. Edison is hoping to help failing schools and failing students while making a return for investors. Proponents argue that market-driven education will result in a better "product"; critics, however, are ideologically opposed to companies making a profit from public school children.

It Is Your Turn Now to Take a Stand!

Do you think public schools would be more efficiently run by private corporations? Why or why not?

more hours per year. Finally, in 2001 almost 25% of 18- to 29-year-olds in the United States were enrolled in colleges or universities—the highest enrollment rate of the nations studied.

A U.S. Department of Education report also attests to the differences between countries in annual per pupil expenditure. The mean expenditure for 30 countries in 2000 was $5,162. The disparities, however, are significant. Mexico had an annual per pupil expenditure rate of $1,414 compared to Switzerland's $8,187 and Austria's $7,851. The U.S. rate of $7,397 per student per year was the fifth highest rate of the countries studied (National Center for Educational Statistics 2004).

Differences also exist in the everyday operations of a school. Some countries empower professionals to organize and operate their school systems. Japan, for example, hires professionals to develop and implement a national curriculum and to administer nationwide financing for its schools. In contrast, school systems in the United States are often run at the local level by school boards and PTAs composed of laypeople. In effect, local communities raise their own funds and develop their own policies for operating the school system in their area; the result is a lack of uniformity from district to district. Thus parents who move from one state to another often find that the quality of education available to their children differs radically. In countries with professionally operated and institutionally coordinated schools, such as Japan, teaching has traditionally been a prestigious and respected profession. Consequently, Japanese students are attentive and obedient to their teachers. In the United States the phrase "Those who can, do; those who can't, teach" reflects the lack of esteem for the teaching profession. The insolence and defiance among students in American classrooms are evidence of the disrespect that many American students have for their teachers.

ThomsonNOW™

Learn more about the **Global Context: Cross-Cultural Variations in Education** by going through the High School Graduates Map Exercise.

Sociological Theories of Education

The three major sociological perspectives—structural functionalism, conflict theory, and symbolic interactionism—are important in explaining different aspects of American education.

Structural-Functionalist Perspective

According to structural functionalism, the educational institution serves important tasks for society, including instruction, socialization, the sorting of individuals into various statuses, and the provision of custodial care (Sadovnik 2004). Many social problems, such as unemployment, crime and delinquency, and poverty, can be linked to the failure of the educational institution to fulfill these basic functions (see Chapters 4, 6, and 7). Structural functionalists also examine the reciprocal influences of the educational institution and other social institutions, including the family, political institution, and economic institution.

Instruction. A major function of education is to teach students the knowledge and skills that are necessary for future occupational roles, self-development, and social functioning. Although some parents teach their children basic knowledge and skills at home, most parents rely on schools to teach their children to read, spell, write, tell time, count money, and use computers. As discussed later, many U.S. students display a low level of academic achievement. The failure of schools to instruct students in basic knowledge and skills both causes and results from many other social problems.

Socialization. The socialization function of education involves teaching students to respect authority—behavior that is essential for social organization (Merton 1968). Students learn to respond to authority by asking permission to leave the classroom, sitting quietly at their desks, and raising their hands before asking a question. Students who do not learn to respect and obey teachers may later disrespect and disobey employers, police officers, and judges.

The educational institution also socializes youth into the dominant culture. Schools attempt to instill and maintain the norms, values, traditions, and symbols of the culture in a variety of ways, such as celebrating holidays (e.g., Martin Luther King Jr. Day, Thanksgiving), requiring students to speak and write in standard English, displaying the American flag, and discouraging violence, drug use, and cheating.

As the number and size of racial and ethnic minority groups have increased, American schools are faced with a dilemma: Should public schools promote only one common culture, or should they emphasize the cultural diversity reflected in the U.S. population? Some evidence suggests that most Americans believe that schools should do both—they should promote one common culture and emphasize diverse cultural traditions (Elam et al. 1994).

Multicultural education—that is, education that includes all racial and ethnic groups in the school curriculum—promotes awareness and appreciation for cultural diversity (see Chapter 9). In 2003, 80% of a sample of U.S. adults responded that "teach[ing] students to get along with people from different backgrounds" is an important role for colleges to perform (Selingo 2003).

Sorting Individuals into Statuses. Schools sort individuals into statuses by providing credentials for individuals who achieve various levels of education at various schools within the system. These credentials sort people into different statuses—for example, "high school graduate," "Harvard alumna," and "English major." In addition, schools sort individuals into professional statuses by awarding degrees in such fields as medicine, engineering, and law. The significance of such statuses lies in their association with occupational prestige and income—the higher one's education, the higher one's income. Furthermore, unemployment rates are tied to educational status (see Figure 8.2).

ThomsonNOW

Learn more about **Structural-Functionalist Perspective** by going through the Functions of Education Learning Module.

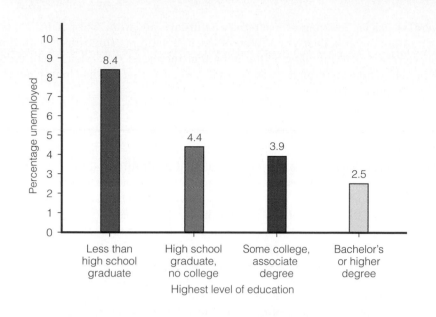

Figure 8.2
Unemployment rate of individuals age 25 years or older by highest level of education, 2005.

Source: U.S. Department of Labor, Bureau of Labor Statistics, Office of Employment and Unemployment Statistics.

Per-pupil expenditure, averaging $7,397 a year, varies dramatically between school districts and states.

Custodial Care. The educational system also serves the function of providing custodial care (Merton 1968), which is particularly valuable to single-parent and dual-earner families and the likely reason for the increase in enrollments of 3- to 5-year-olds. In 1970, 37% of 3- to 5-year-olds were enrolled in formal classes. As noted earlier, in 2001, 64% of this age group were enrolled in either preschool or kindergarten. Ironically, Yale researchers found that state-supported preschools are expelling youngsters at three times the rate of public elementary and secondary schools, primarily for aggressive behavior (Dobbs 2005b).

The school system provides supervision and care for children and adolescents until they are 18 years old—12 years of school, totaling almost 13,000 hours per pupil! Yet some school districts are increasing class hours. For example, the Knowledge Is Power program in Houston, Texas, requires students to attend school several weeks in the summer, on alternate Saturdays, and from 7:25 a.m. to 5:00 p.m. during the regular school year. Some of the motivations behind the "more time" movement are working parents, the hope that increased supervision will reduce delinquency rates, and higher educational standards that require longer hours of study (Wilgoren 2001).

Conflict Perspective

Conflict theorists emphasize that the educational institution solidifies the class positions of groups and allows the elite to control the masses. Although the official goal of education in society is to provide a universal mechanism for achievement, in reality educational opportunities and the quality of education are not equally distributed.

Conflict theorists point out that the socialization function of education is really indoctrination into a capitalist ideology (Sadovnik 2004). In essence, students are socialized to value the interests of the state and to function to sustain it. Such indoctrination begins in kindergarten. Rosabeth Moss Kanter (1972) coined the term "the organization child" to refer to the child in nursery school who is most comfortable with supervision, guidance, and adult control. Teachers cultivate the organization child by providing daily routines and rewarding those who conform. In essence, teachers train future bureaucrats to be obedient to authority.

In addition, to conflict theorists education serves as a mechanism for **cultural imperialism**, or the indoctrination into the dominant culture of a society. When cultural imperialism exists, the norms, values, traditions, and languages of minorities are systematically ignored. A Mexican American student recalls his feelings about being required to speak English (Rodriguez 1990):

> When I became a student, I was literally "remade"; neither I nor my teachers considered anything I had known before as relevant. I had to forget most of what my culture had provided, because to remember it was a disadvantage. The past and its cultural values became detachable, like a piece of clothing grown heavy on a warm day and finally put away. (p. 203)

Conflict theorists are also quick to note that learning is increasingly a commercial enterprise as necessary financial support for equipment, laboratories, and technological upgrades are funded by corporations anxious to bombard students with advertising and other procapitalist messages. For example, Snapple is the "exclusive beverage vendor" for New York's 1,200 public schools. In return for this arrangement, the school system receives a minimum of $8 million a year from Snapple "in commissions and sponsorship money for athletic programs" (Day 2003).

Finally, the conflict perspective focuses on what Kozol (1991) calls the "savage inequalities" in education that perpetuate racial disparities. Kozol documents gross inequities in the quality of education in poorer districts, largely composed of minorities, compared with districts that serve predominantly white middle-class and upper-middle-class families. Kozol reveals that schools in poor districts tend to receive less funding and to have inadequate facilities, books, materials, equipment, and personnel. For example, schools with 70% or more low-income students are more than twice as likely to be overcrowded than schools with 20% or less low-income students (CDF 2004a).

Symbolic Interactionist Perspective

Whereas structural functionalism and conflict theory focus on macrolevel issues, such as institutional influences and power relations, symbolic interactionism examines education from a microlevel perspective. This perspective is concerned with individual and small-group issues, such as teacher-student interactions and the self-fulfilling prophecy.

Teacher-Student Interactions. Symbolic interactionists have examined the ways that students and teachers view and relate to each other. For example, children from economically advantaged homes may be more likely to bring to the classroom social and verbal skills that elicit approval from teachers. From the teachers' point of view middle-class children are easy and fun to teach: They grasp the material quickly, do their homework, and are more likely to "value" the educational process. Children from economically disadvantaged homes often bring fewer social and verbal skills to those same middle-class teachers, who may, inadvertently, hold up social mirrors of disapproval. Teacher disapproval contributes to lower self-esteem among disadvantaged youth.

Self-Fulfilling Prophecy. The **self-fulfilling prophecy** occurs when people act in a manner consistent with the expectations of others. For example, a teacher who defines a student as a slow learner may be less likely to call on that student or to encourage the student to pursue difficult subjects. He or she may also be more likely to assign the student to lower ability groups or curriculum tracks (Riehl 2004). As a consequence of the teacher's behavior, the student is more likely to perform at a lower level.

A classic study by Rosenthal and Jacobson (1968) provides empirical evidence of the self-fulfilling prophecy in the public school system. Five elementary school students in a San Francisco school were selected at random and identified for their teachers as "spurters." Such a label implied that they had superior intelligence and academic ability. In reality, they were no different from the other students in their classes. At the end of the school year, however, these five students scored higher on their intelligence quotient (IQ) tests and made higher grades than their classmates who were not labeled as spurters. In addition, the teachers rated the spurters as more curious, interesting, and happy and more likely to succeed than the non-spurters. Because the teachers expected the spurters to do well, they treated the students in a way that encouraged better school performance.

Who Succeeds? The Inequality of Educational Attainment

Figure 8.3 shows the extent of the variation in highest level of education attained by individuals 25 years of age and older in the United States. As noted earlier, conflict theory focuses on such variations in discussions of education inequalities. Educational inequality is based on social class and family background, race and ethnicity, and gender. Each of these factors influences who succeeds in school.

Social Class and Family Background

One of the best predictors of educational success and attainment is socioeconomic status. Children whose families are in middle and upper socioeconomic brackets are more likely to perform better in school and to complete more years of education

ThomsonNOW™

Learn more about **Self-Fulfilling Prophecy** by going through the Tracking in Education Animation.

"Educational reform measures alone can have only modest success in raising the educational achievements of children from low-income families. The problems of poverty must be attacked directly."

Richard J. Murnane
Harvard University

Figure 8.3

Highest level of education attained by individuals age 25 years old and older, 2002.

Percentages do not sum to 100% because of rounding.

Source: National Center for Educational Statistics, *Digest of Educational Statistics,* 2005.

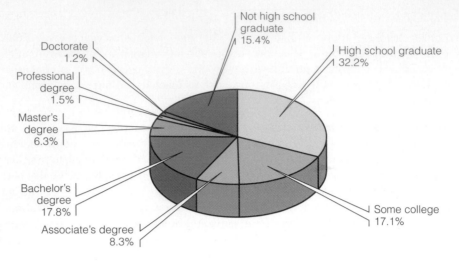

Total individuals age 25 and older = 181.6 million

than children from lower socioeconomic class families. Muller and Schiller (2000) report that students from higher socioeconomic backgrounds are more likely to enroll in advanced mathematics course credits and to graduate from high school—two indicators of future educational and occupational success. In addition, compared to low-income students, high-income students are six times more likely to graduate with a bachelor's degree in five years (Toppo 2004).

Families with low incomes have fewer resources to commit to educational purposes. Low-income families have less money to buy books, computers, tutors, and lessons in activities such as dance and music and are less likely to take their children to museums and zoos. Parents in low-income brackets are also less likely to expect their children to go to college, and their behavior may lead to a self-fulfilling prophecy. Disproportionately, children from low-income families do not go to college (Levinson 2000).

Disadvantaged parents are also less involved in their children's education. For example, 87% of nonpoor children are read to frequently by a family member, compared to 74% of poor children (NCES 2003c). Although working-class parents may value the education of their children, in contrast to middle- and upper-class parents, they are intimidated by their child's schools and teachers and are less likely to attend teacher conferences (Lareau 1989). Furthermore, new teachers in low-income schools are more likely to report having difficulty engaging parents in the educational process, compared to new teachers in higher-income schools (MetLife 2005). Because low-income parents are often low academic achievers themselves, their children are exposed to parents who have limited language and academic skills. Children learn the limited language skills of their parents, which restricts their ability to do well academically. Low-income parents may be unable to help their children with their math, science, and English homework because they often do not have the academic skills to do the assignments. Interestingly, Call and colleagues (1997) report that even among impoverished youths, parental education is one of the best predictors of a child's academic success.

Children from poor families also have more health problems and nutritional deficiencies. In 1965 Project **Head Start** began to help preschool children from the most disadvantaged homes. Head Start provides an integrated program of health care, parental involvement, education, and social services. Today, nearly

910,000 3- to 5-year-olds are enrolled in Head Start (Head Start 2004). Graduates of Head Start "score better on intelligence and achievement tests, their health status is better, and they have the socio-emotional traits to help them adjust to school" (Zigler et al. 2004, p. 341).

Despite the apparent success of Head Start, in 2003 President Bush mandated that all 4-year-olds in the Head Start program participate in a nationwide testing assessment called the National Reporting System (NRS) (CDF 2003a; NABE 2005). The evaluation was designed to measure whether and how English- and Spanish-speaking children were learning and compared facilities between local Head Start programs. Some feared that the assessment was a precursor to dismantling the program. The Children's Defense Fund, a child advocacy group, in response to such fears, aired television advertisements that stated, "Call Congress . . . Tell Congress Head Start's not broken, so don't break it" (CDF 2003b).

However, a 2005 report released by the Government Accountability Office (GAO) found that the Bush administration's testing initiative, the NRS, failed to meet "professional standards." Specifically, the GAO found that the Head Start Bureau failed to follow the rigorous measures necessary to ensure the validity and reliability of the assessment tool, concluding that the NRS could not be counted on to "provide reliable information on children's progress . . . especially the Spanish-speaking children" (NABE 2005).

Assessments of Early Head Start, a program for infants and toddlers from low-income families, have already been conducted. One evaluation found that "after a year or more of program services, when compared with a randomly assigned control group, 2-year-old Early Head Start children performed significantly better on a range of measures of cognitive, language, and socioemotional development" (U.S. Department of Health and Human Services 2001).

Lack of adequate funding for Head Start has long been a problem, as is equality of educational funding in general. Children who live under lower socioeconomic conditions receive fewer public educational resources. Schools that serve low socioeconomic districts are largely overcrowded and understaffed, are less likely to have teachers with advanced degrees and more likely to have higher rates of teacher turnover, and are lacking adequate building space and learning materials (CDF 2004a).

The U.S. tradition of decentralized funding means that local schools depend on local taxes, usually property taxes (Figure 8.4). Although the amount varies by state and municipality, about 43% of school funding comes from local sources. The amount of money available in each district varies by the socioeconomic status of the district. For example, in New York, schools teaching the poorest students receive $2,152 per student less than schools teaching the wealthiest students (Schemo 2003). This system of depending on local communities for financing has several consequences:

- Low socioeconomic status school districts are poorer because less valuable housing means lower property values; in the inner city houses are older and more dilapidated; less desirable neighborhoods are hurt by "white flight," with the result that the tax base for local schools is lower in deprived areas.
- Low socioeconomic status school districts are less likely to have businesses or retail outlets where revenues are generated; such businesses have closed or moved away.
- Because of their proximity to the downtown area, low socioeconomic status school districts are more likely to include hospitals, museums, and art

Figure 8.4

Revenues for public elementary and secondary education by source for 2002–2003 school year.

Source: Data reported by states to the U.S. Department of Education, National Center for Education Statistics, Common Core of Data (CCD), "National Public Education Financial Survey," 2002–2003.

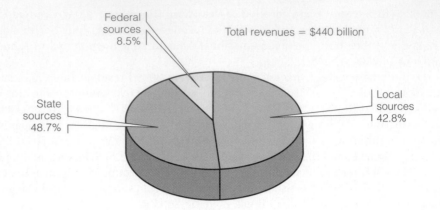

galleries, all of which are tax-free facilities. These properties do not generate revenues.

- Low socioeconomic status neighborhoods are often in need of the greatest share of city services; fire and police protection, sanitation, and public housing consume the bulk of the available revenues. Precious little is left over for education in these districts.

- In low socioeconomic status school districts a disproportionate amount of the money has to be spent on maintaining the school facilities, which are old and in need of repair, so less is available for the children themselves.

Although the state provides additional funding to supplement local taxes, this funding is not enough to lift schools in poorer districts to a level that even approximates the funding available to schools in wealthier districts. *Leandro* v. *State* (1997), a landmark case in North Carolina's Supreme Court, addressed the argument that the quality of a child's education should not be dependent on the wealth of the family and the community into which the child is born or resides. While acknowledging the high costs associated with educating disadvantaged children, *Leandro* also alleged that the state had not done enough to equalize funding. Urban school districts seized the opportunity to also point out the burden they shouldered financially in providing educational services to the large numbers and heavy concentrations of disadvantaged students in their areas. The North Carolina Supreme Court ultimately ruled in two different decisions that the state constitution required North Carolina to provide schools anywhere in the state with adequate resources to fully educate disadvantaged (poor, special education, and limited English proficiency) students and that the state had not met that requirement (NCCAI 2005).

Race and Ethnicity

In comparison to whites, Hispanics and blacks are less likely to succeed in school at almost every level. For example, in 2003, 41% of white fourth graders were reading at grade level, compared to 15% of Hispanics and 13% of blacks. In addition, 37% of white eighth graders performed at grade level in math, compared to 12% of Hispanics and 7% of blacks (CDF 2004b). It is important to note, however, that socioeconomic status interacts with race and ethnicity (Lareau & Horvat 2004). Because race and ethnicity are so closely tied to socioeconomic status, it appears that race or ethnicity alone can determine school success. Although race and ethnicity also have independent effects on educational achievement or lack thereof

The debate over bilingual education is likely to grow. By 2040 less than half of all school-age children will be non-Hispanic whites.

© Michael Newman/PhotoEdit

(Bankston and Caldas 1997; Jencks & Phillips 1998), their relationship is largely a result of the association between race and ethnicity and socioeconomic status.

As Table 8.1 indicates, educational attainment has increased over time and varies by race and ethnicity. In general, the high school graduation gap between racial and ethnic groups is narrowing; the college graduation gap, however, is getting wider—whites and Asians on one side and Hispanics and African Americans on the other.

One reason that some minority students have academic difficulty is that they did not learn English as their native language. For example, in 2004, 9.9 million children in U.S. schools spoke a language other than English in their homes; for most of these children that language was Spanish. This number represents nearly one in five children between the ages of 5 and 17 (U.S. Census Bureau 2005), many of whom live in "linguistically isolated households" where no one over the age of 14 speaks English (AERA 2004). Because of high birthrates, increased immigration, and low levels of private school enrollment, the Hispanic school population continues to grow dramatically and has tripled since 1968 (Frankenberg & Lee 2002).

To help American children who do not speak English as their native language, some educators advocate **bilingual education**—teaching children in both English and their non-English native language. In order to facilitate bilingual education, many school systems across the nation have implemented a recently developed program called **total immersion.** In this program elementary students in particular receive literacy and communication instruction totally in Spanish, for example, thus enabling students to communicate with one another in both languages in and out of the classroom. Advocates claim that bilingual education results in better academic performance of minority students, enriches all students by exposing them to different languages and cultures, and enhances the self-esteem of minority students. Critics argue that bilingual education limits minority students and places them at a disadvantage when they compete outside the classroom, reduces the English skills of minorities, costs money, and leads to hostility with other minorities who are also competing for scarce resources.

Table 8.1

Educational Attainment by Race, Ethnicity, and Sex, 1970 and 2003

	1970		2003	
	Males	Females	Males	Females
Four years of high school or more				
White	54.0	55.0	84.5	85.7
Black	30.1	32.5	79.6	80.3
Hispanic	37.9	34.2	56.3	57.8
Asian	NA	NA	89.5	86.0
Total	51.9	52.8	84.1	85.0
Four years of college or more				
White	14.4	8.4	29.4	25.9
Black	4.2	4.6	16.7	17.8
Hispanic	7.8	4.3	11.2	11.6
Asian	NA	NA	53.9	46.1
Total	13.5	8.1	28.9	25.7

NA = Not available.

Source: U.S. Census Bureau (2004, Table 213).

ThomsonNOW™

Learn more about **Bilingual Education** by going through the English Speakers Map Exercise.

Another factor that hurts minority students academically is that many tests used to assess academic achievement and ability are biased against minorities. Questions on standardized tests often require students to have knowledge that is specific to the white middle-class majority culture, and students for whom English is not their native language are seriously disadvantaged.

In addition to being hindered by speaking a different language and being from a different cultural background, minority students in white school systems are also disadvantaged by overt racism and discrimination. Much of the educational inequality experienced by poor children results because a high percentage of them are also nonwhite or Hispanic. Discrimination against minority students takes the form of unequal funding, as discussed earlier, as well racial profiling and school segregation.

Studies indicate that minority students, and specifically black students, may be the victims of what is being called "learning while black" (Morse 2002b). The allegation is

> not unlike police who stop people on the basis of race; teachers and school officials discipline black students more often—and more harshly—than whites. The result: black students are more likely to slip behind in their studies and abandon school all together—if they're not kicked out first. (Morse 2002b, p. 50)

One study of 11,000 middle school students found that black students were more than twice as likely as whites to be sent to the principal's office or to be suspended. Even in preschool African Americans are twice as likely to be expelled than their white counterparts (Dobbs 2005b). Although the debate is likely to continue, it

Table 8.2

Racial Composition of Schools Attended by the Average Student of Each Race, 2002–2003

Race in Each School	Racial Composition (%) of Schools Attended by Average:				
	White Student	Black Student	Latino Student	Asian Student	Native American Student
White	78	30	28	45	44
Black	9	54	12	12	7
Latino	8	13	54	20	11
Asian	3	3	5	22	2
Native American	1	1	1	1	36
Total	**100**	**100**	**100**	**100**	**100**

Source: Orfield, Gary, and Chungmei Lee, 2005. "Why Segregation Matters: Poverty and Educational Inequality." Cambridge, MA: Harvard University, The Civil Rights Project.

must be noted that differences in discipline patterns are not necessarily a consequence of racism. They may reflect, for example, differences in behavior.

In 1954 the U.S. Supreme Court ruled in *Brown* v. *Board of Education* that segregated education was unconstitutional because it was inherently unequal. Despite this ruling, many schools today are racially segregated. In 1966 a landmark study titled *Equality of Educational Opportunity* (Coleman et al. 1966) revealed the extent of segregation in U.S. schools. In this study of 570,000 students and 60,000 teachers in 4,000 schools, the researchers found that almost 80% of all schools attended by whites contained 10% or less blacks and that whites outperformed minorities (excluding Asian Americans) on academic tests. Coleman and colleagues emphasized that the only way to achieve quality education for all racial groups was to desegregate the schools. This recommendation, known as the **integration hypothesis**, advocated busing to achieve racial balance.

Despite the Coleman report, court-ordered busing, and an emphasis on the equality of education, public schools remain largely segregated. As shown in Table 8.2, most black and Hispanic U.S. students attend schools that are predominantly minority in enrollment. An ongoing study by the Civil Rights Project at Harvard University has found that since 1986 there has been a trend toward "resegregation" (Frankenberg et al. 2003; Orfield & Lee 2005). Research documents the harmful effects of this continued practice. After examining the reading and mathematics achievement levels of a nationally representative sample of high school students, Roscigno (1998) concludes that "school racial composition matters . . . in the direction one would expect, even with class composition and other familial and educational attributes accounted for. Attending a black segregated school continues to have a negative influence on achievement" (p. 1051). Nonetheless, for financial as well as political reasons busing has essentially been abandoned.

Gender

Worldwide, as in the United States, women receive less education than men. An estimated 800 million adults in the world are illiterate, and two-thirds of them are women (see Figure 8.1) (UNESCO 2005). In addition, "although progress has been made toward gender parity in enrollment in primary and secondary levels in all regions over the last ten years . . . the record differs quite strongly between countries [and] at current rates of progress a large minority of countries will not achieve gender parity at primary and secondary levels by 2005" (EFA Global Monitoring Report Team 2003).

Historically, U.S. schools have discriminated against women. Before the 1830s U.S. colleges accepted only male students. In 1833 Oberlin College in Ohio became the first college to admit women. Even so, in 1833 female students at Oberlin were required to wash male students' clothes, clean their rooms, and serve their meals and were forbidden to speak at public assemblies (Fletcher 1943; Flexner 1972).

In the 1960s the women's movement sought to end sexism in education. Title IX of the Education Amendments of 1972 states that no person shall be discriminated against on the basis of sex in any educational program receiving federal funds. These guidelines were designed to end sexism in the hiring and promoting of teachers and administrators. Title IX also sought to end sex discrimination in granting admission to college and awarding financial aid. Finally, the guidelines called for an increase in opportunities for female athletes by making more funds available to their programs. Although gender inequality in education continues to be a problem worldwide, the push toward equality has had some effect. For example, in 1970 nearly twice as many men as women had four years of college or more—8.1% compared to 13.5%. By 2003, 25.7% of women and 28.9% of men had four years of college or more (see Table 8.1).

Traditional gender roles account for many of the differences in educational achievement and attainment between women and men. As noted in Chapter 10, schools, teachers, and educational materials reinforce traditional gender roles in several ways. Some evidence suggests, for example, that teachers provide less attention and encouragement to girls than to boys and that textbooks tend to stereotype females and males in traditional roles (Evans & Davies 2000; Spade 2004).

Studies of academic performance indicate that females tend to lag behind males in math and science. One explanation is that women experience workplace discrimination in these areas, and this restricts their occupational and salary opportunities. The perception of restricted opportunities, in turn, negatively affects academic motivation and performance among girls and women (Baker and Jones 1993).

Most of the research on gender inequality in the schools focuses on how female students are disadvantaged in the educational system. But what about male students? For example, schools fail to provide boys with adequate numbers of male teachers to serve as positive role models. To remedy this, some school systems actively recruit male teachers, especially in the elementary grades where female teachers are in the majority.

The problems that boys bring to school may indeed require schools to devote more resources and attention to them. More than 70% of students with learning disabilities such as dyslexia are male, as are 75% of students identified as having serious emotional problems. Boys are also more likely than girls to have speech impairments, to be labeled as mentally retarded, to exhibit discipline problems, to drop out of school or be expelled, and to feel alienated from the learning process (this chapter's *Self and Society* feature assesses student alienation) (Bushweller 1995; Dobbs 2005b; Goldberg 1999; Sommers 2000). As discussed in Chapter 10, the argument that girls have been educationally shortchanged has recently come under attack; some academicians charge that it is boys, not girls, who have been left behind (Sommers 2000).

Student Alienation Scale

Indicate your agreement with each statement by selecting one of the responses provided:

1. It is hard to know what is right and wrong because the world is changing so fast.
___ Strongly agree ___ Agree ___ Disagree ___ Strongly disagree

2. I am pretty sure my life will work out the way I want it to.
___ Strongly agree ___ Agree ___ Disagree ___ Strongly disagree

3. I like the rules of my school because I know what to expect.
___ Strongly agree ___ Agree ___ Disagree ___ Strongly disagree

4. School is important in building social relationships.
___ Strongly agree ___ Agree ___ Disagree ___ Strongly disagree

5. School will get me a good job.
___ Strongly agree ___ Agree ___ Disagree ___ Strongly disagree

6. It is all right to break the law as long as you do not get caught.
___ Strongly agree ___ Agree ___ Disagree ___ Strongly disagree

7. I go to ball games and other sports activities at school.
___ Always ___ Most of the time ___ Some of the time ___ Never

8. School is teaching me what I want to learn.
___ Strongly agree ___ Agree ___ Disagree ___ Strongly disagree

9. I go to school parties, dances, and other school activities.
___ Always ___ Most of the time ___ Some of the time ___ Never

10. A student has the right to cheat if it will keep him or her from failing.
___ Strongly agree ___ Agree ___ Disagree ___ Strongly disagree

11. I feel like I do not have anyone to reach out to.
___ Always ___ Most of the time ___ Some of the time ___ Never

12. I feel that I am wasting my time in school.
___ Always ___ Most of the time ___ Some of the time ___ Never

13. I do not know anyone that I can confide in.
___ Strongly agree ___ Agree ___ Disagree ___ Strongly disagree

14. It is important to act and dress for the occasion.
___ Always ___ Most of the time ___ Some of the time ___ Never

15. It is no use to vote because one vote does not count very much.
___ Strongly agree ___ Agree ___ Disagree ___ Strongly disagree

16. When I am unhappy, there are people I can turn to for support.
___ Always ___ Most of the time ___ Some of the time ___ Never

17. School is helping me get ready for what I want to do after college.
___ Strongly agree ___ Agree ___ Disagree ___ Strongly disagree

18. When I am troubled, I keep things to myself.
___ Always ___ Most of the time ___ Some of the time ___ Never

19. I am not interested in adjusting to American society.
___ Strongly agree ___ Agree ___ Disagree ___ Strongly disagree

20. I feel close to my family.
___ Always ___ Most of the time ___ Some of the time ___ Never

21. Everything is relative and there just aren't any rules to live by.
___ Strongly agree ___ Agree ___ Disagree ___ Strongly disagree

22. The problems of life are sometimes too big for me.
___ Always ___ Most of the time ___ Some of the time ___ Never

23. I have lots of friends.
___ Always ___ Most of the time ___ Some of the time ___ Never

24. I belong to different social groups.
___ Strongly agree ___ Agree ___ Disagree ___ Strongly disagree

Interpretation

This scale measures four aspects of alienation: powerlessness, or the sense that high goals (e.g., straight A's) are unattainable; meaninglessness, or lack of connectedness between the present (e.g., school) and the future (e.g., job); normlessness, or the feeling that socially disapproved behavior (e.g., cheating) is necessary to achieve goals (e.g., high grades); and social estrangement, or lack of connectedness to others (e.g., being a "loner"). For items 1, 6, 10, 11, 12, 13, 15, 18, 19, 21, and 22, the response indicating the greatest degree of alienation is "strongly agree" or "always." For all other items the response indicating the greatest degree of alienation is "strongly disagree" or "never."

Source: Mau, Rosalind Y., 1992. "The Validity and Devolution of a Concept: Student Alienation." *Adolescence,* 27(107): 739–740. Used by permission of Libra Publishers, Inc., 3089 Clairemont Drive, Suite 383, San Diego, California 92117.

Problems in the American Educational System

When a random sample of Americans were asked, "How satisfied are you with the quality of education students receive in kindergarten through grade twelve in the U.S.?" nearly half responded "somewhat" or "completely dissatisfied" (Gallup 2005a). This is particularly troublesome, given that federal funding for all schools has increased dramatically since 1990 to more than $780 billion in 2003 (NCES 2004b). Consistent with the public's concerns is President Bush's emphasis on educational reform and the problems his administration hopes to address—low academic achievement, high dropout rates, questionable teacher training, and school violence. These and other problems contribute to the widespread concern over the quality of education in the United States.

Low Levels of Academic Achievement

The most recent national data available indicate that 44 million adults in this country cannot "fill out an application, read a food label, or read a simple story to a child"—that is, they are **functionally illiterate** (Literacy Volunteers of America 2003, p. 1). Functionally illiterate adults are disproportionately poor, older than age 55, uneducated, and members of racial or ethnic minority groups (NAAL 2005).

Among children illiteracy tends to be highest among students who attend the poorest schools, although apathy and ignorance can be found among students at more affluent schools as well. For example, an ABC News special titled *Burning Questions: America's Kids—Why They Flunk* began with the following interview with students from middle-class high schools:

> *Interviewer:* Do you know who's running for president?
> *First Student:* Who, run? Ooh. I don't watch the news.
> *Interviewer:* Do you know when the Vietnam War was?
> *Second Student:* Don't even ask me that. I don't know.
> *Interviewer:* Which side won the Civil War?
> *Third Student:* I have no idea.
> *Interviewer:* Do you know when the American Civil War was?
> *Fourth Student:* 1970.

However, results from the National Assessment of Educational Progress, a nationwide testing effort, have improved over time. For example, mathematics proficiency of fourth and eighth graders has increased since 1990 (National Center for Educational Statistics 2004). Writing scores for fourth and eighth graders have also improved, although changes for twelfth graders are not statistically significant (National Center for Educational Statistics 2004). A report issued by the National Commission on Writing in America's Schools and Colleges recommends that the time spent on writing, currently about 15% of the time spent on watching television, be doubled (Lewin 2003).

Another measure of academic success is the extent to which students move forward toward completion of degree requirements. For example, among students entering college in fall 2000, 28% were required to take remedial courses, a variable negatively associated with graduation rates. In general, completion of bachelor's degrees has remained stable over time with 53% of students earning a bachelor's degree within five years. However, the likelihood of working toward a bachelor's degree after five years (i.e., still being enrolled in a degree program) has increased (National Center for Educational Statistics 2004).

Although some international comparisons are improving, American students are still outperformed by many of their foreign counterparts. An international report examining student performances in 30 industrialized nations found that even though the United States is one of the leaders in educational spending, U.S. 15-year-olds scored only "average" in math, reading, and science. Other results from the report include the following (U.S. Department of Education 2003):

- Average class size varies dramatically—from 36 in Korea to 20 in, among other countries, Australia, Denmark, Greece, Italy, and Norway.
- On average, a teacher works 792 hours a year; the range is from 650 hours in Japan to 950 hours in the United States, Scotland, and New Zealand.
- Of all foreign students in the countries studied, more attend colleges and universities in the United States than in any other country.

In addition, the 2003 Trends in International Mathematics and Science Study, which measures math and science scores of eighth graders in 38 countries, indicates that American youth were outperformed by their Asian counterparts in Hong Kong, Singapore, Korea, and Japan. They were also outperformed by students in four European countries—Belgium, Estonia, Hungary, and the Netherlands. The highest international mathematics benchmark includes the ability of students to "solve multi-step word problems involving addition, multiplication, and division" and to "use their understanding of place value and simple fractions to solve problems." More than three-fourths of U.S. eighth graders were unable to achieve this standard. Not surprisingly, when asked about the most difficult course in school, 37% of 13- to 17-year-olds respond math (Saad 2005).

School Dropouts

Globally, dropout rates vary considerably. In the United States 13% of 16- to 24-year-olds are dropouts—that is, they are not presently enrolled in and did not graduate from high school. This is a 16% decrease since 1986 (National Center for Educational Statistics 2004). In some U.S. cities the dropout rate is as high as 55%. However, in the Netherlands the dropout rate is near 0% (BBC 2003).

For event dropout rates—that is, the percentage of students who drop out of high school each year (Figure 8.5)—low-income students are more than twice as

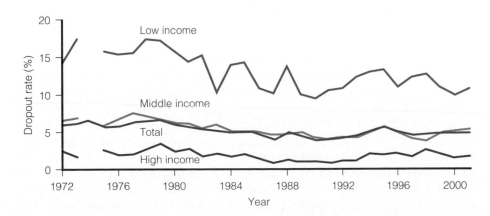

Figure 8.5
Event dropout rates for 15- to 24-year-olds who dropped out of grades 10–12, by family income: October 1972 to 2001.

Source: National Center on Educational Statistics, U.S. Department of Education.

likely to drop out as their middle-class counterparts and are six times as likely to drop out than their peers from high-income families. Event dropout rates also vary by race and ethnicity, with whites and Asian/Pacific Islander youth having lower dropout rates than African Americans or Hispanics (National Center for Educational Statistics 2004). Minorities and low-income students are also more likely to drop out of four-year colleges than their white middle- and upper-class counterparts (Leonhardt 2005).

Why do students drop out of high school? A recent report by the Educational Testing Service, titled *One-Third of a Nation*, found that three factors in combination explain more than half the variation in dropout rates. First are socioeconomic indicators—parent's income, parent's occupation, and the like. Second is the number of parents in the home. Having only one parent in the home increases the likelihood of a child dropping out of school. Finally, the number of times a student has changed schools is positively related to dropping out; that is, the higher the number of different schools attended, the higher the likelihood of dropping out (Barton 2005).

The economic and social consequences of dropping out of school are significant. Dropouts are more likely than those who complete high school to be unemployed (see Figure 8.2) and to earn less when they are employed (NCES 2004b). Individuals who do not complete high school are also more likely to engage in criminal activity, to have poorer health and lower rates of political participation, and to require more government services such as welfare and health care assistance (Natriello 1995; NCES 2002a; Rumberger 1987). Although federal funding for "second chance" opportunities has decreased over the years, General Educational Development (GED) certification is still one of the most commonly sought alternative credentials. Nationally, more than 40% of students who drop out of school complete their high school degree at some point in time (Entwisle et al. 2004).

A relatively new initiative for dropouts is called early or middle school college. Typically, students who have dropped out or are at risk of dropping out are admitted to community colleges or, in some cases, four-year degree programs. They receive a secondary school education, earn a high school degree, and often accrue college credits. Students are offered tutoring services, private teacher conferences, a low student-teacher ratio, individualized attention, and child care if needed. Such programs have been credited with lowering dropout rates of high-risk students (e.g., students who have failing grades, emotional or family problems, or high absenteeism) (Manzo 2005).

Violence in the Schools

Students at school between the ages of 12 and 18 were the victims of 1.9 million crimes in 2002. Most school crime, however, is not serious, with theft accounting for 64% of the total crimes against students. Despite the public outrage over school violence in the wake of Columbine and other school shootings (see this chapter's vignette), the chance of a child being killed at school is quite rare—about one in a million (NCJRS 2003).

Certain school characteristics are associated with an increase in the probability of a student being the victim of a crime. For example, 21% of all 12- to 18-year-olds surveyed in the School Crime Supplement to the National Crime Victimization Survey report attending schools in neighborhoods with gang activity. These students are more likely to be victimized than students in neighborhoods without gang activity (NCES 2005b). In addition, students who report knowing another student who brought a gun to school or who actually saw a student with a gun have higher rates of victimization. Finally, the prevalence of alcohol and other drugs in a school is indirectly related to school violence. Schools in neighborhoods with higher gang activity have students who are more likely to report alcohol and drug availability (Addington et al. 2004).

The Department of Education and the Secret Service combined efforts to study 37 school shootings and their attackers. The final results of the study were published in the *Safe School Initiative* (2004), which found, among other things, that

- School shootings are rarely sudden or impulsive.
- Before the attack many shooters felt that they had been bullied, persecuted, or injured by others.
- The shooter revealed his idea and/or plan of attack to others before the shooting.
- The targets of the shootings were not threatened directly by the shooter before the occurrence.
- Most attackers had access to weapons and had used them before carrying out the shooting.

Furthermore, a study by the National Research Council concludes that school shootings more closely resemble the rampage killings that occur in workplaces than they do urban shootings: "The urban [shooting] cases tended to be classic disputes that spilled into school territory, but the shooters in the rural and suburban cases consciously picked the schools as a place where a general grievance might be resolved" (Bowman 2002, p. 1). Both the *Safe School Initiative* report and the National Research Council conclude that it will be difficult to identify specific assailants in advance because there is no consistent profile of a school shooter. Not surprisingly, only 36% of Americans believe that the government and society can do anything about school shootings (Mason 2005).

Of late, bullying has become one focus of the efforts to stop school violence. **Bullying** is characterized by an "imbalance of power that exists over a long period of time between two individuals, two groups, or a group and an individual in which the more powerful intimidate or belittle others" (Hurst 2005, p. 1). Although many countries have had antibullying policies in place for decades, U.S. policies and research on the topic are comparatively new. One recent research report concluded that television viewing is associated with bullying behavior. At age 4 each hour of television viewed daily is associated with a 9% increased risk of bullying by age 7. Moreover, increased rates of bullying were also associated with decreased cognitive stimulation. For example, grade-school bullies were less likely to be read to by a parent than their nonbullying peers (Zimmerman et al. 2005). A number of websites have cropped up to provide support for children who are being bullied. One such website is Stop Bullying Now, which is maintained by the Department of Health and Human Services.

Students are not the only people victimized in schools. Annually, on the basis of a five-year average, teachers were the victims of approximately 90,000 violent crimes, including rape, sexual assault, aggravated and simple assault, and robbery. Urban rather than rural or suburban teachers are the most likely to be victimized. In addition, teachers at secondary schools are more likely to be the victim of a violent crime than teachers at elementary schools (NCES 2005b).

In response to violence many schools throughout the country have police officers patrolling the halls, require students to pass through metal detectors before entering school, and conduct random locker searches. Video cameras set up in classrooms, cafeterias, halls, and buses purportedly deter some student violence.

Seventy-five percent of the new schools that opened in 2002 were equipped with surveillance cameras (Dillon 2003). More than 2,000 schools nationwide conduct peer mediation and conflict resolution programs to help youth resolve conflict in nonviolent ways.

A relatively new way to deal with school "troublemakers" is the alternative school, which houses students who have committed a variety of offenses while allowing them to continue with their education. In the 2000–2001 school year 39% of all U.S. school districts had an alternative school—66% of urban districts, 41% of suburban districts, and 35% of rural districts (U.S. Census Bureau 2004). In 2003 a class action suit was filed by a group of inner-city teenagers seeking to overturn a state law. The law required students who have committed a wide range of offenses—in or out of school—to be placed in an alternative school facility upon completing their sentences in juvenile detention (Steptoe 2003).

Inadequate School Facilities and Programs

A report titled *America's Crumbling Infrastructure Eroding the Quality of Life* (ASCE 2005) documents the troubling conditions that exist in U.S. schools. According to the report, the American Society of Civil Engineers' report card on American's infrastructure resulted in an overall grade of D for U.S. schools (ASCE 2005). Although the grade came up from the D minus earned in 2001, more financial resources still need to be allocated to provide the best possible conditions for public schools. For the most part this burden falls on the states and their individual localities. For example, infrastructure funding needs for Arkansas are estimated to be between $2.9 billion and $4.5 billion over the next five years, whereas New Jersey funded $8.7 billion to upgrade its poorest districts with estimates now reaching $15 billion. Ohio instituted a $23 billion school building project, whereas Hawaii, the only state-run public school system in the nation, faces a 10-year backlog of maintenance and repair with a $600 million price tag. Although the federal government initially estimated a need for $127 billion in 1999 to address school facilities and their condition, newer sources indicate a need for funding as high as $268 billion (ASCE 2005). Furthermore, many schools present environmental risks.

> A significant number of schoolchildren and teachers in the United States are exposed on an almost daily basis to environmental hazards including volatile chemicals, airborne lead and asbestos, and noise pollution while they are at school. Some school hazards are linked to the aging of many of the nation's schools, to the ongoing siting of schools in close proximity to contaminated waste sites, and to the burgeoning population of school-aged children that has forced financially constrained school districts to use portable classrooms to increase their classroom space. (Wakefield 2002, p. 1)

School lunch programs have also come under attack. Incidents of large-scale food-borne illnesses have increased 10% since 1990, with students reporting a wide range of illnesses from food poisoning to hepatitis A to salmonella (Morse 2002a). There are also concerns about the sale of "junk food" in school cafeterias in light of increasing rates of child obesity (Nash 2003). Less than 20% of school lunches meet the government's requirement that fewer than 10% of calories come from saturated fat. With 15% of today's schoolchildren considered severely overweight or obese, one Florida school district teamed up with Dr. Arthur Agatson to conduct a study using his South Beach Diet guidelines as the foundation for the dietary menu served in several of their schools (CBS 2005). Other efforts to promote healthy eating have been initiated around the nation by school districts that rely on local grocers to supply fresh produce for the school menus (Schneider 2005).

Special education programs pose yet another problem for school systems. The National Council on Disability reports that schools in the United States educated only one in five students with disabilities in 1970. More than 1 million students were excluded from public schools, and 3.5 million did not receive appropriate services. Many states had laws that excluded certain students from public education, including those who were blind, deaf, or labeled "emotionally disturbed" or "mentally retarded." Almost 200,000 school-age children were institutionalized because they had emotional disabilities or suffered from mental retardation. In addition, the likelihood of being excluded from receiving an education was greater for children with disabilities living in low-income, ethnic and racial minority, or rural communities (NCD 2000).

In 1997 the Individuals with Disabilities Education Act (IDEA) was adopted; it was reaffirmed in 2004. As of 2000 more than 6 million children and young people with disabilities, ages 3 through 21, qualified for educational interventions under the IDEA (NCD 2000). Pursuant to IDEA mandates, public school special education programs are now required to provide guaranteed access to education for disabled children. These specially designed instructional programs are structured to meet the unique needs of each child at no cost to the parent. However, a long-term study of the IDEA indicates that states are failing in their efforts to enforce the law and that every state was negligent in complying with at least some component of the act (Sack 2000). For the program to succeed, adequate funding and resources must be readily available. However, with budget cuts, persistent teacher shortages, and insufficient educational resources available to support the special education programs, it is commonly understood that school systems are routinely unable to meet all the compliance standards. As a result, parents have filed complaints with federal agencies to seek full compliance with the law (NCD 2000). Furthermore, given the often volatile, aggressive, and dangerous behaviors exhibited by many of the special education students in the more high-maintenance special education classes, teachers are opting to leave the special education classrooms for traditional classrooms or to pursue different fields entirely.

Recruitment and Retention of Quality Teachers

School districts with inadequate funding and facilities, low salaries, lack of community support, and minimal professional development often have difficulty attracting and retaining qualified school personnel. This has placed the nation's schools at odds with a growing problem in the face of higher accountability standards. With a national average turnover rate of 15.7% in 2004, each year school systems open the new year without enough teachers to cover all their classes (WEP 2005).

In addition to the growing concerns related to recruitment and retention, poorer districts are also faced with the demand for talented teachers who can meet the needs of children from diverse backgrounds and of varying abilities. The number of minority teachers who can serve as role models, have similar life experiences, and have similar language and cultural backgrounds is far too few for the number of minority students (NCES 2000). Furthermore, several studies have shown that poorer communities do not attract quality teachers. Not only are these low-income, high-minority districts likely to employ beginning teachers, they are also 77% more likely to assign these teachers to areas outside their specialty.

"Education is the key to prosperity and the wisest investment we can make in our children's and our nation's future."

Richard Riley
Former U.S. secretary of education

In addition, these high-poverty schools experience turnover rates that are nearly one third higher than the national rate, leading to a shortage of qualified teachers and contributing significantly to the high rate of attrition nationally (CDF 2004a).

Even those who remain in the teaching profession are not necessarily competent or effective. For example, there is evidence that those who go into teaching, on average, have lower college entrance exams than the average college student (NCES 2003a). In addition, a recent survey indicates that nine out of ten principals believe that U.S. colleges and universities are doing such a poor job of training teachers that many of the graduates are not properly prepared to go into the classroom (Winter 2005). Moreover, a Department of Education survey found that less than 36% of current teachers report feeling "very well prepared" to initiate curriculum and performance standards, and less than 20% feel ready to meet the needs of the diverse student population (*A Quality Teacher,* 2002).

In an effort to place "highly qualified" teachers in the classroom, many states have implemented mandatory competency testing as part of state licensure requirements. Eighty percent of those states require prospective teachers to pass the Praxis Series, which tests general knowledge as well as specialty area competency (ETS 2005). The need for teachers who are officially classified as "highly qualified" is tied to national mandates that place an emphasis on the importance of having licensed teachers in the classroom. To be considered "highly qualified," a teacher "must hold at least a bachelor's degree, have full state certification or licensure, and have demonstrated competence in their subject areas" (U.S. Department of Education 2004, p. 2). Some research suggests that teacher licensure has a direct impact on student achievement, with students learning more from certified teachers than from uncertified teachers (Viadero 2005).

Stating that "the single most important action this country can take to improve schools and student learning is to strengthen teaching" (NBPTS 2005, p. 1), the National Board for Professional Teaching Standards created and implemented a plan designed to "retain, reward, and advance" those teachers accomplished in their respective fields. Teachers who have a bachelor's degree and have been in the classroom for three or more years are eligible to seek this higher level of certification, a level that certifies them nationwide to teach in their specialty area.

In an effort to meet the demands of placing teachers in classrooms while facing teacher shortages, states are now allowing skilled professionals who have an interest in teaching but did not receive a teaching degree to enter the teaching profession. Called lateral entry by some states, the program allows the person to obtain a lateral entry teaching license while actually teaching in the classroom. With this approach the teacher is provided with orientation training by the school system as well as assistance in obtaining the license needed to remain in the classroom.

In addition, more than half of the states have adopted **alternative certification programs**, whereby college graduates with degrees in fields other than education can become certified if they have "life experience" in industry, the military, or other relevant jobs. Teach for America, a program originally conceived by a Princeton University student in an honors thesis, is an alternative teacher education program aimed at recruiting liberal arts graduates into teaching positions in economically deprived and socially disadvantaged schools. After completing an eight-week training program, recruits are placed as full-time teachers in rural and inner-city schools. Critics argue that these programs may place unprepared personnel in schools.

Strategies for Action: Trends and Innovations in American Education

Americans rank improving education as one of their top priorities (Harris Poll 2003). Recent attempts to improve schools include raising graduation requirements, barring students from participating in extracurricular activities if they are failing academic subjects, lengthening the school year, and prohibiting dropouts from obtaining driver's licenses. In addition, there is a nationwide movement to eliminate **social promotion**, the passing of students from grade to grade even if they are failing. However, educational reformers are calling for changes that go beyond get-tough policies that maintain the status quo.

National Educational Policy

President Bush's education plan, No Child Left Behind, was signed into law in January 2002. The plan is organized around four principles. The first principle is *accountability.* Each year every state will be required to test the math and reading abilities of third- through eighth-grade students. Parents, administrators, and others will have access to the data, and states will issue a report card rating the performance of each school, teacher, and student.

The second principle is *flexibility.* The new law will permit federal funds to be transferred between programs; local schools will also have more input into how federal funds are used. *Expanding options for parents* is the third principle. After a school is identified as failing, parents will be able to transfer their child to a better-performing or charter school. Also, supplemental funds that can be used for a tutor, summer school programs, or any other school service will be provided to children in failing schools.

AP/Wide World Photos

Margaret Spellings, the eighth U.S. secretary of education, helped draft the No Child Left Behind Act of 2001 while a domestic adviser to the Bush administration.

The Pendulum of Change

Although the primary goal of a teacher is to educate, a teacher's impact on a student can be life changing. For example, in a recent survey of secondary school students, 95% report that they had a teacher who has made a positive difference in their lives. Furthermore, the students indicate that these teachers helped them to do better in school, introduced them to new ideas, and helped them to pursue their individual interests (MetLife 2005). The following narrative poignantly describes a classroom teacher's efforts to make a difference in an environment driven by accountability standards.

It has been about twelve years now. So many things have changed, but I will never forget the first day of administering a new state mandated science experiment as part of a week-long assessment. I am not a scientist, but as an elementary school teacher at the time, science was one of the many topics I had to cover in my classroom. So, here we were, about to embark on this great new scientific experiment, recording for the history books of the state's Department of Education just how effectively my students, and other students across the state, could log this amazing phenomenon they were about to witness.

According to the detailed instructions, some raisins were supposed to "dance" on the top of the surface of a container as two unknown ingredients were mixed together and poured over the raisins sitting in the bottom of a clear cup. Now, the instructions indicated that for ten minutes, I was to record the time on the chalkboard in 15 second intervals. The purpose was to allow the students to have the most accurate information possible for recording in their group journal.

So, with "I Heard It Through the Grapevine" playing through my head, for 10 minutes I dutifully recorded the time in fifteen second intervals. During that 10 minutes, the student interest went from rapt attention to doodling on the test papers to taking out books to read. During that excruciating 10 minutes, absolutely nothing happened. No dancing, not even a flutter. I later learned it was not just in my room that this new-fangled test failed. No one in my building had any success either. Further, no one in the entire county had dancing raisins. As a matter of fact, the raisins refused to dance at every school in the entire state! And it was only Monday morning at 9:00, with two more tasks that day plus three each day for the next four days.

Up until about two years ago, I was sworn to secrecy and wouldn't have been allowed to share the above secure testing information with anyone. However, in a field that operates on a perpetual pendulum, one must accept that bureaucracy will forever be dabbling with ways that some believe to be on the cutting edge of meeting the educational needs of children. Hence, the testing format that started a little over ten years ago has now become obsolete, and a newer state-mandated testing format under No Child Left Behind has been put into motion.

Despite the gradual wariness and loss of confidence that many veteran teachers have about state assessments, we must keep our focus elsewhere. As a seasoned, and oft-times battle-weary veteran teacher, I have come to realize that an educator must maintain an unwavering focus on why he or she has entered this field. What a teacher does has to be done for the sake of the student and no one else. Those of us who have the heart and passion to teach have to remember the motto, "To teach is to touch lives forever." A teacher must realize that he or she will have a tremendous impact on the life of a child, and vice versa.

You will be sobered when you attend the funeral of a student who has died in an auto accident and you find yourself attempting to comfort the parents who are younger than you are, trying to remind them of the joy that their child brought to your classroom. You will be angered and, perhaps, stunned when you read the police blotter section of the local newspaper, and see the names of your former

---

The last principle is using *teaching methods* that are known to work. Money will be provided to "reading first"—a presidential plan to help children read. Teacher quality will be improved because local schools will be better able to recruit and retain excellent teachers. Last, local schools will be able to use federal funds "for hiring new teachers, increasing teacher pay, improving teacher training and development and other uses" (*Fact Sheet* 2003). By 2006, according to the plan, only "highly qualified" teachers will be hired (see this Chapter's *Human Side* feature).

Critics of the law argue that it unfairly burdens the states, which must absorb the financial cost of its provisions. Thus, for example, a Connecticut Department of

students listed under the arrests for DUI, drug possession, and assault and battery. Good kids; challenging kids; it matters not. They were your kids, in your classrooms, and here they are.

For many students, we *become* the parent. Imagine my shock in my second year of teaching when a parent of a problem child I was on the phone with hung up on me right after saying, *"Mr. Jenkins, when my child is home with me she is my problem. When she's in that school with you, she's yours."* Since when did parents stop being parents during school hours?

What we do, we must do for the children. But it has become more and more difficult in recent years, especially in light of the pressures of No Child Left Behind. Never before have I seen such a cloud hanging over doom-and-gloom staff meetings as we compare test scores and study where we must be by this time next year. After all, funding and school recognition are at stake—all because of test results. In some circles the NCLB (No Child Left Behind) program has been coined the NTLTT (No Teachers Left To Teach) program, as we witness increased numbers of teachers opting to take early retirement instead of trying to keep up with the rigors and requirements of becoming and remaining "highly qualified" in our respective fields.

I get more and more frustrated when I hear of test results and "Annual Yearly Progress" reports as we blindly graduate unprepared students. After all, if we graduate those who can't make the grade, once they're gone, our scores are sure to increase. It troubles me to hear the constant threats of "teacher accountability" based on the success of their assigned students; accountability that is measured by test performance. It is especially disturbing when I have over 180 students with as many as 40 in a class. It makes me angry when I feel pressured to teach within a prescribed curriculum based on desired testing outcomes. What ever happened to teaching students to be successful in life; to helping them develop life skills instead of testing skills?

I teach American History. About 95% of the people we talk about in my classroom are dead. My job is to help 13-year-olds understand the significance of the events that played out in the lives of these dead people—the Founding Fathers of our nation; to help them see how those who lived immediately before and after our Founding Fathers served to mold and make this country great—the idol and envy of so many others. I want to teach with my God-given gifts and talents and try to show these youngsters why I believe it is important to know something about their amazing past. They say that history

repeats itself and I have seen proof of it over and over again. I have seen state testing come and go by the wayside and am now facing the pressures and uncertainty of another attempt to find the most appropriate way to teach children. While my school, county, state, and even country attempt to find the answer to how to educate most effectively, I just want to be left alone to teach and reach the children in my own way. Forget the administrative pressure, state tests and government programs. Raisins may dance, or they may fall flat as they did twelve years ago, but, in the long run, it doesn't really matter. Kids will always be kids despite the drastic changes we may see in them. They need role models in their lives—especially when your 68-minute class may be more time spent with that kid in one day than they spend with their parent(s) in a whole week.

Yes, the pendulum is changing directions again, but the kids will not be. If I can hang in there another eight years or so, I just might see it swing back again the other way. In the mean time, I only have 180 days to make an impact on some needy kids' lives before I must send them away and prepare for the next batch. Time waits for no one; especially kids. Reach them while you can, as they won't always be there for you.

Source: Jenkins, Robert F., 2006. *The Pendulum of Change.*

Education report concludes that the state would have to spend $41.6 million by 2008 to carry out all the testing requirements mandated by No Child Left Behind (Archer 2005a). As a result, in August 2005 the Connecticut attorney general filed a lawsuit against the U.S. Department of Education, arguing that the act itself contains prohibitions against unfunded federal mandates (Official Press Release 2005). Other legal maneuvers have also taken place. The governor of Utah signed a bill that gives state legislation priority over the federal act. In doing so, Utah risks more than $76 million in annual education funds. At present, 15 states are in the process of considering anti–No Child Left Behind legislation, and the National Education

Association, the largest organization of teachers in the nation, has joined schools in several districts to bring the first federal lawsuit against the U.S. Department of Education for failing to provide funding for No Child Left Behind initiatives (Dobbs 2005a; Sack 2005).

Proponents of the law hold that such legal maneuvers mask the intent of the legislation—to improve education in the United States. For example, in response to the Connecticut lawsuit, one U.S. Department of Education official noted that Connecticut's students "are suffering from one of the largest achievement gaps in the nation" (Archer 2005b, p. 2). Although Connecticut students are at the top of the performance ladder nationwide, low socioeconomic students and minority students score significantly lower than their high socioeconomic counterparts.

Results of empirical tests of No Child Left Behind are mixed. For example, a recent report by the Northwestern Evaluation Association, a nonprofit independent research group, compared reading and math scores of thousands of students in multiple states and school districts. Two dependent variables were measured: *achievement level* (i.e., how is this year's fourth-grade class doing compared to last year's fourth-grade class?) and *achievement growth* (i.e., how is this child doing in the fourth grade compared to how he or she was doing in the third grade?) (Cronin et al. 2005). Results indicate that, although achievement levels in both math and reading have increased under No Child Left Behind, achievement growth, particularly for minorities, has actually decreased. However, a study by the Center on Education Policy found less conflicting results, with 36 of 49 states surveyed responding that student achievement has increased since No Child Left Behind was instituted (Ripley & Steptoe 2005).

Character Education

An ABC News poll of a national sample of 12- to 17-year-olds found that nearly one-third reported cheating on their schoolwork (Associated Press 2004). In addition, 75% of more than 4,500 students surveyed by the Education Center at Rutgers University reported cheating at least once in their academic careers, and more than half the respondents reported plagiarizing information from the Internet (Slobogin 2002). Particularly worrisome was the finding that more than half the students surveyed did not think that "copying questions and answers from a test" was cheating. As one student put it:

> What's important is getting ahead. The better grades you have, the better school you get into, the better you're going to do in life. And if you learn to cut corners to do that, you're going to be saving yourself time and energy. In the real world, that's what's going to be going on. The better you do, that's what shows. It's not how moral you were in getting there (Slobogin 2002, p. 1).

To many educators and the general public, statements like this signify the need for **character education.**

Despite a national survey of adults that indicates that concerns with values (e.g., ethics, dishonesty, integrity) are one of the nation's top priorities, most school curricula neglect this side of education—the moral and interpersonal aspects of developing as an individual and as a member of society (Gallup 2005b). President Bush's educational reform policy includes support for character education. Proponents of character education argue that "with intentional, thoughtful character education, schools can become communities in which virtues such as responsibility, hard work, honesty, and kindness are taught, expected, celebrated, and continually

"Our society does not need to make its children first in the world in mathematics and science. It needs to care for its children— to reduce violence, to respect honest work of every kind, to reward excellence at every level, to ensure a place for every child.**"**

Nel Noddings
Educational reformer

practiced" (*What Is Character Education?* 2003). For example, service learning programs are increasingly popular at universities and colleges nationwide. Service learning programs are community-based initiatives in which students volunteer in the community and receive academic credit for doing so. Studies on student outcomes have linked service learning to enhanced civic responsibility and moral reasoning, a reduction of risky behaviors, and higher levels of self-esteem (Independent Sector 2002; Jacobs 1999; Ramierz-Valles & Brown 2003). How does the public feel about character education? In a national poll 76% of respondents favored requiring schools to teach about values and morality (NPR 2003). Character education also occurs to some extent in schools that have peer mediation and conflict resolution programs. Such programs teach the value of nonviolence, collaboration, and helping others as well as skills in interpersonal communication and conflict resolution.

Computer Technology in Education

Computers in the classroom allow students to access large amounts of information (see this chapter's *Focus on Technology* feature). The proliferation of computers both in school and at home may mean that teachers will become facilitators and coaches rather than sole providers of information. Not only do computers enable students to access enormous amounts of information, including that from the World Wide Web, but they also allow students to progress at their own pace. However, computer technology is not equally accessible to all students. Although students in poorer school districts are less likely to have access to computers in school or at home, in general, access to computers has increased dramatically over the years. In 1999, 50% of all schools had Internet access; today, that number is 99%. Furthermore, in the 2004–2005 academic year the ratio of students to instructional computers was 4:1; in 1983 the ratio was 125:1 (National Science Board 2003; U.S. Census Bureau 2004).

Interestingly, the conclusion of one of the largest studies of school computers was that students who use computers often scored lower on math tests than their low-use counterparts. An Educational Testing Service study of 14,000 fourth and eighth graders concluded that how the computers were used—repetitive math drills versus real-life simulation—was responsible for the test variations. Also of note, students in classrooms where teachers were trained in computer use did better than students in classrooms where teachers were less skilled (Weiner 2000). Minority and low-income students were less likely to have teachers highly skilled in computer technology than their middle- and upper-income counterparts.

The Enhancing Education Through Technology program is part of the No Child Left Behind Act of 2001. The goals of the program are threefold: (1) to improve student achievement through the use of technology resources, (2) to ensure that teachers integrate technology into the curriculum in such a way as to improve student achievement, and (3) to help students become technically literate by the eighth grade. With the added funding that the program provides, schools will be able to purchase additional technology in support of these goals (NCES 2003e).

School Choice

Traditionally, children have gone to school in the district where they live. School vouchers, charter schools, home schooling, and private schools provide parents with alternative school choices for their children. **School vouchers** are tax credits

Distance Learning and the New Education

Imagine never having an 8 o'clock class or walking into the lecture room late. Imagine no room and board bills or not having to eat your roommate's cooking. Imagine going to class when you want, even 3 o'clock in the morning. Imagine not worrying about parking! The future of higher education? Maybe. It is possible that the World Wide Web and other information technologies have so revolutionized education that the above scenarios are faits accomplis.

What is distance learning? **Distance learning** separates, by time or place, the teacher from the student. They are, however, linked by some communication technology: videoconferencing, satellite, computer, audiotape or videotape, real-time chat room, closed-circuit television, electronic mail, or the like.

Today, it is possible to earn a bachelor's degree, graduate degree, and/or professional degree by means of the Internet. The latest statistics indicate that the number of distance education courses offered has increased dramatically over the years. In 1994, 33% of all two- and four-year degree-granting institutions had distance education courses. By 2001 that number had increased to 56%. Public institutions are more likely to offer distance education courses than private institutions, a higher percentage of students at two-year institutions were enrolled in distance education courses than those at four-year institutions, and the most common medium for distance education was the Internet, followed by live audio or television (PEQIS 2003).

The use of distance education is not limited to higher education. The National Center for Educational Statistics conducted a distance education survey of public schools for the 2002–2003 academic year. The results indicate that 36% of public school districts—more than 8,200 schools—offered distance education courses. Of the total number of students enrolled in distance education, 68% were in high school, 29% were in combined or ungraded schools, 2% were in middle schools, and 1% were in elementary schools (Setzer & Lewis 2005).

The benefits of distance learning are clear. Distance learning provides a less expensive, accessible, and often more convenient way to complete a degree. There are even pedagogical benefits. Research suggests that "students of all ages learn better when they are actively engaged in a process, whether that process comes in the form of a sophisticated multimedia package or a low-tech classroom debate on current events" (Carvin 1997). Distance education also benefits those who have historically been disadvantaged in the classroom. A review of research on gender differences suggests that females outperform males in distance learning environments; they are also more likely to enroll in distance education courses (Koch 1998; NCES 2002b).

---

that are transferred to the public or private school that parents select for their child. President Bush's plan calls for federally funded vouchers for students who are attending failing schools that do not improve test scores for two consecutive years. The vouchers can be used for tuition at a private school, out-of-class tutoring, or other supplemental services.

Proponents of the voucher system argue that it reduces segregation and increases the quality of schools because the schools must compete for students to survive. Opponents argue that vouchers increase segregation because white parents use the vouchers to send their children to private schools with few minorities. Research by Saporito (2003) supports this contention:

> Findings show that white families avoid schools with higher percentages of non-white students. The tendency of white families to avoid schools with higher percentages of non-whites cannot be accounted for by other school characteristics such as test scores, safety, or poverty rates. I also find that wealthier families avoid schools with higher poverty rates. The choices of whites and wealthier students lead to increased racial and economic segregation in the neighborhood schools that these children leave. (p. 181)

But all that glitters is not gold. Evidence suggests that students feel more estranged from their distance learning instructors than from teachers in conventional classrooms. Among distance education users, a higher proportion of them "were less satisfied than more satisfied with the quality of instruction they received in their distance education classes compared with their regular classes" (NCES 2003b).

In addition, the proliferation of "virtual degrees" is problematic. London's Strassford University has an impressive brochure complete with ivy-trimmed buildings and an enviable history dating back to the reign of Queen Victoria. But in fact, it does not exist. It's a fake, and fake degrees in this era of electronic diplomas come with bogus transcripts, letters of recommendation, and a "backup" telephone number if anyone should want to verify the diploma's legitimacy (CBS 2003). Most alarmingly, these fake degrees can be used to get student visas similar to those purchased by the 9/11 terrorists.

Furthermore, teachers, particularly in higher education, are concerned about the quality of distance education. Several regulatory and advisory bodies, including the congressional Web-based Education Commission, are presently establishing quality standards and guidelines (Carnevale 2001). And although a committee of the American Association of University Professors (AAUP) acknowledged that distance learning may be a "valuable pedagogical tool," it also questioned whether "academic quality, academic freedom, intellectual property rights and instructor's workloads and compensation" would be compromised (Arenson 1998, p. A14; AAUP 2003).

Nonetheless, distance education continues to grow, in part, because it is a money-maker—a multibillion-dollar industry. Even one-time critic William Bennett, former U.S. secretary of education, recently threw his hat into the corporate ring by opening K12, a company that markets elementary and secondary school courses to home schoolers and to parents who want their children to have academic help outside the classroom (Wildavsky 2001). In addition, many commercial sites now offer "educational" courses and, alternatively, educational sites increasingly carry advertising banners, consumer discounts, product photographs, and the like (Guernsey 2000).

Will distance learning solve all the problems facing education today? The answer is clearly no. Although distance education—from digital libraries to "virtual" charter schools—does provide a provocative and financially lucrative alternative to traditional education providers, it is not the technological fix some are looking for.

Vouchers, opponents argue, are also unfair to economically disadvantaged students who are not able to attend private schools because of the high tuition. In response to such criticisms, proponents note that the targeted value of a voucher—about $3,000—would be more than sufficient. For example, Education Department statistics indicate that the average cost of tuition for private schools, both primary and secondary, is $3,116 (Boaz and Barrett 2004).

Opponents also argue that the use of vouchers for religious schools violates the constitutional guarantee of separation of church and state. However, the U.S. Supreme Court, in reviewing the voucher program in Cleveland, Ohio, held that the use of tax dollars for enrollment in religious schools is not unconstitutional (Morse 2002c). Still, the debate continues. A recent Florida case wrestles with whether or not "opportunity scholarships" for students in low-rated schools can be used for religious institutions (Richard 2005).

Vouchers can also be used for charter schools. **Charter schools** originate in contracts, or charters, which articulate a plan of instruction that must be approved by local or state authorities. Although charter schools can be funded by foundations, universities, private benefactors, and entrepreneurs, many are supported by tax dollars (Mollison 2001). It is estimated that more than 3,000 charter schools are operating in 39 states and that they are serving 700,000 students (U.S. Charter Schools 2005). Charter schools, like school vouchers, were designed to expand schooling options and to increase the quality of education through competition. Recent evidence suggests that "charter schools produce student learning gains comparable to those of conventional schools, despite resource limitations" (Rand 2003, p. 1).

Some parents are choosing not to send their children to school at all but to teach them at home. In 2003 more than 1 million students in the United States were home-schooled, a significant increase from 850,000 in 1999 (NCES 2004c). For some parents, **home-schooling** is part of a fundamentalist movement to protect children from perceived non-Christian values in the public schools. Other parents are concerned about the quality of their children's education and their safety. How does being schooled at home instead of attending public school affect children? Some evidence suggests that home-schooled children perform as well as or better than their institutionally schooled counterparts (Webb 1989; Winters 2001).

Another choice parents can make is to send their children to a private school. About 11.5% of all students are enrolled in private elementary or private high schools (CAPE 2005). The primary reason parents send their children to private schools is for religious instruction. The second most common reason is the belief that private schools are superior to public schools in terms of academic achievement. Evidence may support this belief. Students in private schools generally perform higher on achievement tests than their public school counterparts (NCES 2003d). Parents also choose private schools for their children to have greater control over school policy, to avoid busing, or to obtain a specific course of instruction, such as dance or music.

Understanding Problems in Education

What can we conclude about the educational crisis in the United States? Any criticism of education must take into account that just over a century ago the United States had no systematic public education system at all. Many American children did not receive even a primary school education. Instead, they worked in factories and on farms to help support their families. Whatever education they received came from the family or the religious institution. In the mid-1800s educational reformer Horace Mann advocated at least five years of mandatory education for all U.S. children. Mann believed that mass education would function as the "balanced wheel of social machinery" to equalize social differences among members of an immigrant nation. His efforts resulted in the first compulsory education law in 1852, which required 12 weeks of attendance by school-age children each year. By World War I every state mandated primary school education, and by World War II secondary education was compulsory as well.

Public schools are supposed to provide all U.S. children with the academic and social foundations necessary to participate in society in a productive and meaningful way. But as conflict theorists note, for many children the educational institution perpetuates an endless downward cycle of failure, alienation, and hopelessness. Legally required protections can help. For example, in 2004, *Williams* v. *California,* a class action suit against the state of California on behalf of more than 1 million low-income students, was settled. Lawyers for the plaintiffs had argued that the students were permanently disadvantaged as a result of the schools they attended. The settlement, which is funded by a $1 billion appropriation, held that

"In order for all children to enter elementary school prepared to learn, we will have to eliminate child poverty in this country and provide every child who needs them not only with adequate health and social services but also with an early childhood education program well beyond anything envisioned by Head Start."

Evans Clinchy
Educational reformer

(1) all students must have needed textbooks and materials, (2) every classroom must be clean, safe, and in good condition, and (3) every teacher must meet standards set down by the state of California and federal law (UCLA 2005).

Breaking the downward cycle, however, is likely to become increasingly difficult as student enrollments swell and diversify, a consequence of the "baby-boom echo" and changing immigration patterns. Between 1989 and 2009 elementary school enrollment will increase 12%, high school enrollment will increase 19%, and full-time college student enrollment will increase 11% (U.S. Department of Education 2000).

The public, however, is supportive of government spending on education. In a national survey 77% of respondents favored paying teachers more, 81% favored placing more computers in classrooms, and 92% favored fixing run-down schools (NPR 2003). Legislation such as the 1998 amendments to the Higher Education Act (which allocated $300 billion to teacher preparation and recruitment, reduced interest on college student loans, and increased the maximum in Pell grants) is essential. But even with financial support, as structural functionalists argue, education alone cannot bear the burden of improving our schools.

Jobs must be provided for those who successfully complete their education. If not, students will have few incentives to exert any effort. Rosenbaum (2002) explains:

> What is missing from current practices is a mechanism for creating and conveying signals that tell students the value of their present actions in achieving desirable career goals. Other countries produce such signals with linkage mechanisms. . . . The German system provides a clear mechanism that makes the relationship between school performance and career option totally obvious [through an apprenticeship program]. . . . Students know that apprenticeships lead to respected occupations, and that school grades affect selection into apprenticeships. (p. 488)

Finally, "if we are to improve the skills and attitudes of future generations of workers, we must also focus attention and resources on the quality of the lives children lead outside the school" (Murnane 1994, p. 290). We must provide support to families so that children grow up in healthy, safe, and nurturing environments. Children are the future of our nation and of the world. Whatever resources we provide to improve the lives and education of children are sure to be wise investments in our collective future.

Chapter Review

ThomsonNOW™

Reviewing is as easy as ① ② ③

1. Before you do your final review, take the ThomsonNOW diagnostic quiz to help you identify the areas on which you should concentrate. You will find information on ThomsonNOW and instructions on how to access all of its great resources on the foldout at the beginning of the text.

2. As you review, take advantage of ThomsonNOW's study videos and interactive Map the Stats exercises to help you master the chapter topics.

3. When you are finished with your review, take ThomsonNOW's posttest to confirm you are ready to move on to the next chapter.

• **Do all countries educate their citizens?**

No. Many societies have no formal mechanism for educating the masses. As a result, worldwide, there are more than 860 million illiterate adults and 100 million children who have no access to formal education. The problem of illiteracy is greater in developing countries than in developed nations, and illiteracy rates are higher for women than for men (see Figure 8.1).

• **According to the structural-functionalist perspective, what are the functions of education?**

There are four major functions. The first is instruction—that is, teaching students knowledge and skills. The second is socialization that, for example, teaches students to respect authority. The third is sorting individuals into statuses by providing them with credentials. The fourth function is custodial care—a babysitting agency of sorts.

- **What is a self-fulfilling prophecy?**

  A self-fulfilling prophecy occurs when people act in a manner consistent with the expectations of others.

- **What variables predict school success?**

  Three variables tend to predict school success. Socioeconomic status predicts school success: the higher the socioeconomic status, the higher the likelihood of school success. Race predicts school success, with nonwhites and Hispanics having more academic difficulty than whites and non-Hispanics. Gender also predicts success, although it varies by grade level.

- **What were the conclusions of the study summarized in the *Social Problems Research Up Close* feature?**

  Roundtree (2000) concluded that (1) carrying a weapon to school is a rare event; (2) age, race, and sex were unrelated to carrying a weapon; (3) prior victimization and fear of crime were unrelated to gun carrying; (4) drug dealing or use was related to carrying a weapon to school; and (5) having a best friend who carried a gun to school was a strong predictor of weapon carrying.

- **What are some of the problems associated with the American school system?**

  One of the main problems is the lack of student achievement in our schools—particularly when U.S. data are compared with data from other industrialized countries. Minority dropout rates are high, and school violence continues to be a threat. School facilities are in need of repair and renovations, and personnel, including teachers, have been found to be deficient.

- **What are some of the solutions being considered to address problems in the schools?**

  The No Child Left Behind Act of 2001 is the primary means by which educational problems in the United States are being addressed. This act focuses on accountability through testing, flexibility of funding, expanding options for parents, and, perhaps most important, improving the quality of teaching in the schools.

- **What are the arguments for and against school choice?**

  Proponents of school choice programs argue that they reduce segregation and that schools, which have to compete with one another, will be of a higher quality. Opponents argue that school choice programs increase segregation and treat disadvantaged students unfairly. Low-income students cannot afford to go to private schools, even with vouchers. Furthermore, those opposed to school choice are quick to note that using government vouchers to help pay for religious schools is unconstitutional.

Critical Thinking

1. Clearly, home schooling has both advantages and disadvantages. After making a list of each, would you want your child to be homeschooled? Why or why not?
2. As discussed in Chapter 12, the proportion of elderly in the United States is increasing dramatically as we move into the 21st century. Because the elderly are unlikely to have children in public schools, how will the allocation of necessary school funds be affected by this demographic trend?
3. Student violence is one of the most pressing problems in U.S. public schools. One response is defensive— that is, having police patrol halls, using metal detectors, and so on. Other than such defensive tactics, what violence prevention techniques should be instituted?

Key Terms

alternative certification programs	Head Start
bilingual education	home schooling
bullying	integration hypothesis
character education	multicultural education
charter schools	privatization
cultural imperialism	school vouchers
distance learning	self-fulfilling prophecy
functionally illiterate	social promotion
	total immersion

Media Resources

The Companion Website for *Understanding Social Problems*, Fifth Edition

http://sociology.wadsworth.com/mooney_knox_schacht5e

Supplement your review of this chapter by going to the companion website to take one of the Tutorial Quizzes, use the flash cards to master key terms, and check out the many other study aids you'll find there. You'll also find special features such as *Wadsworth's Sociology Online Resources and Writing Companion,* GSS Data, and Census 2000 information, data, and resources at your fingertips to help you complete that special project or do some research on your own.

> " The 21st century will be the century in which we redefine ourselves as the first country in world history which is literally made up of every part of the world. " *Kenneth Prewitt, Census Bureau director*

Race, Ethnicity, and Immigration

I n July 2005 British Muslim leader Sheikh Dr. Zaki Badawi flew from London to New York to give a lecture titled "The Law and Religion in Society" at the Chautauqua Institution. Badawi, an Egyptian-born scholar in his 80s, holds a doctorate from the University of London and is the principal of the Muslim College in London. He has been knighted by Queen Elizabeth, has served as an adviser to Tony Blair, and is coeditor of an interfaith magazine. Upon arriving at JFK airport in New York, U.S. border officials detained Badawi for six hours and, without explanation, told Badawi that they had information indicating that Badawi was "inadmissible." So Badawi got on a plane and flew back to London. A customs spokesperson said, "We cannot disclose the information which led to the application being inadmissible because of privacy rules" (BBC 2005, n.p.) Writing about this event, columnist James Carroll (2005a) observed, "If a Muslim of Zaki Badawi's stature can be treated so contemptuously, imagine what the legion of anonymous Muslims face at the burgeoning network of checkpoints, security barriers, and borders that now define daily life" (n.p.).

From the perspective of U.S. customs officials, the barring of Muslim leader Zaki Badawi from entering the United States was justifiable, but from another point of view it is an example of the post–9/11 backlash of prejudice and discrimination against individuals who are (or who are perceived to be) Muslim or Middle Eastern. In the United States Muslims and people of Middle Eastern descent are minorities. A **minority group** is a category of people who have unequal access to positions of power, prestige, and wealth in a society and who tend to be targets of prejudice and discrimination. Minority status is not based on numerical representation in society but rather on social status. For example, before Nelson Mandela was elected president of South Africa, South African blacks suffered the disadvantages of a minority, even though they were a numerical majority of the population.

In this chapter we focus on prejudice and discrimination, their consequences for racial and ethnic minorities, and the strategies designed to reduce these problems.

Was the barring of Muslim leader Zaki Badawi from entering the United States justifiable, or was it an example of discrimination against individuals who are Muslim?

AP/Wide World Photos

We also examine issues related to U.S. immigration, because immigrants often bear the double burden of being minorities *and* foreigners who are not welcomed by many native-born Americans. We begin by examining racial and ethnic diversity worldwide and in the United States, emphasizing first that the concept of race is based on social rather than biological definitions.

The Global Context: Diversity Worldwide

A first-grade teacher asked the class, "What is the color of apples?" Most of the children answered red. A few said green. One boy raised his hand and said "white." The teacher tried to explain that apples could be red, green, or sometimes golden, but never white. The boy insisted his answer was right and finally said, "Look inside" (Goldstein 1999). Like apples, human beings may be similar on the "inside," but they are often classified into categories according to external appearance. After examining the social construction of racial categories, we review patterns of interaction among racial and ethnic groups and examine racial and ethnic diversity in the United States.

The Social Construction of Race

The concept of **race** refers to a category of people who are believed to share distinct physical characteristics that are deemed socially significant. Racial groups are sometimes distinguished on the basis of such physical characteristics as skin color, hair texture, facial features, and body shape and size. Some physical variations among people are the result of living for thousands of years in different geographic regions. For example, humans living in regions with more exposure to ultraviolet radiation from the sun developed darker skin from the natural skin pigment, melanin, which protects the skin from the sun's rays (Brace 2005). In regions with moderate or colder climates, people had no need for protection from the sun and thus developed lighter skin.

Cultural definitions of race have taught us to view race as a scientific categorization of people based on biological differences between groups of individuals. Yet racial categories are based more on social definitions than on biological differences.

Anthropologists note that distinctions among human populations are graded, not abrupt. Skin color is not black or white but rather ranges from dark to light with many gradations of shades. Noses are not either broad or narrow but come in a range of shapes. Physical traits such as these, as well as hair color and other characteristics, come in an infinite number of combinations. For example, a person with dark skin can have any blood type and can have a broad nose (a common combination in West Africa), a narrow nose (a common combination in East Africa), or even blond hair (a combination found in Australia and New Guinea) (Cohen 1998).

The science of genetics also challenges the notion of race. Geneticists have discovered that the genes of any two unrelated persons, chosen at random from around the globe, are 99.9% alike (Ossorio & Duster 2005). Furthermore, "most human genetic variation—approximately 85%—can be found between any two individuals from the same group (racial, ethnic, religious, etc.). Thus, the vast majority of variation is within-group variation" (Ossorio & Duster 2005, p. 117). Classifying people into different races fails to recognize that over the course of human history, migration and intermarriage have resulted in the blending of genetically transmitted traits. Thus there are no "pure" races; people in virtually all societies have genetically mixed backgrounds (Keita & Kittles 1997). And contrary to what some people believe, all humans belong to the same species.

The American Anthropological Association has passed a resolution stating that "differentiating species into biologically defined 'races' has proven meaningless and unscientific" (Etzioni 1997, p. 39). Scientists who reject the race concept now speak of *populations* when referring to groups that most people would call races (Zack 1998).

As clear evidence that race is a social rather than a biological concept, different societies construct different systems of racial classification, and these systems change over time. For example, "at one time in the not too distant past in the United States, Italians, Greeks, Jews, the Irish, and other 'white' ethnic groups were not considered to be white. Over time . . . the category of 'white' was reshaped to include them" (Rothenberg 2002, p. 3).

The significance of race is not biological but social and political, because race is used to separate "us" from "them" and becomes a basis for unequal treatment of one group by another. Despite the increasing acceptance that "there is no biological justification for the concept of 'race'" (Brace, 2005, p. 4), its social significance continues to be evident throughout the world.

Patterns of Racial and Ethnic Group Interaction

When two or more racial or ethnic groups come into contact, one of several patterns of interaction occurs, including genocide, expulsion or population transfer, colonialism, segregation, acculturation, pluralism, assimilation, and amalgamation. These patterns of interaction occur when two or more groups exist in the same society or when different groups from different societies come into contact. Although not all patterns of interaction between racial and ethnic groups are destructive, author and Mayan shaman Martin Prechtel reminds us, "Every human on this earth, whether from Africa, Asia, Europe, or the Americas, has ancestors whose stories, rituals, ingenuity, language, and life ways were taken away, enslaved, banned, exploited, twisted, or destroyed" (quoted by Jensen 2001, p. 13).

Genocide refers to the deliberate, systematic annihilation of an entire nation or people. The European invasion of the Americas, beginning in the 16th century, resulted in the decimation of most of the original inhabitants of North and South America. Some native groups were intentionally killed, whereas others fell victim to diseases brought by the Europeans. In the 20th century Hitler led the Nazi extermination of 12 million people, including 6 million Jews, in what is known as the Holocaust. More recently, in the early 1990s ethnic Serbs attempted to eliminate Muslims from parts of Bosnia—a process they called "ethnic cleansing." In 1994 genocide took pace in Rwanda when Hutus slaughtered hundreds of thousands of Tutsis (called "cockroaches" by the Hutus)—an event depicted in the 2004 film *Hotel Rwanda*. And as this book goes to press, genocide is occurring in the Darfur region of Sudan, where the Sudanese government, using Arab *Janjaweed* militias, its air force, and organized starvation, is systematically killing the black Sudanese population (see also Chapter 16).

Expulsion or **population transfer** occurs when a dominant group forces a subordinate group to leave the country or to live only in designated areas of the country. The 1830 Indian Removal Act called for the relocation of eastern tribes to land west of the Mississippi River. The movement, lasting more than a decade, has been

ThomsonNOW™

Learn more about **Patterns of Racial and Ethnic Group Interaction** by going through the Race and Ethnic Relations Data Experiment.

In 1994 Hutus in Rwanda committed genocide against the Tutsis, resulting in 800,000 deaths.

called the Trail of Tears because tribes were forced to leave their ancestral lands and endure harsh conditions of inadequate supplies and epidemics that caused illness and death. After Japan's attack on Pearl Harbor in 1941, President Franklin Roosevelt authorized the removal of any people considered threats to national security. All people on the West Coast with at least one-eighth Japanese ancestry were transferred to evacuation camps surrounded by barbed wire, where 120,000 Japanese Americans experienced economic and psychological devastation. In 1979 Vietnam expelled nearly 1 million Chinese from the country as a result of long-standing hostilities between China and Vietnam.

Colonialism occurs when a racial or ethnic group from one society takes over and dominates the racial or ethnic group(s) of another society. The European invasion of North America, the British occupation of India, and the Dutch presence in South Africa before the end of apartheid are examples of outsiders taking over a country and controlling the native population. As a territory of the United States, Puerto Rico is essentially a colony whose residents are U.S. citizens, but they cannot vote in presidential elections unless they move to the mainland.

Segregation refers to the physical separation of two groups in residence, workplace, and social functions. Segregation can be **de jure** (Latin meaning "by law") or **de facto** ("in fact"). Between 1890 and 1910 a series of U.S. laws, which came to be known as **Jim Crow laws,** were enacted to separate blacks from whites by prohibiting blacks from using "white" buses, hotels, restaurants, and drinking fountains. In 1896 the U.S. Supreme Court (in *Plessy* v. *Ferguson*) supported de jure segregation of blacks and whites by declaring that "separate but equal" facilities were constitutional. Blacks were forced to live in separate neighborhoods and attend separate schools. Beginning in the 1950s various rulings overturned these Jim Crow laws, making it illegal to enforce racial segregation. Although de jure segregation is illegal in the United States, de facto segregation still exists in the tendency for racial and ethnic groups to live and go to school in segregated neighborhoods.

Acculturation refers to adopting the culture of a group different from the one in which a person was originally raised. Acculturation may involve learning the dominant language, adopting new values and behaviors, and changing the spelling of the family name. In some instances acculturation may be forced, as in the California decision to discontinue bilingual education and force students to learn English in school.

Pluralism refers to a state in which racial and ethnic groups maintain their distinctness but respect each other and have equal access to social resources. In Switzerland, for example, four ethnic groups—French, Italians, Germans, and Swiss Germans—maintain their distinct cultural heritage and group identity in an atmosphere of mutual respect and social equality. In the United States the political and educational recognition of multiculturalism reflects efforts to promote pluralism.

Assimilation is the process by which formerly distinct and separate groups merge and become integrated as one. Assimilation is sometimes referred to as the "melting pot," whereby different groups come together and contribute equally to a new, common culture. Although the United States has been referred to as a melting pot, in reality, many minorities have been excluded or limited in their cultural contributions to the predominant white Anglo-Saxon Protestant tradition.

Assimilation can be of two types: secondary and primary. **Secondary assimilation** occurs when different groups become integrated in public areas and in social institutions, such as neighborhoods, schools, the workplace, and in government. **Primary assimilation** occurs when members of different groups are integrated in personal, intimate associations, as with friends, family, and spouses.

Amalgamation, also known as **marital assimilation,** occurs when different ethnic or racial groups become married or pair-bonded and produce children. Nineteen

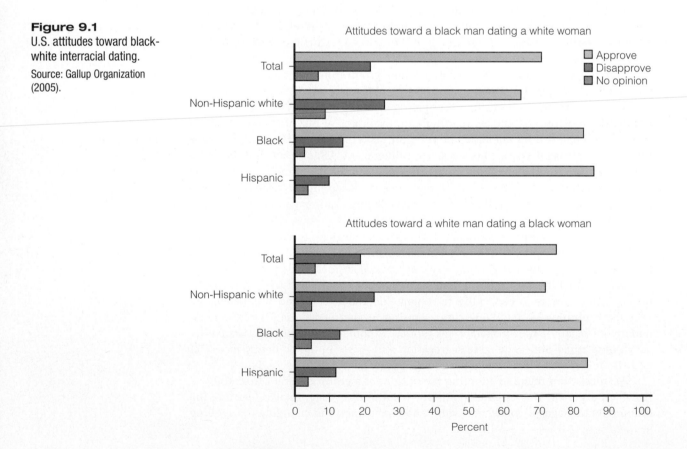

Figure 9.1
U.S. attitudes toward black-white interracial dating.
Source: Gallup Organization (2005).

states had **antimiscegenation laws** banning interracial marriage until 1967, when the Supreme Court (in *Loving* v. *Virginia*) declared these laws unconstitutional.

Since 1960 the number of black-white married couples has increased fivefold; the number of Asian-white married couples has increased more than tenfold; and the number of Hispanics married to non-Hispanics has tripled. This rise in non-traditional unions is related to the increased independence that adult children have from their parents and communities of origin as well as to increased societal acceptance of mixed marriages (Rosenfeld & Kim 2005). Although interracial marriages are more common today than in previous generations, less than 5% of U.S. married couples in 2003 were interracial (Fields 2004). As shown in Figure 9.1, about one in five U.S. adults disapproves of black-white interracial dating.

Racial and Ethnic Group Diversity and Relations in the United States

The racial and ethnic characteristics of the U.S. population are becoming increasingly diversified. In four states—California, New Mexico, Hawaii, and Texas—minority populations outnumber non-Hispanic whites. In these states Hispanics are the largest minority group, except for in Hawaii where the largest minority group is Asian Americans ("National Briefing," 2005).

Racial Diversity in the United States

The first census in 1790 divided the U.S. population into four groups: free white males, free white females, slaves, and other persons (including free blacks and Indians). To increase the size of the slave population, the **one-drop rule** appeared, which specified that even one drop of "Negroid" blood defined a person as black and therefore eligible for slavery. In 1960 the census recognized only two categories: white and nonwhite. In 1970 the census categories consisted of white, black, and "other" (Hodgkinson 1995). In 1990 the U.S. Bureau of the Census recognized four racial classifications: (1) white, (2) black, (3) American Indian, Aleut, or Eskimo, and (4) Asian or Pacific Islander. The 1990 census also included the category of "other." Beginning with the 2000 census the Office of Management and Budget required federal agencies to use a minimum of five race categories: (1) white, (2) black or African American, (3) American Indian or Alaska Native, (4) Asian, and (5) Native Hawaiian or other Pacific Islander (Grieco & Cassidy 2001). In addition, respondents to federal surveys and the census now have the option of officially identifying themselves as being more than one race rather than checking only one racial category (see Figure 9.2). Figure 9.3 presents the 2000 census data on the racial composition of the United States.

Mixed-Race Identity. About 4.5% of U.S. male-female married couples and 10% of unmarried couples are interracial (Fields 2004). The new census option for identifying as "mixed race" avoids putting children of mixed-race parents in the difficult position of choosing the race of one parent over the other when filling out race data on school and other forms. It also avoids impairment of children's self-esteem and social functioning that comes from choosing the racial category of "other."

Golf pro Tiger Woods has referred to himself as "Cablinasian"—reflecting his mixed heritage, which includes Caucasian, black, Asian, and Indian.

Reuters NewMedia Inc./Corbis

Figure 9.2

Reproduction of questions on race and Hispanic origin from the 2000 census.

Source: U.S. Census Bureau, Census 2000 questionnaire.

⟶ NOTE: Please answer BOTH questions 5 and 6.

5. Is this person Spanish/Hispanic/Latino? Mark ☒ the "No" box if **not** Spanish/Hispanic/Latino.

☐ **No,** not Spanish/Hispanic/Latino   ☐ Yes, Puerto Rican
☐ Yes, Mexican, Mexican Am., Chicano   ☐ Yes, Cuban
☐ Yes, other Spanish/Hispanic/Latino — *Print group.*↗

6. What is this person's race? Mark ☒ **one or more races** to indicate what this person considers himself/herself to be.

☐ White
☐ Black, African Am., or Negro
☐ American Indian or Alaska Native — *Print name of enrolled or principal tribe.*↗

☐ Asian Indian   ☐ Japanese   ☐ Native Hawaiian
☐ Chinese   ☐ Korean   ☐ Guamanian or Chamorro
☐ Filipino   ☐ Vietnamese   ☐ Samoan
☐ Other Asian — *Print race.*↗   ☐ Other Pacific Islander — *Print race.*↗

☐ Some other race — *Print race.*↗

Figure 9.3

Race composition of the United States, 2000.

Source: Grieco & Cassidy (2001).

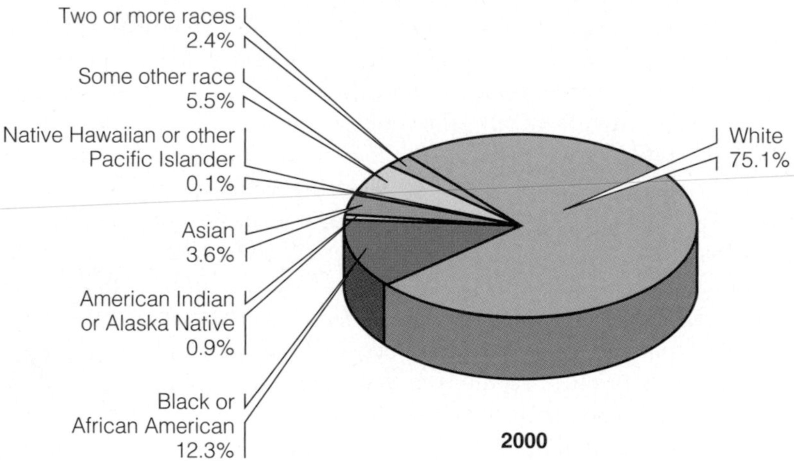

Two or more races 2.4%
Some other race 5.5%
Native Hawaiian or other Pacific Islander 0.1%
Asian 3.6%
American Indian or Alaska Native 0.9%
Black or African American 12.3%
White 75.1%

2000

Such a category implies that the society does not recognize and respect mixed-race individuals, and thus "children growing up within mixed families may feel ashamed of their 'irregular' racial makeup and may experience rejection and alienation in the wider social community" (Zack 1998, p. 23).

Some critics of the new mixed-race option are concerned that the wide-scale recognition of mixed-race identity will decrease the numbers within minority groups and disrupt the solidarity and loyalty based on racial identification. For example, what will happen to organizations and movements devoted to equal rights for blacks if much of the "black" population acquires a new mixed-race identity? However, the 2000 census data suggest that the mixed-race option will not have the large national impact that critics fear. As shown in Figure 9.3, only 2.4% of the U.S. population identified themselves as being of more than one race in the 2000 census. About 5% of U.S. black people said that they were multiracial (Schmitt 2001b).

Ethnic Diversity in the United States

Ethnicity refers to a shared cultural heritage or nationality. Ethnic groups can be distinguished on the basis of language, forms of family structures and roles of family members, religious beliefs and practices, dietary customs, forms of artistic expression such as music and dance, and national origin.

Two individuals with the same racial identity may have different ethnicities. For example, a black American and a black Jamaican have different cultural, or ethnic, backgrounds. Conversely, two individuals with the same ethnic background may identify with different races. Although most Hispanics and Latinos are white, in the 2000 census 2% of Latinos identified themselves as black (Navarro 2003).

The current Census Bureau classification system does not allow people of mixed Hispanic or Latino ethnicity to identify themselves as such. Individuals with one Hispanic and one non-Hispanic parent still must say that they are either Hispanic or not Hispanic. In addition, Hispanics must select one country of origin, even if their parents are from different countries.

U.S. citizens come from a variety of ethnic backgrounds. The largest ethnic population in the United States is of Hispanic origin. (The terms "Hispanic" and "Latino" are used interchangeably here.) More than one in eight (13.3%) people in the United States are Hispanic or Latino and two-thirds (66.9%) of all U.S. Hispanics or Latinos are of Mexican origin (Ramirez & de la Cruz 2003). Other Hispanic Americans have origins in Central and South America (14.3%), Puerto Rico (8.6%), and Cuba (3.7%).

The use of racial and ethnic labels is often misleading. The ethnic classification of "Hispanic/Latino," for example, lumps together such disparate groups as Puerto Ricans, Mexicans, Cubans, Venezuelans, Colombians, and others from Latin American countries. The racial term *American Indian* includes more than 300 separate tribal groups that differ enormously in language, tradition, and social structure. The racial label *Asian American* includes individuals with ancestors from China, Japan, Korea, India, the Philippines, or one of the countries of Southeast Asia.

Race and Ethnic Group Relations in the United States

Despite significant improvements over the last two centuries, race and ethnic group relations continue to be problematic. The racial divide in the United States sharpened in 2005 in the wake of Hurricane Katrina, which left victims—who were predominantly black and poor—waiting for days to be rescued from their flooded attics or rooftops or to be evacuated from overcrowded "shelters" where there was no food, water, medical supplies, or working

Following Hurricane Katrina in 2005, a national survey found that the majority of blacks believe that the government's response to the crisis would have been faster if most of Katrina's victims had been white.

AP/Wide World Photos

Figure 9.4

Perceptions of race and ethnic relations in the United States, 2005. *A national sample of U.S. adults was asked to rate relations between various groups in the United States. The results are depicted in this graph.*

Source: Gallup Organization (2005).

*Non-Hispanic whites.

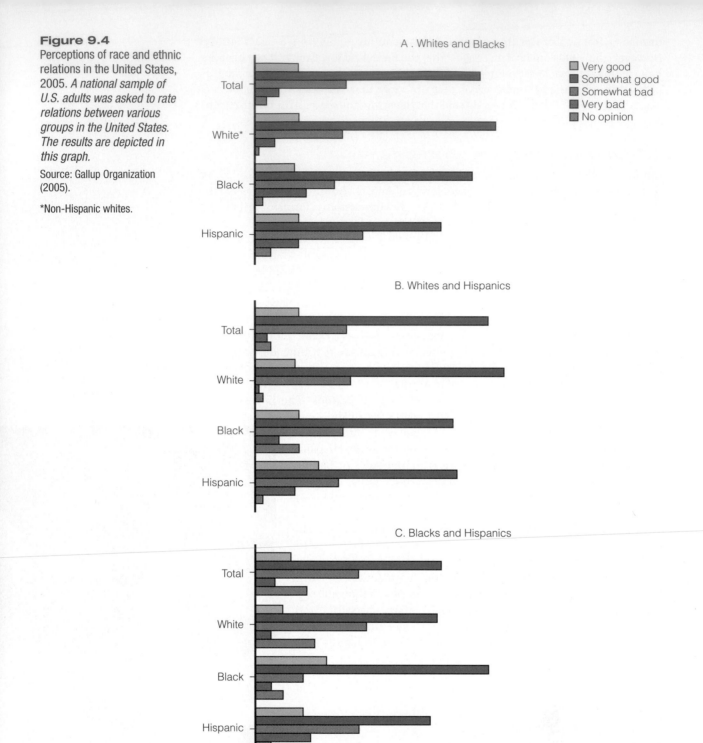

toilets. Following this crisis, a national survey found that 7 in 10 blacks (71%) said that the Katrina disaster shows that racial inequality remains a major problem in the United States, whereas a majority of whites (56%) said that this was not a particularly important lesson of Katrina. The same survey found that most blacks (77%), compared to only 17% of whites, believe that the government's response to the disaster would have been faster if most of Katrina's victims had been white (Pew Research Center 2005).

In response to a question asking whether relations between blacks and whites will always be a problem for the United States or whether a solution will eventually be worked out, 46% of non-Hispanic whites and 57% of blacks said that race relations will always be a problem (Gallup Organization 2005). A Gallup Poll survey asked a national sample of U.S. adults to rate relations between various groups in the United States. Results of this survey are presented in Figure 9.4. Relations between various racial and ethnic groups are influenced by prejudice and discrimination (discussed later in this chapter). Race and ethnic relations are also complicated by issues concerning immigration—the topic we turn to next.

Immigrants in America

The growing racial and ethnic diversity of the United States is largely the result of immigration as well as the higher average birthrates among many minority groups. The many hardships experienced by poor people throughout the world "push" them to leave their home countries, whereas the economic opportunities that exist in more affluent countries "pull" them to those countries. Before reading further, you may want to assess your attitudes toward U.S. immigrants and immigration in this chapter's *Self and Society* feature.

U.S. Immigration: A Historical Perspective

For the first 100 years of U.S. history all immigrants were allowed to enter and become permanent residents. The continuing influx of immigrants, especially those coming from nonwhite, non-European countries, created fear and resentment among native-born Americans, who competed with immigrants for jobs and who held racist views toward some racial and ethnic immigrant populations. Increasing pressures from U.S. citizens to restrict or halt entirely the immigration of various national groups led to legislation that did just that. America's open door policy on immigration ended in 1882 with the Chinese Exclusion Act, which suspended for 10 years the entrance of the Chinese to the United States and declared Chinese ineligible for U.S. citizenship. The Immigration Act of 1917 required all immigrants to pass a literacy test before entering the United States. And in 1921 the Johnson Act introduced a limit on the number of immigrants who could enter the country in a single year, with stricter limitations for certain countries (including those in Africa and the Near East). The 1924 Immigration Act further limited the number of immigrants allowed into the United States and completely excluded the Japanese. Other federal immigration laws include the 1943 Repeal of the Chinese Exclusion Act, the 1948 Displaced Persons Act (which permitted refugees from Europe), and the 1952 Immigration and Naturalization Act (which permitted a quota of Japanese immigrants).

In the 1960s most immigrants were from Europe, but today most immigrants are from Central America (predominantly Mexico) or Asia (see Figure 9.6). In 2003 more than 1 in 10 U.S. residents (11.7%) were born in a foreign country (Larsen

Attitudes Toward U.S. Immigrants and Immigration

For each of the following questions, choose the answer that best reflects your attitudes. After answering all the questions, compare your answers with the results of a national Gallup poll sample of U.S. adults.

1. On the whole, do you think immigration is a good thing or a bad thing for this country?
   (a) Good thing
   (b) Bad thing
2. Do you think that immigrants (a) mostly help the economy by providing low-cost labor or (b) mostly hurt the economy by driving wages down for many Americans?
   (a) Mostly help
   (b) Mostly hurt
3. Which comes closer to your point of view: (a) Immigrants in the long run become productive citizens and pay their fair share of taxes, or (b) immigrants cost the taxpayers too much by using government services like public education and medical services?
   (a) Pay fair share
   (b) Cost too much
4. Do you think the United States should or should not make it easier for illegal immigrants to become citizens of the United States?
   (a) Should make it easier
   (b) Should not make it easier

Comparison Data

A Gallup poll found the following attitudes toward U.S. immigrants and immigration (see Figure 9.5).

Source: Gallup Organization (2005)

2004) (see Table 9.1), and 1 in 5 U.S. children, age 18 and younger, is the child of an immigrant (Fix & Capps 2002).

Being a U.S. Immigrant: Challenges and Achievements

Many immigrants struggle to adjust to life in the United States. Despite the prejudice, discrimination, and lack of social support they experience, many foreign-born U.S. residents work hard to succeed educationally and occupationally. Although immigrants are less likely than U.S. natives to graduate from high school, the percentage of foreign-born residents with a bachelor's degree or more education (27.3%) is nearly equivalent to that of the native-born population (27.2%) (Larsen 2004).

Immigrant employment is relatively concentrated in a small number of sectors, including building and grounds maintenance, food preparation, and construction (Holzer 2005). In a national survey about half (51%) of nonimmigrants said that recent immigrants take jobs away from Americans who want them; 15% said that they did not get a job because it was given to an immigrant instead (NPR et al. 2005). Although immigrant labor displaces some U.S. workers, over the long term immigrants have modest negative effects on the employment of less educated U.S. workers and generate significant benefits for the U.S. economy. Immigrants provide labor in sectors where shortages might occur otherwise, and immigrant labor helps

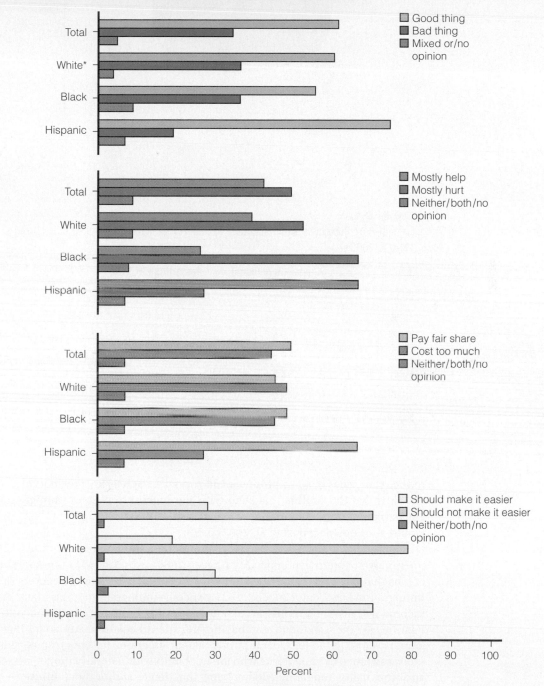

Figure 9.5

Results from a Gallup poll on attitudes toward immigrants and immigration.

Source: Gallup Organization (2005).

*"White" refers to non-Hispanic whites.

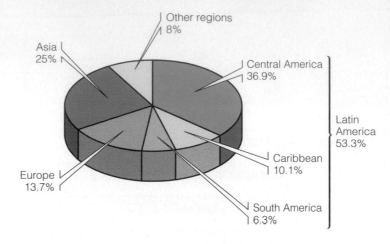

Table 9.1

Foreign-Born Population and Percentage of Total Population for the United States from 1890 to 2003

Year	Number (in Millions)	Percentage of Total
2003	33.5	11.7
1990	19.8	7.9
1970	9.6	4.7
1950	10.3	6.9
1930	14.2	11.6
1890	9.2	14.8

Source: Larsen (2004) and Lollock (2001).

reduce the prices of some products and services, such as housing and food (Holzer 2005). In North Carolina, the state with the highest growth of immigrants in the 1990s, one county commissioner noted that without immigration "our economy would shut down" (Foust et al. 2002, p. 82).

Compared to the native-born U.S. population, immigrants are more likely to be unemployed and to live in poverty (Larsen 2004). Almost 43% of immigrants work at jobs that pay less than $7.50 an hour, compared to 28% of all workers (National Immigration Law Center 2005b). Yet low-income immigrants are less likely than low-income natives to use benefits such as Medicaid, Temporary Assistance to Needy Families (TANF), and food stamps. In part, this is because many noncitizen immigrants are ineligible for federal public benefit programs (although some states provide assistance to immigrants who are not eligible for federally funded services). In addition, many immigrants do not know that they—and/or their children—may be eligible to receive welfare benefits, or they are afraid that seeking benefits will have a negative effect on their legal status and citizenship (Capps et al. 2005).

Table 9.2

Undocumented U.S. Immigrants by Country of Origin

Country of Origin	Percentage of Undocumented Immigrants
Mexico	57
Other Latin American countries	23
Asia	10
Europe and Canada	5
Other parts of the world	5

Source: Passel et al. (2004).

Illegal Immigration

Illegal immigration is an ongoing concern in the United States. There are 9.3 million undocumented immigrants (i.e., "illegal immigrants") in the United States, representing about one-fourth of the total foreign-born population (Passel et al. 2004). More than half of undocumented immigrants (57%) are Mexican (see Table 9.2).

Nearly 7 out of 10 adults—72% of nonimmigrants and 48% of immigrants—say that they are concerned about illegal immigration (NPR et al. 2005). More than half of nonimmigrants are concerned that (1) taxpayers pay for services such as schools and health care for illegal immigrants, (2) too many people are coming into this country, (3) illegal immigration increases the likelihood of terrorism, and (4) "the wrong kinds of people are coming into the country." The biggest concern among immigrants (43%) is that illegal immigration is dangerous for immigrants.

Border Crossing. The U.S. Border Patrol is charged with deterring people from illegally crossing the border into the United States and apprehending those that do. In fiscal year 2004 the U.S. Border Patrol apprehended 1 million people at the Southwest border—an increase of 26% over the previous year (Office of Immigration Statistics 2005). In addition, some groups of U.S. citizens have taken action to try to prevent illegal border crossing. For example, in 2003 a Texas rancher invited a vigilante group called Ranch Rescue to his property to stop undocumented Latinos from crossing his land. Fatima Leiva and Edwin Mancia, Salvadorans, were among a group of immigrants traveling on foot when members of Ranch Rescue detained them. During the detention, Mancia was struck on the back of the head and attacked by a Rottweiler dog owned by a Ranch Rescue member. Although Ranch Rescue purports to be protecting the United States from

Immigrant men commonly work in construction jobs, as well as in landscaping and agriculture.

Courtesy of Caroline Schacht and David Knox

illegal immigration, the underlying motive for their anti-immigrant actions is hate. The president and spokesperson for Ranch Rescue has described Mexicans as "dog turds" who are "ignorant, uneducated and desperate for a life in a decent nation because the one [they] live in is nothing but a pile of dog [excrement] made up of millions of little dog turds" (quoted in SPLC 2005a, p. 3).

Despite efforts to seal the U.S.–Mexican border, illegal border crossings occur every day. Some people cross (or attempt to cross) the U.S.–Mexican border with the help of a *coyote*—a hired guide who typically charges $200–$300 to lead people across the border. Crossing the border illegally involves a number of risks, including death from drowning (e.g., while trying to cross the Rio Grande) or dehydration. In 2003, 19 undocumented immigrants died of dehydration, suffocation, and overheating in the back of a tractor-trailer that had reached 173°. Truck driver Tyrone Williams, who was paid $7,500 to transport the immigrants, was convicted of transporting illegal immigrants but was spared the death penalty when jurors could not determine whether he was directly responsible for the deaths (Nichols 2005).

Undocumented Immigrants in the Workforce. An estimated 5% of U.S. workers are undocumented immigrants (Passel et al. 2004). Virtually all undocumented men are in the labor force. Their labor force participation exceeds that of men who are legal immigrants or who are U.S. citizens because undocumented men are less likely to be disabled, retired, or in school. Undocumented women are less likely to be in the labor force (62%) than undocumented men because they are more likely to be stay-at-home mothers (Passel et al. 2004). "The vast majority of illegal immigrants leave their home countries to work hard, save their money, then return to their homeland. . . . These individuals do not travel their difficult and dangerous journeys searching for a welfare handout; they immigrate to work" (Maril 2004, pp. 11–12).

Undocumented workers often do work that U.S. workers are unwilling to do. Workers routinely work 60 or more hours per week and earn less than the minimum wage of $5.15 per hour. They are not paid overtime and have no benefits. Illegal workers who cleaned Wal-Mart stores generally worked 7 nights a week, 364 days a year, and were often locked in stores (Greenhouse 2005). Immigrant workers who are legally admitted to the United States to work under the temporary foreign worker visa program also endure abusive labor conditions. "Because of language barriers and their vulnerable status under immigration laws, these workers may be the most exploited in the nation" (Bauer, quoted in SPLC 2005c) (see also Chapter 7).

In 1986 Congress approved the Immigration Reform and Control Act, which made hiring illegal immigrants an illegal act punishable by fines and even prison sentences. In 2005 Wal-Mart agreed to pay a record $11 million to settle charges that it used hundreds of illegal immigrants to clean its stores (Greenhouse 2005). This act also prohibits employers from discriminating against legal immigrants who are not U.S. citizens.

Becoming a U.S. Citizen

Foreign-born residents of the United States may or may not apply for and be granted U.S. citizenship. In 2000, of all foreign-born U.S. residents, 35% were **naturalized citizens** (immigrants who applied and met the requirements for U.S. citizenship); 65% were not U.S. citizens (Lollock 2001).

To become a U.S. citizen, immigrants must have been lawfully admitted for permanent residence; must have resided continuously as a lawful permanent U.S. resident for at least five years; must be able to read, write, speak, and understand

Should Undocumented Immigrants Qualify for In-State College Tuition?

An estimated 50,000 to 65,000 undocumented immigrants earned high school diplomas in the United States in spring 2005 (Chu 2005). Although the Supreme Court ruled in 1982 that states must educate illegal immigrants through the twelfth grade, states are divided in their laws regarding supporting illegal immigrants through in-state tuition. Since 2001 nine states have passed laws permitting undocumented students to receive in-state tuition if they (1) attend a school in the state for a certain number of years, (2) graduate from high school, and (3) sign an affidavit stating that they have either applied to legalize their status or will do so as soon as eligible (National Immigration Law Center 2005a). The states with these laws are Texas, California, Utah, Kansas, Washington, New York, Oklahoma, Illinois, and New Mexico. But the Kansas law has been challenged in court, and legislators in 18 other states have denied immigrants in-state tuition. Immigrants' rights advocates argue that it is in the best interest of the United States to help illegal immigrant youth achieve a college education. But opponents worry that granting in-state tuition to undocumented immigrants would (1) be a financial burden to taxpayers, (2) make it harder for their children to get accepted into state universities, and (3) open the door to granting other benefits now reserved to state residents.

It Is Your Turn Now to Take a Stand!

Do you think that undocumented immigrants should qualify for in-state college tuition? Why or why not?

basic English (certain exemptions apply); and must show that they have "good moral character" (U.S. Citizenship and Immigration Services 2003). Applicants who have been convicted of murder or an aggravated felony are permanently denied U.S. citizenship. In addition, applicants are denied if in the last five years they have engaged in any one of a variety of offenses, including prostitution, illegal gambling, controlled substance law violation (except for a single offense of possession of 30 grams or less of marijuana), habitual drunkenness, willful failure or refusal to support dependents, and criminal behavior involving "moral turpitude."

To become a U.S. citizen, one must take the oath of allegiance and swear to support the Constitution and obey U.S. laws, renounce any foreign allegiance, and bear arms for the U.S. military or perform services for the U.S. government when required. Finally, applicants for U.S. citizenship must pass an examination on U.S. government and history administered by the U.S. Citizenship and Immigration Services.

Sociological Theories of Race and Ethnic Relations

Some theories of race and ethnic relations suggest that individuals with certain personality types are more likely to be prejudiced or to direct hostility toward minority group members. Sociologists, however, concentrate on the impact of the structure and culture of society on race and ethnic relations. Three major sociological theories lend insight into the continued subordination of minorities.

Structural-Functionalist Perspective

Structural functionalists emphasize that each component of society contributes to the stability of the whole. In the past, inequality between majority and minority groups was functional for some groups in society. For example, the belief in the superiority of one group over another provided moral justification for slavery, supplying the South with the means to develop an agricultural economy based on cotton. Furthermore, Southern whites perpetuated the belief that emancipation would be detrimental for blacks, who were highly dependent on their "white masters" for survival (Nash 1962).

Structural functionalists recognize, however, that racial and ethnic inequality is also dysfunctional for society (Schaefer 1998; Williams & Morris 1993). A society that practices discrimination fails to develop and utilize the resources of minority members. Prejudice and discrimination aggravate social problems, such as crime and violence, war, unemployment and poverty, health problems, family problems, urban decay, and drug use—problems that cause human suffering as well as impose financial burdens on individuals and society.

The structural-functionalist analysis of manifest and latent functions also sheds light on issues of race and ethnic relations. For example, the manifest function of the civil rights legislation in the 1960s was to improve conditions for racial minorities. However, civil rights legislation produced an unexpected negative consequence, or latent dysfunction. Because civil rights legislation supposedly ended racial discrimination, whites were more likely to blame blacks for their social disadvantages and thus perpetuate negative stereotypes such as "blacks lack motivation" and "blacks have less ability" (Schuman & Krysan 1999).

Conflict Perspective

The conflict perspective examines how competition over wealth, power, and prestige contributes to racial and ethnic group tensions. Consistent with this perspective, the "racial threat" hypothesis views white racism as a response to perceived or actual threats to whites' economic well-being or cultural dominance by minorities. For example, between 1840 and 1870 large numbers of Chinese immigrants came to the United States to work in mining (the California Gold Rush of 1848), railroads (the transcontinental railroad, completed in 1860), and construction. As Chinese workers displaced whites, anti-Chinese sentiment rose, resulting in increased prejudice and discrimination and the eventual passage of the Chinese Exclusion Act of 1882, which restricted Chinese immigration until 1924. More recently, white support for Proposition 209—a 1996 resolution passed in California that ended state affirmative action programs—was higher in areas with larger Latino, African American, or Asian American populations, even after controlling for other factors (Tolbert & Grummel 2003). In other words, opposition to affirmative action programs that help minorities was higher in areas with greater racial and ethnic diversity, suggesting that whites living in diverse areas felt more threatened by the minorities.

In another study researchers interviewed individuals in white racist Internet chat rooms to examine the extent to which people would advocate interracial violence in response to alleged economic and cultural threats (Glaser et al. 2002). The researchers posed three scenarios that might be perceived as threatening: interracial marriage, minority in-migration (i.e., blacks moving into one's neighborhood), and job competition (i.e., competing with a black person for a job). Respondents' reactions to interracial marriage were the most volatile, followed by in-migration. The researchers concluded that violent ideation among white racists stems from perceived threats to white cultural dominance and separateness rather than from perceived economic threats.

Furthermore, conflict theorists suggest that capitalists profit by maintaining a surplus labor force, that is, by having more workers than are needed. A surplus labor force ensures that wages will remain low because someone is always available to take a disgruntled worker's place. Minorities who are disproportionately unemployed serve the interests of the business owners by providing surplus labor, keeping wages low, and, consequently, enabling them to maximize profits. Some of immigration's biggest supporters are business leaders who want to keep wages low. Sociology professor Stephen Klineberg noted that "the accelerating immigration of Hispanics has meant massively downward pressure on their wages, negotiating power, and ability to confront their employers over violations of their rights" (quoted by Greenhouse 2003, n.p.).

Conflict theorists also argue that the wealthy and powerful elite fosters negative attitudes toward minorities to maintain racial and ethnic tensions among workers. So long as workers are divided along racial and ethnic lines, they are less likely to join forces to advance their own interests at the expense of the capitalists. In addition, the "haves" perpetuate racial and ethnic tensions among the "have-nots" to deflect attention away from their own greed and exploitation of workers.

Symbolic Interactionist Perspective

The symbolic interactionist perspective focuses on the social construction of race and ethnicity—how we learn conceptions and meanings of racial and ethnic distinctions through interaction with others—and how meanings, labels, and definitions affect racial and ethnic groups. We have already explained that contemporary race scholars agree that there is no scientific, biological basis for racial categorizations. However, people have learned to think of racial categories as real, and, as the *Thomas Theorem* suggests, if things are defined as real, they are real in their consequences. Ossorio and Duster (2005) explain:

> People often interact with each other on the basis of their beliefs that race reflects physical, intellectual, moral, or spiritual superiority or inferiority. . . . By acting on their beliefs about race, people create a society in which individuals of one group have greater access to the goods of society— such as high-status jobs, good schooling, good housing, and good medical care—than do individuals of another group. (p. 119)

The labeling perspective directs us to consider the role that negative stereotypes play in race and ethnicity. **Stereotypes** are exaggerations or generalizations about the characteristics and behavior of a particular group. When Americans in a 1990 National Opinion Research Center poll were asked to evaluate various racial and ethnic groups, blacks were rated least favorably (Shipler 1998). Most of the respondents labeled blacks as less intelligent than whites (53%), lazier than whites (62%), and more likely than whites to prefer being on welfare to being self-supporting (78%). Negative stereotyping of minorities leads to a self-fulfilling prophecy. As Schaefer (1998) explains:

> Self-fulfilling prophecies can be devastating for minority groups. Such groups often find that they are allowed to hold only low-paying jobs with little prestige or opportunity for advancement. The rationale of the dominant society is that these minority individuals lack the ability to perform in more important and lucrative positions. Training to become scientists, executives, or physicians is denied to many subordinate

group individuals, who are then locked into society's inferior jobs. As a result, the false definition becomes real. The subordinate group has become inferior because it was defined at the start as inferior and was therefore prevented from achieving the levels attained by the majority. (p. 17)

The symbolic interactionist perspective is concerned with how individuals learn negative stereotypes and prejudicial attitudes through language. Different connotations of the colors white and black, for example, may contribute to negative attitudes toward people of color. The white knight is good, and the black knight is evil; angel food cake is white, and devil's food cake is black. Other negative terms associated with black include black sheep, black plague, black magic, black mass, blackballed, and blacklisted. The continued use of such derogatory terms as *Jap, Gook, Spic, Frog, Kraut, Coon, Chink, Wop, Towel-head, Kike,* and *Mick* also confirms the power of language in perpetuating negative attitudes toward minority group members. Similarly, advocates for immigrant rights suggest that the term *illegal aliens* is derogatory; they prefer the term *undocumented worker* as a more neutral term.

In the next section we explore the concepts of prejudice and racism in more depth and discuss ways in which socialization and media perpetuate negative stereotypes and prejudicial attitudes toward racial and ethnic groups.

Prejudice and Racism

Prejudice refers to negative attitudes and feelings toward or about an entire category of people. Prejudice can be directed toward individuals of a particular religion, sexual orientation, political affiliation, age, social class, sex, race, or ethnicity. **Racism** is the belief that race accounts for differences in human character and ability and that a particular race is superior to others.

Aversive and Modern Racism

Compared to traditional, "old-fashioned" prejudice, which is blatant, direct, and conscious, contemporary forms of prejudice are often subtle, indirect, and unconscious. Two variants of these more subtle forms of prejudice are aversive racism and modern racism.

Aversive Racism. Aversive racism represents a subtle, often unintentional form of prejudice exhibited by many well-intentioned white Americans who possess strong egalitarian values and who view themselves as nonprejudiced. The negative feelings that aversive racists have toward blacks and other minority groups are not feelings of hostility or hate but rather feelings of discomfort, uneasiness, disgust, and sometimes fear (Gaertner & Dovidio 2000). Aversive racists may not be fully aware that they harbor these negative racial feelings; indeed, they disapprove of individuals who are prejudiced and would feel falsely accused if they were labeled prejudiced. "Aversive racists find Blacks 'aversive,' while at the same time find any suggestion that they might be prejudiced 'aversive' as well" (Gaertner & Dovidio 2000, p. 14).

Another aspect of aversive racism is the presence of pro-white attitudes, as opposed to antiblack attitudes. In several studies respondents did not indicate that blacks were worse than whites, only that whites were better than blacks (Gaertner & Dovidio 2000). For example, blacks were not rated as being lazier than whites, but whites were rated as being more ambitious than blacks. Gaertner and Dovidio (2000) explain that "aversive racists would not characterize blacks more negatively than whites because that response could readily be interpreted by others or oneself to reflect racial prejudice" (p. 27). Compared with antiblack attitudes, pro-white attitudes reflect a more subtle prejudice that, although less overtly negative, is still racial bias.

Modern Racism. Like aversive racism, **modern racism** involves the rejection of traditional racist beliefs, but a modern racist displaces negative racial feelings onto more abstract social and political issues. The modern racist believes that serious discrimination in the United States no longer exists, that any continuing racial inequality is the fault of minority group members, and that demands for affirmative action for minorities are unfair and unjustified. "Modern racism tends to 'blame the victim' and places the responsibility for change and improvements on the minority groups, not on the larger society" (Healey 1997, p. 55). Like aversive racists, modern racists tend to be unaware of their negative racial feelings and do not view themselves as prejudiced.

Learning to Be Prejudiced: The Role of Socialization and the Media

Psychological theories of prejudice focus on forces within the individual that give rise to prejudice. For example, the frustration-aggression theory of prejudice (also known as the scapegoating theory) suggests that prejudice is a form of hostility that results from frustration. According to this theory, minority groups serve as convenient targets of displaced aggression. The authoritarian-personality theory of prejudice suggests that prejudice arises in people with a certain personality type. According to this theory, people with an authoritarian personality—who are highly conformist, intolerant, cynical, and preoccupied with power—are prone to being prejudiced.

Rather than focus on the individual, sociologists focus on social forces that contribute to prejudice. Earlier we explained how intergroup conflict over wealth, power, and prestige gives rise to negative feelings and attitudes that serve to protect and enhance dominant group interests. In the following discussion we explain how prejudice is learned through socialization and the media.

Learning Prejudice Through Socialization. Although most researchers agree that the majority of children learn conceptions of racial and ethnic distinctions by the time they are about 6 years old, Van Ausdale and Feagin (2001) suggest that children as young as 3 years old have acquired prejudicial attitudes:

> Well before they can speak clearly, children are exposed to racial and ethnic ideas through their immersion in and observation of the large social world. Since racism exists at all levels of society and is interwoven in all aspects of American social life, it is virtually impossible for alert young children either to miss or ignore it. . . . Children are inundated with it from the moment they enter society. (pp. 189–190)

In the socialization process individuals adopt the values, beliefs, and perceptions of their family, peers, culture, and social groups. Prejudice is taught and learned through socialization, although it need not be taught directly and intentionally. Parents who teach their children to not be prejudiced yet live in an

Hate on the Web

Many websites promote hate and intolerance toward various minority groups. The Southern Poverty Law Center, a nonprofit organization that combats hate, intolerance, and discrimination, identified 497 hate websites in 2004 (Potok 2004). Hate websites use sophisticated graphics, music, and entertaining games to lure children, teenagers, and adults (Nemes 2002). Once individuals are hooked on racist ideology, they use Internet discussion groups or chat rooms to connect with other like-minded individuals with whom they can share their racist views and have those views reinforced.

After Hurricane Katrina in 2005, white supremacists posted hundreds of messages on the Internet expressing hopes that blacks in New Orleans would be wiped out. One message suggested that "they pile up all the niggers and use them as human sand bags against the rising storm surge" (Potok 2005, n.p.).

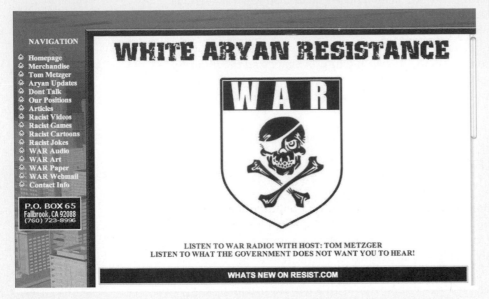

White supremacist groups, such as White Aryan Resistance, spread their message of racial hate through their website.

all-white neighborhood, attend an all-white church, and have only white friends may be indirectly teaching negative racial attitudes to their children. Socialization can also be direct, as in the case of a parent who uses racial slurs in the presence of her children or who forbids her children from playing with children from a certain racial or ethnic background. Children can also learn prejudicial attitudes from their peers. The telling of racial and ethnic jokes among friends, for example, perpetuates stereotypes that foster negative racial and ethnic attitudes.

Prejudice and the Media. The media contribute to prejudice by portraying minorities in negative and stereotypical ways—or by not portraying them at all. In the 2003–2004 television season 73% of characters on prime-time television were white, and 16% were African American (Glaubke & Heintz-Krowles 2004). The percentage of Latino characters was 6.5%—a notable increase from 3% in 1999. Only 3% were Asian, and less than 1% were Arab, Middle Eastern, Indian, or Pakistani. There were no Native American characters. An analysis of character portrayals in the 2003–2004 prime-time television season revealed that Latinos were more likely than any other group to be cast in low-status occupations, such as domestic worker,

Although hate groups use the Internet as a tool for communication, organization, and recruitment, the Internet can also be used to fight hate. For example, the Southern Poverty Law Center's website, Tolerance.org, provides daily news about groups and individuals working for tolerance, online guides to fight hate in communities and on college campuses, and tolerance tips for parents and teachers. In addition, it is possible that the prevalence of racist groups on the Internet may reduce hate crimes by providing outlets for hate that are nonphysical. Furthermore, the presence of hate groups on the Internet has made them more visible to the public, which in turn facilitates monitoring by government and watchdog groups (Glaser et al. 2002).

A number of countries, including Germany and France, have passed laws prohibiting hate speech on the Internet and have brought civil and criminal penalties against defendants. In 2001 Yahoo announced that it would actively try to keep hateful material out of its auctions, classified sections, and shopping areas after a French court ordered Yahoo to pay fines of $13,000 a day if the company did not install technology that would shield French Web users from seeing Nazi-related memorabilia on its auction site (French law prohibits the display of such material). However, in most cases the defendants continued to publish their hate material from the United States, where free speech is protected under the First Amendment. For example, many German neo-Nazi white supremacist groups use U.S.-based Internet servers to avoid prosecution under German law (Kaplan & Kim 2000). One U.S.-based Web page posted in German offered a $7,500 reward for the murder of a young left-wing activist, giving his home address, job, and phone number.

A number of international efforts have attempted to criminalize hate material online and to establish an international legal framework for prosecuting cross-jurisdictional hate speech on the Internet. But unless there is universal ratification, such measures are not likely to be effective. The United States, holding firmly to the value of freedom of speech, is unlikely to participate in international efforts to criminalize hate speech in cyberspace. Even in the unlikely event of international agreement to prohibit online hate speech, how would laws be enforced against websites transmitted from satellites?

When civil liberties groups have challenged legislation against hate speech on the Web, the Supreme Court has sided with them, ruling that such legislation violates free speech. Whether the Supreme Court continues this position in light of anti-American hate speech on the Internet remains to be seen.

and as criminals. Nearly half (46%) of Arabs and Middle Easterners were cast as criminals.

Another media form that is used to promote hatred toward minorities and to recruit young people to the white power movement is "white power music," which contains anti-Semitic, racist, and homophobic lyrics. A CD that was distributed to thousands of middle and high school students by the neo-Nazi record label Panzerfaust Records contains the following lyrics from the group the Bully Boys: "Whiskey bottles/baseball bats/pickup trucks/and rebel flags/we're going on the town tonight/hit and run/let's have some fun/we've got jigaboos on the run." And in a song called "Wrecking Ball," the band H8Machine advises kids to "destroy all your enemies," promising that "the best things come to those who hate" ("Neo-Nazi Label Woos Teens," 2004). The Southern Poverty Law Center has identified 123 domestic and 227 international white power bands that promote hate and intolerance through their music ("White Power Bands," 2002).

The Internet also spreads messages of hate toward minority groups through the websites of various white supremacist and hate group organizations (see this chapter's *Focus on Technology* feature).

ThomsonNOW™

Learn more about
**Discrimination Against
Racial and Ethnic Minorities**
by going through the
Prejudice and Discrimination
Learning Module.

Discrimination Against Racial and Ethnic Minorities

Whereas prejudice refers to attitudes, **discrimination** refers to actions or practices that result in differential treatment of categories of individuals. Although prejudicial attitudes often accompany discriminatory behavior or practices, one can be evident without the other.

Individual Versus Institutional Discrimination

Individual discrimination occurs when individuals treat other individuals unfairly or unequally because of their group membership. Individual discrimination can be overt or adaptive. In **overt discrimination** the individual discriminates because of his or her own prejudicial attitudes. For example, a white landlord may refuse to rent to a Mexican American family because of her own prejudice against Mexican Americans. Or a Taiwanese American college student who shares a dorm room with an African American student may request a roommate reassignment from the student housing office because he is prejudiced against blacks.

Suppose that a Cuban American family wants to rent an apartment in a predominantly non-Hispanic neighborhood. If the landlord is prejudiced against Cubans and does not allow the family to rent the apartment, that landlord has engaged in overt discrimination. But what if the landlord is not prejudiced against Cubans but still refuses to rent to a Cuban family? Perhaps that landlord is engaging in **adaptive discrimination,** or discrimination that is based on the prejudice of others. In this example the landlord may fear that if he or she rents to a Cuban American family, other renters who are prejudiced against Cubans may move out of the building or neighborhood and leave the landlord with unrented apartments. Overt and adaptive individual discrimination can coexist. For example, a landlord may not rent an apartment to a Cuban family because of his or her own prejudices and the fear that other tenants may move out.

Institutional discrimination occurs when normal operations and procedures of social institutions result in unequal treatment of and opportunities for minorities. Institutional discrimination is covert and insidious and maintains the subordinate position of minorities in society. For example, the practice of businesses moving out of inner-city areas results in reduced employment opportunities for America's highly urbanized minority groups. In the retail industry traditional warehouses designed to house inventory are being replaced by large distribution centers that require more space, thus encouraging location away from central cities to lower-priced land in less populated areas. A study by the U.S. Equal Employment Opportunity Commission (EEOC) found that as areas become less populated, the percentage of minority workers declines (EEOC 2004).

Institutional discrimination also occurs in education. When schools use standard intelligence tests to decide which children will be placed in college preparatory tracks, they are limiting the educational advancement of minorities whose intelligence is not fairly measured by culturally biased tests developed from white middle-class experiences.

Institutional discrimination is also found in the criminal justice system, which more heavily penalizes crimes that are more likely to be committed by minorities. For example, the penalties for crack cocaine, more often used by minorities, have traditionally been higher than those for other forms of cocaine use, even though the same prohibited chemical substance is involved. As conflict theorists emphasize, majority group members make rules that favor their own group.

In a national survey of college freshmen, one in four (24.9%) non-Hispanic white students agreed with the statement "Racial discrimination is no longer a major problem in America" (Sax et al. 2004). Yet only 12.5% of African American students, 17.7% of Asian American students, and 18.3% of Hispanic and Latino students agreed with that statement. Perhaps white students are not aware of the degree to which racial and ethnic minorities experience discrimination and its effects in almost every sphere of social life. Next, we look at discrimination in employment, housing, education, and politics. Finally, we expose the extent and brutality of physical and verbal violence against minorities.

Employment Discrimination

Despite laws against it, discrimination against minorities occurs today in all phases of the employment process, from recruitment to interview, job offer, salary, promotion, and firing decisions. Georgetown Place, a Fort Wayne, Indiana, senior community, refused to hire African Americans. A former manager for Georgetowne Place stated that residents at the facility preferred white employees and did not want minorities to come into their rooms (EEOC 2005b). The nationwide chain Abercrombie & Fitch settled a lawsuit alleging that it excluded minorities in its recruiting, hiring, and marketing practices (EEOC 2005a).

A sociologist at Northwestern University studied employers' treatment of job applicants in Milwaukee, Wisconsin, by dividing job applicant "testers" into four groups: blacks with a criminal record, blacks without a criminal record, whites with a criminal record, and whites without a criminal record (Pager 2003). Applicant testers, none of whom actually had a criminal record, were trained to behave similarly in the application process and were sent with comparable résumés to the same set of employers. The study found that white applicants with no criminal record were the most likely to be called back for an interview (34%) and that black applicants with a criminal record were the least likely to be called back (5%). But surprisingly, white applicants *with* a criminal record (17%) were more likely to be called back for an interview than were black applicants *without* a criminal record (14%)! The researcher concluded that "the powerful effects of race thus continue to direct employment decisions in ways that contribute to persisting racial inequality" (Pager 2003, p. 960).

Discrimination in hiring may be unintended. For example, many businesses rely on their existing employees to refer new recruits when a position opens up. Word-of-mouth recruitment is inexpensive and efficient; some companies offer bonuses to employees who bring in new recruits. But this traditional recruitment practice tends to exclude minority workers, because they often do not have a network of friends and family members in higher positions of employment who can recruit them (Schiller 2004).

Employment discrimination contributes to the higher rates of unemployment and lower incomes of blacks and Hispanics compared with whites (see Chapters 6 and 7). Lower levels of educational attainment among minority groups account for some, but not all, of the disadvantages they experience in employment and income. As shown in Figure 9.7, average lifetime earnings of whites are higher than for blacks and Hispanics at the same level of educational attainment (Day & Newburger 2002). Although the lifetime earnings of Asians and Pacific Islanders with advanced

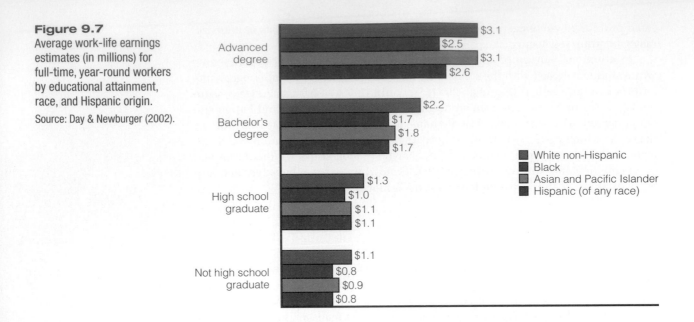

Figure 9.7
Average work-life earnings estimates (in millions) for full-time, year-round workers by educational attainment, race, and Hispanic origin.

Source: Day & Newburger (2002).

Advanced degree
$3.1
$2.5
$3.1
$2.6

Bachelor's degree
$2.2
$1.7
$1.8
$1.7

High school graduate
$1.3
$1.0
$1.1
$1.1

Not high school graduate
$1.1
$0.8
$0.9
$0.8

■ White non-Hispanic
■ Black
■ Asian and Pacific Islander
■ Hispanic (of any race)

degrees are equivalent to those of whites, at every other level of educational attainment, whites earn more than Asians and Pacific Islanders.

Workplace discrimination also includes unfair treatment. In one workplace with a large Hispanic workforce, Hispanic workers were selected each week to clean the lunchroom without being paid for that work (Greenhouse 2003). One Hispanic worker said that this treatment was a matter of dignity and that it made the workers feel humiliated.

Since the terrorist attacks of September 11, 2001, workplace discrimination against individuals who are (or who are perceived to be) Muslim, Arab, Middle Eastern, South Asian, or Sikh has increased. The most common complaints of post—9/11 backlash reported to the EEOC are alleged harassment and firing (EEOC 2003).

Housing Discrimination and Segregation

Before the 1968 Federal Fair Housing Act and the 1974 Equal Credit Opportunity Act, discrimination against minorities in housing and mortgage lending was as rampant as it was blatant. Banks and mortgage companies commonly engaged in "redlining"—the practice of denying mortgage loans in minority neighborhoods on the premise that the financial risk was too great, and the ethical standards of the National Association of Real Estate Boards prohibited its members from introducing minorities into white neighborhoods. Instead, realtors practiced "geographic steering," whereby realtors discouraged minorities from moving into certain areas by showing them homes only in minority neighborhoods.

Although housing discrimination is illegal today, it is not uncommon. To assess discrimination in housing, researchers use a method called "paired testing." In a paired test two individuals—one minority and the other nonminority—are trained to pose as home seekers, and they interact with real estate agents, landlords, rental agents, and mortgage lenders to see how they are treated. The testers are assigned comparable or identical income, assets, and debt as well as comparable or identical housing preferences, family circumstances, education, and job characteristics. A paired testing study of housing discrimination in 23 metropolitan areas found that in the rental market whites were more likely to receive information about available housing units and had more opportunities to inspect available units than did blacks and Hispanics (Turner et al. 2002). The incidence of discrimination was greater for Hispanic renters than for black renters. The same study found that in the home sales market, white home buyers were more likely to be able to inspect available homes and to be shown homes in more predominantly non-Hispanic white neighborhoods than were comparable blacks and Hispanics. Whites were also more likely to receive information and assistance with financing. In other research on mortgage lending discrimination, minorities were less likely to receive information about loan products, they received less time and information from loan officers, they were often quoted higher interest rates, and they had higher loan denial rates than whites, other things being equal (Turner & Skidmore 1999).

Despite continued housing discrimination, homeownership rates among minorities and low-income groups increased substantially in the 1990s, reaching record rates in many central cities. However, minority and low-income homeowner rates still lag behind the overall homeownership rate. Also, many of the gains in minority and low-income homeownership rates are due to increases in *subprime lending*—higher-fee, higher-interest-rate loans offered to borrowers who have poor (or nonexistent) credit records (Williams et al. 2005). This chapter's *Social Problems Research Up Close* feature presents research on housing discrimination conducted by college students.

Although homeownership rates among racial minorities have increased, residential segregation of racial and ethnic groups persists. Analysis of the 2000 census data revealed that although blacks and whites lived in neighborhoods that were slightly more integrated than they were in 1990, people still lived in largely segregated neighborhoods. In 2000 an average white person living in a metropolitan area (which includes city dwellers and suburban residents) lived in a neighborhood that is about 80% white and 7% black (Schmitt 2001a). In contrast, the average black person lived in a neighborhood that is 33% white and 51% black. A 2005 Gallup poll also found that Americans tend to live in neighborhoods largely populated by people of similar racial or ethnic backgrounds: 86% of non-Hispanic whites, 66% of blacks, and 61% of Hispanics reported living in areas where there are many people from their own backgrounds (Carroll 2005b).

For years sociologists have known that U.S. minorities, who are disproportionately represented among the poor, tend to be segregated in concentrated areas of low-income housing, often in inner-city areas of concentrated poverty (Massey & Denton 1993). After Hurricane Katrina in 2005 citizens across the United States and the world were shocked by the degree of poor, segregated communities such as the Lower Ninth Ward of New Orleans, where 98% of residents were black and 36% lived below the poverty line.

Educational Discrimination and Segregation

Both institutional and individual discrimination in education negatively affect racial and ethnic minorities and help to explain why minorities (with the exception of Asian Americans) tend to achieve lower levels of academic attainment and success (see also Chapter 8). Institutional discrimination is evidenced by inequalities in school funding—a practice that disproportionately hurts minority students (Kozol 1991). Nearly half of school funding comes from local taxes. In 2002 the federal government supplied 8% of educational expenditures and state government

ThomsonNOW™

Learn more about **Housing Discrimination and Segregation** by going through the Race Map Exercise.

"The winds of Katrina blew away the flimsy veil that long has shielded most Americans from the ugly reality of our nation's continuing problems with race, class, and poverty."

Mark Potok
Southern Poverty Law Center

An Undergraduate Sociology Class Uncovers Racial Discrimination in Housing

Previous research has indicated that Americans can infer the race of a speaker through the speaker's accent, grammar, and diction, thus offering rental agents an opportunity to discriminate over the phone. Under the guidance of their sociology professor (Douglas Massey) and a postdoctoral fellow (Garvey Lundy), students in an undergraduate sociology research methods class at the University of Pennsylvania designed a study to determine whether rental agents discriminated over the phone.

Sample and Methods

Students involved in the study (Massey & Lundy 2001) developed a data collection instrument that consisted of a standard script that auditors (students posing as renters) followed in their telephone interactions with rental agents. For example, if a machine answered the call, the auditor said, "Hello. My name is _____. I'm interested in the apartment you advertised in _____. Please call me back at _____." If a rental agent

answered the phone, the auditor said, "Hello. My name is _____. I'm interested in the apartment you advertised in _____. Are any apartments still available?" Other questions that auditors asked included "How much do I have to put down?" and "Are there any other fees?"

After devising the data collection instrument, students chose 79 rental listings from three sources: *Apartments for Rent* magazine; *The Apartment Hunter,* published by the *Philadelphia Inquirer;* and the Sunday real estate section of the *Philadelphia Inquirer*. The listings covered all zones of the Philadelphia metropolitan area. Four male and nine female auditors telephoned the selected rental agents, following the script and presenting a profile of a recent college graduate in his or her early to mid-20s with an annual income of $25,000 to $30,000. The auditors spoke White Middle-Class English (WME), Black Accented English (BAE), or Black English Vernacular (BEV). Black Accented English and Black English Vernacular are widely spoken by African Americans in

the United States. Massey and Lundy (2001) explained that "although BEV and BAE may both be identified as 'black sounding,' we suspect that most listeners can tell the difference between the two dialects and that they attach different class labels to each style of speech" (p. 456). Specifically, when an African American speaks standard English with a black pronunciation of certain words (BAE), listeners conclude that the speaker is a middle-class black person, whereas the combination of nonstandard grammar with a black accent (BEV) signals lower-class status.

Based on this assumption, Massey and Lundy (2001) were able to use six independent variables to test for a three-way interaction between race (black or white), sex (male or female), and class (lower or middle). The six independent variables were (1) male BEV, (2) female BEV, (3) male BAE, (4) female BAE, (5) male WME, and (6) female WME. The dependent variables included the various responses of the rental agents, such as whether the agent returned

contributed 48%; local governments provided the remainder (Schiller 2004). Because minorities are more likely than whites to live in economically disadvantaged areas, they are more likely to go to schools that receive inadequate funding. Inner-city schools, which primarily serve minority students, receive less funding per student than do schools in more affluent, primarily white areas.

For example, inequalities in school funding resulted in New York City receiving $2,000 less per pupil than Buffalo, Rochester, Syracuse, and Yonkers. New York State Supreme Court judge Leland DeGrasse found that New York State's system of school funding violated federal civil rights laws because it disproportionately hurt minority students (more than 70% of the state's Asian, black, and Hispanic students live in New York City) (Goodnough 2001).

Another institutional education policy that is advantageous to whites is the policy that gives preference to college applicants whose parents or grandparents are alumni. The overwhelming majority of alumni at the highest ranked universities and colleges are white. Thus white college applicants are the primary beneficiaries

the auditor's phone call, whether the rental agent indicated that a housing unit was still available, and whether the agent indicated that an application fee was required (and if so, how much).

Findings and Conclusions

Massey and Lundy (2001) found "clear and often dramatic evidence of phone-based racial discrimination" (p. 466). Compared to whites, African Americans were less likely to speak to a rental agent (their calls were less likely to be returned). African Americans were also less likely to be told that a unit was available, more likely to pay application fees, and more likely to have credit mentioned as a potential problem in qualifying for a lease.

These racial effects were exacerbated by sex and class. Lower-class blacks experienced less access to rental housing than middle-class blacks, and black females experienced less access than black males. Lower-class black females were the most disadvantaged group. They experienced the lowest probability of contacting and speaking to a rental agent and, even if they did make contact, they faced the lowest probability of being told of a housing unit's availability. Lower-class black females also faced the highest chance of paying an application fee. On average, lower-class black females were assessed $32 more per application than white middle-class males.

The share of auditors reaching an agent and the share being told a unit was available were combined to indicate an overall measure of access to rental units in the Philadelphia housing market. Whereas more than three-fourths (76%) of white middle-class males gained access to a potential rental unit, the figure dropped to 63% for middle-class black men (those speaking BAE), 60% for white middle-class females (those speaking WME), 57% for black middle-class females (those speaking BAE), 44% for lower-class black men (those speaking BEV), and only 38% for lower-class black women (those speaking BEV). "In other words, for every call a white male makes to find out about a rental unit in the Philadelphia housing market, a poor black female must make two calls to achieve the same level of access, roughly doubling her time and effort compared with his" (Massey & Lundy 2001, p. 461). In sum, "being identified as black on the basis of one's speech pattern clearly reduces access to rental housing, but being black and female lowers it further, and being black, female, and poor lowers it further still" (p. 467). Massey and Lundy's findings suggest that much housing discrimination probably occurs over the phone. "Through technology, a racist landlord may discriminate without actually having to experience the inconvenience or discomfort of personal contact with his or her victim" (Massey & Lundy 2001, pp. 454–455). The investigators also concluded that "telephone audit studies offer social scientists a cheap, effective, and timely way to measure the incidence and severity of racial discrimination in urban housing markets" (p. 455).

of these so-called legacy admissions policies. About 10–15% of students in most Ivy League colleges and universities are children of alumni. Harvard University accepts about 11% of its overall applicant pool, but for legacy applicants the admission rate is 40% (Schmidt 2004). As a result of pressure from state lawmakers and minority rights activists, in 2004 Texas A&M University became the first public college to abandon its legacy admittance policy.

Minorities also experience individual discrimination in the schools as a result of continuing prejudice among teachers. One college student completing a teaching practicum reported that some of the teachers in her school often spoke about children of color as "wild kids who slam doors in your face" (Lawrence 1997, p. 111). In a survey conducted by the Southern Poverty Law Center, 1,100 educators were asked whether they had heard racist comments from their colleagues in the past year. More than one-fourth of survey respondents answered yes ("Hear and Now," 2000). It is likely that teachers who are prejudiced against minorities discriminate against them, giving them less teaching attention and less encouragement.

Racial and ethnic minorities are also treated unfairly in educational materials, such as textbooks, which often distort the history and heritages of people of color (King 2000). For example, Zinn (1993) observes, "To emphasize the heroism of Columbus and his successors as navigators and discoverers, and to de-emphasize their genocide, is not a technical necessity but an ideological choice. It serves, unwittingly—to justify what was done" (p. 355).

Finally, racial and ethnic minorities are largely isolated from whites in an increasingly segregated school system. A study by the Civil Rights Project at Harvard University finds that U.S. schools in the 2000–2001 school year were more segregated than they were in 1970 (Orfield 2001). The upward trend in school segregation is due to large increases in minority student enrollment, continuing white flight from urban areas, the persistence of housing segregation, and the termination of court-ordered desegregation plans. Court-mandated busing became a means to achieve equality of education and school integration in the early 1970s after the Supreme Court (in *Swann* v. *Charlotte-Mecklenberg*) endorsed busing to desegregate schools. But in the 1990s lower courts lifted desegregation orders in dozens of school districts (Winter 2003a). In the words of Bradley Schiller (2004), "Racial isolation in the schools is still the hallmark of the American educational system" (p. 178).

Political Discrimination

Historically, African Americans have been discouraged from political involvement by segregated primaries, poll taxes, literacy tests, and threats of violence. However, tremendous strides have been made since the passage of the 1965 Voting Rights Act, which prohibited literacy tests and provided for poll observers. Blacks have won mayoral elections in Atlanta, Cleveland, Washington, D.C., and New York City, and the governorship in Virginia. However, racial minorities and Hispanics continue to be underrepresented in political positions and voting participation.

Discrimination against racial and ethnic minorities in the U.S. political process persists. After "voting irregularities" in the 2000 national elections, the National Association for the Advancement of Colored People (NAACP) and several civil rights groups filed a lawsuit in Florida to eliminate unfair voting practices (NAACP 2001). In the 2000 election thousands of black voters complained that they were wrongfully turned away from the polls or had trouble casting their ballots. Complaints were not limited to Florida; black voters in about a dozen other states reported similar unfair treatment (NAACP 2000). Numerous incidents of voter intimidation and suppression in predominantly African American communities were also reported in the 2004 presidential election (Fitrakis & Wasserman 2004).

Hate Crimes

In June 1998 James Byrd Jr., a 49-year-old father of three, was walking home from a niece's bridal shower in the small town of Jasper, Texas. According to police reports, three white men riding in a gray pickup truck saw Byrd, a black man, walking down the road and offered him a ride. The men reportedly drove down a dirt lane and, after beating Byrd, chained him to the back of the pickup truck and dragged him for two miles down a winding, narrow road. The next day, police found Byrd's mangled and dismembered body. The three men who were arrested had ties to white supremacist groups. James Byrd had been brutally murdered simply because he was black.

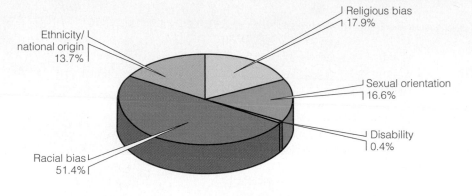

Figure 9.8
Hate crime incidence by
category of bias, 2003.
Source: FBI (2004).

Religious bias
17.9%

Ethnicity/
national origin
13.7%

Sexual orientation
16.6%

Disability
0.4%

Racial bias
51.4%

Total number of single-bias hate crimes: 7,405

The murder of James Byrd exemplifies a **hate crime**—an unlawful act of violence motivated by prejudice or bias. Examples of hate crimes, also known as "bias-motivated crimes" and "ethnoviolence," include intimidation (e.g., threats), destruction of or damage to property, physical assault, and murder.

From the first year that FBI hate crime data were published in 1992, most hate crimes have been based on racial bias (see Figure 9.8). Keep in mind that FBI hate crime data undercount the actual number of hate crimes because (1) not all U.S. jurisdictions report hate crimes to the FBI (reporting is voluntary), (2) it is difficult to prove that crimes are motivated by hate or prejudice, (3) law enforcement agencies shy away from classifying crimes as hate crimes because it makes their community "look bad," and (4) victims are often reluctant to report hate crimes to the authorities.

After the terrorist attacks of September 11, 2001, hate crimes against individuals perceived to be Muslim and/or Middle Eastern increased significantly. Perpetrators of these hate crimes may have felt that they were acting in defense of a perceived threat. A poll by the Council on American-Islamic Relations showed that more than half of America's 7 million Muslims (57%) said they experienced bias or discrimination after September 11 (Morrison 2002). On the positive side the same poll found that most American Muslims (79%) also experienced special kindness or support from friends or colleagues of other faiths. Acts of kindness include verbal assurances, support, and even offers to help guard local mosques and Islamic schools.

More recently, violence against U.S. immigrants has become a growing national problem. This chapter's *Human Side* feature describes the hate crime victimization of a Hispanic immigrant in Georgia—a state with one of the fastest growing immigrant populations.

Motivations for Hate Crimes. Levin and McDevitt (1995) found that the motivations for hate crimes were of three distinct types: thrill, defensive, and mission. Thrill hate crimes are committed by offenders who are looking for excitement and attack victims for the "fun of it." Defensive hate crimes involve offenders who view their attacks as necessary to protect their community, workplace, or college campus from "outsiders" or to protect their racial and cultural purity from being "contaminated" by interracial marriage and childbearing. A study of white racists in Internet chat rooms found that the topic of interracial marriage was more likely than other topics (such as blacks moving into one's neighborhood and competing for one's job) to elicit advocacy of violence (Glaser et al. 2002). For example, one

"The vast majority of the world's 1.2 billion Muslims are as offended by a violent act carried out in the name of Islam as most Christians are horrified by atrocities perpetrated by Serbian Christians or the Real Irish Republican Army.**"**

**Charles Kimball
Wake Forest University**

Anti-Immigrant Hate: One Immigrant's Experience

Domingo Lopez Vargas left his dirt-poor Guatemalan farm village to come to the United States, where he hoped to earn decent money for his wife and nine children. After picking oranges in Florida, he moved to Georgia, where the booming construction business lured immigrant workers. Unlike many of his compadres, Lopez had legal status, which helped him find steady work hanging doors and windows. When work dried up, Lopez joined the more than 100,000 *jornaleros*—day laborers—who wait for landscaping and construction jobs on street corners and in front of convenience stores all across Georgia. Usually there are plenty of pickup trucks that swing by, offering $8 to $12 an hour for digging, planting, painting, or ham-

mering. But this day, *nada*. By late afternoon, Lopez had tired of waiting in the cold, so he walked up the street to pick up a few things at a grocery store.

"I got milk, shampoo and toothpaste," Lopez recalled. "When I was leaving the store, this truck stopped right in front of me and said, 'Do you want to work?' . . . I said, yes, how much? They said nine dollars an hour. I didn't ask what kind of job. I just wanted to work, so I said yes."

Until that afternoon, Lopez said, "Americans had always been very nice to me"—which might explain why he wasn't concerned that the four guys in the pickup truck looked awfully young to be contractors. Or why he didn't think

AP/Wide World Photos

Domingo Lopez Vargas, an immigrant day laborer, was brutally beaten in a hate crime in Canton, Georgia.

respondent said, "better kill her. kill him and her. pull a oj. . . . im not kidding. I would do it if it was my sister. i would gladly go to prison then live a free life knowing some mud babies were calling me uncle whitey" (p. 184).

Mission hate crimes are perpetrated by white supremacist group members or other offenders who have dedicated their lives to bigotry. The Ku Klux Klan, the first major racist white supremacist group in the United States, began in Tennessee shortly after the Civil War. Klansmen have threatened, beaten, mutilated, and lynched blacks as well as whites who dared to oppose them. Other racist groups known to engage in hate crimes are the Identity Church Movement, neo-Nazis, and the skinheads. The Southern Poverty Law Center identified 762 hate groups in the United States in 2004 (SPLC 2005b).

Hate on Campus. Karl Nichols, a white residence hall director at the University of Mississippi, learned about racism in college, but not in the classroom and not from a textbook. Two chunks of asphalt were hurled through Nichols's dor-

twice about being picked up so close to sunset. "I took the offer because I know sometimes people don't stop working until 9 at night," he says.

The four young men, all high school students, drove Lopez to a remote spot strewn with trash. "They told me to pick up some plastic bags that were on the ground. I thought that was my job, to clean up the trash. But when I bent over to pick it up, I felt somebody hit me from behind with a piece of wood, on my back." It was just the start of a 30-minute pummeling that left Lopez bruised and bloody from his thighs to his neck. "I thought I was dying," he said. "I tried to stand up but I couldn't." Finally, after he handed over all the cash in his wallet, $260, along with his Virgin Mary pendant, the teenagers sped away.

As a result of injuries incurred by the beating, Lopez could not work for four months, and he was left with $4,500 in medical bills. Sometimes he still puzzles over his attackers' motives. "They were young," he speculates, "and maybe they didn't have enough education. Or maybe their families . . . taught them to kill people, and that is what they have learned."

Having sworn off day labor, Lopez works night shifts now, cutting up chickens at the nearby Tyson plant, wincing through the pain that shoots up his right arm when he lowers the boom on a bird. But it's only temporary, he says. "I called my wife and told her what happened. She told me to move back to Guatemala. I wanted to, but I didn't have enough money to go back and the police officers told me not to move out of the country because they will still need me to work on the case. After the case is finished, I want to go back to my family."

Source: Adapted from Moser (2004).

mitory room window along with a note warning, "You're going to get it, you God-forsaken nigger-lover" ("Hate on Campus," 2000, p. 10). The next night, someone attempted to set Nichols's door on fire. According to the university's investigation of these incidents, Nichols "may have violated racist taboos . . . by openly displaying his affinity for African American individuals and black culture, by dating black women, by playing black music . . . and by promoting diversity" in his dormitory (p. 10).

Karl Nichols is not alone. At Brown University in Rhode Island, a black student was beaten by three white students who told her she was a "quota" who did not belong at a university. At the State University of New York in Binghamton, an Asian American student was left with a fractured skull after a racially motivated assault by three students. Two students at the University of Kentucky—one white, one black—were crossing the street just off campus when they were attacked by 10 white men. The attackers yelled racist slurs at the black student and choked him until he could not speak or move. The assailants called the white student a "nigger lover" as they broke his hand and nose. "I definitely thought I was going to lose my life," the black student said later. The white student was shocked by the incident, commenting, "I didn't know that much hate existed" ("Hate on Campus," 2000, p. 7).

According to the FBI, more than 1 in 10 hate crimes occurs at schools or colleges (FBI 2004). Far more common than hate crimes are "bias incidents," which are events that are not crimes but still can have the same negative and divisive effects. Howard J. Ehrilich, director of the Prejudice Institute in Baltimore, estimates that each year one-fourth of racial and ethnic minority college students and up to 5% of white college students are targets of bias-motivated name-calling, e-mails, telephone calls, verbal aggression, and other forms of psychological intimidation (Willoughby 2003).

Strategies for Action: Responding to Prejudice, Racism, and Discrimination

> "The problem is . . . we don't believe we are as much alike as we are. Whites and blacks, Catholics and Protestants, men and women. If we saw each other as more alike, we might be very eager to join in one big human family in this world, and to care about that family the way we care about our own."
>
> **Morrie Schwartz**
> **Sociologist (from *Tuesdays with Morrie*)**

Because racial and ethnic tensions exist worldwide, strategies for combating prejudice, racism, and discrimination globally require international cooperation and commitment. The World Conference Against Racism, Racial Discrimination, Xenophobia, and Related Intolerance, held in Durban, South Africa, in 2001 exemplifies international efforts to reduce racial and ethnic tensions and inequalities and to increase harmony among the various racial and ethnic populations of the world. Unfortunately, the U.S. delegation to this conference withdrew because of the expectation that hateful language would be used against Israel (because of the Israeli-Palestinian conflict). Although international strategies to combat racism and discrimination are vital, in this section we focus on legal and political and educational strategies within the United States.

Legal and Political Strategies

Legal and political strategies to reduce prejudice, racism, and discrimination include increasing minority participation and representation in government, eliminating inequalities in school funding, and creating equal opportunities through affirmative action. (Because affirmative action includes voluntary programs and policies in addition to legally mandated ones, we consider affirmative action in a separate section.) Although it is readily apparent that such strategies can prevent or reduce discriminatory practices, the effects of legal and political strategies on prejudice are more complex. Legal policies that prohibit discrimination can actually increase modern forms of prejudice, as in the case of individuals who conclude that because laws and policies prohibit discrimination, any social disadvantages of minorities must be their own fault. On the other hand, any improvement in the socioeconomic status of minorities that results from legal and political policies may help to replace negative images of minorities with positive images.

Increasing Minority Representation and Participation in Government. Increasing minority representation and participation in government promises to increase minorities' voices in influencing public policy that addresses their interests. In addition, minority representation in government affects race relations. One study found a high degree of mistrust of U.S. government among African Americans; however, those who believed that blacks could influence the political process were less likely to distrust the government (Parsons et al. 1999). The researchers concluded that African Americans' "distrust of government will not be reduced until African Americans perceive that they have more of a role to play in their government" (p. 218).

Various national, state, and local minority groups and organizations encourage minorities to register to vote and to vote in governmental elections. Such efforts contributed to an upsurge in black voter participation in recent presidential elections. Efforts to increase voting participation have also targeted Asian Americans and Hispanic populations.

Representation of racial and ethnic minorities in elected government positions has also increased. Members of the 109th U.S. Congress include a record number of Hispanic (29) and black (43) members, as well as 8 Asians/Pacific Islanders, 1 Native American, and 9 foreign-born members (Amer 2005).

Reducing Disparities in Education. Legal remedies have also sought to address institutional discrimination in education by reducing or eliminating disparities in school funding. As noted earlier, schools in poor districts—which predominantly serve minority students—have traditionally received less funding per pupil than do schools in middle- and upper-class districts (which predominantly serve white students). In recent years more than two dozen states have been forced by the courts to come up with a new system of financing schools to increase inadequate funding of schools in poor districts (Goodnough 2001).

Affirmative Action

Affirmative action refers to a broad range of policies and practices in the workplace and educational institutions to promote equal opportunity as well as diversity. Affirmative action is an attempt to compensate for the effects of past discrimination and prevent current discrimination against women and racial and ethnic minorities. Vietnam veterans and people with disabilities may also qualify under affirmative action policies. Although the largest category of affirmative action beneficiaries is women, the majority of students in two sociology classes did not know that women were covered by affirmative action (Beeman et al. 2000).

Federal Affirmative Action. Affirmative action policies developed in the 1960s from federal legislation that required any employer (universities as well as businesses) who received contracts from the federal government to make "good faith efforts" to increase the pool of qualified minorities and women (U.S. Department of Labor 2002). Such efforts can be made by expanding recruitment and training programs. Hiring decisions are to be made on a nondiscriminatory basis.

Affirmative Action in Higher Education. The Supreme Court's 1974 ruling in *University of California Board of Regents* v. *Bakke* marked the beginning of the decline of affirmative action. Alan Bakke, a white male, had applied to the University of California at Davis medical school and was rejected, even though his grade point average and score on the medical school admissions test were higher than those of several minority applicants who had been admitted. The medical school had established fixed racial quotas, guaranteeing admission to 16 minority applicants regardless of their qualifications. Bakke claimed that such quotas discriminated against him as a white male and that the University of California had violated his Fourteenth Amendment right to equal protection under the law. The Supreme Court ruled in Bakke's favor by a 5 to 4 vote (showing how split the court was), concluding that the University of California unwittingly engaged in "reverse discrimination," which was unconstitutional. Affirmative action programs, the court ruled, could not use fixed quotas in admission, hiring, or promotion policies. However, the court affirmed the right for universities and employers to consider race as a factor in admission, hiring, and promotion in order to achieve diversity.

Since the *Bakke* case, numerous legal battles have challenged affirmative action (Olson 2003). In *Hopwood* v. *Texas* Cheryl Hopwood and three other white individuals who had been denied admission to the University of Texas Law School claimed that their Fourteenth Amendment rights to equal protection under the law had been violated by the university's affirmative action admission policies. The Fifth Circuit Court of Appeals decided in 1997 that the University of Texas could no longer use race as a factor in awarding financial aid, admitting students, and hiring and promoting faculty. Higher courts refused to review the decision. But in 2003 the U.S. Supreme Court, after hearing the appeals from two white applicants who applied but were not accepted to the University of Michigan, affirmed the right of colleges to consider race in admissions, but the Court rejected Michigan's use of a point system to do so. According to the University of Michigan undergraduate admissions procedure, minority status provided 20 points on a 150-point scale for admission. The Court found that the problem with the point system was that for some applicants it turned race into the decisive factor instead of just one of many factors (Winter 2003b). In response, the University of Michigan abandoned its point system and created an undergraduate admissions policy similar to its law school admissions policy, which may serve as a model for how other universities can achieve a diverse student body while following court guidelines. In the new "holistic review" approach the university considers the unique circumstances of each student, prioritizing academics and treating all other factors, including race, equally. Applicants are now required to write more essays, including one on cultural diversity. Other colleges and universities that are committed to having a diverse student body but want to avoid charges of "reverse discrimination" ask students to write essays about "cultural traditions," ask if they speak English as a second language, and increase efforts to recruit at high schools with large minority populations.

Attitudes Toward Affirmative Action. Affirmative action remains a divisive issue among Americans. Not surprisingly, blacks and Hispanics are more likely than non-Hispanic whites to favor affirmative action (see Figure 9.8). Women are more likely than men to support affirmative action (Pew Research Center 2003)—not a surprising finding given that women are the largest category targeted to benefit from affirmative action.

Among first-year college students, half (50.4%) agreed that "affirmative action in college admissions should be abolished" (Sax et al. 2004). However, a Pew Research Center (2003) survey asked, "Do you think affirmative action programs designed to increase the number of black and minority students on college campuses is a good thing or a bad thing?" Sixty percent of respondents said it was a good thing.

Public opinion poll results are influenced by how survey questions are worded and framed. Survey questions that ask whether respondents favor "affirmative action programs for women and minorities" elicit more favorable responses than questions that ask about affirmative action for minorities only (Paul 2003). In addition, terms such as "affirmative action," "equal," and "opportunity" in survey questions yield more support for affirmative action policies, whereas terms such as "special preferences," "preferential treatment," and "quotas" tend to lessen support. As shown

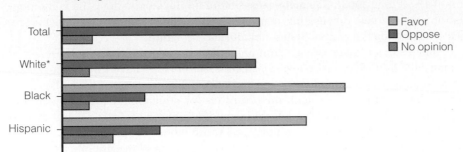

Do you generally favor or oppose affirmative action programs for racial minorities?

- Favor
- Oppose
- No opinion

Total
White*
Black
Hispanic

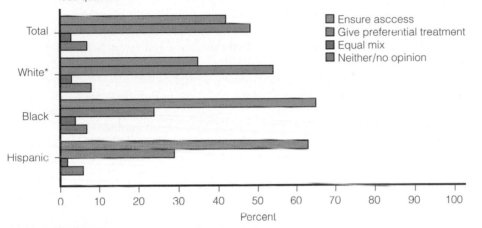

Do affirmative action programs ensure that well-qualified minorities get access to the schools and jobs that they deserve, or do these programs give preferential treatment to minorities in school admission and job hiring—even when those minorities are less qualified?

- Ensure asccess
- Give preferential treatment
- Equal mix
- Neither/no opinion

Total
White*
Black
Hispanic

0 10 20 30 40 50 60 70 80 90 100
Percent

Figure 9.9
Attitudes toward and perceptions of affirmative action.
Source: Gallup Organization (2005).

*Non-Hispanic whites.

in Figure 9.9, blacks and Hispanics are less likely than non-Hispanic whites to view affirmative action as a form of "preferential treatment."

Supporters of affirmative action suggest that such policies have many social benefits. In a review of more than 200 scientific studies of affirmative action, Holzer and Neumark (2000) concluded that these policies produce benefits for women, minorities, and the overall economy. Holzer and Neumark (2000) found that employers that adopt affirmative action increase the relative number of women and minorities in the workplace by an average of 10–15%. Since the early 1960s affirmative action in education has contributed to an increase in the percentage of blacks attending college by a factor of three and the percentage of blacks enrolled in medical school by a factor of four. Black doctors choose more often to practice medicine in inner cities and rural areas serving poor or minority patients than their white medical school classmates do (Holzer & Neumark 2000). Increasing the numbers of minorities in educational and professional positions also provides positive role models for other, especially younger, minorities "who can identify with them and form realistic goals to occupy the same roles themselves" (Zack 1998, p. 51).

Opponents of affirmative action suggest that such programs constitute "reverse discrimination," which hurts whites. However, 2000 data show that there are 1.3 million unemployed blacks and 112 million employed whites; thus if every unemployed black worker in the United States were to displace a white worker, only 1% of white workers would be affected. Because affirmative action pertains only to job-qualified applicants, the actual percentage of affected whites would be a fraction of 1% (Plous 2003). The main causes of unemployment among the white population are corporate downsizing, computerization and automation, and factory relocations outside the United States, not affirmative action.

Some critics of affirmative action argue that it undermines the self-esteem of women and minorities. Although affirmative action may have this effect in rare cases, in many cases affirmative action can raise the self-esteem of women and minorities by providing them with opportunities for educational advancement and employment (Plous 2003). Another criticism of affirmative action is that it fails to help the most impoverished of minorities—those whose deep and persistent poverty impairs their ability to compete not only with whites but with other more advantaged minorities (Wilson 1987).

Opposition to affirmative action threatens the future of such policies and programs and the future educational and occupational opportunities of minorities. After California passed Proposition 209—an initiative to abolish affirmative action—black student enrollment in the UCLA law school declined from 10.3% in 1996 to 1.4% in 2000 (Greenberg 2003). Black and Latino enrollment has also significantly declined in the UC Berkeley undergraduate program and law school, the UC San Diego School of Medicine, and the University of Texas Law School.

It is ironic that President George W. Bush, who does not support affirmative action, was himself a beneficiary of a long-standing policy that gives preferential treatment to college applicants. When President Bush applied to highly selective Yale University in 1964, he had a C average in high school, but he was nevertheless admitted to Yale because he was a "legacy" applicant—that is, the son and grandson of distinguished alumni.

Educational Strategies

Educational strategies to reduce prejudice, racism, and discrimination have been implemented in schools (primary, secondary, and higher education levels), in communities and community organizations, and in the workplace.

Multicultural Education in Schools and Communities. In schools across the nation **multicultural education,** which encompasses a broad range of programs and strategies, works to dispel myths, stereotypes, and ignorance about minorities, to promote tolerance and appreciation of diversity, and to include minority groups in the school curriculum (see also Chapter 8). With multicultural education the school curriculum reflects the diversity of U.S. society and fosters an awareness and appreciation of the contributions of different racial and ethnic groups to U.S. culture. The Southern Poverty Law Center's program Teaching Tolerance publishes and distributes materials and videos designed to promote better human relations among diverse groups. These materials are sent to schools, colleges, religious organizations, and a variety of community groups across the nation.

Many colleges and universities have made efforts to promote awareness and appreciation of diversity by offering courses and degree programs in racial and ethnic studies and by sponsoring multicultural events and student organizations. A national survey by the Association of American Colleges and Universities found that 54% of colleges and universities required students to take at least one course that emphasizes diversity and another 8% were in the process of developing such a requirement (Humphreys 2000). Evidence points to a number of positive outcomes for

More than 600,000 teachers receive *Teaching Tolerance* magazine—a free resource for tolerance education in the classroom (Dees 2000). Other free materials available from the Southern Poverty Law Center (http://www.spl-center.org) include *101 Tools for Tolerance, Responding to Hate at School,* and *Ten Ways to Fight Hate.*

both minority and majority students who take college diversity courses, including increased racial understanding and cultural awareness, increased social interaction with students who have backgrounds different from their own, improved cognitive development, increased support for efforts to achieve educational equity, and higher satisfaction with their college experience (Humphreys 1999).

Whiteness Studies. Courses in "whiteness studies" are also emerging in college curricula. As explained by Professor Gregory Jay (2005):

> Whiteness Studies attempts to trace the economic and political history behind the invention of "whiteness," to challenge the privileges given to so-called "whites," and to analyze the cultural practices (in art, music, literature, and popular media) that create and perpetuate the fiction of "whiteness." . . . "Whiteness Studies" is an attempt to think critically about how white skin preference has operated systematically, structurally, and sometimes unconsciously as a dominant force in American—and indeed in global—society and culture. (n.p.)

Sociologist Dalton Conley (2002) explained that whiteness studies "serve to rectify something wrong with the way we study race in America: By traditionally focusing on minority groups, the implicit message that scholarship projects is that nonwhites are 'deviant,' that's why they are studied" (n.p.)

Diversification of College Student Populations. Recruiting and admitting racial and ethnic minorities in institutions of higher education can foster positive relationships among diverse groups and enrich the educational experience of all students—minority and nonminority alike (American Council on Education & American Association of University Professors 2000). Psychologist Gordon Allport's (1954) "contact hypothesis" suggests that contact between groups is necessary for the reduction of prejudice between group members. For example, native-born Americans with higher levels of contact with immigrants tend to have more positive views of immigrants and immigration than those with less contact (NPR et al. 2005). Ensuring a diverse student population can provide students with opportunities for contact with different groups and can thereby reduce prejudice. One study found that students with the most exposure to diverse populations during college had the most cross-racial interactions five years after leaving college (Gurin 1999).

Understanding Race, Ethnicity, and Immigration

After considering the material presented in this chapter, what understanding about race and ethnic relations are we left with? First, we have seen that racial and ethnic categories are socially constructed; they are largely arbitrary, imprecise, and misleading. Although some scholars suggest that we abandon racial and ethnic labels, others advocate adding new categories—multiethnic and multiracial—to reflect the identities of a growing segment of the U.S. and world population.

Conflict theorists and structural functionalists agree that prejudice, discrimination, and racism have benefited certain groups in society. But racial and ethnic disharmony has created tensions that disrupt social equilibrium. Symbolic interactionists note that negative labeling of minority group members, which is learned through interaction with others, contributes to the subordinate position of minorities.

Prejudice, racism, and discrimination are debilitating forces in the lives of minorities and immigrants. In spite of these negative forces, many minority group members succeed in living productive, meaningful, and prosperous lives. But many others cannot overcome the social disadvantages associated with their minority status and become victims of a cycle of poverty (see Chapter 6). Minorities are disproportionately poor, receive inferior education and health care, and, with continued discrimination in the workplace, have difficulty improving their standard of living.

Achieving racial and ethnic equality requires alterations in the structure of society that increase opportunities for minorities—in education, employment and income, and political participation. In addition, policy makers concerned with racial and ethnic equality must find ways to reduce the racial and ethnic wealth gap and to foster wealth accumulation among minorities (Conley 1999). Social class is a central issue in race and ethnic relations. Professor and activist bell hooks (2000) warns that focusing on issues of race and gender can deflect attention away from the larger issue of class division that increasingly separates the haves from the have-nots. Addressing class inequality must, suggests hooks, be part of any meaningful strategy to reduce inequalities suffered by minority groups. Civil rights activist Lani Guinier, in an interview with Paula Zahn on the *CBS Evening News* on July 18, 1998, suggests that "the real challenge is to . . . use race as a window on issues of class, issues of gender, and issues of fundamental fairness, not just to talk about race as if it's a question of individual bigotry or individual prejudice. The issue is more than about making friends—it's about making change." But, as Shipler (1998) notes, making change requires members of society to recognize that change is necessary, that there is a problem that needs rectifying:

> One has to perceive the problem to embrace the solutions. If you think racism isn't harmful unless it wears sheets or burns crosses or bars blacks from motels and restaurants, you will support only the crudest anti-discrimination laws and not the more refined methods of affirmative action and diversity training. (p. 2)

Public awareness of inequality—racial and economic—became heightened after the tragic events of Hurricane Katrina and the government's slow response to the predominantly poor and black victims of the disaster. U.S. Representative Jesse

"I have a dream that my four little children will one day live in a nation where they will not be judged by the color of their skin, but by the content of their character."

Martin Luther King Jr.
Civil rights leader

Jackson Jr. described what he witnessed in the wake of Katrina: "I saw 5,000 African Americans on the I-10 causeway, desperate, perishing, dehydrated, babies dying. It looked like Africans in the hull of a slave ship. It was so ugly and so obvious. . . . We have great tolerance for black suffering and black marginalization" (quoted by Jackson 2005, n.p.). If Katrina results in decreased tolerance for racial and economic inequality and the suffering it brings, perhaps the more than 1,000 lives lost in the disaster will not have been in vain.

Chapter Review

ThomsonNOW™

Reviewing is as easy as ① ② ③

1. Before you do your final review, take the ThomsonNOW diagnostic quiz to help you identify the areas on which you should concentrate. You will find information on ThomsonNOW and instructions on how to access all of its great resources on the foldout at the beginning of the text.

2. As you review, take advantage of ThomsonNOW's study videos and interactive Map the Stats exercises to help you master the chapter topics.

3. When you are finished with your review, take ThomsonNOW's posttest to confirm you are ready to move on to the next chapter.

• **What is meant by the idea that race is socially constructed?**

Racial categories are based more on social definitions than on biological differences. Genotically, the genes of black and white Americans are 99.9% alike, and there are no "pure" races; people in virtually all societies have genetically mixed backgrounds. Different societies construct different systems of racial classification, and these systems change over time. The significance of race is not biological but social and political, because race is used to separate "us" from "them" and becomes a basis for unequal treatment of one group by another.

• **What are the two types of assimilation?**

Assimilation is the process by which formerly distinct and separate groups merge and become integrated as one. Secondary assimilation occurs when different groups become integrated in public areas and in social institutions, such as neighborhoods, schools, the workplace, and in government. Primary assimilation occurs when members of different groups are integrated in personal, intimate associations, as with friends, family, and spouses.

• **Beginning with the 2000 census, what are the five race categories used to identify the race composition of the United States?**

Beginning with the 2000 census, the five race categories are (1) white, (2) black or African American, (3) American Indian or Alaska Native, (4) Asian, and (5) Native Hawaiian or other Pacific Islander. In addition, respondents to federal surveys and the census now have the option of officially identifying themselves as being of more than one race, rather than checking only one racial category.

• **What is an ethnic group?**

An ethnic group is a population that has a shared cultural heritage or nationality. Ethnic groups can be distinguished on the basis of language, forms of family structures and roles of family members, religious beliefs and practices, dietary customs, forms of artistic expression such as music and dance, and national origin. The largest ethnic population in the United States is Hispanics or Latinos.

• **What percentage of the U.S. population (in 2003) was born outside the United States?**

In 2003 more than 1 in 10 U.S. residents (11.7%) were born in a foreign country, and 1 in 5 U.S. children, ages 18 and younger, was the child of an immigrant.

• **What were the manifest function and latent dysfunction of the civil rights movement?**

The manifest function of the civil rights legislation in the 1900s was to improve conditions for racial minorities. However, civil rights legislation produced an unexpected consequence, or latent dysfunction. Because civil rights legislation supposedly ended racial discrimination, whites were more likely to blame blacks for their social disadvantages and thus perpetuate negative stereotypes such as "blacks lack motivation" and "blacks have less ability."

• **How does contemporary prejudice differ from more traditional, "old-fashioned" prejudice?**

Traditional, old-fashioned prejudice is easy to recognize, because it is blatant, direct, and conscious. More contemporary forms of prejudice are often subtle, indirect, and unconscious.

• **Is it possible for an individual to discriminate without being prejudiced?**

Yes. In overt discrimination an individual discriminates because of his or her own prejudicial attitudes. But sometimes individuals who are not prejudiced discriminate because of someone else's prejudice. For example, a store clerk may watch black customers more closely because the store manager is prejudiced against blacks and has instructed the employee to follow black customers in the store closely. Discrimination based on someone else's prejudice is called adaptive discrimination.

- **Are U.S. schools segregated?**

Racial and ethnic minorities are largely isolated from whites in an increasingly segregated school system. A study by the Civil Rights Project at Harvard University found that U.S. schools in 2000–2001 were more segregated than they were in 1970. The upward trend in school segregation is due to large increases in minority student enrollment, continuing white flight from urban areas, the persistence of housing segregation, and the termination of court-ordered desegregation plans.

- **According to FBI data, the majority of hate crimes are motivated by what kind of bias?**

Since the FBI began publishing hate crime data in 1992, the majority of hate crimes have been based on racial bias.

- **What group constitutes the largest beneficiary of affirmative action policies?**

Affirmative action policies are designed to benefit racial and ethnic minorities, women, and, in some cases, Vietnam veterans and people with disabilities. The largest category of affirmative action beneficiaries is women.

Critical Thinking

1. In the 39th annual survey of college freshmen, the percentage of students indicating that "helping to promote racial understanding" is an "essential" or "very important" personal goal decreased from nearly half (46.4%) in 1992 to close to one-third (29.7%) in 2004 (Sax et al. 2004). In fact, fewer freshmen today indicate that "helping to promote racial understanding" is an "essential" or "very important" personal goal than any other entering class in the history of the survey. Why do you think this is so?
2. Women, most of whom are white, are the largest category designated to benefit from affirmative action. Yet a survey of 35 introductory sociology texts published in the 1990s found that nearly 90% of the texts did not mention affirmative action in their sections on gender inequality, and only 20% of texts included women in their definitions of affirmative action (Beeman et al. 2000). Why do you think many textbooks overlook or

minimize the benefits women may receive from affirmative action?
3. Do you think the time will ever come when a racial classification system will no longer be used? Why or why not? What arguments can be made for discontinuing racial classification? What arguments can be made for continuing it?

Key Terms

acculturation	Jim Crow laws
adaptive discrimination	marital assimilation
affirmative action	minority group
amalgamation	modern racism
antimiscegenation laws	multicultural education
assimilation	naturalized citizen
aversive racism	one-drop rule
colonialism	overt discrimination
de facto segregation	pluralism
de jure segregation	population transfer
discrimination	prejudice
ethnicity	primary assimilation
expulsion	race
genocide	racism
hate crime	secondary assimilation
individual discrimination	segregation
institutional discrimination	stereotype

Media Resources

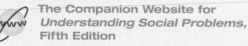

The Companion Website for
Understanding Social Problems,
Fifth Edition

http://sociology.wadsworth.com/mooney_knox_schacht5e

Supplement your review of this chapter by going to the companion website to take one of the Tutorial Quizzes, use the flash cards to master key terms, and check out the many other study aids you'll find there. You'll also find special features such as *Wadsworth's Sociology Online Resources and Writing Companion,* GSS Data, and Census 2000 information, data, and resources at your fingertips to help you complete that special project or do some research on your own.

> " Only a radical transformation of the relationship between women and men to one of full and equal partnership will enable the world to meet the challenges of the 21st century. " *Beijing Declaration and Platform for Action*

Gender Inequality

Since she was little, all Sharon Fullilove wanted to do was fly like a bird. For her, graduation from high school meant she was headed off to the Air Force Academy, the only college she had ever wanted to attend, indeed the only one she had applied to. She had "the right stuff"—she was a cheerleader, a dance champion, and a track star; she sang and did comedy; and she was a straight A student. Her mother, an air force colonel stationed at the Academy, couldn't have been prouder (CNN 2003; Thomas 2003).

One evening, after watching a movie, Sharon and two friends accepted the offer of a ride home by a male upperclassman they all knew. She was the last to be dropped off, and it was then that the senior cadet forced himself on her. Knowing full well the "good-ole boy" culture, she did not report the rape. However, unable to keep her secret and on the advice of her mother, four months after the crime had occurred, she reported it. The case was closed with no arrests or punishment, and Fullilove, disillusioned, quit the academy (CNN 2003; Thomas 2003).

Over the years there have been several accusations of sexual assault against air force cadets but few could have anticipated the results of a Department of Defense study. Of the 579 female cadets surveyed, 1 in 10 seniors reported being the victim of rape or attempted rape; more than 80% of the attacks went unreported (Smith 2003). Female cadets spoke of sexual harassment, accusations of promiscuity, and allegations of "violating rules against drinking, fraternization with upperclassmen, and having sex in the dormitories" (CBS News 2003a, p. 1). First-year Air Force Academy cadets are now required to take a course called Street Smarts, whereby cadets learn to protect themselves and to avoid potentially dangerous situations. In addition, the Department of Defense recently instituted a policy whereby sexual assault victims may receive confidential services (e.g., health care and counseling) whether or not they decide to officially report the crime. This policy replaced an earlier requirement that all sexual assault victims, in order to receive medical treatment or counseling, had to make an official report (U.S. Department of Defense 2005).

ThomsonNOW™

Reviewing is as easy as
① ② ③

Use ThomsonNOW to help you make the grade on your next exam. When you are finished reading this chapter, go to the chapter review for instructions on how to make ThomsonNOW work for you.

Some have argued that sexual assault is a reflection of gender inequality. The term *gender inequality,* however, raises the question: unequal in what way? Depending on the issue, both women and men are victims of inequality. When income, career advancement, household work, and sexual harassment are the focus, women are most often disadvantaged. But when life expectancy, mental and physical illness, and access to one's children following divorce are considered, it is often men who are disadvantaged. In this chapter we seek to understand inequalities for *both* genders.

This chapter looks at **sexism**—the belief that innate psychological, behavioral, and/or intellectual differences exist between women and men and that these differences connote the superiority of one group and the inferiority of the other. As with race and ethnicity, such attitudes often result in prejudice and discrimination at both the individual and institutional levels. Individual discrimination is reflected by the physician who will not hire a male nurse because he or she believes that women are more nurturing and empathetic and are therefore better nurses. Institutional discrimination—that is, discrimination built into the fabric of society—is exemplified by the difficulty many women experience in finding employment; they may have no work history and few job skills as a consequence of living in traditionally defined marriages.

Discerning the basis for discrimination is often difficult because the different types of minority status may intersect. For example, elderly African American and Hispanic women are more likely to receive lower wages and to work in fewer prestigious jobs than younger white women. They may also experience discrimination if they are "out" as homosexuals. Such **double or triple (multiple) jeopardy** occurs when a person is a member of two or more minority groups. In this chapter, however, we emphasize the impact of gender inequality.

Gender refers to the social definitions and expectations associated with being female or male and should be distinguished from **sex,** which refers to one's biological identity. In most Western cultures we take for granted that there are two categories of gender. However, in many other societies three and four genders have been recognized. For example, many Polynesian cultures recognize a third gender called the *mahū*—individuals who take on the work roles of members of the opposite sex (Nanda 2000). Other societies recognize hermaphrodites (individuals born with ambiguous genitalia) as a third gender. In the United States the majority of these babies have surgery to "correct" their genitalia, thus keeping them within the traditional two-gender system.

ThomsonNOW

Learn more about **Sex and Gender** by going through the Sex versus Gender Animation.

The Global Context: The Status of Women and Men

There is no country in the world in which women and men have equal status. Although much progress has been made in closing the gender gap in areas such as education, health care, employment, and government, gender inequality is still prevalent throughout the world.

The World Economic Forum assessed the gender gap in 58 countries by measuring the extent to which women have achieved equality with men in five areas: economic participation, economic opportunity, political empowerment, educational attainment, and health and well-being (Lopez-Claros & Zahidi 2005). Many countries were excluded from the analysis because reliable data were not available for those countries. In reporting the findings of this assessment of the global gender gap, the investigators noted that "even in light of heightened international awareness of gender issues, it is a disturbing reality that no country has yet managed to eliminate the gender gap" (Lopez-Claros & Zahidi 2005, p. 1). Table 10.1 presents the overall scores and rankings of (1) the 10 countries with the smallest gender gap (i.e., the least gender inequality), (2) the 10 countries with the largest gender gap (i.e., the most gender inequality), and (3) the United States, which did not rank in the top 10 or the bottom 10 but ranked 17 out of the 58 countries studied.

Gender inequality varies across cultures, not only in its extent or degree but also in its forms. For example, in the United States gender inequality in family roles commonly takes the form of an unequal division of household labor and child care, with women bearing the heavier responsibility for these tasks. In other countries forms of gender inequality in the family include the expectation that wives ask their husbands for permission to use birth control (see Chapter 13), the practice of aborting female fetuses in cultures that value male children over female children (see also this chapter's *Focus on Technology* feature), and unequal penalties for spouses who commit adultery, with wives receiving harsher punishment. For example, in Afghanistan in 2005 a 29-year-old woman was stoned to death for committing adultery, whereas the man accused of committing adultery with her was allegedly whipped 100 times and freed (Amnesty International 2005).

A global perspective on gender inequality must take into account the different ways in which such inequality is viewed. For example, the practice of Muslim women wearing a headscarf in public is viewed by many non-Muslims as a symbol of female subordination and oppression. To Muslims who embrace this practice (and

Table 10.1

Gender Gap Rankings: Top and Bottom 10 Countries and the United States

Country	Overall Rank	Overall Score[a]	Economic Participation Rank	Economic Opportunity Rank	Political Empowerment Rank	Educational Attainment Rank	Health and Well-being Rank
Top 10 countries							
Sweden	1	5.53	5	12	8	1	1
Norway	2	5.39	13	2	3	6	9
Iceland	3	5.32	17	7	2	7	6
Denmark	4	5.27	6	1	20	5	2
Finland	5	5.19	12	17	4	10	4
New Zealand	6	4.89	16	47	1	11	26
Canada	7	4.87	7	27	11	12	14
United Kingdom	8	4.75	21	41	5	4	28
Germany	9	4.61	20	28	6	34	10
Australia	10	4.61	15	25	22	17	18
United States	**17**	**4.40**	**19**	**46**	**19**	**8**	**42**
Bottom 10 countries							
Venezuela	49	3.42	38	13	52	33	58
Greece	50	3.41	44	48	50	45	22
Brazil	51	3.29	46	21	57	27	53
Mexico	52	3.28	47	45	41	44	51
India	53	3.27	54	35	24	57	34
Korea	54	3.18	34	55	56	48	27
Jordan	55	2.96	58	32	58	43	43
Pakistan	56	2.90	53	54	37	58	33
Turkey	57	2.67	22	58	53	55	50
Egypt	58	2.38	57	50	55	56	49

[a] All scores are reported on a scale of 1 to 7, with 7 representing maximum gender equality.

Source: Lopez-Claros and Zahidi (2005).

not all Muslims do), wearing a headscarf reflects the high status of women and represents the view that women should be respected and not treated as sexual objects.

Similarly, cultures differ in how they view the practice of female genital cutting (FGC), also known as female genital mutilation or female circumcision. There are several forms of FGC, ranging from a symbolic nicking of the clitoris to removal of the clitoris and labia and partial closure of the vaginal opening by stitching the two sides of the vulva together, leaving only a small opening for the passage of urine and menstrual blood. After marriage the sealed opening is reopened to permit intercourse and childbearing. After childbirth the woman's vulva is often stitched back together.

In traditional Muslim societies, women are forbidden to show their faces or other parts of their bodies when in public. As pictured, Muslim women wear a veil to cover their faces and a *chador,* a floor-length loose-fitting garment, to cover themselves from head to toe. Although some women adhere to this norm out of fear of repercussions, many others believe veiling was first imposed on Muhammad's wives out of respect for women and the desire to protect them from unwanted advances. There are more than half a million Muslims living in the United States.

Most FGC procedures are done by nonmedical personnel using unsterilized blades or string. Health risks associated with FGC include pain, hemorrhage, infection, shock, scarring, and infertility. Worldwide, 130 million girls and young women have experienced FGC and an additional 2 million are at risk each year (United Nations Population Fund 2003a). FGC is a common practice in many countries in the northern half of sub-Saharan Africa as well as in Egypt and Yemen. The prevalence of FGC among women ranges from 5% in Niger to 99% in Guinea (Yoder et al. 2004).

People from countries where FGC is not the norm generally view this practice as a barbaric form of violence against women. However, in countries where it commonly occurs, FGC is viewed as an important and useful practice. In some countries it is considered a rite of passage that enhances a woman's status. In other countries it is aesthetically pleasing. For others FGC is a moral imperative based on religious beliefs (Yoder et al. 2004).

Inequality in the United States

Although attitudes toward gender equality are becoming increasingly liberal, the United States has a long history of gender inequality. (You can assess your own beliefs about gender equality in this chapter's *Self and Society* feature.) Women have had to fight for equality: the right to vote, equal pay for comparable work, quality education, entrance into male-dominated occupations, and legal equality. As shown in Table 10.1, the World Economic Forum (2005)—based on its assessment of women's economic participation and opportunities, political empowerment, educational attainment, and health and well-being—ranks the United States only 17th in the world in terms of gender equality. Most U.S. citizens agree that U.S. society does not treat women and men equally: Women have lower incomes, hold fewer prestigious jobs, earn fewer graduate degrees, and are more likely than men to live in poverty.

ThomsonNOW™

Learn more about **Inequality in the United States** by going through the Women-Owned Firms Map Exercise.

The Beliefs about Women Scale (BAWS)

The statements listed here describe different attitudes toward men and women. There are no right or wrong answers, only opinions. Indicate how much you agree or disagree with each statement, using the following scale: (A) strongly disagree, (B) slightly disagree, (C) neither agree nor disagree, (D) slightly agree, (E) strongly agree.

1. Women are more passive than men.
2. Women are less career motivated than men.
3. Women don't generally like to be active in their sexual relationships.
4. Women are more concerned about their physical appearance than men are.
5. Women comply more often than men do.
6. Women care as much as men do about developing a job/career.
7. Most women don't like to express their sexuality.
8. Men are as conceited about their appearance as women are.
9. Men are as submissive as women are.
10. Women are as skillful in business-related activities as men are.
11. Most women want their partner to take the initiative in their sexual relationships.
12. Women spend more time attending to their physical appearance than men do.
13. Women tend to give up more easily than men do.
14. Women dislike being in leadership positions more than men do.
15. Women are as interested in sex as men are.
16. Women pay more attention to their looks than most men do.
17. Women are more easily influenced than men are.
18. Women don't like responsibility as much as men do.
19. Women's sexual desires are less intense than men's.

20. Women gain more status from their physical appearance than men do.

The Beliefs about Women Scale (BAWS) consists of 15 separate subscales; only four are used here. The items for these four subscales and coding instructions are as follows:

Subscales

1. Women are more passive than men (items 1, 5, 9, 13, 17).
2. Women are interested in careers less than men (items 2, 6, 10, 14, 18).
3. Women are less sexual than men (items 3, 7, 11, 15, 19).
4. Women are more appearance conscious than men (items 4, 8, 12, 16, 20).

Items 1–5, 7, 11–14, and 16–20 should be scored as follows: strongly agree, 2; slightly agree, 1; neither agree nor disagree, 0; slightly disagree, 1; and strongly disagree, 2. Items 6, 8, 9, 10, and 15 should be scored as follows: strongly agree, 2; slightly agree, 1; neither agree nor disagree, 0; slightly disagree, 1; and strongly disagree, 2. Scores range from 0 to 40; subscale scores range from 0 to 10. The higher your score, the more traditional your gender beliefs about men and women.

Source: William E. Snell, Jr., Ph.D. (1997). College of Liberal Arts, Department of Psychology, Southeast Missouri State University. Reprinted with permission.

Men are also victims of gender inequality. In 1963 sociologist Erving Goffman wrote that in the United States there is only:

> one complete unblushing male . . . a young, married, white, urban, northern heterosexual, Protestant father of college education, fully employed, of good complexion, weight and height, and a recent record in sports. . . . Any male who fails to qualify in one of these ways is likely to view himself . . . as unworthy, incomplete, and inferior. (Goffman 1963, p. 128)

Although standards of masculinity have relaxed, Williams (2000) argues that masculinity is still based on "success"—at work, on the athletic field, on the streets, and at home—which must be constantly maintained and proven, therefore placing enormous pressure on boys and men.

When U.S. college students were asked to list the best and worst things about being the opposite sex, the same qualities, although in opposite categories, emerged

(Cohen 2001). For example, what males list as the best thing about being female (e.g., free to be emotional), females list as the worst thing about being male (e.g., not free to be emotional). Similarly, what females list as the best thing about being male (e.g., higher pay), males listed as the worst thing about being female (e.g., lower pay). As Cohen notes (2001), although "some differences are exaggerated or over-simplified, . . . we identif[ied] a host of ways in which we 'win' or 'lose' simply because we are male or female" (p. 3).

ThomsonNOW™

Learn more about **Inequality in the United States** by going through the Gender Inequality Data Experiment.

Sociological Theories of Gender Inequality

Both structural functionalism and conflict theory concentrate on how the structure of society and, specifically, its institutions contribute to gender inequality. However, these two theoretical perspectives offer opposing views of the development and maintenance of gender inequality. Symbolic interactionism, on the other hand, focuses on the culture of society and how gender roles are learned through the socialization process.

Structural-Functionalist Perspective

Structural functionalists argue that preindustrial society required a division of labor based on gender. Women, out of biological necessity, remained in the home performing such functions as bearing, nursing, and caring for children. Men, who were physically stronger and could be away from home for long periods of time, were responsible for providing food, clothing, and shelter for their families. This division of labor was functional for society and over time became defined as both normal and natural.

Industrialization rendered the traditional division of labor less functional, although remnants of the supporting belief system still persist. Today, because of day-care facilities, lower fertility rates, and the less physically demanding and dangerous nature of jobs, the traditional division of labor is no longer as functional. Thus modern conceptions of the family have, to some extent, replaced traditional ones—families have evolved from extended to nuclear, authority is more egalitarian, more women work outside the home, and greater role variation exists in the division of labor. Structural functionalists argue, therefore, that as the needs of society change, the associated institutional arrangements also change.

"People call me a feminist whenever I express sentiments that differentiate me from a doormat or a prostitute."

Rebecca West
Author and feminist

Conflict Perspective

Many conflict theorists hold that male dominance and female subordination are shaped by the relationship men and women have to the production process. During the hunting and gathering stage of development, males and females were economic equals, each controlling their own labor and producing needed subsistence. As society evolved to agricultural and industrial modes of production, private property developed and men gained control of the modes of production while women remained in the home to bear and care for children. Male domination was furthered by inheritance laws that ensured that ownership would remain in their hands. Laws that regarded women as property ensured that women would remain confined to the home.

As industrialization continued and the production of goods and services moved away from the home, the male-female gaps continued to grow—women had less education, lower incomes, and fewer occupational skills and were rarely owners. World War II necessitated the entry of a large number of women into the labor force, but in contrast to previous periods, many of them did not return to the home at the end of the war. They had established their own place in the workforce and, facilitated by the changing nature of work and technological advances, now competed directly with men for jobs and wages.

Conflict theorists also argue that continued domination by males requires a belief system that supports gender inequality. Two such beliefs are (1) that women are inferior outside the home (e.g., they are less intelligent, less reliable, and less rational) and (2) that women are more valuable in the home (e.g., they have maternal instincts and are naturally nurturing). Thus, unlike structural functionalists, conflict theorists hold that the subordinate position of women in society is a consequence of social inducement rather than biological differences that led to the traditional division of labor.

Symbolic Interactionist Perspective

Although some scientists argue that gender differences are innate, symbolic interactionists emphasize that through the socialization process both females and males are taught the meanings associated with being feminine and masculine. Gender assignment begins at birth as a child is classified as either female or male. However, the learning of gender roles is a lifelong process whereby individuals acquire society's definitions of appropriate and inappropriate gender behavior.

Gender roles are taught by the family, in the school, in peer groups, and by media presentations of girls and boys and women and men (see the discussion on the social construction of gender roles later in this chapter). Most important, however, gender roles are learned through symbolic interaction as the messages that

Women as well as girls are often portrayed provocatively as a means of selling a product or service. This billboard is a good example of the cultural emphasis placed on women's physical appearance.

others send us reaffirm or challenge our gender performances. As Lorber (1998) notes:

> Gender is so pervasive that in our society we assume it is bred into our genes. Most people find it hard to believe that gender is constantly created and recreated out of human interaction, out of social life, and is the texture and order of social life. Yet gender, like culture, is a human production that depends on everyone constantly "doing gender." (p. 213)

Feminist theory, although also consistent with a conflict perspective, incorporates many aspects of symbolic interactionism. Feminists argue that conceptions of gender are socially constructed as societal expectations dictate what it means to be female or what it means to be male. Thus, for example, women are generally socialized into **expressive roles** (i.e., nurturing and emotionally supportive roles) and males are more often socialized into **instrumental roles** (i.e., task-oriented roles). These roles are then acted out in countless daily interactions as boss and secretary, doctor and nurse, football player and cheerleader "do gender." Feminists also hold that gender "is a central organizing factor in the social world and so must be included as a fundamental category of analysis in sociological research" (Renzetti & Curran 2003, p. 8). Feminists, noting that the impact of the structure and culture of society is not the same for different groups of women and men, encourage research on gender that takes into consideration the differential effects of age, race and ethnicity, and sexual orientation.

Gender Stratification: Structural Sexism

As structural functionalists and conflict theorists agree, the social structure underlies and perpetuates much of the sexism in society. **Structural sexism**, also known as institutional sexism, refers to the ways the organization of society, and specifically its institutions, subordinate individuals and groups based on their sex classification. Structural sexism has resulted in significant differences in the education and income levels, occupational and political involvement, and civil rights of women and men.

Education and Structural Sexism

Literacy rates worldwide indicate that women are less likely than men to be able to read and write, with millions of women being denied access to even the most basic education. For example, on average, women in South Asia have only half as many years of education as their male counterparts. In addition, of the 150 million children age 6 to 11 who are not in school, 90 million are girls (World Bank 2003).

In 2003 few differences existed between men and women in their completion rates of high school and college degrees (U.S. Bureau of the Census 2005). In fact, in recent years most U.S. colleges and universities have had a higher percentage of women than men enrolling directly from high school. This trend is causing some concern that many young American men may not have the education to compete in today's economy. Yet men are still more likely to go on to complete a graduate or professional degree than are women. As Figure 10.1 indicates, dramatic differences appear in the types of advanced degrees that men and women earn. For example, women receive 68% of all psychology doctorates, but they earn only 17% of all doctorates in engineering. Only in the fields of psychology, education, and health (e.g., nursing) are women more likely than men to earn advanced degrees (National

ThomsonNOW

Learn more about **Gender Roles** by going through the Gender Roles and Videos Animation.

"The price of inequality is just too high."

Nafis Sadik
United Nations Population Fund

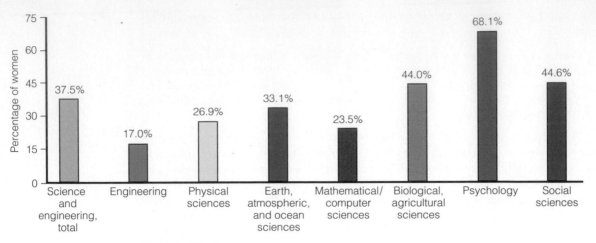

Figure 10.1

Science and engineering doctorates awarded to women, 2003.

Source: National Science Foundation (2003).

Science Foundation 2003). Although the general trend is for men to have higher levels of education than women, in recent years African American women's educational levels have increased at a faster rate than African American men's educational levels. Black women receive twice as many college degrees and twice as many master's degrees as black men. They also earn 50% more PhDs, JDs, MDs, and DDSs (Urban Institute 2003).

One explanation for why women earn fewer advanced degrees than men is that women are socialized to choose marriage and motherhood over long-term career preparation. From an early age women are exposed to images and models of femininity that stress the importance of domestic family life. In a study of college women who were asked about their future plans, Hoffnung (2004) found that the majority wanted a career, marriage, and children. Many expected to go on for an advanced degree. However, seven years after the study those who had become mothers—often unexpectedly—were less likely to have pursued their education than those who did not have children.

Structural limitations also discourage women from advancing in higher education. For example, women seeking academic careers may find that securing a tenure-track position is more difficult than it is for men. In addition, McBrier (2003), in examining sex inequality in law schools, found that women progress through the professorial ranks at a slower pace than men do. She attributes the differences to a variety of variables, including family and geographic constraints, social capital differences, degree prestige, and prior work history. Similarly, a recent study of sociologists found that among those who earned their PhDs in 1997 both males and females were equally as likely to acquire a tenure-track position. However, those with children were less likely than sociologists without children to have done so (American Sociological Association 2004).

Long and colleagues (1993) examined the promotions of 556 men and 450 women with PhDs in biochemistry. They found that women were less likely to be promoted to associate or full professor, were held to a higher standard than men, and were particularly disadvantaged in more prestigious departments. In addition, a committee at the Massachusetts Institute of Technology found that women scientists were allocated less laboratory space, were excluded from search committees, and were encouraged to teach more than their male counterparts at the same rank— all factors that adversely affect promotion (Lawler 1999). Finally, even in public schools, women make up 75% of all classroom teachers but only about 50% of principals and assistant principals (U.S. Bureau of the Census 2005).

Work and Structural Sexism

Women now make up one-third of the world's labor force. Worldwide, women tend to work in jobs that have little prestige and low or no pay, where no product is produced, and where women are the facilitators for others. Women are also more likely to hold positions of little or no authority within the work environment and to have more frequent and longer periods of unemployment (Athreya 2003; United Nations 2000). Although improvements are being made, women of color are even less likely to hold positions of power (Equal Employment Opportunity Commission 2003). In an investigation of female and male African American and white firefighters, black women were the most subordinated group, as black males and white females relied on their superordinate gender and race statuses, respectively (Yoder & Aniakudo 1997).

No matter what the job, if a woman does it, it is likely to be valued less than if a man does it (Barko 2003). For example, in the early 1800s 90% of all clerks were men and being a clerk was a prestigious profession. As the job became more routine, in part because of the advent of the typewriter, the pay and prestige of the job declined and the number of female clerks increased. Today, female clerks predominate, and the position is one of relatively low pay and prestige.

The concentration of women in certain occupations and men in other occupations is referred to as **occupational sex segregation** (see Table 10.2). For example, women are overrepresented in semiskilled and unskilled occupations, and men are disproportionately concentrated in professional, administrative, and managerial positions. In some occupations sex segregation has decreased in recent years. For example, between 1983 and 2003 the percentage of female physicians increased from 16% to 30%, female dentists increased from 7% to 22%, and female clergy increased from 6% to 14% (U.S. Bureau of the Census 2005). Although the pace is slow, increasingly men are applying for jobs traditionally held by women, leading to such terms as *mannies* (male nannies) and *murses* (male nurses) (Cullen 2003). However in some fields the trend has slowed or even reversed; public school teaching is even more dominated by women today than it was two decades ago (U.S. Bureau of the Census 2005). Women are still heavily represented in low-prestige, low-wage **pink-collar jobs** that offer few benefits.

Sex segregation in occupations continues for several reasons (Martin 1992; Renzetti & Curran 2003; Williams 1995). First, cultural beliefs about what is an "appropriate" job for a man or a woman still exist. Cejka and Eagly (1999) report that the more college students believe that an occupation is male or female dominated, the more they attribute success in that occupation to masculine or feminine characteristics. In addition, males and females continue to be socialized to learn different skills and acquire different aspirations. A government report documents that work at age 12 is sex segregated; girls babysit and boys do lawn work (Bureau of Labor Statistics 2000). Furthermore, a study of 14- to 16-year-olds found that, although both males and females held gender equitable ideas about men's and women's roles in society, they enrolled in gender-typical subjects in school and aspired to gender-typical occupations (Tinklin et al. 2005). Finally, Skuratowicz and Hunter (2004), in studying personnel practices in a U.S. bank, found that employees were encouraged to apply

ThomsonNOW™

Learn more about **Work and Structural Sexism** by going through the Women's Employment Patterns Animation.

Table 10.2

Highly Sex-Segregated Occupations 2003

Female-Dominated Occupations	Percentage of Female Workers
Child-care workers	95
Dental hygienists	99
Dietitians	91
Elementary and middle school teachers	82
Librarians	84
Paralegals and legal assistants	84
Prekindergarten and kindergarten teachers	98
Receptionists	93
Registered nurses	92
Secretaries	97
Speech therapists	95
Teacher assistants	92
Travel agents	83

Male-Dominated Occupations	Percentage of Male Workers
Airplane pilots and navigators	97
Architects	78
Automobile mechanics	98
Civil engineers	91
Clergy	86
Construction workers	97
Dentists	78
Firefighters	96
Grounds maintenance workers	94
Lawyers	76
Mechanical engineers	96
Physicians	70
Police officers	88

Source: U.S. Bureau of the Census (2005).

for newly created customer resource manager (CRM) positions with pamphlets picturing women. In contrast, customers were informed of new personal banker (PB) positions with video displays picturing men in business suits. Managers assumed that men would have little interest in the CRM position, which had no direct authority over other employees but included increased customer contact, and would be more attracted to the PB positions, which involved selling financial products and working on commission. Not surprisingly, after the employment process was completed, the CRMs were predominately female and the PBs were mostly male.

Opportunity structures differ as well. For example, women and men, upon career entry, are often channeled by employers into gender-specific jobs that carry different wages and promotion opportunities. However, even women in higher-paying jobs may be victimized by a **glass ceiling**—an often invisible barrier that prevents women and other minorities from moving into top corporate positions. Women and minorities have different social networks than do white men, which contributes to this barrier. White men in high-paying jobs are more likely to have interpersonal connections with individuals in positions of authority (Padavic and Reskin 2002). Women also may be excluded by male employers and by those employees who fear that the prestige of their profession will be compromised by the entrance of women or by those who simply believe that "the ideal worker is normatively masculine" (Martin 1992, p. 220). In addition, women often find that their opportunities for career advancement are adversely affected after returning from family leave. For example, female lawyers returning from maternity leave found their career mobility stalled after being reassigned to less prestigious cases (Williams 2000). Not surprisingly, one study found that in the United States 49% of women in high-level managerial positions had no children, compared to 19% of men (Hewlett 2002).

Finally, because family responsibilities primarily remain with women, working mothers may feel pressure to choose professions that permit flexible hours and career paths, sometimes known as mommy tracks (Moen & Yu 2000). Thus, for example, women dominate the field of elementary education, which permits them to be home when their children are not in school. Nursing, also dominated by women, often offers flexible hours. Although the type of career pursued may be the woman's choice, it is a **structured choice**—a choice among limited options as a result of the structure of society. This chapter's *Social Problems Research Up Close* feature examines the impact that women's and men's different work and family responsibilities have on how they spend their free time.

Income and Structural Sexism

Women are twice as likely as men to be minimum-wage earners. Furthermore, in 2004 women had median weekly earnings of $580, compared to $713 for men (U.S. Bureau of the Census 2005). As Figure 10.2 illustrates, the gender gap in pay varies by state (see Figure 10.2). For example, in the District of Columbia full-time working women earn, on average, 92% of what full-time men earn, whereas in Wyoming the average full-time working woman earns 66% of what the average full-time man earns (Institute for Women's Policy Research 2004). In general, the higher one's education, the higher one's income. Yet, even when men and women have identical levels of educational achievement and both work full-time, women, on average, earn significantly less than men (see Table 10.3). Racial differences also exist. Although white women earned 80% as much as white men in 2003, African American and Hispanic American women earned just 68% and 57%, respectively, of white men's salaries (U.S. Bureau of the Census 2005). Even among celebrities a significant income gap exists. According to *Forbes* magazine's Celebrity 100 issue, the top female athlete (Maria Sharapova) earned 20% of what the top male athlete (Tiger Woods) earned (Burk 2005).

In an investigation of the gender pay gap, Kilbourne and colleagues (1994) analyzed data from the National Longitudinal Survey that included more than 5,000 women and 5,000 men. They concluded that occupational pay is gendered and that "occupations lose pay if they have a higher percentage of female workers

ThomsonNOW

Learn more about **The Glass Ceiling** by going through the Glass Ceiling Animation.

ThomsonNOW

Learn more about **Work and Structural Sexism** by going through the Gender Inequality and Work Learning Module.

"We know that we can do what men can do, but we still don't know that men can do what women can do. That's absolutely crucial. We can't go on doing two jobs."

Gertrude Stein
Author and feminist

Gender and the Quantity and Quality of Free Time

A considerable body of research has been conducted on gender differences in the workplace and in the division of household labor. However, little research has examined differences in how men and women spend their free time. Also known as leisure time, free time is time not spent in paid employment, domestic caregiving, or personal care. Free time includes activities such as socializing with friends, watching television, reading, going for a walk, and other pursuits for fun and relaxation.

In the research study presented here, Mattingly and Bianchi (2003) examined the quantity and quality of free time experienced by men and women in the United States. In doing so, they attempted to answer the following questions:

1. Are there differences in the *quantity* of free time enjoyed by men and women in the United States?
2. Are there gender differences in the *quality* of free-time experiences?
3. Do work and family roles—spouse, parent, worker—affect the quantity and quality of free-time experiences?
4. Are work-family role constraints on free time greater for women than for men?

Sample and Methods

Data for this investigation came from the 1998–1999 Family Interaction, Social Capital, and Trends in Time Use Study conducted by the Survey Research Center at the University of Maryland. Data were collected by telephone interviews of a randomly selected national sample of 1,132 adults (481 men, 651 women). The interviewers first ascertained demographic, family, and socioeconomic characteristics of each respondent, including the respondent's age, race, sex, marital status, family income, number of hours spent in paid employment, and number of children in the household. Interviewers then asked questions to obtain a detailed description of how the respondent spent his or her time on the day before the interview. To construct this "yesterday time diary," the interviewer asked research participants the following questions: "What were you doing on [day of week before diary day] at midnight?" "What time did you finish?" "At any time while you were (REPEAT ACTIVITY) did you do anything else? (like talking, reading, watching TV, listening to the radio, eating, or caring for children)." "While you were (REPEAT ACTIVITY), who was with

you?" "What did you do next?" This cycle of questions continued until the full 24-hour day was recorded. During the interview respondents were also asked "Would you say you always feel rushed even to do the things you have to do, only sometimes feel rushed, or almost never feel rushed?"

Findings and Conclusions

The study produced a number of findings that shed light on the ways in which women and men experience free time.

- *Amount of free time.* On average, men in this study reported having nearly 30 minutes more free time per day than women (5 hours 36 minutes versus 5 hours 8 minutes). A difference of 30 minutes might seem small, unless you consider that it translates into men having an additional 164 hours of free time per year than women, which is equivalent to 4 weeks of paid vacation (40 hours per week) or nearly seven 24-hour days per year.

- *Quality of free time.* One way to characterize the quality of free time is to determine whether it is "pure" or "contaminated." Pure free time occurs without

or require nurturant skills" (p. 708). Cohen and Huffman (2003) also found that in occupations with high female representation, men performing the same job as women earned higher salaries. Budig (2003) reports that regardless of the male-female distribution in the job, men are paid more than women.

Two hypotheses are frequently cited in the literature for why the income gender gap continues to exist. One hypothesis is called the **devaluation hypothesis.** It argues that women are paid less because the work they perform is socially defined as less valuable than the work performed by men. Guy and Newman (2004) argue that these jobs are undervalued in part because they include a significant amount of **emotion work**—that is, work that involves caring, negotiating, and empathizing with people, which is rarely specified in job descriptions or performance evaluations. The other hypothesis, the **human capital hypothesis,** argues that female-male

the contamination of another nonleisure activity, such as cooking, cleaning, or watching children. For example, a woman watching TV alone or with another adult is experiencing pure free time. A woman watching TV while she is cooking or taking care of children is not experiencing pure free time; rather, her free time is contaminated by other nonleisure tasks.

In this study both women and men experienced most of their free time (87%) as pure free time. But because men had more free time overall, they had more pure free time than women. Women's free time was more likely than men's to be contaminated by other tasks or by the presence of children.

- *Effects of marriage on free time.* Married men experienced about the same amount and quality of free time as single men. However, married women had fewer episodes of free time than their single counterparts. Married women also experienced nearly an hour less of pure free time per day than unmarried woman did.
- *Effects of children on free time.* Having children in the home was associated

with less free time for both women and men. Having a preschool-age child significantly decreased free time for mothers but not for fathers. Women living with a young child had more than an hour less of free time per day than did other women.

- *Effects of employment on free time.* Mattingly and Bianchi (2003) found that for employed women and men the more hours a person worked, the less free time he or she had. Surprisingly, employed men experienced more free time than their unemployed counterparts. The researchers explain that unemployed men might be fulfilling traditional homemaker roles or spending their time seeking employment. Although being employed was associated with more free time for men, the same was not true for women. The amount of free time for employed and unemployed women was similar—not surprising, given that many women who are unemployed are likely engaged in household work and child care.
- *Free time and feeling rushed.* Finally, women (34.4%) in this study were more likely than men (28.8%) to report that

they always feel rushed. Not surprisingly, the more free time a respondent had, the less he or she reported feeling rushed. However, this effect was greater for men than for women.

In conclusion, Mattingly and Bianchi (2003) found that women experience less free time and lower-quality free time than men, likely as a result of heavier caregiving responsibilities associated with women's traditional roles. Marriage and having preschool-age children result in less free time for women but not for men. Women, especially those who are wives and/or mothers of young children, have less time than men to relax and refresh, and the free time they do have is often contaminated by other nonleisure activities or the presence of children. In addition, women's free time is not as beneficial to them as men's free time is in reducing feelings of time pressure. Although gender roles may be evolving toward greater equality, this study provides strong evidence that a leisure gap favoring men remains.

---

pay differences are a function of differences in women's and men's levels of education, skills, training, and work experience.

Although these explanations are difficult to test, most researchers agree that there is more support for the devaluation hypothesis than for the human capital hypothesis (Padavic and Reskin 2002). One way researchers test these hypotheses is to compare the earnings of individuals in occupations with comparable worth but different gender compositions. **Comparable worth** refers to the belief that individuals in occupations, even in different occupations, should be paid equally if the job requires "comparable" levels of education, training, and responsibility. In a comparable worth lawsuit, nurses successfully sued the City of Denver for paying them less than other employees (e.g., tree trimmers, sign painters) who had less education (Nelson & Bridges 1999).

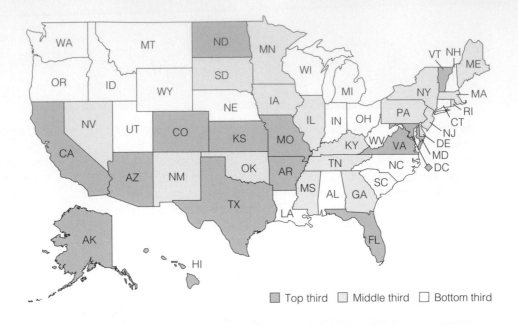

Figure 10.2
Ratio of women's to men's earnings across the United States, 2002.

Source: Institute for Women's Policy Research (2004).

Legend: ■ Top third □ Middle third □ Bottom third

Table 10.3
Effects of Education and Sex on Income of Full-Time Workers, 2004

Educational Attainment	Median Annual Income	
	Men	Women
Total, 25 years old and older	39,624	31,148
Less than 9th grade	20,748	19,292
9th to 12th grade (no diploma)	26,000	23,712
High school graduate (includes equivalency)	33,540	29,848
Some college, no degree	38,948	33,384
Associate degree	40,976	36,088
Bachelor's degree or more	59,436	51,272
Bachelor's degree	54,288	47,632
Master's degree	66,924	57,304
Professional degree	84,656	71,604
Doctorate degree	80,288	72,696

Source: U.S. Department of Labor (2005).

Politics and Structural Sexism

Women received the right to vote in the United States in 1920 with the passage of the Nineteenth Amendment. Even though this amendment went into effect almost 80 years ago, women still play a rather minor role in the political arena. In general, the more important the political office, the lower the probability that a woman will

Table 10.4

Percentage of Women Elected by Level and Type of Government Position, 2005

Level of Government/Position	Number of Seats	Number of Women	Percentage Held by Women
U.S. president	1	0	0.0
U.S. vice president	1	0	0.0
U.S. Congress	535	80	15.0
House	435	66	14.0
Senate	100	14	15.2
Governors	50	8	16.0
State legislators	7,382	1,662	22.5

Source: Center for American Women in Politics (2005).

hold it. Although women are 52% of the population, the United States has never had a woman president or vice president, and until 1993, when a second woman was appointed, the United States had only one female Supreme Court justice. The highest-ranking women ever to serve in U.S. government have been former secretary of state Madeleine Albright and current secretary of state Condoleezza Rice. Appointed in 2005, Rice is also the first African American woman to serve in such a capacity. In 2005 women represented only 16% of all governors and held only 15% of all U.S. congressional seats (see Table 10.4). Worldwide, the percentage of national and local legislative seats held by women is 16% (Population Reference Bureau 2005).

In response to the underrepresentation of women in the political arena, some countries have instituted quotas. In India a 1993 amendment held one-third of all seats in local contests for women. Eight thousand women were elected. A 1996 law in Brazil required that a minimum of 20% of each party's candidates be women. Countries with similar policies include Finland, Germany, Mexico, South Africa, and Spain (Sheehan 2000). In addition, in 2005 the Transitional Administrative Law in Iraq identified a target goal of 25% female representation in the Iraqi parliament. Exceeding that goal, 31% of subsequently elected members of parliament were women.

The relative absence of women in politics, as in higher education and in high-paying, high-prestige jobs in general, is a consequence of structural limitations. Running for office requires large sums of money, the political backing of powerful individuals and interest groups, and a willingness of the voting public to elect women. Thus minority women have even greater structural barriers to election and, not surprisingly, represent an even smaller percentage of elected officials. Women's lack of representation among political leaders begins early. In the last five years less than one-third of student government offices at midwestern universities were held by women (Miller 2004). Nonetheless, 72% of Americans surveyed in 2005 stated that they would vote for a woman president, and 71% of them stated that it was at least "somewhat likely" that the United States would have a woman president by 2030 (Rasmussen Reports 2005).

Would an increase in female participation make a difference in politics? Voting patterns among women suggest that it might. Women tend to be more likely to vote Democratic than men and to express more concern over health care, child care,

> "Women weren't born Democrat, we weren't born Republican, and we weren't born yesterday."
>
> **Patricia Ireland**
> **President, National Organization for Women, 1991–2001**

education, poverty, and homelessness (Renzetti and Curran 2003). Men focus more on the federal deficit, taxes, energy, defense, and foreign policy. Women are less supportive of the death penalty than men are and more supportive of gun control laws. Furthermore, women politicians report that they feel a responsibility to represent both their elected constituency and women in general. As Congresswoman Marge Roukema notes (Hawkesworth et al. 2001):

> I didn't really want to be stereotyped as the woman legislator. . . . I wanted to deal with things like banking and finance. But I learned very quickly that if the women like me in Congress were not going to attend to some of these family concerns, whether it was for jobs or children, pension equity, or whatever, then they weren't going to be attended to. So I quickly shed those biases that I had and said, "Well nobody else is going to do it; I'm going to do it." (p. 6)

Civil Rights, the Law, and Structural Sexism

The 1963 Equal Pay Act and Title VII of the 1964 Civil Rights Act made it illegal for employers to discriminate in wages or employment on the basis of sex. Nevertheless, such discrimination still occurs, as evidenced by the thousands of grievances filed each year with the Equal Employment Opportunity Commission (EEOC). One technique used to justify differences in pay is the use of different job titles for the same work. The courts have ruled, however, that jobs that are "substantially equal," regardless of title, must result in equal pay.

Women are also discriminated against in employment. Discrimination, although illegal, takes place at both the institutional and the individual level (see Chapter 9). Institutional discrimination includes screening devices designed for men, hiring preferences for veterans, the practice of promoting from within an organization based on seniority, and male-dominated recruiting networks (Reskin & McBrier 2000). For example, the Augusta National Golf Club, home of the Masters Golf Tournament and a virtual "who's who of the corporate world," has refused to change its policy forbidding women from joining, despite years of political pressure from women's groups and negative publicity (National Council of Women's Organizations 2005). One of the most blatant forms of individual discrimination is sexual harassment, discussed later in this chapter.

Discrimination takes place in other forms as well. In the United States women often have difficulty obtaining home mortgages or rental property because they have lower incomes, shorter work histories, and less collateral. Until fairly recently, husbands who raped their wives were exempt from prosecution. Even today, some states require a legal separation agreement and/or separate residences for a raped wife to receive full protection under the law. Women in the military have traditionally been restricted in the duties they can perform, and finally, since the U.S. Supreme Court's 1973 *Roe* v. *Wade* decision, which made abortion legal, the right of a woman to obtain an abortion has steadily been limited and narrowed by subsequent legislative acts and judicial decisions (Connolly 2005). The debate has continued with several recent court decisions (see Chapter 15).

The Social Construction of Gender Roles: Cultural Sexism

As social constructionists note, structural sexism is supported by a system of cultural sexism that perpetuates beliefs about the differences between women and men. **Cultural sexism** refers to the ways the culture of society—its norms, values, beliefs, and symbols—perpetuates the subordination of an individual or group because of the sex classification of that individual or group.

For example, the *belief* that females are less valuable than males has serious consequences. China has 20% fewer girls than boys in the birth to 4-years-old age group. China's "missing girl" phenomenon, a growing problem, is a consequence of selective abortions of female fetuses and premature death of female infants as a result of withholding nourishment and health care (Banister 2003). This chapter's *Focus on Technology* feature looks at the issue of sex selection of babies. Cultural sexism takes place in a variety of settings, including the family, the school, and the media as well as in everyday interactions.

Family Relations and Cultural Sexism

From birth, males and females are treated differently. The toys that male and female children receive convey different messages about appropriate gender behavior. For example, 90% of girls age 3 to 11 have a Barbie doll (Dittrich 2002). Recently, retail giant Toys 'R' Us, after much criticism, removed store directories labeled "Boy's World" and "Girl's World." Similarly, toy manufacturer Mattel came under fire after producing a pink, flowered Barbie computer for girls and a blue Hot Wheels computer for boys. The social significance of the gender-specific computers and the public criticism came after it was revealed that the accompanying software packages were different—the boys' package had more educational titles (Bannon 2000).

Household Division of Labor. Little girls and boys work within the home in approximately equal amounts until the age of 18, when the female-to-male ratio begins to change (Robinson & Bianchi 1997; United Nations Population Fund 2003b). In a study of household labor in 10 Western countries, Bittman and Wajcman (2000) report that "women continue to be responsible for the majority of hours of unpaid labor," ranging from a low of 70% in Sweden to a high of 88% in Italy (p. 173). In the United States married women report performing between two-thirds and three-quarters of the household work (United Nations 2000). The fact that women, even when working full-time, contribute significantly more hours to home care than men is known as the "second shift" (Hochschild 1989). In a recent University of Michigan study of 50 pregnant couples, researcher Marlena Studer asked spouses to identify how much child care they each expected to perform after the baby was born. Compared to their husbands' projections, working and stay-at-home mothers anticipated that they would perform almost three times as many hours of child care per week than their husbands (University of Michigan News Service 2005). To the extent that women work the second shift more than men, they convey traditional conceptions of gender to their children that are often carried into adulthood. For example, Cunningham (2001) found that boys are particularly more likely to perform housework as teenagers when they are exposed to nontraditional divisions of household labor.

Three explanations for the continued traditional division of labor emerge from the literature. The first explanation is the *time-availability approach*. Consistent with the structural-functionalist perspective, this position claims that role performance is a function of who has the time to accomplish certain tasks. Because women are more likely to be at home, they are more likely to perform domestic chores.

A second explanation is the *relative resources approach*. This explanation, consistent with a conflict perspective, suggests that the spouse with the least power is relegated the most unrewarding tasks. Because men have more education, higher incomes, and more prestigious occupations, they are less responsible for domestic

Sex Selection of Children

Kristen and John Magill have three daughters whom they adore—11-year-old twins and a 5-year-old. When they began to plan for their next—and last—child, the Magills decided they wanted a boy. In an interview with the *Washington Post,* Kristen Magill explained that she and her husband chose to use sex-selection technology at a Los Angeles fertility clinic to ensure that their next child was a boy. Kristen subsequently became pregnant with twin boys. "I'm excited," she said. "We always wanted a boy. We really wanted just one, but we'll be happy with two" (quoted by Stein 2004, p. A1).

The Magills are part of a small but growing number of Americans who are turning to sex selection—the use of reproductive technology to choose the sex of offspring. Some couples have a social or economic reason for preferring a child of one sex rather than the other. Other couples, such as the Magills, may already have children only or predominantly of one sex and want a child of the other sex to "balance" their existing family (called family balancing).

Couples have been using various techniques to choose the sex of their offspring for centuries. Female infanticide—the killing of a female infant through drowning, strangling, poisoning, or abandonment—is an ancient practice that still occurs today in some parts of Asia. In China and India, where male children are highly valued over female children, many couples choose the sex of their offspring through selective abortion, whereby ultrasound or amniocentesis is used to identify female fetuses, who are then aborted, with the hope that a subsequent pregnancy will produce a male. The effect of selective abortion and female infanticide is evidenced by the imbalanced ratio of boys to girls. In China the ratio of boys to girls in 2000 was 116.9 boys to 100 girls; as many as 40 million Chinese girls are "missing" ("Chinese Look to Their Daughters," 2004). When these boys and girls become adults, there will be millions of single Chinese men who are unable to find a marriage partner because of the shortage of women. A similar problem of an imbalanced sex ratio exists in India, despite the fact that sex selection is not permitted there by law ("India's Disappearing Females," 2004). Concern over the sex imbalance has prompted some Chinese provinces to offer benefits in housing and employment to families with female children. In one Indian province the government instituted a program in 2005 that provides $700 for every female child who reaches the age of 18.

In the United States infanticide obviously is illegal as well as highly morally objectionable, and it is not considered as a method of sex selection. When selective abortion is performed in the United States, it is typically done under one of two conditions: (1) in a multifetus pregnancy to reduce the risk of complications to the remaining fetus(es) and to decrease the health risks to the mother, or (2) when a fetus is diagnosed with a serious birth defect or impairment. Selective abortion is rarely used in the United States for the purpose of choosing the sex of one's offspring.

Two techniques currently available in the United States—sperm separation and preimplantation genetic diagnosis (PGD)—are most often used for preventing sex-linked genetic disorders in children, but they are increasingly being used by couples who, for nonmedical reasons, want to choose the sex of their baby.

In PGD, which was originally designed for detecting genetic diseases, doctors remove eggs from the woman and fertilize them with sperm in the lab, creating embryos. After three days, a cell is extracted from each embryo and the cell chromosomes are examined to identify the sex of the embryo. Then the embryos of the desired sex are implanted in the woman. This method of sex selection is 100% effective in determining the sex of a child, but it is also invasive, costly (up to $20,000), and morally objectionable on the grounds that it involves destroying embryos of the "wrong" sex.

Sex is determined by the sperm that fertilizes the egg. If the sperm has a Y chromosome, it will produce a boy; if it has an X chromosome, it will produce a girl. Thus another method of sex selection involves sorting sperm according to whether they carry male or female chromosomes and then using the sperm carrying the desired chromosome to inseminate a woman or to create an embryo in a laboratory. The most

labor. Thus, for example, because women earn less money than men do, on average, they turn down overtime and other work opportunities to take care of children and household responsibilities, which subsequently reduces their earnings potential even further (Williams 2000).

Gender role ideology, the final explanation, is consistent with a symbolic interactionist perspective. It argues that the division of labor is a consequence of traditional socialization and the accompanying attitudes and beliefs. Women

advanced type of sperm separation technique is called the MicroSort Method, which involves adding fluorescent dye, which binds to X and Y chromosomes, to a sperm sample. Because X-chromosome-bearing sperm are bigger than Y-chromosome-bearing sperm, they can be distinguished because they soak up more dye. The sperm are then separated by a machine that channels them into different receptacles. Either sample can be used to fertilize a woman's eggs, depending on the parents' preference for a boy or a girl. This technique, available only in the United States and in Great Britain, is still experimental. The success rate is 90% for girls and 73% for boys (Hollingsworth 2005). The cost is $2,800 to $4,000 (Stein 2004).

Sex selection using PGD and sperm sorting are highly controversial. As we noted earlier, PGD involves discarding embryos, which is objectionable to many people. A 2003 survey found that, although 78% of Americans approved of embryo testing (PGD) for detecting genetic diseases, only 28% approved of this procedure as a method of sex selection (Ulick 2004). Many U.S. fertility doctors refuse to do the procedure except for medical problems. "My job is to help people make healthy babies, not help people design their babies. Gender is not a disease," said Ralph R. Kazer, a Northwestern University fertility doctor (quoted by Stein 2004).

Opponents of sex selection warn that sex selection could lead to sex imbalances, such as those that already exist in China and In-

dia. Critics also argue that sex selection fuels gender discrimination and uses expensive medical care for frivolous purposes. Michael J. Sandel, a political philosopher at Harvard University, warns that sex selection "runs the risk of turning procreation and parenting into an extension of the consumer society. . . . Sex selection is one step down the road to designer children, in which parents would choose not only the sex of their child but also conceivably the height, hair color, eye color, and ultimately, perhaps, IQ, athletic prowess and musical ability. It's troubling" (quoted by Stein 2004, p. A1).

The American College of Obstetricians and Gynecologists Committee on Ethics (2004) accepts as ethically permissible the practice of sex selection to prevent sex-linked genetic disorders but opposes sex selection for nonmedical reasons because it may reflect and encourage sexism and sex discrimination. Similarly, the President's Commission for the Study of Ethical Problems in Medicine and Biomedical and Behavioral Research, the International Federation of Gynecology and Obstetrics, the Council of Europe's Convention on Human Rights and Biomedicine, and the United Nations oppose the use of sex-selection techniques for nonmedical reasons. The United Nations urges governments of all nations "to take necessary measures to prevent . . . prenatal sex selection" (quoted by American College of Obstetricians and Gynecologists Committee on Ethics 2004, p. 38). The United Kingdom's

Human Fertilisation and Embryology Authority opposes nonmedical use of PGD.

Supporters of sex selection argue that parents should have the right to choose the sex of their baby. They contend that without the option of sex selection, parents might have more children than they desire in order to have at least one child of a particular sex, thus leading to the possibility of unwanted children and financial hardship. Furthermore, proponents argue that a sex imbalance in the United States is unlikely because U.S. couples are just as likely to want a girl as a boy.

However, if we take the position that sex selection in the United States is acceptable, what message do we send to other countries where selective abortion is commonly used to abort female fetuses? Are we setting up a double standard? Marcy Darnovsky, of the Center for Genetics and Society, suggests that "the increased use and acceptance of sex selection in the U.S. would legitimize its practice in other countries, while undermining opposition by human rights and women's rights groups there" (Darnovsky 2004). As the debate over sex selection continues, sex selection continues to be available to couples in the United States, who, like Kristen and John Magill, may decide that any moral or ethical considerations are outweighed by the joy of having a baby . . . girl? Or boy?

and men have been socialized to perform various roles and to expect their partners to perform other complementary roles. Women typically take care of the house, men the yard. These patterns begin with household chores assigned to girls and boys and are learned through the media, schools, books, and toys. A test of the three positions found that although all three had some support, gender role ideology was the weakest of the three in predicting work allocation (Bianchi et al. 2000).

The School Experience and Cultural Sexism

Sexism is also evident in the schools. It can be found in the books students read, the curricula and tests they are exposed to, their extracurricular activities, and the different ways teachers interact with students.

Textbooks. The bulk of research on gender images in textbooks and other instructional materials documents the way males and females are portrayed stereotypically. For example, Purcell and Stewart (1990) analyzed 1,883 storybooks used in schools and found that they tended to depict males as clever, brave, adventurous, and income producing and females as passive and as victims. Females were more likely to be in need of rescue and were also depicted in fewer occupational roles than males. Witt (1996), in a study of third-grade textbooks from six publishers, reports that little girls were more likely to be portrayed as having both traditionally masculine *and* feminine traits, whereas little boys were more likely to be pictured as having masculine characteristics only. These results are consistent with research that suggests that boys are much less free to explore gender differences than females and with Purcell and Stewart's conclusion that boys are often depicted as having "to deny their feelings to show their manhood" (1990, p. 184). Finally, in a recent study of 200 popular children's picture books, researchers found that fathers were largely absent and, when they did appear, they were withdrawn and inept as parents (Anderson & Hamilton 2005). Although some evidence suggests that the frequency of male and female textbook characters is increasingly equal, portrayals of girls and boys largely remain stereotypical (Evans & Davies 2000).

Curricula and Testing. The differing expectations and/or encouragement that females and males receive contribute to their varying abilities, as measured by standardized tests, in such disciplines as math and science. Boys and girls have the same mathematics and science proficiency at age 9; by age 13 boys outperform girls in science but not in math. By age 17 boys outperform girls in both math and science. Are such differences a matter of aptitude? The possibility was recently raised by the president of Harvard University in discussing why women do not excel in science and mathematic fields at the same rate as men. Women's groups and academics around the country criticized him for not recognizing structural and cultural factors, including his own administration, during which tenure offers to women decreased by two-thirds (Winters 2005).

Interestingly, a recent poll reveals that a majority of Americans believe that women and men are equally good at math and science. Even those who believe that men are better at these subjects are split as to whether differences are the result of innate or societal factors (Jones 2005). There is, however, little doubt that social factors play a role. In an experiment at the University of Waterloo, male and female college students, all of whom said they were good in math, were shown either gender-stereotyped or gender-neutral advertisements. When, subsequently, female students who had seen the female-stereotyped advertisements took a math test, they performed not only lower than women who had seen the gender-neutral advertisements but lower than their male counterparts (Begley 2000). Research also indicates that standardized tests themselves are biased. The format of the tests—almost exclusively timed and multiple-choice—favors males, according to some advocates (Smolken 2000). Finally, women faculty at the Massachusetts Institute of Technology (MIT) and Harvard report that their female students were discouraged from entering science fields by the hostile environment in which they saw their female faculty working (Lawler 1999).

Figure 10.3

As this picture from the Bureau of Labor Statistics' Kid's Careers website illustrates, males are less likely than females to be encouraged to choose gender-atypical occupations.

Source: Bureau of Labor Statistics (2005).

Extracurricular Activities. Encouragement to participate in sports and extracurricular activities is gender biased, despite Title IX of the 1972 Educational Amendments Act, which prohibits officials from "tracking" students by sex (Orecklin 2003). Although women's and girls' participation is now at an all-time high with, for example, more than 155,000 women participating in NCAA sporting events, differences remain in the sports that males and females play (U.S. Bureau of the Census 2005). Men are more likely to participate in competitive sports that emphasize traditional male characteristics such as winning, aggression, physical strength, and dominance. Women are more likely to participate in sports that emphasize individual achievement (e.g., figure skating) or cooperation (e.g., synchronized swimming). This may in part be a result of peer pressure. Alley and Hicks (2005) found that adolescents, when asked to read descriptions of male and female peers, rated these individuals as less masculine or less feminine when they participated in cross-gender sports (e.g., ballet for boys and karate among girls). Boys and girls differ in other extracurricular activities as well. For example, girls are more likely to participate in cultural activities than are boys (Dumais 2002).

Teacher-Student Interactions. Sexism is also reflected in the way that teachers treat their students. Millions of young girls are subjected to sexual harassment by male teachers who then fail them when they refuse the teachers' sexual advances (Quist-Areton 2003). After interviewing 800 adolescents, parents, and teachers in three school districts in Kenya, Mensch and Lloyd (1997) report that teachers were more likely to describe girls as lazy and unintelligent. "And when the girls do badly," the researchers remark, "it undoubtedly reinforces teachers' prejudices, becoming a vicious cycle." Similarly, in the United States Sadker and Sadker (1990) observed that elementary and secondary school teachers pay more attention to boys than to girls. Teachers talk to boys more, ask them more questions, listen to them more, counsel them more, give them more extended directions, and criticize and reward them more frequently. However, a book by philosopher Christina Sommers (2000), titled *The War Against Boys,* argues that it is "boys, not girls, on the weak side of the educational gender gap" (p. 14). Noting that boys are at a higher risk for learning disabilities and that they lag in reading and writing scores, Sommers argues that the belief that females are educationally shortchanged is untrue (see Chapter 8). Furthermore, as one researcher notes, the "problem with boys" is not just in the United States but exists throughout the developed world (Poe 2004). Recent patterns also indicate that boys are less likely to go directly to college than girls are (U.S. Bureau of the Census 2005).

Media, Language, and Cultural Sexism

Another concern voiced by social scientists is the extent to which the media portray females and males in a limited and stereotypical fashion and the impact of such portrayals. For example, in the acclaimed documentary *Killing Us Softly 3: Advertising's Image of Women,* Jean Kilbourne (2000) demonstrates how advertisements in the United States routinely turn women into objects, sexualize young girls, portray women as solely responsible for child care (and men as incompetent caregivers), and promote violence against women. Perhaps most surprisingly, the patterns Kilbourne detected have not changed since her first documentary in 1979. Similarly, a study of six months of Sunday morning television talk shows in the United States found that women were almost nonexistent as panelists (Common Dreams 2005). Finally, *Essence* magazine launched a campaign in 2005 against sexism in rap music. This campaign was in response to such inflammatory images as rapper Nelly swiping a credit card down a woman's buttocks and the rapper Ludacris, shown on a compact disk cover, about to bite a woman's leg.

Men are also victimized by media images. A recent study of 1,000 adults found that two-thirds of the respondents thought that women in television advertisements were pictured as "intelligent, assertive and caring," whereas men were portrayed as "pathetic and silly" (Abernathy 2003). In a study of beer and liquor ads, Messner and Montez de Oca (2005) conclude that, although beer ads of the 1950s and 1960s focused on men in their work roles and only depicted women as a reflection of men's home lives, present-day ads portray young men as "bumblers" and "losers" and women as "hotties" and "bitches." In addition, a study of 1,200 children between the ages of 10 and 17 (Dittrich 2002) found that, when asked to name their most admired TV characters, boys' and girls' top five slots were filled by men. The children associated "worrying about appearance and weight, crying, whining and weakness" with female TV characters; playing sports, wanting to be kissed or to have sex, and being a leader were associated with being a male character.

Like media images, both the words we use and the way we use them can reflect gender inequality. The term *nurse* carries the meaning of "a woman who . . . " and the term *engineer* suggests "a man who . . . " Terms such as *broad, old maid,* and *spinster* have no male counterpart. Sexually active teenage females are described by terms carrying negative connotations, whereas terms for equally sexually active male teenagers are considered complimentary among today's youth. Language is so gender stereotyped that the placement of male or female before titles is sometimes necessary, as in the case of "female police officer" or "male prostitute." Furthermore, as symbolic interactionists note, the embedded meanings of words carry expectations of behavior.

Virginia Sapiro (1994) has shown how male-female differences in communication style reflect the structure of power and authority relations between men and women. For example, women are more likely to use disclaimers ("I could be wrong but . . . ") and self-qualifying tags ("That was a good movie, wasn't it?"), reflecting less certainty about their opinions. Communication differences between women and men also reflect different socialization experiences. Women are more often passive and polite in conversation; men are less polite, interrupt more often, and talk more (Renzetti & Curran 2003).

"As I talked to boys across America, I'm struck by how trapped they feel. Our culture puts boys in a gender straightjacket."

William Pollack
Psychologist

Religion and Cultural Sexism

Most Americans claim membership in a church, synagogue, or mosque. Research indicates that women attend religious services more often, rate religion as more important to their lives, and are more likely to believe in an afterlife than are men (Davis et al. 2002). In general, religious teachings have tended to promote traditional conceptions of gender. In 1998 the Southern Baptist Convention *officially* took the stance that a wife should "submit graciously" to her husband, assume her "God-given responsibility to respect her husband, and [to] serve as his helper." Former president Jimmy Carter, a lifelong, active Southern Baptist, left the church in protest over this statement (White 2000). Although Gallagher (2004) found that most evangelical Christians believe that marriage should be considered an equal partnership, she also found that these same Evangelicals continue to believe that the male is the head of the household.

Orthodox Jewish women are not counted as part of the *minyan,* or quorum, required at prayer services, are not allowed to read from the Torah, and are required to sit separately from men at religious services. In addition, Catholic Church doctrine forbids the use of artificial forms of contraception, and Muslim women are required to be veiled in public at all times (Renzetti & Curran 2003). Women cannot serve as ordained religious leaders in the Catholic Church, in Orthodox Jewish synagogues, or in Islamic temples. Even among mainline religious denominations that do allow for the ordination of women, these female clergy often do not hold the same status as their male counterparts and are often limited in their duties (Renzetti & Curran 2003).

However, religious teachings are not all traditional in their beliefs about women and men. Quaker women have been referred to as the "mothers of feminism" because of their active role in the early feminist movement. Reform Judaism has ordained women as rabbis for more than 25 years and gays and lesbians for more than 15 years. In addition, the Women-Church movement—a coalition of feminist faith-sharing groups composed primarily of Catholic women—offers feminist interpretations of Christian teachings. Within many other religious denominations, individual congregations choose to interpret their religious teachings from an inclusive perspective, by replacing masculine pronouns in hymns, the Bible, and other religious readings (Anderson 1997; Renzetti & Curran 2003).

Trends indicate that the number of female clergy in the United States has increased in recent years, with some denominations ordaining women as bishops. Pictured is Reverend Margaret G. Payne, Bishop of the New England Synod of the Evangelical Lutheran Church in America. In contrast, in 2005 the Vatican warned seven Roman Catholic women who were "ordained" as priests to apply for a pardon from the Church under threat of excommunication.

Social Problems and Traditional Gender Role Socialization

Cultural sexism, transmitted through the family, school, media, and language, perpetuates traditional gender role socialization. Gender roles, however slowly, are changing. As one commentator observed (Fitzpatrick 2000), "The hard lines that once helped to define masculine [and feminine] identity are blurring. Women serve in the military, play pro basketball, run corporations and govern. Men diet, undergo cosmetic surgery, bare their souls in support groups and cook" (p. 1).

Despite this **gender tourism** (Fitzpatrick 2000), most research indicates that traditional gender roles remain dominant, particularly for males who, in general, have less freedom to explore the gender continuum. Social problems that result from traditional gender socialization include the feminization of poverty, social-psychological and health costs, and conflict in relationships.

The Feminization of Poverty

Globally, the percentage of female households is increasing dramatically, with one-third of households in developing nations headed by a woman (World Global Issues 2003). Often at the poverty level, many of these households are headed by young women with dependent children and older women who have outlived their spouses. Worldwide, women "work twice as many hours as men, . . . receive only one-tenth of the world's income, and own less than a hundredth of the world's property" (Palmberg 2005, p. 1).

A report card of U.S. efforts to reduce poverty among women was released by U.S. Women Connect, a nonprofit activist group. Although the United States received a grade of B for placing women in decision-making positions, it received an F for efforts to reduce female poverty. Citing federal statistics, the group reports that although the overall poverty rate in the United States has decreased, female poverty has increased over the last five years (Winfield 2000). In 2003 the poverty rate of U.S. females was 13.9%, compared to 11.5% for males (U.S. Bureau of the Census 2005). As noted earlier, both individual and institutional discrimination contribute to the economic plight of women.

Traditional gender role socialization also contributes to poverty among women. Women are often socialized to put family ahead of their education and careers. Women are expected to take primary responsibility for child care, which contributes to the alarming rate of single-parent poor families in the United States. Hispanic and black female-headed households are the poorest of all families headed by a single woman (U.S. Bureau of the Census 2005). In addition, a study of the relationship between marital status, gender, and poverty in the United States, Australia, Canada, and France indicates that in all four countries never-married women are more likely to live in poverty than ever-married women (Nichols-Casebolt & Krysik 1997).

Social-Psychological and Other Health Costs

Many of the costs of traditional gender socialization are social-psychological in nature. Reid and Comas-Diaz (1990) note that the cultural subordination of women results in women having low self-esteem and being dissatisfied with their roles as spouses, homemakers/workers, mothers, and friends. In a study of self-esteem among more than 1,160 students in grades 6 through 10, girls were significantly more likely to have "steadily decreasing self-esteem," whereas boys were more likely to fall into the "moderate and rising" self-esteem group (Zimmerman et al. 1997).

The disparity between boys and girls, men and women, seems to be growing. A study comparing college women in 1966 to those in 1996 found that women's attitudes about their bodies and their overall self-esteem were markedly more negative in 1996 than in 1966, whereas men's levels were the same at both points in time (Sondhaus et al. 2001). It is not surprising, then, that 15% of males, compared to 24% of females, have been found to be clinically depressed at some point in their life (Mazure et al. 2002). Females are also much more likely to be anorexic or bulimic—7 million girls suffer from eating disorders (Renzetti & Curran 2003).

However, not all researchers have found that women have a more negative self-concept than men. Summarizing their research on the self-concepts of women and men in the United States, Williams and Best (1990) found "no evidence of an appreciable difference" (p. 153). They also found no consistency in the self-concepts of women and men in 14 countries: "In some of the countries the men's perceived self was noticeably more favorable than the women's, whereas in others the reverse was found" (p. 152). Furthermore, Burger and Solano (1994) report that women are becoming more assertive and desirous of controlling their own lives rather than merely responding to the wishes of others or the limitations of the social structure.

Men also suffer from traditional gender socialization (Gupta 2003). Men experience enormous cultural pressure to be successful in their work and to earn a high income. Sanchez and Crocker (2005) found that among college-age women *and* men, the more participants were invested in traditional ideals of gender, the lower their self-concept and psychological well-being. Furthermore, among working-class men and boys, "being tough" compensates for the lack of financial success (Williams 2000). In addition, Hochschild (2001) identified a group of self-proclaimed "overtime hounds" who worked long hours for both the money and to prove they were "man enough" to do it. Traditional male socialization also discourages males from expressing emotion—part of what Pollack (2000) calls the boy code. Not surprisingly, males are more likely than females to value materialism and competition over compassion and self-actualization (Beutel & Marini 1995; Cohen 2001; McCammon et al. 1998). This chapter's *Human Side* feature highlights the problems and pressures of being male in American society.

On average, men in the United States die about six years earlier than women, although gender differences in life expectancy have been shrinking (U.S. Bureau of the Census 2005). Traditional male gender socialization is linked to higher rates of cirrhosis of the liver, most cancers, homicide, drug- and alcohol-induced deaths, suicide, and firearm and motor vehicle accidents (U.S. Bureau of the Census 2005). At every stage of life "American males have poorer health and a higher risk of mortality than females" (Gupta 2003, p. 84). Men engage in risky behaviors more often than women do—from smoking, drinking, and abusing drugs to not wearing a seat belt and working in dangerous environments (Gupta 2003). Being married improves men's health more so than it does women's (Williams & Umberson 2004), in large part because wives encourage their husbands to take better care of themselves.

Although men have higher rates of HIV/AIDS worldwide, the disease disproportionately affects women in many areas of the world (see Chapter 2). For example, in sub-Saharan Africa 58% of those infected are women. Women's inequality contributes to the spread of the disease (Heyzer 2003). First, in many of these societies "women lack the power in relationships to refuse sex or negotiate protected sex" (Heyzer 2003, p. 1). Second, women are often the victims of rape and sexual assault, with little social or legal recourse. Finally, some women turn to prostitution as a means of supporting themselves and their children. With little education and no training or work history, they have few options. Ironically, it is women who care for those who are ill, often dropping out of school or quitting jobs to do so, furthering their subordinate role (McGregor 2003).

Are gender differences in morbidity and mortality a consequence of socialization differentials or physiological differences? Although both nature and nurture are likely to be involved, social rather than biological factors may be dominant. As

Real Boys' Voices

Psychologist and author William Pollack traveled from coast to coast talking to boys about the "boy code" (Pollack 1998).

Guys aren't supposed to be weak or vulnerable. Guys aren't supposed to be sweet. A friend of mine died in the hospital. . . . I knew that, as a guy, I was supposed to be strong and I wasn't supposed to show any emotion. . . . I was supposed to be tough . . . when I went home, I just sat by myself and let myself cry. (p.17)
Brad, 14, from a suburb in the Northwest

I think most of the macho stuff guys do is stupid . . . like the kids who do wrestling moves in the hall. At the same time, there are things that I wouldn't do because I'm a guy. I've never gone to a guy friend, for example, and said, "I'm feeling hurt right now and let's talk about it." (p. 31)
Sam, 16, from a city in New England

All the men in my family . . . have been the epitome of negativity. Some have become wrapped up in infidelity, some abuse, some alcoholism. I don't want to become a man, because I don't want to become this. (p. 53)
Gordon, 18, from a small town in the South

Your virginity is what determines whether you're a man or a boy in the eyes of every teenage male. Teenage men see sex as a race: the first one to the finish line wins. (p. 69)
Jeff, 16, from a small town in New England

I think most guys are kind of isolated because it's thought of as weird if you have any really close guy friends. To get around it, guys will go fishing or hunting or bowling or something else "masculine," and then talk about personal or serious things while they're doing that activity. (p. 116)
Brett, 17, from a city in the South

From the girls I've spoken to about relationships, one of their biggest complaints is that they're doing all the giving and the guy is doing all the taking. Girls also tend to be better able to understand social situations. Girls can look at someone and tell what they're feeling. They have more social intuitiveness, more than we clueless guys do. I think that makes them more aware of what's happening in a relationship than we are. (p. 253)
Jesse, 17, from a suburb in New England

We would live in a better society if guys could share their feelings more easily. But guys still hear mixed messages from our society. On the one hand they hear that it's OK now to talk about their feelings, but on the other hand they still hear that they have to be tough and that only girls get emotional. My friend who talked to me and cried about his girlfriend was on the football team. His teammates would laugh at him if he tried to talk to them about that sort of stuff. (p. 272)
Graham, 17, from a suburb in the West

Ever since I've played Little League the word "win" has been forced into my mind. When I was eight years old, the coach would tell us at the beginning of the season that we were just out there for fun, but I knew that it wasn't true. Every day that there was a game, my day would be ruined. (pp. 280–281)
Jake, 16, from a suburb in southern New England

If I get in shape, if I develop a more attractive body, I'd be more popular. It's like the way life is around here, what society shows you. It's a problem to be naturally skinny like me. You're not as athletic or muscular or attractive; you're not as good as the other kids are. (p. 302)
Dylan, 17, from a suburb of Chicago

I think it's hard growing up in the year 2000. It's definitely hard for a guy. Going through high school is tough. I have pressures in sports, school, life all rolled into one. My parents pressure me to do well in school, do well in sports, and I pressure myself to do well in life. . . . I worry about life a lot. I feel that everything is going to work out for everybody except me, that I'll be left in the dust. (p. 341)
Kirk, 18, from a suburb in the Northwest

Source: From *Real Boys: Rescuing Our Sons from the Myths of Boyhood*, by William Pollack. Copyright © 1998 by William Pollack. Used by permission of Random House Inc.

part of the "masculine mystique," men tend to engage in self-destructive behaviors—heavy drinking and smoking, poor diets, lack of exercise, stress-related activities, higher drug use, and a refusal to ask for help. Men are also more likely to work in hazardous occupations than women are. Women's higher rates of depression are also likely to be rooted in traditional gender roles. The heavy burden of child care and household responsibilities, the gender pay gap and occupation gap, and fewer socially acceptable reactions to stress (e.g., it is more acceptable for males than females to drink alcohol) contribute to gender differences in depression.

Are Gender Roles a Matter of Biology or Environment?

The argument is an old one—nature versus nurture. Physically, men and women are different from birth. Men, in general, are stronger, taller, and heavier and have more facial hair. Women develop breasts, have higher-pitched voices, menstruate, and bear children. But are these physical characteristics related to behavioral differences? Are women *innately* nurturers? Are men *innately* aggressive? Noting, for example, that most societies are (were) patriarchal, most would answer yes. Others, however, would be quick to note the role of socialization in traditional gender role assignment. Cross-cultural evidence is mixed. Some societies are characterized by traditional gender roles; in other societies androgyny dominates, and in still other societies traditional gender roles are reversed.

It Is Your Turn Now to Take a Stand!

What do you think? Are gender roles a matter of nature or nurture?

Conflict in Relationships

Worldwide, gender inequality influences relationships. Research indicates that the distribution of resources affects the decision-making process. For example, a study by the International Food Policy Research Institute found that "assets brought to marriage have an impact on bargaining power within the marriage": The more assets, the more power (United Nations Population Fund 2003b). In five of the developing countries studied—Bangladesh, Ethiopia, Ghana, the Philippines, and South Africa—men brought more land and other assets to the marriage and thus had more power in the decision-making process. It should be noted, however, that more than half of women's work is unpaid and thus is not viewed as household income. When the value of household work is included in income totals, women contribute between 40% and 60% of household income (United Nations Population Fund 2003b).

Gender inequality also has an effect on relationships in the United States. For example, negotiating work and home life can be a source of relationship problems. Whereas men in traditional versus dual-income relationships are more likely to report being satisfied with household task arrangements, women in dual-income families are the most likely to be dissatisfied with household task arrangements (Baker et al. 1996). Furthermore, performing an inequitable proportion of the housework and child care results in higher levels of anger among married women than among men (Ross and Van Willigen 1996).

We must also consider, of course, the practical difficulties of raising a family, having a career, and maintaining a happy relationship with a significant other. In one survey more than 80% of both men and women responded that changing gender roles makes it more difficult to have a successful marriage (Morin & Rosenfeld 2000). Successfully balancing work, marriage, and children may require a number of strategies, including (1) a mutually satisfying distribution of household labor, (2) rejection of such stereotypical roles as "supermom" and "breadwinner dad," (3) seeking outside help from others (e.g., child-care providers, domestic workers), and (4) a strong commitment to the family unit. Finally, violence in relationships is gender specific (see Chapters 4 and 5). Although men are more likely to be victims

REUTERS/NASA/HO/Landov

Ret. Col. Eileen Collins became the first woman pilot of a space shuttle in 1995 and has since established herself as the first woman to ever command a space shuttle mission—once in 1999 and then again in July 2005. As a former math teacher who encourages students, especially girls, to study science and mathematics, Col. Collins has successfully balanced her career as an astronaut, wife, and mother.

of violent crime, women are more likely to be victims of rape and domestic violence. Violence against women reflects male socialization that emphasizes aggression and dominance over women. Male violence is a consequence of gender socialization and a definition of masculinity that holds that "as long as nobody is seriously hurt, no lethal weapons are employed, and especially within the framework of sports and games—football, soccer, boxing, wrestling—aggression and violence are widely accepted and even encouraged in boys" (Pollack 2000, p. 40).

Strategies for Action: Toward Gender Equality

In recent decades there has been an increasing awareness of the need to increase gender equality throughout the world. Strategies to achieve this end have focused on empowering women in social, educational, economic, and political spheres and improving women's access to education, nutrition and health care, and basic human rights. But as we will see in the following section on grassroots movements, there is also a men's movement that is concerned with gender inequities and issues facing men.

Grassroots Movements

Efforts to achieve gender equality in the United States have been largely fueled by the feminist movement. Despite a conservative backlash, feminists, and to a lesser extent men's activists groups, have made some gains in reducing structural and cultural sexism in the workplace and in the political arena.

Feminism and the Women's Movement. Feminism is the belief that women and men should have equal rights and re-

"Freedom, especially freedom for women, is more than the absence of oppression. It's the right to speak and vote and worship freely. Human rights require the rights of women."

Laura Bush
First lady of the United States

sponsibilities. The U.S. feminist movement began in Seneca Falls, New York, in 1848 when a group of women wrote and adopted a women's rights manifesto modeled after the Declaration of Independence. Although many of the early feminists were primarily concerned with suffrage, feminism has its "political origins . . . in the abolitionist movement of the 1830s," when women learned to question the assumption of "natural superiority" (Anderson 1997, p. 305). Early feminists were also involved in the temperance movement, which advocated restricting the sale and consumption of alcohol, although their greatest success was the passing of the Nineteenth Amendment in 1920, which recognized women's right to vote.

The rebirth of feminism almost 50 years later was facilitated by a number of interacting forces: an increase in the number of women in the labor force, the publication of Betty Friedan's book *The Feminine Mystique,* an escalating divorce rate, the socially and politically liberal climate of the 1960s, student activism, and the establishment of the Commission on the Status of Women by John F. Kennedy. The National Organization for Women (NOW) was established in 1966 and remains the largest feminist organization in the United States, with more than 500,000 members in 550 chapters across the country. One of NOW's hardest fought battles is the struggle to win ratification of the equal rights amendment (ERA), which states that "equality of rights under the law shall not be denied or abridged by the United States, or by any state, on account of sex." The proposed amendment to the Constitution passed both the House of Representatives and the Senate in 1972 but failed to be ratified by the required 38 states by the 1979 deadline, later extended to 1982. Several lawmakers, however, have introduced legislation that would remove time limits on the ratification process. Should this legislation pass, only 3 more states would be needed to ratify the amendment, which, to date, has been ratified in 35 states.

Supporters of the ERA argue that its opponents used scare tactics—saying that the ERA would lead to unisex bathrooms, mothers losing custody of their children, and mandatory military service for women—to create a conservative backlash. Susan Faludi, in *Backlash: The Undeclared War Against American Women* (1991), contends that contemporary arguments against feminism are the same as those levied against the movement a hundred years ago and that the negative consequences predicted by opponents of feminism (e.g., women unfulfilled and children suffering) have no empirical support.

Today, a new wave of feminism is being led by young women and men who grew up with the benefits won by their mothers but who are shocked by the stoning to death of a woman accused of adultery in Afghanistan, the continuing practice of female genital cutting in at least 25 countries, the fact that millions of women throughout the world lack access to modern contraception, and the fact that, even in one of the "freest" countries in the world, U.S. women face increasing restrictions on abortion, can be denied prescription contraception by a pharmacist who morally objects, earn less than men earn for doing the same job, and experience alarming rates of date rape and intimate partner abuse. Young feminists also grapple with many of the issues their mothers faced, in particular, how to balance work and family life (Orenstein 2001). These young feminists are more inclusive than their predecessors, welcoming all who champion the cause of global equality. Not surprisingly, the new feminists are likely to attract a more diverse group of supporters than their predecessors because future feminist efforts focus on gender equality rather than "gender sameness" (Parker 2000). Some observers, however, note that the new diversity of the women's movement may contribute to "a tension within feminism between the felt urgency to present concerns and grievances as a single unified group of women, and the need to give voice to the variations in concerns and grievances that exist among feminists on the basis of race and ethnicity, social class, sexual orientation, age, physical ability/disability, and a host of other factors" (Renzetti & Curran 2003, p. 25).

The Men's Movement. As a consequence of the women's rights movement, men began to reevaluate their own gender status. As with any grassroots movement, the men's movement has a variety of factions. One of the early branches of the men's movement is known as the mythopoetic men's movement, which began after the publication of Robert Bly's (1990) *Iron John*—a fairy tale about men's wounded masculinity that was on the *New York Times* bestseller list for more than 60 weeks (Zakrzewski 2005). Participants in the mythopoetic men's movement meet in men-only workshops and retreats to explore their internal masculine nature, male identity, and emotional experiences through the use of stories, drumming, dance, music, and discussion. These groups are concerned with improving father-son and male friendship relationships, with overcoming homophobia that interferes with such relationships, and with exploring male forms of intimacy and emotion. A participant observation study of a mythopoetic men's group found that participants, in general, are white middle-class men who feel that they have little emotional support, question relationships with their fathers and sons, and are overburdened by responsibilities, unsatisfactory careers, and the demands of an overly competitive society (Schwalbe 1996).

The men's movement also includes men's organizations that advocate gender equality and work to make men more accountable for sexism, violence, and homophobia. For example, the National Organization of Men Against Sexism (NOMAS) was founded in 1975 to promote a perspective for enhancing men's lives that is profeminist, gay-affirmative, and antiracist and that is committed to justice on a broad range of social issues including class, age, religion, and physical abilities. Other profeminist men's groups besides NOMAS include the national network of Men's Resource Centers.

Other men's groups oppose feminism and view the feminist agenda as an organized form of male bashing. The agenda of antifeminist men's groups is to maintain and promote traditional gender ideology and roles. For example, the Promise Keepers, part of a Christian men's movement, and Louis Farrakhan's Nation of Islam, have often been criticized as patriarchal and antifeminist (Renzetti & Curran 2003).

Some men's groups focus on issues concerning children and fathers' rights. Groups such as the American Coalition for Fathers and Children, Dads Against Discrimination, the National Coalition of Free Men (NCFM), and Fathers for Justice are attempting to change the social and legal bias against men in divorce and child custody decisions, which tend to favor women. Fathers for Justice, a fathers' rights group in Great Britain and Canada, has received international media attention by engaging in publicity stunts and acts of civil disobedience. In one such incident member Jason Hatch dressed as Batman, climbed over the fence around Buckingham Palace, and perched on a palace balcony.

Other concerns on the agenda of some men's rights groups include the domestic violence committed against men by women, false allegations of child sexual abuse, wrongful paternity suits, and the oppressive nature of restrictive masculine gender norms. Just as women have fought against being oppressed by expectations to conform to traditional gender stereotypes, men are beginning to want the same freedom from traditional gender expectations. For example, men who enter nontraditional work roles, such as nurse and primary school teacher, are often stigmatized for participating in "feminine" work. A study of men in nontraditional work roles found that these men commonly experience embarrassment, discomfort, shame, and disapproval from friends and peers (Simpson 2005). Just as the women's movement has fought for women's right to work in male-dominated professions, some men's rights activists want men to be able to work as a nurse, stay home with the kids, order quiche in a restaurant, and drink a frozen piña colada in a bar without being ridiculed for violating male gender norms.

Public Policy

A number of statutes have been passed to help reduce gender inequality. They include the 1963 Equal Pay Act, Title VII of the Civil Rights Act of 1964, Title IX of the Educational Amendments Act of 1972, the Pregnancy Discrimination Act of 1978, the Family and Medical Leave Act of 1993, the 1994 Violence Against Women Act, and the Victims of Trafficking and Violence Protection Act of 2000. Recently, public policy has focused on two issues—sexual harassment and affirmative action.

Sexual Harassment. Sexual harassment is a form of sex discrimination that violates Title VII of the 1964 Civil Rights Act. The U.S. Equal Employment Opportunity Commission (EEOC) (2005) defines **sexual harassment** as unwelcome sexual advances, requests for sexual favors, and other verbal or physical conduct of a sexual nature when this conduct affects an individual's employment, interferes with an individual's work performance, or creates an intimidating, hostile, or offensive work environment. Sexual harassment can be of two types: (1) *quid pro quo,* in which an employer requires sexual favors in exchange for a promotion, salary increase, or any other employee benefit, and (2) the existence of a hostile environment that unreasonably interferes with job performance, as in the case of sexually explicit comments or insults being made to an employee. Common examples of sexual harassment include unwanted touching, the invasion of personal space, making sexual comments about a person's body or attire, and sexual joke telling (Uggen & Blackstone 2004). A newer form of sexual harassment is Internet harassment, in which sexually explicit e-mails are sent to coworkers or images of an individual are posted on a website (Barak 2005).

The victim as well as the harasser may be a woman or a man, although adult women are the most frequent targets of sexual harassment (Uggen & Blackstone 2004). Not surprisingly, women who work in male-dominated occupations and blue-collar jobs are more likely to experience sexual harassment (Jackson & Newman 2004).

In 1998 the U.S. Supreme Court extended protection to victims of same-sex harassment. The victim of sexual harassment does not have to be the person harassed but could be anyone affected by the offensive conduct. In 2005 the California Supreme Court ruled (in *Miller* v. *Department of Corrections*) that a supervisor who has a consensual affair with a subordinate can be sued for sexual harassment by an employee who is not involved in the affair if the employee can demonstrate that widespread sexual favoritism was severe or pervasive enough to alter his or her working conditions and create a hostile working environment (Ison 2005).

Sexual harassment occurs in a variety of settings, including the workplace, public schools, military academies, and college campuses. For example, an independent commission found that university officials knew of accusations of sexual harassment by members of the University of Colorado football program since 1997 and did little about them until victims went public in 2004 (National Women's Law Center 2004).

The Equal Employment Opportunity Commission (2005) suggests that prevention is the best tool to eliminate sexual harassment. Schools and workplaces are encouraged to (1) adopt the policy that sexual harassment will not be tolerated, (2) provide sexual harassment training to administrators and employees, and (3) establish complaint or grievance procedures and take immediate action when an employee or student complains of sexual harassment. Any person who experiences sexual harassment on the job can file a charge of discrimination at any EEOC office. In fiscal year 2004 the EEOC received 13,136 charges of sexual harassment; 15% were filed by males (Equal Employment Opportunity Commission 2005).

Affirmative Action. As discussed in Chapter 9, **affirmative action** refers to a broad range of policies and practices in the workplace and educational institutions to promote equal opportunity as well as diversity in the workplace and on campuses. Affirmative action policies, developed in the 1960s from federal legislation, require that any employer (universities as well as businesses) that receives contracts from the federal government must make "good faith efforts" to increase the number of female and other minority applicants. Such efforts can be made through expanding recruitment and training programs and by making hiring decisions on a nondiscriminatory basis.

People commonly associate affirmative action with racial and ethnic minorities; some people are not aware that affirmative action benefits women. Indeed, the largest category of affirmative action beneficiaries is women (Beeman et al. 2000).

As discussed in Chapter 9, numerous legal battles have challenged affirmative action. In 1996 a California ballot initiative abolished racial and sexual preferences in government programs, which included state colleges and universities (Chavez 2000). Washington state voters passed a similar initiative in 1998, and in 1999 the governor of Florida signed an executive order ending that state's affirmative action program. However, in 2003 the U.S. Supreme Court held that universities have a "compelling interest" in a diverse student population and therefore may take minority status into consideration when making admissions decisions ("Major Rulings of the 2002–2003 Term," 2003) (see Chapter 9). Furthermore, in 2004 attempts to outlaw affirmative action in Michigan and Colorado were defeated (National Organization for Women 2004).

International Efforts

International efforts to address problems of gender inequality date back to the 1979 Convention on the Elimination of All Forms of Discrimination Against Women (CEDAW), often referred to as the International Women's Bill of Rights, adopted by the United Nations in 1979. More than 170 countries ratified the Women's Bill of Rights, including every country in Europe and South and Central America. The United States is the only industrialized country that has not ratified the document (Human Rights Watch 2005).

Another significant international effort occurred in 1995 when representatives from 189 countries adopted the Beijing Declaration and Platform for Action at the Fourth World Conference on Women sponsored by the United Nations. The platform reflects an international commitment to the goals of equality, development, and peace for women everywhere. The platform identifies strategies to address critical areas of concern related to women and girls, including poverty, education, health, violence, armed conflict, and human rights. Every five years the United Nations has sponsored a conference to monitor progress in addressing these concerns. At its 2005 conference delegates voted to encourage member nations to pay particular attention to problems of AIDS among women and girls, the status of women and girls in the West Bank and the Gaza Strip and Afghanistan, the trafficking of women and girls for prostitution, and the economic advancement of women (United Nations 2005).

In addition to international efforts, individual countries have instituted programs or policies designed to combat sexism and gender inequality. For example, Japan implemented the Basic Law for a Gender-Equal Society, a "blueprint for gender equality in the home and workplace" (Yumiko 2000, p. 41). Wage differences remain high, however, with women earning just 65% of men's salaries (CEDAW 2003). A South African Bill of Rights prohibits discrimination on the basis of, among other things, gender, pregnancy, and marital status (International Women's Right's Project 2000), and China has established the Program for the Development of Chinese Women, which focuses on empowering women in the areas of education, human rights, health, child care, employment, and political power (Women's International Network 2000).

Understanding Gender Inequality

Gender roles and the social inequality they create are ingrained in our social and cultural ideologies and institutions and are therefore difficult to alter. For example, in almost all societies women are primarily responsible for child care and men are

Definitions of appropriate gender roles change over time. Fifty years ago, women playing professional basketball would have been unimaginable. Men's roles are also changing, although more slowly.

primarily responsible for military service and national defense (World Bank 2003). Nevertheless, as we have seen in this chapter, growing attention to gender issues in social life has spurred some change. Women who have traditionally been expected to give domestic life first priority are now finding it acceptable to be more ambitious in seeking a career outside the home. Men's roles are also changing. Men who have traditionally been expected to be aggressive and task oriented are now expected to be more caring and nurturing. Increasingly, both women and men are embracing a concept known as **androgyny**—the blending of both masculine and feminine characteristics. The concept of androgyny implies that both masculine and feminine characteristics and roles are equally valued. However, "achieving gender equality . . . is a grindingly slow process, since it challenges one of the most deeply entrenched of all human attitudes" (Lopez-Claros & Zahidi 2005, p. 1). In a recent survey of college freshmen, one in five (27%) agreed with the statement "The activities of a married woman are best confined to the home and family" (Sax et al. 2004).

Although traditionally it has been women who have fought for gender equality, men also have much to gain. They too are victims of discrimination and gender stereotyping. Eliminating gender stereotypes and redefining gender in terms of equality does not mean simply liberating women but liberating men and our society as well. "What we have been talking about is allowing people to be more fully human and creating a society that will reflect that humanity. Surely that is a goal worth striving for" (Basow 1992, p. 359). Regardless of whether traditional gender roles emerged out of biological necessity, as the structural functionalists argue, or out of economic oppression, as the conflict theorists hold, or both, it is clear that, today, gender inequality carries a high price: poverty, loss of human capital, feelings of worthlessness, violence, physical and mental illness, and death. Surely, the costs are too high to continue to pay.

Chapter Review

ThomsonNOW™

Reviewing is as easy as ① ② ③

1. Before you do your final review, take the ThomsonNOW diagnostic quiz to help you identify the areas on which you should concentrate. You will find information on ThomsonNOW and instructions on how to access all of its great resources on the foldout at the beginning of the text.

2. As you review, take advantage of ThomsonNOW's study videos and interactive Map the Stats exercises to help you master the chapter topics.

3. When you are finished with your review, take ThomsonNOW's posttest to confirm you are ready to move on to the next chapter.

- **Does gender inequality exist worldwide?**

There is no country in the world where men and women are treated equally. Although women suffer in terms of income, education, and occupational prestige, men are more likely to suffer in terms of mental and physical health, mortality, and the quality of their relationships.

- **How do the three major sociological theories view gender inequality?**

Structural functionalists argue that the traditional division of labor was functional for preindustrial society and over time has become defined as both normal and natural. Today, however, modern conceptions of the family have replaced traditional ones to some extent. Conflict theorists hold that male dominance and female subordination evolved in relation to the means of production—from hunting and gathering societies in which females and males were economic equals to industrial societies in which females were subordinate to males. Symbolic interactionists emphasize that through the socialization process both females and males are taught the meanings associated with being feminine and masculine.

- **What is meant by the terms** *structural sexism* **and** *cultural sexism*?

Structural sexism refers to the ways in which the organization of society, and specifically its institutions, subordinate individuals and groups based on their sex classification. Structural sexism has resulted in significant differences between education and income levels, occupational and political involvement, and civil rights of women and men. Structural sexism is supported by a system of cultural sexism that perpetuates beliefs about the differences between women and men. Cultural sexism refers to the ways the culture of society—its norms, values, beliefs, and symbols—perpetuates the subordination of an individual or group because of the sex classification of that individual or group.

- **What are some of the problems caused by traditional gender roles?**

First is the feminization of poverty. Women are socialized to put family ahead of education and careers, a belief that is reflected in their less prestigious occupations and lower incomes. Second are social-psychological and health costs. Women tend to have lower self-esteem and higher rates of depression and eating disorders than men. Men, on average, die about six years earlier than women and have higher rates of cirrhosis of the liver, most cancers, homicide, drug- and alcohol-induced deaths, suicide, and firearm and motor vehicle accidents. Last, gender inequality influences the nature of relationships. For example, although men are more likely to be victims of violent crime, women are more likely to be victims of domestic violence.

- **What strategies can be employed to end gender inequality?**

Grassroots movements, such as feminism and the women's rights movement and the men's rights movement, have made significant inroads in the fight against gender inequality. Their accomplishments, in part, have been the result of successful lobbying for passage of laws concerning sex discrimination, sexual harassment, and affirmative action. Besides these national efforts, international efforts continue as well. One of the most important is the Convention to Eliminate All Forms of Discrimination Against Women (CEDAW), also known as the International Women's Bill of Rights, which was adopted by the United Nations in 1979.

Critical Thinking

1. Some research suggests that "men and women with more androgynous gender orientations—that is to say, those having a balance of masculine and feminine personality characteristics—show signs of greater mental

health and more positive self-images" (Anderson 1997, p. 34). Do you agree or disagree? Why or why not?

2. Recent evidence suggests that a "gender gap" exists in the number of men and women entering college, particularly among African Americans, with women attending at higher rates than men. Although the number of females in the population is slightly higher, the difference does not explain the projected gap in enrollments. Why are black women entering college at a higher rate than black men?

3. Why are women more likely to work in traditionally male occupations than men are to work in traditionally female occupations? Are the barriers that prevent men from doing "women's work" cultural, structural, or both? Explain.

Key Terms

affirmative action
androgyny
comparable worth
cultural sexism
devaluation hypothesis
double or triple
  (multiple) jeopardy

emotion work
expressive roles
feminism
gender
gender tourism
glass ceiling

human capital hypothesis
instrumental roles
occupational sex
  segregation
pink-collar jobs

sex
sexism
sexual harassment
structural sexism
structured choice

Media Resources

The Companion Website for *Understanding Social Problems*, Fifth Edition

http://sociology.wadsworth.com/mooney_knox_schacht5e

Supplement your review of this chapter by going to the companion website to take one of the Tutorial Quizzes, use the flash cards to master key terms, and check out the many other study aids you'll find there. You'll also find special features such as Wadsworth's *Sociology Online Resources and Writing Companion,* GSS data, and Census 2000 information, data, and resources at your fingertips to help you complete that special project or do some research on your own.

> " Homophobia alienates mothers and fathers from sons and daughters, friend from friend, neighbor from neighbor, Americans from one another. So long as it is legitimated by society, religion, and politics, homophobia will spawn hatred, contempt, and violence, and it will remain our last acceptable prejudice. " *Byrne Fone*

Issues in Sexual Orientation

The Global Context: A World View of Laws Pertaining to Homosexuality

Homosexuality and Bisexuality in the United States: A Demographic Overview

The Origins of Sexual Orientation Diversity

Sociological Theories of Sexual Orientation Issues

Heterosexism, Homophobia, and Biphobia

Discrimination Against Sexual Orientation Minorities

Effects of Homophobia, Heterosexism, and Antigay Discrimination on Heterosexuals

Strategies for Action: Reducing Antigay Prejudice and Discrimination

Understanding Issues in Sexual Orientation

Chapter Review

as a junior and senior at Homewood-Flossmoor High School in the suburbs of Chicago, Myka Held played a key role in leading a campaign to promote tolerance of gay and lesbian students. The campaign involved selling gay-friendly t-shirts to students and teachers and having as many people as possible wear the t-shirts to school on a designated day. The t-shirts, made by Duke University, say, "gay? fine by me." "I think it's really important for gay people out there to know that there are straight people who support them," Ms. Held said (quoted in Puccinelli 2005). "I have always supported equal rights for every person and have been disgusted by discrimination and prejudice. As a young Jewish woman, I believe it is my duty to stand up and support minority groups. . . . In my mind, fighting for gay rights is a proxy for fighting for every person's rights" (Held 2005). More than 100 students and 3 brave teachers wore the t-shirts in the first year of the campaign, and in the second year more than 200 students showed their support by buying and wearing the t-shirts. "Since shirt day, I have become more interested in politics and seeing how just one or two people can make a difference," said Held. "What I did was on a small scale, it's true, but the difference . . . I made in opening the minds of some students is important" (Held 2005).

As the opening vignette illustrates, fighting prejudice and discrimination against sexual orientation minorities is an issue not just for lesbians, gays, and bisexuals but for all those who value fairness and respect for human beings in all their diversity. In this chapter we examine prejudice and discrimination toward sexual orientation minorities—nonheterosexual individuals: homosexual (or gay) women (also known as lesbians), homosexual (or gay) men, and bisexual individuals.

It is beyond the scope of this chapter to explore how sexual diversity and its cultural meanings vary throughout the world. Rather, in this chapter we focus on Western conceptions of diversity in sexual orientation. The term **sexual orientation** refers to the classification of individuals as heterosexual, bisexual, or homosexual, based on their emotional and sexual attractions, relationships, self-identity, and lifestyle. **Heterosexuality** refers to the predominance of emotional and sexual attraction to individuals of the other sex. **Homosexuality** refers to the predominance of emotional and sexual attraction to individuals of the same sex, and **bisexuality** is emotional and sexual attraction to members of both sexes. Lesbians, gays, and bisexuals, sometimes referred to collectively as the **lesbigay population,** are considered part of a larger population referred to as the transgendered community. **Transgendered individuals** include "a range

ThomsonNOW™

Reviewing is as easy as
①②③

Use ThomsonNOW to help you make the grade on your next exam. When you are finished reading this chapter, go to the chapter review for instructions on how to make ThomsonNOW work for you.

At the Homewood-Flossmoor high school in the suburbs of Chicago, Myka Held was involved in a campaign to promote gay tolerance by selling t-shirts that say, "gay? fine by me" and having students and teachers wear them on the same day at school. The t-shirts are made by Duke University.

of people whose gender identities do not conform to traditional notions of masculinity and femininity" (Cahill et al. 2002, p. 10). Transgendered individuals include not only homosexuals and bisexuals but also cross-dressers, transvestites, and transsexuals. Cross-dressers are individuals, usually heterosexual men, who occasionally dress in the clothing of the other sex; transvestites are homosexual men who occasionally dress as women; and transsexuals are individuals who have undergone hormone treatment and sex-reassignment surgery to change their gender identity (McCammon et al. 2004). Much of the current literature on the treatment and political and social agendas of the lesbigay population includes other members of the transgendered community; hence the term **LGBT** or **GLBT** is often used to refer collectively to lesbians, gays, bisexuals, and transgendered individuals.

ThomsonNOW

Learn more about **Gender Identity** by going through the Sex: The Biological Dimension Learning Module.

After summarizing the legal status of lesbians and gay men around the world, we discuss the prevalence of homosexuality, heterosexuality, and bisexuality in the United States, review explanations for sexual orientation diversity, and apply sociological theories to better understand societal reactions to sexual diversity. Then, after detailing the ways in which nonheterosexuals are victimized by prejudice and discrimination, we end the chapter with a discussion of strategies to reduce antigay prejudice and discrimination.

The Global Context: A World View of Laws Pertaining to Homosexuality

Homosexual behavior has existed throughout human history and in most, perhaps all, human societies (Kirkpatrick 2000). A global perspective on laws and social attitudes regarding homosexuality reveals that countries vary tremendously in their treatment of same-sex sexual behavior—from intolerance and criminalization to acceptance and legal protection. In 84 countries sexual activity between consenting adults of the same sex is illegal (International Gay and Lesbian Human Rights Commission 2003a). In 55 of these countries laws criminalizing same-sex sexual behavior apply to both female and male homosexuality; in 29 of these countries the laws apply to male homosexuality only. Legal penalties vary for violating laws that prohibit homosexual sexual acts. In nine countries individuals found guilty of engaging in same-sex sexual behavior may receive the death penalty (see Table 11.1).

In general, countries throughout the world are moving toward increased legal protection of sexual orientation minorities. The latest available data from the International Gay and Lesbian Human Rights Commission (1999) finds that 22 countries have national laws that ban various forms of discrimination against gays, lesbians, and bisexuals. In 1996 South Africa became the first country in the world to include in its constitution a clause banning discrimination based on sexual orientation. Fiji, Canada, and Ecuador also have constitutions that ban discrimination based on sexual orientation ("Constitutional Protection," 1999).

In recent years legal recognition of same-sex relationships has become more widespread. In 2001 the Netherlands became the first country in the world to offer full legal marriage to same-sex couples. Same-sex married couples and opposite-sex married couples in the Netherlands are treated identically, with two exceptions. Unlike opposite-sex marriages, same-sex couples married in the Netherlands are unlikely to have their marriages recognized as fully legal abroad. Regarding children, parental rights will not automatically be granted to the nonbiological spouse in gay couples. To become a fully legal parent, the spouse of the biological parent

Table 11.1

Countries in Which Homosexual Acts Are Subject to the Death Penalty

- Afghanistan
- Iran
- Mauritania
- Pakistan
- Saudi Arabia
- Somalia
- Sudan
- United Arab Emirates
- Yemen

Sources: "Jail, Death Sentences in Africa" (2001) and Mackay (2001).

must adopt the child. In 2003 Belgium legalized same-sex marriages. In 2005 Spain, Canada, and South Africa became the third, fourth, and fifth countries, respectively, to legalize same-sex marriage.

Other countries recognize same-sex **registered partnerships** or "civil unions," which are federally recognized relationships that convey most but not all the rights of marriage (some countries also offer registered partnerships or civil unions to opposite-sex couples). Federally recognized registered partnerships or civil unions for same-sex couples are available in Australia, Belgium, Brazil, Canada, Denmark, Finland, France, Germany, Great Britain, Greenland, Hungary, Iceland, Israel, Italy, New Zealand, Norway, Portugal, Spain, and Sweden (International Gay and Lesbian Human Rights Commission 2005b; LAWbriefs 2005). As we discuss later in this chapter, a number of U.S. states give legal recognition to same-sex couples.

Human rights treaties and transnational social movement organizations have increasingly asserted the rights of people to engage in same-sex relations. International organizations such as Amnesty International, which resolved in 1991 to defend those imprisoned for homosexuality, the International Lesbian and Gay Association (founded in 1978), and the International Gay and Lesbian Human Rights Commission (founded in 1990) continue to fight antigay prejudice and discrimination. The United Nations Human Rights Commission has proposed the Resolution on Sexual Orientation and Human Rights. This landmark resolution recognizes the existence of sexual orientation–based discrimination around the world, affirms that such discrimination is a violation of human rights, and calls on all governments to promote and protect the human rights of all people, regardless of their sexual orientation (International Gay and Lesbian Human Rights Commission 2004). Despite the worldwide movement toward increased acceptance and protection of homosexual individuals, the status and rights of lesbians and gays in the United States continues to be one of the most divisive issues in U.S. society.

Gert Kasteel, left, and Dolf Pasker were among the world's first same-sex couples to marry legally under Dutch law after the Netherlands became the first country to allow same-sex marriages.

© AP/Wide World Photos

Homosexuality and Bisexuality in the United States: A Demographic Overview

Before looking at demographic data concerning homosexuality and bisexuality in the United States, it is important to understand the ways in which identifying or classifying individuals as homosexual, gay, lesbian, and bisexual is problematic.

Sexual Orientation: Problems Associated with Identification and Classification

The classification of individuals into sexual orientation categories (e.g., gay, straight, bisexual, lesbian, homosexual, heterosexual) is problematic for a number of reasons. First, because of the social stigma associated with nonheterosexual identities, many individuals conceal or falsely portray their sexual orientation identities to protect themselves against prejudice and discrimination. But more important, distinctions among sexual orientation categories are simply not as clear-cut as many people would believe.

Consider the early research on sexual behavior by Kinsey and his colleagues (1948, 1953), which found that although 37% of men and 13% of women had had at least one same-sex sexual experience since adolescence, few of the individuals reported exclusive homosexual behavior. These data led Kinsey to conclude that most people are not exclusively heterosexual or homosexual. Rather, Kinsey suggested that an individual's sexual orientation may have both heterosexual and homosexual elements. In other words, Kinsey suggested that heterosexuality and homosexuality represent two ends of a sexual orientation continuum and that most individuals are neither entirely homosexual nor entirely heterosexual but fall somewhere along this continuum. In other words, most people are, to some degree, bisexual.

Sexual orientation classification is also complicated by the fact that sexual behavior, attraction, love, desire, and sexual orientation identity do not always match. For example, "research conducted across different cultures and historical periods (including present-day Western culture) has found that many individuals develop passionate infatuations with same-gender partners in the absence of same-gender sexual desires . . . whereas others experience same-gender sexual desires that never manifest themselves in romantic passion or attachment" (Diamond 2003, p. 173).

Consider the findings of a national study of U.S. adults that investigated (1) sexual attraction to individuals of the same sex, (2) sexual behavior with people of the same sex, and (3) homosexual self-identification (Michael et al. 1994). This survey found that 4% of women and 6% of men said that they are sexually attracted to individuals of the same sex, and 4% of women and 5% of men reported that they had had sexual relations with a same-sex partner after age 18. Yet less than 3% of the men and less than 2% of women in this study identified themselves as homosexual or bisexual (Michael et al. 1994). What these data tell us is that, first, "those who acknowledge homosexual desires may be far more numerous than those who actually act on those desires" (Black et al. 2000, p. 140). Second, not all people who are sexually attracted to or have had sexual relations with individuals of the same sex view themselves as homosexual or bisexual. A final difficulty in labeling a person's sexual orientation is that an individual's sexual attractions, behavior, and identity may change across time.

> "Although it is typically presumed that heterosexual individuals only fall in love with other-gender partners and gay-lesbian individuals only fall in love with same-gender partners, this is not always so."
>
> **Lisa M. Diamond**
> **Psychologist**

The Prevalence of Nonheterosexual Adults in the United States

Despite the difficulties inherent in categorizing individuals into sexual orientation categories, recent data reveal the prevalence of individuals who identify as lesbian, gay, or bisexual and of cohabiting same-sex couples living in the United States. As noted earlier, in the national survey by Michael et al. (1994), less than 3% of the men and less than 2% of women in this study identified themselves as homosexual or bisexual. More recently, Smith and Gates (2001) estimated that there are more than 10 million gay and lesbian adults in the United States, which represents between 4% and 5% of the total U.S. adult population. In addition, a 2004 national poll found that about 5% of U.S. high school students identify as lesbian or gay (Curtis 2004). An estimated 1 to 3 million Americans older than age 65 are gay, lesbian, bisexual, or transgender (National Gay and Lesbian Task Force 2005c).

Prevalence of Same-Sex Unmarried Couple Households in the United States

The 2000 census found that about 1 in 9 (or 594,000) unmarried-partner households in the United States involve partners of the same sex (Simmons & O'Connell 2003). Nationally, 51% of same-sex couples had male partners. About one-fifth (22.3%) of gay male couples and one-third (34.3%) of lesbian couples have children present in the home (Simmons & O'Connell 2003).

Census 2000 data also revealed that 99.3% of U.S. counties reported same-sex cohabiting partners, compared to 52% of counties in 1990 (Bradford et al. 2002). Same-sex couples are more likely to live in metropolitan areas than in rural areas. However, the largest proportional increases in the number of same-sex couples self-reporting in 2000 compared to 1990 came in rural, sparsely populated states. States with the highest percentage of same-sex unmarried partners of all coupled

Residents of America's first gay and lesbian retirement community, Palms of Manasota, in Palmetto, Florida.

Palms of Manasota

households include California, Massachusetts, Vermont, and New York (Bradford et al. 2002; Simmons & O'Connell 2003).

Why are data on the numbers of U.S. GLBT adults and couples relevant? Primarily because census numbers on the prevalence of GLBT adults and couples can influence laws and policies that affect gay individuals and their families. In anticipation of the 2000 census, the National Gay and Lesbian Task Force Policy Institute and the Institute for Gay and Lesbian Strategic Studies conducted a public education campaign urging people to "out" themselves on the 2000 census. The slogan was, "The more we are counted, the more we count" (Bradford et al. 2002, p. 3). "The fact that the Census documents the actual presence of same-sex couples in nearly every state legislative and U.S. Congressional district means anti-gay legislators can no longer assert that they have no gay and lesbian constituents" (Bradford et al. 2002, p. 8).

The Origins of Sexual Orientation Diversity

One of the prevailing questions regarding sexual orientation centers on its origin or "cause." Questions about the causes of sexual orientation are typically concerned with the origins of homosexuality and bisexuality. Because heterosexuality is considered normative and "natural," causes of heterosexuality are rarely considered.

Much of the biomedical and psychological research on sexual orientation attempts to identify one or more causes of sexual orientation diversity. The driving question behind this research is, Is sexual orientation inborn or is it learned or acquired from environmental influences? Although a number of factors have been correlated with sexual orientation, including genetic factors, gender role behavior in childhood, and fraternal birth order, there is no single theory that can explain diversity in sexual orientation (McCammon et al. 2004).

Beliefs About What "Causes" Homosexuality

Aside from what "causes" homosexuality, sociologists are interested in what people *believe* about the "causes" of homosexuality. Most gays believe that homosexuality is an inherited, inborn trait. In a national study of homosexual men 90% believe that they were born with their homosexual orientation; only 4% believe that environmental factors are the sole cause (Lever 1994). The percentage of Americans who believe that homosexuality is something a person is born with increased from 13% in 1977 to 40% in 2002 (Newport 2002).

Individuals who believe that homosexuality is biologically based tend to be more accepting of homosexuality. In contrast, "those who believe homosexuals choose their sexual orientation are far less tolerant of gays and lesbians and more likely to conclude homosexuality should be illegal than those who think sexual orientation is not a matter of personal choice" (Rosin & Morin 1999, p. 8).

Can Homosexuals Change Their Sexual Orientation?

Individuals who believe that homosexuals choose their sexual orientation tend to think that homosexuals can and should change their sexual orientation. Various forms of **reparative therapy** or **conversion therapy** are dedicated to changing homosexuals' sexual orientation. Some religious organizations sponsor "exgay ministries," which claim to "cure" homosexuals and transform them into heterosexuals through prayer and other forms of "therapy." Consider the following examples:

- One church counselor told a gay woman that God could heal her and make her "whole again." The counselor laid hands on her to rid her of the demonic "spirit of homosexuality," gave her Bible verses to memorize, and worked on her "femininity" by teaching her how to apply makeup, encouraging her to grow her hair long, and instructing her to replace her old jeans, sweatpants, and gym shorts with skirts and dresses (Human Rights Campaign 2000).
- A 21-year-old gay man struggling with his sexual orientation reported that his strict religious family, upon learning of his orientation, threatened to disown him and have him excommunicated unless he changed. He went into an "exgay" program voluntarily because he could not bear to lose his family, but other young men, he said, were forced into the program after being kidnapped. The program "counselors" strapped him to a chair with electrodes and sensors and showed him pictures of nude men, shocking him when he became aroused; they continued this treatment until he did not respond. He finally fled the program after being sexually abused by a male orderly. The experience was traumatizing and left him with feelings of self-hatred and fear and thoughts of suicide.
- Sixteen-year-old Zach was enrolled by his parents in a Christian camplike program to change him into a heterosexual. The program, called Refuge, discourages homosexual behavior by imposing the following rules on its participants: no secular music, no more than 15 minutes per day behind a closed bathroom door, no contact with any practicing homosexual, no masturbation, and (no joke) no Calvin Klein underwear (Buhl 2005).

Critics of reparative therapy and ex-gay ministries take a different approach: "It is not gay men and lesbians who need to change . . . but negative attitudes and discrimination against gay people that need to be abolished" (Besen 2000, p. 7). The American Psychiatric Association, the American Psychological Association, the American Academy of Pediatrics, the American Counseling Association, the National Association of School Psychologists, the National Association of Social Workers, and the American Medical Association agree that homosexuality is not a mental disorder and needs no cure—that sexual orientation *cannot* be changed and that efforts to change sexual orientation do not work and may, in fact, be harmful (Human Rights Campaign 2000; Potok 2005). According to the American Psychiatric Association, "clinical experience suggests that any person who seeks conversion therapy may be doing so because of social bias that has resulted in internalized homophobia, and that gay men and lesbians who have accepted their sexual orientation are better adjusted than those who have not done so" (quoted by Holthouse 2005, p. 14). Close scrutiny of reports of "successful" reparative therapy reveal that (1) many claims come from organizations with an ideological perspective on sexual orientation rather than from unbiased researchers, (2) the treatments and their outcomes are poorly documented, and (3) the length of time that clients are followed after treatment is too short for definitive claims to be made about treatment success (Human Rights Campaign 2000).

Indeed, at least 13 ministries of Exodus International—the largest ex-gay ministry network—have closed because their directors reverted to homosexuality (Fone 2000). Michael Bussee, who in 1976 helped start Exodus International, said,

"After dealing with hundreds of people, I have not met one who went from gay to straight. Even if you manage to alter someone's sexual behavior, you cannot change their true sexual orientation" (quoted by Holthouse 2005, p. 14). Bussee worked to help "convert" gay people for three years, until he and another male Exodus employee fell in love and left the organization.

Sociological Theories of Sexual Orientation Issues

Sociological theories do not explain the origin or "cause" of sexual orientation diversity; rather, they help to explain societal reactions to homosexuality and bisexuality and ways in which sexual identities are socially constructed.

Structural-Functionalist Perspective

Structural functionalists, consistent with their emphasis on institutions and the functions they fulfill, emphasize the importance of monogamous heterosexual relationships for the reproduction, nurturance, and socialization of children. From a structural-functionalist perspective homosexual relations, as well as heterosexual nonmarital relations, are defined as "deviant" because they do not fulfill the family institution's main function of producing and rearing children. Clearly, however, this argument is less salient in a society in which (1) other institutions, most notably schools, have supplemented the traditional functions of the family; (2) reducing (rather than increasing) population is a societal goal; and (3) same-sex couples can and do raise children.

Some structural functionalists argue that antagonisms between heterosexuals and homosexuals disrupt the natural state, or equilibrium, of society. Durkheim, however, recognized that deviation from society's norms can also be functional. As Durkheim observed, deviation "may be useful as a prelude to reforms which daily become more necessary" (Durkheim 1993, p. 66). Specifically, the gay rights movement has motivated many people to reexamine their treatment of sexual orientation minorities and has produced a sense of cohesion and solidarity among members of the gay population (although bisexuals have often been excluded from gay and lesbian communities and organizations). Gay activism has been instrumental in advocating more research on HIV and AIDS, more and better health services for HIV and AIDS patients, protection of the rights of HIV-infected individuals, and HIV/AIDS public education. Such HIV/AIDS prevention strategies and health services benefit the society as a whole.

Finally, the structural-functionalist perspective is concerned with how changes in one part of society affect other aspects. With this focus on the interconnectedness of society we note that urbanization has contributed to the formation of strong social networks of gays and bisexuals. Cities "acted as magnets, drawing in gay migrants who felt isolated and threatened in smaller towns and rural areas" (Button et al. 1997, p. 15). Given the formation of gay communities in large cities, it is not surprising that the gay rights movement first emerged in large urban centers.

Other research has demonstrated that the worldwide increase in liberalized national policies on same-sex relations and the lesbian and gay rights social movement has been influenced by three cultural changes: the rise of individualism, increasing gender equality, and the emergence of a global society in which nations are influenced by international pressures (Frank & McEneaney 1999).

ThomsonNOW

Learn more about **HIV/AIDS** by going through the AIDS Map Exercise.

Individualism "appears to loosen the tie between sex and procreation, allowing more personal modes of sexual expression" (p. 930).

> Whereas once sex was approved strictly for the purpose of family reproduction, sex increasingly serves to pleasure individualized men and women in society. This shift has involved the casting off of many traditional regulations on sexual behavior, including prohibitions of male-male and female-female sex. (Frank & McEneaney 1999, p. 936)

Gender equality involves the breakdown of sharply differentiated sex roles, thereby supporting the varied expressions of male and female sexuality. Globalization permits the international community to influence individual nations. For example, when Zimbabwe president Robert Mugabe pursued antihomosexual policies in 1995, 70 members of the U.S. Congress signed a letter asking him to halt his antihomosexual campaign. Many international organizations and human rights associations joined the protest. The pressure of international opinion led Zimbabwe's Supreme Court to rule in favor of lesbian and gay groups' right to organize.

Conflict Perspective

Conflict theorists, particularly those who do not emphasize a purely economic perspective, note that the antagonisms between heterosexuals and nonheterosexuals represent a basic division in society between those with power and those without power. When one group has control of society's institutions and resources, as in the case of heterosexuals, they have the authority to dominate other groups. The recent battle over gay rights is just one example of the political struggle between those with power and those without it.

A classic example of the power struggle between gays and straights took place in 1973 when the American Psychiatric Association (APA) met to revise its classification scheme of mental disorders. Homosexual activists had been appealing to the APA for years to remove homosexuality from its list of mental illnesses but with little success. The view of homosexuals as mentally ill contributed to their low social prestige in the eyes of the heterosexual majority. In 1973 the APA's board of directors voted to remove homosexuality from its official list of mental disorders. The board's move encountered a great deal of resistance from conservative APA members and was put to a referendum, which reaffirmed the board's decision (Bayer 1987).

More recently, gays and lesbians are waging a political battle to win civil rights protections in the form of laws prohibiting discrimination on the basis of sexual orientation (discussed later in this chapter). Conflict theory helps to explain why many business owners and corporate leaders oppose civil rights protection for gays and lesbians. Employers fear that such protection would result in costly lawsuits if they refused to hire homosexuals, regardless of the reason for their decision. Business owners also fear that granting civil rights protections to homosexual employees would undermine the economic health of a community by discouraging the development of new businesses and even driving out some established firms (Button et al. 1997).

However, many companies are recognizing that implementing antidiscrimination policies that include sexual orientation is good for the "bottom line." The majority (82%) of Fortune 500 companies have included sexual orientation in their nondiscrimination policies, and employers are increasingly offering benefits to domestic partners of LGBT employees (Human Rights Campaign 2005b). Gay-friendly work policies help employers maintain a competitive edge in recruiting and maintaining a talented and productive workforce. Companies are also competing for the gay and lesbian dollar. LGBT consumers—who had an estimated total buying power of $610 billion in 2005—report that they trust brands more if they are made by companies that have progressive policies toward gay and lesbian employees (Human Rights Campaign 2005b).

In summary, conflict theory frames the gay rights movement and the opposition to it as a struggle over power, prestige, and economic resources. Recent trends toward increased social acceptance of homosexuality may, in part, reflect the corporate world's competition over the gay and lesbian consumer dollar.

Symbolic Interactionist Perspective

Symbolic interactionism focuses on the meanings of heterosexuality, homosexuality, and bisexuality, how these meanings are socially constructed, and how they influence the social status and self-concepts of nonheterosexual individuals. The meanings we associate with same-sex relations are learned from society—from family, peers, religion, and the media. The negative meanings associated with homosexuality are reflected in the current slang use of the phrase "That's so gay" or "You're so gay," which is meant to convey something that is considered bad or valueless, as a synonym for "dumb" or "stupid" (Kosciw & Cullen 2002).

Historical and cross-cultural research on homosexuality reveals the socially constructed nature of homosexuality and its meaning. Although many Americans assume that same-sex romantic relationships have always been taboo in our society, during the 19th century "romantic friendships" between women were encouraged and regarded as preparation for a successful marriage. The nature of these friendships bordered on lesbianism. President Grover Cleveland's sister Rose wrote to her friend Evangeline Whipple in 1890: "It makes me heavy with emotion . . . all my whole being leans out to you. . . . I dare not think of your arms" (Goode & Wagner 1993, p. 49).

The symbolic interactionist perspective also points to the effects of labeling on individuals. Once individuals become identified or labeled as lesbian, gay, or bisexual, that label tends to become their **master status.** In other words, the dominant heterosexual community tends to view "gay," "lesbian," and "bisexual" as the most socially significant statuses of individuals who are identified as such. Esterberg (1997) notes that "unlike heterosexuals, who are defined by their family structures, communities, occupations, or other aspects of their lives, lesbians, gay men, and bisexuals are often defined primarily by what they do in bed. Many lesbians, gay men, and bisexuals, however, view their identity as social and political as well as sexual" (p. 377).

Negative social meanings associated with homosexuality also affect the self-concepts of nonheterosexual individuals. **Internalized homophobia**—a sense of personal failure and self-hatred among lesbians and gay men resulting from social rejection and stigmatization—has been linked to increased risk for depression, substance abuse and addiction, anxiety, and suicidal thoughts (Bobbe 2002; Gilman et al. 2001). Unlike other marginalized minorities who have the support of their families and churches, lesbian and gay individuals are often rejected by their religion and sometimes by their families as well. When Reverend Mel White, a closeted gay Christian man who was nearly driven to suicide after two decades of struggling to save his marriage and his soul with "reparative therapies" finally came out to his mother, her response was, "I'd rather see you at the bottom of that swimming pool, drowned, than to hear this" (White 2005, p. 28).

Labels also affect heterosexuals' attitudes. When Gallup tested alternative terms for referring to gay Americans, it found that using the label "gays and

ThomsonNOW™

Learn more about **Master Status** by going through the Roles and Status Learning Module.

lesbians" results in somewhat more favorable, pro-gay responses than does using the term "homosexual" (Saad 2005). The negative meanings associated with homosexuality are explored in the following section.

Heterosexism, Homophobia, and Biphobia

The United States, along with many other countries throughout the world, is predominantly heterosexist. **Heterosexism** refers to "the institutional and societal reinforcement of heterosexuality as the privileged and powerful norm" (SIECUS 2000). Heterosexism is based on the belief that heterosexuality is superior to homosexuality; it results in prejudice and discrimination against homosexuals and bisexuals. Prejudice refers to negative attitudes, whereas discrimination refers to behavior that denies individuals or groups equality of treatment. Before reading further, you may wish to complete this chapter's *Self and Society* feature, which assesses your behaviors toward individuals you perceive to be homosexual.

Homophobia

SIECUS, the Sex Information and Education Council of the United States, "strongly supports the right of each individual to accept, acknowledge, and live in accordance with his or her orientation . . . and deplores all forms of prejudice and discrimination against people based on sexual orientation" (SIECUS 2000, p. 1). Nevertheless, negative attitudes toward homosexuality are reflected in the high percentage of the U.S. population who disapprove of homosexuality. According to national surveys by the Gallup Organization, the percentage of Americans saying that homosexuality should be considered an acceptable alternative lifestyle was only 34% in 1982 and 38% in 1992, but the percentage increased to just over half (51%) in 2005 (Saad 2005).

The term **homophobia** is commonly used to refer to negative attitudes and emotions toward homosexuality and those who engage in it. Homophobia is not necessarily a clinical phobia (i.e., one involving a compelling desire to avoid the feared object despite recognizing that the fear is unreasonable). Other terms that refer to negative attitudes and emotions toward homosexuality include *homonegativity* and *antigay bias*.

In general, individuals who are more likely to have negative attitudes toward homosexuality and to oppose gay rights are older, attend religious services, are less educated, live in the South or Midwest, and reside in small rural towns (Curtis 2003; Loftus 2001; Page 2003). In general, heterosexuals have more favorable attitudes toward gay men and lesbian women if they have had prior contact with or know someone who is gay or lesbian (Mohipp & Morry 2004).

Public opinion surveys also indicate that men are more likely than women to have negative attitudes toward gays (Moore 1993). But many studies on attitudes toward homosexuality do not distinguish between attitudes toward gay men and attitudes toward lesbians. Research that has assessed attitudes toward male versus female homosexuality has found that heterosexual women and men hold similar attitudes toward lesbians but that men are more negative toward gay men (Louderback & Whitley 1997; Price & Dalecki 1998).

A study of undergraduates (110 men, 98 women) attending a Canadian university found that attitudes toward gay men were more negative than those toward lesbians (Schellenberg et al. 1999). Compared with science or business majors, students in the arts and social sciences had more positive attitudes toward gay men, and women were more positive than men.

The Self-Report of Behavior Scale (Revised)

This questionnaire is designed to examine which of the following statements most closely describes your behavior during past encounters with people you thought were homosexuals. Rate each of the following self-statements as honestly as possible by choosing the frequency that best describes your behavior: (1) never, (2) rarely, (3) occasionally, (4) frequently, or (5) always.

1. I have spread negative talk about someone because I suspected that he or she was gay. _____

2. I have participated in playing jokes on someone because I suspected that he or she was gay. _____

3. I have changed roommates and/or rooms because I suspected my roommate to be gay. _____

4. I have warned people who I thought were gay and who were a little too friendly with me to keep away from me. _____

5. I have attended antigay protests. _____

6. I have been rude to someone because I thought that he or she was gay. _____

7. I have changed seat locations because I suspected the person sitting next to me to be gay. _____

8. I have had to force myself to stop from hitting someone because he or she was gay and very near me. _____

9. When someone I thought to be gay has walked toward me as if to start a conversation, I have deliberately changed directions and walked away to avoid him or her. _____

10. I have stared at a gay person in such a manner as to convey to him or her my disapproval of his or her being too close to me. _____

11. I have been with a group in which one (or more) person(s) yelled insulting comments to a gay person or group of gay people. _____

12. I have changed my normal behavior in a restroom because a person I believed to be gay was in there at the same time. _____

13. When a gay person has "checked" me out, I have verbally threatened him or her. _____

14. I have participated in damaging someone's property because he or she was gay. _____

15. I have physically hit or pushed someone I thought was gay because he or she brushed his or her body against me when passing by. _____

16. Within the past few months, I have told a joke that made fun of gay people. _____

17. I have gotten into a physical fight with a gay person because I thought he or she had been making moves on me. _____

18. I have refused to work on school and/or work projects with a partner I thought was gay. _____

19. I have written graffiti about gay people or homosexuality. _____

20. When a gay person has been near me, I have moved away to put more distance between us. _____

Scoring: The revised Self-Report of Behavior Scale is scored by totaling the number of points endorsed on all items (never, 1; rarely, 2; occasionally, 3; frequently, 4; always, 5), yielding a range from 20 to 100 total points. The higher the score, the more negative the attitudes toward homosexuals.

Comparison Data

The Self-Report of Behavior Scale was originally developed by Sunita Patel (1989) in her psychology master's thesis research at East Carolina University. College men (from a university campus and from a military base) were the original participants (Patel et al. 1995). The scale was revised by Shartra Sylivant (1992), who used it with a coed high school student population, and by Tristan Roderick (1994), who involved college students to assess the scale's psychometric properties. The scale was found to have high internal consistency. Two factors were identified: (1) a passive avoidance toward homosexuals and (2) active or aggressive reactions.

In Roderick's study (Roderick et al. 1998) the mean score for 182 college women was 24.76. The mean score for 84 men was significantly higher, 31.60. In Sylivant's (1992) high school sample, the mean scores were 33.74 for young women and 44.40 for young men.

The revised Self-Report of Behavior Scale is reprinted by the permission of the students and faculty who participated in its development: S. Patel, S. L. McCammon, T. E. Long, L. J. Allred, K. Wuensch, T. Roderick, and S. Sylivant.

Cultural Origins of Homophobia. Why do many Americans disapprove of homosexuality? Antigay bias has its roots in various aspects of U.S. culture.

Religion. Most Americans who view homosexuality as unacceptable say that they object on religious grounds (Rosin & Morin 1999). Indeed, conservative Christian ideology has been identified as the best predictor of homophobia (Plugge-Foust & Strickland 2000). Many religious leaders teach that homosexuality is sinful and prohibited by God. The Roman Catholic Church rejects all homosexual expression and resists any attempt to validate or sanction the homosexual orientation. Some fundamentalist churches have endorsed the death penalty for homosexual people and teach the view that AIDS is God's punishment for engaging in homosexual sex. The Westboro Baptist Church (Topeka, Kansas), headed by the antigay Reverend Fred Phelps, maintains a website called godhatesfags.com. Members of this church have held antigay demonstrations near the funerals of people who have died from AIDS, carrying signs reading "Gays Deserve to Die." Furthermore, an organization of Christian fundamentalists claimed that the destruction brought on by Hurricane Katrina in 2005 was God's judgment against New Orleans for holding the annual gay Southern Decadence party (Curtis 2005).

Theologians and religious scholars have different viewpoints on the Bible's position on homosexuality. Many scholars believe that key passages actually are denouncing orgies and prostitution—or in the case of the town of Sodom, inhospitality—and not homosexuality. Two Old Testament passages that do condemn homosexual acts are found amid a long list of religious prohibitions, including eating pork and wearing mixed fabrics—rules that have been abandoned by most contemporary Christians (Potok 2005).

Some religious groups, such as the Quakers and the United Church of Christ (UCC), are accepting of homosexuality, and other groups have made reforms toward increased acceptance of lesbians and gays. In the early 1970s the UCC became the first major Christian church to ordain an openly gay minister, and in 2005 the UCC became the largest Christian denomination to endorse same-sex marriages. Some Episcopal priests perform "ceremonies of union" between same-sex couples; some Reform Jewish groups sponsor gay synagogues (Fone 2000). In June 2003 the Reverend V. Gene Robinson was elected bishop of the Episcopal diocese of New Hampshire, becoming the first openly gay bishop in the church's history. Although the official position of the United Methodist Church is one that condemns homosexuality, some Methodist ministers advocate acceptance of and equal rights for lesbians and gay men (*The United Methodist Church and Homosexuality,* 1999). Acceptance of gays and lesbians is even found among Southern Baptists, one of the more conservative Christian religions. The Association of Welcoming and Affirming Baptists, organized in 1993, is a network of nearly 50 congregations and other church groups that advocate for the inclusion of LGBT individuals within the Baptist community of faith.

Marital and Procreative Bias. Many societies have traditionally condoned sex only when it occurs in a marital context that provides for the possibility of producing and rearing children. Although a handful of several states have granted legal recognition to same-sex couples, only one—Massachusetts—allows same-sex couples to be legally married. Furthermore, even though assisted reproductive technologies make it possible for gay individuals and couples to have children, many people believe that these advances should be used only by heterosexual married couples.

© AP/Wide World Photos

When the Reverend V. Gene Robinson was elected bishop of the Episcopal diocese in New Hampshire in 2003, he became the first openly gay bishop in the church's history.

Concern About HIV and AIDS. Although most cases of HIV and AIDS world-wide are attributed to heterosexual transmission, the rates of HIV and AIDS in the United States are much higher among gay and bisexual men than among other groups. Because of this, many people associate HIV and AIDS with homosexuality and bisexuality. This association between AIDS and homosexuality has fueled antigay sentiments. Lesbians, incidentally, have a very low risk for sexually transmitted HIV—a lower risk than heterosexual women.

Rigid Gender Roles. Antigay sentiments also stem from rigid gender roles. When Cooper Thompson (1995) was asked to give a guest presentation on male roles at a suburban high school, male students told him that the most humiliating put-down was being called a fag. The boys in this school gave Thompson the impression that they were expected to conform to rigid, narrow standards of masculinity in order to avoid being labeled in this way.

From a conflict perspective heterosexual men's subordination and devaluation of gay men reinforces gender inequality. "By devaluing gay men . . . heterosexual men devalue the feminine and anything associated with it" (Price & Dalecki 1998, pp. 155–156). Negative views toward lesbians also reinforce the patriarchal system of male dominance. Social disapproval of lesbians is a form of punishment for women who relinquish traditional female sexual and economic dependence on men. Not surprisingly, research findings suggest that individuals with traditional gender role attitudes tend to hold more negative views toward homosexuality (Louderback & Whitley 1997).

Myths and Negative Stereotypes. Prejudice toward homosexuals can also stem from some of the myths and negative stereotypes regarding homosexuality. One negative myth about homosexuals is that they are sexually promiscuous and lack "family values," such as monogamy and commitment to relationships. Although some homosexuals do engage in casual sex, as do some heterosexuals, many homosexual couples develop and maintain long-term committed relationships. Between 64% and 80% of lesbians report that they are in a committed relationship at any given time, and between 46% and 60% of gay men report being in a committed relationship (Cahill et al. 2002).

Another myth that is not supported by data is that homosexuals, as a group, are child molesters. In fact, 95% of all reported incidents of child sexual abuse are committed by heterosexual men (SIECUS 2000). Most often the abuser is a father, stepfather, or heterosexual relative of the family. Research cited by Cahill and Jones (2003) finds that a child's risk of being molested by his or her relative's heterosexual partner is more than 100 times greater than the risk of being molested by someone who is homosexual or bisexual. Furthermore, "when a man abuses a young girl, the problem is not heterosexuality. . . . Similarly, when a priest sexually abuses a boy or under-age teen, the problem is not homosexuality. The problem is child abuse" (Cahill & Jones 2003, p. 1).

Some of the negative myths and stereotypes about lesbians and gays can be traced to the antigay propaganda of Paul Cameron in the 1970s and 1980s. A former psychology instructor at University of Nebraska, Cameron began publishing pseudoscientific pamphlets "proving" that gay people commit more serial murders, molest more children, and intentionally spread diseases. Cameron's antigay pamphlets, largely distributed in fundamentalist churches, also claimed that gay people "were obsessed with consuming human excrement, allowing them to spread deadly diseases simply by shaking hands with unsuspecting strangers or using public restrooms" (Moser 2005, p. 14).

Biphobia

Just as the term *homophobia* is used to refer to negative attitudes toward gay men and lesbians, **biphobia** refers to "the parallel set of negative beliefs about and stigmatization of bisexuality and those identified as bisexual" (Paul 1996, p. 449). Although both homosexual- and bisexual-identified individuals are often rejected by heterosexuals, bisexual-identified women and men also face rejection from many homosexual individuals. Thus bisexuals experience "double discrimination."

Biphobia includes negative stereotyping of bisexuals, the exclusion of bisexuals from social and political organizations of lesbians and gay men, and fear and distrust of, as well as anger and hostility toward, people who identify themselves as bisexual (Firestein 1996). Individuals with negative attitudes toward bisexual individuals often believe that bisexuals are actually homosexuals who are in denial about their true sexual orientation or are trying to maintain heterosexual privilege (Israel & Mohr 2004). Bisexual individuals are sometimes viewed as heterosexuals who are looking for exotic sexual experiences. One negative stereotype that encourages biphobia is the belief that bisexuals are, by definition, nonmonogamous. However, many bisexual women and men prefer and have long-term committed monogamous relationships.

Lesbians seem to exhibit greater levels of biphobia than do gay men. This is because many lesbian women associate their identity with a political stance against sexism and patriarchy. Some lesbians view heterosexual and bisexual women who "sleep with the enemy" as traitors to the feminist movement.

Changing Attitudes Toward Homosexuality

Over the last few decades attitudes toward the morality of homosexuality have become more liberal, and support for protecting civil rights of gays and lesbians has increased (Loftus 2001). Part of the explanation for these changing attitudes is the increasing levels of education in the U.S. population, because individuals with more education tend to be more liberal in their attitudes toward homosexuality. Increased contact between homosexuals and heterosexuals and positive depictions of gay and lesbian individuals in the media also contribute to increased acceptance of homosexuality.

Increased Contact Between Homosexuals and Heterosexuals. Greater acceptance of homosexuality may also be due to increased personal contact between heterosexuals and openly gay individuals as more gay and lesbian Americans are coming out to their family, friends, and coworkers. Psychologist Gordon Allport's (1954) "contact hypothesis" asserts that contact between groups is necessary for the reduction of prejudice. Recent research has found that, in general, heterosexuals have more favorable attitudes toward gay men and lesbian women if they have had prior contact with or know someone who is gay or lesbian (Mohipp & Morry 2004). Contact with openly gay individuals reduces negative stereotypes and ignorance and increases support for gay and lesbian equality (Wilcox & Wolpert 2000). Polls of U.S. adults find that more than half of respondents (56%) say they have a friend or acquaintance who is gay or lesbian, nearly one-third (32%) say they work with someone who is gay or lesbian, and nearly one-fourth (23%) say that someone in their family is gay or lesbian (Newport 2002). A national poll of U.S. high school students found that 16% of students have a gay or lesbian person in their family and 72% know someone who is gay or lesbian (Curtis 2004). One woman describes how her brother's coming out of the closet changed her views on homosexuality (Yvonne 2004):

> I was raised in a devout born-again Christian family. I was raised as a child to believe that homosexuality was evil and a perversion. When I was growing up, I used to wonder if any of the kids at my school could be gay. I couldn't imagine it could be so. As it happens, there was a gay individual even closer than I imagined. My brother Tommy came out of the closet in the early 1990s. After he came out, I had to confront my own denial about the fact that, in my heart of hearts, I had always known that Tommy was gay. Over the years, his partner Rod has come to be a loved and cherished member of our family, and we have all had to confront the prejudices and stereotypes we have held onto for so long about sexual orientation. It seems to me that coming out of the closet is the greatest weapon that gays and lesbians have. If my own brother had never come out, my family would never have been forced to confront the deep-seated prejudices we were raised with. . . . I am still a Christian, but my husband (who also has a gay brother) and I attend a church that truly puts the teachings of Christ into practice—teachings about love, tolerance and inclusivity. As a Christian who grew up with a gay family member, I know that the propaganda put forth by the religious right on this issue is founded in fear, hatred and prejudice. None of these are values taught by Jesus!

Gays and Lesbians in the Media.   Another explanation for the increasing acceptance of gays and lesbians is the positive depiction of gays and lesbians in the popular media. In 1998 Ellen DeGeneres came out on her sitcom *Ellen,* and by 2005 many television viewers had seen gay and lesbian characters in television shows such as *Will and Grace, Buffy the Vampire Slayer, Queer Eye for the Straight Guy,* and *Six Feet Under.* "Honest, non-stereotyped and diverse portrayals of gays and lesbians in prime time can offer youth a realistic representation of the gay community . . . [and] can offer positive role models for gay and lesbian youth" (Miller et al. 2002, p. 21).

One study found that college students reported lower levels of antigay prejudice after watching television shows with prominent homosexual characters (e.g., *Six Feet Under* and *Queer Eye for the Straight Guy*) (Schiappa et al. 2005). The researchers propose that just as Allport's contact hypothesis suggests that intergroup contact may reduce prejudice, the "parasocial contact hypothesis" suggests that contact with individuals through the media has similar effects.

After the Supreme Court invalidated antisodomy laws in June 2003, lesbians, gays, bisexuals, and their allies gathered throughout the country to celebrate the landmark ruling.

© Reuters NewMedia Inc. /Corbis

Discrimination Against Sexual Orientation Minorities

In June 2003 a Supreme Court decision in *Lawrence* v. *Texas* invalidated state laws that criminalize sodomy—oral and anal sexual acts. This historic decision overruled a 1986 Supreme Court case (*Bowers* v. *Hardwick*), which upheld a Georgia **sodomy law** as constitutional. The 2003 ruling, which found that sodomy laws were discriminatory and unconstitutional, removes the stigma and criminal branding that sodomy laws have long placed on GLBT individuals. Before this historic ruling was made, sodomy was illegal in 13 states: Alabama, Florida, Idaho, Kansas, Louisiana, Mississippi, Missouri, North

Carolina, Oklahoma, South Carolina, Texas, Utah, and Virginia, and four states (Kansas, Missouri, Oklahoma, and Texas) targeted only same-sex acts. Penalties for engaging in sodomy ranged from a $200 fine to 20 years' imprisonment. In states that criminalized both same- and opposite-sex sodomy, sodomy laws were usually not used against heterosexuals but were used primarily against gay men and lesbians.

Like other minority groups in U.S. society, homosexuals and bisexuals experience various forms of discrimination. Next, we look at sexual orientation discrimination in the workplace and in housing, in the military, in marriage and parenting, in violent expressions of hate, and in treatment by police. This chapter's *Focus on Technology* feature discusses the discriminatory effects of Internet filtering and monitoring technology on sexual orientation minorities.

Discrimination in the Workplace and in Housing

The percentage of Americans saying that homosexuals should have equal job opportunities grew from 56% in 1977 to 74% in 1992 to 88% in 2003 (Saad 2005). Yet as of June 2005 it was legal in 34 states to fire, decline to hire or promote, or otherwise discriminate against an employee because of his or her sexual orientation (National Gay and Lesbian Task Force 2005e).

In a survey of gay, lesbian, and bisexual residents of Topeka, Kansas, 16% of respondents reported that they had been denied employment because of their sexual orientation; 11% reported being denied a promotion; 15% reported that they had been fired; and 35% reported that they had received harassing letters, e-mail, or faxes at work (Colvin 2004). Two respondents in the study reported the following experiences (p. 3):

> My job found out that I was a lesbian and my "friend" that comes in every night was my girlfriend. She was told not to come in anymore or I would be fired. And later because she came in I was fired.
>
> As soon as my . . . boss suspected I was gay, he harassed me until I took a job with another state agency. Prior to that I had three outstanding employee evaluations, but he couldn't find anything I did right.

Like other minority group members, LGBT individuals are sometimes denied the option to rent or buy residential property because of prejudice against them. Only the District of Columbia and 12 states ban sexual orientation discrimination in housing: California, Connecticut, Illinois, Massachusetts, Minnesota, New Hampshire, New Jersey, New York, New Mexico, Rhode Island, Vermont, and Wisconsin, (Cahill 2005). In other states lesbian and gay individuals and couples are vulnerable to discrimination in housing. In addition, same-sex families are not allowed to qualify as a "family" when applying for public housing, which decreases their chances of being eligible for public housing assistance.

Discrimination in the Military

A majority of Americans (76%) believe that homosexuals should be hired in the armed forces (Saad 2005). Nevertheless, sexual orientation discrimination occurs in the military. In 1993 President Clinton instituted a "don't ask, don't tell" policy in which recruiting officers are not allowed to ask about sexual orientation and homosexuals are encouraged not to volunteer such information. Thousands of service members have been discharged under the policy since it took effect, with more

Effect of Internet Filtering and Monitoring on Sexual Orientation Minorities

When Darrel and Estella, gay and lesbian high school students in Southern California, were spit on and otherwise harassed by their antigay classmates, they found refuge in Will's classroom. Will is one of the few out gay teachers at their school. Wanting to help Darrel and Estella, Will sat at his classroom computer and typed in "gay and lesbian youth resources." "Access denied," the screen said. "Objectionable material." "I couldn't believe it," Will said. "Here I was trying to help students in dire need, and all I was able to offer them was that the terms they used to name themselves were 'objectionable'" (Sapp 2005).

As a result of the Children's Internet Protection Act of 2000, Internet filtering devices limit the types of material an Internet user can access in schools and libraries. Filtering technologies are promoted as tools to help parents, schools, libraries, and communities to prevent children's access to sexually explicit and pornographic material on the Internet. However, some filtering software also denies users access to several lesbian,

gay, and bisexual youth resource sites as well as to health (especially sexual health) sites. Filters block websites by screening the keywords that index the sites and blocking any sites with keywords in "objectionable" categories. Filtering products can be customized by selecting topics or categories for blocking or can be used at a default setting.

The words *gay* and *lesbian* and their slang synonyms are often among those categories that are blocked by Internet filtering products. For example, the filtering software CyberSitter automatically filters out words and phrases such as *gay, lesbian, gay rights,* and *gay community* (Javier 1999). In a study of seven filtering products nearly one in three "safe sex" sites was blocked by at least one of the filters at the *least restrictive* setting, and at the most restrictive setting one-fourth (24%) of 3,053 health websites were blocked (Rideout et al. 2002).

Mandated Internet filters are an example of institutional homophobia—an institutional mandate that silences and marginalizes gay

and lesbian individuals (Sapp 2005). Given the impact of Internet filtering on the gay community, the Gay and Lesbian Alliance Against Defamation does not support the use of filtering software. Instead, they advocate parental oversight, school supervision, and training of young Internet users (Bowes 1999). However, the Children's Internet Protection Act (CIPA), passed by Congress in 2000, requires schools and libraries that receive certain types of federal funding to use Internet filtering devices. The CIPA requirement for libraries was struck down in spring 2002 by a circuit court on the grounds that it violates the First Amendment. But in June 2003 the U.S. Supreme Court, in *United States* v. *American Library Association,* reversed the lower court's decision and upheld the CIPA requirement, stipulating that public libraries must turn off the filter upon request by an adult patron.

Another concern of sexual orientation minorities is the use of monitoring software that allows parents, teachers, and other authority figures to track sites that Web surfers

women than men being discharged. The "don't ask, don't tell" policy also provided an additional means for servicemen to harass lesbian service members by threatening to "out" those who refused their advances or threatening to report them, thus ending their careers (Human Rights Watch 2001).

A Government Accounting Office report found that the "don't ask, don't tell" policy has cost nearly $200 million for the replacement and training of personnel who were discharged as a result of the policy (Human Rights Campaign 2005a). The report also found that nearly 800 specialists with critical military skills have been discharged, including 54 linguists who specialize in Arabic. As of this writing, gay rights advocates are pushing for the passage of the Military Readiness Enhancement Act, which would repeal the "don't ask, don't tell" policy.

Most European countries, including France, Germany, Spain, Switzerland, and Denmark, have lifted their bans on gays in the military. Great Britain has gone a step further by actively encouraging gays to enlist (Lyall 2005).

try to access. The monitoring software industry markets its products by claiming that they allow for parental awareness without censorship and that parental use of monitoring software encourages open family communication. But for youth who are not ready to reveal their sexual orientation to their family, "such software could potentially 'out' them before they are ready, leading to strained family relations and deeper isolation" (Javier 1999, p. 8).

The best-known story of a gay person affected by an "online outing" is that of a naval officer who was outed In 1998 after an America Online employee divulged his personal data and online history to the navy. This invasion of privacy cost this officer his career.

Websites also record information about your Internet usage. When you visit a free news website, you may be required to "subscribe" by giving them your name and e-mail address. After subscribing, you read a few articles. But unknown to you, the news website has put a "cookie," or piece

of computer code, on the hard drive of your computer. This cookie contains a unique identifier permitting the news site to recognize you when you return to that site. Cookies can do helpful things, such as remember your password for accessing that site and tailor Web pages for preset preferences, but they can also contain a record of what you did on that site, including what searches you made and what articles you read (Aravosis 1999).

Although some websites have strict privacy policies, others sell information about site visitors to corporate America. One company boasts on its website that it is "the world's oldest and largest mailing list manager and broker for Gay, Lesbian, and HIV-related names, currently managing almost two million names, which we estimate to be about 65 percent of all those commercially available in this segment" (Aravosis 1999, p. 33). Databases containing the names of gay consumers are valuable because gays are perceived as a "wealthy and wired market."

The Internet has been a useful tool for gays, lesbians, and bisexuals. Going online has allowed the gay community to create safe places for support and information. However, Internet filtering and monitoring software that has been installed on computers in homes, schools, libraries, and workplaces throughout the country represents a threat to the gay community. Filtering and monitoring technologies make it impossible or dangerous for closeted gays or lesbians to seek out support and information about their community.

We live in an age in which the Internet has become an extremely important part of the coming out process for many gay and lesbian youth. In many cases, it can be a lifeline to those in geographically isolated areas. To deny basic educational and support resources to lesbian and gay youth could seriously endanger their physical and emotional well-being. (Gay and Lesbian Alliance Against Defamation 1999, p 46)

Discrimination in Marriage

Before the 2003 Massachusetts Supreme Court ruling in *Goodridge* v. *Department of Public Health,* no state had declared that same-sex couples have a constitutional right to be legally married. In response to growing efforts to secure legal recognition of same-sex couples, opponents of same-sex marriage have prompted antigay marriage legislation. In 1996 Congress passed and President Clinton signed the **Defense of Marriage Act,** which states that marriage is a "legal union between one man and one woman" and which denies federal recognition of same-sex marriage. In effect, this law allows states to either recognize or not recognize same-sex marriages performed in other states. As of May 2005, 39 states had banned gay marriage either through statute or a state constitutional amendment, and 15 states had passed broader antigay family measures that ban other forms of partner recognition in addition to marriage, such as domestic partnerships and civil unions (National

Should Religious Organizations That Receive Federal Funds to Provide Social Services Under the Faith-Based Initiative Be Allowed to Discriminate in Hiring?

The Faith-Based Initiative, issued as an executive order by President Bush in 2002, allows religious organizations to receive federal funding to pay for the delivery of a wide range of social services, including services for the poor, homeless, and drug addicted (see also Chapter 6). In 2003 the White House sent a memo to Congress that advocated allowing faith-based service providers to ignore local and state nondiscrimination laws that include sexual orientation when hiring for positions paid for with federal funds. In 2005 the U.S. House of Representatives passed an amendment to the School Readiness Act that will allow taxpayer-funded faith-based organizations that provide preschool Head-Start programs to discriminate in employment. Under this amendment Head-Start teachers working for faith-based organizations can be fired for being lesbian or gay. Conservative groups argue that such discrimination is necessary for religious freedom and to maintain the religious character of a faith-based organization. Although such discrimination is allowed with private funds, the legality of allowing faith-based groups to discriminate in employment funded by taxpayer monies is highly controversial.

It Is Your Turn Now to Take a Stand!

Do you think that religious organizations that receive federal funds to provide social services under the Faith-Based Initiative should be allowed to discriminate in hiring?

Gay and Lesbian Task Force 2005a). These broader measures, known as "Super DOMAs," potentially endanger employer-provided domestic partner benefits, joint and second-parent adoptions, health care decision-making proxies, or any policy or document that recognizes the existence of a same-sex partnership (Cahill & Slater 2004).

A group of conservative Republicans introduced the Federal Marriage Amendment, which would amend the U.S. Constitution to define marriage as being between a man and a woman. The Federal Marriage Amendment did not pass in 2004 in the Senate or the House, but supporters vowed to continue the fight. The Federal Marriage Amendment was reintroduced in the 109th Congress in 2005 as the Marriage Protection Amendment. If passed, the constitutional amendment would deny marriage and likely civil union and domestic partnership rights to same-sex couples (LAWbriefs 2005a). This chapter's *Human Side* feature presents one woman's plea to protect same-sex families and to oppose a federal marriage constitutional amendment.

Arguments in Favor of Same-Sex Marriage.   Advocates of same-sex marriage argue that banning same-sex marriages or refusing to recognize same-sex marriages granted in other states is a violation of civil rights that denies same-sex couples the many legal and financial benefits that are granted to heterosexual married couples. For example, married couples have the right to inherit from a spouse who dies without a will, to avoid inheritance taxes between spouses, to make crucial medical decisions for a partner and to take family leave to care for a partner in the event of the partner's critical injury or illness, to receive Social Security survivor benefits, and to include a partner in his or her health insurance coverage. Other rights bestowed on married (or once married) partners include assumption of spouse's pension, bereavement leave, burial determination, domestic violence protection, reduced-rate memberships, divorce protections (such as equitable division of assets and visitation of partner's children), automatic housing lease transfer, and immunity from testifying against a spouse (Sullivan 2003). Finally, unlike 17 other countries that recognize same-sex couples for immigration purposes, the United States does not recognize same-sex couples in granting immigration status because such couples are not considered "spouses."

Another argument for same-sex marriage is that it would promote relationship stability among gay and lesbian couples. "To the extent that marriage provides status, institutional support, and legitimacy, gay and lesbian couples, if allowed to marry, would likely experience greater relationship stability" (Amato 2004, p. 963). This outcome, argues Amato, would be beneficial to the children of same-sex parents. Without legal recognition of same-sex families, children living in gay- and lesbian-headed households are denied a range of securities that protect children of heterosexual married couples. These include the right to get health insurance coverage and Social Security survivor benefits from a nonbiological parent. In some cases children in same-sex households lack the automatic right to continue living with their nonbiological parent should their biological mother or father die (Tobias & Cahill 2003). It is ironic that the same pro-marriage groups that stress that children are better off in married couple families disregard the benefits of same-sex marriage to children.

Finally, there are religious-based arguments in support of same-sex marriage. In a sermon titled "The Christian Case for Gay Marriage," Jack McKinney (2004) interprets Luke 4: "Jesus is saying that one of the most fundamental religious tasks is to stand with those who have been excluded and marginalized. . . . [Jesus] is determined to stand with them, to name them beloved of God, and to dedicate his life to seeing them empowered." McKinney goes on to ask, "Since when has it been immoral for two people to commit themselves to a relationship of mutual love and caring? No, the true immorality around gay marriage rests with the heterosexual majority that denies gays and lesbians more than 1,000 federal rights that come with marriage." As noted earlier, in 2005 the United Church of Christ became the largest Christian denomination to endorse same-sex marriages.

Arguments Against Same-Sex Marriage. Whereas advocates of same-sex marriage argue that so long as same-sex couples cannot be legally married, they will not be regarded as legitimate families by the larger society, opponents do not want to legitimize homosexuality as a socially acceptable lifestyle. Opponents of same-sex marriage who view homosexuality as unnatural, sick, and/or immoral do not want their children to view homosexuality as an accepted "normal" lifestyle.

Opponents of same-sex marriage commonly argue that such marriages would subvert the stability and integrity of the heterosexual family. However, Sullivan (1997) suggests that homosexuals are already part of heterosexual families:

> [Homosexuals] are sons and daughters, brothers and sisters, even mothers and fathers, of heterosexuals. The distinction between "families" and "homosexuals" is, to begin with, empirically false; and the stability of existing families is closely linked to how homosexuals are treated within them. (p. 147)

Many opponents of same-sex marriage base their opposition on their religious views. In a Pew Research Center national poll, the majority of Catholics and

Effects of a Federal Marriage Amendment on Gay and Lesbian Families: One Woman's Plea to "Do No Harm"

In April 2005 Dr. Kathleen Moltz, an assistant professor at Wayne State University School of Medicine, testified against the passage of the Federal Marriage Amendment before a United States Senate Judiciary Committee. The following paragraphs are excerpts from her testimony.

Members of the Subcommittee:

. . . I am here as the mother of two beautiful children whose welfare I am trying desperately to protect, as the partner of the wonderful woman with whom I share my life, and as a pediatrician who has taken the oath "first, do no harm." In 1990 . . . my partner Dahlia Schwartz and I became a couple. After several years together, we were married in a traditional Jewish wedding ceremony in 1996. . . . This was years before any state recognized marriage rights for same-sex couples. Dahlia and I were together for several more years before we decided to have children. We are now proud parents to our daughter Aliana, and our son Itamar. Dahlia carried Itamar through a difficult pregnancy, a precipitous labor and an emergency C-section. At several times during

the delivery, I was asked to leave the room— something that a different-sex spouse would not have to go through. I cannot put into the words the agonizing emotions of not being able to be present when my partner and child were in medical distress.

Immediately after he was born, Itamar experienced temperature regulation problems, rapid breathing and hypoglycemia. . . . The pediatrician on staff refused to discuss Itamar's condition with me because I was not his "real mother." What is a "real mother" if not the person who would lay down her life for her child? I am a real mother to my children, and so is Dahlia. Fortunately, since then I have adopted Itamar through second-parent adoption, and Dahlia has adopted Aliana. Both children now have the benefit of a legal relationship with two parents.

On May 21, 2004, Dahlia and I were legally married in Massachusetts, where we had lived for eight years. . . . In June 2004 we moved to Michigan, where I took a job as a pediatric endocrinologist at Wayne State University. . . . I moved my family to Michigan so that I could take a job that would allow Dahlia to stay home with the kids. The domestic partner health benefits that Wayne State provided made this possible, particularly because Dahlia has a continuing medical

condition that makes health insurance a necessity. I would not have taken the job without the domestic partnership benefits. The move also allowed me to be near my parents—our children's grandparents—who live eight houses down from us. . . .

Not long after we moved to Michigan, the state became embroiled in a campaign to pass Proposal 2, an amendment to the state constitution that would ban marriage rights for same-sex couples. When our daughter asked what it was all about, we told her that there were people who believed that we couldn't really be a family. We told her that we thought this was silly because, obviously, we are a family—we share love, children and a commitment to raising healthy, happy kids. When the results of the election came in, Aliana asked about the outcome. We told her that the amendment had passed. With tears in her eyes, she asked, "Does this mean our family has to split up?" Like children do, our 4-year-old went straight to the heart of the issue. The voters had sent us a message that day: you are not a family.

We were dismayed and stunned by the results of the Michigan election and spent days wondering which of our neighbors and colleagues thought that our family should not have equal rights. . . . When anti-gay groups

Protestants oppose legalizing same-sex marriage, whereas the majority of secular respondents favor it (Green 2004). However, churches have the right to say no to marriage for gays in their congregations, but marriage is still a *civil* option that does not require religious sanctioning.

Finally, opponents of gay marriage point to public opinion polls that suggest that the majority of Americans are against same-sex marriage. In a March 2005 national Gallup poll most adults (68%) said that marriages between homosexuals should not be recognized by the law as valid (Newport 2005). In the same poll more than half (57%) of U.S. adults said that they favored a constitutional amendment

from outside our state tried to use the amendment to take away the health benefits insurance I obtain through my work, I could not sit idly by. Throughout the campaign, supporters of the amendment insisted that the amendment had nothing to do with the health benefits that families like mine receive. In fact, their brochure even claimed that it was "only about marriage." But as soon as the amendment passed, it became a weapon to take away the health insurance upon which many families—including my own—rely.

. . . I am concerned that the Federal Marriage Amendment, which is very similar to Michigan's amendment, will be used to deny equal benefits nationwide. . . . Ironically, the same people who promoted this amendment favor policies that permit or encourage one parent to stay at home with children. But under the false pretense of protecting marriage, this law might force us either to move away from grandparents and family or to deprive our kids of precious time with their parents. No one has been able to explain to me how even one marriage is protected by this unfair, discriminatory law.

I have also heard that the Michigan amendment—and the [proposed] federal amendment . . . is necessary to "protect" marriage as a sacred institution. . . . As an American with great respect for our Constitution, I don't understand why federal law should play a role in defining for the various religions which marriages are "sacred." . . . [N]o religious denomination can ever be forced to perform marriages that don't meet its standards. For instance, rabbis cannot be compelled to perform interfaith marriages even though the laws of every state allow them.

. . . Finally, I have heard that marriage must be "protected" from families like mine for the good of children. As a pediatrician, I know that this is completely unsupported by any scientific fact. Every piece of creditable medical evidence I can find, every study, indicates that children with lesbian and gay parents do just as well as their peers. Every major medical, psychiatric and psychological association that has issued an opinion on the subject endorses increasing, not removing, legal protection of gay and lesbian families. Their endorsement is based on a commitment to protecting the health and welfare of children and their families. In short . . . it is of the utmost importance to extend, not to remove, legal protection to the children and to both parents in gay and lesbian families.

. . . I don't know what harm . . . a constitutional amendment might cause. I fear that families like mine, with young children, will lose health benefits; will be denied common decencies like hospital visitation when tragedy strikes; will lack the ability to provide support for one another in old-age. I fear that my loving, innocent children will face hatred and insults implicitly sanctioned by a law that brands their family as unequal. I know that these sweet children have already been shunned and excluded by people claiming to represent values of decency and compassion. I also know what such an amendment will not do. It will not help couples who are struggling to stay married. It will not assist any impoverished families struggling to make ends meet or to obtain health care for sick children. It will not keep children with their parents when their parents see divorce as their only option. It will not help any single American citizen to live life with more decency, compassion or morality. In the coming months and debates, I urge you to consider both the medical evidence and the experiences of families like mine with an open heart and an open mind. . . . And I pray, and my family prays, that in dealing with our precious Constitution, you will follow the dictates of the oath that binds my profession: first, do no harm.

Source: Testimony from Dr. Kathleen Moltz, April 2005, U.S. Senate Judiciary Committee.

that would define marriage as being between a man and a woman, thus barring marriages between gay and lesbian couples.

Discrimination in Child Custody and Visitation

Several respected national organizations—including the Child Welfare League of America, the American Psychological Association, the American Psychiatric Association, and the National Association of Social Workers—have taken the position that a parent's sexual orientation is irrelevant in determining child custody (Landis 1999). In a review of research on family relationships of lesbians and gay men, Patterson (2001) concluded that "the greater majority of children with lesbian or gay parents grow up to identify themselves as heterosexual" and that "concerns about possible difficulties in personal development among children of lesbian and gay parents have not been sustained by the results of research" (p. 279). Patterson (2001) notes that the "home environments provided by lesbian and gay parents are just as likely as those provided by heterosexual parents to enable psychosocial growth among family members" (p. 283). Nevertheless, some court judges are biased against lesbian and gay parents in custody and visitation disputes. For example, in 1999 the Mississippi Supreme Court denied custody of a teenage boy to his gay father and instead awarded custody to his heterosexual mother who remarried into a home "wracked with domestic violence and excessive drinking" (*Custody and Visitation,* 2000, p. 1).

Discrimination in Adoption and Foster Care

A 2003 poll of U.S. adults reveals that Americans are almost evenly split on the issue of the right of gay couples to adopt children: 49% support allowing gay couples to adopt and 48% oppose it (Page 2003). Gay and lesbian people can adopt children in at least 22 states and the District of Columbia; however, Florida and Mississippi forbid adoption by gay and lesbian people, Utah forbids adoption by any unmarried couple (which includes all same-sex couples), and in Arkansas lesbians and gay men are prohibited from serving as foster parents (National Gay and Lesbian Task Force 2004).

Most adoptions by gay people have been by individual gay men or lesbians who adopt the biological children of their partners. Such second-parent or stepparent adoptions ensure that the children can enjoy the benefits of having two legal parents, especially if one parent dies or becomes incapacitated. However, in four states (Colorado, Nebraska, Ohio, and Wisconsin) court rulings have decided that the state adoption law does not allow for second-parent or stepparent adoption by members of same-sex couples, and in 22 states it is unclear whether the state adoption law permits second-parent adoption (National Gay and Lesbian Task Force 2005d).

Hate Crimes Against Sexual Orientation Minorities

On October 6, 1998, Matthew Shepard, a 21-year-old student at the University of Wyoming, was abducted and brutally beaten. He was found tied to a wooden ranch fence by two motorcyclists who had initially thought he was a scarecrow. His skull had been smashed, and his head and face had been slashed. The only apparent reason for the attack: Matthew Shepard was gay. On October 12 Shepard died of his injuries. Media coverage of his brutal attack and subsequent death focused nationwide attention on hate crimes against sexual orientation minorities.

Other incidents of brutal hate crimes against gays include the murder of Billy Jack Gaither in Alabama in February 1999. Two men who claimed to be angry over a sexual advance made by Gaither plotted his murder, beat Gaither to death with an ax handle, and then burned his body on a pyre of old tires. In July 1999 at Fort Campbell, Kentucky, Private First Class Barry Winchell was beaten to death by another soldier with a baseball bat because Winchell was perceived to be homosexual.

In eighteenth-century America, when laws against homosexuality often included the death penalty, violence against gays and lesbians was widespread and included beatings, burnings, various kinds of torture, and execution (Button et al. 1997). Although such treatment of sexual orientation minorities is no longer legally condoned, gays, lesbians, and bisexuals continue to be victimized by hate crimes. Anti-LGBT hate crimes are crimes against individuals or their property that are based on bias against the victim because of his or her perceived sexual orientation or gender identity. Such crimes include verbal threats and intimidation, vandalism, sexual assault and rape, physical assault, and murder.

According to the FBI Uniform Crime Reports, there were 1,479 victims of sexual orientation hate crimes in 2003, representing 16.3% of all hate crime victims reported to the FBI (FBI 2004). But as discussed in Chapter 9, FBI hate crime statistics underestimate the incidence of hate crimes. The National Coalition of Anti-Violence Programs (2005) reported that there were 2,131 victims of antigay hate crimes in 2004—a 7% increase from 2003—and 20 anti-LGBT murders. And because it is not uncommon for heterosexual men and women to be mistaken for gay men and lesbians, nearly 1 in 10 victims (9%) of anti-LGBT violence in 2004 identified as being heterosexual.

Antigay Hate in Schools and on Campuses

America's schools are not safe places for gay, lesbian, and bisexual youth. More than two-thirds (69%) of gay and lesbian students have been verbally, physically, or sexually harassed at school (Chase 2000). In a survey of 904 LGBT youth (grades 6–12) from 48 states, Kosciw and Cullen (2002) found that homophobic language is pervasive in U.S. schools and that harassment is a common experience for LGBT youth. Other findings from this study include the following (Kosciw & Cullen 2002):

- More than 80% of students had heard antigay words such as *faggot* or *dyke* frequently or often, and more than 90% had heard the phrase "That's so gay" or "You're so gay." Almost one-fourth of the youth (23.6%) reported hearing homophobic remarks from faculty or school staff. The frequency of homophobic remarks was higher than that of sexist and racist remarks.
- More than one-third of the youth reported experiencing physical harassment in the past year (being shoved, pushed) because of their sexual orientation; nearly 10% reported that physical harassment occurred frequently.
- More than 20% of the youth reported having been physically assaulted (being punched, kicked, injured with a weapon) in the past year because of their sexual orientation.
- Nearly one-third of youth (30.8%) had missed a day of school in the past month because they felt unsafe.

Given the harsh treatment of LGBT youth in school settings, it is not surprising that 40% of gay youth report that their schoolwork is negatively affected by conflicts over their sexual orientation (Gay, Lesbian, and Straight Education Network 2000). More than one-fourth of gay youth drop out of school—usually to escape the harassment, violence, and alienation they endure there (Chase 2000). Lesbian, gay, and bisexual youth who report high levels of victimization at school also have higher levels of substance use, suicidal thoughts, and sexual risk behaviors than heterosexual peers who report high levels of at-school victimization (Bontempo & D'Augelli 2002). A survey of youths' risk behavior conducted by the Massachusetts Department of Education in 1999 found that 30% of gay teens attempted suicide in the previous year, compared to 7% of their straight peers (reported in Platt 2001).

Antigay hate is also common among college students (see this chapter's *Social Problems Research Up Close* feature). In a survey of 484 young adults at six community colleges in California, 10% reported physically assaulting or threatening people whom they believed to be homosexual and 24% reported calling homosexuals insulting names (Franklin 2000). The researcher concluded from the findings of the study that, overall, many young adults believe that antigay harassment and violence is socially acceptable.

Police Mistreatment of Sexual Orientation Minorities

One reason that LGBT victims of hate crimes do not report these crimes to the police is the perception that the police will not be helpful and may even inflict further abuse. A recent report by Amnesty International (2005) found that police mistreatment and abuse of LGBT people are widespread in the United States. The report focused on 4 cities (Chicago, Los Angeles, New York, and San Antonio), surveyed the 50 largest police departments in the country, and included hundreds of interviews and testimonies.

Types of police mistreatment revealed in the report include targeted and discriminatory enforcement of laws against LGBT people, verbal abuse, inappropriate pat-down and strip searches, failure to protect LGBT people in holding cells, inappropriate response or failure to respond to hate crimes or domestic violence calls, sexual harassment and abuse, and physical abuse. In one case a 31-year-old gay African American man who was arrested by Chicago police following an altercation with his landlord alleged that officers handcuffed him and raped him with a billy club that had been dipped in cleaning fluid. The victim received an out-of-court settlement of $20,000.

Some police departments have taken initiatives to improve their practices regarding the treatment of LGBT individuals. For example, Washington, D.C., created the Gay and Lesbian Liaison Unit, and many police departments provide training on hate crimes against LGBT individuals. However, 28% of police departments in the Amnesty International (2005) study do not provide any training in LGBT issues.

Effects of Homophobia, Heterosexism, and Antigay Discrimination on Heterosexuals

The homophobic and heterosexist social climate of our society is often viewed in terms of how it victimizes the gay population. However, heterosexuals are also victimized by homophobia, heterosexism, and antigay discrimination. We touch on some of these effects in the following list.

1. *Restriction of male gender expression.* Because of the antigay climate, heterosexuals, especially males, are hindered in their own self-expression and

"Injustice anywhere is a threat to justice everywhere. We are caught in an inescapable network of mutuality, tied in a single garment of destiny. Whatever affects one directly affects all indirectly.**"**

Martin Luther King Jr.
Civil rights leader

intimacy in same-sex relationships. "The threat of victimization (i.e., antigay violence) probably also causes many heterosexuals to conform to gender roles and to restrict their expressions of (nonsexual) physical affection for members of their own sex" (Garnets et al. 1990, p. 380). Homophobic epithets frighten youth who do not conform to gender role expectations, leading some youth to avoid activities that they might otherwise enjoy and benefit from (e.g., arts for boys, athletics for girls) (Gay, Lesbian, and Straight Education Network 2000).

2. *Dysfunctional sexual behavior.* Some cases of rape and sexual assault are related to homophobia and compulsory heterosexuality. For example, college men who participate in gang rape, also known as "pulling train," entice each other into the act "by implying that those who do not participate are unmanly or homosexual" (Sanday 1995, p. 399). Homonegativity also encourages early sexual activity among adolescent men. Adolescent male virgins are often teased by their male peers, who say things like "You mean you don't do it with girls yet? What are you, a fag or something?" Not wanting to be labeled and stigmatized as a "fag," some adolescent boys "prove" their heterosexuality by having sex with girls.

3. *Loss of rights for individuals in unmarried relationships.* Homophobia has contributed to the passage of state constitutional amendments that prohibit same-sex marriage, which can also result in denying rights and protections to opposite-sex unmarried couples. For example, in 2005 Judge Stuart Friedman of Cuyahoga County (Ohio) agreed that a man who was charged with assaulting his girlfriend could not be charged with a domestic violence felony because the Ohio state constitutional amendment granted no such protections to unmarried couples (Human Rights Campaign 2005b). As we mentioned earlier, some antigay marriage measures also threaten the provision of domestic partnership benefits to unmarried heterosexual couples.

4. *Heterosexual victims of hate crimes.* As we discussed earlier in this chapter, extreme homophobia contributes to instances of violence against homosexuals—acts known as hate crimes. But hate crimes are crimes of perception, meaning that victims of antigay hate crimes may not be homosexual; they may just be perceived as being homosexual. In 2004 the National Coalition of Anti-Violence Programs (2005) reported that in the United States, 192 heterosexual individuals were victims of antigay hate crimes—9% of all antigay hate crime victims.

5. *Fear and grief.* Many heterosexual family members and friends of homosexuals live with the fear that their lesbian or gay friend or family member could be victimized by antigay prejudice and discrimination. They may also be afraid of being mistreated themselves simply for having a gay or lesbian friend or family member. Youth with gay and lesbian family members are often taunted by their peers. And finally, imagine the grief experienced by the family and friends of victims of antigay hate, such as the mother of Matthew Shepard.

6. *School shootings.* Antigay harassment has also been a factor in many of the school shootings in recent years. In March 2001, 15-year-old Charles Andrew Williams fired more than 30 rounds in a San Diego suburban high school, killing 2 and injuring 13 others. A woman who knew Williams reported that the students had teased Williams and called him gay (Dozetos 2001a). According to the Gay, Lesbian, and Straight Education Network (GLSEN), Williams's story is not unusual. Referring to a study of harassment of U.S. students that was commissioned by the American Association of University Women, a GLSEN report concluded, "For boys, no other type of harassment provoked as

Campus Climate for GLBT People: A National Perspective

Since the mid-1980s, numerous studies have documented the hostile climate that GLBT students, staff, faculty, and administrators experience on college campuses. A recent survey by the National Gay and Lesbian Task Force Policy Institute (Rankin 2003) is a significant contribution to the research literature in this area.

Sample and Methods

Because of the difficulty in identifying lesbian, gay, bisexual, and transgender individuals, Rankin (2003) used purposeful sampling of GLBT individuals and snowball sampling. Contacts were made with "out" GLBT individuals on campus, and they were asked to share the survey with other members of the GLBT community who were not so open about their sexual/gender identity.

Surveys, both paper-and-pencil and online versions, were administered to students, faculty, staff, and administrators at 14 colleges and universities from around the country. The surveys, which contained 35 questions and an additional space for re-

spondents to provide commentary, were designed to collect data about (1) respondents' personal campus experiences as members of the GLBT community, (2) their perception of the climate for GLBT members of the academic community, and (3) their perceptions of institutional actions, including administrative policies and academic initiatives regarding GLBT issues and concerns on campus. A total of 1,669 usable surveys were returned, representing the following groups:

- 1,000 students (undergraduate and graduate), 150 faculty, and 467 staff/administrators
- 720 men, 848 women
- 326 people of color
- 572 gay people (mostly male), 458 lesbians, 334 bisexuals, and 68 transgender individuals
- 825 "closeted" people

Findings and Discussion

Some of the findings of the Campus Climate Assessment are discussed in the following sections.

LIVED OPPRESSIVE EXPERIENCES

Nineteen percent of the respondents reported that within the past year they had feared for their physical safety because of their sexual orientation or gender identity, and 51% had concealed their sexual or gender identity to avoid intimidation. GLBT people of color were more likely than white GLBT people to conceal their sexual or gender identity to avoid harassment. Many nonwhite respondents reported feeling out of place at predominantly white GLBT settings. In the words of one student, "As a chicana, I felt ostracized even more. Forget about feeling a sense of community when you're a member of two minority groups" (Rankin 2003, p. 25). The study also found that 27% of faculty, staff, and administrators and 40% of students indicated that they had concealed their sexual identity to avoid discrimination in the past year.

Respondents were also asked if they had experienced harassment in the past year. Harassment was defined as "conduct that has interfered unreasonably with your

strong a reaction on average; boys in this study would be less upset about physical abuse than they would be if someone called them gay" (Dozetos 2001a).

Strategies for Action: Reducing Antigay Prejudice and Discrimination

"Resistance begins with people confronting pain, whether it's theirs or somebody else's, and wanting to do something to change it."

bell hooks
Author

Many of the efforts to change policies and attitudes regarding sexual orientation minorities have been spearheaded by organizations such as the Human Rights Campaign (HRC), the National Gay and Lesbian Task Force (NGLTF), the Gay and Lesbian Alliance Against Defamation (GLAAD), the International Lesbian and Gay Association (ILGA), the Lambda Legal Defense and Education Fund, the National Center for Lesbian Rights, and the Gay, Lesbian, and Straight Education Network

ability to work or learn on this campus or has created an offensive, hostile, intimidating working or learning environment." Undergraduate students were the most likely to have experienced harassment (36%), followed by faculty (27%), graduate students (23%), and staff (19%). Derogatory remarks were the most common form of harassment (89%). Other types of harassment included verbal harassment or threats, anti-GLBT graffiti, threats of physical violence, denial of services, and physical assault.

PERCEPTIONS OF ANTI-GLBT OPPRESSION ON CAMPUS

About three-fourths of faculty, students, administrators, and staff rated the campus climate as homophobic. In contrast, most respondents rated the campus generally (not specific to GLBT people) as friendly, concerned, and respectful. Thus "even though respondents feel that the overall campus climate is hospitable, heterosexism and homophobia are still prevalent" (Rankin 2003, p. 31).

INSTITUTIONAL ACTIONS

Participants were asked to respond to several questions about institutional actions regarding GLBT concerns on campus. Less than half (37%) agreed that "the college/university thoroughly addresses campus issues related to sexual orientation/gender identity" and only 22% agreed that "the curriculum adequately represents the contributions of GLBT persons." However, the majority of respondents agreed that their classrooms or their job sites were accepting of GLBT persons (63%) and that the college or university provided resources on GLBT issues and concerns (71%). These positive responses may be due in part to the inclusion of sexual orientation in the nondiscrimination policies of all but one of the participating colleges and universities. Several of the campuses also provided domestic partner benefits, had GLBT resource centers, and offered safe-space programs.

These findings indicate that intolerance and harassment of GLBT students, staff,

faculty, and administrators continue to be prevalent on U.S. campuses. It is important to note that the colleges and universities that participated in this study are not representative of most institutions of higher learning in the United States and "may be among the most gay-friendly campuses in the country" (Rankin 2003, p. 3). Rankin explains that "all of the institutions who participated in this survey had a visible GLBT presence on campus, including, in most cases, a GLBT campus center. Most had sexual orientation nondiscrimination policies. As only 100 of the 5,500 U.S. colleges and universities have GLBT student centers, the 14 universities surveyed here are not representative of most institutions of higher education in the U.S." (p. 3). Thus, Rankin suggests that the findings of this study "may significantly understate the problems facing GLBT students and staff at U.S. colleges and universities" (p. 3).

(GLSEN). But the effort to reduce antigay prejudice and discrimination is not just a "gay agenda"; many heterosexuals and mainstream organizations (such as the National Education Association) have worked on this agenda as well.

Many of the advancements in gay rights have been the result of political action and legislation. Barney Frank (1997), an openly gay U.S. representative, emphasized the importance of political participation in influencing social outcomes. He noted that demonstrative and cultural expressions of gay activism, such as "gay pride" celebrations, marches, demonstrations, or other cultural activities promoting gay rights, are important in organizing gay activists. However, he warned:

> Too many people have seen the cultural activity as a substitute for democratic political participation. In too many cases over the past decades we have left the political arena to our most dedicated opponents [of gay rights], whose letter writing, phone calling, and lobbying have easily triumphed over our marching, demonstrating, and dancing.

The most important lesson . . . for people who want to make America a fairer place is that politics—conventional, boring, but essential politics—will ultimately have a major impact on the extent to which we can rid our lives of prejudice. (Frank 1997, p. xi)

Next, we look at efforts to reduce employment discrimination against sexual orientation minorities, provide recognition and support to lesbian and gay families, and include sexual orientation in hate crime legislation. We also provide an overview of educational policies and programs designed to reduce intolerance and support nonheterosexual students in schools and on campuses.

Reducing Employment Discrimination Against Sexual Orientation Minorities

Most Americans are in favor of equal rights for homosexuals in the workplace (Saad 2005). Yet, as of this writing, federal law protects individuals from discrimination only on the basis of race, religion, national origin, sex, age, and disability. A bill called the **Employment Nondiscrimination Act (ENDA)** would make it illegal to discriminate on the basis of sexual orientation. ENDA would not apply to religious organizations, businesses with fewer than 15 employees, or the military. ENDA was introduced in Congress in 1994. As of this writing, each time the bill has come up before Congress, it has not passed. In its current form, ENDA, if passed, would *not* protect transgender employees against discrimination based on gender expression or identity.

With the absence of federal legislation prohibiting antigay discrimination, some state and local governments and some private employers have taken measures to prohibit employment discrimination based on sexual orientation. The scope of these measures varies, from prohibiting discrimination only in public employment to comprehensive protection against discrimination in public and private employment, education, housing, public accommodations, credit, and union practices.

Local and State Bans on Antigay Employment Discrimination.   In 1974 Minneapolis became the first municipality to ban antigay job discrimination. By early 2005, 157 cities and counties prohibited sexual orientation discrimination in the private workplace (Human Rights Campaign 2005b). In 1982 Wisconsin became the first state to prohibit antigay employment discrimination. By early 2005, 16 states (and the District of Columbia) had laws banning sexual orientation discrimination in the workplace (both public and private sectors), and nearly half (47%) of the U.S. population was covered by either state or local sexual orientation nondiscrimination laws, up from 34% in 1995 (Cahill 2005) (see Table 11.2 and Figure 11.1).

Nondiscrimination Policies in the Workplace.   In 1975 AT&T became the first employer to add sexual orientation to its nondiscrimination policy. By the end of 2004, 2,867 employers, including more than 80% of Fortune 500 companies, included sexual orientation in their equal employment opportunity or nondiscrimination policies (Human Rights Campaign 2005b). Of the top 50 U.S. four-year colleges and universities, only one—the University of Notre Dame—does not include sexual orientation in its nondiscrimination policies.

Effects of Nondiscrimination Laws and Policies.   A survey of 126 U.S. cities and counties that had implemented laws or policies prohibiting discrimination on the basis of sexual orientation found that such laws and policies sometimes had

Table 11.2

States with Laws[a] Banning Discrimination Based on Sexual Orientation

- California (1992)
- Connecticut (1991)
- Hawaii (1991)
- Illinois (2005)
- Maine (2005)
- Maryland (2001)
- Massachusetts (1989)
- Minnesota (1993)
- Nevada (1999)
- New Hampshire (1997)
- New Jersey (1992)
- New Mexico (2003)
- New York (2002)
- Rhode Island (2001)
- Vermont (1992)
- Wisconsin (1982)

[a]Laws covering both public and private sector workplaces.

Source: Human Rights Campaign (2005b).

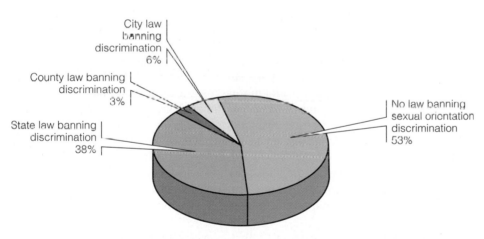

Figure 11.1

Proportion of total U.S. population covered by a sexual orientation nondiscrimination law as of January 2005.

Source: Cahill (2005).

negative effects and "resulted in divisions in the community, greater controversy or tension, or the increased mobilization of those opposed to gay rights" (Button et al. 1997, p. 120). However, the more commonly cited effects were positive and included the following:

- Reduced discrimination against lesbians and gays in public employment and other institutions covered by the law.

- Increased feelings of security, comfort, and acceptance by lesbians and gay men, resulting in lesbians and gay men being more likely to "come out of the closet"—to reveal their sexual orientation to others (or to at least stop hiding it).
- An increased sense of legitimacy among gays and lesbians that helped them overcome "internalized homophobia."

Providing Legal Recognition and Support to Lesbian and Gay Couples

In 2003 the Massachusetts Supreme Court ruled (in *Goodridge* v. *Department of Public Health*) that denying same-sex couples the protections, benefits, and obligations of civil marriage violated the basic premises of individual liberty and equality in the Massachusetts Constitution (Cahill & Slater 2004). As of this writing, no other state has affirmed the right of same-sex couples to be legally married. Although the California state legislature approved a bill in 2005 that would have allowed same-sex couples to marry, Governor Arnold Schwarzenegger vetoed the bill.

Some states have passed legislation to allow same-sex couples legal status that entitles them to many of the same rights and responsibilities as married heterosexual couples (see Table 11.3). For example, in Vermont and Connecticut same-sex couples can apply for a civil union license. A **civil union** is a legal status parallel to civil marriage under state law. Civil union status entitles same-sex couples to all the rights and responsibilities available under state law to married couples. Unlike marriage for heterosexual couples, the rights of partners in same-sex civil unions are not recognized by federal law, so they do not have the more than 1,000 federal protections that go along with civil marriage, nor is their legal status recognized in other states.

Some states, counties, cities, and workplaces allow unmarried couples, including gay couples, to register as **domestic partners.** The rights and responsibilities granted to domestic partners vary from place to place but may include coverage under a partner's health and pension plan, rights of inheritance and community property, tax benefits, access to married student housing, child custody and child and spousal support obligations, and mutual responsibility for debts.

In 1991 the Lotus Development Corporation became the first major American firm to extend domestic partner recognition to gay and lesbian employees. By the end of 2004 the Human Rights Campaign (2005b) identified 8,250 employers that provided domestic partner health insurance benefits to their employees—an increase of 13% from the previous year. However, even when companies offer domestic partner benefits to same-sex partners of employees, these benefits are usually taxed as income by the federal government, whereas spousal benefits are not.

In 10 states and the District of Colombia as well as in several dozen municipalities, same-sex partners of public employees are granted domestic partner benefits. Three states—Hawaii, New Jersey, and California—have enacted laws that provide varying degrees of protection for domestic partners. The California law (the Domestic Partner Rights and Responsibilities Act of 2003) is the most comprehensive, granting same-sex domestic partners nearly all the rights and responsibilities granted to married couples in the state (Cahill & Slater 2004).

Another attempt to provide equal benefits to same-sex couples is the Uniting American Families Act—a bill introduced in Congress in 2005 that would enable foreign same-sex partners of U.S. citizens or permanent residents to seek

Table 11.3

States That Recognize Same-Sex Relationships

State	Same-Sex Relationship Recognition
California	California expanded its domestic partner registry, effective January 2005, to confer almost all the state-level spousal rights and responsibilities on registered couples.
Connecticut	In April 2005 Governor Jodi Rell signed a bill allowing same-sex couples to enter into civil unions that offer almost all the benefits of marriage under state law. Connecticut is the first state to establish civil unions voluntarily, without having been ordered to do so by a court.
Hawaii	Hawaii offers "reciprocal beneficiary" status to same-sex couples that provides eligible registered couples with certain rights and obligations associated with survivorship, inheritance, property ownership, and insurance.
Maine	In 2004 the Maine legislature passed a bill that established a domestic partner registry and provides registered couples with inheritance rights, next-of-kin status, victim's compensation, and priority in guardian and conservator rights.
Massachusetts	Massachusetts began offering civil union marriage licenses to same-sex couples in May 2004.
New Jersey	New Jersey began offering "domestic partner" status to same-sex and some opposite-sex couples in July 2004. The status allows the partner to be treated as a dependent for the purposes of administering certain retirement and health benefits.
Vermont	Vermont enacted a civil union law in 2000 that entitles same-sex couples to the more than 300 state-level rights and responsibilities extended to opposite-sex spouses.

Source: Human Rights Campaign (2005b).

immigration status. As noted earlier, 17 other countries throughout the world recognize same-sex couples for immigration purposes.

Protecting Gay and Lesbian Parental Rights

As noted earlier, 2000 census data revealed that one-third (33%) of female same-sex householders and 22% of male same-sex householders were living with their children (under 18 years of age) (Simmons & O'Connell 2003). A number of policies and rulings reflect the increasing provision of parental rights and responsibilities to gay and lesbian parents and co-parents.

- In an unprecedented ruling the California Board of Equalization granted head-of-household tax status to a nonbiological lesbian co-parent ("NCLR Wins Equal Tax Benefits for Nonbiological Lesbian Mother," 2000). This is the first ruling by any state tax board that provides equitable tax status for gay and lesbian families.

ThomsonNOW™

Learn more about **Protecting Gay and Lesbian Parental Rights** by going through the Diversity among Families Animation.

- A Massachusetts court ruled that two women may be listed as "mother" on a birth certificate when one of the women donated the egg for the child and the other carried the child (LAWbriefs 2000).
- A Pennsylvania court ordered a nonbiological, nonadoptive parent to pay child support for the five children she jointly parented with her former partner (LAWbriefs 2003).
- In 2005 the Supreme Court of Pennsylvania overturned a lower court decision that found it to be in the best interest of a child not to have any contact with her nonbiological lesbian mother because the child's biological mother had alienated the child from her. The court noted that a heterosexual parent who alienated the child would not be rewarded for such behavior and that "this scenario is equally applicable" (LAWbriefs 2005a).
- The West Virginia Supreme Court awarded a woman custody of the child she and her deceased lesbian partner raised together. The deceased partner's parents had fought for custody of the child, and, after being denied by a trial court, an appellate court reversed the decision and awarded custody to the grandparents. The supreme court overruled the appellate court decision, recognizing the parental rights of the woman who parented the child for his entire life (LAWbriefs 2005b).
- In Tennessee Christy Berry had primary custody of her young son after her divorce from the child's father. The father sought a change in custody, claiming that Ms. Berry's sexual orientation would harm their child. A trial court granted the father's request for custody, but an appeals court reversed the decision, noting that the boy was doing well with his mother and that there was no evidence her sexual orientation had any adverse effect on him (LAWbriefs 2005b).

Antigay Hate Crimes Legislation

Hate crime laws call for tougher sentencing when prosecutors can prove that the crime committed was a hate crime. As of May 2005, 32 states and the District of Columbia had hate crime laws that include sexual orientation, 13 states had hate crime laws that did not include sexual orientation, 1 state had hate crime laws that addressed crimes motivated by bias or prejudice but did not list categories, and 4 states had no hate crime laws (National Gay and Lesbian Task Force 2005b).

At the federal level, the hate crimes law passed in 1968 protects citizens against hate crimes based on race, religion, and national origin. For years gay rights advocates lobbied Congress to add federal hate crime law protections based on sexual orientation, gender, gender identity, and disability. In 2005 the Local Law Enforcement Hate Crimes Prevention Act was finally passed, adding sexual orientation, gender, gender identity, and disability to the federal hate crimes law. Under this new law states can receive federal grants and assistance to help them prosecute hate crimes against individuals based on sexual orientation as well as the other covered categories.

Educational Strategies: Policies and Programs in the Public Schools

If schools are to promote the health and well-being of all students, they must address the needs of gay, lesbian, and bisexual youth and promote acceptance of sexual orientation diversity within the school setting. In a national survey by the Kaiser Family Foundation (2000), 76% of parents of youths in grades 7–12 indicated that sex education should include discussion of homosexuality, yet only 41% of students said that "homosexuality" was covered in sex education class.

One strategy for promoting tolerance for diversity among students involves establishing and enforcing rules that prohibit antigay behavior in the schools. But most U.S. K–12 students (more than 75%) go to schools with no legal protections against anti-LGBT harassment and bullying. Only eight states and the District of Columbia have statewide legal protections for students based on sexual orientation (Snorton 2005). Seven states (Alabama, Arizona, Mississippi, Oklahoma, South Carolina, Texas, and Utah) have laws that stigmatize LGBT people by specifically prohibiting any positive portrayal of LGBT issues or people in schools. In the absence of state laws prohibiting antigay harassment in schools, some schools have their own policies against such behavior. Schools that do not protect students against harassment can face legal challenges. A number of court rulings have held school districts responsible for failing to protect LGBT students from discrimination, violence, and harassment and have ordered school districts to pay between $40,000 and $1 million in damages (Cianciotto & Cahill 2003).

One resource for creating a "harassment-free" climate is the Gay, Lesbian, and Straight Education Network (GLSEN)—the leading national organization fighting against harassment and discrimination in K–12 schools. GLSEN conducts training for school staffs around the country and has developed the faculty training program of the Massachusetts Department of Education: the Safe Schools for Gay and Lesbian Students program, the first statewide effort aimed at ending homophobia in schools (Gay, Lesbian, and Straight Education Network 2000). The National Education Association has also implemented national training programs to educate teachers in every state about the role they can and must play to stop antigay harassment in their schools (Chase 2000).

Some schools have established school-based support groups for LGBT students. Such groups can help students increase their self-esteem and overcome their sense of isolation, provide information and resources, and serve as a resource for parents. School counselors can be trained to work with gay, lesbian, and bisexual youth and their parents. Education about sexual orientation can be implemented in sex education or health education classes or in conflict resolution or diversity curricula. In-service training for teachers and other staff is important and can include examining the effects of antigay bias, dispelling myths about homosexuality, and brainstorming on ways to create a more inclusive environment (Mathison 1998).

Gay-straight alliances (GSAs) are school-sponsored clubs for gay teens and their straight peers. The first GSA was established in a Los Angeles high school in 1984, and today there are more than 1,000 GSAs in U.S. schools (Kosciw & Cullen 2002). However, most public schools offer little or no support and education regarding sexual orientation diversity. Most schools have no support groups or special counseling services for gay and lesbian youth.

Campus Policies and Programs Dealing with Sexual Orientation

D'Emilio (1990) suggests that colleges and universities have the ability and the responsibility to promote gay rights and social acceptance of homosexual people:

> For reasons that I cannot quite fathom, I still expect the academy to embrace higher standards of civility, decency, and justice than the society around it. Having been granted the extraordinary privilege of thinking critically as a way of life, we should be

Logos of Safe Zone programs at four universities and colleges. From left to right: Pennsylvania State University, University of North Carolina at Chapel Hill, Purdue University, and Cornell College.

(from left to right) Courtesy of Pennsylvania State University, Courtesy of University of North Carolina at Chapel Hill, Courtesy of Purdue University, Courtesy of Cornell College

astute enough to recognize when a group of people is being systematically mistreated. We have the intelligence to devise solutions to problems that appear in our community. I expect us also to have the courage to lead rather than follow. (p. 18)

Student groups have been active in the gay liberation movement since the 1960s. Because of the activism of students and the faculty and administrators who support them, nearly 400 U.S. colleges and universities have nondiscrimination policies that include sexual orientation (Singh & Wathington 2003). Other measures to support the LGBT college student population include gay and lesbian studies programs, social centers, and support groups, as well as campus events and activities that celebrate diversity. Many campuses have Safe Zone or Ally programs designed to visibly identify students, staff, and faculty who support the LGBT population. The Safe Zone and Ally programs may require a training session that provides a foundation of knowledge needed to be an effective ally to LGBT students and those questioning their sexuality. Participants in Safe Zone or Ally programs display some type of sign or placard outside their office or residence hall room that identifies them as individuals who are willing to provide a safe haven, a listening ear, and support for LGBT people and those struggling with sexual orientation issues (Safe Zone Resources 2003).

Understanding Issues in Sexual Orientation

Recent years have witnessed a growing acceptance of lesbians, gay men, and bisexuals as well as increased legal protection and recognition of these marginalized populations. The advancements in gay rights are notable and include the Supreme

Court's 2003 decriminalization of sodomy, the growing adoption of nondiscrimination policies covering sexual orientation, the increased recognition of domestic partnerships, the state civil union rights and responsibilities offered to same-sex couples in Vermont, Connecticut, and California, the Massachusetts ruling that same-sex marriage is allowable under the state constitution, and the election of the first openly gay bishop in the New Hampshire Episcopal diocese. But the winning of these battles in no way signifies that the war is over. Gay, lesbian, and bisexual individuals continue to be frequent victims of discrimination, harassment, and violence. In some countries homosexuality is formally condemned, with penalties ranging from fines to imprisonment and even death.

As both structural functionalists and conflict theorists note, nonheterosexuality challenges traditional definitions of family, child rearing, and gender roles. Every victory in achieving legal protection and social recognition for sexual orientation minorities fuels the backlash against them by groups who are determined to maintain traditional notions of family and gender. Often, this determination is rooted in and derives its strength from uncompromising religious ideology.

As symbolic interactionists note, meanings associated with homosexuality are learned. Powerful individuals and groups opposed to gay rights focus their efforts on maintaining the negative meanings of homosexuality to keep the gay, lesbian, and bisexual population marginalized. Since taking office, President George W. Bush has issued more than 250 proclamations, including those honoring National School Lunch Week and Wright Brothers Day, but he has refused to recognize Gay Pride Month with an official proclamation (Tobias & Cahill 2003). Bush did issue a presidential proclamation for Marriage Protection Week as part of an effort to define marriage as being between a woman and a man, thus denying marriage rights to same-sex couples. Another direct attack on gay rights is the proposal of the Marriage Protection Amendment, which would amend the Constitution to define marriage as being between one woman and one man. New York senator Hillary Rodham Clinton notes that the push for the amendment represents the first time an attempt has been made to amend the Constitution to specifically deny rights to any group of individuals. She also suggests that the current campaign against gay rights is a political strategy to drive wedges between Americans and to divert the public's attention away from social problems. "They'd rather talk about taking away rights and undermining the ability of Americans to live their own lives, to have their own families," said Clinton, "than to talk about the miserable economy, to talk about their miserable foreign policy, to talk about their rollback of environmental laws and workers' rights, education and health care!" (quoted in Johnston 2003).

Political efforts to undermine gay rights and recognition must contend with the fact that "gay is everywhere": 2000 census data reveal that 99.3% of U.S. counties reported same-sex cohabiting partners, compared to 52% of counties in 1990. With the gay population being more "out" than ever before, more heterosexual individuals are reporting having family, work-related, or friendship or acquaintance ties to lesbian and gay individuals. This increased contact, as well as gay-straight alliances in schools and more frequent portrayals of lesbians and gays in the media, have inspired many heterosexually identified individuals to support the gay rights movement.

Referring to politicians who support a federal marriage amendment that would deny same-sex couples the right to marry, Senator Hillary Rodham Clinton suggests that politicians who support the amendment are driving a wedge between Americans and diverting the public's attention away from social problems. "They'd rather talk about taking away rights and undermining the ability of Americans to live their own lives, to have their own families," said Clinton, "than to talk about the miserable economy, to talk about their miserable foreign policy, to talk about their rollback of environmental laws and workers' rights, education and health care!"

AP/Wide World Photos

But, as one scholar noted, "The new confidence and social visibility of homosexuals in American life have by no means conquered homophobia. Indeed it stands as the last acceptable prejudice" (Fone 2000, p. 411). True, the American public is becoming increasingly supportive of gay rights. But as Yang (1999) points out, as the antigay minority diminishes in size, "it often becomes more dedicated and impassioned" (p. ii).

Our task in the coming years is to get the heterosexual Americans who support our cause to feel as passionately outraged by the injustices we face and to be as strongly motivated to act in support of our rights as our adversaries are in their opposition to our rights. (Yang 1999, p. iii)

Chapter Review

ThomsonNOW™

Reviewing is as easy as ① ② ③

1. Before you do your final review, take the ThomsonNOW diagnostic quiz to help you identify the areas on which you should concentrate. You will find information on ThomsonNOW and instructions on how to access all of its great resources on the foldout at the beginning of the text.

2. As you review, take advantage of ThomsonNOW's study videos and interactive Map the Stats exercises to help you master the chapter topics.

3. When you are finished with your review, take ThomsonNOW's posttest to confirm you are ready to move on to the next chapter.

- **Are there any countries in the world that have national laws that ban discrimination against gays, lesbians, and bisexuals?**

Yes, more than 20 countries have national laws that ban discrimination based on sexual orientation.

- **Is there any country where same-sex couples can be legally married?**

Yes. In 2001 the Netherlands became the first country in the world to offer legal marriage to same-sex couples. Belgium, Spain, Canada, and South Africa as well as the state of Massachusetts, have also legalized same-sex marriage.

- **In what ways is the classification of individuals into sexual orientation categories problematic?**

Classifying individuals into sexual orientation categories is problematic for a number of reasons: (1) Some people conceal or falsely portray their sexual orientation identity; (2) attractions, love, behavior, and self-identity do not always match; and (3) sexual orientation can change over time.

- **What is the relationship between beliefs about what "causes" homosexuality and attitudes toward homosexuality?**

Individuals who believe that homosexuality is biologically based or inborn tend to be more accepting of homosexuality. In contrast, individuals who believe that homosexuals choose their sexual orientation are less tolerant of gays and lesbians.

- **What is the position of the American Psychiatric Association, the American Psychological Association, the American Academy of Pediatrics, and the American Medical Association on conversion or reparative therapy for gays and lesbians?**

These organizations agree that sexual orientation cannot be changed and that efforts to change sexual orientation (conversion or reparative therapy) do not work and may, in fact, be harmful.

- **What is internalized homophobia? What are the effects of internalized homophobia on lesbian and gay individuals?**

Internalized homophobia is a sense of personal failure and self-hatred among lesbian and gay individuals resulting from social rejection and stigmatization. Internalized homophobia has been linked to increased risk for depression, substance abuse and addiction, anxiety, and suicidal thoughts.

- **What factor has been identified as the best predictor of homophobia?**

Conservative Christian ideology has been identified as the best predictor of homophobia.

- **What factors help to explain the increased acceptance of homosexuality and support for gay rights in the last decade?**

Factors that help explain the increased acceptance of homosexuality and support of gay rights include the increasing levels of education among U.S. adults, the positive depiction of gays and lesbians in the popular media, and increased contact between heterosexuals and openly gay and lesbian individuals.

- **Is employment discrimination based on sexual orientation illegal in all 50 states?**

No. As of June 2005, it is legal in 34 states to discriminate against someone on the basis of sexual orientation.

- **What is the Marriage Protection Amendment?**

  The Marriage Protection Amendment, previously known as the Federal Marriage Amendment, is a proposal to amend the U.S. Constitution to define marriage as being between a man and a woman.

- **What percentage of the population is covered by sexual orientation nondiscrimination laws?**

  In early 2005 nearly half (47%) of the U.S. population was covered by either state or local sexual orientation nondiscrimination laws, up from 34% in 1995.

- **What are Safe Zone or Ally programs?**

  Safe Zone or Ally programs are designed to visibly identify students, staff, and faculty who support the LGBT population. Participants in Safe Zone or Ally programs display a sign or placard outside their office or residence hall room that identifies them as individuals who are willing to provide a safe haven and support for LGBT people and those struggling with sexual orientation issues.

Critical Thinking

1. How is the homosexual population similar to and different from other minority groups?
2. Do you think that social acceptance of homosexuality leads to the creation of laws that protect lesbians and gays? Or does the enactment of laws that protect lesbians and gays help to create more social acceptance of gays?
3. In Massachusetts, of the more than 5,000 same-sex couples who got married in the first year after a court order legalizing such marriages went into effect, 63% were lesbian couples and 36% were gay male couples (Johnston 2005). Why do you think there was a significantly higher percentage of lesbian marriages than gay male marriages?

Key Terms

biphobia	homophobia
bisexuality	homosexuality
civil union	internalized homophobia
conversion therapy	lesbigay population
Defense of Marriage Act	LGBT
domestic partners	master status
Employment	registered partnerships
Nondiscrimination	reparative therapy
Act (ENDA)	sexual orientation
GLBT	sodomy laws
heterosexism	transgendered individuals
heterosexuality	

Media Resources

The Companion Website for *Understanding Social Problems*, Fifth Edition

http://sociology.wadsworth.com/mooney_knox_schacht5e

Supplement your review of this chapter by going to the companion website to take one of the Tutorial Quizzes, use the flash cards to master key terms, and check out the many other study aids you'll find there. You'll also find special features such as *Wadsworth's Sociology Online Resources and Writing Companion*, GSS data, and Census 2000 information, data, and resources at your fingertips to help you complete that special project or do some research on your own.

> "The quality of a nation is reflected in the way it recognizes that its strength lies in its ability to integrate the wisdom of elders with the spirit and vitality of its children and youth." *Margaret Mead, Anthropologist*

12

Problems of Youth and Aging

W ho am I?" asked Nathan Grieco. "That's a question I haven't asked myself for quite some time. Many things (mostly bad) have happened in my life. There are so many it would take me two whole lifetimes to type about it" (quoted by Carpenter & Kopas 1999, p. 1).

There is no doubt that Nathan Grieco was depressed. Things had not been easy for him. His girlfriend had broken up with him, he was socially awkward and not very popular at school, and although his grades were good, he had been diagnosed with attention-deficit disorder and was on medication.

The greatest "torture" in his life, however, was the ongoing custody battle between his feuding parents. After eight years there was no end in sight for him and his two younger brothers. After several occasions of being forced to see their father, a man they described as abusive and overbearing, the most recent court order held that failure to visit their father would result in contempt-of-court charges against their mother and her possible incarceration. Having no legal rights of their own, they had few choices. Within a year of the order, at age 16, Nathan Grieco was dead; found by his mother, kneeling next to his bed, with a leather belt around his neck. The coroner ruled that there was "insufficient evidence of either a suicide or an accident" (Carpenter & Kopas 1999, p. 1).

Like too many other youths, Nathan Grieco defined his life as one of "endless torment." Interestingly, some research suggests that depression is "curvilinear" with age—that is, highest at the extremes of the age continuum (DeAngelis 1997). Depression is not the only characteristic shared by the young and the old. Both groups are often the victims of stereotyping, physical abuse, age discrimination, and poverty; both groups are also major population segments of American society. In this chapter we examine the problems and potential solutions associated with youth and aging. We begin by looking at age in a cross-cultural context.

The Global Context: Youth and Aging Around the World

The young and the old receive different treatment in different societies. Differences in the treatment of the dependent young and old have traditionally been associated with whether the country is developed or less developed. Although proportionately more elderly live in developed countries than in less-developed ones, these societies have fewer statuses for the elderly to occupy. Their positions as caretakers, home owners, employees, and producers are often usurped by those ages 18 to 64. Paradoxically, the more primitive the society, the more likely that society is to practice senilicide—the killing of the elderly. In some societies the elderly are considered a burden and are left to die or, in some cases, actually killed. For example, in Malawi the elderly are killed for their body parts, which bring a high price on the black market ("Elderly Become 'Muti' Targets," 2003).

Not all societies treat the elderly as a burden. Scandinavian countries provide government support for in-home care workers for the elderly who can no longer perform such tasks as cooking and cleaning. Eastern cultures such as Japan revere the elderly, in part, because of their presumed proximity to honored ancestors. Japan has the fastest growing proportion of the elderly of any country in the world: 26% of the population by 2015 and 30% of the population by 2030 (Kakuchi 2003).

Societies also differ in the way they treat children. In less developed societies children work as adults, marry at a young age, and pass from childhood directly to adulthood with no recognized period of adolescence. In contrast, in industrialized nations children are often expected to attend school for 12 to 16 years and, during this time, to remain financially and emotionally dependent on their families.

Because of this extended period of dependence, the United States treats "minors" and adults differently. There is a separate justice system for juveniles, and there are age limits for driving, drinking alcohol, joining the military, entering into a contract, marrying, dropping out of school, and voting. These limitations would not be tolerated if placed on individuals on the basis of sex or race. Hence **ageism**— the belief that age is associated with certain psychological, behavioral, and/or intellectual traits, at least with reference to children—is significantly more tolerated than sexism or racism in the United States.

Despite this differential treatment, people in the United States are fascinated with youth and being young. This was not always the case. The elderly were once highly valued in the United States, particularly older men who headed families and businesses. Younger men even powdered their hair, wore wigs, and dressed in a way that made them look older. It should be remembered, however, that in 1900, the average life expectancy in the United States was 47 and more than half the population was under the age of 16. Being old was rare and respected; to some it was a sign that God looked upon the individual favorably.

One theory argues that the shift from valuing the old to valuing the young took place during the transition from an agriculturally based society to an industrial one. Land, which was often owned by elders, became less important, as did their knowledge and skills about land-based economies. With industrialization, technological skills, training, and education became more important than landownership. Called **modernization theory**, this position argues that as a society becomes more technologically advanced, the position of the elderly declines (Quadagno 2005).

Youth and Aging

Age is largely socially defined. Cultural definitions of "old" and "young" vary from society to society, from time to time, and from person to person. For example, in ancient Greece or Rome, where the average life expectancy was 20 years, one was old at 18; similarly, one was old at age 30 in medieval Europe and at age 40 in the United States in 1850.

Age is also a variable that has a dramatic impact on one's life (Matras 1990):

- Age determines one's life experiences because the date of birth determines the historical period in which a person lives. Twenty years ago cell phones and palm-sized computers could not have been imagined.
- Different ages are associated with different developmental stages (physiological, psychological, and social) and abilities. Ben Franklin observed, "At 20 years of age the will reigns; at 30 the wit; at 40 judgment."
- Age defines roles and expectations of behavior. The expression "act your age" implies that some behaviors are not considered appropriate for people of certain ages.
- Age influences the social groups to which one belongs. Whether one is part of a sixth-grade class, a labor union, or a senior's bridge club depends on one's age.
- Age defines one's legal status. Sixteen-year-olds can get a driver's license, 18-year-olds can vote and get married without their parents' permission, and 65-year-olds are eligible for Social Security benefits.

Table 12.1
Percentage of U.S. Population in Three Age Groups: 1950, 2000, and 2050

Year	All Ages	Percentage Younger Than 18 Years	Percentage Age 18–64 Years	Percentage Age 65 Years and Older
1950	100.0	31.3	60.6	8.2
2000	100.0	25.7	61.9	12.4
2050	100.0	23.7	56.0	20.3

Data are for the resident population. Data for 1950 exclude Alaska and Hawaii.

Sources: U.S. Bureau of the Census (1980 [includes data for 1950], 2000a, 2000b).

Childhood, Adulthood, and Elderhood

Every society assigns different social roles to different age groups. **Age grading** is the assignment of social roles to given chronological ages (Matras 1990). Although the number of age grades varies by society, most societies make at least three distinctions: childhood, adulthood, and elderhood (see Table 12.1).

Childhood. The period of childhood in our society is from birth through age 17 and is often subdivided into infancy, childhood, and adolescence. Infancy has always been recognized as a stage of life, but the social category of childhood developed only after industrialization, urbanization, and modernization took place. Before industrialization infant mortality was high because of the lack of adequate health care and proper nutrition. Once infants could be expected to survive infancy, the concept of childhood emerged, and society began to develop norms in reference to children. In the United States child labor laws prohibit children from being used as cheap labor, educational mandates require that children attend school until the age of 16, and federal child pornography laws impose severe penalties for the sexual exploitation of children.

Adulthood. The period from age 18 through age 64 is generally subdivided into young adulthood, adulthood, and middle age. Each of these statuses involves dramatic role changes related to entering the workforce, getting married, and having children. The concept of middle age is relatively recent and developed as life expectancy extended. Some people in this phase are known as members of the **sandwich generation** because they are often emotionally and economically responsible for both their young children and their aging parents.

Elderhood. At age 65 one is likely to be considered elderly, a category that is often subdivided into the young-old, the old, and the old-old. Membership in one of these categories does not necessarily depend on chronological age. The healthy, active, independent elderly—increasing in number—are often considered to be the young-old, whereas the old-old are less healthy, less active, and more dependent.

"I'm a little older now. I've moved from the sandwich generation to the club sandwich generation."

Ellen Goodman
Syndicated columnist

Little did members of this age cohort know that they were to become the middle-aged "sandwich generation," emotionally and economically responsible for both their children and their aging parents. The stress and pressures from these caregiver roles are often part of what is called the "midlife crisis."

© Tom Miner / The Image Works

Sociological Theories of Age Inequality

Three sociological theories help to explain age inequality and the continued existence of ageism in the United States. These theories—structural functionalism, conflict theory, and symbolic interactionism—are discussed in the following sections.

Structural-Functionalist Perspective

Structural functionalism emphasizes the interdependence of society—how one part of a social system interacts with other parts to benefit the whole. From a structural-functionalist perspective the elderly must gradually relinquish their roles to younger members of society. This transition is viewed as natural and necessary to maintain the integrity of the social system. The elderly gradually withdraw as they prepare for death, and society withdraws from the elderly by segregating them in housing such as retirement villages and nursing homes. In the interim the young have learned through the educational institution how to function in the roles surrendered by the elderly. In essence, a balance in society is achieved whereby the various age groups perform their respective functions: The young go to school, adults fill occupational roles, and the elderly, with obsolete skills and knowledge, disengage. As this process continues, each new group moves up and replaces another, benefiting society and all its members.

This theory is known as **disengagement theory** (Quadagno 2005). Some researchers no longer accept this position as valid, however, given the increased number of elderly who remain active throughout life (Riley 1987). In contrast to disengagement theory, **activity theory** emphasizes that the elderly disengage in part because they are structurally segregated and isolated, not because they have a natural tendency to do so (Quadagno 2005). For those elderly who remain active, role

Despite age grading, many singers such as Stevie Nicks, Mick Jagger, and Cher continue to perform and are commercially successful. Tina Turner, almost 60 in this picture, defies age stereotypes.

loss may be minimal. In studying 1,720 respondents who reported using a senior center in the previous year, Miner and colleagues (1993) found that those who attended the center were less disengaged and more socially active than those who did not.

Conflict Perspective

The conflict perspective focuses on age grading as another form of inequality because both the young and the old occupy subordinate statuses. Some conflict theorists emphasize that individuals at both ends of the age continuum are superfluous to a capitalist economy. Children are untrained, inexperienced, and neither actively producing nor consuming in an economy that requires both. Similarly, the elderly, although once working, are no longer productive and often lack required skills and levels of education. Both the young and the old are considered part of what is called the dependent population; that is, they are an economic drain on society. Hence children are required to go to school in preparation for entry into a capitalist economy, and the elderly are forced to retire.

Other conflict theorists focus on how different age strata represent different interest groups that compete with one another for scarce resources. Debates about funding for public schools, child health programs, Social Security, and Medicare largely represent conflicting interests of the young versus the old.

Symbolic Interactionist Perspective

The symbolic interactionist perspective emphasizes the importance of examining the social meaning and definitions associated with age. Teenagers are often portrayed as lazy, aimless, and awkward. The elderly are also defined in a number of stereotypical ways, contributing to a host of myths surrounding the inevitability of physical and mental decline. Table 12.2 identifies some of these myths.

Table 12.2

Myths and Facts About the Elderly

Category	Myth	Fact
Health	The elderly are always sick; most are in nursing homes.	More than 85% of the elderly are healthy enough to engage in their normal activities. Less than 5% are confined to a nursing home.
Mental status	The elderly are senile.	Although some of the elderly learn more slowly and forget more quickly, most remain oriented and mentally intact. Only 20–25% develop Alzheimer's disease or some other incurable form of brain disease. Senility is not inevitable as one ages.
Isolation	The elderly are alone and isolated from family members	Most older people have regular contact with friends and family; many see at least one of their children once a week.
Employment	The elderly are inefficient employees.	Although only 18.3% of men and 10.7% of women, age 65 and older, are still employed, those who continue to work are efficient workers. Compared to younger workers, the elderly have lower job turnover, fewer accidents, and less absenteeism. Older workers also report higher satisfaction in their work.
Politics	The elderly are not politically active.	In 2004 individuals age 65 and older were more likely to be registered to vote and/or to vote than any other age group.
Sexuality	Sexual satisfaction disappears with age.	Many elderly people report active and satisfying sex lives. For example, of couples age 75 years old and older, more than 25% reported having sexual intercourse once a week.
Adaptability	The elderly cannot adapt to new working conditions.	A high proportion of the elderly are flexible in accepting change in their occupations and earnings. Adaptability depends on the individual: Many young people are set in their ways, and many older people adapt to change readily.

Source: AOA (2004), Binstock (1986), Federal Elections Commission (2005), Palmore (1984), Quadagno (2005), Seeman & Adler (1998), Toner (1999), and U.S. Bureau of the Census (2004).

Media portrayals of the elderly contribute to their negative image. The young are typically portrayed in active, vital roles and are often overrepresented in commercials. In contrast, the elderly are portrayed as difficult, complaining, and burdensome and are often underrepresented in commercials. A study of the elderly in popular 1940s through 1980s films concluded that "older individuals of both genders were portrayed as less friendly, having less romantic activity, and enjoying fewer positive outcomes than younger characters at a movie's conclusion" (Brazzini et al. 1997, p. 541). Media images are powerful; a recent study found that children as young as 5 years old have already developed negative stereotypes of the elderly (EurekAlert 2003) (see this chapter's *Social Problems Research Up Close* feature).

The elderly are also portrayed as childlike in terms of clothes, facial expressions, temperament, and activities—a phenomenon known as **infantilizing elders**. For example, young and old are often paired together. A promotional advertisement for the movie *Just You and Me, Kid,* with Brooke Shields and George Burns, described it as "the story of two juvenile delinquents." In *Grumpy Old Men* the characters played by Jack Lemmon and Walter Matthau get "cranky" when they get tired, and the media focus on images of Santa visiting nursing homes and local elementary school children teaching residents arts and crafts. Finally, the elderly are often depicted in role reversal, cared for by their adult children, as in the television situation comedies *Golden Girls, Frasier,* and *King of Queens.*

Negative stereotypes and media images of the elderly engender **gerontophobia**, a shared fear or dread of the elderly, which may create a self-fulfilling prophecy. For example, in a 20-year study of 600 respondents, researchers found that elderly people with positive perceptions of aging (e.g., seeing the elderly as wise) lived, on average, seven and a half years longer than those who had a negative image of aging (e.g., considering most elderly to be senile) (Ramirez 2002).

> "Since it is the Other within us who is old, it is natural that the revelation of our age should come to us from outside, from others."
>
> **Simone de Beauvoir**
> **Feminist author**

Problems of Youth

The number of people under the age of 18—73.6 million in 2005—is the largest in history and will grow to 80 million by the year 2020 (U.S. Bureau of the Census 2004). Although recent evidence indicates that conditions for children are improving, numerous problems remain. Indeed, some of our most pressing social problems can be traced to early childhood experiences and adolescent behavioral problems. Table 12.3 provides a summary of problems associated with youth in America.

Child Labor

Child labor involves children performing work that is hazardous, that interferes with a child's education, or that harms a child's health or physical, mental, spiritual, or moral development (U.S. Department of Labor 1995). Even though virtually every country in the world has laws that limit or prohibit the extent to which children can be employed, child labor persists throughout the world. An estimated 250 million school-age children are child laborers—16 out of every 100 children worldwide (Human Rights Watch 2005a; UNICEF 2005a). To grasp the scale of child labor, imagine a country as populous as the United States in which the entire population consists of child laborers.

Child laborers work in factories, workshops, construction sites, mines, quarries, and fields, on deep-sea fishing boats, at home, on the street, and on the battlefield where, globally, 300,000 child soldiers endure the rigors of armed conflict (Becker 2004). Child laborers make bricks, shoes, soccer balls, fireworks and matches, furniture, toys, rugs, and clothing. They work in the manufacturing of brass, leather goods, and glass. They tend livestock and pick crops. In Egypt more than 1 million children, ages 7 to 12, work each year in cotton-pest management.

Television and the Portrayal of the Elderly

The extent to which minorities, including the elderly, are stereotyped in the media is often the subject of social science research. However, the following study by Donlan and colleagues (2005) goes a step further than simply documenting the existence of such stereotypes. In their study the researchers ask important questions regarding the impact of lifetime television exposure and the effects of maintaining a TV character impression diary on, among other things, views of the elderly and projected future media consumption.

Sample and Methods

Seventy respondents between the ages of 60 and 92 were recruited from a cross-section of New England communities. The average age was 73, and, reflecting the general trend for women to live longer than men, 64% of the sample were women. Seventy percent of the participants were retired, and, of those who were employed, two-thirds worked part-time.

Participants were randomly assigned to one of two conditions. Members of the control group were asked to keep a "media diary" of television viewing for a week's time. For seven days respondents recorded the name of each show they watched and the channel, time, and type of program (e.g., drama, comedy). The intervention group, that is, the experimental group, in addition to recording the same information as the control group, also responded to several questions about the portrayal of the elderly. Specifically, each day they were asked to select "one older character that made the strongest impression" on them and answer questions about that character's portrayed level of health, role in the program (e.g., major or minor part), and estimated age. After completing the viewing diaries, all respondents were asked to answer questions concerning their views on aging and estimated frequency of future television viewing.

There are two independent variables in this study. The first is what the researchers call lifetime television exposure, as measured by asking respondents the age at which they began watching television and the number of hours, in general, they watched television per week. The second independent variable is the experimental manipulation, that is, having respondents in the intervention group keep an "impression" diary of selected older television characters.

Attitude toward old age is the first dependent variable. Respondents were asked the extent to which negative aging words (e.g., wrinkled, grumpy, senile) matched or did not match their views of the elderly. This was rated on a seven-point scale from "does not match at all" to "completely matches." Donlan and colleagues (2005) hypothesized that "elders with higher exposure to television will harbor more negative images of aging than those with lower exposure to television" (p. 309).

The second dependent variable was measured by asking respondents about the portrayal of the elderly on television. For example, respondents were asked, "When older persons appear on television as characters, how are they most often portrayed?" Responses ranged on a 5-point scale from "very negatively" to "very positively." Donlan and coworkers hypothesized that "an intervention based on maintaining a diary of viewing impressions will increase elders' awareness that television programming

They endure routine beatings by their foremen as well as exposure to heat and pesticides (Human Rights Watch 2001). Children typically earn the equivalent of $1 per day and work from 7:00 a.m. to 6:00 p.m. daily, with one midday break, seven days a week.

Illegal and oppressive employment of children also occurs in the United States in restaurants, grocery stores, meat-packing plants, sweatshops in urban garment districts, and in agriculture. Between 300,000 and 800,000 U.S. child workers labor on commercial farms alone, frequently under dangerous and grueling conditions (Human Rights Watch 2003). Child farmworkers in the United States often work 12-hour days, sometimes beginning at 3:00 or 4:00 a.m. Many are exposed to dangerous pesticides that cause cancer and brain damage, with short-term symptoms including rashes, headaches, dizziness, nausea, and vomiting. They often work in

presents the old in a stereotypical and infrequent manner" (p. 309).

Findings and Conclusions

Supporting the first hypothesis, Donlan and colleagues (2005) found that lifetime television exposure was significantly and directly associated with negative age stereotypes; that is, the higher the lifetime television exposure, the greater the negative attitudes toward the elderly. In fact, lifetime television exposure was a better predictor of negative age stereotypes than the respondent's age, education, or health. As Donlan and co-workers conclude, "The impersonal programming of television seemed to have a greater impact on images-of-aging formation than did more personal factors, such as depression and self-rated health" (p. 314).

The second hypothesis was also supported. Participants in the intervention group, compared to the control group participants, were more likely to be aware of the infrequency of older characters on television. They were also more likely to report their intention to decrease television viewing in the future. However, there was no

significant difference between participants in the intervention group and participants in the control group with regard to their views on aging. This is not surprising, however. As Donlan and colleagues note, it would be unrealistic to expect significant changes in the way respondents view the elderly as a result of one week of heightened awareness.

Furthermore, the analysis revealed that participants in the intervention group were more likely to acknowledge the negative portrayal of the elderly on television after the intervention than before the intervention. This finding held only for comedies, not dramas. In explaining this result, the investigators note that comedy programming has long stereotyped the elderly. In fact, one intervention group participant stated that, in reference to a popular situation comedy character, "The portrayal of this character is negative—but it's a comedy"(Donlan et al. 2005, p. 314).

In addition to allowing hypotheses testing, the media diaries provide a qualitative glimpse into perceptions of the elderly by older television viewers. For example, the

omission of the elderly was noted in a variety of program types.

A 71-year-old male viewer of television news noted that "When people are interviewed about different matters, older people were left out." Similarly, a 60-year-old game show viewer indicated that, "They should try and get older Americans as contestants." [And] according to a 68-year-old homemaker who watched more than 45 hours of television a week, "I feel like we've been ignored. I feel like we're non-existent." (Donlan et al. 2005, pp. 314–315)

Finally, the researchers acknowledge that the results may not be applicable to other countries. For example, several of the participants noted that BBC (British Broadcasting Corporation) programming portrays the elderly in a more positive light. As one participant commented in reference to a British comedy shown on public television in the United States, "I wish America could get a show going like this that has some intelligence to it—not some idiotic nonsense. In America older parents are always portrayed as the outlaw in-laws."

100° temperatures without adequate access to drinking water and are sometimes forced to work without access to toilets or hand-washing facilities (Human Rights Watch 2003). Agriculture is also one of the most dangerous occupations, and children (as well as adults) sustain high rates of injury from work with knives, other sharp tools, and heavy equipment. For example, between 1992 and 2000 more than 40% of work-related fatalities of young people were related to agriculture, and half the victims were under the age of 15 (National Consumers League 2005).

Child Prostitution and Trafficking. One of the worst forms of child labor is child prostitution and child trafficking. Worldwide, it is estimated that there are 1 million child prostitutes; in the United States the estimate is 300,000 (Dorman 2001; United Nations 2003a). In poor countries the sexual services of children are often sold by their families in an attempt to get money. Some children are kidnapped or

> "Child labor is a scar on the world's conscience."
>
> **UNICEF Report**

Table 12.3
Each Day In America

1	mother dies in childbirth.
4	children are killed by abuse or neglect.
5	children or teens commit suicide.
8	children or teens are killed by firearms.
76	babies die before their first birthdays.
182	children are arrested for violent crimes.
366	children are arrested for drug abuse.
390	babies are born to mothers who received late or no prenatal care.
860	babies are born at low birth weight.
1,186	babies are born to teen mothers.
1,707	babies are born without health insurance.
1,887	public school students are corporally punished.[a]
2,171	babies are born into poverty.
2,341	babies are born to mothers who are not high school graduates.
2,455	children are confirmed as abused or neglected.
2,539	high school students drop out.[a]
3,742	babies are born to unmarried mothers.
4,440	children are arrested.
17,072	public school students are suspended.

[a] Based on calculations per school day (180 days of 7 hours each).

Source: CDF (2005).

lured by traffickers with promises of employment, only to end up in a brothel. For example, promised a job as a waitress, this 13-year-old girl was brought to the United States and forced to work as a "sex slave" (Allen 2002):

> She was a virgin, and she didn't know what they were talking about, but she knew it was bad, so she refused. She was then brutally gang-raped to induct her into the business. For the next six months, she was forced to service 10 to 15 men a day. Twice she was impregnated, twice forced to have an abortion, and twice she was back in the brothel the next day. The traffickers circulated her through trailer brothels and private parties, where she was passed around. She was pistol-whipped and raped if she resisted. The girl was rescued when two girls ran to neighbors who called police. She had multiple sexually transmitted diseases, scar tissue from the forced abortions, and was addicted to drugs and alcohol. She had posttraumatic stress syndrome, including severe depression and suicidal thoughts. She was physically, mentally, emotionally and spiritually broken. (p. 1)

The U.S. Immigration and Naturalization Service identified 250 brothels in 26 American cities where forced prostitutes, including children, are taken. Americans also engage in child-sex tourism abroad, particularly in Southeast Asia. Of

Street children are often subject to police brutality and violence. Many are held for long periods of time without due process. It is estimated that there are 30–170 million street children worldwide. In this picture, two plainclothes policemen take a youth into custody.

240 identified cases in which legal action was taken against foreigners for sexually abusing children in this region, researchers found that about one-fourth of the violators were from the United States (Dorman 2001).

Orphaned and Street Children

Worldwide, there are millions of orphans, many kept in institutions, some living on the streets, and still others, with no means of supporting themselves, turning to sex work, crime, or drugs. In many of these institutions children are simply warehoused with little food, clean water, or medical care. Even in institutions that do provide such necessities, many of the children are emotionally neglected or otherwise subject to inhumane treatment (Human Rights Watch 2005b). Recently, the number of orphans has increased dramatically for at least two reasons. First is the devastation of the HIV/AIDS pandemic. Globally, the number of children orphaned by the disease, predominantly in developing nations, is 13 million, and this number is expected to grow to 25 million by 2010 ("AIDS Creating Global 'Orphans Crisis,'" 2003). More than half of children 0 to 14 years old in Zimbabwe, Botswana, Zambia, Kenya, and Uganda have lost one or both parents to HIV/AIDS (United Nations 2003d). Second, children are increasingly orphaned from armed conflicts. For example, the civil war in Rwanda is responsible for hundreds of thousands of orphans, many of whom are now street children. Yet a third reason is the result of natural disasters, such as the December 2004 tsunami, which left hundreds of thousands of children orphaned and homeless.

It is estimated that between 30 million and 170 million children live on the streets (Street Children 2005). Many of the children are "addicted to inhalants, such as cobbler's glue, which offers them an escape from reality in exchange for a host of physical and psychological problems" (CASA Alianza 2003). Often resorting to prostitution, many of the street children have HIV/AIDS. For example, of Mexico's

© TOMAS BRAVO/Reuters/Corbis

2 million street children at least 7% are HIV-positive (CASA Alianza 2003). The number of orphaned and street children is predicted to grow dramatically over the next decade.

Children's Rights

Historically, children have had little control over their lives. They have been "double dependent" on both their parents and the state. Indeed, colonists in America regarded children as property. Beginning in the 1950s, however, the view that children should have greater autonomy became popular and was codified in several legal decisions and international treaties. In 1959 the United Nations General Assembly approved the Declaration on the Rights of the Child, which held that health care, housing, and education, as well as freedom from abuse, neglect, and exploitation, are fundamental children's rights.

A second measure, the Convention on the Rights of the Child, further articulated the rights of children and was adopted by the United Nations in 1989. Countries ratifying the document have made significant improvements in the lives of children. For example, the draft constitution of the Democratic Republic of the Congo now prohibits conscription into the army before age 18. Only two countries, the United States and Somalia, have failed to ratify the Convention (UNICEF 2005b). One reason the United States has not signed the pact may be Article 11, which requires that children be assured "the highest attainable standard of health." Some people are concerned that accepting such a position would result in cases being brought to court that the United States is simply unwilling, at present, to hear. The two most recent provisions of the Convention deal with the involvement of children in armed conflicts, the sale of children, child prostitution, and child pornography (United Nations 2003c).

Children are both discriminated against and granted special protections under the law. Although legal mandates require that children go to school until age 16, other laws provide a separate justice system whereby children have limited legal responsibility based on their age status. For example, in 2005 the U.S. Supreme Court prohibited execution of convicted criminals who committed their crimes before the age of 18. In a 5–4 decision Justice Kennedy, writing for the majority, argued that "the age of 18 is the point where society draws the line for many purposes between childhood and adulthood. It is, we conclude, the age at which the line for death eligibility ought to rest" (Totenberg 2005, p. 1).

Finally, in what is being called one of the most important court decisions on children's rights in recent years, a federal court has held that abused and neglected children have the constitutional right to legal representation. The decision, resulting from a class action suit against Georgia's child welfare agency, will amend Georgia law, which previously held that abused and neglected children had a right to an attorney only when the state had entered a plea to terminate parental rights (Dewan 2005).

Foster Care and Adoption

According to the Child Welfare League of America, as of September 2003 there were more than 500,000 children in the U.S. foster care system. Most children are placed in foster care because of parental abuse or neglect (CWLA 2005). Although black and Hispanic children are disproportionately represented in foster care, we cannot conclude that child abuse and neglect are more likely in black and Hispanic households (see Chapter 5). For example, because of the tendency for minority families to have lower incomes than white families, relatives may be less able to provide a home for the abused or neglected child.

Because older children are less likely to be adopted, the average age of a child in foster care is 10.2 years, with children younger than 1 year representing less than 5% of the total number of foster care children. Many of these children have special physical or emotional needs or are part of a sibling group, which makes adoption difficult. In 2002 the average stay in the foster care system was 32 months (CWLA 2005).

The foster care system has grown dramatically over the years, with federal reimbursement to states increasing from $1.2 billion in 1989 to $6.2 billion in 2004 (PBS 2005). Despite increased funding, stories of abused, neglected, and missing children in the foster care system remain. Furthermore, a survey conducted by Casey Family Programs (2003) indicate that, compared to the general population, adults who were in the foster care system as children are more likely to be homeless, to have higher rates of unemployment, to have lower incomes, and to be arrested.

Poverty and Economic Discrimination

There are 2.2 billion children in the world, and nearly half of them live in poverty (SOTWC 2005). Worldwide it is estimated that more than one-third of poor children live in developing countries. The highest rates are in sub-Saharan Africa, where 65% of the children are poor, and in South Asia, where 59% of the children are poor. Most of these children live in rural areas (World Youth Report 2005).

Of the major industrialized nations of the world, the United States leads in child poverty, with a rate of 17.6%, or 1 out of every 6 children living in poverty in 2003. Further, the number of American children in *extreme* poverty—that is, those living in families with incomes one half the poverty level or less—grew twice as fast as child poverty in general. In 2003, 34.1% of black children, 29.7% of Hispanic children, 12.5% of Asian children, and 9.8% of non-Hispanic white children were considered poor (CDF 2004a).

The effects of childhood poverty are far-reaching. Studies have shown that child poverty is related to school failure (CDF 2004a), negative involvement with parents (Harris & Marmor 1996), stunted growth, reduced cognitive abilities, limited emotional development (Brooks-Gunn & Duncan 1997; CDF 2004a), and a higher likelihood of dropping out (Duncan et al. 1998) or being held back from progressing to the next grade level in school (CDF 2004a).

Three decades ago the elderly were the poorest age group in the United States, but today it is children ("Snapshots of the Elderly," 2000). Although Social Security, Medicare, housing subsidies, and Supplemental Security Income (SSI) keep millions of elderly out of poverty, the United States has no universal government policy to protect children. In addition, recent welfare reform has led to cutbacks in the few programs that do benefit children. For example, more than half of food stamp recipients are under the age of 18 (CDF 2003b).

Children are also the victims of discrimination in terms of employment, age restrictions, wages, training programs, and health benefits. Traditionally, children worked on farms and in factories but were displaced during the Industrial Revolution. In 1938 Congress passed the Fair Labor Standards Act, which required factory workers to be at least 16 years old. Although the law was designed to protect children, it was also discriminatory in that it prohibited minors from having free access to jobs and economic independence. Today, 14- and 15-year-olds are restricted in the number of hours per day and hours per week they can work; those younger than 14 are, in general, prohibited from working.

Children, Violence, and the Media

Gang violence, child abuse, and crime are all-too-common childhood experiences (see this chapter's *Human Side* feature). For example, the results of a survey of ninth through twelfth graders conducted by the Centers for Disease Control and Prevention reveals that as much as 6% of students stay home from school on one or more days because they feel unsafe at school and/or on their way to or from school (YRBSS 2004).

According to the Children's Defense Fund, U.S. children are 12 times more likely to die from gunfire, 16 times more likely to be murdered by a gun, and 9 times more likely to die from a firearm accident than children in 25 other industrialized nations *combined* (CDF 2003a). Child abuse and neglect remain at epidemic levels, with an estimated 3 million children victimized annually; children younger than age 18 account for 16.3% of all arrests, excluding traffic violations (FBI 2004).

School violence, and particularly the tragic deaths of 12 students and 1 teacher at Columbine High School, has focused attention on the relationship between youth violence, guns, and the media. In fact, when a random sample of Americans were asked the causes of school violence, 68% responded that "the portrayal of violence and use of guns in today's entertainment and music" was an extremely or very important factor (Gallup 2004). The typical child watches 28 hours of television a week. By the time the child is 18, he or she will have seen 16,000 murders and 200,000 other acts of violence on television. Ironically, the Center for Media and Public Affairs reports that commercial television for children is 50 to 60 times more violent than prime-time TV with, for example, cartoons averaging 80 violent acts an hour (CDF 2003c).

"New media"—computers, the Internet, video games, and the like—have also come under attack. For example, with children playing video games, on average, 50 minutes a day (Kaiser Family Foundation 2005a), there is some fear that violent video games may lead to aggressive behavior. In 2005 the brother of a murder victim filed a multimillion-dollar wrongful death suit against the makers and marketers of *Grand Theft Auto,* arguing that the alleged offender was programmed for violence as a result of playing the video game. Several states are now considering banning the sale of violent video games to those under the age of 17 (Bradley 2005).

Children's Health

In addition to the problems already discussed, children suffer from a variety of health problems, both mental and physical. One in five children, for example, has a diagnosable mental disorder, such as depression or schizophrenia (see Chapter 2). Many children, particularly girls, suffer from eating disorders, such as anorexia nervosa or bulimia. It is estimated that 1% of young girls will develop anorexia and that of that number, 10% will die from the disease. Suicide is the third most common cause of death of 15- to 24-year-olds, the sixth most common cause of death of 5- to 15-year-olds, and the third most common cause of death of college students. Suicide rates among the young have tripled since 1960. More youth die from suicide than from heart disease, cancer, HIV/AIDS, stroke, birth defects, and chronic lung disease combined (NMHA 2003).

Growing Up in the Other America

Author Alex Kotlowitz conducted a two-year participant observation research study of children in a Chicago housing project (Kotlowitz 1991). The following description captures the horrific living conditions endured by Lafeyette, Pharoah, and Dede—three children living in the "jects."

The children called home "Hornets" or, more frequently, "the projects" or, simply, the "jects" (pronounced *jets*). Pharoah called it "the graveyard." But they never referred to it by its full name: the Governor Henry Horner Homes.

Nothing here, the children would tell you, was as it should be. Lafeyette and Pharoah lived at 1920 West Washington Boulevard, even though their high-rise sat on Lake Street. Their building had no enclosed lobby; a dark tunnel cut through the middle of the building, and the wind and strangers passed freely along it. Those tenants who received public aid had their checks sent to the local currency exchange, since the building's first-floor mailboxes had all been broken into. And since darkness engulfed the building's corridors, even in the daytime, the residents always carried flashlights, some of which had been handed out by a local politician during her campaign.

Summer, too, was never as it should be. It had become a season of duplicity.

On June 13, a couple of weeks after their peaceful afternoon on the railroad tracks, Lafeyette celebrated his twelfth birthday. Under the gentle afternoon sun, yellow daisies poked through the cracks in the sidewalk as children's bright faces peered out from behind their windows. Green leaves clothed the cottonwoods, and pastel cotton shirts and shorts, which had sat for months in layaway, clothed the children. And like the fresh buds on the crabapple trees, the children's spirits blossomed with the onset of summer.

Lafeyette and his nine-year-old cousin Dede danced across the worn lawn outside their building, singing the lyrics of an L. L. Cool J rap, their small hips and spindly legs moving in rhythm. The boy and girl were on their way to a nearby shopping strip, where Lafeyette planned to buy radio headphones with $8.00 he had received as a birthday gift.

Suddenly, gunfire erupted. The frightened children fell to the ground. "Hold your head down!" Lafeyette snapped, as he covered Dede's head with her pink nylon jacket. If he hadn't physically restrained her, she might have sprinted for home, a dangerous action when the gangs started warring. "Stay down," he ordered the trembling girl.

The two lay pressed to the beaten grass for half a minute, until the shooting subsided. Lafeyette held Dede's hand as they cautiously crawled through the dirt toward home. When they finally made it inside, all but fifty cents of Lafeyette's birthday money had trickled from his pockets.

Source: Alex Kotlowitz, 1991. *There Are No Children Here: The Story of Two . . .* by Alex Kotlowitz. Copyright © 1991 by Alex Kotlowitz. Used by permission of Doubleday, a division of Random House, Inc.

One particular health problem is child obesity. According to the American Heart Association, "childhood obesity is one of the most critical public health problems today and threatens to reverse the last half century's gains in reducing cardiovascular (CVD) disease and death" (AHA 2005). The prevalence of overweight children has increased from 5% in 1980 to 16% in 2005, with rates as high as 20% for African American and Hispanic children. In addition to an increased risk of a variety of health problems, overweight children are less likely to have friends, have more difficulty networking, and, if teased about their weight, are more likely to commit suicide.

Kids in Crisis

Childhood is a stage of life that is socially constructed by structural and cultural forces of the past and present. The old roles for children as laborers and farm helpers are disappearing, yet no new roles have emerged. While being bombarded by the media, children must face the challenges of an uncertain future, peer culture, music videos, divorce, poverty, and crime. Parents and public alike fear that children are becoming increasingly involved with sex, drugs, alcohol, and violence. Some even argue that childhood as a stage of life is disappearing.

Despite the many problems associated with childhood, the condition of children in the United States is better than the public's perception of it. For example, 12% of children lack health insurance, but 93% of a national sample of adults believed that 20–30% of children lacked health insurance. In addition, almost half the adults sampled believed that 30% of children live in poverty when in fact 17% live in poverty. The significance of these discrepancies lies in the public's influence on social policy. If the public has an inaccurate picture of the state of children in the United States, it may incorrectly impact policy makers (Peterson 2003).

Despite some gains, conditions remain intolerable for millions of children in the United States and around the world: homelessness, sexual exploitation, low birth weight, hunger, childhood depression, absent or inadequate health care, and dangerous living conditions (e.g., foster care, child abuse). A report from the National Research Council called for a "new national dialogue focused on rethinking the meaning of both shared responsibility for children and a strategic investment in their future" (Jacobson 2000, p. 1). Such an investment, particularly in preventive initiatives, is not only humane but also cost-effective, because millions of dollars that would be spent on solving child-related problems after they occur could be saved.

Demographics: The Graying of America

In recent years the ratio of children to old people has changed dramatically (see Table 12.1 and Figure 12.1). In 2003 one in every eight Americans was older than age 65; in 2030 one in five people will be over the age of 65 (AOA 2004).

These statistics reflect three significant demographic changes. First, 76 million baby boomers, born between 1946 and 1964, are getting older. Second, life expectancy has increased as a result of a general trend toward modernization, including better medical care, sanitation, and nutrition. Finally, decreasing birthrates mean fewer children and a higher percentage of the elderly. For example, in Japan low birthrates and rising life expectancies have contributed to the highest proportion of elderly in the world (Zhong Xin Net 2003).

ThomsonNOW™

Learn more about **Demographics: The "Graying of America"** by going through the Aging Population Learning Module.

Age and Race and Ethnicity

In 2003, 17.6% of the elderly were racial or ethnic minorities. Minority populations are projected to represent 26.4% of the U.S. elderly population by 2030, up from 13% in 1990. Although the elderly white population is projected to increase 77% between 2000 and 2030, the elderly minority population is expected to increase 223% in the same time period (AOA 2004). The growth of the elderly minority population is a consequence of higher fertility rates and immigration patterns, particularly among Hispanics. Given their higher rates of diabetes, arthritis, and cardiovascular disease, the increased numbers of older minorities present a unique challenge to health care providers.

Age and Sex

In the United States elderly women outnumber elderly men. The sex ratio for individuals age 60 and older is 77 men for every 100 women (United Nations 2003b). Men die at an earlier age than women for both biological and sociological reasons: heart disease, stress, and occupational risk (see Chapters 2 and 10). At age 65 men can expect to live an additional 16.6 years, whereas women can expect to live an

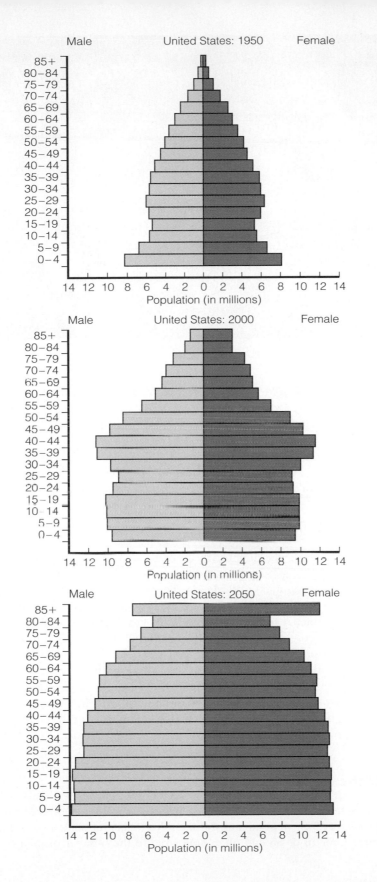

Figure 12.1
U.S. population pyramids.

Source: U.S. Bureau of the Census (2002).

Women, as they get older, are often portrayed as sexually unattractive, something men are rarely subjected to. Advertising contributes to this image by idealizing youth in advertisements that advocate the use of wrinkle creams, hair dyes, and getting one's "girlish figure" back.

additional 19.5 years (AOA 2004). The fact that women live longer results in a sizable number of elderly women who are poor. Not only do women, in general, make less money than men, but also older women may have spent their savings on their husband's illness, and, as homemakers, they may receive fewer Social Security benefits. In addition, retirement benefits and other major sources of income may be lost with a husband's death. Seventy percent of all elderly poor are women, half of whom were not poor before the death of their husbands.

Age and Social Class

How long a person lives is influenced by his or her social class. In general, the higher the social class, the longer one lives, the fewer the debilitating illnesses, the greater the number of social contacts and friends, the less likely one is to define oneself as "old," and the greater the likelihood of success in adapting to retirement. Higher social class is also related to fewer residential moves, higher life satisfaction, more leisure time, and more positive self-rated health. Of those individuals age 65 and older, 26% of those with annual incomes of $35,000 or higher reported that their health was "excellent," whereas only 10% of those with incomes under $10,000 rated their health as such. Functional limitations such as problems with walking, dressing, and bathing were also less common among higher income groups (Seeman & Adler 1998). In short, the higher one's socioeconomic status, the longer, happier, and healthier one's life (Quadagno 2005).

Problems of the Elderly

The increase in the number of the elderly worldwide presents a number of institutional problems. The **dependency ratio**—the number of societal members who are younger than 18 or 65 and older compared to the number of people who are

Facts on Aging Quiz

Answer the following questions about the elderly, and assess your knowledge of the world's fastest-growing age group.

	True	False
1. Lung capacity tends to decline in old age.	_____	_____
2. The majority of old people say they are seldom bored.	_____	_____
3. Old people tend to become more religious as they age.	_____	_____
4. The health and economic status of old people will be about the same or worse in the year 2010 (compared with younger people).	_____	_____
5. A person's height tends to decline in old age.	_____	_____
6. The aged are more fearful of crime than are younger persons.	_____	_____
7. The proportion of blacks among the aged is growing.	_____	_____
8. The majority of old people live alone.	_____	_____
9. The five senses all tend to weaken in old age.	_____	_____
10. Medicare pays over half of the medical expenses of the elderly.	_____	_____
11. Older persons who reduce their activity tend to be happier than those who do not.	_____	_____
12. Older persons have more injuries in the home than younger persons.	_____	_____
13. Physical strength tends to decline with age.	_____	_____
14. The aged are the most law abiding of all adult age groups.	_____	_____
15. Older persons have more acute Illnesses than do younger persons.	_____	_____

Answers: Items 1, 2, 5–7, 9, 13, and 14 are true. The remainder are false.

Source: Erdman B. Palmore, 1999. *Ageism: Negative and Positive*. New York: Springer Publishing Co. Used by permission.

between 18 and 64—is increasing. In 2000 there were 63 "dependents" for every 100 people between the ages of 18 and 64. By 2050 the estimated ratio will be 78 to 100 (Quadagno 2005). This dramatic increase, and the general movement toward global aging, may lead to a shortage of workers and military personnel, foundering pension plans, and declining consumer markets. It may also lead to increased taxes as governments struggle to finance elder-care programs and services, heightening intergenerational tensions as societal members compete for scarce resources. In addition to these macrolevel concerns, the elderly face a number of challenges of their own. This chapter's *Self and Society* feature tests your knowledge of the elderly and some of these concerns.

Work and Retirement

What one does (occupation), for how long (work history), and for how much (wages) are important determinants of retirement income. Indeed, employment is important because it provides the foundation for economic resources later in life. Yet for the elderly who want to work, entering and remaining in the labor force may be difficult because of negative stereotypes, lower levels of education, reduced geographic mobility, fewer employable skills, and discrimination. Nonetheless, as retiree benefits decline and recession increases, older Americans are working longer than ever before. Since the mid-1990s, older Americans have become the fastest-growing group in the labor force, with those over age 55 expected to make up 19.1% of all workers by 2012 (Porter & Walsh 2005).

One way that older employees remain in the workforce is through phased retirement. **Phased retirement** allows workers to ease into retirement by reducing hours worked a day, days worked a week, or months worked a year. For example, many universities allow faculty to work part-time for half salary while collecting partial pension benefits.

Phased retirement is not only beneficial for retirees, who may not want to continue to work full-time, but also good for employers, who benefit from the skills and knowledge of older employees.

In 1967 Congress passed the Age Discrimination in Employment Act (ADEA), which was designed to ensure continued employment for people between the ages of 40 and 65. In 1986 the upper limit was removed, making mandatory retirement illegal in most occupations. Nevertheless, thousands of cases of age discrimination occur annually. Although employers cannot advertise a position by age, they can state that the position is an "entry level" one or that "2 to 3 years' experience" is required. Despite strong evidence to the contrary, a study conducted by the National Council on the Aging revealed that "50 percent of employers surveyed believed that older workers cannot perform as well as younger workers" (Reio and Sanders-Reto 1999, p. 12). In addition, if displaced, older workers remain unemployed longer than younger workers; they often have to accept lower salaries than they earned earlier and may be more likely to give up looking for work (Goldberg 2000).

Retirement is a relatively recent phenomenon. Before passage of the Social Security laws, individuals continued to work into old age. Today, Social Security payments are limited to those older than 65 years of age—age 67 by the year 2022. However, in a recent survey, when a sample of retired respondents were asked at what age they left the workforce, 68% responded that they had retired before the age of 65 (Carroll 2005).

Retirement is difficult in the United States because "work" is often equated with "worth." A job structures one's life and provides an identity; retirement often culturally signifies the end of one's productivity. Retirement may also involve a dramatic decrease in personal income. Six in 10 older Americans are worried about "not having enough money for retirement" (Jones 2005). The desire to remain financially independent, a lack of confidence in the Social Security system, increased educational levels and technological skills, and the desire to continue working have led to a recent reduction in early retirements. In addition, as noted, the number of people who stop working once they retire is also decreasing as retirees open their own businesses, work part-time, continue their educations, volunteer in the community, and begin second careers (Porter & Walsh 2005).

Poverty

In 2003 the median income for individuals 65 years old and older was $14,664, and 29.3% of the elderly reported incomes of $10,000 or less (AOA 2004) (see Figure 12.2). Poverty among the elderly varies dramatically by sex, race, ethnicity, marital status, and age: Women, minorities, those who are single or widowed, and the old-old are most likely to be poor (Quadagno 2005). Nine percent of elderly whites

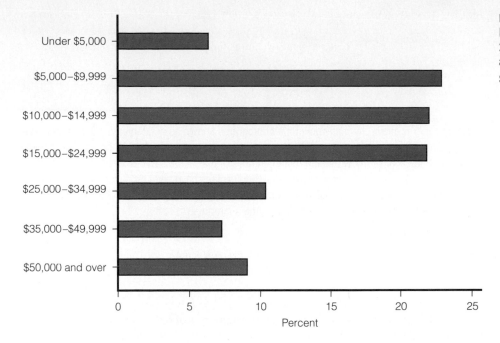

Figure 12.2
Distribution of income (%) in 2003 for individuals age 65 and older.
Source: AOA (2004).

are poor, compared to 23.7% of elderly blacks and 19.5% of elderly Hispanics (AOA 2004).

Elderly women are more likely to be poor than elderly men. Nearly half of elderly Hispanic women, and 4 out of 10 elderly black women are living in poverty. The comparable percentage for white women is half that of minorities. Minority women are more likely to have been working in low-paying jobs with no retirement plan, to have little or no savings, and to have fewer resources to fall back on. For example, older black households have an estimated median net worth of $13,000 compared to older white households, which have an estimated median net worth of $181,000 (Hounsell & Humphlett 2003). For many minority women Social Security is the sole source of support. The highest poverty rate, over 40%, is among elderly Hispanic women who live alone or with nonrelatives (AOA 2004).

Social Security, actually titled "Old Age, Survivors, Disability, and Health Insurance," is a major source of income for many elderly. When Social Security was established in 1935, it was never intended to be a person's sole economic support in old age; rather, it was meant to supplement other savings and assets. A Social Security Administration survey reveals that when asked about "major sources of income," 90% of the elderly reported Social Security, 55% reported income from assets, 43% reported public and private pensions, and 22% reported earnings (AOA 2004). Spending on the elderly has increased over time, as have Social Security benefits. Today, 92% of Americans age 65 and older receive Social Security payments, and another 3% will be eligible once they retire. The average monthly check for a retiree in 2005 was $955 (Goozner 2005).

Although Social Security may keep as many as 4 out of 10 elderly from being poor (NCSC 2000), the Social Security system has been criticized for being based on the number of years of paid work and preretirement earnings. Hence women and minorities, who often earn less during their employment years, receive less in retirement benefits. Another concern is whether funding for the Social Security system will be adequate to provide benefits for the increased numbers of elderly in the next

Should Social Security Be Privatized?

Social Security is one of the many government programs known as entitlements. Funded by a payroll tax, Social Security in fiscal year 2004 cost the federal government $496 billion. The growing number of baby boomers now nearing retirement means that in the coming decades the Social Security system may become bankrupt. Currently, various plans have been proposed to fix the system. President Bush's proposal would allow workers to divert up to 4% of their Social Security contributions to private investment accounts while reducing guaranteed benefits.

It Is Your Turn Now to Take a Stand!

Do you think Social Security should be privatized? Why or why not?

decades. In the next 25 years the number of Americans of retirement age will *increase* by 100%, whereas the ratio of workers who contribute to the Social Security system will *decrease* by an estimated 60% (Sloan 2005). Fear of Social Security's demise has led some economists and the Bush administration to recommend privatization. Such a system would allow workers to put the money they now have deducted from their payroll for Social Security (FICA) into a personally owned and invested retirement account. Despite some claims that the system is collapsing, surveys indicate that most Americans are opposed to such a plan (Gallup 2005; Morin & VandeHei 2005).

Health Issues

The biology of aging is called **senescence**. Senescence follows a universal pattern but does not have universal consequences. "Massive research evidence demonstrates that the aging process is neither fixed nor immutable. Biologists are now showing that many symptoms that were formerly attributed to aging—for example, certain disturbances in cardiac function or in glucose metabolism in the brain—are instead produced by disease" (Riley & Riley 1992, p. 221). Biological functioning is also intricately related to social variables. Altering lifestyles, activities, and social contacts affects mortality and morbidity. For example, a longitudinal study of men and women between the ages of 70 and 79 found that regular physical activity, higher levels of ongoing positive social relationships, and a sense of self-efficacy enhanced physical and cognitive functioning (Seeman & Adler 1998).

Biological changes are consequences of either **primary aging,** caused by physiological variables such as cellular and/or molecular variation (e.g., gray hair), or secondary aging. **Secondary aging** entails changes attributable to poor diet, lack of exercise, increased stress, and the like. Secondary aging exacerbates and accelerates primary aging.

Alzheimer's disease, an example of primary aging, received national attention when former president Ronald Reagan announced that he had the disease. Named for German neurologist Alois Alzheimer, the debilitating disease affects both the mental and the physical condition of 4.5 million Americans and will affect a projected 13 million by 2050 (NIH 2003).

"If exercise could be put in a pill, it would be the number one anti-aging medicine.**"**

Robert Butler
Gerontologist

In 2004 more than one-third (37.4%) of the elderly rated their health as excellent or very good; 57.7% of blacks and 60.1% of Hispanics rated their health as excellent or very good, compared to 76.9% of older whites (AOA 2004). Although elderly Americans are in better health than in previous generations, health often declines with age, with most older Americans having at least one chronic condition. Older people account for 36% of all hospital stays, averaging 5.8 days per year compared with 5 days for people between the ages of 45 and 64 (AOA 2004). Health is a major quality-of-life issue for the elderly, especially because they face higher medical bills with reduced incomes. Older Americans spend three times as much on health care as their younger counterparts, and more than half of the amount is spent on insurance. The poor elderly, often women and/or minorities, spend an even higher proportion of their resources on health care.

Medicare was established in 1966 to provide medical coverage to those over the age of 65, and it insures more than 40 million people (NCHS 2003). Although it is widely assumed that the medical bills of the elderly are paid by the government, the elderly are responsible for as much as 25% of their total health costs. Medicare, for example, does not pay for routine physical examinations, most immunizations, dental care, glasses, and hearing aids (Quadagno 2005). In 2002 people older than age 65 spent $3,741 on out-of-pocket health care expenditures, compared with $2,416 for the general population (AOA 2004). The difference between Medicare benefits and the actual cost of medical care is called the **medigap**. Almost three-fourths of Medicare recipients have some type of medigap coverage. In addition, a prescription drug benefit was added to Medicare as part of a 2003 legislative package designed to improve and modernize the program (Quadagno 2005).

Because health is associated with income, the poorest old are often the most ill: They receive less preventive medicine, have less knowledge about health care issues, and have limited access to health care delivery systems. **Medicaid** is a federally and state-funded program for those who cannot afford to pay for medical care. However, eligibility requirements often disqualify many of the elderly poor, often minorities and women.

Living Arrangements

The elderly live in a variety of contexts, depending on their health and financial status. Most elderly do not want to be institutionalized, preferring to remain in their own homes or in other private households with friends and relatives (see Figure 12.3). Of the noninstitutionalized elderly population, 53.9% live at home with their spouse, although that number decreases with age (AOA 2004). The homes of the elderly, however, are usually older, located in inner-city neighborhoods, in need of repair, and often too large to be cared for easily. For example, in 2003 the median year of construction of homes occupied by older residents was 1965, and 5.2% of the homes were in need of repair (AOA 2004).

Although many of the elderly poor live in government housing or apartments with subsidized monthly payments, the wealthier elderly often live in retirement communities. These are often planned communities, located in states with warmer climates, and are often expensive. These communities offer various amenities and activities, have special security, and are restricted by age. One criticism of these communities is that they segregate the elderly from the young and discriminate against younger people by prohibiting them from living in certain areas.

Those who cannot afford retirement communities or who may not be eligible for subsidized housing often live with relatives in their own home or in the homes of

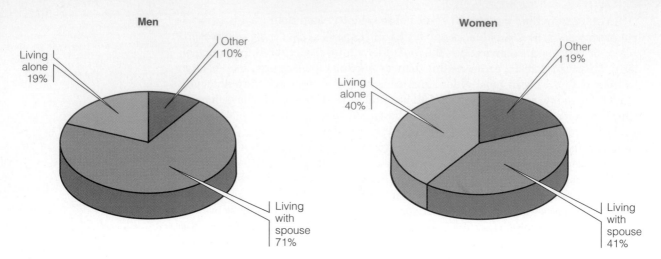

Figure 12.3
Living arrangements of individuals age 65 and older in 2003.
Source: AOA (2004).

others. It is estimated that more than 22 million people provide care for aging family members (Jackson 2003). Almost three-fourths are women, many of whom are daughters who care not only for their elderly parents but also for their own children.

Other living arrangements include shared housing, modified independent living arrangements, and nursing homes. With shared housing people of different ages live together in the same house or apartment; they have separate bedrooms but share a common kitchen and dining area. They share chores and financial responsibilities. In modified independent living arrangements the elderly live in their own house, apartment, or condominium within a planned community where special services, such as meals, transportation, and home repairs, are provided. Skilled or semiskilled health care professionals are available on the premises, and the residences have call buttons so help can be summoned in case of an emergency.

Nursing homes are residential facilities that provide full-time nursing care for residents. Nursing homes may be private or public. Private facilities are expensive and are operated for profit by an individual or a corporation. Public facilities are nonprofit and are operated by a government agency, religious organization, or the like. The probability of being in such an extended care facility is associated with race, age, and sex: Whites, the old-old, and women are more likely to be in residence. The elderly with chronic health problems are also more likely to be admitted to nursing homes. Nursing homes vary dramatically in cost, services provided, and quality of care. One study found that of the 17,000 nursing homes in the United States, 90% were understaffed (Pear 2002).

The fastest growing facilities for the elderly are assisted-living quarters—36,399 in 2002, a 48% increase from 1998 (McCoy 2003). Assisted-living facilities, although not as "full service" as nursing homes, offer private living units with the confidence of an around-the-clock staff. In 2003 the government proposed federal guidelines that require such facilities to reveal all costs, services, and policies and to have an adequately trained staff on duty at all times. In addition, the proposal suggests the creation of a National Center for Excellence in Assisted Living and increases state and federal funding for the Long-Term Care Ombudsman Program (McCoy 2003).

ThomsonNOW™

Learn more about **Living Arrangements** by going through the Nursing Homes Map Exercise.

Victimization and Abuse

Elder abuse refers to physical or psychological abuse, financial exploitation, medical abuse, and/or neglect of the elderly (Green 2003; Hooyman & Kiyak 1999). In one California case an adult protective services (APS) worker found a

> 70-year-old diabetic woman hungry and shoeless. A pan of rotten chicken fat sat on the stove. The smell of spent crack cocaine permeated the air. A man and a disheveled woman emerged from a filthy bedroom and quickly left. Alone with the APS worker, the frightened woman wept and said, "Thank God. I knew he would send somebody to help me. I've prayed and prayed for so long." She had been repeatedly hit and kept in her room, she said, and described being walked around the neighborhood and portrayed as the caregiver's sick grandmother in a ploy to beg for money. (Flaherty 2004, p. 1)

In the United States alone the number of older Americans who are abused in some way is estimated to be over 2 million (APA 2005). Even in Japan, where the elderly are traditionally revered, incidents of elder abuse are on the rise. Although elder abuse can take place in private homes by family members, the elderly, like children, are particularly vulnerable to abuse when they are institutionalized. In 1990 the Nursing Home Reform Act was passed, establishing various rights for nursing care residents: the right to be free of mental and physical abuse, the right not to be restrained unless necessary as a safety precaution, the right to choose one's physician, and the right to receive mail and telephone communication (Harris 1990, p. 362).

Whether the abuse occurs in the home or in an institution, the victim is most likely to be female, widowed, white, on a limited income, and in her mid-70s. The abuser tends to be an adult child or spouse of the victim who misuses alcohol. Some research suggests that the perpetrator of the abuse is more often an adult child who is financially dependent on the elderly victim (Mack & Jones 2003). Whether the abuser is an adult child or a spouse may simply depend on the person with whom the elder victim lives.

Many of the problems of the elderly are compounded by their lack of interaction with others, loneliness, and inactivity. This is particularly true for the old-old. The elderly are also segregated in nursing homes and retirement communities, separated from family and friends, and isolated from the flow of work and school. As with most problems of the elderly, the problems of isolation, loneliness, and inactivity are not randomly distributed. They are higher among the elderly poor, women, and minorities. A cycle is perpetuated—being poor and old results in being isolated and engaging in fewer activities. Such withdrawal affects health, which makes the individual less able to establish relationships or participate in activities.

Quality of Life

Although some elderly do suffer from declining mental and physical functioning, many others do not. Being old does not mean being depressed, poor, and sick. Less than 6% of those age 65 and older suffer from any type of depression (NIMH 2003). Interestingly, Blumenthal and colleagues report that depression in the elderly is better treated by exercise than by medication alone or by medication and exercise combined (Livni 2000).

ThomsonNOW™

Learn more about **Physician-Assisted Suicide and Social Policy** by going through the Euthanasia Map Exercise.

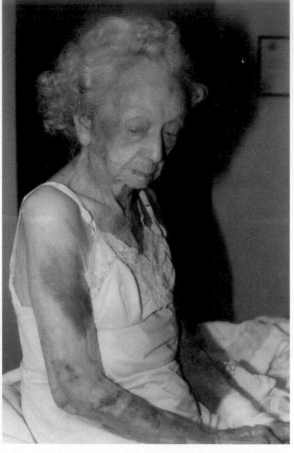

Courtesy of National Center on Elder Abuse

If you suspect elder abuse or are concerned about the well-being of an older person, call your state abuse hotline immediately.

Among the elderly who are depressed, two social factors tend to be in operation. One is society's negative attitude toward the elderly. Words and phrases such as "old," "useless," and "a has-been" reflect cultural connotations of the elderly that influence feelings of self-worth. The roles of the elderly also lose their clarity. How is a retiree supposed to feel or act? What does a retiree do? As a result, the elderly become dependent on external validation that may be weak or absent.

The second factor contributing to depression among the elderly is the process of growing old. This process carries with it a barrage of stressful life events, all converging in a relatively short time period. These include health concerns, retirement, economic instability, loss of significant other(s), physical isolation, job displacement, and increased awareness of the inevitability of death as a result of physiological decline. All these events converge on the elderly and increase the incidence of depression and anxiety and may affect the decision not to prolong life (see the *Focus on Technology* feature in this chapter).

Strategies for Action: Growing Up and Growing Old

Activism by or on behalf of children or the elderly has been increasing in recent years, and, as the numbers of children and the elderly grow, such activism is likely to escalate and to be increasingly successful. For example, global attention on the elderly led to 1999 being declared the International Year of Older Persons, and "the first nearly universally ratified human rights treaty in history" deals with children's rights (United Nations 1998, p. 1). Such activism takes several forms, including collective action through established organizations and the exercise of political and economic power.

Collective Action

Countless organizations are working on behalf of children, including the Children's Defense Fund, UNICEF, Children's Partnership, Children Now, Save the Children, and the Children's Action Network. Many successes take place at the local level, where parents, teachers, corporate officials, politicians, and citizens join together in the interest of children. In Kentucky AmeriCorp volunteers raised the reading competency of underachieving youths 116% in just six months; in Dayton, Ohio, behavioral problems of students at an elementary school were significantly reduced once "character education" was introduced; and in the Los Angeles Crenshaw High School student entrepreneurs established "Food from the Hood," a garden project that is expected to earn $50,000 in profit.

Some programs combine the interests of both the young and the old. The Adopt-a-Grandparent Program arranges for children to visit nursing home residents, and

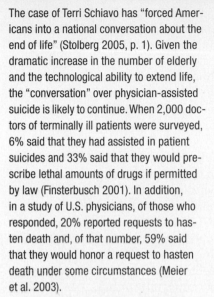

Physician-Assisted Suicide and Social Policy

The case of Terri Schiavo has "forced Americans into a national conversation about the end of life" (Stolberg 2005, p. 1). Given the dramatic increase in the number of elderly and the technological ability to extend life, the "conversation" over physician-assisted suicide is likely to continue. When 2,000 doctors of terminally ill patients were surveyed, 6% said that they had assisted in patient suicides and 33% said that they would prescribe lethal amounts of drugs if permitted by law (Finsterbusch 2001). In addition, in a study of U.S. physicians, of those who responded, 20% reported requests to hasten death and, of that number, 59% said that they would honor a request to hasten death under some circumstances (Meier et al. 2003).

Americans, in general, support physician-assisted suicide. When a random sample of Americans were asked, "When a person has a disease that cannot be cured, do you think that doctors should be allowed by law to end the patient's life by some painless means if the patient and the family request it?" 75% responded yes (Moore 2005). In addition, 63% of Americans disagreed with the 1997 U.S. Supreme Court decision that individuals do not have a constitutional right to doctor-assisted suicide (Taylor 2002). This decision, however, did validate the concept of double effect. **Double effect** refers to the use of medical interventions that relieve pain and suffering but that may also hasten death.

One of the most important factors influencing a physician's decision to restrict technological interventions is "family preference" (Randolph et al. 1997). The family's decision, however, is most often based on the physician's recommendations to limit care, whether it be withdrawal of life support (food, water, or mechanical ventilation), administering medications to end life (intravenous vasopressors), or withholding certain procedures that would prolong life (cardiopulmonary resuscitation) (Luce 1997).

As of 2005 only Oregon allows physician-assisted suicide through its Death with Dignity Act. Two physicians must agree that the patient is terminally ill and is expected to die within six months, the patient must ask three times for death, both orally and in writing, and the patient must swallow the barbiturates him- or herself rather than be injected with a drug by the physician. In 2002, 38 physician-assisted suicides occurred in Oregon—71% were male, 97% were white, 53% were married, and 84% had cancer. The median age was 69 (Vollmar 2003). Loss of autonomy and a "diminished ability to participate in activities that make life enjoyable" were the two most commonly cited "end-of-life concerns" (Center for Health Statistics 2003, p. 20).

In 2001 the U.S. attorney general legally challenged Oregon's Death with Dignity Act by instructing the Drug Enforcement Administration to "prosecute physicians and pharmacists who prescribe and dispense controlled substances under Oregon law" (Center for Health Statistics 2003). Oregon then brought suit against the attorney general, resulting in a finding favorable to the state; the federal government could not overturn Oregon's death with dignity law. In 2005 the U.S. Supreme Court agreed to hear the Bush administration's final challenge (Associated Press 2005).

The national debate over physician-assisted suicide was fueled by images of Dr. Jack Kevorkian administering a deadly dose of drugs, at the request of a terminally ill patient, on the CBS evening show *60 Minutes.* Although Kevorkian was convicted of second-degree murder for the *60 Minutes* death, advocates of physician-assisted suicide have tried to pass Oregon-like provisions in other states. As of 2005, Vermont, Hawaii, and California have pending physician-assisted suicide legislation.

Despite what some say is a movement toward a greater acceptance of physician-assisted suicide, the official position of the American Medical Association is that physicians must respect the patient's decision to forgo life-sustaining treatment but should not participate in physician-assisted suicide. One argument against physician-assisted suicide is that the practice is subject to abuses—a spouse with self-serving interests, a depressed patient making a hasty decision, or an overburdened family pressuring a vulnerable loved one. Concern also exists that legalizing physician-assisted suicide may disproportionately end the lives of minority, ethnic, or psychiatrically disturbed individuals (Allen 1998).

Some would argue, however, that ultimately the decision should reside with the patient. As one elderly person said:

> I came into this world as a human-being and I wish to leave in the same manner. Being able to walk, to communicate, to take care of my own needs, to think, to feel. . . . There is no need for me to re-experience my first few months of life through my last months in this world. . . . I do not wish to be once again in a diaper. Just the thought of me losing control over my body frightens me. (Leichtenritt & Rettig 2000, p. 3)

the Foster Grandparent Program pairs children with special needs with low-income elderly. In addition, the Children's Defense Fund, the Child Welfare League of America, the AARP, and the National Council on the Aging joined together in 1986 to create Generations United, a "national organization that focuses solely on promoting inter-generational strategies, programs and policies" (Generations United 2003, p. 1). It is one of the few national organizations that acts as an advocate on behalf of the young and the old.

More than a thousand organizations are directed toward realizing political power, economic security, and better living conditions for the elderly. One of the earliest and most radical groups is the Gray Panthers, founded in 1970 by Margaret Kuhn. The Gray Panthers were responsible for revealing the unscrupulous practices of the hearing aid industry, persuading the National Association of Broadcasters to add "age" to "sex" and "race" in the Television Code of Ethics statement on media images, and eliminating the mandatory retirement age. In view of these successes, it is interesting that the Gray Panthers, with only 40,000 members, is a relatively small organization compared to the AARP.

The AARP has more than 35 million members (10 times that of the National Rifle Association), age 50 and older, and has an annual budget of $800 million and nearly 2,000 employees (Birnbaum 2005). Services of the AARP include discounted mail-order drugs, investment opportunities, travel information, volunteer opportunities, and health insurance. The AARP is the largest volunteer organization in the United States with the exception of the Roman Catholic Church. Not surprisingly, it is one of the most powerful lobbying groups in Washington. Some observers note that the AARP is such a powerful lobbying group that "it holds the key to how or whether Social Security will be restructured" (Birnbaum 2005, p. 1).

Political Power

Children are unable to hold office, to vote, or to lobby political leaders. Child advocates, however, acting on behalf of children, have wielded considerable political influence in such areas as child care, education, health care reform, and crime prevention. In addition, funding of such programs is supported by most Americans. A Children's Defense Fund study found that the majority of those surveyed favored federally funded after-school programs and subsidized child care even if it meant raising taxes (CDF 2000).

As conflict theorists emphasize, the elderly compete with the young for limited resources. They have more political power than the young and more political power in some states than in others. In Florida there is concern that the elderly may eventually wield too much political power and act as a voting bloc, demanding excessive services at the cost of other needy groups. For example, if the elderly were concentrated in a particular district, they could block tax increases for local schools. To the extent that future political issues are age based and that the elderly are able to band together, their political power may increase as their numbers grow over time. By 2030 almost half of all adults in developed countries and two-thirds of all voters will be near or at retirement age. The growing political power of the elderly is already becoming evident; the Netherlands has a political party called the Pension Party.

Economic Power

Although children have little economic power, the economic power of the elderly has grown considerably in recent years, leading one economist to refer to the elderly as a "revolutionary class" (Thurow 1996). The 2003 median income for males

Protests are not limited to the young. The elderly have already been active in protesting changes to Social Security and other government policies that they believe are detrimental. As the number of elderly grows, social activism is likely to increase.

age 65 and older was $20,363; for females age 65 and older, $11,845. Households headed by individuals age 65 and older had a median income of $35,310 (AOA 2004). Although these incomes are significantly lower than the incomes for men and women between 45 and 54 years of age, income is only one source of economic power. The fact that many elderly own their homes, have substantial savings and investments, enjoy high levels of disposable income, and are growing in number contributes to their increased importance as consumers.

Finally, in many parts of the country the elderly have become a major economic power as part of what is called the **mailbox economy**. The mailbox economy refers to the tendency for a substantial portion of local economies to be dependent on pension and Social Security checks received in the mail by older residents (Atchley 2000). States with the highest proportion of the elderly—as a percentage of each state's population—include Florida (17%), Pennsylvania (15.4%), West Virginia (15.3%), North Dakota (14.8%), and Iowa (14.7%) (AOA 2004).

Fighting Discrimination

The number of age-based complaints to the U.S. Equal Employment Opportunity Commission (EEOC) has risen as the population has aged and the economy has worsened. In 1999 there were 14,141 complaints; in 2002, there were 19,921 complaints—a 41% increase. Of those filing complaints, 64% were between the ages of 40 and 59 (Nicholson 2003).

Modeled after Title VII (see Chapter 10), the Age Discrimination in Employment Act was passed in 1967 and protects workers between the ages of 40 and 65. In 1978 the upper age limit was extended to 70. In 1996 the U.S. Supreme Court heard *O'Connor* v. *Consolidated Coin Corp.,* which held that workers do not have to prove that they were replaced by someone younger than 40 to have a finding of age discrimination. One of the most significant cases was heard in 2003—*EEOC and Arnett et al.* v. *CalPERS* (California Public Employees' Retirement System). In this case the EEOC recovered $250 million for California public safety officers who had been discriminated against on the basis of age. Age discrimination suits have the lowest success rate of any of the eight protected classes, including sex, race, and religious discrimination (AARP 2003). However, in 2005 the U.S. Supreme Court made winning age bias suits a little easier. The Court ruled that, for an age discrimination lawsuit to be successful, workers who bring the suits against their employers do not have to prove that the discrimination was intentional (Greenhouse 2005).

Government Policy

When registered voters were asked what they thought was the single most important issue facing the government, three of the four top answers dealt with issues confronting the elderly—health care (discussed in Chapter 2), Social Security, and Medicare and Medicaid (CBS & New York Times 2000). As mentioned earlier, Medicaid provides health care for the poor; **Medicare** is a national health care insurance program designed for people over the age of 65. In 1988 Congress passed the Medicare Catastrophic Coverage Act (MCCA), which was the most significant change in Medicare since its establishment in 1966. The new benefits included unlimited hospitalization, an upper limit on the amount of money recipients would pay for physicians' services, home health care and nursing home services, and unlimited hospice care. These changes were particularly significant because many of the illnesses of the elderly are chronic in nature.

The reforms were financed by increasing monthly medical premiums by $4 a month and by imposing an annual fee based on a person's federal income tax bracket. The maximum premium paid was $800 per person and $1,600 per couple (Harris 1990; Torres-Gil 1990). The AARP initially supported the reforms but later withdrew its support, as did many other organizations. The additional monies paid were simply not worth the new benefits, they contended. The AARP also argued that the elderly should not have to bear the burden of reforms necessary for the general public. In 1989, under pressure from the AARP and other organizations of the elderly, Congress repealed the MCCA. Pressured by public demands and fear of bankruptcy, Congress is currently in the process of reforming health care policy (see Chapter 2), including Medicare. For example, the Medicare Prescription Drug, Improvement, and Modernization Act of 2003 extends prescription drug coverage to the elderly and the disabled beginning in 2006. The net federal cost of the new benefit is projected to be $724 billion from 2006 to 2015 (Kaiser Family Foundation 2005b). Congress is also considering the Elder Justice Act, which, if passed, will oversee services and prevention related to elder abuse, including monitoring long-term care facilities (OLPA 2005).

Other government initiatives for the elderly include the Elderly Nutrition Program and the Older Americans Act Amendments of 2000 (AOA 2001). The Elderly Nutrition Program provides delivered meals to locations where the elderly congregate—homes, senior centers, and schools—and provides nutritional training, counseling, and education. The Older Americans Act Amendments of 2000 created the National Family Caregiver Support Program. This program will "help hundreds of thousands of family caregivers of older loved ones who are ill or who have disabilities" (AOA 2003). The National Family Caregiver Support Program has been funded at $125 million a year through fiscal year 2005. Among other things the program will provide respite care, supplemental services, and individual counseling and training for caregivers (AOA 2003).

In 1997 Congress passed the State Children's Health Insurance Program (SCHIP), which "is the largest single expansion of health insurance coverage for children in more than 30 years" (U.S. Department of Health and Human Services 1999, p. 1). SCHIP was designed for families who cannot afford health insurance but whose incomes are too high to qualify for Medicaid. The program currently provides health care for 6 million children. However, in 2004 Congress failed to allocate the more than $1 billion in available federal funds to SCHIP, threatening the health coverage of more than 200,000 children (CDF 2004b).

Finally, there are several important child-centered legislative initiatives before Congress, including the Child Protective Services Improvement Act, the Give a Kid a Chance Omnibus Mental Health Services Act, and, perhaps most important, the MediKids Insurance Act of 2003. This act, in amending Social Security, would ensure that all children born after December 31, 2004, are guaranteed comprehensive health care insurance (Legislative Agenda 2005). There are presently 10 million children without health insurance coverage.

> "No man or woman stands as tall as one who stoops to help a child."
>
> **Teddy Roosevelt**
> **Former U.S. president**

Understanding Youth and Aging

What can we conclude about youth and aging in American society? Age is an ascribed status and, as such, is culturally defined by role expectations and implied personality traits. Society regards both the young and the old as dependent and in need of the care and protection of others. Society also defines the young and the old as physically, emotionally, and intellectually inferior. As a consequence of these and other attributions, both age groups are sociologically a minority with limited opportunity to obtain some or all of society's resources.

Although both the young and the old are treated as minority groups, different meanings are assigned to each group. In general, in the United States the young are more highly valued than the old. Structural functionalists argue that this priority on youth reflects the fact that the young are preparing to take over important statuses, whereas the elderly are relinquishing them. Conflict theorists emphasize that in a capitalist society, both the young and the old are less valued than more productive members of society. Conflict theorists also point out the importance of propagation, that is, the reproduction of workers, which may account for the greater value placed on the young than the old. Finally, symbolic interactionists describe the way images of the young and the old intersect and are socially constructed.

The collective concerns for the elderly and the significance of defining ageism as a social problem have resulted in improved economic conditions for the elderly. Currently, the elderly are one of society's more powerful minorities. Research indicates, however, that despite their increased economic status, the elderly are still subject to discrimination in such areas as housing, employment, and medical care and are victimized by systematic patterns of stereotyping, abuse, and prejudice.

In contrast, the position of children in the United States, although improving, remains tragic, with one in six children living in poverty. Wherever there are poor families, there are poor children, many of whom are educated in inner-city schools, live in dangerous environments, and lack basic nutrition and medical care. In addition, age-based restrictions limit the entry of these children into certain roles (e.g., employee) and demand others (e.g., student). Although most of society's members would agree that children require special protections, concerns regarding quality-of-life issues and rights of self-determination are only recently being debated.

Age-based decisions are potentially harmful. If budget allocations were based on indigence rather than age, more resources would be available for those truly in need. Furthermore, age-based decisions could eventually lead to intergenerational conflict. Government assistance is a zero-sum relationship—the more resources one group gets, the fewer resources another group receives.

Social policies that allocate resources on the basis of need rather than age would shift the attention of policy makers to remedying social problems rather than serving the needs of special interest groups. Age should not be used to cause negative effects on an individual's life any more than race, ethnicity, gender, or sexual orientation is. Although eliminating all age barriers or requirements is unrealistic, a movement toward assessing the needs of individuals and their abilities would be more consistent with the American ideal of equal opportunity for all.

Chapter Review

ThomsonNOW™

Reviewing is as easy as ① ② ③

1. Before you do your final review, take the ThomsonNOW diagnostic quiz to help you identify the areas on which you should concentrate. You will find information on ThomsonNOW and instructions on how to access all of its great resources on the foldout at the beginning of the text.

2. As you review, take advantage of ThomsonNOW's study videos and interactive Map the Stats exercises to help you master the chapter topics.

3. When you are finished with your review, take ThomsonNOW's posttest to confirm you are ready to move on to the next chapter.

- **What problems do the young and old have in common?**

  Among others, both the young and the old are victims of stereotypes, physical abuse, age discrimination, and poverty.

- **What age distinctions are commonly made in most societies?**

  Most societies make a distinction between childhood, adulthood, and elderhood. Childhood, often subdivided into infancy, childhood, and adolescence, is usually thought of as extending from birth to 17 years old; adulthood extends from 18 to 64 years old, and elderhood begins at 65 years of age.

- **What is disengagement theory?**

  According to disengagement theory, to achieve a balanced society, the elderly must relinquish their roles to younger members. Thus the young go to school, adults fill occupational roles, and the elderly, with obsolete skills and knowledge, disengage.

- **What are some of the problems of youth?**

  Problems of the young include (1) child labor, (2) orphaned children, (3) street children, (4) limited civil rights, (5) poverty, (6) foster care and adoption, (7) economic discrimination, (8) violence by and against children (e.g., gang violence, child abuse, and crime), and (9) health concerns.

- **What three independent variables affect the consequences of aging?**

  Race, sex, and social class affect the consequences of aging. Racial minorities, women, and the lower classes are more likely to suffer adversely from the aging process.

- **What are some of the problems of the elderly?**

  For the elderly who want to work, entering and remaining in the labor force may be difficult because of negative stereotypes, lower levels of education, reduced geographic mobility, fewer employable skills, and discrimination. Retirement may also be difficult because of role transition and lowered income. Poverty is a problem as well, with 30% of the elderly reporting incomes of less than $10,000. Some elderly also suffer from chronic health problems, including depression. Finally, elder abuse, although also present in private homes, is particularly problematic in institutionalized settings.

- **In terms of strategies for action, what are some of the similarities between the young and the old?**

  Both the young and the old have organizations working on their behalf—for example, the Children's Defense Fund and UNICEF, and the AARP and the Gray Panthers. Both groups have advocates working for their political and economic interests, although the elderly, according to conflict theorists, already wield considerable political and economic power. Finally, both the young and the old have benefited from government policies, such as the Older Americans Act, Medicare, and the State Children's Health Insurance Program (SCHIP).

Critical Thinking

1. In many ways American society discriminates against children. Children are segregated in schools, in a separate justice system, and in the workplace. Identify everyday examples of the ways in which children are treated like "second-class" citizens in the United States.

2. Age pyramids pictorially display the distribution of people by age (see Figure 12.1). How do different age pyramids influence the treatment of the elderly?
3. Regarding children and the elderly, what public policies or programs from other countries might be beneficial if they were adopted in the United States? Do you think that policies from other countries would necessarily be successful here?

Key Terms

activity theory	mailbox economy
age grading	Medicaid
ageism	Medicare
child labor	medigap
dependency ratio	modernization theory
disengagement theory	phased retirement
double effect	primary aging
elder abuse	sandwich generation
gerontophobia	secondary aging
infantilizing elders	senescence

Media Resources

The Companion Website for *Understanding Social Problems*, Fifth Edition

http://sociology.wadsworth.com/mooney_knox_schacht5e

Supplement your review of this chapter by going to the companion website to take one of the Tutorial Quizzes, use the flash cards to master key terms, and check out the many other study aids you'll find there. You'll also find special features such as *Wadsworth's Sociology Online Resources and Writing Companion,* GSS data, and Census 2000 information, data, and resources at your fingertips to help you complete that special project or do some research on your own.

> "Population may be the key to all the issues that will shape the future: economic growth; environmental security; and the health and well-being of countries, communities, and families."
>
> *Nafis Sadik, Executive Director, UN Population Fund*

Population Growth and Urbanization

The Global Context: A World View of Population Growth and Urbanization

Sociological Theories of Population Growth and Urbanization

Social Problems Related to Population Growth and Urbanization

Strategies for Action: Responding to Problems of Population Growth, Population Decline, and Urbanization

Understanding Problems of Population Growth and Urbanization

Chapter Review

W hen I think about population growth in the world, I conjure up an image of a bus hurtling down the highway toward what appears to be a cliff. The bus is semi-automatic and has no driver in charge of its progress. Some of the passengers on the bus are ignorant of what seems to lie ahead and are more worried about whether or not the air conditioning is turned up high enough or wondering how many snacks they have left for the journey. Other more alert passengers are looking down the road, but some of them think that what seems like a cliff is really just an optical illusion and is nothing to worry about; some think it may just be a dip, not really a cliff. Those who think it is a cliff are trying to figure out how to apply the brakes, knowing that a big bus takes a long time to slow down even after the brakes are put on. . . . The population bus is causing damage and creating vortexes of change as it charges down the highway, whether or not we are on the cliff route; and the better we understand its speed and direction, the better we will be at steering it and managing it successfully. (Weeks 2005, p. xxi)

Although thousands of years passed before the world's population reached 1 billion, in less than 300 years the population exploded from 1 billion to 6 billion. Population growth affects virtually every aspect of social life and "is the single most important set of events ever to occur in human history" (Weeks 2005, p. 4). In this chapter we provide an overview of the world's population situation, noting trends in population size and growth. Because nearly half of the world's population lives in urban areas, in this chapter we tie together the overlapping concerns of population growth and urbanization, focusing on problems of population growth and urbanization and strategies to alleviate these problems.

The Global Context: A World View of Population Growth and Urbanization

We begin this chapter with a brief overview of the history of world population growth, current population trends, and future population projections. We also review the development of urbanization and describe the current state of urbanization throughout the world.

World Population: History, Current Trends, and Future Projections

Humans have existed on this planet for at least 200,000 years. For 99% of human history population growth was restricted by disease and limited food supplies. Around 8000 B.C. the development of agriculture and the domestication of animals led to increased food supplies and population growth, but even then harsh living conditions and disease still put limits on the rate of growth. This pattern continued until the mid-18th century, when the Industrial Revolution improved the standard of living for much of the world's population. The improvements included better food, cleaner drinking water, improved housing and sanitation, and advances in medical technology, such as antibiotics and vaccinations against infectious diseases—all contributed to rapid increases in population.

Population Doubling Time. Population **doubling time** is the time required for a population to double from a given base year if the current rate of growth continues. It took several thousand years for the world's population to double from 4 million

Figure 13.1
World's 10 largest countries in population: 2005 and 2050 (projected).

Source: Based on data from Population Reference Bureau (2005).

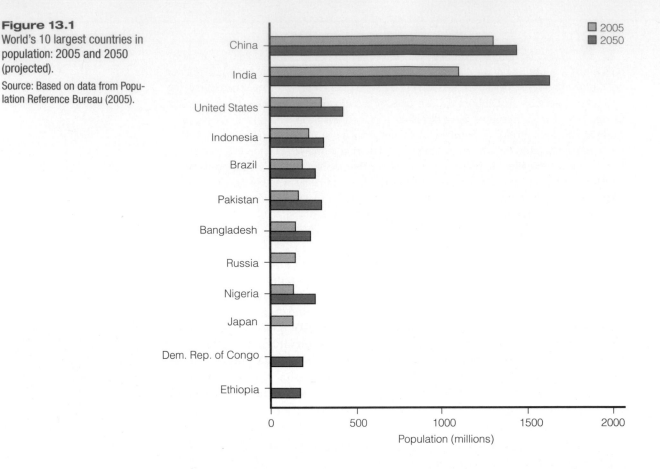

to 8 million, a few thousand years to double from 8 million to 16 million, about 1,000 years to double from 16 million to 32 million, and less than 1,000 years to double to 64 million. The next doubling of the world's population occurred between the European Renaissance and the Industrial Revolution in a time span of about 400 years. But after 1750 the doubling of world population took little more than 100 years, and the next doubling took less time. The most recent doubling (from 3 billion in 1960 to 6 billion in 1999) took only about 40 years (Weeks 2005).

A look at current population trends and future projections suggests that although world population will continue to grow in the coming decades, it may never double in size again. In fact, because fertility rates have dropped around the world, "we're finally at a point where it's possible that a child born today will live to see the stabilization of world population" (Nelson 2004, p. 17).

Current Population Trends and Future Projections. According to the United Nations (2005b), the world's population is growing at an annual rate of 1.14%, resulting in the addition of 76 million people per year. Projections of future population growth suggest that world population will grow from 6.5 billion in 2005 to 9.1 billion in 2050. Although China is the most populated country in the world today, India will become the most populated country by 2050 (see Figure 13.1).

Ninety-nine percent of world population growth is in developing countries, mostly in Africa and Asia. As the size of a country's population grows, so does its

Table 13.1

Population Density

Area	Population Density (People per Square Mile)
World	125
More-developed countries	61
Less-developed countries	165

Source: Population Reference Bureau (2005).

population density, or the number of people per unit of land area. Overall, population density is higher in less-developed countries (see Table 13.1), which means that people in less-developed countries live in more-crowded conditions. The population density of India, for example, is 869 people per square mile, compared with 80 people per square mile in the United States. To get an idea of how crowded the population is in India, consider that India has one-third the land area of the United States but nearly four times the population. Imagine if the U.S. population quadrupled. Then imagine that this quadrupled population all lived in the eastern third of the United States. That will give you an idea of how crowded living conditions are in India. And many countries have even more crowded living conditions compared with India. In 18 countries the population density is more than 1,000 people per square mile (Population Reference Bureau 2005).

Higher population growth in developing countries is largely due to higher **total fertility rates**—the average lifetime number of births per woman in a population. As shown in Figure 13.2, the least developed countries of the world have the highest rates of fertility and population growth. Fertility rates range from 8 births to women in the African country of Niger to 1.2 births per woman in Central and Eastern European countries and South Korea.

Population size is affected not only by fertility rates but also by immigration. Much of the population growth occurring in the United States is due to high immigration rates. In other countries populations are changing as a result of HIV/AIDS. For example, despite their high fertility rates, the four African countries most affected by HIV/AIDS—Botswana, Lesotho, Namibia, and Swaziland—are expected to experience a decline in population over the coming decades as deaths (resulting from AIDS) outnumber births.

Will there be an end to the rapid population growth that has occurred in recent decades? Will the population of the world stabilize? Although some predict that population will stabilize around the middle of the 21st century, no one knows for sure. There has been a significant reduction in fertility rates around the world, from a global average of 5 children per woman in the 1950s to 2.65 children in 2005 (United Nations 2005b). To reach population stabilization, fertility rates throughout the world would need to achieve what is called "replacement level," whereby births would replace, but not outnumber, deaths. **Replacement-level fertility** is 2.1 births per woman, that is, slightly more than 2 because not all female children will live long enough to reach their reproductive years. More than 50 countries have already achieved below-replacement fertility rates, and by 2050 the average fertility rate worldwide is projected to be below replacement level (see Figure 13.2). However,

ThomsonNOW™

Learn more about **Population Density** by going through the Population Map Exercise.

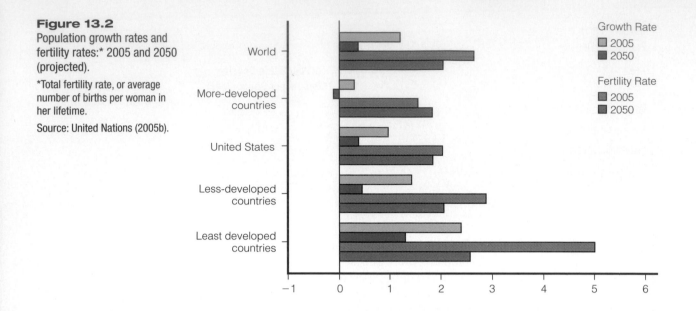

Figure 13.2
Population growth rates and fertility rates:* 2005 and 2050 (projected).

*Total fertility rate, or average number of births per woman in her lifetime.

Source: United Nations (2005b).

Growth Rate
□ 2005
■ 2050

Fertility Rate
■ 2005
■ 2050

"Are we experiencing a population explosion or birth dearth? The answer may be both."

Population Reference Bureau

ThomsonNOW™

Learn more about **Population** by going through the Three Basic Demographic Processes Animation.

even if every country in the world achieved replacement-level fertility rates, populations would continue to grow for several decades because of **population momentum**—continued population growth as a result of past high fertility rates that have resulted in a large number of young women who are currently entering their childbearing years. So despite the below-replacement fertility rates in more developed regions, population in these regions is expected to continue to grow until about 2030 and then to begin to decline. The U.S. population, however, will continue to increase through 2050 because of immigration.

In sum, there are two population trends occurring simultaneously that, on the surface, appear to be contradictory: (1) The total number of people on this planet is rising and is expected to continue to increase over the coming decades; and (2) about 40% of the world's population lives in countries in which couples have so few children that the countries' populations are likely to decline over the coming years (Population Reference Bureau 2004b). These countries include China, Japan, and most of Europe. As we discuss later in this chapter, each of these trends presents a set of problems and challenges.

An Overview of Urbanization Worldwide and in the United States

As early as 5000 B.C., cities of 7,000 to 20,000 people existed along the Nile, Tigris-Euphrates, and Indus River valleys. But not until the Industrial Revolution in the 19th century did **urbanization,** the transformation of a society from a rural to an urban one, spread rapidly.

As population has increased, so has the proportion of people living in urban areas. An **urban area** is a spatial concentration of people whose lives are centered around nonagricultural activities. Although countries differ in their definitions of "urban," most countries designate places with 2,000 people or more as being urbanized. According to the U.S. census definition, an "urban population" consists of individuals living in cities or towns of 2,500 or more inhabitants. An "urbanized

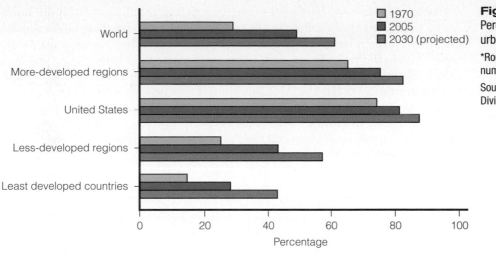

Figure 13.3
Percentage* of population in urban areas, by year.

*Rounded to nearest whole number.

Source: United Nations Population Division (2004).

area" refers to one or more places and the adjacent densely populated surrounding territory that together have a minimum population of 50,000.

The share of the global population living in urban areas has increased from 29% in 1950 to 49% in 2005 and is expected to reach 61% by 2030 (United Nations Population Division 2004). In more developed regions 76% of the population already lives in urban areas (see Figure 13.3). Virtually all the population growth expected between 2000 and 2030 will be in urban areas, primarily in less-developed regions. Although the urban population is expected to rise from 3 billion to 5 billion between 2003 and 2030, the rural population is expected to decline slightly from 3.3 billion to 3.2 billion in the same time period (United Nations Population Division 2004).

The number of **megacities**—urban areas with 10 million residents or more—is also increasing. In 1950 there was only one city with more than 10 million inhabitants: New York City. By 2015, 23 cities are projected to have more than 10 million people; all but four will be in less developed countries (Population Reference Bureau 2004a).

Increasing urbanization results about equally from births in urban areas and from migration of people from rural areas to the cities (United Nations Population Fund 1999). Rural dwellers migrate to urban areas to flee war or natural disasters or to find employment. As foreign corporate-controlled commercial agriculture displaces traditional subsistence farming in poor rural areas, peasant farmers flock to the city looking for employment. Some rural dwellers migrate to urban areas in search of a better job—one that has higher wages and better working conditions. Governments have also stimulated urban growth by spending more to improve urban infrastructures and services while neglecting the needs of rural areas (Clark 1998).

History of Urbanization in the United States. Urbanization of the United States began as early as the 1700s, when most major industries were located in the most populated areas, including New York City, Philadelphia, and Boston. Unskilled laborers, seeking manufacturing jobs, moved into urban areas as industrialization accelerated in the 19th century. The "pull" of the city was not the only reason for urbanization, however. Technological advances were making it possible for fewer farmers to work the same amount of land. Thus "push" factors were also involved—making a living as a farmer became more and more difficult as technology, even then, replaced workers.

Urban populations continued to multiply as a large influx of European immigrants in the late 1800s and early 1900s settled in U.S. cities. This influx was followed by a major migration of southern rural blacks to northern urban areas. People were lured to the cities by the promise of employment and better wages and such urban amenities as museums, libraries, and entertainment. Immigrants are also attracted to cities with large ethnic communities that provide a familiar cultural environment in which to live and work.

Over the last two centuries the growth of the U.S. urban population—which includes residents of inner cities and their surrounding suburbs—has been dramatic. Between 1800 and 2005 the percentage of the U.S. population that is urban grew from 6.1% to 81%.

Suburbanization. In the late 19th century railroad and trolley lines enabled people to live outside the city and still commute into the city to work. As more and more people moved to the **suburbs**—urban areas surrounding central cities—the United States underwent **suburbanization.** As city residents left the city to live in the suburbs, cities lost population and experienced **deconcentration,** or the redistribution of the population from cities to suburbs and surrounding areas.

Many factors have contributed to suburbanization and deconcentration. After World War II many U.S. city dwellers moved to the suburbs out of concern for the declining quality of city life and the desire to own a home on a spacious lot. Suburbanization was also spurred by racial and ethnic prejudice, as the white majority moved away from cities that, because of immigration, were becoming increasingly diverse. Mass movement into suburbia was encouraged by the federal interstate highway system (financed by the government under the guise of ensuring national defense), the affordability of the automobile, and the dismantling of metropolitan mass transit systems (Lindstrom & Bartling 2003). In the 1950s Veterans Administration and Federal Housing Administration loans made housing more affordable, enabling many city dwellers to move to the suburbs. Suburb dwellers who worked in the central city could commute to work or work in a satellite branch in suburbia that was connected to the main downtown office. As increasing numbers of people moved to the suburbs, so did businesses and jobs. Without a strong economic base, city services and the quality of city public schools declined, which furthered the exodus from the city.

U.S. Metropolitan Growth and Urban Sprawl. Simply defined, a **metropolitan area** is a densely populated core area together with adjacent communities. The largest city in each metropolitan area is designated the **central city.** Another term for metropolitan area is **metropolis,** from the Greek roots meaning "mother city."

Metropolises have grown rapidly in the United States. As new areas reach the minimum required city or urbanized area population and as adjacent towns, cities, and counties satisfy the requirements for inclusion in metropolitan areas, both the number and size of metropolitan areas have grown. Most Americans live in one of the more than 360 metropolitan areas in the nation. One U.S. state—New Jersey—is entirely occupied by metropolitan areas as designated by the U.S. census.

The growth of metropolitan areas is often referred to as **urban sprawl**—the ever-increasing outward growth of urban areas. Urban sprawl results in the loss of green open spaces, the displacement and endangerment of wildlife, traffic congestion and noise, and pollution liabilities—problems that we discuss later in this chapter.

Those who enjoy the conveniences and amenities of urban life but who find large metropolitan areas undesirable may choose to live in a **micropolitan area**—

a small city (between 10,000 and 50,000 people) located beyond congested metropolitan areas. These areas are large enough to attract jobs, restaurants, community organizations, and other benefits yet are small enough to elude traffic jams, high crime rates, and high costs of housing and other living expenses associated with larger cities.

Sociological Theories of Population Growth and Urbanization

The three main sociological perspectives—structural functionalism, conflict theory, and symbolic interactionism—can be applied to the study of population and urbanization.

Structural-Functionalist Perspective

Structural functionalism focuses on how changes in one aspect of the social system affect other aspects of society. For example, the **demographic transition theory** of population describes how industrialization has affected population growth. According to this theory, in traditional agricultural societies high fertility rates are necessary to offset high mortality and to ensure continued survival of the population. As a society becomes industrialized and urbanized, improved sanitation, health, and education lead to a decline in mortality. The increased survival rate of infants and children along with the declining economic value of children leads to a decline in fertility rates. About one-third of the world's countries have completed the demographic transition—the progression from a population with short lives and large families to one in which people live longer and have smaller families (Cincotta et al. 2003). Many low-fertility countries have entered what is known as a "second demographic transition," in which fertility falls below the two-child replacement level. This second demographic transition has been linked to greater educational and job opportunities for women, increased availability of effective contraception, and the rise of individualism and materialism (Population Reference Bureau 2004b).

Urbanization plays a significant role in the demographic transition. Because health care delivery is more cost-effective in cities than in rural areas, governments prioritize urban health clinics. With greater access to health care urban dwellers are the first to experience declines in infant mortality and fertility. A study of contraceptive use in Kenya found that women who lived in urban areas were more likely to have used contraception (Kimuna & Adamchak 2001). Recent declines in fertility throughout Africa are mostly an urban phenomenon (Cincotta et al. 2003).

Structural functionalists view the development of urban areas as functional for societal development. Although cities initially served as centers of production and distribution, today they are centers of finance, administration, education, health care, and information.

Urbanization is also dysfunctional, because it leads to increased rates of anomie, or normlessness, as the bonds between individuals and social groups become weak (see also Chapters 1 and 4). Whereas in rural areas social cohesion is based on shared values and beliefs, in urban areas social cohesion is based on interdependence created by the specialization and social diversity of the urban population. Anomie is linked to higher rates of deviant behavior, including crime, drug addiction, and alcoholism. Overcrowding, poverty, the rapid spread of infectious disease, and environmental destruction are also considered dysfunctions associated with urbanization.

Conflict Perspective

The conflict perspective focuses on how wealth and power, or the lack thereof, affect population problems. In 1798 Thomas Malthus predicted that the population would grow faster than the food supply and that masses of people were destined to be poor and hungry. According to Malthusian theory, food shortages would lead to war, disease, and starvation, which would eventually slow population growth. However, conflict theorists argue that food shortages result primarily from inequitable distribution of power and resources (Livernash & Rodenburg 1998).

Conflict theorists also note that population growth results from pervasive poverty and the subordinate position of women in many less developed countries. Poor countries have high infant and child mortality rates. Hence women in many poor countries feel compelled to have many children to increase the chances that some will survive into adulthood. Their subordinate position prevents many women from limiting their fertility. For example, in 14 countries around the world a woman must get her husband's consent before she can receive any contraceptive services (United Nations Population Fund 1997). Thus, according to conflict theorists, population problems result from continued economic and gender inequality.

Power and wealth also affect the development and operations of urban areas. The capitalistic pursuit of wealth contributed to the development of cities, because capitalism requires that the production and distribution of goods and services be centrally located, thus, at least initially, leading to urbanization. Today, global capitalism and corporate multinationalism, in search of new markets, cheap labor, and raw materials, have largely spurred urbanization of the developing world. Capitalism also contributes to migration from rural areas into cities because peasant farmers who have traditionally produced goods for local consumption are being displaced by commercial agriculture that is geared to producing fruits, flowers, and vegetables for export to the developed world. Displaced from their traditional occupations, peasant farmers are flocking to cities to find employment (Clark 1998).

"Cities have been ignored because they are blacker, browner, poorer, and more female than the rest of the nation."

Julianne Malveaux
Writer and scholar

The conflict perspective also focuses on how individuals and groups with wealth and power influence decisions that affect urban populations. For example, according to citizens' groups working to stop urban sprawl in Central and Eastern Europe, city officials may be bribed to approve a new shopping mall or other development project (Sheehan 2001). In addition, deteriorating conditions in U.S. inner cities are often ignored because the residents of inner cities lack the wealth, power, and status to solicit government spending on needed infrastructure and services, such as sidewalk and road repairs, street cleaning, and beautification projects.

Symbolic Interactionist Perspective

The symbolic interactionist perspective focuses on how meanings, labels, and definitions learned through interaction affect population problems. For example, many societies are characterized by **pronatalism**—a cultural value that promotes having children. Throughout history many religions have worshiped fertility and recognized it as being necessary for the continuation of the human race. In many countries religions prohibit or discourage birth control, contraceptives, and abortion. Women in pronatalistic societies learn through interaction with others that deliberate control of fertility is socially unacceptable. Women who use contraception in communities where family planning is not socially accepted face ostracism by their community, disdain from relatives and friends, and even divorce and abandonment by their husbands (Women's Studies Project 2003). However, once some women learn new definitions of fertility control, they become role models and influence the attitudes and behaviors of others in their personal networks (Bongaarts & Watkins 1996). This chapter's *Human Side* feature presents quotes from women and men around the world that relate to family planning and what it means to them.

The symbolic interaction perspective can also be applied to understanding how urban life affects interaction patterns and social relationships. The classical and modern views represent different observations of how urban living affects social relationships.

Classical Theoretical View. Cities have the reputation of being cold and impersonal. George Simmel observed that urban living involved an overemphasis on punctuality, individuality, and a detached attitude toward interpersonal relationships (Wolff 1978). It is not difficult to find evidence to support Simmel's observations: New Yorkers pushing each other to get onto the subway during rush hour; motorists cursing each other in Los Angeles traffic jams; Chicago residents ignoring the homeless man asleep on the sidewalk.

Louis Wirth (1938), a second-generation student of Simmel, argued that urban life is disruptive for both families and friendships. He believed that because of the heterogeneity, density, and size of urban populations, interactions become segmented and transitory, resulting in weakened social bonds. Wirth held that as social solidarity weakens, people exhibit loneliness, depression, stress, and antisocial behavior.

Modern Theoretical View. In contrast to Wirth's pessimistic view of urban areas, Herbert Gans (1984) argued that cities do not interfere with the development and maintenance of functional and positive interpersonal relationships. Among other communities Gans studied an Italian urban neighborhood in Boston and found such neighborhoods to be community oriented and marked by close interpersonal ties. Rather than finding the social disorganization described by Wirth, Gans observed that kinship and ethnicity helped bind people together. Intimate small groups with strong social bonds characterized these enclaves. Thus Gans saw the city as a patchwork quilt of different neighborhoods or urban villages, each of which helped individuals deal with the pressures of urban living.

A Theoretical Synthesis. Fisher (1982) interviewed more than 1,000 respondents in various urban areas and found evidence for both the classical and the modern theoretical views of urbanism. From the classical perspective he found that heterogeneity (the diversity among urban residents) does make community integration and consensus difficult—community cohesion is less tight. Ties that do exist are less often kin related than in nonurban areas and are more often based on work relationships, memberships in voluntary and professional organizations, and proximity to neighbors.

Fisher also found, however, that the diversity of urban populations facilitates the development of subcultures that have a sense of community ties. For example, large urban areas include such diverse groups as gays, ethnic and racial minorities, and artists. These individuals find each other and develop their own unique subcultures.

Voices from Around the World: Women and Men Talk About Family Planning

The Women's Studies Project involves numerous studies in 14 countries that assess the impact of family planning on women's lives. The following excerpts from focus group discussions and in-depth interviews provide glimpses into the family planning experiences and attitudes of women and men around the world (Women's Studies Project 2003). Comments made by respondents in the Women's Studies Project suggest that women's family planning experiences—their contraceptive use and nonuse, their pregnancies and childbearing, and their experiences with family planning and reproductive health programs—affect other aspects of their lives, including their roles as individuals, mothers, and wives; their participation in the workforce; their self-esteem; and their marital and sexual satisfaction.

If family planning had been available earlier, my future would have been different. That is my life-long regret. Because I had too many children, I had to quit [teaching].

Woman in South Jiangsu, China

Yes, people are happy with family planning. They see that their family is in harmony, their children are big enough to take care of themselves, while the mother can take care of herself.

Woman in rural northern Sumatra, Indonesia

If I had [had] access to the method of preventing pregnancy, I wouldn't have been pregnant . . . and I would be working somewhere in town, and maybe I would be having a better life than this one.

Zimbabwean woman

Without family planning and the consequent child spacing and limitation, there is no quality of life. As a woman, you cannot get enough time to give love to your children and your husband if you have many children.

Zimbabwean woman

My parents had eight children. My father died when I was 20—there was no money for the doctor. Some siblings were given to other families. Having too many children—not only do the parents suffer, but also the children, with bad nutrition and bad housing conditions.

Woman in China

Having four children nearly made me crazy. I couldn't give them food and clothes. They wandered from door to door and were driven away like dogs. One day my son asked, "Why did you give me birth if you can't feed me?"

Woman in Bangladesh

Because you have free time to take care of your husband, you can see the affection is reborn.

Contraceptive user in Mali

The man of the house never likes it if the woman can't work. He says, "Did I marry you to keep you as a pet? I married you to work

---

Other research has also found support for a synthesis of the classical and modern views of urban life. Tittle (1989) found that the larger the size of the community, the weaker a respondent's social bonds and the higher his or her anonymity, tolerance, alienation, and reported incidence of deviant behavior. Tittle also found that racial and ethnic groups created their own sense of community in urban neighborhoods.

Social Problems Related to Population Growth and Urbanization

Next, we examine some of the social problems related to population growth and urbanization, including environmental problems, poverty and unemployment, global insecurity, poor maternal and infant health, transportation and traffic problems,

in my house! If you sit around, who will look after the children, and who will do all of the chores?" This is why I stopped taking the [birth control] pills.

Woman in Bangladesh who experienced
negative side effects from using the Pill

A woman who lived with us, she used family planning. She fell ill and even had two operations. She has not had any more children. . . . When I saw her experience, I was afraid.

Mali woman, explaining why she
does not use contraception

People said many things about my having the operation [sterilization]. . . . "Don't you stand next to us. Stay away! Even to look at you is a sin!" I would just weep when people said those things to me.

Woman from rural Bangladesh

I talked secretly with eight or ten women about this. Some of the women said, "If the elders find out about anyone having this op-

eration, they will not let her live in the village anymore. No one will eat food cooked by a woman who has been operated on."

Woman from rural Bangladesh
who sought sterilization

The husband always has the final say. What happens is that women are limited in their thinking, and if you do not show your dominance, you will have problems.

Zimbabwean man who believes that men
should control family planning decisions

If my wife makes the decision to use family planning without my consent, I would divorce her.

Malian man

If I want four children and my wife wants six, she has to listen to me because I am the one who supports the family financially. If I decide to have five children, this is because I know I can look after them. The husband is the head of the family, and the wife can never

tell me the number of children she wants to have.

A husband in Zimbabwe

The phrase "many children, more economic future" is out-of-date. Today, many children means lots of problems, lots of responsibility.

Man from Ujung Pandang, Indonesia

From *Women's Voices, Women's Lives: The Impact of Family Planning.* Family Health International, 2003, Women's Studies Project.

and the effects of sprawl on wildlife and human health. First, we consider the problems faced by countries where fertility rates are low and population is declining.

Problems Associated with Below-Replacement Fertility

In more than one-third of the world's countries—including China, Japan, and all of Europe—fertility rates have fallen well below the 2.1 children replacement level (Butz 2005). Because these low fertility rates will eventually lead to population decline, some reports have sounded an alarm about the possibility of a "birth dearth." Low fertility rates lead not only to a decline in population size but also to an increasing proportion of elderly members. A birth dearth eventually results in fewer workers to support the pension, social security, and health care systems for the elderly. Below-replacement fertility rates also raise concern about a country's ability to maintain a productive economy, because there may not be enough future workers to replace current workers as they age and retire.

Environmental Problems and Resource Scarcity

As we discuss in Chapter 14 (on environmental problems), population growth places increased demands on natural resources, such as forests, water, cropland, and oil, and results in increased waste and pollution. Over the last 50 years the earth's ecosystems have been degraded more rapidly and extensively than in any other comparable period of time in human history (Millennium Ecosystem Assessment 2005).

The countries that suffer most from shortages of water, farmland, and food are developing countries with the highest population growth rates. For example, about one-third of the developing world's population (1.7 billion people) live in countries with severe water stress (Weiland 2005). However, countries with the largest populations do not necessarily have the largest impact on the environment. This is because the impact that each person makes on the environment—each person's **environmental footprint**—is determined by the patterns of production and consumption in that person's culture. The environmental footprint of an average person in a high-income country is about six times bigger than that of someone in a low-income country and many more times bigger than in the least developed nations (United Nations Population Fund 2004). For example, even though the U.S. population is only one-fourth the size of India's, its environmental footprint is more than three times bigger—the United States releases 15.7 million tons of carbon into the atmosphere each year, compared to India's 4.9 million tons. Hence, although population growth is a contributing factor in environmental problems, patterns of production and consumption are at least as important in influencing the effects of population on the environment.

Poverty and Unemployment

Poverty and unemployment are problems that plague countries with high population growth as well as urban areas in both rich and poor countries. Less-developed, poor countries with high birthrates do not have enough jobs for a rapidly growing population, and land for subsistence farming becomes increasingly scarce as populations grow. In some ways poverty leads to high fertility, because poor women are less likely to have access to contraception and are more likely to have large families in the hope that some children will survive to adulthood and support them in old age. But high fertility also exacerbates poverty, because families have more children to support and national budgets for education and health care are stretched thin.

In many countries in developing regions one in four urban residents lives in absolute poverty (National Research Council 2003). In some of the world's poorest developing countries half the urban population lives in conditions of extreme deprivation. In the United States the highest rates of poverty are in the central cities (see Figure 13.4), in part because many central cities face unemployment rates that are much higher than the national average.

In the United States and other industrialized countries urban unemployment and poverty are partly the results of deindustrialization, or the loss and/or relocation of manufacturing industries. Since the 1970s many urban factories have closed or relocated, forcing blue-collar workers into unemployment. When prospects of finding decent employment are low, the resulting feelings of frustration and worthlessness can lead to drug use, crime, and violence (Rodriquez 2000).

Urban Housing and Sanitation Problems. In both developed and less-developed countries the urban poor struggle to find affordable decent housing. Many cities are experiencing a housing crisis, because the number of low-income renters has in-

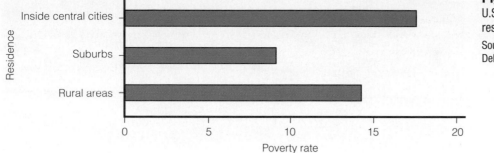

Figure 13.4
U.S. poverty rates by
residence: 2003.

Source: Based on data from
DeNavas-Walt et al. (2004).

creased while the number of low-cost rental units has dropped. Housing that is available and affordable is often substandard, characterized by outdated plumbing and wiring, overcrowding, rat infestations, toxic lead paint, and fire hazards (see also Chapter 6).

Low-income housing tends to be concentrated in inner-city areas of extreme poverty, but jobs are increasingly moving to the suburbs, and central-city residents often lack transportation to get to these jobs. The more affluent suburbs restrict development of affordable housing to keep out "undesirables" and maintain their high property values. Suburban zoning regulations that require large lot sizes, minimum room sizes, and single-family dwellings serve as barriers to low-income development in suburban areas (Orfield 1997).

Concentrated areas of poverty and poor housing in urban neighborhoods are called **slums.** In the United States slums that are occupied primarily by African Americans are known as **ghettos,** and those occupied primarily by Latinos are called **barrios.** Nearly one in three city dwellers worldwide—almost 1 billion people—live in slums characterized by overcrowding, little employment, and poor water, sanitation, and health care services (United Nations 2005a). Half the urban population in Africa, Asia, Latin America, and the Caribbean suffers from disease associated with inadequate water and sanitation (Millennium Ecosystem Assessment 2005).

Global Insecurity

A Population Institute report titled *Breeding Insecurity: Global Security Implications of Rapid Population Growth,* warns that rapid population growth is a contributing factor to global insecurity, including civil unrest, war, and terrorism (Weiland 2005). The report points out that many developing countries are characterized by a "youth bulge"—a high proportion of 15- to 29-year-olds relative to the adult population. Youth bulges result from high fertility rates and declining infant mortality rates, a common pattern in developing countries today. Youth bulges are growing rapidly throughout Africa and the Middle East. For example, in Iraq young adults make up nearly half (47%) of the adult population. The combination of a youth bulge with other characteristics of rapidly growing populations, such as resource scarcity, high unemployment rates, poverty, and rapid urbanization, sets the stage for political unrest. "Large groups of unemployed young people, combined with overcrowded cities and lack of access to farmland and water create a population that is angry and frustrated with the status quo, and thus more likely to resort to violence to bring about change" (Weiland 2005, p. 3).

Nearly one in three city dwellers—almost 1 billion people—live in slums characterized by overcrowding, little employment, and poor water, sanitation, and health care services.

© David Turnley/Corbis

Poor Maternal, Infant, and Child Health

As noted in Chapter 2, maternal deaths (deaths related to pregnancy and childbirth) are the leading cause of mortality for reproductive-age women in the developing world. Having several children at short intervals increases the chances of premature birth, infectious disease, and death for the mother or the baby. Childbearing at young ages (teens) has been associated with anemia and hemorrhage, obstructed and prolonged labor, infection, and higher rates of infant mortality (Zabin & Kiragu 1998). In developing countries one in four children is born unwanted, increasing the risk of neglect and abuse. In addition, the more children a woman has, the fewer the parental resources (parental income and time and maternal nutrition) and social resources (health care and education) available to each child. The adverse health effects of high fertility on women and children are, in themselves, compelling reasons for providing women with family planning services. "Reproductive health and choice are often the key to a woman's ability to stay alive, to protect the health of her children and to provide for herself and her family" (Catley-Carlson & Outlaw 1998, p. 241).

Transportation and Traffic Problems

Urban areas are often plagued with transportation and traffic problems. A study of 85 U.S. urban areas found that in 2003 traffic congestion caused 3.7 billion hours of traffic delay and wasted 2.3 billion gallons of fuel. The average annual delay per traveler increased from 16 hours in 1982 to 40 hours in 1993 and 47 hours in 2003 (Schrank and Lomax 2005).

Many public roads in urban areas are afflicted with what some call *autosclerosis*—"clogged vehicular arteries that slow rush hour traffic to a crawl or a stop, even when there are no accidents or construction crews ahead" ("Bridge to the

21st Century," 1997, p. 1). According to Jan Lundberg, director of the Alliance for a Paving Moratorium, "the average vehicle speed for crosstown traffic in New York City is less than six miles per hour—slower than it was in the days of horse-drawn buggies" (quoted in Jensen 2001, p. 6). And "traffic jams in Atlanta have been so entangled that babies have been born in traffic standstills, and some desperate drivers have had to leave their cars to relieve themselves behind roadside bushes" (Shevis 1999, p. 2). Traffic congestion creates stress on drivers, which sometimes leads to aggressive driving and violent reactions to other drivers—a phenomenon known as **road rage.**

Cars and light trucks are the largest single source of air pollution (Union of Concerned Scientists 2003). Air pollution and traffic congestion have been major forces that drive residents and businesses away from densely populated urban areas (Warren 1998). Health problems associated with congested traffic include stress, respiratory problems, and death. More than 20 million people are severely injured or killed on the world's roads each year (World Health Organization 2003). Indeed, far more people are killed and injured in automobile accidents than by violent crime. Therefore it has been argued that, despite higher crime rates in the inner city, the suburbs are the more dangerous place to live, because suburbanites "drive three times as much, and twice as fast, as urban dwellers" (Durning 1996, p. 24).

In addition to the day-to-day concerns of traffic congestion and air pollution from vehicles, cities are faced with protecting their public transportation systems from terrorist attacks, such as the 2004 bombing of a commuter train in Madrid and the 2005 bombings of London's public transit system. Finally, Hurricanes Katrina and Rita in 2005 reminded city officials that they must have a transportation plan in the event of an evacuation.

© Peter Johnson/Corbis

Worldwide, pregnancy is the leading cause of death for young women, ages 15 to 19. Most (95%) maternal deaths occur in Africa and Asia. This woman in sub-Saharan Africa has a 1 in 16 risk of dying in pregnancy or childbirth, compared to a 1 in 2,800 chance for a woman in a developed country.

Effects of Sprawl on Wildlife and Human Health

The spread of urban and suburban areas increasingly is replacing natural habitats with pavement, buildings, and human communities. The loss of open green space, trees, and plant life affects animals whose homes are turned into parking lots, shopping centers, office buildings, and housing developments. According to the U.S. Fish and Wildlife Service, habitat loss resulting from urban and suburban sprawl is the number one reason that wildlife species are becoming increasingly endangered (Shevis 1999) (see also Chapter 14).

Evidence of wildlife displacement resulting from sprawl is found across the nation. Coyotes, normally found only in the West and in Appalachia, are now being sighted in every state (Shevis 1999). Other displaced species include bear, Canada geese, and deer.

Sprawl also poses health hazards to humans. As more wooded areas are cleared for development, deer are displaced. "With no place to go, they bound into suburban backyards in search of food and water and across highways, frequently injuring

❝Everybody says that living in the inner city is dangerous, but the truth is that, if you take car crashes into account, the suburbs are statistically far more dangerous places to live.❞

Jan Lundberg
Director of the Alliance for a Paving Moratorium

themselves and causing harm to drivers. . . . Deer cause an estimated half-million vehicle accidents a year, killing 100 people and injuring thousands more" (Shevis 1999, pp. 2–3).

Suburban sprawl has also brought humans into greater contact with ticks that carry Lyme disease (UNEP 2005). If bitten by an infected tick, humans can develop various symptoms, including skin rash, neurological problems, fatigue, abdominal and joint pain, headache, and heart damage. Other health problems associated with sprawl include lack of physical activity and obesity. As discussed in this chapter's *Social Problems Research Up Close* feature, people who live in high-sprawl areas tend to walk less than people who live in more compact communities. You can assess your attitudes toward walking and proposals to create more walkable communities in this chapter's *Self and Society* feature.

Suburban sprawl brings humans into greater contact with ticks that carry Lyme disease. If bitten by an infected tick, humans can develop various symptoms, including skin rash, neurological problems, abdominal and joint pain, fatigue, headache, and heart damage. Most cases of Lyme disease occur in the northeastern United States, the northern Midwest, and the West.

Strategies for Action: Responding to Problems of Population Growth, Population Decline, and Urbanization

As shown in Figure 13.5, although many countries are satisfied with their population growth, the majority of national governments view their population growth rates as either too high or too low. While some countries are struggling to slow population growth, others are challenged with maintaining or even increasing their populations. After describing efforts to maintain or increase population in low-fertility countries, we look at strategies to slow population growth, including providing access to family planning services and contraceptive methods, improving the status of women, increasing economic development and improving health status, and imposing governmental regulations and policies. Because populations increasingly live in urban areas, strategies that address population problems also address, indirectly, problems of urbanization. Additional strategies that seek to alleviate urban problems include those designed to alleviate poverty and stimulate economic development in inner

Figure 13.5
Governments' views on population growth rate.

Source: Based on data from Population Reference Bureau (2005).

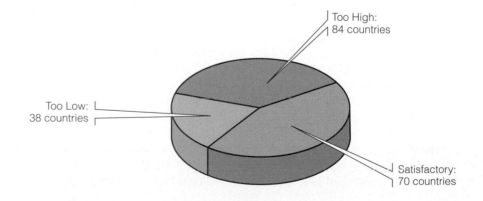

Too High: 84 countries

Too Low: 38 countries

Satisfactory: 70 countries

Relationship Between Urban Sprawl and Physical Activity, Obesity, and Morbidity

Health experts agree that most Americans are overweight and do not get enough exercise. As discussed in Chapter 2, obesity has reached epidemic proportions and is the second leading cause of preventable deaths in the United States. Although eating too much of the wrong kinds of foods is a major cause of obesity, lack of physical activity also plays a major role.

Here, we describe the first national study to investigate whether our physical activity levels, weight, and health are related to the type of place in which we live (Ewing et al. 2003). This study assesses how sprawl development—where homes are far from workplaces, shops, restaurants, and other destinations—affects physical activity, weight, and health of community residents.

Sample and Methods

Ewing and colleagues (2003) used the Centers for Disease Control and Prevention Behavioral Risk Factor Surveillance System (BRFSS) surveys (1998–2000) to obtain data on physical activity levels, body mass index and obesity, hypertension, diabetes, and heart disease of respondents. The BRFSS is a random telephone survey administered to U.S. civilian noninstitutionalized adults. More than 200,000 respondents from 448 counties and more than 175,000 respondents from 83 metropolitan areas were selected from the larger BRFSS samples because they lived in areas for which urban sprawl indexes were available.

Ewing and co-workers (2003) measured the degree of sprawl in 448 counties by using a "sprawl index" based on data available from the U.S. Census Bureau and other federal sources. The bigger the sprawl index, the more compact the metropolitan area; the smaller the index, the more sprawling the region.

Findings and Conclusions

Sprawling areas are not conducive to walking or biking, and people who live in sprawling communities have fewer opportunities to walk or bicycle as part of their daily routine. Rather, residents in sprawling areas depend on driving as the most convenient form of transportation. Therefore it is not surprising that Ewing and colleagues found that people who live in counties with sprawling development are likely to walk less and weigh more than people who live in less-sprawling counties. The people living in the most sprawling areas were likely to weigh six pounds more than people in the most compact county.

In addition, after controlling for factors such as age, education, sex, and race and ethnicity, people in more sprawling counties were more likely to suffer from hypertension (high blood pressure). The researchers did not find a statistically significant relationship between level of sprawl and diabetes or cardiovascular disease.

The study by Ewing and colleagues (2003) is the first national study to establish a direct association between the level of sprawl development in a community and the health of the people who live there. The study's findings suggest that we can improve public health by creating more compact communities that offer opportunities and design features (e.g., sidewalks) to include walking in our daily routines.

cities, implement "smart growth" and "new urbanism" in development, improve transportation and alleviate traffic congestion, and curb urban growth in developing countries.

Efforts to Maintain or Increase Population in Low-Fertility Countries

In some countries with below-replacement fertility levels, population strategies have focused on *increasing* rather than decreasing the population. For example, Australia's total fertility rate hit a record low of 1.73 in 2001, prompting the government to begin paying a $3,000 bonus in 2004 (which increased to $4,000 in 2005) to families who have babies (Lalasz 2005). The town of Yamatsuri, Japan,

Attitudes Toward Walking and Creating Better Walking Communities

For each of the following items, select the answer that best represents your attitudes. You can compare your answers with those of a national random sample of U.S. adults who participated in a 2002 telephone survey conducted by Belden Russonello & Stewart (*Americans' Attitudes Toward Walking*, 2003).

1. Which of the following statements describes you more: (A) If it were possible, I would like to walk more throughout the day either to get to specific places or for exercise, or (B) I prefer to drive my car wherever I go?

2. How much of a factor is each of the following in why you do not walk more right now (A, a major reason; B, somewhat of a reason; C, not much of a reason; D, not a reason at all):

   2a. Things are too far to get to and it is not convenient to walk.

   2b. Not enough time to walk.

   2c. Laziness.

   2d. It is hard to walk where I live because of traffic and lack of places to walk.

   2e. It is hard to walk where I live because there are not enough sidewalks or crosswalks.

   2f. Physically I am unable to walk more.

   2g. I do not like to walk.

   2h. There is too much crime to walk where I live.

3. For each of the following proposals to create more walkable communities, indicate your views using the following key: (A) strongly favor, (B) somewhat favor, (C) somewhat oppose, (D) strongly oppose.

   3a. Better enforce traffic laws such as speed limit.

   3b. Use part of the transportation budget to design streets with sidewalks, safe crossings, and other devices to reduce speeding in residential areas and make it safer to walk, even if this means driving more slowly.

   3c. Use part of the state transportation budget to create more sidewalks and stop signs in communities, to make it safer and easier for children to walk to school, even if this means less money to build new highways.

   3d. Increase federal spending on making sure people can safely walk across the street, even if this means less tax dollars go to building roads.

   3e. Have your state government use more of its transportation budget for improvements in public transportation, such as trains, buses and light rail, even if this means less money to build new highways.

   3f. Design communities so that more stores and other places are within walking distance of homes, even if this means building homes closer together.

Comparison Data

1. Walk more: 55%; drive: 41%; don't know: 5%

---

offers a $9,200 monetary reward (over a 10-year period) to persuade women who have at least two children to have more (Wiseman 2005). Aside from monetary rewards, many countries encourage childbearing by implementing policies designed to help women combine child rearing with employment. For example, as noted in Chapter 5, many European countries have generous family leave policies and universal child care.

Another way to increase population is to increase immigration. Spain, for example, has eased restrictions on immigration as a way to gain population.

Provide Access to Family Planning Services

Since the 1950s governments and nongovernmental organizations such as the International Planned Parenthood Federation have sought to lower fertility through family planning programs that provide reproductive health services and access to

2. Reasons for not walking more:

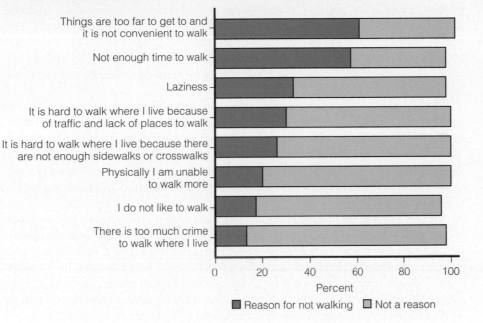

3. Proposals to create more walkable communities:

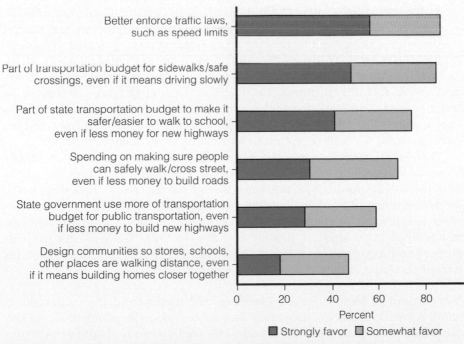

Source: Adapted from *Americans' Attitudes Toward Walking and Creating Better Walking Communities* (2003).

Strategies for Action: Responding to Problems of Population Growth, Population Decline, and Urbanization

contraceptive information and methods. Such programs, along with developments in contraceptive technology, have achieved the desired result: Globally, the average number of children born to each woman has fallen from 5 in the 1950s to 2.65 in 2005, because more women today want to limit family size and are using modern methods of birth control to control the number and spacing of births (United Nations 2005b). Today, about 60% of married women worldwide use some form of modern contraception, compared to 10% in 1960 (Population Reference Bureau 2004b).

Yet there is still an unmet need for contraception; more than 100 million women in less developed countries—about 17% of married women—would prefer to avoid pregnancy but are not using any form of family planning (Population Reference Bureau 2004b). Although 68% of married women in the United States use modern contraception, in 25 countries the figure is less than 10%. The highest rate is in China, where 86% of married women use modern contraception (Population Reference Bureau 2005).

Family planning programs are effective in increasing use of contraceptives. For example, in rural Pakistan married women living in households served by a "lady health worker," who provides doorstep delivery of contraceptive supplies, are more likely than other married women to use modern contraceptives (Douthwaite & Ward 2005). The provision of doorstep services through female workers is especially valuable in Pakistan, where women's mobility is limited and female modesty is highly valued.

In some countries family planning personnel often refuse or are forbidden by law or policy to make referrals for contraceptive and abortion services for unmarried women. Furthermore, many women throughout the world do not have access to legal, safe abortion. Without access to contraceptives many women who experience unwanted pregnancy resort to abortion—even under illegal and unsafe conditions. More than half of the nearly 80 million unintended pregnancies that occur worldwide every year end in abortion. Research in 12 countries of Central Asia and Eastern Europe has found that increased contraception use has resulted in significant declines in the rate of abortion (Leahy 2003).

In 1994 delegations from 179 UN member countries met in Cairo, Egypt, to develop a plan for the future of population and development. This meeting resulted in a declaration called the Programme of Action that stressed four goals to achieve by 2015: (1) universal education, (2) reduction in infant and child mortality, (3) reduction in maternal mortality, and (4) access to reproductive and sexual health services, including family planning. Two-thirds of the funding for the Programme of Action was to come from developing countries, and developed donor countries were to supply one-third. Only three donor countries, Denmark, Norway, and the Netherlands, have met their promised contribution levels (Goodrich 2004).

Cuts in U.S. assistance to international family planning programs are largely the result of opposition to abortion practices in some countries. On his first day in office in 2001, George W. Bush reinstated the "global gag rule," which denies U.S. international family planning assistance to organizations that use their own privately raised funds to counsel women on the availability of abortion, advocate change in abortion laws, or provide abortion services. The gag rule has cost women's health programs millions of dollars, resulting in the closure of thousands of family planning clinics in developing countries.

Involvement of Men in Family Planning. Although men play a central role in family planning decisions, they often do not have access to information and services that would empower them to make informed decisions about contraceptive use (Women's Studies Project 2003). Therefore family planning programs need to direct educational programs and health services to men. For example, men need education about the health risks to women when pregnancies are spaced too closely or when pregnancies occur before age 20 and after age 40.

Another important component of family planning and reproductive health programs involves changing traditional male attitudes toward women. According to traditional male gender attitudes, (1) a woman's most important role is being a wife and mother; (2) it is a husband's right to have sex with his wife at his demand; and (3) it is a husband's right to refuse to use condoms and to forbid his wife to use any other form of contraception. A number of programs around the world work with groups of boys and young men to change such traditional male gender attitudes (Schueller 2005).

Improve the Status of Women: The Importance of Education and Employment

Throughout the developing world the primary status of women is that of wife and mother. Women in developing countries traditionally have not been encouraged to seek education or employment; rather, they are encouraged to marry early and have children.

Improving the status of women by providing educational and occupational opportunities is vital to curbing population growth. Educated women are more likely to marry later, want smaller families, and use contraception. In countries where primary school enrollment is widespread or nearly universal, fertility declines more rapidly because (1) schools help spread attitudes about the benefits of family planning and (2) universal education increases the cost of having children, because parents sometimes are required to pay school fees for each child and because they lose potential labor that children could provide (Population Reference Bureau 2004b).

In addition, "education can result in smaller family size when the education provides access to a job that offers a promising alternative to early marriage and childbearing" (Population Reference Bureau 2004b, p. 18). Providing employment opportunities for women is also important to slow population growth, because high levels of female labor force participation and higher wages for women are associated with smaller family size. Low fertility rates in developed countries are largely due to women postponing having children until they have completed their education and established their careers.

Increase Economic Development and Improve Health

Although fertility reduction can be achieved without industrialization, economic development may play an important role in slowing population growth. Families in poor countries often rely on having many children to provide enough labor and

Recognizing that men play a crucial role in family planning decisions, family planning programs are making efforts to include men in family planning education and services.

Do You Think Safe Abortion Should Be Part of Reproductive Health Care in Developing Countries?

In developing countries hundreds of thousands of women die each year as a result of unsafe abortions. The majority of U.S. adults favor U.S. economic aid for family planning programs in developing countries, but only half favor U.S. economic aid to provide voluntary, safe abortions as part of reproductive health care in developing countries that request it (DaVanzo 2000).

It Is Your Turn Now to Take a Stand!

Do you think that the United States should provide funding to developing countries for medical facilities and equipment and training to health care personnel for the purpose of providing women with access to safe abortion?

How would you vote on this issue?

income to support the family. Economic development decreases the economic value of children and is also associated with more education for women and greater gender equality. As previously noted, women's education and status are related to fertility levels.

Economic development tends to result in improved health status of populations. Reductions in infant and child mortality are important for fertility decline, because couples no longer need to have many pregnancies to ensure that some children survive into adulthood. Finally, the more developed a country is, the more likely women are to be exposed to meanings and values that promote fertility control through their interaction in educational settings and through media and information technologies (Bongaarts & Watkins 1996).

> "Increasing urbanization has the potential for improving human life or increasing human misery. The cities can provide opportunities or frustrate their attainment; promote health or cause disease; empower people to realize their needs and desires or impose on them a simple struggle for basic survival."
>
> **United Nations**
> *1996 State of the World Population Report*

Restore Urban Prosperity

Nearly 9 in 10 Americans (86%) want their states to fund improvements in existing communities rather than provide incentives for new development (*American Community Survey*, 2004). A number of strategies have been proposed and implemented to restore prosperity to U.S. cities and well-being to their residents, businesses, and workers, including strategies to attract new businesses, create jobs, and repopulate cities. The economic development and revitalization of cities also involves improving affordable housing options (see Chapter 6), alleviating urban problems related to HIV/AIDS, addiction, and crime (discussed in Chapters 2–4), and reducing problems of traffic and transportation, which we address later in this chapter.

Empowerment Zone/Enterprise Community Program. The federal Empowerment Zone/Enterprise Community Initiative, or EZ/EC program, provides tax incentives, grants, and loans to businesses to create jobs for residents living within various designated zones or communities, many of which are in urban areas. Federal money provided to empowerment zones and enterprise communities is also used to train and educate youth and families and to improve child care, health care, and transportation. The EZ/EC program provides grant funding so that communities can

Funding from the U.S. Department of Housing and Urban Development's Brownfields Economic Development Initiative was used to transform an old abandoned factory building in Wheeling, West Virginia, into a new, usable office facility.

design local solutions that empower residents to participate in the revitalization of their neighborhoods.

Infrastructure Improvements. Urban revitalization often involves making improvements in the **infrastructure**—the underlying foundation that enables a city to function. Infrastructure includes such things as water and sewer lines, phone lines, electricity cables, sidewalks, streets, curbs, lighting, and storm drainage systems. Improving infrastructure may help to attract business to an area. Infrastructure improvements in urban areas also increase property values and renew residents' sense of pride in their neighborhood (Cowherd 2001).

Brownfield Redevelopment. Brownfields are abandoned or undeveloped sites that are located on contaminated land. There are more than 400,000 brownfields throughout the United States and an estimated 5 million acres of abandoned industrial sites in U.S. cities—roughly the same amount of land occupied by 60 of the nation's largest cities (U.S. Department of Housing and Urban Development 2004).

Cleaning up and redeveloping brownfields is not only an important environmental measure but also a key component of urban revitalization because it provides jobs, increases tax revenues, attracts more businesses, residents, and tourists, and helps to curb urban sprawl. Hundreds of urban brownfield sites across the United States have been successfully redeveloped into residential, business, and recreational property. Recognizing that a major obstacle to brownfield redevelopment is lack of funding, the U.S. Department of Housing and Urban Development developed a funding program known as the Brownfields Economic Development Initiative, which has provided millions of dollars to communities across the country for the purpose of brownfield redevelopment (U.S. Department of Housing and Urban Development 2004).

Gentrification and Incumbent Upgrading. Gentrification is a type of neighborhood revitalization in which middle- and upper-income individuals buy and rehabilitate older homes in an economically depressed neighborhood. The city provides tax incentives for investing in old housing with the goal of attracting wealthier residents back into these neighborhoods and increasing the tax base. After the house is renovated, the owner may live there, rent, or sell the house. A downside to gentrification is that low-income city residents are often forced into substandard housing because less affordable housing is available. In effect, gentrification often displaces the poor and the elderly (Johnson 1997).

An alternative to gentrification is **incumbent upgrading,** in which aid programs help residents of depressed neighborhoods buy or improve their homes and stay in the community. Both gentrification and incumbent upgrading improve decaying neighborhoods, attracting residents as well as businesses.

Community Development Corporations. In many low- and moderate-income urban neighborhoods, community development corporations (CDCs)—nonprofit groups formed by residents, small business owners, congregations, and other local stakeholders—work to create jobs and affordable housing and renovate parks and other community facilities. An analysis of CDCs found that they have produced dramatic improvements and raised property values in neighborhoods (Galster et al. 2005). A major strength of CDCs is that they involve community residents in planning and implementing urban renewal projects, giving residents a sense of empowerment in their communities.

Improve Transportation and Alleviate Traffic Congestion

"Adding highway capacity to solve traffic congestion is like buying larger pants to deal with your weight problem."

Michael Replogle
Transportation director for
Environmental Defense

An important strategy for reducing traffic congestion involves increasing the use of public transit, such as buses, trains, and subways. A national survey found that the majority of Americans (80%) support building more rail systems serving cities, suburbs, and entire regions to give them the option of not driving their cars (U.S. Conference of Mayors 2001).

Transportation planners increasingly recognize that building more roads does not necessarily ease traffic problems. The U.S. public agrees: In a poll of randomly selected registered voters the majority (66%) said that they do not think traffic congestion will be eased if more roads are built (U.S. Conference of Mayors 2001). More important, concern is growing over the social and environmental problems related to road building and motor vehicle use. According to Jan Lundberg, director of a grassroots group called the Alliance for a Paving Moratorium, nearly half of all urban space is paved; more land is devoted to cars than to housing, and every year nearly 100,000 people are displaced by highway construction (Jensen 2001). The Alliance for a Paving Moratorium advocates a halt to road building. "In the Alliance's view, a paving moratorium would limit the spread of population, redirect investment from suburbs to inner cities, and free up funding for mass transportation and maintenance of existing roads" (Jensen 2001, p. 6).

Another way to ease traffic congestion is to encourage means of transportation other than motor vehicles. In some cities motorists must pay a "congestion charge" for driving in a "congestion charge zone" during weekday high-traffic periods. Congestion charges encourage travelers to us public transport, bicycles, motorcycles, or alternative fuel vehicles, which are exempt from the charge. In 1998 Singapore became the first city to use a congestion charge. Other cities that levy congestion charges include Oslo, Bergen, Trondheim, and London.

Finally, the development of communities that enable residents to walk or ride a bicycle to schools, shops, and other locations can help relieve traffic congestion.

This strategy is part of the "smart growth" and "new urbanism" movement discussed in the next section. This chapter's *Focus on Technology* feature looks at the use of bicycles as transportation vehicles.

Responding to Urban Sprawl: Growth Boundaries, Smart Growth, and New Urbanism

In a national survey nearly half of Americans (46%) said that slowing development of open space was an "extremely high" or "high" priority (*American Community Survey*, 2004). Some cities have tried to manage urban sprawl by establishing growth boundaries. Rather than simply put a limit on urban growth, another approach to managing urban sprawl is to develop land according to principles known as **smart growth.** A smart-growth urban development plan entails the following principles (Froehlich 1998; Rees 2003; Smart Growth Network 2004):

- **Mixed-use land,** which allows homes, jobs, schools, shops, workplaces, and parks to be located within close proximity of each other.
- Ample sidewalks, encouraging residents to walk to jobs and shops.
- Compact building design.
- Housing and transportation choices.
- Distinctive and attractive community design.
- Preservation of open space, farmland, natural beauty, and critical environmental areas.
- Redevelopment of existing communities, rather than letting them decay and building new communities around them.
- Regional planning and collaboration among businesses, private residents, community groups, and policy makers on development and redevelopment issues.

Smart growth is similar to another movement in urban planning called **New Urbanism.** The goals and methods of New Urbanism are similar to those of smart growth, but the impetus for these movements is slightly different. Smart growth approaches the idea of sustainable urban communities with the primary goal of stopping sprawl. The New Urbanism approach is to raise the quality of life for all those in the community by creating compact communities with a sustainable infrastructure. Smart growth and New Urbanism are often impeded by local zoning codes that mandate large housing setbacks, wide streets, and separation of residential and commercial areas (Pelley 1999).

Regionalism

The various social problems that face urban areas may best be addressed through **regionalism**—a form of collaboration among central cities and suburbs that encourages local governments to share common responsibility for common problems. Central cities, declining inner suburbs, and developing suburbs are often in conflict over the distribution of government-funded resources, zoning and land use plans, transportation and transit reform, and development plans. Rather than compete with each other, regional government provides a mechanism for achieving the interests of an entire region. A metropolitan-wide government would handle the inequities and concerns of both suburban and urban areas. As might be expected, suburban officials resist regionalization because they believe it will hurt their neighborhoods economically by draining off money for the cities.

A Return to Simpler Technology: Bicycles as a Solution to Urban Problems

The application of technology to solving modern social problems often involves relatively current, state-of-the-art technologies. However, sometimes an effective solution to a social problem can be found in an older, simpler technology.

In cities around the world bicycles are emerging as a solution to some of today's urban problems. "For safer streets, less congestion, and cleaner air, the bicycle is poised to become an integral part of urban transportation systems in the 21st century" (Gardner 1999, p. 23). The use of bicycles can also result in improved health and lower health care costs—benefits associated with cleaner air, reduced noise, and more exercise. In the United States cyclists are 12 times more likely than people in cars to die en route (Gardner 2005). However, studies in the United Kingdom have found that the health benefits of cycling, which include decreased risk of heart disease and diabetes, far outweigh the risks of bicycle accidents (Sheehan 2001). Increased bicycle use would also result in fewer people being injured and killed by cars. Nearly 1 million people are killed on the world's roads each year; most of them are pedestrians. In addition, replacing motor vehicle trips with bike trips could improve health by reducing vehicle emissions and improving air quality. In the United States air pollution kills more people than traffic accidents do (Blatt 2005). Increased bicycle use could also help alleviate the problem of noise pollution and its associated adverse health effects. Noise, which is perceived by many urban residents as one of the greatest problems associated with road traffic, contributes to stress disturbances, cardiovascular disease, and hearing loss (Sheehan 2001).

Although bicycles could contribute to improving urban life by reducing traffic congestion, pollution-causing vehicle emissions, and noise pollution and by improving public health, their use in urban areas is not widespread. A number of factors discourage widespread bicycle use, including unsafe roads and a lack of safe bike parking. Ironically, air pollution, largely caused by vehicle emissions, discourages the use of bicycles, thus hindering a mode of transportation that would help to alleviate the air pollution.

However, bicycles have become a major mode of transportation in some European countries. In the past 20 years the Netherlands has doubled the length of its bikeways and Germany has tripled its bikeway networks. Cycling accounts for 12% of all trips in Germany and for 27% in the Netherlands, compared to less than 1% in the United States (Gardner 2003).

Encouraging the use of bicycles requires changes in urban design and policy to make cycling a safer, more viable, and/or necessary option. Some European cities, such as Munich, Vienna, and Copenhagen, have commercial centers that restrict vehicle traffic to ambulances, delivery trucks, and cars owned by local residents. About 20 car-free communities are in various stages of development in Germany (Sheehan 2001). Programs in Lima, Peru, help low-income residents buy bicycles, and a program in Copenhagen, Denmark, provides bikes for public use (Gardner 1999). The United Kingdom has built an 8,000-kilometer National Cycle Network that will pass within 4 kilometers of half of the country's population (Brown 2000). It is hoped that the accessibility of safe cycling paths will induce people to shift from cars to bicycles on short trips.

To encourage bike use for longer trips, cities can establish convenient connections between cycling and public transit. "Bicycles and transit can complement each other when people are able to carry their bikes aboard buses or trains, or park them at stations" (Sheehan 2001, p. 17). Urban planning based on "mixed use" designs can also facilitate bike use by combining housing, public facilities (such as schools, parks, and libraries), and commercial sites (such as grocery stores and banks) in the same neighborhood.

Even if cities are designed to promote safe bicycling as a mode of transportation, people in the United States and other industrialized countries are not likely to trade their car keys for a bicycle helmet. "In the United States, where 95 percent of parking is free and where gas prices, vehicle taxes, and other driving-related costs are among the lowest in the industrial world, using a car is a rational choice and a key reason that biking remains marginalized" (Gardner 2005, p. 58). People throughout the developed world have acquired a love for the automobile and its images of freedom, power, adventure, and sexiness. These images are perpetuated by the automobile industry, which spends more money on advertising than any other industry in the United States and worldwide (Sheehan 2001). Until our love for cleaner air and better health outweighs our love of the automobile, most Americans will continue to leave their bicycles (if they own them) in the closet or garage.

Strategies for Reducing Urban Growth in Developing Countries

In developing countries limiting population growth is essential for alleviating social problems associated with rapidly growing urban populations. Another strategy for minimizing urban growth in less-developed countries involves redistributing the population from urban to rural areas. Such redistribution strategies include the following: (1) promoting agricultural development in rural areas, (2) providing incentives to industries and businesses to relocate from urban to rural areas, (3) providing incentives to encourage new businesses and industries to develop in rural areas, and (4) developing the infrastructure of rural areas, including transportation and communication systems, clean water supplies, sanitary waste disposal systems, and social services. Of course, these strategies require economic and material resources, which are in short supply in less-developed countries.

Understanding Problems of Population Growth and Urbanization

What can we conclude from our analysis of population growth and urbanization? First, although fertility rates have declined significantly in recent years and although some countries are experiencing a decline in their population size, world population will continue to grow for several decades. This growth will largely occur in urban areas in developing regions. Given the problems associated with population growth, such as environmental problems and resource depletion, global insecurity, poverty and unemployment, and poor maternal and infant health, most governments recognize the value of controlling population size and support family planning programs. However, efforts to control population must go beyond providing safe, effective, and affordable methods of birth control. Slowing population growth necessitates interventions that change the cultural and structural bases for high fertility rates. Two of these interventions are increasing economic development and improving the status of women, which includes raising their levels of education, their economic position, and their (and their children's) health. Addressing problems associated with population growth also requires the willingness of wealthier countries to commit funds to providing reproductive health care to women, improving the health of populations, and providing universal education for people throughout the world (see Table 13.2).

Attention to urban problems and issues is increasingly important because the United States and the rest of the world are rapidly becoming urbanized. Aside from population concerns, problems affecting urban residents include poverty and unemployment, inferior or unaffordable housing, and traffic and transportation problems.

The social forces affecting urbanization in industrialized countries are different from those in developing countries. Countries such as the United States have experienced urban decline as a result of deindustrialization, deconcentration, and the shift to a service economy in which jobs that pay well and come with full benefits are scarce. At the same time, developing countries have experienced rapid urban growth as a result of industrialization, fueled in part by a global economy in which transnational corporations locate industry in developing countries to gain access to cheap labor, raw materials, and new markets.

One of the shadows lingering over cities throughout the world is cast by environmental problems, because urban populations consume the largest share of the world's natural resources and contribute most of the pollution and waste that

Table 13.2

Annual Expenditures on Luxury Items Compared with Funding Needed to Meet Selected Basic Needs

Product	Annual Expenditure
Makeup	$18 billion
Pet food in Europe and the United States	$17 billion
Perfume	$15 billion
Ocean cruises	$14 billion
Ice cream in Europe	$11 billion

Social or Economic Goal	Additional Annual Investment Needed to Achieve Goal
Reproductive health care for all women	$12 billion
Elimination of hunger and malnutrition	$19 billion
Universal literacy	$5 billion
Clean drinking water for all	$10 billion
Immunizing every child	$1.3 billion

Source: World Watch Institute Press Release, 2004 (January 1). "State of the World 2004: Consumption by the Numbers." www.worldwatch.org

compromise the health of our planet. Global environmental problems, discussed in Chapter 14, are linked to both population growth and urban life. "Key global environmental problems have their roots in cities—from the vehicular exhaust that pollutes and warms the atmosphere, to the urban demand for timber that denudes forests and threatens biodiversity, to the municipal thirst that heightens tensions over water" (Sheehan 2003, p. 131). As we ponder ways to improve cities, and as we strive to meet the needs of growing populations, we must include the larger environment in the equation.

Chapter Review

ThomsonNOW™

Reviewing is as easy as ❶ ❷ ❸

1. Before you do your final review, take the ThomsonNOW diagnostic quiz to help you identify the areas on which you should concentrate. You will find information on ThomsonNOW and instructions on how to access all of its great resources on the foldout at the beginning of the text.

2. As you review, take advantage of ThomsonNOW's study videos and interactive Map the Stats exercises to help you master the chapter topics.

3. When you are finished with your review, take ThomsonNOW's posttest to confirm you are ready to move on to the next chapter.

- **Where is most of the world's population growth occurring?**

  Ninety-nine percent of world population growth is in developing countries. Most of this growth is occurring in urban areas.

- **What is "urban sprawl," and why is it a problem?**

  Urban sprawl refers to the ever-increasing outward growth of urban areas. Urban sprawl results in the loss of green, open spaces, the displacement and endangerment of wildlife, traffic congestion and noise, and pollution liabilities. Sprawl is also linked to negative health effects in humans, because people in high-sprawl areas walk less and are more overweight than people who live in low-sprawl areas.

- **What is the demographic transition?**

  The demographic transition is the progression from a population with short lives and large families to one in which people live longer and have smaller families. About one-third of countries have completed the demographic transition.

- **Many countries are experiencing below-replacement fertility (fewer than 2.1 children born to each woman). Why are some countries concerned about a "birth dearth"?**

  In countries with below-replacement fertility, there are or will be fewer workers to support a growing number of elderly retirees and to maintain a productive economy.

- **Why is population growth considered a threat to global security?**

  In developing countries rapid population growth results in a "youth bulge"—a high proportion of 15- to 29-year-olds relative to the adult population. The combination of a youth bulge with other characteristics of rapidly growing populations, such as resource scarcity, high unemployment rates, poverty, and rapid urbanization, sets the stage for civil unrest, war, and terrorism, because large groups of unemployed young people resort to violence in an attempt to improve their living conditions.

- **Globally, what was the average number of children born to each woman in 1960? In 2005?**

  Globally, the average number of children born to each woman has fallen from 5 in 1960 to 2.65 in 2005.

- **What are brownfields?**

  Brownfields are abandoned or undeveloped sites that are located on contaminated land. Cleaning up and redeveloping brownfields is a key component of urban revitalization because it provides jobs, increases tax revenues, and potentially attracts more businesses, residents, and tourists.

- **What does the term "mixed-use land" refer to? What are the benefits of mixed-use land?**

  The use of mixed-use land is a strategy of the smart-growth movement whereby homes, jobs, schools, shops, workplaces, and parks are located within close proximity of each other. Mixed-use land encourages walking as a means of transportation and minimizes the use of cars.

Critical Thinking

1. In cities across the country families with children are leaving the city to move to the suburbs where they can afford a bigger house with more space. Consequently, many cities are experiencing a decline in their population of children (Egan 2005). How might significant reductions in the youth population affect cities? What could cities do to attract more families with children?

2. What could be done in your community to encourage the use of bicycles as an alternative to motor vehicles?

3. One strategy for encouraging childbearing in low-fertility European countries is to provide work-family supports to make it easier for women to combine childbearing with employment. If the United States offered more generous work-family benefits, such as paid parenting leave and government-supported child care, would the U.S. birthrate increase? Would such policies affect the number of children you would want to have?

Key Terms

barrio	mixed-use land
brownfields	New Urbanism
central city	population density
deconcentration	population momentum
demographic transition theory	pronatalism
doubling time	regionalism
environmental footprint	replacement-level fertility
gentrification	road rage
ghetto	slum
incumbent upgrading	smart growth
infrastructure	suburbanization
megacities	suburbs
metropolis	total fertility rate
metropolitan area	urban area
micropolitan area	urbanization
	urban sprawl

Media Resources

The Companion Website for *Understanding Social Problems*, Fifth Edition

http://sociology.wadsworth.com/mooney_knox_schacht5e

Supplement your review of this chapter by going to the companion website to take one of the Tutorial Quizzes, use the flash cards to master key terms, and check out the many other study aids you'll find there. You'll also find special features such as Wadsworth's *Sociology Online Resources and Writing Companion,* GSS data, and Census 2000 information, data, and resources at your fingertips to help you complete that special project or do some research on your own.

> "The natural world is the larger sacred community to which we belong. To be alienated from this community is to become destitute in all that makes us human." *Thomas Berry, Catholic monk and ecotheologian*

Environmental Problems

AP/Wide World Photos

Sister Dorothy Stang, an American nun and environmental
activist, dedicated her life to
protecting the Amazon rain forest
and its poor inhabitants. At age 73
she was shot dead by two gunmen
who were allegedly hired by illegal
loggers and ranchers who were encroaching on a federal peasant farming reserve that she helped to establish in the state of Pará, Brazil (Reel
2005). Ten days after Stang's murder,
Brazilian environmentalist Dionisio
Ribeiro Filho was shot in the head at
the Tingua federal reserve near Rio
de Janeiro; Filho had defended the
reserve for more than 15 years from
poachers and illegal palm tree cut-
ters (Gaier 2005). Edson Bedin, the
head of Brazil's federal environmental
agency, said, "This business of shutting up ecologists and environmentalists with violence, it's not going to

Sister Dorothy Stang, an American
nun and environmentalist, dedicated her life to protecting the
Amazon rain forest in Brazil and
its poor residents. She was shot
and killed in 2005 in a contract
killing by illegal loggers and
ranchers who were encroaching
on a federal peasant farming reserve that she helped establish in
the state of Pará, Brazil.

stop. . . . Threats against agents, workers have become routine" (Gaier 2005, p. 1). The day
after Filho's murder, two men showed up at the house of Bulgarian antinuclear activist Al-
bena Simeonova, threatening to kill her if she did not stop her public opposition to plans to
build a nuclear power plant in Belene ("Bulgarian Green Leader Threatened With Death,"
2005). "This is not only a serious threat against my life," said Simeonova. "It represents a
threat to all who campaign against nuclear plants trying to protect their lives and the local
environment" (n.p.).

A Gallup poll found that two-thirds of U.S. adults say they worry either "a great
deal" (35%) or "a fair amount" (31%) about the quality of the environment (Lyons
2005). As the opening vignette suggests, some people are so concerned about environmental problems that they are willing to put their lives on the line to fight for
protection of the environment.

In this chapter we focus on environmental problems that threaten the lives and
well-being of people, plants, and animals all over the world—today and in future
generations. After examining how globalization affects environmental problems,
we view environmental issues through the lens of structural functionalism, conflict
theory, ecofeminist theory, and symbolic interactionism. We then present an
overview of major environmental problems, examining their social causes and exploring strategies that attempt to reduce or alleviate them.

The Global Context: Globalization and the Environment

In 1992 leaders from across the globe met at the first Earth Summit in Rio de Janeiro
to forge agreements to protect the planet's environment and at the same time alleviate world poverty. When world leaders met a decade later in 2002 for the second

Earth Summit in Johannesburg, South Africa, the overall state of the environment had deteriorated and poverty had deepened. How is it that the combined efforts of leaders who met at the first Earth Summit were so ineffectual in achieving their goals? A large part of the answer lies in the increasing globalization of the last two decades. Three aspects of globalization that have affected the environment are (1) the permeability of international borders to pollution and environmental problems, (2) cultural and social integration spurred by communication and information technology, and (3) growth of free trade and transnational corporations.

Permeability of International Borders

Environmental problems such as global warming and destruction of the ozone layer (discussed later in this chapter) demonstrate that environmental problems extend far beyond their source to affect the entire planet and its inhabitants. A striking example of the permeability of international borders to pollution is the spread of toxic chemicals (such as PCBs) from the Southern Hemisphere into the Arctic. In as little as five days chemicals from the tropics can evaporate from the soil, ride the winds thousands of miles north, condense in the cold air, and fall on the Arctic in the form of toxic snow or rain (French 2000). This phenomenon was discovered in the mid-1980s, when scientists found high levels of PCBs in the breast milk of Inuit women in the Canadian Arctic.

Another environmental problem involving permeability of borders is **bioinvasion**: the emergence of organisms in regions where they are not native. Bioinvasion is largely a product of the growth of global trade and tourism (Chafe 2005). Exotic species travel in the ballast water of ships (water taken in to stabilize empty vessels as they cross waterways), in packing material, in shipments of crops and other goods, and in many other ways. Invasive species may compete with native species for food, start an epidemic, or prey on natives, threatening not only their immediate victims but also the entire ecosystem in which the victims live.

Cultural and Social Integration

As mass media infiltrate the world, people across the globe aspire to consume the products and mimic the materially saturated lifestyles portrayed in movies, television, and advertising. As patterns of consumption in China, India, and other developing countries increasingly follow those in wealthier Western nations, so do the problems associated with overconsumption: depletion of natural resources, pollution, and global warming.

On the positive side the Internet and other forms of mass communication have helped to integrate the efforts of diverse environmental groups across the globe. The globalization of the environmental movement has created new opportunities for environmental groups to join forces, share information, and educate the public through mass communication.

The Growth of Transnational Corporations and Free Trade Agreements

As discussed in Chapter 7, the world's economy is dominated by transnational corporations, many of which have established factories and other operations in developing countries where labor and environmental laws are lax. Many transnational corporations are implicated in environmentally destructive activities—from mining and cutting timber to dumping of toxic waste.

The World Trade Organization (WTO) and free trade agreements such as NAFTA (North American Free Trade Agreement) and the FTAA (Free Trade Area of the Americas) allow transnational corporations to pursue profits, expand markets, use natural resources, and exploit cheap labor in developing countries while weakening the ability of governments to protect natural resources or to implement environmental legislation. Transnational corporations have influenced the world's most powerful nations to institutionalize an international system of governance that values commercialism, corporate rights, and "free" trade over environment, human rights, worker rights, and human health (Bruno & Karliner 2002). Under NAFTA's Chapter 11 provisions corporations can challenge local and state environmental policies, federal controlled substances regulations, and court rulings if such regulatory measures and government actions negatively affect the corporation's profits. Any country that decides, for example, to ban the export of raw logs as a means of conserving its forests or, as another example, to ban the use of carcinogenic pesticides can be charged under the WTO by member states on behalf of their corporations for obstructing the free flow of trade and investment. A secret tribunal of trade officials would then decide whether these laws were "trade restrictive" under the WTO rules and should therefore be struck down. Once the secret tribunal issues its edict, no appeal is possible. The convicted country is obligated to change its laws or face the prospect of perpetual trade sanctions (Clarke 2002, p. 44). As of early 2005, 42 cases had been filed by corporate interests and investors, 11 of which had been finalized. Five corporations that won their claims received $35 million paid by taxpayers in Canada and Mexico. For example, in the late 1990s Ethyl, a U.S. chemical company, used NAFTA rules to challenge Canadian environmental regulation of the toxic gasoline additive MMT. Ethyl won the suit, and Canada paid $13 million in damages and legal fees to Ethyl and reversed the ban on MMT (Public Citizen 2005). Although the United States has not, as of this writing, lost a case, it is only a matter of time before a corporation based in Mexico or Canada wins a NAFTA case against the United States (Public Citizen 2005).

Sociological Theories of Environmental Problems

Each of the three main sociological theories—structural functionalism, conflict theory, and symbolic interactionism—as well as ecofeminist theory provide insights into social causes of and responses to environmental problems.

Structural-Functionalist Perspective

Structural functionalism emphasizes the interdependence between human beings and the natural environment. From this perspective human actions, social patterns, and cultural values affect the environment, and, in turn, the environment affects social life. For example, population growth affects the environment as more people use natural resources and contribute to pollution. However, the environmental impact of population growth varies tremendously according to a society's patterns of economic production and consumption. As discussed in Chapter 13, the impact that each person makes on the environment—each person's **environmental footprint**—is determined by the patterns of production and consumption in that person's culture. The environmental footprint of an average person in a high-income country is about six times bigger than that of someone in a low-income country and many more times larger than in the least developed nations (United Nations Population Fund 2004).

Structural functionalism focuses on how changes in one aspect of the social system affect other aspects of society. For example, in the two years after the terrorist attacks of September 11, 2001, public concern about most environmental problems declined sharply, most likely as a result of increasing concern about the economy and terrorism over the same period (Saad 2002). The effect of oil prices on the economy provides another illustration of how a change in one aspect of the social system affects other aspects of society. When the price of oil skyrocketed after Hurricane Katrina in 2005, businesses suffered and consumers struggled to pay higher prices for food and other goods. Because so much of our economy depends on oil, an oil shortage or price spike affects virtually every aspect of our economy.

The structural-functionalist perspective raises our awareness of latent dysfunctions—negative consequences of social actions that are unintended and not widely recognized. For example, the more than 840,000 dams worldwide provide water to irrigate farmlands and supply 17% of the world's electricity. Yet dam building has had unintended negative consequences for the environment, including the loss of wetlands and wildlife habitat, the emission of methane (a gas that contributes to global warming) from rotting vegetation trapped in reservoirs, and the alteration of river flows downstream, which kills plant and animal life ("A Prescription for Reducing the Damage Caused by Dams," 2001). Dams have also displaced millions of people from their homes. As philosopher Kathleen Moore points out, "Sometimes in maximizing the benefits in one place, you create a greater harm somewhere else. . . . While it might sometimes seem that small acts of cruelty or destruction are justified because they create a greater good, we need to be aware of the hidden systematic costs" (Jensen 2001, p. 11). Being aware of latent dysfunctions means paying attention to the unintended and often hidden environmental consequences of human activities.

Conflict Perspective

The conflict perspective focuses on how wealth, power, and the pursuit of profit underlie many environmental problems. Wealth is related to consumption patterns that cause environmental problems. Wealthy nations have higher per capita consumption of petroleum, wood, metals, cement, and other commodities that deplete the earth's resources, emit pollutants, and generate large volumes of waste. The wealthiest 20% of the world's population is responsible for 86% of total private consumption (Bright 2003). The United States is responsible for 25% of the world's oil consumption, yet the United States produces less than 3% of the world's oil supplies (Pope 2005).

The capitalistic pursuit of profit encourages making money from industry regardless of the damage done to the environment. McDaniel (2005) notes that "our culture tolerates environmentalism only so long as it has minimal impact on big business. . . . In an economically centered culture, jobs come first, not the health of people or the environment" (pp. 22–23).

To maximize sales, manufacturers design products intended to become obsolete. As a result of this **planned obsolescence**, consumers continually throw away used products and purchase replacements. Industry profits at the expense of the environment, which must sustain the constant production and absorb ever-increasing amounts of waste.

Industries also use their power and wealth to influence politicians' environmental and energy policies as well as the public's beliefs about environmental issues. ExxonMobil, the world's largest oil company, has spent millions of dollars on lobbying and has funded numerous organizations that have tried to discredit scientific findings that link fossil fuel burning to global climate change (Mooney 2005). Despite the agreement of hundreds of scientists from around the world that global warming is occurring and is largely due to "greenhouse gases" released by fossil fuel burning, the Bush administration pulled out of the Kyoto Protocol—an international agreement to reduce greenhouse gas emissions—citing "incomplete" science as the reason. The electric utility industry gave President Bush's election campaigns more than $1 million and has spent millions of dollars on lobbying. In return, President Bush pushed for the Clear Skies Initiative, which would save power companies billions of dollars by relaxing pollution emission caps and extending timelines for air pollution reduction, as required by the 1970 Clean Air Act (Clarren 2005). Bush's legislation created a "cap-and-trade" system, whereby a plant can exceed its permitted level of emissions by buying credits from a plant in the same region whose emissions are below what is allowed. Critics of the cap-and-trade approach argue that it fails to achieve the lowest possible emissions because it does not require all plants to use the best available technology to reduce emissions. By allowing some plants to have higher emissions, it also exposes populations living near these high-emissions plants to excessive air pollution.

With both President G. W. Bush and Vice President Dick Cheney having personal and family ties to the oil industry, it is not surprising that the Bush administration supported numerous policy decisions that benefit the oil industry at the expense of the environment and public health. The energy bill passed in 2005 included $85 million in subsidies and tax breaks for most forms of energy, but oil, gas, and nuclear energy received the lion's share. One of the oft-cited criticisms of the war on Iraq is the assertion that it was motivated by the interests of the U.S. oil industry. Finally, the Bush administration has appointed industry-friendly officials to environmental posts. For example, Bush appointed former American Petroleum Institute attorney Philip Cooney, who opposed the Kyoto Protocol, as chief of staff of the White House Council on Environmental Quality. Also, Larisa Dobriansky, a former lobbyist who worked on climate change issues for ExxonMobil, was appointed the deputy assistant secretary for national energy policy at the Department of Energy, where she managed the department's Office of Climate Change Policy (Mooney 2005).

Ecofeminist Perspective

Ecological feminism, or **ecofeminism,** began in 1974 when French feminist Francoise d'Eaubonn coined the term *ecological feminisme* to call attention to women's potential to energize an ecological revolution (Warren 2000). Ecofeminists view environmental problems as resulting from human domination of the environment and see connections between the domination of women, people of color, children, and the poor and the domination of nature. Throughout the world and in developing countries in particular, men are dominant in deciding how natural resources are

"Women must see that there can be no liberation for them and no solution to the ecological crisis within a society whose fundamental model of relationships continues to be one of domination."

Karen J. Warren
Ecofeminist philosopher

used. Men are dominant in positions of government and corporate leadership and own most of the land. By some estimates women around the world hold title to less than 2% of the land that is owned (MacDonald & Nierenberg 2003). In contrast to a male-oriented view of natural resources as a means to an end—a means to profit and power—ecofeminists often embrace a spiritual approach to addressing environmental problems, drawing on pagan, Native American, New Age, and Eastern religious traditions that emphasize the close connection between women and nature (Mother Earth) (Schaeffer 2003).

Symbolic Interactionist Perspective

The symbolic interactionist perspective focuses on how meanings, labels, and definitions learned through interaction and through the media affect environmental problems. Whether an individual recycles, drives an SUV (sport utility vehicle), or joins an environmental activist group is influenced by the meanings and definitions of these behaviors that the individual learns through interaction with others.

Large corporations and industries commonly use marketing and public relations strategies to construct favorable meanings of their corporation or industry. The term **greenwashing** refers to the way in which environmentally and socially damaging companies portray their corporate image and products as being "environmentally friendly" or socially responsible. Greenwashing is commonly used by public relations firms that specialize in damage control for clients whose reputations and profits have been hurt by poor environmental practices. Wal-Mart, for example, has paid millions of dollars in fines and settlements for a number of environmental violations, including excessive storm runoff at construction sites, petroleum storage violations, and Clean Air Act violations (Wal-Mart Watch 2005). When Wal-Mart announced in 2005 that it would donate $35 million (less than 1% of its 2004 profits) to the National Fish and Wildlife Foundation over a 10-year period to purchase 1 acre of conservation land for every acre of land developed by Wal-Mart, many environmentalists accused Wal-Mart of greenwashing. Just before Earth Day (April 22) in 2005, Wal-Mart bought full-page ads in at least 20 newspapers touting its new program, Acres for America (Associated Press 2005). Such publicity suggests that Wal-Mart's Acres for America program is designed to improve Wal-Mart's public image. The comments of Eric Olson of the Sierra Club reflect the skepticism that many environmentalists have toward corporate greenwashing: "Wal-Mart thinks it can paint over its record with a nice shade of green, but that won't hide its true colors" (quoted by Associated Press 2005).

Ford Motor Company was accused of greenwashing in 2004 when it advertised its Escape Hybrid sport utility vehicle as an example of Ford's commitment to the environment when, for the fifth consecutive year, Ford's vehicles had the worst overall fuel economy of all major automakers. In fact, Ford's fleetwide fuel economy was worse in 2004 than it was in 1984! And hybrids accounted for less than 1% of Ford's annual sales (Johnson 2004).

Greenwashing also occurs in the political arena when political leaders attempt to portray an environmentally damaging policy or program as environmentally friendly. Consider, for example, President Bush's Clear Skies Initiative, which allows five times more mercury emissions, one and a half times more sulfur dioxide emissions, and hundreds of thousands more tons of smog-forming nitrogen oxide emissions than those allowed under the 1970 Clean Air Act (Clarren 2005). Naming this proposal the Clear Skies Initiative is a blatant example of greenwashing.

"Despite their eco-friendly rhetoric, for most corporations in the world green is nothing more than the color of money."

Kenny Bruno and Joshua Karlinger
Authors of *Earthsummit.biz: The Corporate Takeover of Sustainable Development*

Although greenwashing involves manipulation of public perception to maximize profits, many corporations make genuine and legitimate efforts to improve their operations, packaging, or overall sense of corporate responsibility toward the environment. For example, in 1990 McDonald's announced that it was phasing out foam packaging and switching to a new, paper-based packaging that is partially degradable. But many environmentalists are not satisfied with what they see as token environmentalism, or as Peter Dykstra of Greenpeace suggests, 5% of environmental virtue to mask 95% of environmental vice (Hager & Burton 2000).

Greenwashing can cause controversy and divisions among environmentalists. Consider Earth Day, which has brought environmentalists together to celebrate the earth and to increase public awareness of environmental issues every April 22 since 1970. In recent years Earth Day events have been sponsored by corporations such as Office Depot, Texas Instruments, Raytheon Missile Systems, and Waste Management (a Houston-based company responsible for numerous hazardous waste sites). Some environmentalists, including Earth Day's founder, former senator Gaylord Nelson, consider the participation of corporations in Earth Day evidence of the celebration's success. But other environmentalists oppose corporate sponsorship of Earth Day events, accusing corporations of using their financial support of Earth Day as a public relations greenwashing strategy. John Stauber, a critic of corporations that hide their damaging environmental practices behind green advertising and marketing, says that "Waste Management sponsoring Earth Day is similar to Enron sponsoring a seminar on corporate responsibility" (quoted by Cappiello 2003). In recent years environmentalists have withdrawn from and protested Earth Day events in cities across the country because the events were sponsored by corporations.

Environmental Problems: An Overview

Over the past 50 years humans have altered **ecosystems**—the complex and dynamic relationships between forms of life and the environments they inhabit—more rapidly and extensively than in any other comparable period of time in history (Millennium Ecosystem Assessment 2005). As a result, humans have created environmental problems, including depletion of natural resources, air, land, and water pollution, global warming, environmental illness, environmental injustice, threats to biodiversity, and disappearing livelihoods. Because many of these environmental problems are related to the ways that humans produce and consume energy, we begin this section with an overview of global energy use. Before reading further, you may want to assess your attitudes toward energy and the environment in this chapter's *Self and Society* feature.

Energy Use Worldwide: An Overview

Being mindful of environmental problems means seeing the connections between energy use and our daily lives.

> Everything we consume or use—our homes, their contents, our cars and the roads we travel, the clothes we wear, and the food we eat—requires energy to produce and package, to distribute to shops or front doors, to operate, and then to get rid of. We rarely consider where this energy comes from or how much of it we use—or how much we truly need. (Sawin 2004, p. 25)

Attitudes Toward Energy and the Environment

For each of the following statements, select the answer that best represents your views on energy and the environment. Then compare your answers with those of a national representative sample of U.S. adults.

1. With which of the following statements do you most agree?
   (a) The development of U.S. energy supplies—such as oil, gas, and coal—should be given priority, even if the environment suffers; or
   (b) Protection of the environment should be given priority, even at the risk of limiting the amount of energy supplies.
2. Do you favor or oppose the construction of a nuclear energy plant in your area as one of the ways to provide electricity for the United States?
   (a) Strongly favor
   (b) Somewhat favor
   (c) Somewhat oppose
   (d) Strongly oppose
3. Do you think the Arctic National Wildlife Refuge in Alaska should or should not be opened up for oil exploration?
   (a) Yes, should
   (b) No, should not
4. Which of the following approaches to solving the nation's energy problems do you think the United States should follow right now?
   (a) Emphasize more conservation by consumers of existing energy supplies.
   (b) Emphasize production of more oil, gas, and coal supplies.
5. Do you favor or oppose the use of nuclear energy as one of the ways to provide electricity for the United States?
   (a) Strongly favor
   (b) Somewhat favor
   (c) Somewhat oppose
   (d) Strongly oppose

Comparison Data from a National Sample

A 2005 Gallup poll survey (Gallup Organization 2005) of a nationally representative sample of U.S. adults revealed the following results in response to the five questions (the questions presented here are slightly modified from the actual questions asked in the poll).

1. Prioritize energy supplies: 39%
   Prioritize environment: 52%
   Both equally: 4%
   Neither or no opinion: 5%
2. Strongly favor: 11%
   Somewhat favor: 24%
   Somewhat oppose: 19%
   Strongly oppose: 44%
3. Yes, should: 42%
   No, should not: 53%
   No opinion: 5%
4. Emphasize more production: 28%
   Emphasize more conservation: 61%
   Both equally: 7%
   Neither or no opinion: 2%
5. Strongly favor: 17%
   Somewhat favor: 37%
   Somewhat oppose: 22%
   Strongly oppose: 21%
   No opinion: 3%

ThomsonNOW™

Learn more about **Energy Use Worldwide** by going through the Share of Global Emissions Map Exercise.

Most of the world's energy—86% in 2003—comes from fossil fuels, which include petroleum (or oil), coal, and natural gas (Energy Information Administration 2005) (see Figure 14.1). As you continue reading this chapter, notice that the major environmental problems facing the world today—air, land, and water pollution, destruction of habitats, biodiversity loss, global warming, and environmental illness—are linked to the production and use of fossil fuels.

The next most common source of energy is hydroelectric power (6.5%), which involves generating electricity from moving water. As water passes through a dam

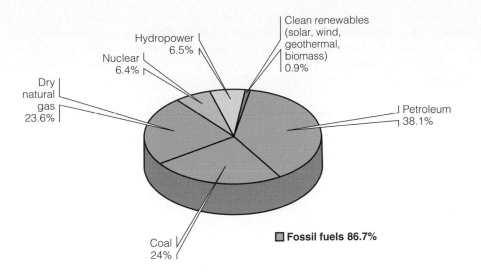

Figure 14.1
World energy production by source: 2003.

Source: Energy Information Administration (2005).

Hydropower 6.5%

Nuclear 6.4%

Dry natural gas 23.6%

Clean renewables (solar, wind, geothermal, biomass) 0.9%

Petroleum 38.1%

Coal 24%

■ Fossil fuels 86.7%

into a river below, energy is produced by a turbine in the dam. Although hydroelectric power is nonpolluting and inexpensive, it is criticized for affecting natural habitats. For example, dams make certain fish unable to swim upstream to reproduce. Canada produces more hydroelectric power than any other country, followed by Brazil, China, and the United States (Energy Information Administration 2005).

Nuclear power, accounting for 6.4% of world energy production in 2003, is associated with a number of problems related to radioactive nuclear waste—problems that are discussed later in this chapter. The United States produces more nuclear power than any other country (Energy Information Administration 2005). Less than 1% of the world's energy comes from other renewable resources, which include geothermal power (from the heat of the earth), solar power, wind power, and biomass (e.g., fuel wood, crops, animal wastes), which are primarily used by poor populations in developing countries. One-fourth of the world's population— 1.6 billion people—lack access to electricity in their homes and use biomass for heating and cooking (Flavin & Aeck 2005).

Depletion of Natural Resources

Population growth, combined with consumption patterns, is depleting natural resources such as forests, water, minerals, and fossil fuels. For example, freshwater resources are being consumed by agriculture, by industry, and for domestic use. More than 1 billion people lack access to clean water (Fornos 2005).

The world's forests are also being depleted. The demand for new land, fuel, and raw materials has resulted in **deforestation**—the conversion of forest land to nonforest land (Intergovernmental Panel on Climate Change 2000). Global forest cover has been reduced by half of what it was 8,000 years ago (Gardner 2005). Between the 1960s and the 1990s one-fifth of the world's tropical forests were cut or burned (Youth 2003). The major causes of deforestation are the expansion of agricultural land, human settlements, wood harvesting, and road building.

Deforestation displaces people and wild species from their habitats; soil erosion caused by deforestation can cause severe flooding; and, as we explain later in this chapter, deforestation contributes to global warming. Deforestation also contributes to **desertification**—the degradation of semiarid land, which results in the expansion of desert land that is unusable for agriculture. Overgrazing by cattle and other herd animals also contributes to desertification. The problem of desertification is most severe in Africa (Reese 2001). As more land turns into desert, populations can no longer sustain a livelihood on the land, and so they migrate to urban areas or other countries, contributing to social and political instability.

Air Pollution

Transportation vehicles, fuel combustion, industrial processes (such as the burning of coal and the processing of minerals from mining), and solid waste disposal have contributed to the growing levels of air pollutants, including carbon monoxide, sulfur dioxide, nitrogen dioxide, mercury, and lead. Air pollution levels are highest in areas with both heavy industry and traffic congestion, such as Los Angeles, New Delhi, Jakarta, Bangkok, Tehran, Beijing, and Mexico City. In the mid-1990s breathing the air in Mexico City was equivalent to smoking two packs of cigarettes a day (Weiner 2001).

In the United States emissions of the six major air pollutants decreased 51% from 1970 to 2003. But, largely because of lax enforcement of the 1970 Clean Air Act, 95 million Americans in 224 counties and the District of Columbia breathe air with levels of toxicity that exceed federal health standards (Clarren 2005).

Air pollution is linked to heart disease, lung cancer, and respiratory ailments, such as emphysema, chronic bronchitis, and asthma. Electric power plants, the largest source of air pollution in the United States, emit soot—tiny particles of toxic chemicals—causing 554,000 asthma attacks and 38,200 heart attacks each year (Clarren 2005). Indoor air pollution from burning wood and biomass for heating and cooking, which we discuss next, is a significant cause of respiratory illness, lung cancer, and blindness in developing countries.

© Ted Spiegel/Corbis

Indoor air pollution is a serious problem in developing countries. As this woman cooks food for her family, she is exposed to harmful air contaminants from the fumes.

Indoor Air Pollution. When we hear the phrase *air pollution,* we typically think of industrial smokestacks and vehicle exhausts pouring gray streams of chemical matter into the air. But indoor air pollution is also a major problem, especially in poor countries.

About half the world's population and up to 90% of rural households in developing countries rely on wood and unprocessed biomass (dung and crop residues) for cooking and heating fuel (Bruce et al. 2000). These fuels are typically burned indoors in open fires or in poorly functioning stoves, producing hazardous emissions such as soot particles, carbon monoxide, nitrous oxides, sulfur oxides, and formaldehyde. Long-term exposure to the smoke contributes to respiratory illness, lung cancer, tuberculosis, and blindness; an estimated 2 million deaths in developing countries each year result from exposure to indoor air pollution. According to the World Health Organization, indoor air pollution ranks fifth as a risk factor for ill health worldwide—behind malnutrition, AIDS, tobacco use, and poor water and sanitation (Bruce et al. 2000; Mishra et al. 2002).

Even in affluent countries much air pollution is invisible to the eye and exists where we least expect it—in our homes, schools, workplaces, and public buildings. Sources of indoor air pollution include lead dust (from old lead-based paint); secondhand tobacco smoke; by-products of combustion (e.g., carbon monoxide) from stoves, furnaces, fireplaces, heaters, and dryers; and other common household, personal, and commercial products (American Lung Association 2005). Some of the most common indoor pollutants include carpeting (which emits more than a dozen toxic chemicals); mattresses, sofas, and pillows (which emit formaldehyde and fire retardants); pressed wood found in kitchen cabinets and furniture (which emits formaldehyde); and wool blankets and dry-cleaned clothing (which emit trichloroethylene). Air fresheners, deodorizers, and disinfectants emit the pesticide paradichlorobenzene. Potentially harmful organic solvents are present in numerous office supplies, including glue, correction fluid, printing ink, carbonless paper, and felt-tip markers. Many homes today contain a cocktail of toxic chemicals. "Styrene (from plastics), benzene (from plastics and rubber), toluene and xylene, trichloroethylene, dichloromethane, trimethylbenzene, hexanes, phenols, pentanes and much more outgas from our everyday furnishings, construction materials, and appliances" (Rogers 2002).

Destruction of the Ozone Layer. The ozone layer of the earth's atmosphere absorbs most of the harmful ultraviolet-B radiation from the sun and completely screens out lethal ultraviolet-C radiation. The ozone layer is thus essential to life on earth. Yet the use of certain chemicals has weakened the ozone layer. The depletion of the ozone layer allows hazardous levels of ultraviolet rays to reach the earth's surface and is linked to increases in skin cancer and cataracts, weakened immune systems, reduced crop yields, damage to ocean ecosystems and reduced fishing yields, and adverse effects on animals.

The ozone hole above Antarctica spanned a record 11 million square miles in 2003, exposing the southern tip of South America. The size of the hole decreased slightly in 2004 but increased in 2005 (Reuters 2005).

Ninety-six chemicals have been identified as being harmful to the ozone layer. These include chlorofluorocarbons (CFCs) (used in refrigerators, air conditioners, spray cans, and other applications), hydrochlorofluorocarbons (HCFCs) (developed as a replacement for CFCs), halons (used in fire extinguishers), and methyl bromide (used as a fumigant for crops, pest control, and quarantine treatment of agricultural exports). These and other ozone-damaging chemicals remain in the atmosphere for various lengths of time, ranging from about 1 year to 1,700 years (UNEP 2003).

Acid Rain. Air pollutants, such as sulfur dioxide and nitrogen oxide, mix with precipitation to form **acid rain.** Polluted rain, snow, and fog contaminate crops, forests, lakes, and rivers. As a result of the effects of acid rain, all the fish have died in a third of the lakes in New York's Adirondack Mountains (Blatt 2005). Because pollutants in the air are carried by winds, industrial pollution in the Midwest falls back to earth as acid rain on southeast Canada and the northeast New England states. Acid rain is not just a problem in North America; it decimates plant and animal species around the globe. Acid rain also deteriorates the surfaces of buildings and statues. "The Parthenon, Taj Mahal, and Michelangelo's statues are dissolving under the onslaught of the acid pouring out of the skies" (Blatt 2005, p. 161).

Greenland's Illulissat glacier, considered one of the wonders of the world, has shrunk by more than 6 miles in 2 or 3 years. If global warming continues, Greenland's ice cap could melt within a few hundred years, raising the level of the earth's oceans by 20–23 feet.

© RH Production / Robert Harding World Imagery/Getty Images

Global Warming and Climate Change

In 2005 scientists reported that the Illulissat glacier in Greenland, considered one of the wonders of the world, shrunk by more than 6 miles in 2 or 3 years (Bohan 2005). In the same year Russian researchers discovered that the world's largest frozen peat bog, located in Siberia, is melting, turning this once frozen area the size of Germany and France combined into a mass of shallow lakes (Pearce 2005). The melting of Greenland's Illulissat glacier and Siberia's peat bog contributes to growing evidence of **global warming**.

Globally, the 1990s was the warmest decade on record since record keeping began in the late 1800s. The hottest year on record was 1998, followed by 2002 and 2003. Global temperatures in 2004 were nearly 1°F above the long-term average (1880–2003), ranking 2004 as the fourth warmest year on record (National Oceanic and Atmospheric Administration 2005). Average global air temperature rose by 0.6°C over the 20th century and, between 1990 and 2100 this temperature is expected to rise another 1.4°–5.8°C (Intergovernmental Panel on Climate Change 2001b). Over the 20th century the average annual U.S. temperature has risen 1°F and is expected to rise by 5°–9°F over the next 100 years (National Assessment Synthesis Team 2000).

Causes of Global Warming. Although the cause of global warming remains a disputed topic, the prevailing scientific view is that **greenhouse gases**—primarily carbon dioxide, methane, and nitrous oxide—accumulate in the atmosphere and act like the glass in a greenhouse, holding heat from the sun close to the earth. Since 1750, atmospheric concentrations of carbon dioxide, the main greenhouse gas, have increased by 31%, methane by 151%, and nitrous oxide by 17% (Intergovernmental Panel on Climate Change 2001b).

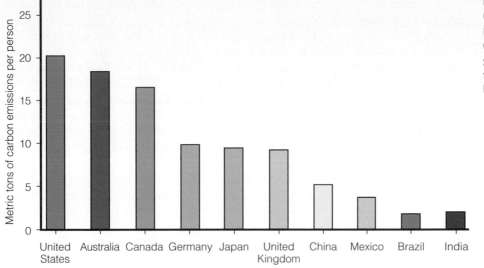

Figure 14.2
Carbon emissions per person in selected countries (2002 data).

Source: Author created using data from UN Human Development Report (2005).

The primary source of carbon dioxide emissions is fossil fuel burning. Total world carbon dioxide emissions from fossil fuels were 25.2 billion metric tons in 2003, an increase of 17.1% from 1993. With less than 5% of the world's population, the United States produces 24% of the world's carbon emissions from fossil fuel burning (Energy Information Administration 2005). U.S. carbon emissions per person are roughly double that of other major industrialized countries and 17 times that of India (see Figure 14.2).

Deforestation also contributes to increasing levels of carbon dioxide in the atmosphere. Trees and other plant life use carbon dioxide and release oxygen into the air. As forests are cut down, fewer trees are available to absorb the carbon dioxide.

Effects of Global Warming and Climate Change. Numerous effects of global warming and climate change have been observed and are anticipated in the future (Intergovernmental Panel on Climate Change 2001a; National Assessment Synthesis Team 2000; UNEP 2005). After Hurricane Katrina devastated the U.S. Gulf Coast, increased attention was given to the role that global warming and climate change play in weather events. "While not all extreme weather events can be attributed directly to climate change, the intensity of such events is likely to increase as a result of global warming" (UNEP 2005, p. 2).

Global warming and climate change are projected to affect regions in different ways. As temperature increases, some areas will experience heavier rain, whereas other regions will get drier. Some regions will experience increased water availability and crop yields, but other regions, particularly tropical and subtropical regions, are expected to experience decreased water availability and a reduction in crop yields. Global warming results in shifts of plant and animal habitats and the increased risk of extinction of some species. Regions that experience increased rainfall as a result of increasing temperatures may face increases in waterborne diseases and diseases transmitted by insects. Over time climate change may produce opposite effects on the same resource. For example, in the short term forest productivity is likely to increase in response to higher levels of carbon dioxide in the

"The uncomfortable truth is that the current scale and character of human activities are decreasing the planet's life-support capacity both in known and in unanticipated ways, and perhaps more rapidly than we suspect."

Carl N. McDaniel
Author of *Wisdom for a Livable Planet*

air, but over the long term forest productivity is likely to decrease because of drought, fire, insects, and disease.

As global warming melts glaciers and permafrost, the sea level will continue to rise. Global average sea levels rose 0.1–0.2 meter during the 20th century and are expected to rise by 0.09–0.88 meter from 1990 to 2100. As sea levels rise, some island countries, as well as some barrier islands off the U.S. coast, are likely to disappear and low-lying coastal areas will become increasingly vulnerable to storm surges and flooding. If global warming continues, Greenland's ice cap could melt within a few hundred years, raising the level of the earth's oceans by 20–23 feet and threatening the lives of the 21.2 billion people who live within 20 miles of the coastline (Bohan 2005).

In urban areas flooding can be a problem where storm drains and waste management systems are inadequate. Increased flooding associated with global warming is expected to result in increases in drownings and in diarrheal and respiratory diseases. Increases in the number of people exposed to insect- and water-related diseases, such as malaria and cholera, are also expected. Even if greenhouse gases are stabilized, global air temperature and sea level are expected to continue to rise for hundreds of years.

Another effect of global warming is that it contributes to further warming of the planet—a process known as a *positive feedback loop*. For example, the melting of Siberia's frozen peat bog—a result of global warming—could release billions of tons of methane, a potent greenhouse gas, into the atmosphere (Pearce 2005). And the melting of ice and snow—another result of global warming—exposes more land and ocean area, which absorbs more heat than ice and snow, further warming the planet.

Land Pollution

About 30% of the world's surface is land, which provides soil to grow the food we eat. Increasingly, humans are polluting the land with nuclear waste, solid waste, and pesticides. In 2003, 1,296 hazardous waste sites were on the National Priority List (also called Superfund sites) (U.S. Bureau of the Census 2004). States with the most Superfund sites include New Jersey (116 sites), California (98 sites), Pennsylvania (95 sites), New York (91 sites), Michigan (69 sites), and Florida (52 sites).

Nuclear Waste. Nuclear waste, resulting from both nuclear weapons production and nuclear reactors or power plants, contains radioactive plutonium, a substance linked to cancer and genetic defects. Radioactive plutonium has a half-life of 24,000 years, meaning that it takes 24,000 years for the radioactivity to be reduced by half (Mead 1998). Thus nuclear waste in the environment remains potentially harmful to human and other life for thousands of years.

Under the 1982 Nuclear Waste Policy Act, since 1998 the Department of Energy has been responsible for handling the storage of nuclear waste from commercial nuclear power plants. Proposals for disposing of nuclear waste have included burying it in rock formations deep below the earth's surface or under Antarctic ice, injecting it into the ocean floor, hurling it into outer space, and storing the waste in huge casks that could be parked in a handful of high-security lots around the country. Each of these options is risky and costly. Because of the government's faltering in its plan to bury nuclear waste at Yucca Mountain in Nevada, every nuclear reactor operator in the country has sued the Energy Department for failure to accept waste on time (Wald 2005).

Nuclear plants have about 52,000 tons of radioactive spent fuel, with about 10,000 tons of that amount sealed in casks (Vedantam 2005b). Because of inadequate oversight and gaps in safety procedures, radioactive spent fuel is missing or unaccounted for at some U.S. nuclear power plants, which raises serious safety concerns (Vedantam 2005a). Accidents at nuclear power plants, such as the 1986 accident at Chernobyl, and the potential for nuclear reactors to be targeted by terrorists add to the actual and potential dangers of nuclear power plants.

Recognizing the hazards of nuclear power plants and their waste, Germany became the first country to order all of its 19 nuclear power plants shut down by 2020 ("Nukes Rebuked," 2000). But other countries are building or planning to build new reactors. In 2004 six new reactors were connected to the grid, and five reactors were permanently shut down, bringing the total number of nuclear reactors worldwide to 114 (Lenssen 2005).

Solid Waste. In 1960 each U.S. citizen generated 2.7 pounds of garbage on average every day. In 2001 this figure had increased to 4.4 pounds (U.S. Bureau of the Census 2004). This figure does not include mining, agricultural, and industrial waste; demolition and construction wastes; junked autos; or obsolete equipment wastes. Some solid waste is converted into energy by incinerators; more than half is taken to landfills. The availability of landfill space is limited, however. Some states have passed laws that limit the amount of solid waste that can be disposed of; instead, they require that bottles and cans be returned for a deposit or that lawn clippings be used in a community composting program. This chapter's *Focus on Technology* feature examines the hazards of disposing of computers in landfills (and other environmental and health hazards of computers).

Pesticides. Pesticides are used worldwide for crops and gardens; outdoor mosquito control; the care of lawns, parks, and golf courses; and indoor pest control. Pesticides contaminate food, water, and air and can be absorbed through the skin, swallowed, or inhaled. At least 53 carcinogenic pesticides are applied to major U.S. food crops (Blatt 2005). Many common pesticides are considered potential carcinogens and neurotoxins. Even when a pesticide is found to be hazardous and is banned in the United States, other countries from which we import food may continue to use it. In an analysis of more than 7,000 domestic and imported food samples, pesticide residues were detected in more than one-third (37%) of the domestic samples and in more than one-fourth (28%) of the imported samples (Food and Drug Administration 2005). Pesticides also contaminate our groundwater supplies.

Water Pollution

According to the Environmental Protection Agency, 40% of all U.S. waters are not safe for fishing or swimming (Smith et al. 2000). Our water is being polluted by a number of harmful substances, including pesticides, vehicle exhaust, acid rain, oil spills, and industrial, military, and agricultural waste.

Water pollution is most severe in developing countries, where more than 1 billion people lack access to clean water. In developing nations as much as 95% of untreated sewage is dumped directly into rivers, lakes, and seas that are also used for drinking and bathing (Pimentel et al. 1998). Mining operations, located primarily in developing countries are notoriously damaging to the environment. Modern gold-mining techniques use cyanide to extract gold from low-grade ore. Cyanide is extremely toxic: One teaspoon of a 2% cyanide solution can kill an adult (cyanide was used to kill Jews in Hitler's gas chambers). In 2000 a dam holding cyanide-laced

Environmental and Health Hazards of Computer Manufacturing, Disposal, and Recycling

Much of the electronic waste from the United States is shipped to India. At this scrap shop in New Delhi, mercury and lead from electronic waste pose environmental and health hazards.

AP/Wide World Photos

More than 1 billion computers worldwide have been sold, and the 2 billion mark may be reached as early as 2008 (Thibodeau 2002). In the United States alone more than 100,000 computers become obsolete each day (Silicon Valley Toxics Coalition 2005). The toxic materials (see table) used in computer manufacturing pose significant hazards for workers in the production of computers, and the toxic **e-waste** (waste from electronic equipment) makes the disposal and recycling of computers a human and environmental hazard.

Toxic Materials Used in Computer Manufacturing

- Lead and cadmium in computer circuit boards
- Lead oxide and barium in cathode ray tubes of computer monitors
- Mercury in switches and flat screens
- Brominated flame retardants on printed circuit boards, cables, and plastic casing
- Polychlorinated biphenyls (PCBs) present in older capacitors and transformers
- Polyvinyl chloride (PVC)-coated copper cables and plastic computer casings

waste at a Romanian gold mine broke, dumping 22 million gallons of cyanide-laced waste into the Tisza River, which flowed into Hungary and Serbia. Some have called this event the worst environmental disaster since the 1986 Chernobyl nuclear explosion. Today, in West Papua, Indonesia, a U.S.-owned gold mine dumps 120,000 tons of cyanide-laced waste into local rivers every day (Ayres 2004). Most gold is used to make jewelry, hardly a necessity warranting the environmental degradation that results from gold mining. To make matters worse, about 300,000 tons of waste are generated for every ton of marketable gold—which translates roughly into 3 tons of waste per gold wedding ring (Sampat 2003).

Hazards of Computer Manufacturing

The manufacturing of semiconductors, printed circuit boards, disk drives, and monitors involves hazardous chemicals. Numerous reports have documented higher-than-normal rates of cancer and birth defects among computer manufacturing workers who are regularly exposed to carcinogenic and other toxic chemicals (Silicon Valley Toxics Coalition 2001).

Hazards of E-Waste Disposal

When consumers upgrade to faster and sleeker PCs with more memory and new features, what happens to their old machines? Most unwanted, obsolete computers are destined for landfills, incinerators, or hazardous waste exports; less than 10% of unwanted computers are recycled (Computer Take Back Campaign 2005).

HAZARDS OF LANDFILL DUMPING AND INCINERATING E-WASTE

The main concern with regard to dumping e-waste in landfills is that hazardous substances in landfills can leach and contaminate the soil and groundwater. In many city landfills cathode ray tubes from computers and TVs are the largest source of lead (Motavalli 2001). Lead has toxic effects on humans as well as on plants, animals, and microorganisms. The leaching of mercury from landfilled e-waste is also a major concern. Just one-seventieth of a teaspoon of mercury can contaminate 20 acres of a lake, making the fish unfit for consumption (Computer Take Back Campaign 2005).

Some computer waste is disposed of through incineration (burning). Incineration of scrap computers generates extremely toxic dioxins that are released in air emissions.

HAZARDS OF RECYCLING COMPUTERS

Computers and printers can be recycled, and the plastic materials and precious metals, such as copper and gold, can be recovered. But recycling computers releases dust containing toxic polybrominated flame retardants into the air, and high concentrations of these substances have been found in the blood of workers in recycling plants (Silicon Valley Toxics Coalition 2001).

THE EXPORTATION OF HAZARDOUS E-WASTE

Fifty to 80% of the United States' electronic waste that is collected for recycling is exported to developing countries where labor costs are cheap and worker safety and environmental regulations are lax compared with U.S. law. One pilot program that collected electronic scrap in San Jose, California, estimated that shipping computer monitors to China for recycling was 10 times cheaper than recycling them in the United States (Silicon Valley Toxics Coalition 2001).

When investigators visited a waste site in Guiyu, China, they saw men, women, and children wearing little or no protective gear smashing, picking apart, and burning computers, exposing themselves and their surroundings to toxic substances (Silicon Valley Toxics Coalition 2002). The groundwater near the site is so polluted that drinking water is trucked in from 18 miles away. A nearby river water sample contained 190 times the pollution levels allowed by the World Health Organization.

A 1989 treaty known as the Basel Convention restricts the export of hazardous waste from rich countries to poor countries—even when the waste is destined for recycling. But the United States has not ratified this treaty and has lobbied Asian governments to establish bilateral trade agreements to allow continued exportation of hazardous waste.

Solutions to the Problem of Toxic E-Waste

LEGISLATIVE SOLUTIONS

At least 26 states have considered legislation concerning the disposal of e-waste. In

---

In the United States one indicator of water pollution is the number of fish advisories issued; these advisories warn against the consumption of certain fish caught in local waters because of contamination with such pollutants as mercury and dioxin. In 2003, 48 states issued more than 3,000 fish advisories. About one-third (35%) of total U.S. lake acreage and one-fourth (24%) of river miles are under advisory (EPA 2004a). The Environmental Protection Agency advises women who may become pregnant, pregnant women, nursing mothers, and young children to avoid eating certain fish altogether (swordfish, shark, king mackerel, and tilefish) because of the high levels of mercury (EPA 2004b).

2000 Massachusetts became the first state to ban dumping of video monitors (computer screens and televisions) (Massachusetts Department of Environmental Protection 2001). Maine and Minnesota also prohibit dumping cathode ray tubes (found in monitors). And in 2003 California passed the Electronic Waste Recycling Act, becoming the first state to pass legislation to require recycling. This act requires retailers to add a $6 to $10 fee to certain electronic products to cover the costs of collection and recycling programs and requires recycling programs to use environmentally sound methods. Electronics manufacturers favor regulating e-waste recycling through federal legislation so that all manufacturers abide by the same rules.

THE EUROPEAN UNION'S WEEE DIRECTIVE
In 2002 the European Union passed legislation on waste from electrical and electronic equipment (the WEEE Directive) that makes producers responsible for taking back their old products and for phasing out the use of certain toxic materials in computer manufacturing; the legislation also encourages cleaner product design and less waste

generation. This strategy, known as *Extended Producer Responsibility,* is based on the principle that producers bear a degree of responsibility for all the environmental impacts of their products. This includes impacts arising from choice of materials and from the manufacturing process as well as the use and disposal of products. When manufacturers are financially responsible for waste management of their products, they will have a financial incentive to design those products with less hazardous and more recyclable materials.

INDUSTRY INITIATIVES
A number of industries have taken steps to alleviate the environmental and health hazards of computers. Some computer manufacturers, such as Dell and Gateway, lease out their products, thereby ensuring that they get them back to further upgrade and lease out again. In 1998 IBM introduced the first computer that uses 100% recycled resin in all its major plastic parts. Hewlett-Packard (HP) has developed a safe cleaning method for chips, using carbon dioxide for cleaning as a substitute for hazardous solvents. To encourage recycling of e-waste,

HP's Planet Partners program allows people to have their old computers and printers picked up at their door for a fee of $17 to $31 to cover transportation and recycling costs. In exchange, they receive a $50 coupon toward the purchase of an HP product. Toshiba is working on a modular upgradable and customizable computer to reduce product obsolescence. Pressures to eliminate halogenated flame retardants and to design products for recycling have led to the use of metal shielding in computer housings, and a range of lead-free solders are now available.

As governments and industries continue to develop strategies for reducing environmental hazards associated with computer manufacturing and disposal, consumers can also make a difference by choosing manufacturers who practice "product stewardship"—making products that are less toxic, conserve natural resources, and reduce waste. *A Guide to Environmentally Preferable Computer Purchasing* is available from the Northwest Product Stewardship Council at www.govlink.org/nwpsc/ (Northwest Product Stewardship Council 2000).

Chemicals, Carcinogens, and Health Problems

When scientists tested journalist Bill Moyers's blood as part of a documentary on the chemical industry, they found traces of 84 of the 150 chemicals they had tested (PBS 2001). Sixty years ago, when Moyers was 6 years old, only one of these chemicals—lead—would have been present in his blood. During a 2004 World Health Organization convention in Budapest, 44 different hazardous chemicals were found in the bloodstream of top EU officials (Schapiro 2004). And in a study of umbilical cord blood of 10 newborns, researchers found an average of 200 industrial chemicals, pesticides, and other pollutants (Environmental Working Group 2005).

In the United States more than 62,000 chemical substances are in commercial use, and 1,500 new chemicals are introduced each year. But complete data on the health and environmental effects are known for only 7% of chemicals produced in high volume (UNEP 2002).

The *11th Report on Carcinogens* (U.S. Department of Health and Human Services 2004) lists 246 chemical substances that are "known to be human carcinogens" or "reasonably anticipated to be human carcinogens," meaning that they are linked to cancer. However, the report suggests that these 246 chemical substances may constitute only a fraction of actual human carcinogens. Many of the chemicals we are exposed to in our daily lives can cause not only cancer but also other health problems, such as infertility, birth defects, and a number of childhood developmental and learning problems (Fisher 1999; Kaplan & Morris 2000; McGinn 2000). Some chemicals, such as persistent organic pollutants, accumulate in the food chain and persist in the environment, taking centuries to degrade.

Chemicals found in common household, personal, and commercial products can result in a variety of temporary acute symptoms, such as drowsiness, disorientation, headache, dizziness, nausea, fatigue, shortness of breath, cramps, diarrhea, and irritation of the eyes, nose, throat, and lungs. Long-term exposure can affect the nervous system, reproductive system, liver, kidneys, heart, and blood. Fragrances, which are found in many consumer products, may produce sensory irritation, pulmonary irritation, decreases in expiratory airflow velocity, and possible neurotoxic effects (Fisher 1998). Fragrance products can cause skin sensitivity, rashes, headache, sneezing, watery eyes, sinus problems, nausea, wheezing, shortness of breath, inability to concentrate, dizziness, sore throat, cough, hyperactivity, fatigue, and drowsiness (DesJardins 1997). More and more businesses are voluntarily limiting fragrances in the workplace or banning them altogether to accommodate employees who experience ill effects from them. This chapter's *Social Problems Research Up Close* feature describes the *Third National Report on Human Exposure to Environmental Chemicals* (Centers for Disease Control and Prevention 2005).

Vulnerability of Children. The World Health Organization estimates that 5,500 children die each day from diseases linked to polluted water, air, and food (Mastny 2003). In the United States an estimated 1 in 200 children suffers developmental or neurological deficits as a result of exposure to toxic substances during pregnancy or after birth (UNEP 2002). Asthma, the number one childhood illness in the United States, is linked to air pollution. Children are more vulnerable than adults to the harmful effects of most pollutants for a number of reasons. For instance, children drink more fluids, eat more food, and inhale more air per unit of body weight than do adults; in addition, crawling and a tendency to put their hands and other things in their mouths provide more opportunities to ingest chemical or heavy metal residues.

Multiple Chemical Sensitivity Disorder. Multiple chemical sensitivity (MCS), also known as environmental illness, is a condition whereby individuals experience adverse reactions when exposed to low levels of chemicals found in everyday substances (vehicle exhaust, fresh paint, housecleaning products, perfume and other fragrances, synthetic building materials, and numerous other petrochemical-based products). Symptoms of MCS include headache, burning eyes, difficulty breathing, stomach distress or nausea, loss of mental concentration, and dizziness. The onset of MCS is often linked to acute exposure to a high level of chemicals or to chronic long-term exposure. Sufferers of MCS often avoid public places and/or wear a protective breathing filter to avoid inhaling the many chemical substances in the environment. Some individuals with MCS build houses made from materials that do not

The Third National Report on Human Exposure to Environmental Chemicals

In 2001 the Centers for Disease Control and Prevention published a landmark study that reported measures of human exposure to 27 environmental chemicals, 24 of which had never been measured before in a nationally representative sample of the U.S. population. The *Second National Report on Human Exposure to Environmental Chemicals* and the *Third National Report on Human Exposure to Environmental Chemicals* (Centers for Disease Control and Prevention 2003, 2005) are continuations of the ongoing assessment of the U.S. population's exposure to environmental chemicals. For the *Second Report* human exposure to 116 chemicals was tested. One of the key findings of the *Second Report* was that phthalates—compounds found in such products as soap, shampoo, hair spray, many types of nail polish, and flexible plastics—are absorbed by humans. For the *Third Report* exposure to 148 chemicals was tested.

Methods and Sample

The *Third National Report on Human Exposure to Environmental Chemicals* is based on 2001–2002 data from the *National Health and Nutrition Examination Survey (NHANES)* conducted by the National Center for Health Statistics. The NHANES is a series of surveys designed to collect data on the health and nutritional status of the U.S. population.

The NHANES is based on a representative sample of the noninstitutionalized, civilian U.S. population, with an oversampling of African Americans, Mexican Americans, adolescents (ages 12–19 years), older Americans (age 60 years and older), and pregnant women to produce more reliable estimates for these groups. The samples did not specifically target people who were believed to have high or unusual exposures to environmental chemicals.

Assessing research participants' exposure to environmental chemicals involved measuring the chemicals or their metabolites (breakdown products) in the participants' blood or urine—a method known as *biomonitoring*. Blood was obtained by venipuncture for participants age 1 year and older, and urine specimens were collected for people age 6 years and older.

The chemicals measured in the research participants' blood and urine were grouped into the following categories:

1. Metals (including barium, cobalt, lead, mercury, uranium, and others)
2. Cotinine (tracks human exposure to tobacco and tobacco smoke)
3. Organophosphate, organochlorine, and other pesticides
4. Phthalates (chemicals commonly used in such consumer products as soap, shampoo, hair spray, nail polish, and flexible plastics)
5. Polycyclic aromatic hydrocarbons (PAHs)
6. Dioxins, furans, and polychlorinated biphenyls (PCBs)
7. Phytoestrogens
8. Herbicides
9. Pyrethroid, carbamate, and other pesticides

How did the Centers for Disease Control and Prevention (CDC) determine which chemicals to test? In 2002 the CDC solicited nominations from the public for candidate chemicals or categories of chemicals for possible inclusion in future reports and received nominations for hundreds of chemicals. The CDC selected the chemicals to be tested on the basis of several factors, including (1) scientific data that suggested exposure in the U.S. population, (2) the seriousness of health effects known or suspected to result from exposure, (3) the need to assess public health efforts to reduce exposure to a chemical, and (4) the cost of analysis for the chemical.

The researchers conducting the study for the CDC had several objectives, including the following:

- To determine which selected environmental chemicals are getting into the bodies of Americans and at what concentrations.
- For chemicals that have a known toxicity level, to determine the prevalence of people in the U.S. population with levels above those toxicity levels.

contain chemicals that are typically found in building materials. Although estimates of the prevalence of chemical sensitivity vary, one recent study reported that in the Atlanta, Georgia, metropolitan area, 12.6% of a population sample reported an unusual sensitivity to common chemicals, and 3.1% had been diagnosed as having MCS or environmental illness (Caress et al. 2002).

- To establish reference ranges that can be used by physicians and scientists to determine whether a person or group has an unusually high level of exposure.
- To assess the effectiveness of public health efforts to reduce the exposure of Americans to specific environmental chemicals.
- To track, over time, trends in the levels of exposure of the U.S. population to environmental chemicals.
- To determine whether exposure levels are higher among minorities, children, women of childbearing age, or other potentially vulnerable groups.

Key Findings and Conclusions
Selected findings from the *Third National Report on Human Exposure to Environmental Chemicals* (Centers for Disease Control and Prevention 2005) include the following:

- *New measures for some widely used insecticides.* The *Third Report* presents first-time exposure information for five commonly used pyrethroid insecticides. The findings suggest widespread exposure to pyrethroid insecticides in the U.S. population.
- *Establishment of reference ranges.* The levels of chemicals found in the blood and urine of the research participants provide *reference ranges* (also known as background exposure levels) for all the

environmental chemicals for which participants were tested. Physicians, researchers, and public health officials use the reference ranges to determine whether individuals or groups are experiencing levels of exposure that are unusually high compared to the level of exposure experienced by the rest of the population. The *Third Report* established reference ranges for 38 of the 148 chemicals tested (reference ranges for the remaining 110 chemicals were established in earlier reports).

- *Reduced exposure of the U.S. population to environmental tobacco smoke.* Cotinine is a metabolite of nicotine that tracks exposure to environmental tobacco smoke (ETS) among nonsmokers; higher cotinine levels reflect more exposure to ETS. Results from the *Third Report* show that from 1988–1991 to 1999–2002, median levels in nonsmokers decreased by 68–69% in children and adolescents and by about 75% percent in adults. Levels in blacks were more than twice the levels in Mexican Americans and non-Hispanic whites. Although efforts to reduce ETS exposure show significant progress, ETS exposure remains a major public health concern.
- *Continued progress in reducing blood-lead levels among children.* For the period 1999–2002, 1.6% of children ages

1–5 had elevated blood-lead levels. This percentage has decreased from 4.4% in the early 1990s. This significant decrease reflects the success of public health efforts to decrease the exposure of children to lead. Nevertheless, children at high risk for lead exposure (e.g., those living in homes containing lead-based paint or lead-contaminated dust) remain a major public health concern.

Because the presence of an environmental chemical in a person's blood or urine does not necessarily mean that the chemical will cause disease, more studies are needed to determine which levels of these chemicals in people result in disease. Research has already documented adverse health effects of lead and environmental tobacco smoke. However, for many environmental chemicals, few studies are available, and more research is needed to assess health risks associated with different blood or urine levels of these chemicals. Future CDC reports, expected to be released every two years, will include more chemicals and will also address the following questions:

1. Are chemical exposure levels increasing or decreasing over time?
2. Are public health efforts to reduce chemical exposure effective?
3. Do certain groups of people have higher levels of chemical exposure than others?

Environmental Injustice

Although environmental pollution and degradation and depletion of natural resources affect us all, some groups are more affected than others. **Environmental injustice,** also referred to as **environmental racism,** refers to the tendency for socially and politically marginalized groups to bear the brunt of environmental ills.

Environmental Injustice in the United States. A Gallup poll of U.S. adults found that nonwhites (43%) are more likely than whites (31%) to worry "a great deal" about the environment (Lyons 2005). These findings may be explained by the fact that in the United States polluting industries, industrial and waste facilities, and transportation arteries (which generate vehicle emissions pollution) are often located in minority communities (Bullard 2000; Bullard & Johnson 1997). African Americans and Hispanics are more likely than whites to live in high-pollution areas and to have higher prevalence of and death rates from respiratory illness such as asthma (American Lung Association 2005).

An area between New Orleans and Baton Rouge, Louisiana—known as Cancer Alley because of the high rates of cancer in this area—has about 140 chemical plants. The people who live and work in Cancer Alley are disproportionately poor and black. One resident of Norco, Louisiana, who lived next to a Shell Oil refinery and chemical plant, reported that nearly everyone in the community suffered from health problems caused by industry pollution (Bullard 2000). When nearly all the chemical plants in Cancer Alley were damaged by Hurricane Katrina in 2005, residents and their homes were exposed to the toxic poisons that were released from these plants into floodwaters.

A study of 370 communities in Massachusetts found that compared with communities with higher median household incomes and lower percentages of people of color, communities with lower household incomes and higher percentages of people of color contained significantly higher levels of chemical emissions from industrial facilities and had more hazardous waste sites (Leutwyler 2001). Another study of air lead concentrations in 3,111 U.S. counties found that counties with the largest proportion of black youth under age 16 have more than 7% more lead in the air than counties with no black youth, and counties with the largest proportion of white youth have nearly 10% less lead in the air than counties with the smallest proportion of white youth (Stretesky 2003). Consequently, nearly 11% of black children suffer from lead poisoning (which is associated with deficits in intellectual and academic functioning), compared to 2% of white children. In addition, in North Carolina, hog industries—and the associated environmental and health risks associated with hog waste—tend to be located in communities with large black populations, low voter registration, and low incomes (Edwards & Ladd 2000).

Native Americans also experience the effects of environmental injustice. The U.S. federal government approved a plan by a private corporation to store tens of thousands of tons of radioactive nuclear waste on a Native American reservation in Utah (Vedantam 2005b). The reservation leaders agreed to the plan with the hope that the nuclear waste storage facility will provide jobs to Native Americans and bring in needed income.

Environmental Injustice Around the World. Environmental injustice affects marginalized populations around the world, including minority groups, indigenous peoples, and other vulnerable and impoverished communities, such as peasants and nomadic tribes (Renner 1996). These groups are often powerless to fight against government and corporate powers that sustain environmentally damaging industries. At the Second International Indigenous Youth Conference in Vancouver, a delegate from the Igorot tribal people of the Philippines explained, "Indigenous people are no longer able to plant fruits or vegetables because the . . . mercury poisoning, produced from massive logging and mining operations, inhibits the growth of any plant life" (Sterritt 2005, n.p.). Indigenous groups in Nigeria, such as the Urhobo, Isoko, Kalabare, and Ogoni, are facing environmental threats caused by oil production operations run by transnational corporations. Oil spills, natural gas fires, and leaks from toxic waste pits have polluted the soil, water, and air and have compromised the health of various local tribes. "Formerly lush agricultural land is now covered by oil slicks, and much vegetation and wildlife has been destroyed. Many Ogoni suffer from respiratory diseases and cancer, and birth defects are frequent" (Renner 1996, p. 57). The environmental injustices experienced by the Ogoni and the Igorot reflect only the tip of the iceberg. Renner (1996) warns that "minority populations and indigenous peoples around the globe are facing massive degradation of their environments that threatens to irreversibly alter, indeed destroy, their ways of life and cultures" (p. 59).

Threats to Biodiversity

About 1.9 million species of life have been identified on earth (Baillie et al. 2004). However, there may be 10 times that number and perhaps many times more if we include microbial diversity (Chivian & Bernstein 2004). This enormous diversity of life, known as **biodiversity,** provides food, medicines, fibers, and fuel; purifies air and freshwater; pollinates crops and vegetation; and makes soils fertile.

In recent decades we have witnessed mass extinction rates of diverse life forms. On average, one species of plant or animal life becomes extinct every 20 minutes (Levin & Levin 2002). Unlike the extinction of the dinosaurs millions of years ago, humans are the primary cause of disappearing species today (Hunter 2001). Biologists expect that at least half of the roughly 10 million species alive today will become extinct over the next few centuries as a result of habitat loss caused by deforestation and urban sprawl, overharvesting, and global warming (Cincotta & Engelman 2000). Another threat to biodiversity is overexploitation of species for their meat, hides, horns, or medicinal or entertainment value. A poacher, for example, can earn several hundred dollars for a single rhinoceros horn—equivalent to a year's salary in many African countries. That same horn, ground up and used as a medicinal product, can sell for half a million dollars in Asia (Schmidt 2004). The International Fund for Animal Welfare (2005) checked the Internet for one week and found more than 9,000 wild animal products and specimens and live wild animals for sale—predominantly from species protected by law—a figure that represents "only the tip of the iceberg" (p. ii).

As shown in Table 14.1, 15,589 species worldwide are threatened with extinction. This figure is an underestimation because it is based on an assessment of less than 3% of the world's 1.9 million known species (Baillie et al. 2004).

Environmental Problems and Disappearing Livelihoods

Environmental problems threaten the ability of the earth to sustain its growing population. Agriculture, forestry, and fishing provide 50% of all jobs worldwide and 70% of jobs in sub-Saharan Africa, East Asia, and the Pacific (World Resources Institute 2000). As croplands become scarce or degraded, as forests shrink, and as marine life dwindles, millions of people who make their living from these natural resources must find alternative livelihoods. In Canada's maritime provinces collapse of the cod fishing industry in the 1990s from overfishing left 30,000 fishers dependent on government welfare payments and decimated the economies of hundreds of communities (World Resources Institute 2000).

"People everywhere must come to understand that they depend completely on biodiversity— for food, much medicine, and many other products; for global sustainability and stability; and for spiritual renewal. Acting on this realization will put human beings in their proper place, as only one among millions of kinds of organisms, with responsibility for all our fellow humans and for the future of the planet as a whole."

Peter Raven
Missouri Botanical Garden

Table 14.1

Threatened Species Worldwide: 2004[a]

Category	Number of Threatened Species in 2004
Mammals	1,101
Birds	1,213
Amphibians	1,856
Reptiles	304
Fishes	800
Invertebrates (insects, mollusks, crustaceans, etc.)	1,992
Plants	8,321
Lichens	2
Total	15,589

[a] Threatened species include those classified as critically endangered, endangered, and vulnerable.

Source: Baillie et al. (2004).

Globally, there are an estimated 30 million **environmental refugees**—individuals who have migrated because they can no longer secure a livelihood as a result of deforestation, desertification, soil erosion, and other environmental problems (Margesson 2005). As individuals lose their source of income, so do nations. In one-fourth of the world's nations, crops, timber, and fish contribute more to the nation's economy than industrial goods do (World Resources Institute 2000).

Social Causes of Environmental Problems

Various structural and cultural factors have contributed to environmental problems. These include population growth, industrialization and economic development, and cultural values and attitudes such as individualism, materialism, and militarism.

Population Growth

As noted in Chapter 13, the world's population doubled from 3 billion in 1960 to 6 billion in 1999, and it is projected to reach more than 9 billion by 2050. Population growth places increased demands on natural resources and results in increased waste. As Hunter (2001) explains:

> Global population size is inherently connected to land, air, and water environments because each and every individual uses environmental resources and contributes to environmental pollution. While the scale of resource use and the level of wastes produced vary across individuals and across cultural contexts, the fact remains that land, water, and air are necessary for human survival. (p. 12)

However, population growth itself is not as critical as the ways in which populations produce, distribute, and consume goods and services. Recall from Chapter 13 that wealthy, industrialized countries have the slowest rates of population growth, yet it is these countries that consume the most natural resources and contribute most to pollution, global warming, and other environmental problems.

ThomsonNOW™

Learn more about **Population Growth** by going through the Population Map Exercise.

Industrialization and Economic Development

Many of the environmental problems confronting the world, including global warming and the depletion of the ozone layer, are associated with industrialization and economic development. Industrialized countries, for example, consume more energy and natural resources and contribute more pollution to the environment than poor countries.

The relationship between level of economic development and environmental pollution is curvilinear rather than linear. For example, industrial emissions are minimal in regions with low levels of economic development and are high in the middle-development range as developing countries move through the early stages of industrialization. However, at more advanced industrial stages industrial emissions ease because heavy-polluting manufacturing industries decline and "cleaner" service industries increase and because rising incomes are associated with a greater demand for environmental quality and cleaner technologies. However, a positive linear correlation has been demonstrated between per capita income and national carbon dioxide emissions (Hunter 2001).

In less developed countries environmental problems are largely the result of poverty and the priority of economic survival over environmental concerns. Vajpeyi (1995) explains:

> Policymakers in the Third World are often in conflict with the ever-increasing demands to satisfy basic human needs—clean air, water, adequate food, shelter, education—and to safeguard the environmental quality. Given the scarce economic and technical resources at their disposal, most of these policymakers have ignored long-range environmental concerns and opted for short-range economic and political gains. (p. 24)

A major environmental concern today is the growing economy of China—the most populated country in the world. The income of China's 1.3 billion people has been growing at a rapid rate, and along with that rising income Chinese consumers are increasingly able to spend money on cars, travel, meat, and other goods and services that use fossil fuels. If China were to burn coal at the U.S. level of 2 tons per person, the country would use 2.8 billion tons per year—more than the current world production of 2.5 billion tons. In addition, if the Chinese were to use oil at the same rate as Americans, by 2031 China would need 99 million barrels of oil a day—more than the current world production of 79 million barrels per day (Aslam 2005). Lester Brown of World Watch Institute says that "China is teaching us that we need a new economic model, one that is based not on fossil fuels but that instead harnesses renewable sources of energy" (quoted by Aslam 2005, n.p.).

In considering the effects of economic development on the environment, note that the ways we measure economic development and the "health" of economies also influence environmental outcomes. Two primary measures of economic development and the health of economies are gross domestic product (GDP) and consumer spending. But these measures overlook the social and environmental costs of the production and consumption of goods and services. Until definitions and measurements of "economic development" and "economic health" reflect these costs, the pursuit of economic development will continue to contribute to environmental problems.

Cultural Values and Attitudes

Cultural values and attitudes that contribute to environmental problems include individualism, materialism, and militarism.

Individualism. Individualism, which is a characteristic of U.S. culture, puts individual interests over collective welfare. Even though recycling is good for our collective environment, many individuals do not recycle because of the personal inconvenience involved in washing and sorting recyclable items. Similarly, individuals often indulge in countless behaviors that provide enjoyment and convenience at the expense of the environment: long showers, use of dish-washing machines, recreational boating, frequent meat eating, the use of air conditioning, and driving large, gas-guzzling SUVs, to name just a few.

Materialism. Materialism, or the emphasis on worldly possessions, also encourages individuals to continually purchase new items and throw away old ones. The media bombard us daily with advertisements that tell us life will be better if we purchase a particular product. After the terrorist attacks of September 11, 2001, President Bush advised Americans that it was their patriotic duty to go to the malls and spend money. Materialism contributes to pollution and environmental degradation by supporting polluting and resource-depleting industries and by contributing to waste.

Militarism. The cultural value of militarism also contributes to environmental degradation (see also Chapter 16). "It is generally agreed that the number one polluter in the United States is the American military. It is responsible each year for the generation of more than one-third of the nation's toxic waste . . . an amount greater than the five largest international chemical companies combined" (Blatt 2005, p. 25). Toxic substances from military vehicles, weapons materials, and munitions pollute the air, land, and groundwater in and around military bases and training areas. The Pentagon has asked Congress to loosen environmental laws for the military, so that training exercises can occur unimpeded (Janofsky 2005). In addition, the Environmental Protection Agency is forbidden to investigate or sue the military (Blatt 2005).

> "In America we have two dominant religions: Christianity and accumulation."
>
> **Brian Swimme**
> **California Institute of Integral Studies**

Strategies for Action: Responding to Environmental Problems

One strategy for alleviating environmental problems is to lower fertility rates and slow population growth, as discussed in detail in Chapter 13. Responses to environmental problems also include environmental activism, environmental education, the use of "green" energy, modifications in consumer products and behavior, and government regulations and legislation. Sustainable economic development and international cooperation and assistance also play important roles in alleviating environmental problems.

> "Just as we have the capacity to hasten the degradation and destruction of our planet, so, too, do we have the capacity to preserve, build, and improve the quality of life on our planet. The choice is ours."
>
> **Hal Burdett**
> **Population Institute**

Environmental Activism

With more than 10,000 environmental organizations with a combined membership of between 19 million and 41 million, "the U.S. environmental movement is now one of the most powerful social movements in the United States" (Faber 2005, p. 51). Environmental organizations exert pressure on government and private industry to initiate or intensify actions related to environmental protection. Environmentalist groups also design and implement their own projects and disseminate information to the public about environmental issues. In a national survey of college freshmen nearly one in five (17.8%) said that "becoming involved in programs to clean up the environment" was personally "essential" or "very important" (Sax et al. 2004). This chapter's *Human Side* feature contains excerpts from an interview with a grandmother who is nationally known for her activism against the environmentally devastating practice of mountaintop-removal mining.

In the United States environmentalist groups date back to 1892 with the establishment of the Sierra Club, followed by the Audubon Society in 1905. Other environmental groups include the National Wildlife Federation, the World Wildlife Fund, the Environmental Defense Fund, Friends of the Earth, the Union of Concerned Scientists, Greenpeace, Environmental Action, the Natural Resources Defense Council, World Watch Institute, World Resources Institute, the Rainforest Alliance, and Conservation International.

Online Activism. The Internet and e-mail provide important tools for environmental activism. For example, Action Network, an online environmental activism community sponsored by Environmental Defense, sends e-mail action alerts to 800,000 members informing them when Congress and other decision makers threaten the health of the environment. These members can then send e-mails and faxes to Congress, the president, and business leaders, urging them to support policies that protect the environment. In 2002, with the help of Environmental Defense's Action Network, California passed a bill to mandate stricter greenhouse gas emissions controls for automobiles (Environmental Defense 2004). Other environmental organizations, such as the Sierra Club, also have similar action alert features on their websites.

Religious Environmentalism. From a religious perspective environmental degradation can be viewed as sacrilegious, sinful, and an offense against God (Gottlieb 2003a). "The world's dominant religions—as well as many people who identify with the 'spiritual' rather than with established faiths—have come to see that the environmental crisis involves much more than assaults on human health, leisure, or convenience. Rather, humanity's war on nature is at the same time a deep affront to one of the essentially divine aspects of existence" (Gottlieb 2003b, p. 489). This view has compelled religious groups to take an active role in environmental activism. For example, the National Association of Evangelicals, an umbrella group of 51 church denominations, adopted a platform called For the Health of the Nation: An Evangelical Call to Civic Responsibility. This platform, which has been signed by nearly 100 evangelical leaders, calls on the government to "protect its citizens from the effects of environmental degradation" (Goodstein 2005). Larry Schweiger, president of the National Wildlife Federation, welcomes evangelicals as allies and explains that conservative lawmakers who might not pay attention to what environmental groups say may be more likely to pay attention to what the faith community is saying. Recently, the faith community has been voicing concerns over global warming and climate change. The magazine *Christianity Today* ran an editorial endorsing a bill (sponsored by Senators McCain and Lieberman) that would curb greenhouse gases (Goodstein 2005).

Ecoterrorism. An extreme form of environmental activism is **ecoterrorism,** defined as any crime intended to protect wildlife or the environment that is violent, puts human life at risk, or results in damages of $10,000 or more (Denson 2000).

An Interview with West Virginia Anti–Mountaintop Removal Activist Julia Bonds

Julia Bonds, a grandmother and native of West Virginia, was awarded one of six Goldman Environmental Prizes for her activism in fighting mountaintop-removal mining—a practice in which mountain peaks are literally sliced off so that coal can be extracted from within the mountain. In the following excerpts from an interview published in Grist Magazine, Julia Bonds describes the devastating effects of mountaintop-removal mining and her activism experiences.

Grist: Can you describe the effects of mountaintop removal mining on communities and the environment?

Julia Bonds: What I've seen happening in the coalfields is the . . . complete annihilation of the communities and culture of Appalachia. . . . The wonderful and valuable hardwood forests are being destroyed and they will not return for over 600 years, if ever. Our beautiful mountain streams have been devastated. . . . The blasts from the mine damage homes . . . and the air quality where they're blasting and mining is the worst anyone can imagine. . . . The worst

This mountaintop removal is just one of the mining sites on Black Mountain in Kentucky.

devastation comes from the flooding. . . . Once you remove soil and vegetation from the top of a very steep mountain valley, you're going to increase runoff from rain. . . .

Grist: . . . You and your family were the last residents to evacuate from your hometown of Marfork Hollow. What happened to the town?

The best-known ecoterrorist groups are the Earth Liberation Front (ELF) and the Animal Liberation Front (ALF). ALF and the ELF are international underground movements consisting of autonomous individuals and small groups who engage in "direct action" to (1) inflict economic damage on those profiting from the destruction and exploitation of the natural environment, (2) save animals from places of abuse (e.g., laboratories, factory farms, fur farms), and (3) reveal information and educate the public on atrocities committed against the earth and all the species that populate it. Direct actions of ALF and ELF have included setting fire to a ski resort in Vail, Colorado, to express opposition to the resort's proposed expansion into the forest habitat of the Canada lynx, setting fire to gas-guzzling Hummers and other SUVs at car dealerships, vandalizing SUVs parked on streets and in driveways, setting fire to construction sites (in opposition to the development), and smashing windows at the homes of fur retailers. The FBI estimates that between 1996 and

Julia Bonds: I'm the seventh generation to live in that hollow, and my grandson is the ninth. Massey Coal [Company] moved in there around 1994. Now, I'm used to coal mining—I'm from a coal-mining family—but I was not prepared for what Massey brought down on our heads in Marfork. The reserves they're mining now are not the clean reserves they were mining in the '40s, '50s, and '60s. These reserves create more waste than coal. The air pollution, the coal dust, is unbearable in that little community. My grandson now has asthma, and my home and my neighbors' homes were damaged by coal dust.

. . . The thing that really sticks in my mind is a 6-year-old child, my grandson, standing in a stream full of dead fish and asking, "What's wrong with these fish?" I looked down at the water and screamed. My family, for generations, has enjoyed that stream, but we never went back in the river again. We also witnessed several black-water spills [of coal waste]. Those are so thick they're like pea soup, with big black chunks in it. I knew people were going to have to drink that crap.

In Marfork, there's a huge earthen dam for coal waste—it's eventually going to be 924 feet tall and will hold 7 billion gallons of waste—that sat three miles above my home. I was sitting out on the front porch with my grandson, and he told me he had picked out an escape route in case the dam failed. I knew in my heart there was really no escape. How do you tell a child that his life is a sacrifice for corporate greed? . . .

Grist: What was the first step you took as an activist?

Julia Bonds: When I saw the fish kill, I called the neighbor that lived above me. . . . The neighbor said, "Here's the Department of Environmental Protection's number, call them." Two weeks later I noticed a flyer on a window that said there was a rally against irresponsible mining. I went to that rally, I went to one meeting, and I never looked back.

Grist: Can you tell me about some of the threats you've experienced because of your work?

Julia Bonds: You really haven't been intimidated until you see a 60-ton coal truck swerve at you on a narrow road, when there's a rock cliff on one side and a 100-foot drop-off on the other. . . . The coal companies pack permit hearings with their men, and they brainwash the men, telling them, "These are the people who are going to take your jobs away." So there's foul language, threats, phone calls. . . . Once I heard someone say at a permit hearing, "If I were these ladies, I'd be afraid to go home tonight."

Grist: What's been your greatest victory so far?

Julia Bonds: I think it's watching or reading when an oppressed Appalachian person stands up with a protest sign or writes a letter to the editor. When people empower themselves, that's the greatest victory.

Source: Michelle Nijhuis, 2003 (April 14). "Coal-Miner's Slaughter," *Grist Magazine.* www.grist.com

2002 ALF and ELF committed more than 600 criminal acts in the United States, resulting in damages in excess of $43 million (Jarboe 2002). Because the direct actions of ELF and ALF are illegal, activists work anonymously in small groups, or individually, and do not have any centralized organization or coordination. Although critics of ecoterrorism cite the damage done by such tactics, members and supporters of ecoterrorist groups claim that the real terrorists are corporations that plunder the earth.

The Role of Corporations in the Environmental Movement. Corporations are major contributors to environmental problems and often fight against environmental efforts that threaten their profits. However, some corporations are joining the environmental movement for a variety of reasons, including pressure from consumers

Wind energy is harnessed by turbines such as those pictured in this photo of a wind farm in Alamont Pass, California.

© Morton Beebe/Corbis

and environmental groups, the desire to improve their public image, genuine concern for the environment, and/or concern for maximizing current or future profits.

In 1994, out of concern for public and environmental health, Ray Anderson, CEO of Interface, set a goal of being a sustainable company by 2020—"a company that will grow by cleaning up the world, not by polluting or degrading it" (McDaniel 2005, p. 33). Anderson envisioned recycling all the materials used, not releasing any toxins into the environment, and using solar energy to power all production and has made significant progress toward these goals.

Car manufacturers can play a role in the environmental movement by developing fuel-efficient and alternative-fuel vehicles. In collaboration with the environmental organization Environmental Defense, FedEx has developed a hybrid electric delivery truck that cuts smog-forming pollution by 65%, reduces soot by 96%, and goes 57% farther on a gallon of fuel (Environmental Defense 2004).

Rather than hope that industry voluntarily engages in eco-friendly practices, Robert Hinkley suggests that corporate law be changed to mandate socially responsible behavior (see this chapter's *Taking a Stand* feature).

Environmental Education

One goal of environmental organizations and activists is to educate the public about environmental issues and the seriousness of environmental problems. Some Americans underestimate the seriousness of environmental concerns because they lack accurate information about environmental issues. In a national survey of U.S. adults most (69%) rated themselves as having either "a lot" (10%) or "a fair amount" (59%) of knowledge about environmental issues and problems (National Environmental Education and Training Foundation & Roper Starch Worldwide 1999). Yet on 7 of 10 questions asked about emerging environmental issues, more

Should Corporate Law Be Changed to Include the Code for Corporate Citizenship?

Corporate attorney Robert Hinkley suggests that corporations pursue profit at the expense of the public good, including the environment, not because corporate executives are greedy but because they are bound by corporate law to try to make a profit for shareholders (Cooper 2004). According to Hinkley, the solution to environmentally destructive corporate practices, as well as other corporate abuses, is to add the **Code for Corporate Citizenship** to corporate law. The code would say, "The duty of directors henceforth shall be to make money for shareholders but not at the expense of the environment, human rights, public health and safety, dignity of employees, and the welfare of the communities in which the company operates" (Cooper 2004, p. 6).

It Is Your Turn Now to Take a Stand!

Should states change their corporate laws to add the Code of Corporate Citizenship?

Americans gave incorrect answers than correct ones. The average was 3.2 correct answers out of 10 questions. Even those who rated themselves as having "a lot" of environmental knowledge, as well as those with a college degree, scored only 4 out of 10 items correctly. The authors of the study concluded:

> Increased knowledge is the key to changing attitudes and behaviors on issues critical to our environmental future. If Americans can answer an average of only 3 of 10 simple knowledge questions, there is a clear need to provide environmental information in a form that the American public can easily digest and act upon. (National Environmental Education and Training Foundation & Roper Starch Worldwide 1999, p. 41)

A main source of information about environmental issues for most Americans is the media. However, because the media are owned by corporations and wealthy individuals with corporate ties, unbiased information about environmental impacts of corporate activities may not readily be found in mainstream media channels. Indeed, the public must consider the source in interpreting information about environmental issues. Propaganda by corporations sometimes comes packaged as "environmental education." Hager and Burton (2000) explain: "Production of materials for schools is a growth area for public relations companies around the world. Corporate interests realize the value of getting their spin into the classrooms of future consumers and voters" (p. 107).

"Green" Energy

As noted earlier, many of the environmental problems facing the world today are linked to energy use, particularly the use of fossil fuels. Increasing the use of "green" energy—energy that is renewable and nonpolluting—can help alleviate environmental problems associated with fossil fuels. Also known as clean energy, green energy sources include solar power, wind power, biofuel, and hydrogen. Although clean energy represents less than 1% of world energy production, it is expected to increase in coming years.

Solar Power. The world's fastest developing clean energy is photovoltaic cells—a form of solar power that converts sunlight to electricity (Sawin 2005b). Japan, which produces half of all photovoltaic cells, aims for photovoltaic cells to generate 10% of Japan's electricity by 2030. Driven by government incentive programs, Germany and other European countries are also increasing their use of solar cells to provide electricity. In the United States, California leads the nation in photovoltaic cell use.

Another form of solar power is the use of solar thermal collectors, which capture the sun's warmth to heat building space and water. China is the world's leader in solar thermal production and use, followed by Japan, Europe, and Turkey (Sawin 2005b). The United States was once the world leader in solar thermal production but has fallen far behind because of low natural gas prices and the lack of government incentives. Another form of solar power comes from "concentrating solar power plants," which use the sun's heat to make steam to turn electricity-producing turbines.

Wind Power. Wind power is the world's second fastest-growing energy source, after solar power. Wind turbines, which turn wind energy into electricity, are operating in more than 65 countries, although nearly three-fourths of wind power is produced in Europe. In Germany, the leading country in wind power, wind energy meets 6.6% of electricity needs. The United States is in third place worldwide (after Spain) in use of wind power (Sawin 2005a).

One disadvantage of wind power is that wind turbines have been known to result in bird mortality. However, this problem has been mitigated in recent years through the use of painted blades, slower rotational speeds, and careful placement of wind turbines.

Biofuel. Biofuels are fuels derived from agricultural crops. Biofuels burn cleaner than fossil fuels, are renewable, and can create agricultural jobs and revenues for countries while reducing dependence on imported oil.

Two types of biofuels are ethanol and biodiesel. Ethanol is an alcohol-based fuel that is produced by fermenting and distilling sugar obtained from agricultural crops such as corn, beets, sugar cane, barley, and wheat. Ethanol is blended with gasoline to create E85 (85% ethanol, 15% gasoline). Vehicles that run on E85, called flexible fuel vehicles, have been used by the government and in private fleets for years and have just recently been available to consumers. However, many owners of flexible fuel vehicles do not know that their vehicle can operate with E85, and there are only 150 gas stations that sell E85 to the public, most of which are in the Midwest (U.S. Department of Energy 2005).

Biodiesel fuel—America's fastest growing alternative fuel—is a cleaner-burning diesel fuel made from vegetable oils and/or animal fats, including recycled cooking oil. Some individuals who make their own biodiesel fuel obtain used cooking oil from restaurants at no charge. The European Union, which is the leading manufacturer of biodiesel, hopes that biofuels will supply 20% of the fuel market in 2020 (Aeck 2005).

Hydrogen Power. Hydrogen, the most plentiful element on earth, is a clean-burning fuel that can be used for electricity production, heating, cooling, and transportation. Many see a movement to a hydrogen economy as a long-term solution to the environmental and political problems associated with fossil fuels. Further research is needed, however, to develop nonpolluting and cost-effective ways to extract and transport hydrogen. In the meantime, all the major auto manufacturers are developing hydrogen fuel cell technology. In addition, in 2004 Governor Schwarzenegger issued an executive order calling for the development of a Hydrogen Highway Network Blueprint Plan to make hydrogen a viable alternative transportation fuel in California.

Modifications in Consumer Behavior

In the United States and other industrialized countries consumers are making "green" choices in their behavior and purchases that reflect concern for the environment. In some cases these choices carry a price tag, such as paying more for organically grown food or for clothing made from organic cotton. In many cases, however, consumers are motivated to make green choices in their purchasing behavior because doing so saves money. For example, after gas prices topped $3 a gallon in 2005, sales of gas-guzzling SUVs dropped, and sales of fuel-efficient cars, such as hybrids, increased. According to Dan Becker, director of the Sierra Club's global warming program, switching to more fuel-efficient vehicles such as hybrids is the single biggest step to reducing dependency on foreign oil (Gartner 2005).

Consumers often consider their utility bill when they choose energy-efficient appliances and electrical equipment. To assist consumers in choosing energy-efficient products, in 1992 the U.S. Environmental Protection Agency introduced the Energy Star label for products that meet EPA energy-efficiency standards. Globally, 37 countries use an energy-efficient labeling system for appliances and electronic equipment (Makower et al. 2005).

Consumers can also choose to purchase "green power"—clean energy from nonpolluting sources (e.g., wind and solar). Green energy is available in every state, but not every utility company offers it. Purchasing green power does not necessarily mean the electricity you actually receive comes directly from a wind turbine or solar panel because the local grid that delivers electricity combines all power—green or not. Instead, the amount of clean power consumers buy is generated on their behalf and added to the larger pool of electricity (Vartan 2005). Buying green power is not widespread because many consumers do not know that it is an option. To find out if it is available in their area, consumers can call their utility company and ask.

Green Building. The proposed design for New York City's Freedom Tower includes a "wind farm" of turbines 60 floors up that will generate 20% of the tower's electricity. The proposed Freedom Tower is an example of a "green building."

The U.S. Green Building Council developed green building standards known as LEED (Leadership in Energy and Environmental Design). These standards consist of 69 criteria to be met by builders in six areas, including energy use and emissions, water use, materials and resource use, and sustainability of the building site. LEED buildings include the Pentagon Athletic Center, the Detroit Lions' football training facility, and the David L. Lawrence Convention Center in Pittsburgh (Makower et al. 2005).

One of the most environmentally friendly green buildings in the world is the Adam Joseph Lewis Center for Environmental Studies at Oberlin College. The building is constructed with either recycled or sustainably produced nontoxic materials; bathroom and other liquid wastes are purified on-site and used again to flush toilets or to water outside vegetation; power and heat are produced by solar and geothermal power (McDaniel 2005).

The Adam Joseph Lewis Center for Environmental Studies, located on the Oberlin College campus, is among the most innovative, environmentally friendly buildings in the world. The building resulted from the vision of David Orr, professor of environmental studies at Oberlin.

Ed Hancock, courtesy of DOE/NREL

Government Policies, Regulations, and Funding

Worldwide, governments spend about $1 trillion (U.S. dollars) per year in subsidies that allow the prices of fuel, timber, metals, and minerals (and products using these materials) to be much lower than they otherwise would be, encouraging greater consumption (Renner 2004). Thus governmental policies can contribute to environmental problems. But government policies, regulations, and funding can also play a role in protecting and restoring the environment. At the local, state, regional, and federal levels governments have implemented policies and regulations affecting the production and use of pollutants, the preservation of natural resources and biodiversity, and the use of renewable energy.

For example, in 2003 the European Union drafted legislation known as REACH (Registration, Evaluation, and Authorization of Chemicals) that requires chemical companies to conduct safety and environmental tests to prove that the chemicals they are producing are safe. If they cannot prove that a chemical is safe, it will be banned from the market (Rifkin 2004). The European Union has become a world leader in environmental stewardship by placing the "precautionary principle" at the center of EU regulatory policy. The precautionary principle requires industry to prove that their products are safe. In contrast, in the United States chemicals are assumed to be safe unless proven otherwise, and the burden is put on the consumer, the public, or the government to prove that a chemical causes harm.

Government policies and regulations also affect energy use. In 2004 more than 20 countries committed to specific targets for the renewable share of total energy use (UNEP 2005). At least 12 U.S. states require utilities to offer customers power generated by wind, biomass, and other green sources (Clayton 2005). In addition, more than 70 mayors and other local leaders from around the world signed the Urban Environmental Accords, pledging to obtain 10% of energy from renewable resources by 2012 and to reduce greenhouse gases by 25% by 2030 (Stoll 2005).

Some environmentalists propose that governments use taxes to discourage environmentally damaging practices and products (Brown & Mitchell 1998). In the

1990s a number of European governments increased taxes on environmentally harmful activities and products (such as gasoline, diesel, and motor vehicles) and decreased taxes on income and labor (Renner 2004).

Sustainable Economic Development

Achieving global cooperation on environmental issues is difficult, in part, because developed countries (primarily in the Northern Hemisphere) have different economic agendas from those of developing countries (primarily in the Southern Hemisphere). The northern agenda emphasizes preserving wealth and affluent lifestyles, whereas the southern agenda focuses on overcoming mass poverty and achieving a higher quality of life (Koenig 1995). Southern countries are concerned that northern industrialized countries—having already achieved economic wealth—will impose international environmental policies that restrict the economic growth of developing countries just as they are beginning to industrialize. Global strategies to preserve the environment must address both wasteful lifestyles in some nations and the need to overcome overpopulation and widespread poverty in others.

Development involves more than economic growth; it involves sustainability—the long-term environmental, social, and economic health of societies. **Sustainable development** involves meeting the needs of the present world without endangering the ability of future generations to meet their own needs. "The aim here is for those alive today to meet their own needs without making it impossible for future generations to meet theirs. . . . This in turn calls for an economic structure within which we consume only as much as the natural environment can produce, and make only as much waste as it can absorb" (McMichael et al. 2000, p. 1067).

Sustainable development requires the use of clean, renewable energy. Renewable energy projects in developing countries have demonstrated that providing affordable access to green energy such as solar, wind, and biofuels helps to alleviate poverty by providing energy for creating business and jobs and by providing power for refrigerating medicine, sterilizing medical equipment, and supplying freshwater and sewer services needed to reduce infectious disease and improve health (Flavin & Aeck 2005).

> "Man is here for only a limited time, and he borrows the natural resources of water, land and air from his children who carry on his cultural heritage to the end of time. . . . One must hand over the stewardship of his natural resources to the future generations in the same condition, if not as close to the one that existed when his generation was entrusted to be the caretaker."
>
> **Delano Saluskin**
> **Yakima Indian Nation**

International Cooperation and Assistance

Global environmental concerns such as global warming, destruction of the ozone layer, and loss of biodiversity call for global solutions forged through international cooperation and assistance. For example, the 1987 Montreal Protocol on Substances That Deplete the Ozone Layer forged an agreement made by 70 nations to curb the production of CFCs (which contribute to ozone depletion and global warming). The 1992 Earth Summit in Rio de Janeiro brought together heads of state, delegates from more than 170 nations and nongovernmental organizations, and participants to discuss an international agenda for both economic development and the environment. The 1992 Earth Summit resulted in the Rio Declaration—"a nonbinding statement of broad principles to guide environmental policy, vaguely committing its signatories not to damage the environment of other nations by activities within their borders and to acknowledge environmental protection as an integral part of development" (Koenig 1995, p. 15).

In 1997 delegates from 160 nations met in Kyoto, Japan, and forged the **Kyoto Protocol**—the first international agreement to place legally binding limits on greenhouse gas emissions from developed countries. The United States, which produces one-fourth of the world's greenhouse gas emissions, rejected Kyoto in 2001. In 2004 the Russian Federation joined more than 120 other countries in ratifying the treaty, which entered into force in early 2005.

In another global environmental treaty known as the Stockholm Convention, representatives from 122 countries drafted an agreement to phase out a group of dangerous chemicals known as persistent organic pollutants. More recently, Germany hosted a major intergovernmental conference, Renewables 2004, that resulted in the adoption of the International Action Programme on Renewables; this program supports more than 200 ongoing or planned policies and projects in over 150 countries (UNEP 2005).

Some countries do not have the technical or economic resources to implement the requirements of environmental treaties. Wealthy industrialized countries can help less-developed countries address environmental concerns through economic aid. Because industrialized countries have more economic and technological resources, they bear primary responsibility for leading the nations of the world toward environmental cooperation. Jan (1995) emphasizes the importance of international environmental cooperation and the role of developed countries in this endeavor:

> Advanced countries must be willing to sacrifice their own economic well-being to help improve the environment of the poor, developing countries. Failing to do this will lead to irreparable damage to our global environment. Environmental protection is no longer the affair of any one country. It has become an urgent global issue. Environmental pollution recognizes no national boundaries. No country, especially a poor country, can solve this problem alone. (pp. 82–83)

Understanding Environmental Problems

Environmental problems are linked to a number of interrelated features of social life, including corporate globalization, rapid and dramatic population growth, expanding world industrialization, patterns of excessive consumption, and reliance on fossil fuels for energy. Growing evidence of the irreversible effects of global warming and loss of biodiversity, increased concerns about the link between global warming and extreme weather events such as Hurricane Katrina, and adverse health effects of toxic waste and other forms of pollution suggest that we cannot afford to ignore environmental problems.

Many Americans believe in a "technological fix" for the environment—that science and technology will solve environmental problems. Paradoxically, the same environmental problems that have been caused by technological progress may be solved by technological innovations designed to clean up pollution, preserve natural resources and habitats, and provide clean forms of energy. But leaders of government and industry must have the will to finance, develop, and use technologies that do not pollute or deplete the environment. When asked how companies can produce products without polluting the environment, corporate attorney Robert Hinkley suggested that, first, it must become a goal to do so:

> I don't have the technological answers for how it can be done, but neither did President John F. Kennedy when he announced a national goal to land a man on the moon by the end of the 1960s. The point is that, to eliminate pollution, we first have to make it our goal. Once we've done that, we will devote the resources necessary to make it happen. We will develop technologies that we never thought possible. But if we don't make it our goal, then we will never devote the resources, never develop the technology, and never solve the problem. (quoted by Cooper 2004, p. 11)

But the direction of technical innovation is largely in the hands of big corporations that place profits over environmental protection. Unless the global community challenges the power of transnational corporations to pursue profits at the expense of environmental and human health, corporate behavior will continue to take a heavy toll on the health of the planet and its inhabitants. Because oil has been implicated in political and military conflicts involving the Middle East (see Chapter 16), such conflicts are likely to continue so long as oil plays the lead role in providing the world's energy.

Global cooperation is also vital to resolving environmental concerns but is difficult to achieve because rich and poor countries have different economic development agendas: Developing poor countries struggle to survive and provide for the basic needs of their citizens; developed wealthy countries struggle to maintain their wealth and relatively high standard of living. Can both agendas be achieved without further pollution and destruction of the environment? Is sustainable economic development an attainable goal? With mounting concern about climate change, the health impacts of air pollution, rising oil prices, and the need to ensure energy access to all, governments worldwide have strengthened their commitment to sustainable, renewable energy policies and projects (UNEP 2005).

Since September 11, 2001, world leaders, the media, and citizens in countries across the globe have been preoccupied with terrorism. Lester Brown, of the Earth Policy Institute, warns against the environmental dangers of such preoccupation:

> Terrorism is certainly a matter of concern, but if it diverts us from the environmental trends that are undermining our future until it is too late to reverse them, Osama bin Laden and his followers will have achieved their goal of bringing down Western civilization in a way they could not have imagined. (Brown 2003, p. 5)

In 2004 the Nobel Peace Prize was given to Wangari Maathai for leading a grassroots environmental campaign to plant 30 million trees across Kenya. This was the first time ever that the Nobel Peace Prize was awarded to someone for his or her accomplishments in restoring the environment. In her acceptance speech, Maathai

The 2004 Nobel Peace Prize was awarded to Wangari Maathai for leading a grassroots environmental campaign called the Green Belt Movement, which is responsible for planting 30 million trees across Kenya. Maathai is the first person to be awarded the Nobel Peace Prize for environmental work.

explained, "A degraded environment leads to a scramble for scarce resources and may culminate in poverty and even conflict" (quoted by Little 2005, p. 2). With ongoing conflict around the globe, it is time for world leaders to recognize the importance of a healthy environment for world peace and to prioritize environmental protection in their political agendas.

Chapter Review

ThomsonNOW™

Reviewing is as easy as ① ② ③

1. Before you do your final review, take the ThomsonNOW diagnostic quiz to help you identify the areas on which you should concentrate. You will find information on ThomsonNOW and instructions on how to access all of its great resources on the foldout at the beginning of the text.

2. As you review, take advantage of ThomsonNOW's study videos and interactive Map the Stats exercises to help you master the chapter topics.

3. When you are finished with your review, take ThomsonNOW's posttest to confirm you are ready to move on to the next chapter.

- **In what ways does corporate globalization pose a threat to environmental protection?**

The rules of corporate globalization, set out by the World Trade Organization (WTO) and free trade agreements such as NAFTA and the FTAA, provide transnational corporations with privileges to pursue profits, expand markets, use natural resources, and exploit cheap labor in developing countries while weakening the ability of governments to protect natural resources or to implement environmental legislation. Corporate globalization is based on values of commercialism, corporate rights, and "free" trade rather than on values of environmental protection, health and safety, and social justice.

- **How did the terrorist attacks of September 11, 2001, affect public concern for environmental issues in the two years following the attacks?**

Gallup poll surveys found that in the two years after the terrorist attacks of September 11, public concern about most environmental problems declined sharply, most likely because of increasing concern about the economy and terrorism over the same period. Perhaps environmental problems look less serious by contrast with the economy and the terrorist threat, or perhaps the environment did not receive as much media attention, because media stories focused on terrorism and war and economic issues.

- **Where does most of the world's energy come from?**

Most of the world's energy comes from fossil fuels, which include oil, coal, and natural gas. This is significant because many of the serious environmental problems in the world today, including pollution, biodiversity loss, and global warming, stem from the use of fossil fuels.

- **What are some examples of common household, personal, and commercial products that contribute to indoor pollution?**

Some common indoor air pollutants include carpeting, mattresses, drain cleaners, oven cleaners, spot removers, shoe polish, dry-cleaned clothes, paints, varnishes, furniture polish, potpourri, mothballs, fabric softener, caulking compounds, air fresheners, deodorizers, disinfectants, glue, correction fluid, printing ink, carbonless paper, and felt-tip markers.

- **What is the primary cause of global warming?**

The prevailing view on what causes global warming is that greenhouse gases—primarily carbon dioxide, methane, and nitrous oxide—accumulate in the atmosphere and act like the glass in a greenhouse, holding heat from the sun close to the earth. The primary greenhouse gas is carbon dioxide, which is released into the atmosphere by burning fossil fuels.

- **How does global warming contribute to further global warming?**

As global warming melts ice and snow, it exposes more land and ocean area, which absorbs more heat than ice and snow, further warming the planet. The melting of Siberia's frozen peat bog—a result of global warming—could release billions of tons of methane, a potent greenhouse gas, into the atmosphere and cause further global warming. This process, whereby the effects of global warming cause further global warming, is known as a positive feedback loop.

- **What is the relationship between level of economic development and environmental pollution?**

There is a curvilinear relationship between level of economic development and environmental pollution. In regions with low levels of economic development, industrial emissions are minimal, but emissions rise in countries that are in the middle economic development range as they move through the early stages of industrialization. However, at more advanced industrial stages, industrial emissions ease because heavy-polluting manufacturing industries decline and "cleaner" service industries increase and because rising incomes are associated with a greater demand for environmental quality and cleaner technologies.

- **How do environmental groups use the Internet as a tool for environmental activism?**

Some environmental groups (including Environmental Defense and the Sierra Club) have an "action alert" feature whereby anyone with an e-mail address can sign up to receive e-mail action alerts. These e-mail alerts inform members when Congress and other decision makers threaten the health of the environment and provide an easy way for members to send e-mails and faxes to Congress, the president, and business leaders, urging them to support policies that protect the environment. Other environmental organizations, such as the Sierra Club, also have similar action alert features on their websites.

- **What is the fastest growing source of "clean" renewable energy?**

Solar power is the fastest developing alternative source of energy, followed by wind power.

- **According to 2004 Nobel Peace Prize winner Wangari Maathai, why is environmental protection important for national and international security?**

In her acceptance speech for the 2004 Nobel Peace Prize, Wangari Maathai explained that "a degraded environment leads to a scramble for scarce resources and may culminate in poverty and even conflict."

Critical Thinking

1. In Chapters 9 and 11, we discussed hate crimes, noting that hate crime laws impose harsher penalties on the perpetrator of a crime if the motive for that crime was hate or bias. Should motives be considered in imposing penalties on individuals who are convicted of acts of ecoterrorism? For example, should a person who sets fire to a business to protest that business's environmentally destructive activities receive a lighter penalty than a person who sets fire to a business for some other reason?
2. Consumers who want to make environmentally friendly purchasing decisions are sometimes faced with difficult choices. For example, suppose that an individual lives in an area where locally grown organic produce is not available. In this case, is it better to purchase (1) organic produce that has been trucked in from a distant state (no pesticides, but the transportation contributes to fossil fuel emissions) or (2) locally grown produce from a farm that uses pesticides?
3. A 2005 Gallup poll (Lyons 2005) found that Americans who seldom or never attend religious services are more likely to worry about the environment than do Americans who attend church weekly (40% versus 28%). Why do you think this is so?

Key Terms

acid rain	environmental racism
biodiversity	environmental refugees
bioinvasion	e-waste
Code for Corporate Citizenship	Extended Producer Responsibility
deforestation	global warming
desertification	greenhouse gases
ecofeminism	greenwashing
ecoterrorism	Kyoto Protocol
ecosystems	multiple chemical sensitivity
environmental footprint	planned obsolescence
environmental injustice	sustainable development

Media Resources

The Companion Website for *Understanding Social Problems,* Fifth Edition

http://sociology.wadsworth.com/mooney_knox_schacht5e

Supplement your review of this chapter by going to the companion website to take one of the Tutorial Quizzes, use the flash cards to master key terms, and check out the many other study aids you'll find there. You'll also find special features such as Wadsworth's *Sociology Online Resources and Writing Companion,* GSS data, and Census 2000 information, data, and resources at your fingertips to help you complete that special project or do some research on your own.

> " Most of the consequences of technology that are causing concern at the present time— pollution of the environment, potential damage to the ecology of the planet, occupational and social dislocations, threats to the privacy and political significance of the individual, social and psychological malaise . . . are with us in large measure because it has not been anybody's explicit business to foresee and anticipate them. "

Emmanuel Mesthene, former director of the Harvard Program on Technology and Society

Science and Technology

at 19, Erica Robinson lived in a household of "fighters." Her brothers were both athletes—one a star basketball player, the other a wrestler. Her father's strength was summed up in a plaque hanging from the wall of their home: "Don't quit." When Erica found out she was pregnant, she was determined to do whatever was best for her baby (Kilen 2003).

But something was terribly wrong. At Erica's six-month checkup the doctor said that the baby wasn't getting enough blood to survive. An emergency caesarean section was performed, and with it E'Maria Robinson-Butler was introduced to the world—10 inches, 11 ounces—one of the tiniest babies ever born. In fact, on any given day in the United States 1,300 premature babies are born. "Preemies" are more likely than full-term infants to have lasting disabilities, such as chronic lung disease, cerebral palsy, mental retardation, and vision and hearing problems. The rate of premature babies has increased 27% since 1981 (March of Dimes 2003).

As soon as E'Maria was born, a tube was placed in her windpipe; she was hooked up to a respirator, fed intravenously, and given medicine to help her tiny lungs develop. During her five-month stay in the neonatal unit, E'Maria had three surgeries to correct her eyesight, heart, and a hernia. Erica was finally able to bring her little girl home, but not without a heart monitor, multiple medications, and an oxygen tank. There were still problems, but the prognosis is good.

Twenty years ago, E'Maria, born at 25 weeks and weighing little more than two sticks of butter, would not have survived. Ten years ago, over half of babies like E'Maria would not have survived. Today, as a result of medical technology, 85% of babies born at 25 weeks' gestation survive. Moreover, the limit of viability is being pushed back one week every five years (Kilen 2003).

"She always keeps her fist balled up," said Grandpa Robinson. "She's a little fighter."

Many of the medical technologies available today seem futuristic. But such technologies, for example, virtual reality, cloning, and teleportation, are no longer just the stuff of popular sci-fi movies. Virtual reality is now used to train workers in occupations as diverse as medicine, engineering, and professional football. The ability to genetically replicate embryos has sparked worldwide debate over the ethics of reproduction, and California Institute of Technology scientists have transported a ray of light from one location to another. Just as the telephone, the automobile, television, and countless other technological innovations have forever altered social life, so will more recent technologies (see this chapter's *Focus on Technology* feature).

Science and technology go hand in hand. **Science** is the process of discovering, explaining, and predicting natural or social phenomena. A scientific approach to understanding AIDS, for example, might include investigating the molecular structure of the virus, the means by which it is transmitted, and public attitudes about AIDS. **Technology,** as "a form of human cultural activity that applies the principles of science and mechanics to the solution of problems," is intended to accomplish a specific task—in this case, the development of an AIDS vaccine.

Societies differ in their level of technological sophistication and development. In agricultural societies, which emphasize the production of raw materials, the use of tools to accomplish tasks previously done by hand, or **mechanization,** dominates. As societies move toward industrialization and become more concerned with the mass production of goods, automation prevails. **Automation** involves the use of self-operating machines, as in an automated factory where autonomous robots

New Technological Inventions

Here are some of the most innovative inventions of recent years:

- *The Earth Simulator.* A group of Japanese engineers has created a supercomputer capable of 35 trillion calculations a second—five times faster than the next fastest computer. At a cost of $350 million, the computer is capable of forecasting the weather for the entire planet far into the future. It has already calculated global ocean temperatures for the next 50 years.

- *Tendon-activated hand.* Prosthetics are generally clumsy and not very responsive to fine-tuned actions. However, a new "tendon-activated hand" connects sensors in the artificial hand to remnant tendons, providing flexible and natural movement in the prosthetic. One recent accident victim has even returned to playing the piano!

- *Lego Mindstorms.* Researchers at the Massachusetts Institute of Technology (MIT) have developed computer-controlled building blocks. Described as half robot and half Lego, the custom-designed software helps children write computer programs by piecing together instructions on a computer screen that are transmitted to a robot. The robot then carries out the instructions—whether it be dealing a hand of gin rummy or playing a game of hide-and-seek. Besides being fun, children learn the elementary principles of programming.

- *Talking lights.* Imagine you are visually impaired and in search of the "Main Street" exit of a building. As you walk through a door, the badge you are wearing says, "This is the Main Street exit." Designed by electrical engineers at MIT, talking lights transform ordinary fluorescent lightbulbs into "global positioning satellites" that can be used to guide people around malls, senior centers, hospitals, and the like. The lights simply emit information to a decoder that then translates it to verbal messages. The unit is wireless and inexpensive and consumes no more energy than a regular lightbulb.

- *Recodable lock.* One of the biggest concerns of the 21st century is computer security—Internet privacy, virus-free computer environments, secure e-mail, and so on. Although software security continues to be a multibillion dollar industry, two researchers may have discovered a way to secure a computer from the inside out. This new microscopic lock is almost impossible to crack and may do to electronic snoopers what software programs have failed to do—put a lock on cyberspace.

- *Microsystems Jini.* All of us have spent hours poring over the manuals of our latest high-tech purchases trying to figure out the operating instructions of each new gadget. Jini, however, is a new software package that, once installed, allows your computer to talk to other "intelligent" appliances installed into the Jini system. From the Jini Web page simply click on an icon to warm your coffee, retrieve pictures from a digital camera, or toast a Pop Tart.

- *Computer shoes.* Feet get tired on that long walk home? Help is in sight. Adidas has just manufactured a new tennis shoe with a computer! The sensors in the bottom of the shoe adjust the firmness of the heel based on the type of terrain you are walking on—for example, rigid on dirt trails, softer on pavement. Cushion adjustment can also be set with buttons on the shoes.

- *Teeth telephone.* Two British developers have created a phone tooth. The device is a miniature telephone that is embedded in a tooth, usually a molar, where phone calls are received. The signals are then transmitted, through vibrations, from the tooth to the skull and then to the inner ear. Unfortunately, at this time, the talking tooth is really a listening tooth—there is no way to talk to the caller.

- *Virtual touch.* Using a computer is a multisensory experience—bright lights, moving images, and real-life audio, but until recently, no tactile stimuli. Researchers at MIT have now created what they call a haptic interface—that is, a joystick that gives the computer user a sense of touch. The advantages of such an innovation are unlimited, allowing doctors to actually feel tissue during cybersurgery or engineers to experience the goodness-of-fit of parts just designed.

- *Breaking through.* That all too familiar sound of a jackhammer may be a thing of the past. In trying to design a quiet way to break up concrete, researchers have developed RAPTOR, a lightweight "gun" that fires penny nails at speeds of 5,000 feet per second. When the nails are fired into the concrete in consecutive lines, stress fractures occur and even the thickest of concrete begins to crumble. Because RAPTOR can be fitted with a silencer, the firing of nails breaks up the concrete quietly and with much less effort on the part of the operator.

- *Remote boat.* For the water-skiing enthusiast who wants to ski without the hassle of finding a driver and/or spotter comes the remote-controlled 8-foot-long fiberglass boat that pulls a skier behind it. The "personless" boat includes a tow handle that contains all the controls, including start/stop, acceleration, and turning mechanisms. A must for the avid water-skier!

Sources: Buechner et al. (2002), D'Agnese (2000), Hamilton et al. (2004), and *1999 Emerging Technology Finalist* (2000).

assemble automobiles. Finally, as a society moves toward post-industrialization, it emphasizes service and information professions (Bell 1973). At this stage technology shifts toward **cybernation,** whereby machines control machines—making production decisions, programming robots, and monitoring assembly performance.

What are the effects of science and technology on humans and their social world? How do science and technology help to remedy social problems, and how do they contribute to social problems? Is technology, as Postman (1992) suggests, both a friend and a foe to humankind? We address each of these questions in this chapter.

ThomsonNOW™

Learn more about the **Effects of Science and Technology** by going through the Technology and Change Learning Module.

The Global Context: The Technological Revolution

Less than 50 years ago, traveling across state lines was an arduous task, a long-distance phone call was a memorable event, and mail carriers brought belated news of friends and relatives from far away. Today, travelers journey between continents in a matter of hours, and for many, e-mail, faxes, videoconferencing, instant messaging, and electronic fund transfers have replaced conventional means of communication.

The world is a much smaller place than it used to be, and it will become even smaller as the technological revolution continues. The Internet has 850 million users in more than 100 countries—nearly 140 million users in the United States alone (Burkholder 2005; Internet World Stats 2005). The most common online language populations are English (35.2%), followed by Chinese (13.7%), Spanish (9.0%), Japanese (8.4%), German (6.9%), and French (4.2%) (Global Reach 2004). Although three-fourths of all Internet users live in industrialized countries, there is some movement toward the Internet's becoming a truly global medium as Africans, Middle Easterners, and Latin Americans increasingly "get online" (Internet World Stats 2005). Table 15.1 displays the Internet activities of the average global user.

The movement toward globalization of technology is, of course, not limited to the use and expansion of the Internet. The world robot market—and the U.S. share of it—continues to expand; Microsoft's Internet platform and support products are sold all over the world; scientists collect skin and blood samples from remote islanders for genetic research; a global treaty regulating trade of genetically altered products has been signed by more than 100 nations; and Intel computer processing units (CPUs) power an estimated four-fifths of the world's personal computers (PCs).

To achieve such scientific and technological innovations, sometimes called research and development (R&D), countries need material and economic resources. Research entails the pursuit of knowledge; development refers to the production of materials, systems, processes, or devices directed toward the solution of practical problems. In 2003 the United States spent $276 billion on R&D. As in most other countries, U.S. funding sources are primarily from four sectors: private industry (66%), the federal government (28%), colleges and universities (3%), and other nonprofit organizations, such as research institutes (3%) (National Science Foundation 2004).

Scientific discoveries and technological developments also require the support of a country's citizens and political leaders. For example, although abortion has been technically possible for years, millions of the world's citizens live in countries where abortion is either prohibited or permitted only when the life of the mother is in danger. Thus the degree to which science and technology are considered good or bad, desirable or undesirable, is, as social constructionists argue, a function of time and place.

Table 15.1

Global Internet Use: April 2005

• Average number of sessions per month	31
• Average number of unique domains visited	62
• Average pages viewed per month	1,151
• Average pages viewed per surfing session	37
• Average time online per month	25 hours, 48 minutes
• Average time spent during surfing session	50 minutes
• Average duration of a page viewed	43 seconds
• Average online population	295,495,299

Source: Nielsen/NetRatings. Available at http://www.nielsennetratings.com

The world was made a smaller place in the mid to late 1800s by the Pony Express. Today, the Internet provides a forum for millions of users worldwide who shop, surf, e-mail, and bank "the net." Globally, the number of Internet users is predicted to grow dramatically over the next several decades.

Postmodernism and the Technological Fix

Many Americans believe that social problems can be resolved through a **technological fix** (Weinberg 1966) rather than through social engineering. For example, a social engineer might approach the problem of water shortages by persuading people to change their lifestyle: Use less water, take shorter showers, and wear clothes more than once before washing. A technologist would avoid the challenge of changing people's habits and motivations and instead concentrate on the development of new technologies that would increase the water supply. Social problems can be tackled through both social engineering and a technological fix. In recent years, for example, social engineering efforts to reduce drunk driving have included imposing stiffer penalties for drunk driving and disseminating public service announcements, such as "Friends don't let friends drive drunk." An example of a technological fix for the same problem is the development of car airbags, which reduce injuries and deaths resulting from car accidents.

Not all individuals, however, agree that science and technology are good for society. **Postmodernism,** an emerging worldview, holds that rational thinking and the scientific perspective have fallen short in providing the "truths" they were once presumed to hold. During the industrial era, science, rationality, and technological innovations were thought to hold the promises of a better, safer, and more humane world. Today, postmodernists question the validity of the scientific enterprise, often pointing to the unforeseen and unwanted consequences of resulting technologies. Automobiles, for example, began to be mass-produced in the 1930s in response to consumer demands. But the proliferation of automobiles has also led to increased air pollution and the deterioration of cities as suburbs developed, and today, traffic fatalities are the number one cause of accident-related deaths.

Sociological Theories of Science and Technology

Each of the three major sociological frameworks helps us to better understand the nature of science and technology in society.

Structural-Functionalist Perspective

Structural functionalists view science and technology as emerging in response to societal needs—that "science was born indicates that society needed it" (Durkheim 1973). As societies become more complex and heterogeneous, finding a common and agreed-on knowledge base becomes more difficult. Science fulfills the need for an assumed objective measure of "truth" and provides a basis for making intelligent and rational decisions. In this regard, science and the resulting technologies are functional for society.

If society changes too rapidly as a result of science and technology, however, problems may emerge. When the material part of culture (i.e., its physical elements) changes at a faster rate than the nonmaterial part (i.e., its beliefs and values), a **cultural lag** may develop (Ogburn 1957). For example, the typewriter, the conveyor belt, and the computer expanded opportunities for women to work outside the home. With the potential for economic independence, women were able to remain single or to leave unsatisfactory relationships and/or establish careers. But although new technologies have created new opportunities for women, beliefs about women's roles, expectations of female behavior, and values concerning equality, marriage, and divorce have lagged behind.

Robert Merton (1973), a structural functionalist and founder of the subdiscipline sociology of science, also argued that scientific discoveries or technological innovations may be dysfunctional for society and may create instability in the social system. For example, the development of time-saving machines increases production, but it also displaces workers and contributes to higher rates of employee alienation. Defective technology can have disastrous effects on society. In 1994 a defective Pentium chip was discovered to exist in 2 million computers in aerospace, medical, scientific, and financial institutions, as well as in schools and government agencies. Replacing the defective chip was a massive undertaking but was necessary to avoid thousands of inaccurate computations and organizational catastrophe.

Conflict Perspective

Conflict theorists, in general, argue that science and technology benefit a select few. For some conflict theorists technological advances occur primarily as a response to capitalist needs for increased efficiency and productivity and thus are motivated by profit. As McDermott (1993) notes, most decisions to increase technology are made by "the immediate practitioners of technology, their managerial cronies, and for the profits accruing to their corporations" (p. 93). In the United States private industry spends more money on research and development than the federal government does. The Dalkon Shield and silicone breast implants are examples of technological advances that promised millions of dollars in profits for their developers. However, the rush to market took precedence over thorough testing of the products' safety. Subsequent lawsuits filed by consumers who argued that both products had compromised the physical well-being of women resulted in large damage awards for the plaintiffs.

Science and technology also further the interests of dominant groups to the detriment of others. The need for scientific research on AIDS was evident in the early 1980s, but the required large-scale funding was not made available so long as the virus was thought to be specific to homosexuals and intravenous drug users. Only when the virus became a threat to mainstream Americans were millions of dollars allocated to AIDS research. Hence conflict theorists argue that granting agencies act as gatekeepers to scientific discoveries and technological innovations. These agencies are influenced by powerful interest groups and the marketability of the product rather than by the needs of society.

When the dominant group feels threatened, it may use technology as a means of social control. For example, the use of the Internet is growing dramatically in China, the world's second largest Internet market. In 2005 the government of China announced that every website must be registered "and provide complete information on its organizers . . . or face being declared illegal" (Reuters 2005, p. 1). Thousands of censors now monitor Web pages, blogs, and chat rooms looking for politically incorrect statements.

Finally, conflict theorists as well as feminists argue that technology is an extension of the patriarchal nature of society that promotes the interests of men and ignores the needs and interests of women. As in other aspects of life, women play a subordinate role in reference to technology in terms of both its creation and its use. For example, washing machines, although time-saving devices, disrupted the communal telling of stories and the resulting friendships among women who gathered together to do their chores. Bush (1993) observed that in a "society characterized by a sex-role division of labor, any tool or technique . . . will have dramatically different effects on men than on women" (p. 204).

© CORBIS

Motivated by profit, private industry spends more money on research and development than the federal government does.

Symbolic Interactionist Perspective

Knowledge is relative. It changes over time, over circumstances, and between societies. We no longer believe that the world is flat or that the earth is the center of the universe, but such beliefs once determined behavior because individuals responded

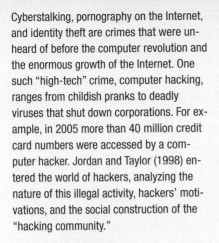

The Social Construction of the Hacking Community

Cyberstalking, pornography on the Internet, and identity theft are crimes that were unheard of before the computer revolution and the enormous growth of the Internet. One such "high-tech" crime, computer hacking, ranges from childish pranks to deadly viruses that shut down corporations. For example, in 2005 more than 40 million credit card numbers were accessed by a computer hacker. Jordan and Taylor (1998) entered the world of hackers, analyzing the nature of this illegal activity, hackers' motivations, and the social construction of the "hacking community."

Sample and Methods

Jordan and Taylor (1998) researched computer hackers and the hacking community through 80 semistructured interviews, 200 questionnaires, and an examination of existing data on the topic. As is often the case in crime, illicit drug use, and other similarly difficult research areas, a random sample of hackers was not possible. **Snowball sampling** is often the preferred method in these cases, that is, one respondent refers the researcher to another respondent, who then refers the researcher to another respondent, and so forth. Through their analysis the investigators provide insight into this increasingly costly social problem and the symbolic interactionist notion of "social construction"—in this case, of an online community.

Findings and Conclusions

Computer hacking, or "unauthorized computer intrusion," is an increasingly serious problem, particularly in a society dominated by information technologies. Unlawful entry into computer networks or databases can be achieved by several means, including (1) guessing someone's password, (2) tricking a computer about the identity of another computer (called "IP spoofing"), or (3) "social engineering," a slang term referring to getting important access information by stealing documents, looking over someone's shoulder, going through their garbage, and so on.

Hacking carries with it certain norms and values, because, according to Jordan and Taylor (1998), the hacking community can be thought of as a culture within a culture. The two researchers identified six elements of this socially constructed community:

- *Technology.* The core of the hacking community is the technology that allows it to occur. As one professor who was interviewed stated, the young today have "lived with computers virtually from the cradle, and therefore have no trace of fear, not even a trace of reverence."
- *Secrecy.* The hacking community must, on the one hand, commit secret acts because their "hacks" are illegal. On the other hand, much of the motivation for hacking requires publicity to achieve the notoriety often sought. In addition, hacking is often a group activity that bonds members together. As one hacker stated, hacking "can give you a real kick some time. But it can give you a lot more satisfaction and recognition if you share your experiences with others."
- *Anonymity.* Whereas secrecy refers to the hacking act, anonymity refers to the importance of the hacker's identity remaining unknown. Thus, for example, hackers and hacking groups take on names such as Legion of Doom, the Inner Circle I, Mercury, and Kaos, Inc.
- *Membership fluidity.* Membership is fluid rather than static, often characterized by high turnover rates, in part, as a response to law enforcement pressures.

Unlike more structured organizations, there are no formal rules or regulations.
- *Male dominance.* Hacking is defined as a male activity; consequently, there are few female hackers. Jordan and Taylor (1998) also note, after recounting an incident of sexual harassment, that "the collective identity hackers share and construct . . . is in part misogynist" (p. 768).
- *Motivation.* Contributing to the articulation of the hacking communities' boundaries are the agreed-upon definitions of acceptable hacking motivations, including (1) addiction to computers, (2) curiosity, (3) excitement, (4) power, (5) acceptance and recognition, and (6) community service through the identification of security risks.

Finally, Jordan and Taylor (1998, p. 770) note that hackers also maintain group boundaries by distinguishing between their community and other social groups, including "an antagonistic bond to the computer security industry (CSI)." Ironically, hackers admit a desire to be hired by the CSI, which would not only legitimize their activities but also give them a steady income.

Jordan and Taylor conclude that the general fear of computers and of those who understand them underlies the common, although inaccurate, portrayal of hackers as pathological, obsessed computer "geeks." When journalist Jon Littman asked hacker Kevin Mitnick if he was being demonized because of increased dependence on and fear of information technologies, Mitnick replied, "Yeah. . . . That's why they're instilling fear of the unknown. That's why they're scared of me. Not because of what I've done, but because I have the capability to wreak havoc" (Jordan & Taylor 1998, p. 776).

to what they thought to be true. The scientific process is a social process in that "truths"—socially constructed truths—result from the interactions between scientists, researchers, and the lay public.

Kuhn (1973) argues that the process of scientific discovery begins with assumptions about a particular phenomenon (e.g., the world is flat). Because unanswered questions always remain about a topic (e.g., why don't the oceans drain?), science works to fill these gaps. When new information suggests that the initial assumptions were incorrect (e.g., the world is not flat), a new set of assumptions or framework emerges to replace the old one (e.g., the world is round). It then becomes the dominant belief or paradigm.

Symbolic interactionists emphasize the importance of this process and the effect that social forces have on it. Conrad (1997), for example, describes the media's contribution in framing societal beliefs that alcoholism, homosexuality, and racial inequality are genetically determined. Technological innovations are also affected by social forces, and their success depends, in part, on the social meaning assigned to any particular product. As social constructionists argue, reality is socially constructed by individuals as they interpret the social world around them, including the meaning assigned to various technologies. If claims makers can successfully define a product as impractical, cumbersome, inefficient, or immoral, the product is unlikely to gain public acceptance. Such is the case with RU486, an oral contraceptive that is widely used in France, Great Britain, and China but whose availability, although legal in the United States, is opposed by a majority of Americans (Gallup 2000; Gottlieb 2000).

Not only are technological innovations subject to social meaning, but who becomes involved in what aspects of science and technology is also socially defined. Men, for example, outnumber women three to one in earning computer science degrees, and although women make up 47% of the general workforce, they make up just 29% of the information technology workforce (CAW 2003; U.S. Bureau of the Census 2004). Societal definitions of men as being rational, mathematical, and scientifically minded and as having greater mechanical aptitude than women are, in part, responsible for these differences. This chapter's *Social Problems Research Up Close* feature highlights one of the consequences of the masculinization of technology, as well as the ways in which computer hacker identities and communities are socially constructed.

Technology and the Transformation of Society

A number of modern technologies are considerably more sophisticated than technological innovations of the past. Nevertheless, older technologies have influenced the social world as profoundly as the most mind-boggling modern inventions. Postman (1992) describes how the clock—a relatively simple innovation that is taken for granted in today's world—profoundly influenced not only the workplace but also the larger economic institution:

> The clock had its origin in the Benedictine monasteries of the twelfth and thirteenth centuries. The impetus behind the invention was to provide a more or less precise regularity to the routines of the monasteries, which required, among other things, seven periods of devotion during the course of the day. The bells of the monastery were to be rung to signal the canonical hours; the mechanical clock was the technology that

could provide precision to these rituals of devotion. . . . What the monks did not foresee was that the clock is a means not merely of keeping track of the hours but also of synchronizing and controlling the actions of men. And thus, by the middle of the fourteenth century, the clock had moved outside the walls of the monastery, and brought a new and precise regularity to the life of the workman and the merchant. . . . In short, without the clock, capitalism would have been quite impossible. The paradox . . . is that the clock was invented by men who wanted to devote themselves more rigorously to God; it ended as the technology of greatest use to men who wished to devote themselves to the accumulation of money. (pp. 14–15)

Technology has far-reaching effects not only on the economy but also on every aspect of social life. In the following sections we discuss societal transformations resulting from various modern technologies, including workplace technology, computers, the information highway, and science and biotechnology.

Technology in the Workplace

All workplaces, from government offices to factories and from supermarkets to real-estate agencies, have felt the impact of technology. The Office of Technology Assessment of the U.S. Congress estimates that 7 million U.S. workers are under some type of computer surveillance, which lessens the need for supervisors and makes control by employers easier. In addition, technology can make workers more accountable by gathering information about their performance. Through such time-saving devices as personal digital assistants (PDAs) and battery-powered store-shelf labels, technology can enhance workers' efficiency. Technology can also contribute to worker error, however. In a recent study of a popular hospital computer system, researchers found several ways that the computerized drug-ordering program endangered the health of patients. For example, the software program warned a doctor of a patient's drug allergy only *after* the drug was ordered, and, rather than showing the usual dose of a particular drug, the program showed the dosage available in the hospital pharmacy (DeNoon 2005).

Technology is also changing the location of work. Worldwide, it is estimated that by 2008, 41 million corporate employees will spend one day a week **teleworking,** that is, completing all or part of their work away from the traditional workplace (see Chapter 7). In addition, more than 100 million people, the majority of them Americans, will work from home at least one day a month (Jones 2004).

Information technologies are also changing the nature of work. Lilly Pharmaceutical employees communicate by means of their own "intranet," on which all work-related notices are posted. Federal Express not only created a FedEx network for their employees but also allowed customers to enter their package-tracking database, saving the company millions of dollars a year. The importance of such technologies is empirically documented. In a survey of British office workers, 25% reported that information technologies were more important than office managers (Hayday 2003).

Robotic technology, sometimes called computer-aided manufacturing (CAM), has also revolutionized work. Ninety percent of robots work in factories, and more than half of these are used in heavy industry, such as automobile manufacturing (Tesler 2003). An employer's decision to use robotics depends on direct costs (e.g., initial investment) and indirect costs (e.g., unemployment compensation), the feasibility and availability of robots to perform the desired tasks, and the resulting increased rate of productivity. Use of robotics may also depend on whether labor unions resist the replacement of workers by machines.

Automation means that machines can now perform the labor originally provided by human workers, such as the robots that perform tasks on automobile assembly lines.

© Adam Lubroth /Stone /Getty Images

The Computer Revolution

Early computers were much larger than the small machines we have today and were thought to have only esoteric uses among members of the scientific and military communities. In 1951 only about a half dozen computers existed (Ceruzzi 1993). The development of the silicon chip and sophisticated microelectronic technology allowed tens of thousands of components to be imprinted on a single chip smaller than a dime. The silicon chip led to the development of laptop computers, cellular phones, digital cameras, the iPod, and portable DVDs. The silicon chip also made computers affordable. Although the first PC was developed only 20 years ago, today 62% of American homes have a computer (U.S. Bureau of the Census 2004). Ownership varies by a number of factors but is greatest among households with higher incomes (U.S. Bureau of the Census 2004) (see Table 15.2).

Seventy-two million employees use a computer at work—more than half the labor force (U.S. Bureau of the Census 2004). The percentage of workers who use computers varies by occupation, with 80% of managers and professionals using a computer but only 20% of, for example, laborers using a computer. Females, whites, and the more educated have higher rates of computer use at work. The most common computer activity is accessing the Internet or e-mail, followed by word processing, working with spreadsheets or databases, and accessing or updating calendars or schedules.

Not surprisingly, computer education has also mushroomed in the last two decades (see Chapter 8). In 1971, 2,388 U.S. college students earned a bachelor's degree in computer and information sciences; by 2002 that number had increased to 47,299 (U.S. Bureau of the Census 2004). Universities are moving toward requiring their students to have laptop computers; wireless corridors are an increasingly common occurrence, and college and university spending on hardware and software is at an all-time high.

ThomsonNOW™

Learn more about the **Computer Revolution** by going through the Mobil Telephones Map Exercise.

"I think there is a world market for maybe five computers."

Thomas Watson
Chairman of IBM, speaking in 1943

Table 15.2
Households with Computers and Internet Access by Selected Characteristic: 2003

Characteristic	Households with Computers	Households with Internet Access
All households	**61.8**	**54.6**
Age of householder		
Under 25 years old	56.5	46.9
25 to 34 years old	68.6	60.2
35 to 44 years old	73.2	65.2
45 to 54 years old	71.9	65.1
55 years old or older	46.6	40.8
Sex		
Male	65.6	58.6
Female	57.4	50.1
Education of householder		
Elementary	20.6	14.0
Some high school	32.7	24.3
High school graduate or GED	51.1	43.0
Some college	70.6	62.4
BA degree or higher	83.3	78.3
Household Income		
Under $5,000	35.6	26.8
$5,000 to $9,000	26.9	20.0
$10,000 to $14,999	31.7	23.7
$15,000 to $19,999	38.2	29.4
$20,000 to $24,999	46.1	36.7
$25,000 to $34,999	55.4	45.6
$35,000 to $49,999	71.1	62.8
$50,000 to $74,999	81.9	76.0
$75,000 to $99,999	88.1	84.1
$100,000 to $149,999	92.9	90.4
$150,000 or more	94.7	92.4

Source: National Science Foundation, Science and Engineering Indicators, 2002.

Computers are also big business, and the United States is one of the most successful producers of computer technology in the world, boasting several of the top companies—Dell, Hewlett-Packard, IBM, and Apple. Retail sales of computers exceed $12.5 billion annually (U.S. Bureau of the Census 2004). Americans spend $400 million on educational software alone, and by 2006 spending on home computers is predicted to grow tenfold as consumers increasingly define PCs as a necessity rather than a luxury (Klein 1998; Tanaka 2000).

Computer software is also big business, and in some states too big. In 2000 a federal judge found that Microsoft Corporation was in violation of antitrust laws, which prohibit unreasonable restraint of trade. At issue were Microsoft's Windows operating system and the vast array of Windows-based applications (e.g., spreadsheets, word processors, tax software)—applications that work *only* with Windows. The court held that the 70,000 programs written exclusively for Windows made "competing against Microsoft impractical" (Markoff 2000). In an agreement reached between the U.S. Department of Justice and Microsoft Corporation, Microsoft is to be divided into two companies—an operating systems company and an applications company. However, several states rejected the government's settlement and are continuing their antitrust suits against Microsoft. Furthermore, in 2005 a federal court held that Norvell Inc., producers of WordPerfect and Quattro Pro, could proceed with two of their four antitrust allegations (Associated Press 2005b).

The Internet

Information technology, or **IT,** refers to any technology that carries information. Most information technologies were developed within a 100-year span: photography and telegraphy (1830s), rotary power printing (1840s), the typewriter (1860s), transatlantic cable (1866), the telephone (1876), motion pictures (1894), wireless telegraphy (1895), magnetic tape recording (1899), radio (1906), and television (1923) (Beniger 1993). The concept of an "information society" dates back to the 1950s, when an economist identified a work sector he called "the production and distribution of knowledge." In 1958, 31% of the labor force was employed in this sector—today more than 50% is. When this figure is combined with those in service occupations, more than 75% of the labor force is involved in the information society.

The development of a national information infrastructure was outlined in the Communications Act of 1994. An information infrastructure performs three functions (Kahin 1993). First, it carries information, just as a transportation system carries passengers. Second, it collects data in digital form that can be understood and used by people. Finally, it permits people to communicate with one another by sharing, monitoring, and exchanging information based on common standards and networks. In short, an information infrastructure facilitates telecommunications, knowledge, and community integration.

The **Internet** is an international information infrastructure—a network of networks—available through universities, research institutes, government agencies, libraries, and businesses. In 2003, 60% of all Americans used the Internet. U.S. users were equally likely to be male or female but were more likely to be Asian, college graduates, full-time employees, and those with annual household incomes of $50,000 or more (U.S. Bureau of the Census 2004). The most active Internet users are connected to broadband—that is, services that provide high-speed (DSL and cable) access rather than dial-up service—and are high-income young males (McGann 2005a).

Despite dramatic growth, there is some evidence that Internet use in the United States may be slowing down. Researchers from the Pew Internet and American Life Project found that 42% of those surveyed were nonusers (Pew 2003). The majority of nonusers (56%) said that they did not intend to go online, citing cost, difficulty of use, and fear (e.g., computer fraud) as reasons for remaining offline. Nonusers were also less likely to use other technologies (e.g., cell phones) and less likely to be socially content. "Those who are socially content—who trust others, have lots of people to draw on for support, and believe that generally others are fair—are more likely to be wired than those who are less content" (Pew 2003, p. 1).

E-commerce, the buying and selling of goods and services over the Internet, continues to grow. For example, online banking is the "fastest growing internet activity in the U.S. over the last five years with 53 million Americans, or 44 percent of all U.S. internet users, now using some form of online banking" (McGann 2005b). One of the largest online retailers in the United States is the federal government, with annual sales in excess of $3.6 billion in 2000 (Hasson & Browning 2002). Amazon.com's reported net sales for the same year were $2.8 billion. The government has at least 164 sites that sell property or products to the public. The most profitable site is the Treasury Department's Treasury Direct Web page, which sells savings bonds, T-bills, and the like. Figure 15.1 displays the most common online activities. Interestingly, online consumerism may actually help the environment. A team of energy experts reports that given present trends, by 2007 "e-commerce could prevent the annual release of 35 million tons of greenhouse gases by reducing the need for up to 3 billion square feet of energy-consuming office buildings and malls in the United States" (Sampat 2000, p. 94).

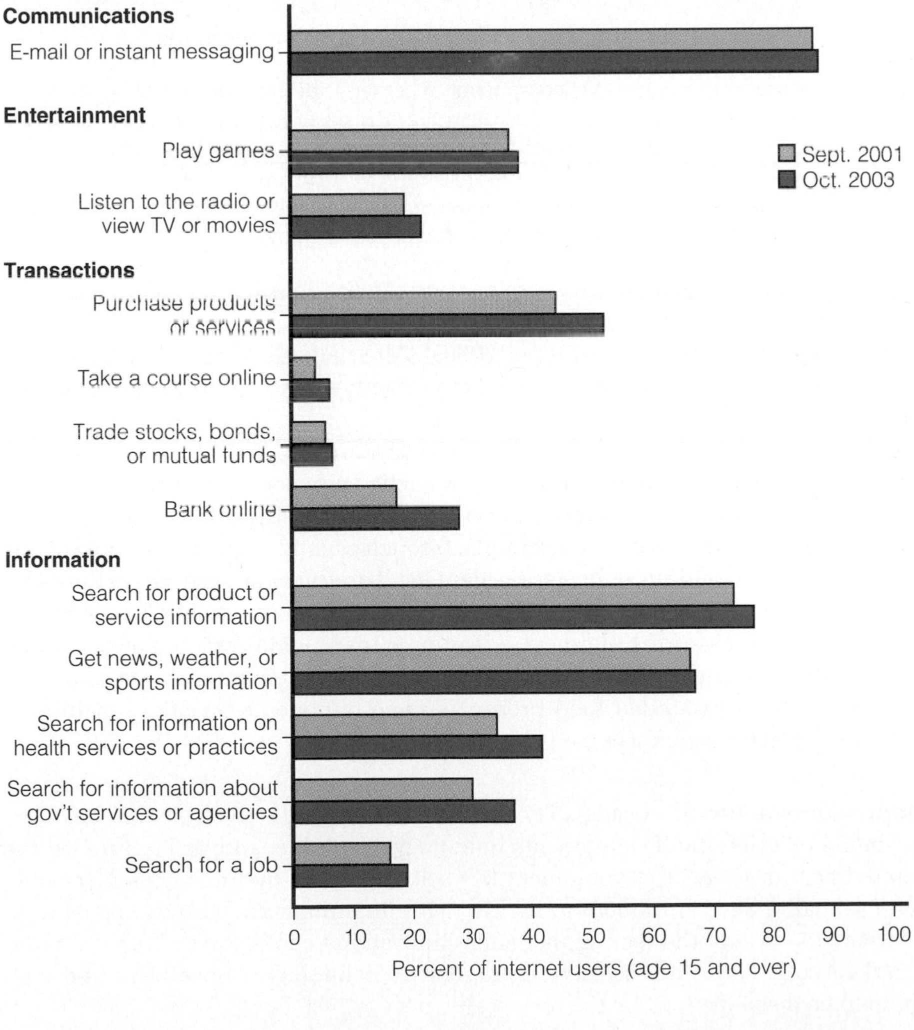

Figure 15.1
Online activities of users, age 15 and older, for 2001 and 2003.

Source: National Science Foundation, Science and Engineering Indicators, 2002.

Science and Biotechnology

Although recent computer innovations and the establishment of an information highway have led to significant cultural and structural changes, science and its resulting biotechnologies have produced not only dramatic changes but also hotly contested issues. In this section we look at some of the issues raised by developments in genetics and reproductive technology.

Genetics. Molecular biology has led to a greater understanding of the genetic material found in all cells—DNA (deoxyribonucleic acid)—and with it the ability for **genetic screening.**

> If you could uncoil a strip of DNA, it would reach 6 feet in length, a code written in words of four chemical letters: A, T, G, and C. Fold it back up, and it shrinks to trillionths of an inch, small enough to fit in any one of our 100 trillion cells, carrying the recipe for how to create human beings from scratch. (Gibbs 2003, p. 42)

Currently, researchers are trying to complete genetic maps that will link DNA to particular traits. Already, specific strands of DNA have been identified as carrying such physical traits as eye color and height as well as such diseases as breast cancer, cystic fibrosis, prostate cancer, depression, and Alzheimer's.

The U.S. Human Genome Project (HGP), a 13-year effort to decode human DNA, is now complete. Conclusion of the project is transforming medicine.

> All diseases have a genetic component whether inherited or resulting from the body's response to environmental stresses like viruses or toxins. The successes of the HGP have . . . enabled researchers to pinpoint errors in genes—the smallest units of heredity—that cause or contribute to disease. The ultimate goal is to use this information to develop new ways to treat, cure, or even prevent the thousands of diseases that afflict humankind. (Human Genome Project 2005c, p. 1)

The hope is that if a defective or missing gene can be identified, possibly a healthy duplicate can be acquired and transplanted into the affected cell. This is known as **gene therapy** (Human Genome Project 2005a). Alternatively, viruses have their own genes that can be targeted for removal. Experiments are now under way to accomplish these biotechnological feats.

Genetic engineering is the ability to manipulate the genes of an organism in such a way that the natural outcome is altered. Genetic engineering is accomplished by splicing the DNA from one organism into the genes of another. Often, however, unwanted consequences ensue. For example, through genetic engineering some plants are now self-insecticiding—that is, the plant itself produces an insect-repelling substance. Ironically, the plant's continual production of the insecticide, in contrast to only sporadic application by farmers, is leading to insecticide-resistant pests (Ehrenfeld 1998). The debate over genetically engineered crops is ongoing as advocates note the prospects for expanded food production and opponents question health and environmental consequences (see Chapter 6).

Reproductive Technologies. The evolution of "reproductive science" has been furthered by scientific developments in biology, medicine, and agriculture. At the same time, however, its development has been hindered by the stigma associated with sexuality and reproduction, its link with unpopular social movements (e.g., contraception), and the feeling that such innovations challenge the natural order (Clarke 1990). Nevertheless, new reproductive technologies have been and continue to be developed.

In **in vitro fertilization** (IVF) an egg and a sperm are united in an artificial setting, such as a laboratory dish or test tube. Although the first successful attempt at IVF occurred in 1944, the first test-tube baby, Louise Brown, was not born until 1978. In 2003 there were 400,000 frozen embryos in U.S. clinics—"88 percent set aside for future family building by patients. Only about 3 percent have been earmarked for medical research and just 2 percent for donation to other couples" (Rosenberg 2003, p. 41). Criticisms of IVF are often based on traditional definitions of the family and the legal complications created when a child can have as many as five potential parental ties—egg donor, sperm donor, surrogate mother, and the two people who raise the child (depending on the situation, IVF may not involve donors and/or a surrogate). Litigation over who the "real" parents are has already occurred.

Perhaps more than any other biotechnology, abortion epitomizes the potentially explosive consequences of new technologies. **Abortion** is the removal of an embryo or fetus from a woman's uterus before it can survive on its own. Since the U.S. Supreme Court's ruling in *Roe* v. *Wade* in 1973, abortion has been legal in the United States. However, recent Supreme Court decisions have limited the *Roe* v. *Wade* decision. For example, in *Planned Parenthood of Southeastern Pennsylvania* v. *Casey,* the Court ruled that a state may restrict the conditions under which an abortion is granted, such as requiring a 24-hour waiting period or parental consent for minors. In 2005 the Supreme Court agreed to hear a challenge to parental notification laws. Specifically, the Court will consider "whether laws requiring parental notification before a minor can get an abortion must make an explicit exception when the minor's health is at stake" (Barbash 2005, p. 1). In recent years the number of abortions has decreased in the United States (U.S. Bureau of the Census 2004), even though 78% of respondents in a national survey support abortion under any or some circumstances (Gallup 2005).

Several laws that advocate and protect "fetal rights" have also been passed. The Unborn Victims of Violence Act of 2004 protects all children in utero, regardless of stage of development. To date, 28 states have criminalized harm to a fetus. For example, because California defines an 8-week-old fetus as a person, Scott Peterson was convicted in the deaths of his wife, Laci, *and* his unborn son, Conner. Observers note that if in law a fetus is defined as a human being with the same rights as women and men, then it may in effect overturn *Roe* v. *Wade* (Rosenberg 2003).

Most recent debates concern intact dilation and extraction (D&X) abortions. Opponents refer to such abortions as **partial birth abortions** because the limbs and the torso are typically delivered before the fetus has expired. D&X abortions are performed because the fetus has a serious defect, the woman's health is jeopardized by the pregnancy, or both. In 2003 a federal ban on partial birth abortions was signed into law (White House 2003). However, several constitutional challenges to the ban have occurred and, in 2004 a federal judge ruled that the ban was unconstitutional because it imposes an "undue burden on a woman's right to choose an abortion" (Willing 2005).

Feminists strongly oppose the ban, arguing that it is just one step closer to making abortions illegal. They are also quick to note that the ban was not supported by "the American Medical Association, the American College of Obstetricians and Gynecologists, the American Medical Women's Association, the American Nurses Association, or the American Public Health Association" (U.S. Newswire 2003, p. 1). Says National Organization for Women president Kim Gandy, "Try as you might, you won't find the term 'partial birth abortion' in any medical dictionary. That's because it doesn't exist in the medical world—it's a fabrication of the anti-choice machine" (U.S. Newswire 2003, p. 1).

Abortion is a complex issue for everyone, but especially for women, whose lives are most affected by pregnancy and childbearing. Women who have abortions are disproportionately poor, unmarried minority women who say that they intend to have children in the future. Abortion is also a complex issue for societies, which must respond to the pressures of conflicting attitudes toward abortion and the reality of high rates of unintended and unwanted pregnancy. Figure 15.2 illustrates the percentage of Americans who support legal abortions under various circumstances.

Attitudes toward abortion tend to be polarized between two opposing groups of abortion activists—pro-choice and pro-life. Advocates of the pro-choice movement hold that freedom of choice is a central human value, that procreation choices must be free of government interference, and that because the woman must bear the burden of moral choices, she should have the right to make such decisions. Alternatively, pro-lifers hold that the unborn fetus has a right to live and be protected, that abortion is immoral, and that alternative means of resolving an unwanted pregnancy should be found. Assess your attitudes toward abortion in this chapter's *Self and Society* feature.

In July 1996 scientist Ian Wilmut of Scotland successfully cloned an adult sheep named Dolly. To date, cattle, goats, mice, pigs, cats, rabbits, and horses have also been cloned (Associated Press 2005a; Weiss 2003). This technological breakthrough has caused worldwide concern about the possibility of human cloning, leading the United Nations to adopt a declaration that calls for governments to ban all forms of cloning that are at odds with human dignity and the preservation of human life (Lynch 2005). One argument in favor of developing human cloning technology is its medical value; it may potentially allow everyone to have "their own reserve of therapeutic cells that would increase their chance of being cured of various diseases, such as cancer, degenerative disorders and viral or inflammatory diseases" (Kahn 1997, p. 54). Human cloning could also provide an alternative reproductive route for couples who are infertile and for those in which one partner is at risk for transmitting a genetic disease.

Arguments against cloning are largely based on moral and ethical considerations. Critics of human cloning suggest that whether used for medical therapeutic purposes or as a means of reproduction, human cloning is a threat to human dignity. For example, cloned humans would be deprived of their individuality, and as

Figure 15.2
Support for legal abortions under specific circumstances: 2003.

Source: Gallup (2003).

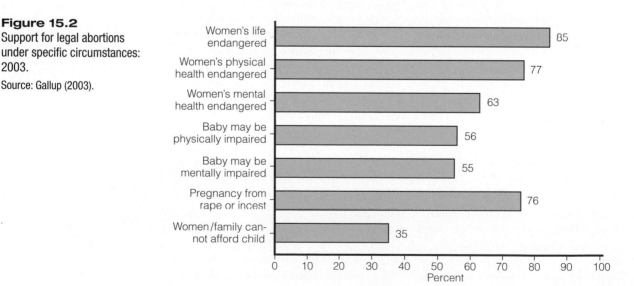

Abortion Attitude Scale

This is not a test. There are no wrong or right answers to any of the statements, so just answer as honestly as you can. The statements ask you to tell how you feel about legal abortion (the voluntary removal of a human fetus from the mother during the first three months of pregnancy by a qualified medical person). Tell how you feel about each statement by circling one of the choices beside each sentence. Respond to each statement and circle only one response. Use the following scale for your answers:

5   Strongly agree
4   Agree
3   Slightly agree
2   Slightly disagree
1   Disagree
0   Strongly disagree

1. The Supreme Court should strike down legal abortions in the United States.   5 4 3 2 1 0
2. Abortion is a good way of solving an unwanted pregnancy.   5 4 3 2 1 0
3. A mother should feel obligated to bear a child she has conceived.   5 4 3 2 1 0
4. Abortion is wrong no matter what the circumstances are.   5 4 3 2 1 0
5. A fetus is not a person until it can live outside its mother's body.   5 4 3 2 1 0

6. The decision to have an abortion should be the pregnant mother's.   5 4 3 2 1 0
7. Every conceived child has the right to be born.   5 4 3 2 1 0
8. A pregnant female not wanting to have a child should be encouraged to have an abortion.   5 4 3 2 1 0
9. Abortion should be considered killing a person.   5 4 3 2 1 0
10. People should not look down on those who choose to have abortions.   5 4 3 2 1 0
11. Abortion should be an available alternative for unmarried pregnant teenagers.   5 4 3 2 1 0
12. Persons should not have the power over the life or death of a fetus.   5 4 3 2 1 0
13. Unwanted children should not be brought into the world.   5 4 3 2 1 0
14. A fetus should be considered a person at the moment of conception.   5 4 3 2 1 0

Scoring and Interpretation

As its name indicates, this scale was developed to measure attitudes toward abortion. It was developed by Sloan (1983) for use with high school and college students. To compute your score, first reverse the point scale for items 1, 3, 4, 7, 9, 12, and 14. Total the point responses for all items. Sloan provided the following categories for interpreting the results:

70–56 Strong pro-abortion
55–44 Moderate pro-abortion
43–27 Unsure
26–16 Moderate pro-life
15–0 Strong pro-life

Reliability and Validity

The Abortion Attitude Scale was administered to high school and college students, Right to Life group members, and abortion service personnel. Sloan (1983) reported a high total test estimate of reliability (0.92). Construct validity was supported in that Right to Life members' mean scores were 16.2; abortion service personnel mean scores were 55.6; and other groups' scores fell between these values.

Source: *Abortion Attitude Scale,* by L. A. Sloan. Reprinted with permission from the *Journal of Health Education* Vol. 14, No. 3, May/June 1983. The Journal of Health Education is a publication of the American Allegiance for Health, Physical Education, Recreation and Dance, 1900 Association Drive, Reston, VA 20191.

Christopher Reeve was a long-time advocate of federally funded research on embryonic stem cells. He died on October 10, 2004, at the age of 52.

Kahn (1997, p. 119) points out, "creating human life for the sole purpose of preparing therapeutic material would clearly not be for the dignity of the life created." **Therapeutic cloning** uses stem cells from human embryos. **Stem cells** can produce any type of cell in the human body and thus can be "modeled into replacement parts for people suffering from spinal cord injuries or degenerative diseases, including Parkinson's and diabetes" (Eilperin & Weiss 2003, p. A6). For example, stem cells have been used for repairing spinal cord injuries in mice, allowing them to walk normally (Weiss 2005b). Because the use of stem cells can entail the destruction of human embryos, many conservatives, including President Bush, are opposed to the practice (see Table 15.3).

The Bush administration is also critical of a process known as SCNT, or somatic cell nuclear transfer, a "complex technique that merges eggs (whose nuclei have been removed) with adult cells to create specialized embryonic stem cell lines" (Kalb 2005, p. 52). Although pioneered by South Korean researchers, U.S. scientists believe that SCNT "will allow them to study the origins of disease, hunt for cures, and create genetically matched repair cells for patients" (Kalb 2005, p. 52).

Congress has pledged to relax restrictions imposed on stem cell research by President Bush in 2001, but the president has vowed to veto any such legislation (Weiss 2005a). Interestingly, a recent survey found that only 20% of Americans support the president's position of restricting federal funding for embryonic stem cell research (Tumulty 2005). This chapter's *Taking a Stand* feature asks your opinion on whether federal funds should be used to support stem cell research.

Despite what appears to be a universal race to the future and the indisputable benefits of such scientific discoveries as the workings of DNA and the technology of IVF, some people are concerned about the duality of science and technology. Science and the resulting technological innovations are often life assisting and life giving (see *The Human Side* feature in this chapter); they are also potentially destructive and life-threatening. The same scientific knowledge that led to the discovery of nuclear fission, for example, led to the development of both nuclear power plants and the potential for nuclear destruction. Thus we now turn our attention to the problems associated with science and technology.

Societal Consequences of Science and Technology

Scientific discoveries and technological innovations have implications for all social actors and social groups. As such, they also have consequences for society as a whole.

Alienation, Deskilling, and Upskilling

As technology continues to play an important role in the workplace, workers may feel that there is no creativity in what they do—they feel alienated (see chapter 7). For example, a study of California's high-tech "white-collar factories" found that

Table 15.3

Demographics of the Stem Cell Issue (Percentages), 2004

	Which Is More Important?		
	New Cures from Stem Cell Research	Protecting Human Embryos	Don't Know
Total	56	32	12
Age			
18–29	61	32	7
30–49	58	31	11
50–64	55	34	11
64 and older	50	32	18
Political affiliation			
Republican	45	45	10
Democrat	68	22	10
Independent	58	30	12
Self-described			
Conservative	44	45	11
Moderate	61	27	12
Liberal	77	16	7
Religion			
Protestants	52	38	10
White evangelical	33	58	9
White mainline	60	10	12
Black Protestant	47	36	17
Catholics	63	28	9
Secular	70	16	14

Source: Adapted from Pew Research Center (2005).

employees were suffering from isolation, job insecurity, and pressure to update skills (BBC 2003). The movement from mechanization to automation to cybernation increasingly removes individuals from the production process, often relegating them to flipping a switch, staring at a computer monitor, or entering data at a keyboard. Many low-paid employees, often women, sit at computer terminals for hours entering data and keeping records for thousands of businesses, corporations, and agencies. The work that takes place in these "electronic sweatshops" is monotonous, solitary, and almost completely devoid of autonomy for the worker.

Not only are these activities routine, boring, and meaningless, but they also promote **deskilling**—that is, "labor requires less thought than before and gives the workers fewer decisions to make" (Perrolle 1990, p. 338). Deskilling stifles development of alternative skills and limits opportunities for advancement and creativity as

"For a list of all the ways technology has failed to improve the quality of life, please press 3."

Alice Kahn
Humorist

Should Federal Funds Be Used to Support Embryonic Stem Cell Research?

Although some believe that an embryonic stem cell is a living being with fundamental human rights, others argue that it is too early in the embryo's development to be considered an individual. Many believe that the use of stem cells for research could ultimately benefit those already living who are suffering from serious life-threatening diseases and injuries. Still others argue that it is morally reprehensible to kill one living being in the hopes of saving another. Despite President Bush's promise to veto any bill passed by either of the congressional houses that would expand on his limited research stem cell provisions, in May 2005 the House of Representatives voted to allow surplus frozen embryos from in vitro fertilization clinics to be used for stem cell research with permission from the donors.

It Is Your Turn Now to Take a Stand!

What do you think? Given the controversy over the use of stem cells, should public funds be spent to support embryonic stem cell research?

---

old skill sets become obsolete. Conflict theorists believe that deskilling also provides the basis for increased inequality, for "throughout the world, those who control the means of producing information products are also able to determine the social organization of the 'mental labor' which produces them" (Perrolle 1990, p. 337).

Technology in some work environments, however, may lead to **upskilling.** Unlike deskilling, upskilling reduces alienation because employees find their work more rather than less meaningful and have greater decision-making powers as information becomes decentralized. Futurists argue that upskilling in the workplace could lead to a "horizontal" work environment in which "employees do not so much what they are told to do, but what their expansive knowledge of the entire enterprise suggests to them needs doing" (Global Internet Project 1998).

Social Relationships and Social Interaction

Technology affects social relationships and the nature of social interaction. The development of telephones has led to fewer visits with friends and relatives; with the coming of VCRs and cable television, the number of places where social life occurs (e.g., movie theaters) has declined. Even the nature of dating has changed as computer networks facilitate cyberdates and "private" chat rooms. As technology increases, social relationships and human interaction are transformed.

Technology also makes it easier for individuals to live in a cocoon—to be self-sufficient in terms of finances (e.g., Quicken), entertainment (e.g., movies on demand, DVR), work (e.g., telework), recreation (e.g., virtual reality), shopping (e.g., eBay), communication (e.g., e-mail, instant messaging), and many other aspects of social life. Even "seeing" your doctor may go online! Blue Shield of California is now paying doctors $25 for each e-mail consultation. Contrary to alienating patients, many respond that they feel closer to their doctors, given the often conversational tone of online communication (Freudenheim 2005). However, although technology can bring people together, it can also isolate them from each other (Klotz 2004). For example, children who use a home computer "spend much less time on sports and outdoor activities than non-computer users" (Attewell et al. 2003, p. 277). In addition, a recent study of more than 1,500 U.S. Internet users between the ages of 18 and 64 found that for every hour a respondent was on the Internet there was a corresponding 23.5-minute reduction in face-to-face interaction with family members (Nie et al. 2004). Some technological innovations replace social roles—an answering machine may replace a secretary, a computer-operated vending machine may replace a waitperson, an automatic teller machine may replace a banker, and closed circuit television, a teacher. These technologies may improve efficiency, but they also reduce necessary human contact.

Loss of Privacy and Security

Schools, employers, and the government are increasingly using technology to monitor individuals' performance and behavior. A 2005 study reports that 36% of companies use "keystroke monitoring so they can read what people type as well as track how much time they spend at the computer," and "55 percent retain and review e-mail messages" (MacMillan 2005, p. 1). Today, the legality of monitoring e-mails is under scrutiny—1 in 20 companies has been sued for e-mail-related surveillance. In addition to e-mail monitoring, high-tech machines monitor countless other behaviors, including counting a telephone operator's minutes online, videotaping a citizen walking down a city street, or tracking the whereabouts of a student or faculty member on campus. One group of privacy advocates, using the Freedom of Information Act, has filed for "technical information about a network of video cameras that has been established in their city" (Markoff 2002).

> "People aren't aware that mouse clicks can be traced, packaged, and sold."
>
> **Larry Irving**
> **U.S. Department of Commerce**

Employers and schools may subject individuals to drug-testing technology (see Chapter 3), and in 2002 the number of identity thefts doubled from the previous year, with this crime representing the most frequent complaint to the Federal Trade Commission (Lee 2003) (see Chapter 4). Through computers individuals can obtain access to someone's phone bills, tax returns, medical reports, credit histories, bank account balances, and driving records. In 2005 alone (CNN 2005a; Krim 2005; Nowell 2005a, 2005b).

- Nearly 700,000 customers of four banks were notified that their financial records had been stolen.
- Bank of America Corporation lost computer data tapes with personal information on 1.2 million federal employees.
- A laptop computer with the names and Social Security numbers of current and former MCI employees was stolen from a financial analyst's car.
- Personal information from 175,000 people was mistakenly sold to identity thieves posing as legitimate businesspeople.
- Tapes containing information on 4 million Citigroup customers were lost by United Parcel Service (UPS) en route to a credit bureau.

Although just inconvenient for some, unauthorized disclosure is potentially devastating for others. If a person's medical records indicate that he or she is HIV-positive, for example, that person could be in danger of losing his or her job or health benefits. If DNA testing of hair, blood, or skin samples reveals a condition that could make the person a liability in the insurer's or employer's opinion, the individual could be denied insurance benefits, medical care, or even employment. In

Grow, Cells, Grow: One Child's Fight for Survival

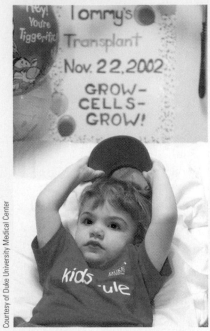

Courtesy of Duke University Medical Center

Tommy Bennett's big brown eyes and sweet demeanor make it that much harder to accept his plight. Just three years old, he blithely endures the constant barrage of drugs, needles, and tests as though he instinctively knows that they are destined to cure him. The development of new technologies has produced new forms of work and new demands for highly skilled workers in certain segments of the labor market.

Born with a rare, degenerative disease called Sanfilippo syndrome, Tommy lacks a critical enzyme needed for proper organ and brain development. Without enzymes, Tommy will die by adolescence. With the enzymes, Tommy's brain may unlock the potential to allow him to talk, dress, and care for himself. Such skills have eluded his two affected siblings, four-year-old Hunter and six-year-old Ciara. Ciara had just been diagnosed with Sanfilippo syndrome when their mom became pregnant with Tommy.

Since that time, they have searched desperately for someone willing to take a chance on helping Ciara, Hunter, and Tommy. The Bennetts found hope at Duke University Medical Center, the only program in the country willing to apply the benefits of stem cells—derived from newborn babies' umbilical cords—to treat this disease.

Proof of a Sanfilippo cure remains elusive, and Tommy is only the sixth Sanfilippo patient ever to have received a stem cell transplant. Yet if the transplant is to help, Tommy is a good candidate. He is young enough that the disease has only just begun to wreak havoc on his brain and organs. His siblings have progressed too far to be helped. Still, the sting of disappointment was palpable when doctors deemed Tommy the only viable candidate.

Thankful as they are for the opportunity, the Bennetts have embarked on a costly gamble—financially, emotionally, and physically. The Bennetts uprooted their kids and moved 900 miles away from family and friends to undergo a series of grueling tests before Tommy's transplant could begin.

response to the possibility of such consequences, Brin (1998), author of *The Transparent Society,* argues that, because it is impossible to prevent such intrusions, "reciprocal transparency," or complete openness, should prevail. If organizations can collect the information, then citizens should have access to it and to its uses.

Our privacy is also disturbed by the intrusion in our e-mail in-boxes of unwanted mail called spam. Between 2002 and 2003 the amount of junk mail sent on the Internet increased by 85% to a total of 4.9 trillion spam messages. Motivated in part by the "promise of an easy profit, spammers have gone from pests to an invasive species of parasite that threatens to clog the inner workings of the Internet" (Taylor 2003). Recently, spammers have targeted cell phones, sending thousands of unwanted text messages to unsuspecting subscribers (CNN 2004). For the average Internet user, about 5 minutes out of every hour online is spent dealing with spam (Nie et al. 2004).

Then came the real test of endurance. Confined to the hospital unit for four straight weeks, Tommy's small body was ravaged by toxic doses of chemotherapy designed to wipe out his immune system and make way for a new one that might provide the crucial enzyme.

Alicia took on hospital duty, caring for Tommy night and day, and catching a few winks of sleep as time permitted on a pull-down cot. John assumed full-time care for Ciara and Hunter at a rented apartment nearby, no easy task given Ciara's penchant for 3 a.m. awakenings.

The process is clearly daunting, yet the transplant itself is deceptively simple. It takes just fifteen minutes for a bag of red liquid to drip intravenously into a child's bloodstream. Nurses literally squeeze every last drop from the bag, lest they lose a single stem cell that floats amidst a billion blood and supporting cells.

Every parent knows that stem cells hold the key to their child's survival. If they grow, the child has a fighting chance to live. If they do not, the child has probably exhausted his or her last resort at a cure.

Then comes the waiting, and the familiar refrain: "Grow, Cells, Grow." The words resonate within the halls, grace the walls of every room, and are sprinkled throughout cards of love and hope. Parents recite them like a battle cry designed to incite soldiers in action.

Indeed, stem cells are like tiny soldiers who descend upon bone marrow and rescue it from near-certain demise. So powerful are stem cells that it takes only ten to a hundred of them to restore a child's entire blood-forming and immune system—in Tommy's case providing the missing enzymes. Moreover, they know exactly where to go and what function to perform.

Yet such remarkable power is not without its drawbacks. Stem cells can attack the last remnants of the child's immune system, a complication called graft-versus-host disease. Stem cells take time to grow and mature, leaving the child's developing immune system vulnerable to minor infections that could prove deadly.

Children also suffer mightily from the dangerously high levels of chemotherapy needed to wipe out their immune system. Often, their mucous linings literally slough off from within, causing severe diarrhea and vomiting. Nausea, painful sores, fatigue, and stomach pains also plague the children as the chemo exerts its effects.

Luckily, Tommy endured far less of the usual symptoms of his transplant, but only time will tell if the new cells have become his own. A year must pass before his new immune system will be running at full force. A lot can happen in that time, but hope, prayer, and a will to overcome will be on their side.

Sadly, Tommy Bennett died on November 24, 2003.

Source: Reprinted from *Phi Kappa Phi Forum* 83(1) (Winter 2003). Copyright by Duke University Medical Center. Reprinted by permission of the publisher.

Technology has created threats not only to the privacy of individuals but also to the security of entire nations. Computers can be used (or misused) in terrorism and warfare to cripple the infrastructure of a society and to tamper with military information and communication operations (see Chapter 16). In a 2004 assessment of efforts made to address computer security standards, the Department of Homeland Security and seven other federal agencies received a failing grade for cybersecurity from a congressional oversight committee (Krebs 2005).

Unemployment

Some technologies replace human workers—robots replace factory workers, word processors displace secretaries and typists, and computer-assisted diagnostics reduce the need for automobile mechanics. Unemployment rates can also increase when companies "outsource" their information technology operations to

The development of new technologies has produced new forms of work and new demands for highly skilled workers in certain segments of the labor market.

© David Sams/Stock Boston

ThomsonNOW

Learn more about **Unemployment** by going through the Unemployment Map Exercise.

lower-wage countries. For example, the U.S. accounting firm Ernst and Young has offices in India that prepare 2% of the firm's total tax returns processed. According to one expert, "more than 300 of the Fortune 500 firms do business with Indian information-technology-services companies" (O'Meara 2003, p. 32).

Not surprisingly, unemployment rates for information technology workers are at an all time high, exceeding the overall jobless rate in 2004 (Schneider 2004). When layoffs occur, older IT workers are more likely to remain unemployed, being less likely than younger workers to take a job outside the field. Downsizing of corporations, business failures, less need for technological support, and an economic recession in addition to "outsourcing" have contributed to disappearing high-tech jobs (McNair 2003; Schneider 2004).

Technology also changes the nature of work and the types of jobs available. For example, fewer semi-skilled workers are needed because many of their jobs are now being done by machines. Of those jobs projected to decline the fastest by 2012, a substantial majority are in manufacturing—textile mill workers, cut-and-sew apparel manufacturers, leather and hide tanners, and so forth (U.S. Bureau of the Census 2004). The jobs that remain, often white-collar jobs, require more education and technological skills. Technology thereby contributes to the split labor market as the pay gulf between skilled and unskilled workers continues to grow. For example, Addison and colleagues (2000) report that employees who use computers at work are at a lower risk of losing their jobs than are nonusers.

The Digital Divide

One of the most significant social problems associated with science and technology is the increased division between the classes. As Welter (1997) notes:

> It is a fundamental truth that people who ultimately gain access to, and who can manipulate, the prevalent technology are enfranchised and flourish. Those individuals (or cultures) that are denied access to the new technologies, or cannot master and pass them on to the largest number of their offspring, suffer and perish. (p. 2)

The fear that technology will produce a "virtual elite" is not uncommon. Several theorists hypothesize that as technology displaces workers, most notably the unskilled and uneducated, certain classes of people will be irreparably disadvantaged—the poor, minorities, and women. There is even concern that biotechnologies will lead to a "genetic stratification," whereby genetic screening, gene therapy, and other types of genetic enhancements are available only to the rich.

The wealthier the family, for example, the more likely the family is to have a computer. Of American families with an annual income of $150,000 or more, 94.7% have at least one computer in the household (see Table 15.2). However, only 26.9% of households with annual incomes between $5,000 and $9,999 own a computer (U.S. Bureau of the Census 2004). Furthermore, 61.1% of white Americans own a computer, compared to 37.1% of African Americans. Whites are also twice as likely as blacks or Hispanics to have broadband service (ANOL 2004).

Racial disparities also exist in Internet access. In 2003, although 65.1% of whites used the Internet from one or more locations, only 45.6% of blacks and 37.2% of Hispanics used the Internet from one or more locations (ANOL 2004). Inner-city neighborhoods are disproportionately populated by racial and ethnic minorities and are simply less likely to be "wired," that is, to have the telecommunications hardware necessary for access to online services. In fact, cable and telephone companies are less likely to lay fiber optic cables in these areas—a practice called "information apartheid" or "electronic redlining." Urban-rural differences also exist. For example, when survey respondents were asked why they did not have high-speed Internet access, 22.1% of the respondents who live in rural areas and 4.7% of those who live in urban areas replied that broadband service was not available (ANOL 2004).

The cost of equalizing such differences is enormous, but the cost of not equalizing them may be even greater. Employees who are technologically skilled have higher incomes than those who are not—up to 15% higher (Hancock 1995; International Labor Organization 2001). Further, technological disparities exacerbate the structural inequities perpetuated by the split labor force and the existence of primary and secondary labor markets.

Mental and Physical Health

Some new technologies have unknown risks. Biotechnology, for example, has promised and, to some extent, has delivered everything from life-saving drugs to hardier pest-free tomatoes. However, biotechnologies have also created **technology-induced diseases,** such as those experienced by Chellis Glendinning (1990). Glendinning, after using the Pill and, later, the Dalkon Shield IUD, became seriously ill.

> Despite my efforts to get help, medical professionals did not seem to know the root of my condition lay in immune dysfunction caused by ingesting artificial hormones and worsened by chronic inflammation. In all, my life was disrupted by illness for twenty years, including six years spent in bed. . . . For most of the years of illness, I lived in isolation with my problem. Doctors and manufacturers of birth control technologies never acknowledged it or its sources. (p. 15)

Other technologies that pose a clear risk to a large number of people include nuclear power plants, the pesticide DDT, automobiles, X-rays, food coloring, and breast implants.

The production of new technologies may also place manufacturing employees in jeopardy. For example, the electronics industry uses thousands of hazardous chemicals:

> The semiconductor industry uses large amounts of toxic chemicals to manufacture the components that make up a computer, including disk drives, circuit boards, video display equipment, and silicon chips themselves, the basic building blocks of computer devices. The toxic materials needed to make the 220 billion silicon chips manufactured annually are staggering in amount and include highly corrosive hydrochloric acid; metals such as arsenic, cadmium, and lead; volatile solvents such as methyl chloroform, toluene, benzene, acetone, and trichloroethylene; and toxic gases such as arsine. Many of these chemicals are known or probable human carcinogens. (Chepesiuk 1999, p. 1)

Finally, technological innovations are, for many, a cause of anguish and stress, particularly when the technological changes are far-reaching (Hormats 2001). Nearly

60% of workers report being "technophobes," that is, fearful of technology, and as many as 10% of Internet users are "addicted" to being online (Boles & Sunoo 1998; Papadakis 2000). Researchers at the University of Florida have developed a list of symptoms associated with Internet addiction, as signified by the acronym MOUSE: "**M**ore than intended time spent online; **O**ther responsibilities neglected; **U**nsuccessful attempts to cut down; **S**ignificant relationship discord; and **E**xcessive thoughts or anxieties when not online" (Deutsche Welle 2003, p. 3).

The Challenge to Traditional Values and Beliefs

Technological innovations and scientific discoveries often challenge traditionally held values and beliefs, in part because they enable people to achieve goals that were previously unobtainable. Before recent advances in reproductive technology, for example, women could not conceive and give birth after menopause. Technology that allows postmenopausal women to give birth challenges societal beliefs about childbearing and the role of older women. Macklin (1991) notes that the techniques of egg retrieval, in vitro fertilization, and gamete intrafallopian transfer make it possible for two different women to each make a biological contribution to the creation of a new life. Such technology requires society to reexamine its beliefs about what a family is and what a mother is. Should family be defined by custom, law, or the intentions of the parties involved?

Medical technologies that sustain life lead us to rethink the issue of when life should end. The increasing use of computers throughout society challenges the traditional value of privacy. New weapons systems make questionable the traditional idea of war as something that can be survived and even won. And cloning causes us to wonder about our traditional notions of family, parenthood, and individuality. Toffler (1970) coined the term **future shock** to describe the confusion resulting from rapid scientific and technological changes that unravel our traditional values and beliefs.

Strategies for Action: Controlling Science and Technology

As technology increases, so does the need for social responsibility. Nuclear power, genetic engineering, cloning, and computer surveillance all increase the need for social responsibility: "Technological change has the effect of enhancing the importance of public decision making in society, because technology is continually creating new possibilities for social action as well as new problems that have to be dealt with" (Mesthene 1993, p. 85). In the following sections we address various aspects of the public debate, including science, ethics, and the law, the role of corporate America, and government policy.

Science, Ethics, and the Law

Science and its resulting technologies alter the culture of society through the challenging of traditional values. Public debate and ethical controversies, however, have led to structural alterations in society as the legal system responds to calls for action. For example, several states now have what are called genetic exception laws.

Genetic exception laws require that genetic information be handled separately from other medical information, leading to what is sometimes called patient shadow files (AMA 2001). The logic of such laws rests with the potentially devastating effects of genetic information being revealed to insurance companies, other family members, employers, and the like. At present, 16 states "require informed consent for a third party either to perform or require a genetic test or to obtain genetic information" and 24 states require informed consent to disclose genetic information (NCSL 2005a).

Are such regulations necessary? In a society characterized by rapid technological and thus social change—a society where custody of frozen embryos is part of the divorce agreement—many would say yes. Cloning, for example, is one of the most hotly debated technologies in recent years. Bioethicists and the public vehemently debate the various costs and benefits of this scientific technique. Despite such controversy, however, the chairman of the National Bioethics Advisory Commission warned more than five years ago that human cloning will be "very difficult to stop" (McFarling 1998). At present, 14 states have laws pertaining to human cloning, some prohibiting cloning for reproductive purposes, some prohibiting therapeutic cloning, and still others, both (NCSL 2005b).

Should the choices that we make, as a society, be dependent on what we can do or what we should do? Whereas scientists and the agencies and corporations who fund them often determine what we *can* do, who should determine what we *should* do? Although such decisions are likely to have a strong legal component—that is, they must be consistent with the rule of law and the constitutional right of scientific inquiry—legality or the lack thereof often fails to answer the question, What should be done? *Roe* v. *Wade* (1973) did little to squash the public debate over abortion and, more specifically, the question of when life begins. Thus it is likely that the issues surrounding the most controversial of technologies will continue into the 21st century with no easy answers (see Figure 15.3).

Technology and Corporate America

As philosopher Jean-Francois Lyotard notes, knowledge is increasingly produced to be sold (Powers 1998). The development of genetically altered crops, the commodification of women as egg donors, and the harvesting of regenerated organ tissues are all examples of potentially market-driven technologies. Like the corporate pursuit of computer technology, profit-motivated biotechnology creates several concerns.

First is the concern that only the rich will have access to such life-saving technologies as genetic screening and cloned organs. Such fears are justified. Companies with obscure names such as Progenitor, Millennium Pharmaceuticals, Darwin Molecular, and Myriad Genetics have been patenting human life. Millennium Pharmaceuticals holds the patent on the melanoma gene and the obesity gene, Darwin Molecular controls the premature aging gene, Progenitor controls the gene for schizophrenia, and Myriad Genetics has nine patents on the breast and ovarian cancer genes (Mayer 2002; Shand 1998).

These patents result in **gene monopolies,** which could lead to astronomical patient costs for genetic screening and treatment. One company's corporate literature candidly states that its patent of the breast cancer gene will limit competition and lead to huge profits (Shand 1998, p. 47). The biotechnology industry argues that such patents are the only way to recoup research costs that, in turn, lead to further innovations. At present, more than 3 million gene-related patent applications

"Prohibiting scientific and medical activities would also raise troubling enforcement issues. . . . Would the FBI raid research laboratories and universities? Seize and read the private medical records of infertility patients? Burst into operating rooms with their guns drawn? Grill new mothers about how their babies were conceived?"

Mark Eibert
Attorney

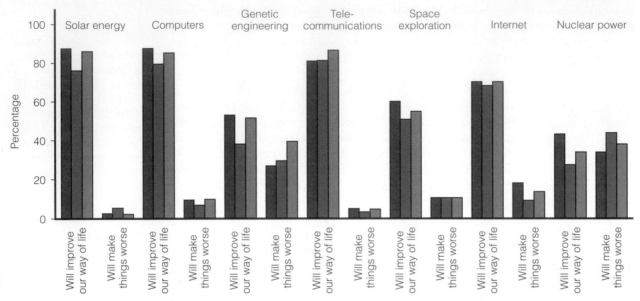

Figure 15.3

Public attitudes toward selected technologies in the United States, Europe, and Canada.

Source: National Science Foundation (2002).

have been filed with the U.S. Patent and Trademark Office (Human Genome Project 2005b).

The commercialization of technology causes several other concerns, including issues of quality control and the tendency for discoveries to remain closely guarded secrets rather than collaborative efforts (Lemonick & Thompson 1999; Mayer 2002; Rabino 1998). In addition, industry involvement has made government control more difficult because researchers depend less and less on federal funding. More than 66% of research and development in the United States is supported by private industry using their own company funds (National Science Foundation 2004).

Finally, although there is little doubt that profit acts as a catalyst for some scientific discoveries, other less commercially profitable but equally important projects may be ignored. As biologist Isaac Rabino states, "Imagine if early chemists had thrown their energies into developing profitable household products before the periodic table was discovered" (Rabino 1998, p. 112).

Runaway Science and Government Policy

Science and technology raise many public policy issues. Policy decisions, for example, address concerns about the safety of nuclear power plants, the privacy of electronic mail, the hazards of chemical warfare, and the legality of cloning. In creating science and technology, have we created a monster that has begun to control us rather than the reverse? What controls, if any, should be placed on science and technology? And are such controls consistent with existing law? Consider the use of Grokster software to download music files (the question of intellectual property rights and copyright infringement); a 2005 Utah law limiting children's access to material on the Internet (free speech issues); and Carnivore (DCS 1000), an FBI surveillance program that can search every message that passes through an Internet service provider (Fourth Amendment privacy issues) (ACLU 2005; Kaplan 2000; Schaefer 2001; White 2000).

The government, often through regulatory agencies and departments, prohibits the use of some technologies (e.g., assisted-suicide devices) and requires others (e.g., seat belts). The Food and Drug Administration, the Environmental Protection Agency, and the Department of Agriculture investigated the use of genetically altered corn—corn that had been approved only for animal feed—in the making of Taco Bell taco shells (Brasher 2000). A treaty has been adopted by 130 nations that permits countries to "bar imports of genetically altered seeds, microbes, animals, and crops that they deem a threat to their environment" (Pollack 2000, p. 1). Despite the agreement, developing countries are under enormous pressure from corporations to permit such imports (Mayer 2002).

The federal government has instituted several initiatives dealing with technology-related crime. A 2005 bill introduced in the U.S. Senate would amend the federal criminal code to make "phishing" illegal—that is, it would criminalize the practice of fraudulently obtaining personal information (U.S. Congress 2005). As a government report notes, "Many of the attributes of this Internet technology—low cost, ease of use, and anonymous nature, among others make it an attractive medium for fraudulent scams, child sexual exploitation, and . . . cyberstalking" (U.S. Department of Justice 1999, p. 1).

Of late, the issue of online pornography has come to the forefront. A Federal Bureau of Investigation (FBI) report states that "computer telecommunications have become one of the most prevalent techniques used by pedophiles to share illegal photographic images of minors and to lure children into illicit sexual relationships" (FBI 2002, p. 1). For example, it is estimated that there are more than 100,000 websites that offer illegal child pornography, an industry that generates over $3 billion a year (*Internet Pornography Statistics*, 2005). A second concern is the ease with which children can access and view online pornography. In response to such availability, the federal government enacted the Children's Internet Protection Act, which requires "public schools and libraries to install Internet filters on their computers so children and adults cannot view 'inappropriate' information on the Internet" (MacMillan 2004). Americans support such initiatives. In a 2005 survey of likely American voters, 71% responded that Congress needs to pass new laws designed to keep the Internet safe (CNN 2005b).

Finally, the government has several science and technology boards, including the National Science and Technology Council, the Office of Science and Technology Policy, and the President's Council of Advisors on Science and Technology. These agencies advise the president on matters of science and technology, including research and development, implementation, national policy, and coordination of different initiatives.

> "As president, I will prohibit genetic discrimination, criminalize identity theft, and guarantee the privacy of medical and sensitive financial records."
>
> **George W. Bush**
> **U.S. president**

Understanding Science and Technology

What are we to understand about science and technology from this chapter? As structural functionalists argue, science and technology evolve as a social process and are a natural part of the evolution of society. As society's needs change, scientific discoveries and technological innovations emerge to meet these needs, thereby serving the functions of the whole. Consistent with conflict theory, however, science and technology also meet the needs of select groups and are characterized by political components. As Winner (1993) notes, the structure of science and technology conveys political messages, including "power is centralized," "there are barriers between social classes," "the world is hierarchically structured," and "the good things are distributed unequally" (Winner 1993, p. 288).

The scientific discoveries and technological innovations that are embraced by society as truth itself are socially determined. Research indicates that science and the resulting technologies have both negative and positive consequences—a **technological dualism.** Technology saves lives and time and money; it also leads to death, unemployment, alienation, and estrangement. Weighing the costs and benefits of technology poses ethical dilemmas, as does science itself. Ethics, however, "is not only concerned with individual choices and acts. It is also and, perhaps, above all concerned with the cultural shifts and trends of which acts are but the symptoms" (McCormick & Richard 1994, p. 16).

Thus society makes a choice by the very direction it follows. These choices should be made on the basis of guiding principles that are both fair and just, such as those listed here (Buchanan et al. 2000; Eibert 1998; Goodman 1993; Murphie & Potts 2003; Winner 1993):

1. Science and technology should be prudent. Adequate testing, safeguards, and impact studies are essential. Impact assessment should include an evaluation of the social, political, environmental, and economic factors.
2. No technology should be developed unless all groups, and particularly those who will be most affected by the technology, have at least some representation "at a very early stage in defining what that technology will be" (Winner 1993, p. 291). Traditionally, the structure of the scientific process and the development of technologies have been centralized (i.e., decisions have been made by a few scientists and engineers); decentralization of the process would increase representation.
3. Means should not exist without ends. Each new innovation should be directed toward fulfilling a societal need rather than the more typical pattern in which a technology is developed first (e.g., high-definition television) and then a market is created (e.g., "You'll never watch a regular TV again!"). Indeed, from the space program to research on artificial intelligence the vested interests of scientists and engineers, whose discoveries and innovations build careers, should be tempered by the demands of society.

What the 21st century will hold, as the technological transformation continues, may be beyond the imagination of most of society's members. Technology empowers; it increases efficiency and productivity, extends life, controls the environment, and expands individual capabilities. According to a National Intelligence Council report, "Life in 2015 will be revolutionized by the growing effort of multi-disciplinary technology across all dimensions of life: social, economic, political, and personal" (NIC 2003, p. 1).

As we proceed into the first computational millennium, one of the great concerns of civilization will be the attempt to reorder society, culture, and government in a manner that exploits the digital bonanza yet prevents it from running roughshod over the checks and balances so delicately constructed in those simpler precomputer years.

Chapter Review

• **What are the three types of technology?**

The three types of technology, escalating in sophistication, are mechanization, automation, and cybernation. Mechanization is the use of tools to accomplish tasks previously done by hand. Automation involves the use of self-operating machines, and cybernation is the use of machines to control machines.

• **What are some Internet global trends?**

Globally, English speakers are the largest language group online, but non-English speakers constitute the fastest growing group on the Internet. The clear majority of Internet users live in industrialized countries, although there is some movement toward the Internet's becoming truly global as those in developing countries "get online."

• **According to Kuhn, what is the scientific process?**

Kuhn describes the process of scientific discovery as occurring in three steps. First are assumptions about a particular phenomenon. Next, because unanswered questions always remain about a topic, science works to start filling in the gaps. Then, when new information suggests that the initial assumptions were incorrect, a new set of assumptions or framework emerges to replace the old one. It then becomes the dominant belief or paradigm until it is questioned and the process repeats.

• **What is meant by the computer revolution?**

The silicon chip made computers affordable. Today, 62% of American homes have a computer. In addition, over half the labor force uses a computer at work. The most common computer activity at work is accessing the Internet or e-mail, followed by word processing, working with spreadsheets or databases, and accessing or updating calendars or schedules.

• **What is the Human Genome Project?**

The U.S. Human Genome Project is an effort to decode human DNA. The 13-year-old project is now complete, allowing scientists to "transform medicine" through early diagnosis and treatment as well as possibly preventing disease through gene therapy. Gene therapy entails identifying a defective or missing gene and then replacing it with a healthy duplicate that is transplanted to the affected area.

• **How are some of the problems of the Industrial Revolution similar to the problems of the technological revolution?**

The most obvious example is in unemployment. Just as the Industrial Revolution replaced many jobs with technological innovations, so too has the technological revolution. Furthermore, research indicates that many of the jobs created by the Industrial Revolution, such as working on a factory assembly line, were characterized by high rates of alienation. Rising rates of alienation are also a consequence of increased estrangement as high-tech employees work in "white-collar factories."

• **What is meant by the acronym MOUSE?**

MOUSE is a reference to Internet addiction and refers to a list of symptoms, as indicated by each letter: "**M**ore than intended time spent online; **O**ther responsibilities neglected; **U**nsuccessful attempts to cut down; **S**ignificant relationship discord; and **E**xcessive thoughts or anxieties when not online" (Deutsche Welle 2003, p. 3).

• **What is the digital divide?**

The digital divide is the tendency for technology to be most accessible to the wealthiest and most educated. For example, some fear that there will be "genetic stratification," whereby the benefits of genetic screening, gene therapy, and other genetic enhancements will be available to only the richest segments of society.

• **What is meant by the commercialization of technology?**

The commercialization of technology refers to profit-motivated technological innovations. Whether it be the isolation of a particular gene, genetically modified organisms, or the regeneration of organ tissues, where there's a possibility for profit, private enterprise will be there.

Critical Thinking

1. Use of the Internet by neo-Nazi and white supremacist groups has recently increased. Despite such increases, the U.S. Supreme Court has strengthened First Amendment protections of Internet material (Whine 1997). Should such groups have the right to disseminate information about their organizations and recruit members through the Internet?
2. What currently existing technologies have had more negative than positive consequences for individuals and for society?
3. Some research suggests that productivity actually declines with the use of computers (Rosenberg 1998). Assuming that this "paradox of productivity" is accurate, what do you think causes the reduction in efficiency?

Key Terms

abortion	mechanization
automation	partial birth abortion
cultural lag	postmodernism
cybernation	science
deskilling	snowball sampling
e-commerce	stem cells
future shock	technological dualism
gene monopolies	technological fix
gene therapy	technology
genetic engineering	technology-induced diseases
genetic screening	telework
Internet	therapeutic cloning
in vitro fertilization	upskilling
IT	

Media Resources

 The Companion Website for
Understanding Social Problems,
Fifth Edition

http://sociology.wadsworth.com/mooney_knox_schacht5e

Supplement your review of this chapter by going to the companion website to take one of the Tutorial Quizzes, use the flash cards to master key terms, and check out the many other study aids you'll find there. You'll also find special features such as Wadsworth's Sociology Online Resources and Writing Companion, GSS data, and Census 2000 information, data, and resources at your fingertips to help you complete that special project or do some research on your own.

"Every gun that is made, every warship launched, every rocket fired, signifies in the final sense a theft from those who hunger and are not fed, those who are cold and not clothed. The world in arms is not spending money alone. It is spending the sweat of its laborers, the genius of its scientists, and the hopes of its children." *Dwight D. Eisenhower, former U.S. president and military leader*

16

Conflict, War, and Terrorism

The Global Context: Conflict in a Changing World

Sociological Theories of War

Causes of War

Terrorism

Social Problems Associated with Conflict, War, and Terrorism

Strategies for Action: In Search of Global Peace

Understanding Conflict, War, and Terrorism

Chapter Review

arwan Abu Ubeida was born in Fallujah, Iraq, where he attended high school and enjoyed all the advantages that having a successful father can bring (Ghosh 2005). Although a Sunni Muslim—as was Saddam Hussein—Marwan decided not to join the resistance as so many others had. Surely, after the fall of the Ba'athist regime and the capture of Saddam the Americans would leave his homeland. But they stayed and stayed, and after U.S. soldiers fired and killed 12 demonstrators and injured many more, Marwan, who had been at the demonstration, joined the Iraqi insurgency.

Now, at only age 20, Marwan is a seasoned soldier. He has experienced dozens of assaults against U.S. troops and is considered an expert with the Russian-made PKC machine gun, his weapon of choice. Marwan is a "jihadi foot soldier," a member of Abu Mousab al-Zarqawi's terrorist group, Al Qaeda (Ghosh 2005, p. 24). But for the last several months Marwan has been training to carry out a suicide mission. "I can't wait," he says, recounting the day that his name was added to the list of volunteers as the happiest day of his life. When he is called to duty, he will spend his final days in seclusion, spiritually and psychologically preparing for his mission. Time permitting, he will call his family to say goodbye.

Marwan has practiced and practiced his final prayer. What is it he is praying for? "First I will ask Allah to bless my mission with a high rate of casualties among the Americans. . . . Then I will ask him to purify my soul so I am fit to see him, and I will ask to see my mujahedin brothers who are already with him. . . . The most important thing is that he should let me kill many Americans" (Ghosh 2005, pp. 23–24).

Religious fanatic or calculated strategy? A new book by Robert Pape, titled *Dying to Win: The Strategic Logic of Suicide Bombers* (2005), suggests that most suicide bombers are, in fact, not Islamic fundamentalists but soldiers in a "coherent campaign" to rid their homeland of foreign military forces. Although religion may play a part, particularly in recruitment, only 40% of the 462 suicide attacks studied were religiously motivated. Pape concludes that suicide attacks, in Iraq and around the world, are on the rise because they are an effective means to accomplish a collective end.

Thus the use of suicide bombers can be seen as a direct result of the continued presence of U.S.-led troops in Iraq. The pivotal events leading to the war in Iraq and, more generally, the war on terrorism, include the bombing of the World Trade Center and the U.S. Pentagon on September 11, 2001. **War**, the most violent form of conflict, refers to organized armed violence aimed at a social group in pursuit of an objective. Wars have existed throughout human history and continue in the contemporary world.

War is one of the great paradoxes of human history. It both protects and annihilates. It creates and defends nations but also destroys them. Whether war is just or unjust, defensive or offensive, it involves the most horrendous atrocities known to humankind. In this chapter we focus on the causes and consequences of conflict, war, and terrorism. Along with population and environmental problems, conflict, war, and terrorism are among the most serious of all social problems in their threat to the human race and life on earth.

The Global Context: Conflict in a Changing World

As societies have evolved and changed throughout history, the nature of war has also changed. Before industrialization and the sophisticated technology that resulted, war occurred primarily between neighboring groups on a relatively small scale. In the modern world war can be waged between nations that are separated by thousands of miles as well as between neighboring nations. In the following sections we examine how war has changed our social world and how our changing social world has affected the nature of war in the industrial and post-industrial information age.

War and Social Change

The very act that now threatens modern civilization—war—is largely responsible for creating the advanced civilization in which we live. Before large political states existed, people lived in small groups and villages. War broke the barriers of autonomy between local groups and permitted small villages to be incorporated into larger political units known as chiefdoms. Centuries of warfare between chiefdoms culminated in the development of the state. The **state** is "an apparatus of power, a set of institutions—the central government, the armed forces, the regulatory and police agencies—whose most important functions involve the use of force, the control of territory and the maintenance of internal order" (Porter 1994, pp. 5–6). The creation of the state in turn led to other profound social and cultural changes:

> And once the state emerged, the gates were flung open to enormous cultural advances, advances undreamed of during—and impossible under—a regimen of small autonomous villages. . . . Only in large political units, far removed in structure from the small autonomous communities from which they sprang, was it possible for great advances to be made in the arts and sciences, in economics and technology, and indeed in every field of culture central to the great industrial civilizations of the world. (Carneiro 1994, pp. 14–15)

Industrialization and technology could not have developed in the small social groups that existed before military action consolidated them into larger states. Thus war contributed indirectly to the industrialization and technological sophistication that characterize the modern world. Industrialization, in turn, has had two major influences on war. Cohen (1986) calculated the number of wars fought per decade in industrial and pre-industrial nations and concluded that "as societies become more industrialized, their proneness to warfare decreases" (p. 265). For example, in 2004 there were 19 major armed conflicts, the majority of which were in less developed countries in Africa and Asia (Stockholm International Peace Research Institute 2005).

Although industrialization may decrease a society's propensity to war, it also increases the potential destruction of war. With industrialization, military technology became more sophisticated and more lethal. Rifles and cannons replaced the clubs, arrows, and swords used in more primitive warfare and, in turn, were replaced by tanks, bombers, and nuclear warheads. This chapter's *Focus on Technology* feature looks at how information technology is transforming war and what it means to have a "secure" nation.

© Bob Mahoney/The Image Works

Industrialization can decrease a society's propensity for war, but it also increases the potential destructiveness of war because, with industrialization, warfare technology becomes more sophisticated and lethal.

Cyberwarfare and Cybersecurity

In the post-industrial information age computer technology has revolutionized the nature of warfare and future warfare capabilities. Today, a "whole range of new technologies are offered for the next generations of weapons and military operations" (Bonn International Center for Conversion 1998, p. 3), including the use of high-performance sensors, information processors, directed-energy technologies, precision-guided munitions, and computer worms and viruses (Bonn International Center for Conversion 1998; O'Prey 1995; PBS 2003). With the increasing proliferation and power of computer technology, military strategists and political leaders are exploring the horizons of information warfare often called cyberwarfare. Essentially, **cyberwarfare** utilizes information technology to attack or manipulate the military and civilian infrastructure and information systems of an enemy. For example, cyberwarfare capabilities include the following, from the least to most serious (National Security 2003):

- *Web vandalism.* This tactic is used to "deactivate or deface" a government Web page.
- *Disinformation campaigns.* These campaigns use misinformation to confuse the enemy.
- *Gathering of secret data.* In this tactic classified information is intercepted and tampered with.
- *Disruption in the field.* This tactic seeks to interfere with military activities by, for example, blocking vital communication and intercepting commands.
- *Attacks on critical infrastructure.* This tactic involves electronic attacks on America's infrastructures, including transportation, electricity, water, fuel, and finances.

Concerns over cyberwarfare are warranted. For example, U.S. officials stumbled across a "two-year pattern of probing of computer systems in the Pentagon, NASA, the Energy Department, and university and research labs" (PBS 2003, p. 2). The attacks were traced to a mainframe in Russia.

At the Naval Postgraduate School students from diverse backgrounds—some military, some civilian—come together for cyberspace maneuvers, that is, mock Internet warfare with a computer, modem, software, and keyboard (Howe 2003). Scientists at Los Alamos National Laboratory, as part of the Department of Homeland Security's effort to protect against terrorists, build and attack computer models of cities "inhabited by millions of virtual individuals who go to work, shopping centers, soccer games and anywhere else their real life counterparts go" (Cha 2005, p. A1). Furthermore, Air Force "battlelabs" have been established across the United States (Sietzen 2000). Because "the ability of American forces to deny access to space by any enemy of the United States or its allies" is paramount, battlelab personnel are conducting research on space control technologies. Today, the concept of space control includes controlling "everyday communications moving through space: voice, e-mail, paging signals, computer data and weather projections, . . . reconnaissance images of enemy forces and basic military communications between forces, fleets and communication centers" (Sietzen 2000, p. 2).

There are two concerns, however, about the use of cyberwarfare technologies. First, although the United States is developing a cyberwarfare strategy, other countries, including Russia, China, and Israel, are further along in their information warfare capabilities (Messmer 2000). In fact, a House committee, reviewing cybersecurity of federal agencies, gave an overall grade of F to the 24 agencies reviewed (McDonald 2002), and a more recent report, "Cybersecurity: A Crisis of Prioritization," concludes that "the federal government is applying short-term remedies to evolving threats that require a long-term vision" (Vlahos 2005, p. 1).

Another problem is that cyberwarfare is relatively inexpensive and readily available (McDonald 2002). With a computer, a modem, and some rudimentary knowledge of computers, anyone could initiate an information attack. For example, after the initial fall of the Taliban and Al Qaeda, computers were seized and prisoners were interrogated. One computer contained

> software and connections to a programming site where the users had been pulling specific information about digital switches on power and water company system infrastructures. It showed how Al Qaeda was doing research through open, available resources to learn about U.S. critical infrastructure and how to exploit it. (McDonald 2002)

The 2006 budget requests $1.69 billion for information technology security—a 7.2% increase from the previous year (Perera 2005). In 2005 a bill was passed by the House of Representatives that would establish an assistant secretary for cybersecurity within the Department of Homeland Security. If the position receives final approval, the assistant secretary would be responsible for "establishing and managing a national cybersecurity response and information sharing system with the capability to detect and prevent attacks on the nation's cybersecurity and to help in the restoration of cybersecurity infrastructure in the wake of such attacks" (McKenna 2005, p. 1).

The Economics of Military Spending

The increasing sophistication of military technology has commanded a large share of resources totaling, worldwide, $1.04 trillion in 2004 (Stockholm International Peace Research Institute 2005). Global military spending has been increasing since 1998, with dramatic increases between 2002 and 2005 as a consequence of U.S.-led post–September 11, 2001, expenditures. For example, according to government figures, the war in Iraq costs an estimated $4.3 billion a month; the war in Afghanistan costs $800 million a month (Lumpkin 2005). When a representative sample of Americans were asked about national defense and military spending, 30% responded "too little" was being spent, 38% responded that the amount was "about right," and 30% responded "too much" was being spent (Gallup 2005a).

The **Cold War**, the state of political tension and military rivalry that existed between the United States and the former Soviet Union, provided justification for large expenditures on military preparedness. However, the end of the Cold War, along with the rising national debt, resulted in cutbacks in the U.S. military budget in the 1990s.

Today, military spending has approached Cold War levels. The United States accounts for nearly half the world's military spending, the largest single percentage of any nation—more than the combined total of the 32 next most powerful nations (Stockholm International Peace Research Institute 2005). The next highest military spenders are England, France, Japan, and China, which, along with the United States, account for 64% of the world's military budget. U.S. national defense outlays, estimated to be $468 billion in 2005, include expenditures for salaries of military personnel, research and development, weapons, veterans' benefits, and other defense-related expenses (Office of Management and Budget 2005).

The U.S. government not only spends money on its own military and defense but also sells military equipment to other countries, either directly or by helping U.S. companies sell weapons abroad. Although the purchasing countries may use these weapons to defend themselves from hostile attack, foreign military sales may pose a threat to the United States by arming potential antagonists. For example, the United States, the world's leading arms-exporting nation, supplied weapons to Iraq to use against Iran. These same weapons were then used against Americans in the Gulf and Iraq wars.

In 2003 the United States transferred arms to several countries in active conflict, including Chad, Angola, Pakistan, Israel, and Colombia. A 2005 report by the World Policy Institute, titled *U.S. Weapons at War: Promoting Freedom or Fueling Conflict?* concludes that far "from serving as a force for security and stability, U.S. weapons sales frequently serve to empower unstable, undemocratic regimes to the detriment of U.S. and global security" (Berrigan & Hartung 2005).

ThomsonNOW™

Learn more about the **Economics of Military Spending** by going through the Annual Military Expenditures Map Exercise.

Sociological Theories of War

Sociological perspectives can help us understand various aspects of war. In this section we describe how structural functionalism, conflict theory, and symbolic interactionism can be applied to the study of war.

Structural-Functionalist Perspective

Structural functionalism focuses on the functions that war serves and suggests that war would not exist unless it had positive outcomes for society. We have already noted that war has served to consolidate small autonomous social groups

Keep us flying!

BUY WAR BONDS

© David Pollack/Corbis

As structural functionalists argue, a major function of war is that it produces unity among societal members. War provides a common cause and a common identity. Societal members feel a sense of cohesion, and they work together to defeat the enemy.

into larger political states. An estimated 600,000 autonomous political units existed in the world at about 1000 B.C. Today, that number has dwindled to fewer than 200.

Another major function of war is that it produces social cohesion and unity among societal members by giving them a "common cause" and a common enemy. For example, in 2005 *Newsweek* began a new feature about everyday American heroes called "Red, White, and Proud." Unless a war is extremely unpopular, military conflict also promotes economic and political cooperation. Internal domestic conflicts between political parties, minority groups, and special interest groups dissolve as they unite to fight the common enemy. During World War II, U.S. citizens worked together as a nation to defeat Germany and Japan.

In the short term war also increases employment and stimulates the economy. The increased production needed to fight World War II helped pull the United States out of the Great Depression. The investments in the manufacturing sector during World War II also had a long-term impact on the U.S. economy. Hooks and Bloomquist (1992) studied the effect of the war on the U.S. economy between 1947 and 1972 and conclude that the U.S. government "directed, and in large measure, paid for a 65 percent expansion of the total investment in plant and equipment" (p. 304). War can also have the opposite effect, however. In a 2005 restructuring of the military, the Pentagon, seeking a "meaner, leaner fighting machine," recommended shutting down or reconfiguring nearly 180 military installations, "ranging from tiny Army reserve centers to sprawling Air Force bases that have been the economic anchors of their communities for generations" (Schmitt 2005). If the plan is approved as initially recommended, thousands of civilian jobs will be lost.

Another function of war is the inspiration of scientific and technological developments that are useful to civilians. Research on laser-based defense systems led to laser surgery, for example, and research in nuclear fission and fusion facilitated the development of nuclear power. The airline industry owes much of its technology to the development of air power by the Department of Defense, and the Internet was created by the Pentagon for military purposes. Other **dual-use technologies**, a term referring to defense-funded innovations that have commercial and civilian applications, include SLICE, a high-speed twin-hull water vessel originally made for the Office of Naval Research. SLICE has a variety of commercial applications, "including its use as a tour or sport fishing boat, oceanographic research vessel, oil spill response ship, and high-speed ferry" (State of Hawaii 2000).

Finally, war serves to encourage social reform. After a major war members of society have a sense of shared sacrifice and a desire to heal wounds and rebuild normal patterns of life. They put political pressure on the state to care for war victims, improve social and political conditions, and reward those who have sacrificed lives, family members, and property in battle. As Porter (1994) explains, "Since . . . the lower economic strata usually contribute more of their blood in battle than the wealthier classes, war often gives impetus to social welfare reforms" (p. 19).

Conflict Perspective

Conflict theorists emphasize that the roots of war are often antagonisms that emerge whenever two or more ethnic groups (e.g., Bosnians and Serbs), countries (United States and Vietnam), or regions within countries (the U.S. North and South) struggle for control of resources or have different political, economic, or religious ideologies. In addition, conflict theory suggests that war benefits the corporate, military, and political elites. Corporate elites benefit because war often results in the victor taking control of the raw materials of the losing nations, thereby creating a bigger supply of raw materials for its own industries. Indeed, many corporations profit from defense spending. Under the Pentagon's bid-and-proposal program, for example, corporations can charge the cost of preparing proposals for new weapons as overhead on their Defense Department contracts. Also, Pentagon contracts often guarantee a profit to the developing corporations. Even if the project's cost exceeds initial estimates, called a cost overrun, the corporation still receives the agreed-on profit. In the late 1950s President Dwight D. Eisenhower referred to this close association between the military and the defense industry as the **military-industrial complex**. Conflict theorists would be quick to note that "many former Republican officials and political associates of the Bush administration are associated with the Carlyle Group, an equity investment firm with billions of dollars in military and aerospace assets" (Knickerbocker 2002, p. 2).

ThomsonNOW™

Learn more about **Conflict Perspective** by going through the Power Elite Model Animation.

The military elite benefit because war and the preparations for it provide prestige and employment for military officials. For example, Military Professional Resources Inc. (MPRI), an organization staffed by former military, defense, law enforcement, and other professionals, serves "the national security needs of the U.S. government, selected foreign governments, international organizations, and the private sector" and lists such capabilities as war gaming, force development and management, and democracy transition assistance (Military Professional Resources Inc. 2005). According to the Center for Public Integrity, in 2004 the total value of MPRI's contracts in Iraq was more than $2.5 million (Center for Public Integrity 2005).

War also benefits the political elite by giving government officials more power. Porter (1994) observed that "throughout modern history, war has been the lever by which . . . governments have imposed increasingly larger tax burdens on increasingly broader segments of society, thus enabling ever-higher levels of spending to be sustained, even in peacetime" (p. 14). Political leaders who lead their country to a military victory also benefit from the prestige and hero status conferred on them.

Finally, feminists argue that war and other conflicts are often justified using the "language of feminism" (Viner 2002). For example, the attack on Afghanistan in 2001 was, in part, to liberate women who had been subjugated by the Taliban regime. Ironically, the position of women in Afghanistan has improved little in the last several years, leading many Muslim women to reject "Western-style feminism" and embrace Muslim feminism.

> Muslim women deplore misogyny just as Western women do, and they know that Islamic societies also oppress them; why wouldn't they? But liberation for them does not encompass destroying their identity, religion, or culture, and many of them want to retain the veil. (Viner 2002, p. 2)

Other differences also exist. Muslim feminism is based on the teachings of Islam, is profamily, and rejects the concept of patriarchy (McElroy 2003).

Generation X and the Military

Research indicates that attitudes and values differ between generations. For example, members of Generation X—that is, Americans born in the post–baby boom years—are often thought to be more materialistic, less civically engaged, and less socially trusting compared to previous generations. Rarely studied, however, are the differences within a generation; that is, are there different mind-sets within the same age cohort? Thus, Franke (2001) compares the values and attitudes of cadets at West Point with the values and attitudes of a sample of college students at Syracuse University. The research addresses an important question: To what extent are the attitudes and values of future military officers representative of the civilian population?

Sample and Methods

In the fall of 1995 the Future Officer Survey was administered anonymously to a representative sample of West Point cadets (N = 1,233) representing 31% of the total student population. The response rate was 48.2%, yielding 594 usable questionnaires. A revised version of the same survey, with specific references to military duties and West Point deleted or modified, was administered to 372 students in upper- and lower-division social science classes at Syracuse University (SU).

Using a five-point Likert scale (from "strongly agree" to "strongly disagree") respondents were asked their level of agreement or disagreement with 36 state-

ments, each one an indicator of a dependent variable. The dependent variables include the following: (1) political conservatism (e.g., "The government should provide health insurance for every American"), (2) patriotism (e.g., "All Americans should be willing to fight for their country"), (3) warriorism (e.g., "Sometimes war is necessary to protect the national interest"), and (4) globalism and global institutions (e.g., "World government is the best way to ensure international peace"). A fifth dependent variable was measured by asking respondents whether they defined themselves as conservative or liberal. The independent variables included sample type (i.e., comparing cadets to SU students) and class level in school (e.g., comparing first-year cadets to fourth-year cadets).

Several hypotheses were generated from the literature. First, Franke (2001) hypothesizes that West Point cadets will have higher levels of political conservatism, patriotism, and warriorism than SU students. Second, Franke hypothesizes that SU students will be more likely to agree with statements that advocate peacekeeping operations and that are supportive of globalism and global institutions. Finally, Franke hypothesizes that such differences will be magnified by years in school. Thus, for example, Franke predicts that senior cadets will be more patriotic than first-year cadets and that SU students in their fourth year will be more liberal than those in their first year.

Findings and Conclusions

Looking at statistical differences between and within samples yields some interesting results.

1. *Political conservatism.* Compared to first-year cadets, senior cadets were significantly more likely to agree with only one of the four indicators of political conservatism ("The United States has gone too far in providing for equal opportunity under the law"). Contrary to predictions, the overall measure of political conservatism did not vary significantly between year in school for cadets or SU students. Between-sample differences were highly significant, however, with cadets, as expected, having higher rates of political conservatism than SU students.

2. *Patriotism.* Contrary to predictions, cadets' patriotism scores *decreased* between class cohorts; that is, seniors were significantly less rather than more patriotic than first-year cadets. Between-sample comparisons were also significant. Cadets on each of the six indicators of patriotism scored significantly higher than SU students. For example, 90% of cadets agreed with the statement that "an American should always feel that his or her primary allegiance is to his or her country," whereas only 75% of SU students felt similarly.

3. *Warriorism.* Warriorism increased across class year for West Point cadets. Alternatively, in general, SU students'

Symbolic Interactionist Perspective

The symbolic interactionist perspective focuses on how meanings and definitions influence attitudes and behaviors regarding conflict and war (see this chapter's *Social Problems Research Up Close* feature). The development of attitudes and behaviors that support war begins in childhood. American children learn to glorify and

levels of warriorism decreased as class increased, although statistical significance was reached for only one of the seven indicators of warriorism. Twice as many SU first-year students (43%) as seniors (21%) agreed with the statement "The most important role of the U.S. military is preparation for and conduct of war." Not surprisingly, between-sample results were significant, with cadets scoring much higher on this dependent variable than SU students.

4. *Globalism and global institutions.* Support for globalism and global institutions decreased between class cohorts for cadets and increased, although not significantly, between classes for SU students. Between-sample differences were not significant. Overall, neither cadets nor SU students were supportive of the ideals of globalism and global institutions such as a multinational military or the United Nations.

Finally, respondents were asked about their political affiliation. Cadets, in general, and senior cadets in particular, most often categorized themselves as conservative. For example, 58.6% of first-year cadets and 67.7% of senior cadets self-identified as conservatives. Interestingly, SU students' political affiliation was essentially the opposite. First-year students (49.6%) as well as seniors (60.8%) defined themselves as liberal.

The results of Franke's study, although mixed, generally support the hypotheses. On average, West Point cadets were more politically conservative, patriotic, and warrioristic than civilian students at a private university. They also were less supportive of globalism and global institutions. Differences also existed within samples, with, for example, self-identified conservatism increasing between cadets' class level—from first to fourth year.

There are two explanations for the results. The first is called self-selection. The self-selection model argues that people who attend military academies are more politically conservative, patriotic, and warrioristic *before* they enter the academy; that is, they select a military academy because it is consistent with their attitudes and values. The fact that first-year cadets had higher rates of warriorism than first-year SU students would support this perspective. The second model is called the socialization model; it holds that the longer an individual is socialized into an institution, the more that institution influences the individual's attitudes and values. Thus, for example, SU students were more liberal than cadets upon entering college (self-selection), but the extent of their liberal thinking increased with years at SU (socialization).

There are, however, several limitations of the study that need to be addressed. First, Syracuse University is a private university, and the students who attend it may not be representative of all "Gen X" college students. Future research needs to look at students from both private and public colleges and universities. Second, variables other than class position may be responsible for the results. As Franke notes, differences in maturity may explain the within-sample findings. Seniors not only have undergone more socialization at their respective institutions but also have had more "life experiences," and it is these life experiences that may be responsible for the findings. Finally, the research is cross-sectional in nature, which means that it looked at only one point in time. It did not follow respondents through their academic careers but measured different respondents in different class cohorts and then compared the results. Future research should be longitudinal in nature; that is, it should measure the same people at different points in time.

Given these limitations, it is difficult to answer the question posed at the beginning of this article. As Franke (2001) notes:

> While the present analysis cannot provide a definitive answer, the data indicate that a gap between West Point cadets and their civilian generational peers exists already prior to military socialization. The observed correlations between value orientation and the general sociopolitical views could be another manifestation of the growing politicization of the U.S. military. The data also suggest that socialization may widen the gap. (p. 116)

Franke (2001) concludes that his research supports the notion of service academies but, at the same time, questions their role in an increasingly "global security environment."

celebrate the Revolutionary War, which created our nation. Movies romanticize war, children play war games with toy weapons, and various video and computer games glorify heroes conquering villains. Indeed, from 1938 to 1942 a series of "Horrors of Wars" cards were manufactured and distributed in the United States and collected by millions of American youth, much like baseball cards (Nelson 1999).

"Why do we kill people who are killing people to show that killing people is wrong?"

Holly Near
Singer-songwriter

The face of patriotism is changing. Although, traditionally, older Americans have been the most patriotic, the spread of patriotism has been extending to all ages. For example, a recent survey of 2005 college graduates found that 83% defined themselves as patriotic.

Symbolic interactionism helps to explain how military recruits and civilians develop a mind-set for war by defining war and its consequences as acceptable and necessary. The word *war* has achieved a positive connotation through its use in various phrases—the war on drugs, the war on poverty, and the war on crime. Positive labels and favorable definitions of military personnel facilitate military recruitment and public support of armed forces. Recently, the Army National Guard launched a $38 million marketing campaign targeting young men and women with advertisements showing "troops with weapons drawn, helicopters streaking and tanks rolling," all "in an attempt to remind people what the Guard has been about since Colonial Days: fighting wars and protecting the homeland." The new slogan? "The most important weapon in the war on terrorism. You" (Davenport 2005, p. 1).

Many government and military officials convince the masses that the way to ensure world peace is to be prepared for war. Most world governments preach peace through strength rather than strength through peace. Governments may use propaganda and appeals to patriotism to generate support for war efforts and to motivate individuals to join armed forces. Salladay (2003), for example, notes that those in favor of the war on Iraq have commandeered the language of patriotism, making it difficult but necessary for peace activists to use the same symbols or phrases.

To legitimize war, the act of killing in war is not regarded as "murder." Deaths that result from war are referred to as casualties. Bombing military and civilian targets appears more acceptable when nuclear missiles are "peacekeepers" that are equipped with multiple "peace heads." Killing the enemy is more acceptable when derogatory and dehumanizing labels such as Gook, Jap, Chink, Kraut, and Haji convey the attitude that the enemy is less than human.

Such labels are often socially constructed as images, often through the media, and are presented to the public. Social constructionists, like symbolic interactionists in general, emphasize the social aspects of "knowing." Thus Li and Izard (2003) used content analysis to analyze newspaper and television coverage of the World Trade Center and Pentagon attacks on September 11, 2001. The researchers examined the first eight hours of coverage of the attacks presented on CNN, ABC, CBS, NBC, and FOX as well as in eight major U.S. newspapers (including the *Los Angeles Times,* the *New York Times,* and the *Washington Post*). Results of the analysis indicate that newspaper articles tended to have a "human interest" emphasis, whereas television coverage was more often "guiding and consoling." Other results suggest that both media relied most heavily on government sources, that newspapers and the networks were equally factual, and that networks were more homogeneous in their presentation than newspapers. One indication of the importance of the media lies in President George W. Bush's creation of the Office of Global Communications— "a huge production company, issuing daily scripts on the Iraq war to U.S. spokesmen around the world, auditioning generals to give media briefings, and booking administration stars on foreign news shows" (Kemper 2003, p. 1).

Causes of War

The causes of war are numerous and complex. Most wars involve more than one cause. The immediate cause of a war may be a border dispute, for example, but religious tensions that have existed between the two combatant countries for decades may also contribute to the war. The following section reviews various causes of war.

Conflict over Land and Other Natural Resources

Nations often go to war in an attempt to acquire or maintain control over natural resources, such as land, water, and oil. Michael Klare, author of *Resource Wars: The New Landscape of Global Conflict* (2001), predicts that wars will increasingly be fought over resources as supplies of the most needed resources diminish. Disputed borders are one of the most common motives for war. Conflicts are most likely to arise when borders are physically easy to cross and are not clearly delineated by natural boundaries, such as major rivers, oceans, or mountain ranges.

Water is another valuable resource that has led to wars. Unlike other resources, water is universally required for survival. At various times the empires of Egypt, Mesopotamia, India, and China all went to war over irrigation rights. In 1998, five years after Eritrea gained independence from Ethiopia, forces clashed over control of the port city Assab and with it, access to the Red Sea.

Not only do the oil-rich countries in the Middle East present a tempting target in themselves, but war in the region can also threaten other nations that are dependent on Middle Eastern oil. Thus, when Iraq seized Kuwait and threatened the supply of oil from the Persian Gulf, the United States and many other nations reacted militarily in the Gulf War. In a document prepared for the Center for Strategic and International Studies, Starr and Stoll (1989) warn that soon

> water, not oil, will be the dominant resource issue of the Middle East. According to World Watch Institute, "despite modern technology and feats of engineering, a secure water future for much of the world remains elusive." The prognosis for Egypt, Jordan, Israel, the West Bank, the Gaza Strip, Syria, and Iraq is especially alarming. If present consumption patterns continue, emerging water shortages, combined with a deterioration in water quality, will lead to more competition and conflict. (p. 1)

Despite such predictions, tensions in the Middle East changed dramatically following the death of Palestine Liberation Organization (PLO) leader Yasser Arafat in 2004. With new leadership in place, Israeli prime minister Ariel Sharon and newly elected Palestinian leader Mahmoud Abbas have agreed to peace talks, which will include discussion about removing Jewish settlers from the Jordan River basin area.

Conflict over Values and Ideologies

Many countries initiate war not over resources but over beliefs. World War II was largely a war over differing political ideologies: democracy versus fascism. The Cold War involved the clash of opposing economic ideologies: capitalism versus communism. Wars over differing religious beliefs have led to some of the worst episodes of bloodshed in history, in part, because some religions are partial to martyrdom—the idea that dying for one's beliefs leads to eternal salvation. The Shiites (one of the two main sects within Islam) in the Middle East represent a classic example of holy warriors who feel divine inspiration to kill the enemy.

Conflicts over values or ideologies are not easily resolved. The conflict between secularism and Islam has lasted for 14 centuries. Conflict over values and ideologies are less likely to end in compromise or negotiation because they are fueled by people's convictions. For example, when a representative sample of American Jews were asked, "Do you agree or disagree with the following statement? 'The goal of Arabs is not the return of occupied territories but rather the destruction of Israel,'" 84% agreed, 13% disagreed, and 3% were unsure (American Jewish Committee 2005).

If ideological differences can contribute to war, do ideological similarities discourage war? The answer seems to be yes; in general, countries with similar ideologies are less likely to engage in war with each other than countries with differing ideological values (Dixon 1994). Democratic nations are particularly disinclined to wage war against one another (Doyle 1986).

Racial, Ethnic, and Religious Hostilities

Racial, ethnic, and religious groups vary in their cultural beliefs, values, and traditions. Thus conflicts between racial, ethnic, and religious groups often stem from conflicting values and ideologies. Such hostilities are also fueled by competition over land and other scarce natural and economic resources. Gioseffi (1993) notes that "experts agree that the depleted world economy, wasted on war efforts, is in great measure the reason for renewed ethnic and religious strife. 'Haves' fight with 'have-nots' for the smaller piece of the pie that must go around" (p. xviii). Racial, ethnic, and religious hostilities are also perpetuated by the wealthy majority to divert attention away from their exploitations and to maintain their own position of power.

As described by Paul (1998), sociologist Daniel Chirot argues that the recent worldwide increase in ethnic hostilities is a consequence of "retribalization," that is, the tendency for groups, lost in a globalized culture, to seek solace in the "extended family of an ethnic group" (p. 56). Chirot identifies five levels of ethnic conflict: (1) multiethnic societies without serious conflict (e.g., Switzerland), (2) multiethnic societies with controlled conflict (e.g., United States, Canada), (3) societies with ethnic conflict that has been resolved (e.g., South Africa), (4) societies with serious ethnic conflict leading to warfare (e.g., Sri Lanka), and (5) societies with genocidal ethnic conflict, including "ethnic cleansing" (e.g., Kosovo).

Religious differences as a source of conflict have recently come to the forefront. An Islamic jihad, or holy war, has been blamed for the September 11 attacks on the World Trade Center and Pentagon as well as for bombings in Kashmir, Sudan, the Philippines, Kenya, Tanzania, Saudi Arabia, Spain, and Great Britain. Some claim that Islamic beliefs in and of themselves have led to recent conflicts (Feder 2003). Others contend that religious fanatics, not the religion itself, are responsible for violent confrontations. For example, Islamic leader Osama bin Laden claims that unjust U.S. Middle East policies are responsible for "dividing the whole world into two sides—the side of believers and the side of infidels" (Williams 2003, p. 18). Despite bin Laden's decree, the terrorist activity in Iraq is increasingly sectarian as Sunni Arabs rebel against the Shiite Arab–dominated government (Tavernise 2005).

Defense Against Hostile Attacks

The threat or fear of being attacked may cause the leaders of a country to declare war on the nation that poses the threat. The threat may come from a foreign country or from a group within the country. After Germany invaded Poland in 1939, Britain and France declared war on Germany out of fear that they would be Germany's next victims. Germany attacked Russia in World War I, in part out of fear that Russia had entered the arms race and would use its weapons against Germany. Japan bombed Pearl Harbor hoping to avoid a later confrontation with the U.S. Pacific fleet, which posed a threat to the Japanese military. In 2001 a U.S.-led coalition bombed Afghanistan in response to the September 11 terrorist attacks. Moreover, in March 2003 the United States, Great Britain, and a loosely coupled "coalition of the willing" invaded Iraq in response to perceived threats of weapons of mass destruction and the reported failure of Saddam Hussein to cooperate with UN weapons inspectors. Yet, in 2005 a presidential commission concluded that the attack on Iraq was based on faulty intelligence and that, in fact, "America's spy agencies were 'dead wrong' in most of their judgments about Iraq's weapons of mass destruction" (Shrader 2005, p. 1). As a result, many Americans, nearly 60%, favor a partial or complete withdrawal from Iraq (Jones 2005).

Revolution

Revolutions involve citizens warring against their own government and often result in significant political, economic, and social change. A revolution may occur when a government is not responsive to the concerns and demands of its citizens and when strong leaders are willing to mount opposition to the government (Barkan & Snowden 2001; Renner 2000).

The birth of the United States resulted from colonists revolting against British control. Contemporary examples of civil war include Sri Lanka, where the Tamils, a separatist group living in the northern region of the country, have been at war with the Sri Lankan government for 22 years. The war has resulted in more than 64,000 deaths and has often been "confined to jungle skirmishes, government air strikes and ambushes by the Liberation of Tamil Tigers (LTTE) which want to carve a separate Tamil state out of Sri Lanka's north and east" (Reuters 2001).

Similarly, Liberia's 14-year on-again off-again civil war pits Liberians United for Reconciliation and Democracy rebels against government officials. In 2003, under heavy pressure from the United States, the president of Liberia resigned. His exile has led to talks between warring factions, the arrival of much-needed U.S. supplies and 15,000 United Nations peacekeepers, and the beginnings of an interim government (Global Security 2005; Sengupta 2003). Civil wars have also erupted in newly independent republics created by the collapse of communism in Eastern Europe, as well as in Rwanda, Sierra Leone, Chile, Uganda, and Sudan.

"Those who make peaceful evolution impossible, make violent revolution inevitable."

John F. Kennedy
Former U.S. president

ThomsonNOW

Learn more about **Revolution** by going through the Revolutions Animation.

Nationalism

Some countries engage in war in an effort to maintain or restore their national pride. For example, Scheff (1994) argues that "Hitler's rise to power was laid by the treatment Germany received at the end of World War I at the hands of the victors" (p. 121). Excluded from the League of Nations, punished by the Treaty of Versailles, and ostracized by the world community, Germany turned to nationalism as a reaction to material and symbolic exclusion. Furthermore, some observers note that despite the *official* end of the war in Iraq and, with it, Saddam Hussein's rule, the dictator has "become a symbol of nationalist resistance to forces opposing the U.S.-led occupation of Iraq" (Morahan 2003, p. 1).

In the late 1970s Iranian fundamentalist groups took hostages from the U.S. embassy in Iran. President Carter's attempt to use military forces to free the hostages was not successful. That failure intensified doubts about America's ability to use military power effectively to achieve its goals. The hostages in Iran were eventually released after President Reagan took office, but doubts about the strength and effectiveness of the U.S. military still called into question America's status as a world power. Subsequently, U.S. military forces invaded the small island of Grenada because the government of Grenada was building an airfield large enough to accommodate major military armaments. U.S. officials feared that this airfield would be used by countries in hostile attacks on the United States. From one point of view, the large-scale "successful" attack on Grenada functioned to restore faith in the power and effectiveness of the U.S. military.

Terrorism

Terrorism is the premeditated use, or threatened use, of violence by an individual or group to gain a political or social objective (Barkan & Snowden 2001; Brauer 2003; Interpol 1998). Terrorism may be used to publicize a cause, promote an ideology, achieve religious freedom, attain the release of a political prisoner, or rebel against a government. Terrorists use a variety of tactics, including assassinations, skyjackings, suicide bombings, armed attacks, kidnapping and hostage taking, threats, and various forms of bombings. Despite President Bush's declaration of a "war on terrorism," unlike war where there is a winner and a loser, terrorism is unlikely to be completely defeated.

> There can be no final victory in the fight against terrorism, for terrorism (rather than full-scale war) is the contemporary manifestation of conflict, and conflict will not disappear from earth as far as one can look ahead and human nature has not undergone a basic change. But it will be in our power to make life for terrorists and potential terrorists much more difficult. (Laqueur 2006, p. 173)

Table 16.1 highlights U.S. responses to terrorist activity.

Types of Terrorism

Terrorism can be either transnational or domestic. **Transnational terrorism** occurs when a terrorist act in one country involves victims, targets, institutions, governments, or citizens of another country. The 1988 bombing of Pan Am Flight 103 over Lockerbie, Scotland, exemplifies transnational terrorism. The incident took the lives of 270 people and, after a 10-year investigation, resulted in the life sentence of a Libyan intelligence agent (CNN 2001a). In 2003 the Libyan government agreed to pay $2.7 billion in compensation to the victims' families (Smith 2004). Other examples

Table 16.1

U.S. Policy Responses to Terrorism

- **Diplomacy and Constructive Engagement**—Using diplomacy, including verbal contact and direct negotiations, to create a "global antiterror coalition."
- **Economic Sanctions**—Banning or threatening to ban, for example, trade and investment relations with a nation as a means of control. Other economic sanctions include freezing bank accounts, imposing trade embargos, and suspending foreign aid.
- **Economic Inducements**—Developing assistance programs to reduce poverty and literacy in countries that are breeding grounds for terrorist activity.
- **Covert Action**—Infiltrating terrorist groups, military operations, and intelligence gathering. Also included in this category are sabotaging weapons facilities and capturing wanted terrorists.
- **Rewards for Information Programs**—Exchanging money for information on or the capture of a wanted terrorist. For example, the reward for capturing Osama bin Laden is $25 million.
- **Extradition/Law Enforcement Cooperation**—Enlisting the worldwide cooperation of law enforcement agencies, including international extradition of terrorists.
- **Military Force**—Using military force in fighting terrorism. For example, military force was used to successfully overthrow the Taliban in Afghanistan.
- **International Conventions**—Entering into agreements with other countries that obligate the signatories to conform to the articles of the convention, including, for example, prosecution and extradition of terrorists.

Source: Perl (2003).

On July 7, 2005, four bombs exploded during a London rush hour, killing more than 50 people and wounding more than 700. All the suspects were British citizens, thus setting off fears of a new brand of terrorists—"home-grown." Ties to Al Qaeda have been confirmed.

© Sion Touhig/Corbis

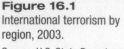

Figure 16.1
International terrorism by
region, 2003.

Source: U.S. State Department
(2004).

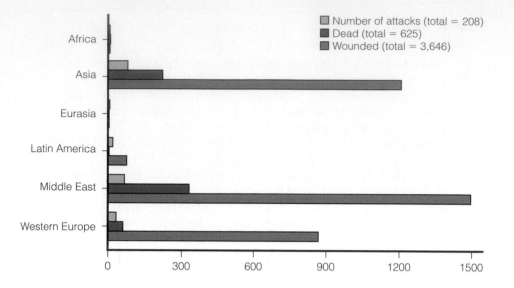

of transnational terrorism include attacks on U.S. embassies in Kenya and Tanzania and the bombing of a naval ship, the USS *Cole,* moored in Aden Harbor, Yemen. The 2001 attacks on the World Trade Center, the Pentagon, and Flight 93 are also examples of transnational terrorism. The September 11 attacks, linked to Al Qaeda and its leader, Osama Bin Laden, are the most devastating acts of terrorism in U.S. history.

Domestic terrorism, sometimes called insurgent terrorism (Barkan & Snowden 2001), is exemplified by the 1995 truck bombing of a nine-story federal office building in Oklahoma City, resulting in 168 deaths and the injury of more than 200 people. Gulf War veteran Timothy McVeigh and Terry Nichols were convicted of the crime. McVeigh is reported to have been a member of a paramilitary group that opposes the U.S. government. In 1997 McVeigh was sentenced to death for his actions, and he was executed in 2001 (Barnes 2004). The 2004 bombing of a Russian school, which killed 324 people, nearly half of them children, is also an act of domestic terrorism, as Chechen rebels continue to fight for an independent state.

Patterns of Global Terrorism

Although estimates vary, a 2004 report by the U.S. State Department describes patterns of terrorism around the world (U.S. State Department 2004). In 2003 (see Figure 16.1):

- There were 208 transnational acts of terrorism—a slight increase since 2002.
- The number of deaths from terrorist acts was 625, a decrease from the previous year's number of 725.
- A total of 3,646 people were wounded, representing "numerous indiscriminate attacks" (e.g., commercial districts, places of worship).
- Thirty-five U.S. citizens were killed in terrorist attacks.
- The Middle East had the highest rate of terrorist attacks, followed by Asia and Western Europe.
- The Middle East also had the highest number of deaths, followed by Asia and Western Europe.

Counterterrorism officials report that the number of terrorist threats against the United States has decreased 20–50% over the last two years, and they are presently at their lowest levels since the September 11, 2001, attacks (Priest & Hsu 2005). Despite the decline, concern with terrorism remains. When a random sample of U.S. adults were asked, "Who do you think is currently winning the war against terrorism?" 37% responded the United States and its allies. However, 42% responded "neither side" and 20% responded "the terrorists" (Gallup 2005b).

The Roots of Terrorism

Walter Laqueur's book, *No End to War: Terrorism in the 21st Century* (2003), dispels the myths that poverty and political oppression give birth to terrorist activity. As Laqueur notes, almost no terrorism has occurred in the world's poorest countries. Further, the most repressive regimes of the 21st century, Hitler's Germany, Russia under Stalin, and Mussolini's fascist Italy, were relatively free of terrorism.

Laqueur's research also suggests that terrorism flourishes in democracies. What, then, are the causes of terrorism? In 2003 a panel of terrorist experts came together in Oslo, Norway, to address the causes of terrorism (Bjorgo 2003). Although not an exhaustive list, several causes emerged from the conference:

- A failed or weak state, which is unable to control terrorist operations.
- Rapid modernization, when, for example, a country's sudden wealth leads to rapid social change.
- Extreme ideologies—religious or secular.
- A history of political violence, civil wars, and revolutions.
- Repression by a foreign occupation (i.e., invaders to the inhabitants).
- Large-scale racial or ethnic discrimination.
- The presence of a charismatic leader.

Note that Iraq, part of what President George W. Bush calls the "axis of evil," has several of the characteristics listed here, including rapid modernization (e.g., oil reserves), extreme ideologies (e.g., Islamic fundamentalism), a history of violence (e.g., invasion of Kuwait), and a weak state that is unable to control terrorist operations (e.g., the newly elected Iraqi government).

The causes of terrorism listed here, however, are macro in nature. What of social-psychological variables? Borum (2003) suggests a four-stage microlevel process. The decision to commit a terrorist act begins with an individual's assessment that *something is not right* (e.g., government-imposed restrictions). Next, individuals *define the situation as unfair* in that the "not right" condition does not apply to everyone (e.g., government-imposed restrictions are imposed on some but not on others). Individuals then begin to *blame specific others for the injustice* (e.g., government leaders) and, finally, to *redefine those who are responsible for the injustice as bad or evil* (e.g., a fascist regime). This process of "ideological development," as portrayed in Figure 16.2, often leads to stereotyping and dehumanizing the enemy, which then facilitates violence. Although the process was developed as a heuristic device, Borum notes that "understanding the mind-set" of a terrorist can help in the fight against terrorism.

America's Response to Terrorism

A government can use both defensive and offensive strategies to fight terrorism. Defensive strategies include using metal detectors and X-ray machines at airports and strengthening security at potential targets, such as embassies and military command posts. The newly created Department of Homeland Security, made up of

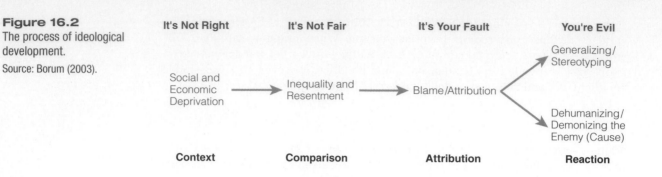

Figure 16.2
The process of ideological development.

Source: Borum (2003).

It's Not Right   It's Not Fair   It's Your Fault   You're Evil

Social and Economic Deprivation → Inequality and Resentment → Blame/Attribution

Generalizing/Stereotyping

Dehumanizing/Demonizing the Enemy (Cause)

Context   Comparison   Attribution   Reaction

22 domestic agencies (e.g., the U.S. Coast Guard, the Immigration and Naturalization Service, the Secret Service) and 180,000 employees, coordinates such defensive tactics. The new department's first priority is to "protect the nation against further terrorist attacks," as "component agencies analyze threats and intelligence, guard our borders and airports, protect our critical infrastructure, and coordinate the response of our nation for future emergencies" (U.S. Department of Homeland Security 2003, p. 1).

Offensive strategies include retaliatory raids, such as the U.S. bombing of terrorist facilities in Afghanistan, the invasion of Iraq, group infiltration, and preemptive strikes. New legislation facilitates such offensive tactics. In October 2001 the USA Patriot Act (Uniting and Strengthening America by Providing Appropriate Tools Required to Intercept and Obstruct Terrorism) was signed into law. The act increases police powers both domestically and abroad. Critics hold that the act poses a danger to civil liberties.

For example, the act provides for the *indefinite* detention of immigrants if the immigrant group is defined as a "danger to national security" (Romero 2003). Thus some 540 prisoners, many without charge, are now being held at the Guantanamo Bay prison camp in Cuba, where at least 10 cases of abuse or mistreatment have been reported and several others are pending (Dodds 2005). Furthermore, abuses of Iraqi prisoners at Abu Ghraib in Baghdad have led to several court-martials and convictions of military personnel. Table 16.2 contains the results of a national survey of 1,464 Americans who were asked whether they thought the reports of mistreatment at Guantanamo Bay were "isolated incidents" or a "wider pattern" of abuse.

Advocates of the USA Patriot Act argue that during war, some restrictions of civil liberties are necessary. Moreover, the legislation is not "a substantive shift in policy but a mere revitalization of already established precedents" (Smith 2003, p. 25). Finally, proponents note that many of the provisions of the Patriot Act are set to expire at the end of 2005, making the critics' predictions of a 1984 Orwellian doom-and-gloom scenario absurd. Nonetheless, there is some evidence that the renewal of the Patriot Act's "sunset provisions" are not a foregone conclusion. For example, the 9/11 Commission report recommends that "a full and informed debate on the Patriot Act would be healthy" (Congressional Research Service 2005). This chapter's *Taking a Stand* feature asks your opinion about antiterrorism legislation.

Combating terrorism is difficult, and recent trends will make it increasingly problematic (Strobel et al. 2001; Zakaria 2000). First, data stored on computers can be easily acquired by hackers who illegally gain access to classified information. Interlopers obtained the fueling and docking schedules of the USS *Cole*. Second, the Internet permits groups with similar interests, once separated by geography, to share plans, fund-raising efforts, recruitment strategies, and other coordinated efforts.

Table 16.2

Reports of Prisoner Mistreatment at Guantanamo Bay

	Isolated Incidents (%)	Wider Pattern (%)	Neither/ Don't Know (%)
Total	54	34	12
Men	56	64	10
Women	52	34	14
18–29 years old	43	46	11
30–49 years old	55	36	9
50–64 years old	59	28	13
65 years old and older	56	25	19
White	57	31	12
Black	35	52	13
Hispanic	44	45	11
Republican	76	14	10
Democrat	43	45	12
Independent	45	44	11

Source: Adapted from Pew Research Center (2005).

AFP/Getty Images

The prisoner abuses at Guantanamo Bay (Cuba) and Abu Ghraib (Iraq) shocked the nation. Several of the photos, transmitted around the world on the Internet, were made into billboards in Arab countries. Army Pfc. Lynndie England, pictured, was convicted on six of seven counts involving prisoner mistreatment and was sentenced to three years in prison and given a dishonorable discharge.

Does Antiterrorism Legislation Threaten Civil Liberties?

After September 11, 2001, the Bush administration passed several laws in the fight against terrorism. Some of these measures allow for covert government surveillance, detention of immigrant groups, tracking of e-mail, and roving wiretaps. Concern over individual liberties has been heightened by the accusations of abuses at Guantanamo Bay and the horrors of Abu Ghraib. Although many civil libertarians believe that the new laws violate citizens' rights, advocates argue that during war, measures must be taken to protect national security.

It Is Your Turn Now to Take a Stand!

What do you think? Under what circumstances, if any, should civil liberties be abridged?

Worldwide, thousands of terrorists keep in touch through Hotmail.com e-mail accounts. Third, globalization contributes to terrorism by providing international markets where the tools of terrorism—explosives, guns, electronic equipment, and the like—can be purchased. Finally, fighting terrorism under guerrilla warfare–like conditions is increasingly a concern. Unlike terrorist activity, which targets civilians and may be committed by lone individuals, **guerrilla warfare** is committed by organized groups opposing a domestic or foreign government and its military forces. Guerrilla warfare often involves small groups who use elaborate camouflage and underground tunnels to hide until they are ready to execute a surprise attack. In December 2004 an estimated 20,000 Iraqi insurgents were fighting U.S. and allied forces ("Iraq: Status Report," 2005).

The possibility of terrorists using weapons of mass destruction is the most frightening scenario of all and, as stated earlier, the motivation for the 2003 war with Iraq. **Weapons of mass destruction (WMD)** include chemical, biological, and nuclear weapons. Anthrax, for example, although usually associated with diseases in animals, is a highly deadly disease in humans, and, although preventable by vaccine, it has a "lethal lag time." In a hypothetical city of 100,000 people, delaying a vaccination program 1 day would result in 5,000 deaths; 6 days, 35,000 deaths. In 2001 trace amounts of anthrax were found in several letters sent to media and political figures, resulting in five deaths and the inspection and closure of several postal facilities (Baliunas 2004).

Other examples of the use of WMD exist. On at least eight occasions Japanese terrorists dispersed aerosols of anthrax and botulism in Tokyo (Inglesby et al. 1999), and in 2000 a religious cult, hoping to disrupt elections in an Oregon county, "contaminated local salad bars with salmonella, infecting hundreds" (Garrett 2001, p. 76). Furthermore, in 2003 the poison ricin was detected on a mail-opening machine in Senate majority leader Bill Frist's Washington, D.C., office (Associated Press 2005). Government officials have initiated a variety of laws, policies, and technological innovations designed to combat WMD. For example, federal funds have been used to develop an experimental medicine tentatively called BCTP. The drug has successfully protected mice from injections of anthrax-like bacteria (Stipp 2004).

Social Problems Associated with Conflict, War, and Terrorism

Social problems associated with conflict, war, and terrorism include death and disability; rape, forced prostitution, and displacement of women and children; social-psychological costs; diversion of economic resources; and destruction of the environment.

Death and Disability

Many American lives have been lost in wars, including 53,000 in World War I, 292,000 in World War II, 34,000 in Korea, and 47,000 in Vietnam (U.S. Bureau of the Census 2004). More recently, between March 2003 and May 2005, 13,000 U.S. troops have been wounded and 1,700 have been killed in Iraq. Furthermore, 24,000 civilian deaths have been recorded during the same time period (Barry et al. 2005). Globalization and sophisticated weapons technology combined with increased population density have made it easier to kill a large number of people in a short amount of time. When the atomic bomb was dropped on the Japanese cities of Hiroshima and Nagasaki during World War II, 250,000 civilians were killed.

The impact of war and terrorism extends far beyond those who are killed. Many of those who survive war incur disabling injuries or contract diseases. For example, 1 million people worldwide have been killed or disabled by landmines—a continuing problem in the aftermath of the 2003 war with Iraq (Renner 2005). In 1997 the Mine Ban Treaty, which requires that governments destroy stockpiles within 4 years and clear landmine fields within 10 years, became international law. To date, 144 countries have signed the agreement; 42 countries remain, including China, India, Israel, Russia, and the United States (International Campaign to Ban Landmines 2005). This chapter's *Self and Society* feature tests your knowledge of the subject.

War-related deaths and disabilities also deplete the labor force, create orphans and single-parent families, and burden taxpayers who must pay for the care of orphans and disabled war veterans.

Individuals who participate in experiments for military research may also suffer physical harm. Representative Edward Markey of Massachusetts identified 31 experiments dating back to 1945 in which U.S. citizens were subjected to harm from participation in military experiments. Markey charged that many of the experiments used human subjects who were captive audiences or populations considered "expendable," such as the elderly, prisoners, and hospital patients. Eda Charlton of New York was injected with plutonium in 1945. She and 17 other patients did not learn of their poisoning until 30 years later. Her son, Fred Shultz, said of his deceased mother:

> I was over there fighting the Germans who were conducting these horrific medical experiments . . . at the same time my own country was conducting them on my own mother. (Miller 1993, p. 17)

Rape, Forced Prostitution, and Displacement of Women and Children

Half a century ago the Geneva Convention prohibited rape and forced prostitution in war. Nevertheless, both continue to occur in modern conflicts.

Before and during World War II the Japanese military forced 100,000 to 200,000 women and teenage girls into prostitution as military "comfort women." These women were forced to have sex with dozens of soldiers every day in "comfort stations." Many of the women died as a result of untreated sexually transmitted diseases, harsh punishment, or indiscriminate acts of torture.

"The use of rape in conflict reflects the inequalities women face in their everyday lives in peacetime. . . . Women are raped because their bodies are seen as the legitimate spoils of war."

Amnesty International

The Landmine Knowledge Quiz

Carefully read each of the following statements, and select the correct answer. When you are finished, compare your answers to those provided and rank your performance using the scale that follows the statements.

1. The most heavily mined regions of the world are in Southeast Asia and Central America. True or False?

2. Worldwide, more than 300 different types of landmines exist. True or False?

3. Each landmine costs
   (a) $1,000–$2,000
   (b) $500–$1,000
   (c) $3–$30
   (d) $100–$500

4. What percentage of landmine victims are civilians?
   (a) 10%
   (b) 60%
   (c) Less than 1%
   (d) Over 80%

5. Landmines remain active long after their intended use, killing or injuring innocent people up to 10 years after they have been deployed. True or False?

6. It takes about a year and a half to train a mine-detection dog. True or False?

7. Metal detectors are effective in identifying buried mines. True or False?

8. Landmines kill or maim over 8,000 children every year. True or False?

9. In Cambodia, 1 out of every 1,236 people is an amputee from a landmine. True or False?

10. Today, approximately 10 million landmines lie in thousands of minefields around the world. True or False?

Answers: (1) False (Africa and the Middle East are the most heavily mined); (2) True; (3) c; (4) d; (5) False (most landmines last forever); (6) True; (7) False (mines are increasingly plastic); (8) True; (9) False (1 out of every 236); (10) False (there are 60–70 million active landmines).

Rank Your Performance: Number Correct

9 or 10,	Excellent
7 or 8,	Good
5 or 6,	Average
3 or 4,	Fair
0, 1, or 2,	Poor

Source: Adapted from information at http://www.landmines.org. Used by permission.

Since 1998, Congolese government forces have fought Uganda and Rwanda rebels. Women have paid a high price for this civil war, in which gang rape is "so violent, so systematic, so common . . . that thousands of women are suffering from vaginal fistula, leaving them unable to control bodily functions and enduring ostracism and the threat of debilitating health problems" (Wax 2003, p. 1). Rapes of Albanian women by Serbian and Yugoslavian paramilitary soldiers were, according to a Human Rights Watch report, "used deliberately as an instrument to terrorize the civilian population, extort money from families, and push people to flee their homes" (quoted in Lorch & Mendenhall 2000, p. 2). Feminist analysis of wartime rape emphasizes that the practice reflects not only a military strategy but also ethnic and gender dominance. In 2001 three Serbian soldiers accused of mass rape and forced prostitution were convicted by a United Nations tribunal and sentenced to a combined total of 60 years in prison (CNN 2001b).

War and terrorism also force women and children to flee to other countries or other regions of their homeland. For example, since 1990, 17 million children have been forced to leave their homeland because of armed conflict (Save the Children 2005). Refugee women and female children are particularly vulnerable to sexual abuse and exploitation by locals, members of security forces, border guards, or other refugees. In refugee camps women and children may also be subjected to sexual violation. A 2003 report by Save the Children examined the treatment of women and children in 40 conflict zones. The use of child soldiers was reported in 70% of the zones, and trafficking of women and girls was reported in 85% of the zones. The most dangerous zones for women and children were Angola, Burundi, Sierra Leone, the Democratic Republic of the Congo, and Afghanistan, where the lives of 4 million women and 6 million children are endangered (Save the Children 2003).

Social-Psychological Costs

Terrorism, war, and living under the threat of war interfere with social-psychological well-being and family functioning. In a study of 269 Israeli adolescents, Klingman and Goldstein (1994) found a significant level of anxiety and fear, particularly among younger females, with regard to the possibility of nuclear and chemical warfare. Similarly, Myers-Brown and colleagues (2000) report that Yugoslavian children suffer from depression, anxiety, and fear as a response to conflicts in that region. Children, as well as adults, had similar emotional responses to the September 11 attacks on the World Trade Center and Pentagon (NASP 2003). Whether it is war or terrorism, "virtually every aspect of a child's development is damaged in such circumstances" (Bellamy 2003).

Guerrilla warfare is particularly costly in terms of its psychological toll. In Iraq, "guerrilla insurgents attack with impunity. Friends and foes are indistinguishable, creating the need for constant vigilance. Death is always lurking, and can come from hand grenades thrown by children, earth-rattling bombs in suicide trucks, or snipers hidden in bombed-out buildings" (Waters 2005, p. 1). As a result, as many as 24 U.S. soldiers have committed suicide in Iraq, and more than 900 have been evacuated for psychiatric reasons. In addition, between 2001 and 2004 the divorce rates of active-duty army officers and enlisted personnel have almost doubled (Crary 2005).

Military personnel who engage in combat and civilians who are victimized by war may experience a form of psychological distress known as **post-traumatic stress disorder** (PTSD), a clinical term referring to a set of symptoms that can result from any traumatic experience, including crime victimization, rape, or war. Symptoms of PTSD include sleep disturbances, recurring nightmares, flashbacks, and poor concentration (National Center for Post Traumatic Stress Disorder 2005). For example, Canadian Lt. General Romeo Dallaire, head of the UN peacekeeping mission in Rwanda, witnessed horrific acts of genocide. Four years after his return he continued to have images of "being in a valley at sunset, waist deep in bodies, covered in blood" (quoted in Rosenberg 2000, p. 14). PTSD is also associated with other personal problems, such as alcoholism, family violence, divorce, and suicide.

One study estimates that about 30% of male veterans of the Vietnam War have experienced PTSD and that about 15% continue to experience it (Hayman & Scaturo

1993). Another study of 215 Army National Guard and Army Reserve troops who served in the Gulf War and who did not seek mental health services upon return to the states reports that 16–24% exhibited symptoms of PTSD (Sutker et al. 1993). Compared with other civilians, refugees have higher rates of PTSD and depression owing to stressors common to refugee camps (De Jong et al. 2000). Finally, research on PTSD in children reveals that females are generally more symptomatic than males and that PTSD may disrupt the normal functioning and psychological development of children (Cauffman et al. 1998; National Center for Post Traumatic Stress Disorder 2003; Pfefferbaum 1997).

Diversion of Economic Resources

As discussed earlier, maintaining the military and engaging in warfare require enormous financial capital and human support. In 2004 worldwide military expenditures approached $1.4 trillion (Stockholm International Peace Research Institute 2005). This amount exceeds the combined government research expenditures on developing new energy technologies, improving human health, raising agricultural productivity, and controlling pollution. Although military spending by rogue states (Cuba, Iraq, Iran, Libya, North Korea, Sudan, and Syria) was just $14.4 billion in 2002, U.S. allies weighed in at $212 billion in the same year (Center for Defense Information 2003).

Money that is spent for military purposes could be allocated to social programs. The decision to spend $567 million for one Trident II D-5 missile, equal to the operating cost of the Smithsonian Institution (Center for Defense Information 2003), is a political choice. Similarly, allocating $2.3 billion for a "Virginia" attack submarine while our schools continue to deteriorate is also a political choice. If just 3% of the proposed military budget for 2006 was redirected toward domestic issues, (1) 218,361 new teachers could be hired, or (2) health care coverage for 2.8 million uninsured could be provided, or (3) the number of public safety officers could be expanded by 283,904, or (4) 113,451 affordable housing units could be built, or (5) scholarships for 2.5 million university students could be provided (National Priorities Project 2005). As Figure 16.3 indicates, estimated expenditures for the 2005 fiscal year include more money for national defense than for justice, transportation, veterans' benefits, and natural resources and the environment combined (Office of Management and Budget 2005).

Destruction of the Environment

Traditional definitions of and approaches to national security have assumed that political states or groups constitute the principal threat to national security and welfare. This assumption implies that national defense and security are best served by being prepared for war against other states. The environmental costs of military preparedness are often overlooked or minimized in such discussions. Annually, for example, the U.S. Navy Public Works Center in San Diego produces 12,000 tons of paint cans. The depleted cans, enough to fill 12 semitrucks, contain volatile chemicals, making their disposal or any proposed recycling method problematic (Lifsher 1999). A seemingly trivial matter—paint cans, in one year, at one facility. Imagine the environmental problems generated by the military worldwide.

Destruction of the Environment During War. The environmental damage that occurs during war continues to devastate human populations long after war ceases. As the casings for landmines erode, poisonous substances—often

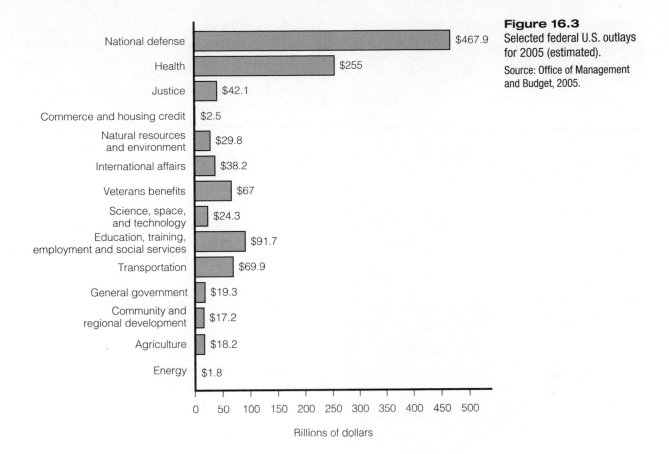

Figure 16.3
Selected federal U.S. outlays for 2005 (estimated).

Source: Office of Management and Budget, 2005.

carcinogenic—leak into the ground (Landmines 2003) In Vietnam 13 million tons of bombs left 25 million bomb craters. Nineteen million gallons of herbicides, including Agent Orange, were spread over the countryside. An estimated 80% of Vietnam's forests and swamplands were destroyed by bulldozing or bombing (Funke 1994).

The Gulf War also illustrates how war destroys the environment (Environmental Media Services 2002; Funke 1994; Renner 1993). In 6 weeks 1,000 air missions were flown, dropping 6,000 bombs. In 1991 Iraqi troops set 650 oil wells on fire, releasing oil, which still covers the surface of the Kuwaiti desert and seeps into the ground, threatening to poison underground water supplies. The 6–8 million barrels of oil that spilled into the Gulf waters are threatening marine life. This spill is by far the largest in history—25–30 times the size of the 1989 Exxon Valdez oil spill in Alaska.

The clouds of smoke that hung over the Gulf region for eight months contained soot, sulfur dioxide, and nitrogen oxides—the major components of acid rain—and a variety of toxic and potentially carcinogenic chemicals and heavy metals. The U.S. Environmental Protection Agency estimates that in March 1991 about 10 times as much air pollution was being emitted in Kuwait as by all U.S. industrial and power-generating plants combined. Acid rain destroys forests and harms crops. It also activates several dangerous metals normally found in soil, including aluminum, cadmium, and mercury. Furthermore, the 900,000 depleted uranium munitions fired at Kuwait and Iraq are thought to have serious environmental and health consequences. This chapter's *Human Side* feature poignantly describes the destructive course of Gulf War syndrome—one of the many diseases thought to be associated with this radioactive material.

Sometimes environmental degradation is indirect. In Afghanistan two decades of conflict have

> put previous environmental management and conservation strategies on hold, brought about a collapse of local and national governance, destroyed infrastructure, hindered agricultural activity, and driven people into cities already lacking the most basic public amenities (Haavisto 2005 p. 158).

In urban areas of Afghanistan, for example, safe water may be reaching less than 12% of the residents.

The ultimate environmental catastrophe facing the planet is a massive exchange thermonuclear war. Aside from the immediate human casualties, poisoned air, poisoned crops, and radioactive rain, many scientists agree that the dust storms and concentrations of particles would block vital sunlight and lower temperatures in the Northern Hemisphere, creating a **nuclear winter**. In the event of large-scale nuclear war, most living things on earth would die. The fear of nuclear war has greatly contributed to the military and arms buildup, which, ironically, also causes environmental destruction even in times of peace.

Destruction of the Environment in Peacetime. Even when military forces are not engaged in active warfare, military activities assault the environment. For example, modern military maneuvers require large amounts of land. The use of land for military purposes prevents other uses, such as agriculture, habitat protection, recreation, and housing. More important, military use of land often harms the land. In practicing military maneuvers, the armed forces demolish natural vegetation, disturb wildlife habitats, erode soil, silt up streams, and cause flooding. Military bombing and shooting ranges leave the land pockmarked with craters and contaminate the soil and groundwater with lead and other toxic residues. Furthermore,

> mined areas can restrict access to large areas of agricultural land, forcing populations to use small tracts of land to earn their livelihoods. The limited productive land that is available is over-cultivated, which contributes to long term underproduction, as minerals are depleted from the soil, and the loss of valuable vegetation. (Landmines 2003, p. 1)

Bombs exploded during peacetime leak radiation into the atmosphere and groundwater. From 1945 to 1990, 1,908 bombs were tested—that is, exploded—at more than 35 sites around the world. Although underground testing has reduced radiation, some radioactive material still escapes into the atmosphere and is suspected of seeping into groundwater. Similarly, in Russia decommissioned submarines have been found to emit radioactive pollution ("Foreign Conservationists Under Siege," 2000).

Finally, although arms control and disarmament treaties of the last decade have called for the disposal of huge stockpiles of weapons, no completely safe means of

Oil smoke from the 650 burning oil wells left in the wake of the Gulf War contains soot, sulfur dioxide, and nitrogen oxides, the major components of acid rain, along with a variety of toxic and potentially carcinogenic chemicals and heavy metals.

"My God . . . what have we done?"

Robert Lewis
Copilot of the Enola Gay, after dropping the atomic bomb on Hiroshima

The Death of a Gulf War Veteran

The following letter to American Legion Magazine *is from the parents of a Gulf War veteran.*

Our son may be unique. We really don't know. We're desperate for information that is extremely hard to come by. We are very frustrated that what happened to us may have already happened to other families—or will—before something is done.

This is for Scott, our son, who said, "Go for it, Mom," when I told him I'd never quit trying to find out what made him so ill.

He was sent to the Persian Gulf with the 1133rd Transportation Co., National Guard, of Mason City, Iowa. He left for Saudi as a very healthy young man and he returned in much the same physical condition. Just tired, but not unusual considering where he'd been and what he'd seen.

. . . Two years after he returned home in the summer of 1993, he developed a rash on his torso. It would erupt, disappear and then come back again. It didn't always look the same. We thought it was a heat rash, something ho ate, new laundry soap, etc.

In the fall of 1993 he developed sores in his mouth. He could eat very little and quickly lost about 40 pounds. He went from specialist to specialist, who sent tests many places to try and find a cause. He was put on steroids and depending on the dosage, it would get a little better and then flare up again. By winter he could only eat pureed foods and liquids. He could not use a straw as it hurt too much. Many doctors, many tests, many different medications. Nothing helped very much. And no concrete diagnosis.

In May 1994, he was examined by the VA hospital in Des Moines. They were not able to find the cause for the terrible sores in his mouth and the rash that now also affected his feet, hands and arms. He made many trips to Des Moines—a three-hour drive.

In early August 1994, the VA Hospital in Des Moines came up with a diagnosis of lupus. My heart just broke when I heard those words, but he was thrilled to know there was finally a name and treatment for his symptoms.

By mid-August he was hardly able to walk, his feet swollen and so extremely sore from the rash that seemed to get worse by the day. He went to the VA Hospital on Friday, August 19 . . . and was admitted. The rash had become blisters about the size of a 50-cent piece and were breaking and bleeding. He was running a fever. . . .

He was transferred to University [hospital] in Iowa City and taken to surgery . . . [where] they removed all his skin and replaced it with what is called pig skin. Out of surgery, bandaged from head to toe, he was given a five percent chance of survival.

Our family gathered together to give him all the love and support we could. Ten days later the pig skin was removed. Infection had been found. It was too risky to take him back to surgery, so it was removed in a sterile room close to his room.

The next seven weeks are a blur. We had many ups—a good day for Scott—and many downs—a bad day for him. He had to endure burn baths every day. Water jets removed sloughed skin. He'd grow a tiny patch of skin only to lose it in a day or so later. He was fed through a tube. Many antibiotics were given in the hopes of warding off infection; morphine for the tremendous pain. We read to him—he'd correctly pronounce words we missed. His sense of humor never left. He worried about all of us.

Scott's life ended at 3 a.m. on October 15, 1994. . . .

We are convinced beyond anything we've ever felt as parents that Persian Gulf Syndrome killed our only son. We want someone to tell us the truth. We won't quit until we have believable answers to our question.

What do we tell his oldest son, now 11, who has nothing but pictures of his daddy . . . ?

What do we say to a three-year-old whose wish is to build a rocket so he can go get daddy and make everyone happy?

What do you say to a son who was born two weeks after his daddy died . . . ?

What do you say to a wife who longs only for her husband's love, strength and support?

If chemicals and gases were used over there, tell us the truth. It kills innocent people and destroys innocent families. We are real people with real feelings and we deserve the truth. This was not easy to write. We did it in hopes it may help someone. Only then will Scott's death make sense to us. He would have wanted it that way. He was that special.

Ardie and Rollie Siefken, Plainfield, Iowa

disposing of weapons and ammunition exist. Many activist groups have called for placing weapons in storage until safe disposal methods are found. Unfortunately, the longer the weapons are stored, the more they deteriorate, increasing the likelihood of dangerous leakage. In 2003 a federal judge gave permission, despite objections by environmentalists, to incinerate 2,000 tons of nerve agents and mustard gas left from the Cold War era. Although the army says that it is safe to dispose of the weapons, they have issued protective gear in case of an "accident" to the nearly 20,000 residents who live nearby (CNN 2003).

Strategies for Action: In Search of Global Peace

Various strategies and policies are aimed at creating and maintaining global peace. These include the redistribution of economic resources, the creation of a world government, peacekeeping activities of the United Nations, mediation and arbitration, and arms control.

Redistribution of Economic Resources

Inequality in economic resources contributes to conflict and war because the increasing disparity in wealth and resources between rich and poor nations fuels hostilities and resentment. Therefore any measures that result in a more equal distribution of economic resources are likely to prevent conflict. John J. Shanahan (1995), retired U.S. Navy vice admiral and director of the Center for Defense Information, suggests that wealthy nations can help reduce social and economic roots of conflict by providing economic assistance to poorer countries. Nevertheless, U.S. military expenditures for national defense far outweigh U.S. economic assistance to foreign countries. For example, the money pledged by the United States for the 2004 Asian tsunami disaster is the equivalent of just one and a half day's spending in Iraq (Monbiot 2005).

As we discussed in Chapter 13, strategies that reduce population growth are likely to result in higher levels of economic well-being. Funke (1994) explains that "rapidly increasing populations in poorer countries will lead to environmental overload and resource depletion in the next century, which will most likely result in political upheaval and violence as well as mass starvation" (p. 326). Although achieving worldwide economic well-being is important for minimizing global conflict, it is important that economic development does not occur at the expense of the environment.

Finally, UN secretary general Kofi Annan in an address to the United Nations concludes that it is not poverty per se that leads to conflict but rather the "inequality among domestic social groups" (Deen 2000). Referencing a research report completed by the Tokyo-based UN University, Annan argues that "inequality . . . based on ethnicity, religion, national identity or economic class . . . tends to be reflected in unequal access to political power that too often forecloses paths to peaceful change" (Deen 2000).

World Government

Some analysts have suggested that world peace might be attained through the establishment of a world government. The idea of a single world government is not new. In 1693 William Penn advocated a political union of all European monarchs,

Millions of Iraqis defied the insurgents and voted in their first free election in half a century. Afghanistan's first ever parliamentary election is scheduled to take place in the fall of 2005.

and in 1712 Jacques-Henri Bernardin de Saint-Pierre of France suggested an all-European union with a "Senate of Peace." Proposals such as these have been made throughout history.

From the 1950s to the early 1990s world power was primarily divided between the United States and the Soviet Union. With the fall of Eastern European communism in the early 1990s the United States became the world's leader, causing observers to debate the proper role of the United States in global politics. Many argued that with the demise of the Soviet Union the world no longer needed a watchdog—a role the United States had historically played. Still others argued that "it is both the moral duty and in the strategic interest of the United States to become involved in regions where there is unrest, and stand as a leading force in international organizations, especially the United Nations" (Social Science Research Council 2003, p. 1).

Although some commentators are pessimistic about the likelihood of a new world order, Lloyd (1998) identifies three global trends that, he contends, signify the "ghost of a world government yet to come" (p. 28). First is the increasing tendency for countries to engage in "ecological good behavior," indicating a concern for a global rather than national well-being. Second, despite several economic crises worldwide, major world players such as the United States and Great Britain have supported rather than abandoned nations in need. And, finally, an International Criminal Court has been created with "powers to pursue, arraign, and condemn those found guilty of war and other crimes against humanity" (p. 28). On the basis of these three trends, Lloyd (1998) concludes, "Global justice, a wraith pursued by peace campaigners for over a century, suddenly seems achievable" (p. 28). Despite Lloyd's optimism, the French overwhelmingly rejected the proposed constitution of the European Union (EU) in a 2005 referendum. The EU is an organization of 25 member states, all of which must ratify the constitution for it to become law. Presently, nine countries have ratified the constitution—Austria, Germany, Greece, Hungary, Italy, Lithuania, Slovakia, Slovenia, and Spain (BBC 2005).

Table 16.3

United Nations Peacekeeping Operations: Summary Data, 2005

Military personnel and civilian police serving in peacekeeping operations	65,973
Countries contributing military personnel and civilian police	105
International civilian personnel	4,378
Local civilian personnel	8,112
UN volunteers	1,763
Total number of fatalities in peacekeeping operations since 1948	1,983
Approved budgets for July 1, 2004, to June 30, 2005	$4.47 billion
Estimated total cost of operations from 1948 to June 2004	$36.0 billion

Source: United Nations (2005).

The United Nations

The United Nations (UN), whose charter begins, "We the people of the United Nations—Determined to save succeeding generations from the scourge of war," has engaged in over 60 peacekeeping operations since 1948 (see Table 16.3) (United Nations 2005). The UN Security Council can use force, when necessary, to restore international peace and security.

> United Nations peacekeepers—military personnel in their distinctive blue helmets or blue berets, civilian police and a range of other civilians—help implement peace agreements, monitor cease fires, create buffer zones, or support complex military and civilian functions essential to maintain peace and begin reconstruction and institution-building in societies devastated by war. (United Nations 2003, p. 1).

Recently, the United Nations has been involved in overseeing multinational peacekeeping forces in Sudan, Burundi, Haiti, Liberia, and Ethiopia (United Nations 2005).

In the last few years, the United Nations has come under heavy criticism. First, in recent missions, developing nations have supplied more than 75% of the troops while developed countries—United States, Japan, and Europe—have contributed 85% of the finances. As one UN official commented, "You can't have a situation where some nations contribute blood and others only money" (quoted by Vesely 2001, p. 8). Second, a recent review of UN peacekeeping operations noted several failed missions, including an intervention in Somalia in which 44 U.S. marines were killed (Lamont 2001). Third, as typified by the debate over the disarming of Iraq, the UN cannot take sides but must wait for a consensus of its members which, if not forthcoming, undermines the strength of the organization (Goure 2003).

Finally, the concept of the United Nations is that its members represent individual nations, not a region or the world. And because nations tend to act in their own best economic and security interests, UN actions performed in the name of world peace may be motivated by nations acting in their own interests.

As a result of such criticisms, Secretary-General Kofi Annan has called on the 191 members of the UN to approve the most far-reaching changes in the 60-year history of the organization (Lederer 2005). One of the most controversial recommendations concerns the composition of the Security Council, the most important decision-making body of the organization. Annan's recommendation that the 15 members of the Security Council—a body dominated by the United States, Great Britain, France, Russia, and China—be changed to include a more representative number of nations could, if approved, shift the global balance of power. Other recommendations include the establishment of a Humans Rights Council and the development of a comprehensive antiterrorism strategy.

Mediation and Arbitration

Most conflicts are resolved through nonviolent means (Worldwatch Institute 2003). Mediation and arbitration are just two of the nonviolent strategies used to resolve conflicts and stop or prevent war. In mediation a neutral third party intervenes and facilitates negotiation between representatives or leaders of conflicting groups. Mediators do not impose solutions but rather help disputing parties generate options for resolving the conflict (Conflict Research Consortium 2003). Ideally, a mediated resolution to a conflict meets at least some of the concerns and interests of each party to the conflict. In other words, mediation attempts to find "win-win" solutions in which each side is satisfied with the solution. Although mediation is used to resolve conflict between individuals, it is also a valuable tool for resolving international conflicts. For example, in 2000 mediators were used to resolve the escalating violence between Israeli and Palestinian troops, and in 2004 mediation talks between the Sudan Liberation Army and the government led to a temporary cease fire. Using mediation as a means of resolving international conflict is often difficult, given the complexity of the issues.

> "War doesn't determine who's right, just who's left."
>
> **George Carlin**
> **Comedian**

Arbitration also involves a neutral third party who listens to evidence and arguments presented by conflicting groups. Unlike mediation, however, in arbitration the neutral third party arrives at a decision or outcome that the two conflicting parties agree to accept. As Sweet and Brunell (1998) note, **triad dispute resolution**—that is, resolution that involves two disputants and a negotiator—performs "profoundly political functions including the construction, consolidation, and maintenance of political regimes" (p. 64).

Arms Control and Disarmament

In the 1960s the United States and the Soviet Union led the world in an arms race, each competing to build a more powerful military arsenal than its adversary. If either superpower were to initiate a full-scale war, the retaliatory powers of the other nation would result in the destruction of both nations. Thus the principle of **mutually assured destruction (MAD)** that developed from nuclear weapons capabilities transformed war from a win-lose proposition to a lose-lose scenario. If both sides would lose in a war, the theory goes, neither side would initiate war.

Because of the end of the Cold War and the growing realization that current levels of weapons literally represented overkill, governments have moved in the direction of arms control, which, at least in the past, involved reducing or limiting defense spending, weapons production, and armed forces. Recent arms control initiatives include SALT (the Strategic Arms Limitation Treaty), START (the Strategic Arms Reduction Treaty), SORT (the Strategic Offensive Reduction Treaty), and NPT (the Nuclear Nonproliferation Treaty).

Strategic Arms Limitation Treaty. Under the 1972 SALT agreement (SALT I), the United States and the Soviet Union agreed to limit both their defensive weapons and their land-based and submarine-based offensive weapons. Also in 1972, Henry

Kissinger drafted the Declaration of Principles, known as **détente**, which means "negotiation rather than confrontation." A further arms-limitation agreement (SALT II) was reached in 1979 but was never ratified by Congress because of the Soviet invasion of Afghanistan. Subsequently, the arms race continued with the development of new technologies and an increase in the number of nuclear warheads.

Strategic Arms Reduction and Strategic Offensive Reduction Treaties. Strategic arms talks resumed in 1982 but made relatively little progress for several years. During this period, President Reagan proposed the Strategic Defense Initiative (SDI), more commonly known as "star wars," which purportedly would be able to block missiles launched by another country against the United States. Although some research was conducted on the system, SDI, which was land based, was never actually built. More recently, a space-based national weapons program has been proposed by the U.S. Air Force. Such a program, although costly, would provide "freedom to attack as well as freedom from attack" from space (Weiner 2005).

In 1991 the international situation changed dramatically. The communist regime in the Soviet Union had fallen, the Berlin Wall had been dismantled, and many Eastern European and Baltic countries were under self-rule. SALT was renamed START and was signed in 1991. A second START agreement, signed in 1993 and ratified by the U.S. Senate, signaled the end of the Cold War. START II called for the reduction of nuclear warheads to 3,500 by the year 2003 but was superseded by SORT, which requires both the United States and Russia to reduce their nuclear arsenals to 1,700–2,200 warheads by December 31, 2012 (Arms Control Association 2003a).

Nuclear Nonproliferation Treaty. The 1970 NPT was renewed in 2000 and adopted by 187 countries. Only India, Israel, and Pakistan have not signed the agreement (Dickey 2005). The treaty holds that countries without nuclear weapons will not to try to get them; in exchange, the countries with nuclear weapons (the United States, United Kingdom, France, China, and Russia) agree that they will not provide nuclear weapons to countries that do not have them. In January 2003 North Korea, under suspicion of secretly producing nuclear weapons, announced that it was withdrawing from the treaty, effective immediately (Arms Control Association 2003b). However, in July 2005, after "more than a year of stalemate, North Korea agreed . . . to return to disarmament talks . . . and pledged to discuss eliminating its nuclear-weapons program" (Brinkley & Sanger 2005, p. 1).

However, even if military superpowers honor agreements to limit arms, the availability of black-market nuclear weapons and materials presents a threat to global security. Kyl and Halperin (1997) note that U.S. security is threatened more by nuclear weapons falling into the hands of a terrorist group than by a nuclear attack from an established government. One of the most successful nuclear weapons brokers was Pakistan's Abdul Qadeer Khan. Khan, the "father" of Pakistan's nuclear weapons program, sold the technology and equipment required to make nuclear weapons to such rogue states as Iran and North Korea. He is also suspected of selling nuclear weapons to Al Qaeda. At present, his activities are under investigation by the United States and the International Atomic Energy Agency (Powell & McGirk 2005).

In 2003 the United States accused Iran of operating a covert nuclear weapons program. Although Iranian officials responded that their nuclear program was solely for the purpose of generating electricity, concerns continue. In 2005 the United States announced that it would freeze the assets of any company doing business with Iran's Atomic Energy Organization or any other organizations thought to be involved in nuclear proliferation (Dickey 2005).

"The equivalent of a nuclear war has already happened. Over the last half century, millions have died as a result of accidents, experiments, lies and cover-ups by the nuclear industry.**"**

Eduardo Goncalves
Journalist and environmentalist

Understanding Conflict, War, and Terrorism

As we come to the close of this chapter, how might we have an informed understanding of conflict, war, and terrorism? Each of the three theoretical positions discussed in this chapter reflects the realities of global conflict. As structural functionalists argue, war offers societal benefits—social cohesion, economic prosperity, scientific and technological developments, and social change. Furthermore, as conflict theorists contend, wars often occur for economic reasons because corporate elites and political leaders benefit from the spoils of war—land and water resources and raw materials. The symbolic interactionist perspective emphasizes the role that meanings, labels, and definitions play in creating conflict and contributing to acts of war.

The September 11 attacks on the World Trade Center and the Pentagon and the aftermath—the battle against terrorism, the bombing of Afghanistan, and the war with Iraq—forever changed the world we live in. For some theorists these events were inevitable. Political scientist Samuel P. Huntington argues that such conflict represents a **clash of civilizations**. In *The Clash of Civilizations and the Remaking of World Order* (1996), Huntington argues that in the new world order

> the most pervasive, important and dangerous conflicts will not be between social classes, rich and poor, or economically defined groups, but between people belonging to different cultural entities . . . the most dangerous cultural conflicts are those along the fault lines between civilizations . . . the line separating peoples of Western Christianity, on the one hand, from Muslim and Orthodox peoples on the other. (p. 28)

Although not without critics, the clash-of-civilizations hypothesis has some support. In an interview of almost 10,000 people from nine Muslim states representing half of all Muslims worldwide, only 22% had favorable opinions toward the United States (CNN 2003). Even more significantly, 67% saw the September 11 attacks as "morally justified," and the majority of respondents found the United States to be overly materialistic, secular, and having a corrupting influence on other nations.

Ultimately, we are all members of one community—Earth—and have a vested interest in staying alive and protecting the resources of our environment for our own and future generations. But, as we have seen, conflict between groups is a feature of social life and human existence that is not likely to disappear. What is at stake—human lives and the ability of our planet to sustain life—merits serious attention. World leaders have traditionally followed the advice of philosopher Karl von Clausewitz: "If you want peace, prepare for war." Thus nations have sought to protect themselves by maintaining large military forces and massive weapons systems. These strategies are associated with serious costs, particularly in hard economic times. In diverting resources away from other social concerns, defense spending undermines a society's ability to improve the overall security and well-being of its citizens. Conversely, defense-spending cutbacks, although unlikely in the present climate, could potentially free up resources for other social agendas, including lowering taxes, reducing the national debt, addressing environmental concerns, eradicating hunger and poverty, improving health care, upgrading educational services, and improving housing and transportation. Therein lies the promise of a "peace dividend." The hope is that future dialogue on the problems of war and terrorism will redefine national and international security to encompass social, economic, and environmental well-being.

> "I hate war as only a soldier who has lived it can, only as one who has seen its brutality, its futility, its stupidity."
>
> **Dwight D. Eisenhower**
> Former U.S. President, military leader, hero

Chapter Review

• **What is the relationship between war and industrialization?**

War indirectly affects industrialization and technological sophistication because military research and development advances civilian-used technologies. Industrialization, in turn, has had two major influences on war: The more industrialized a country, the lower the rate of conflict, and if conflict occurs, the higher the rate of destruction.

• **In general, how do feminists view war?**

Feminists are quick to note that wars are part of the patriarchy of society. Although women and children may be used to justify a conflict (e.g., improving women's lives by removing the repressive Taliban in Afghanistan), the basic principles of male dominance and control are realized through war.

• **What are some of the causes of war?**

The causes of war are numerous and complex. Most wars involve more than one cause. Some of the causes of war are conflict over land and natural resources; values or ideologies; racial, ethnic, and religious hostilities; defense against hostile attacks; revolution; and nationalism.

• **What is terrorism, and what are the different types of terrorism?**

Terrorism is the premeditated use, or threatened use, of violence by an individual or group to gain a political or social objective. Terrorism can be either transnational or domestic. Transnational terrorism occurs when a terrorist act in one country involves victims, targets, institutions, governments, or citizens of another country. Domestic terrorism involves only one nation, such as the 1995 truck bombing of a nine-story federal office building in Oklahoma City.

• **What are some of the macrolevel "roots" of terrorism?**

Although not an exhaustive list, some of the macro-level "roots" of terrorism include (1) a failed or weak state, (2) rapid modernization, (3) extreme ideologies, (4) a history of violence, (5) repression by a foreign occupation, (6) large-scale racial or ethnic discrimination, and (7) the presence of a charismatic leader.

• **How has the United States responded to the threat of terrorism?**

The United States has used both defensive and offensive strategies to fight terrorism. Defensive strategies include using metal detectors and X-ray machines at airports and strengthening security at potential targets, such as embassies and military command posts. The Department of Homeland Security coordinates such defensive tactics. Offensive strategies include retaliatory raids such as the U.S. bombing of terrorist facilities in Afghanistan, group infiltration, and preemptive strikes.

• **What is meant by "diversion of economic resources"?**

Worldwide, the billions of dollars used on defense could be channeled into social programs dealing with, for example, education, health, and poverty. Thus defense monies are economic resources diverted from other needy projects.

• **What are some of the criticisms of the United Nations?**

First, in recent missions, developing nations have supplied more than 75% of the troops. Second, several recent UN peacekeeping operations have failed. Third, the UN cannot take sides but must wait for a consensus of its members, which, if not forthcoming, undermines the strength of the organization. Finally, the concept of the United Nations is that its members represent individual nations, not a region or the world. Because nations tend to act in their own best economic and security interests, UN actions performed in the name of world peace may be motivated by nations acting in their own interests.

Critical Thinking

1. Certain actions constitute "war crimes." Such actions include the use of forbidden munitions, such as biological weapons, purposeless destruction, killing civilians, poisoning waterways, and violation of surrender terms. In addition to these actions, what other actions should constitute war crimes?

2. Selecting each of the five major institutions in society, what part could each play in attaining global peace?

3. Make a list of famous war movies (e.g., *Schindler's List*). With specific movies in mind, list media sounds and images of war. Has the portrayal of war in movies changed over time? If so, how and why?

Key Terms

arbitration
clash of civilizations
Cold War
cyberwarfare
détente
domestic terrorism
dual-use technologies
guerrilla warfare
military-industrial
  complex
mutually assured
  destruction (MAD)

nuclear winter
post-traumatic stress
  disorder
state
terrorism
transnational terrorism
triad dispute resolution
war
weapons of mass
  destruction (WMD)

Media Resources

The Companion Website for
Understanding Social Problems,
Fifth Edition

http://sociology.wadsworth.com/mooney_knox_schacht5e

Supplement your review of this chapter by going to the companion website to take one of the Tutorial Quizzes, use the flash cards to master key terms, and check out the many other study aids you'll find there. You'll also find special features such as *Wadsworth's Sociology Online Resources and Writing Companion,* GSS data, and Census 2000 information, data, and resources at your fingertips to help you complete that special project or do some research on your own.

Epilogue

Today, there is a crisis—a crisis of faith: faith in the ideals of equality and freedom, faith in political leadership, faith in the American dream, and, ultimately, faith in the inherent goodness of humankind and the power of one individual to make a difference.

To some extent, faith is shaken by texts such as this one. The global gap between the rich and the poor is widening; war and conflict around the world continues to take lives and use resources that otherwise could be spent on improving health and education; profit-seeking multinational corporations are increasingly influencing world affairs; political corruption is everywhere; the environment is killing us—if we don't kill it first. Social problems are everywhere, and what's worse, many solutions seem only to create more problems.

The transformation of American society in recent years has been dramatic. With the exception of the Industrial Revolution, no other period in human history has seen such rapid social change. The structure of U.S. society, forever altered by such macrosociological processes as corporate multinationalization, deindustrialization, and globalization, continues to be characterized by social inequities—in our schools, in our homes, in our cities, and in our salaries.

The culture of society has also undergone rapid change, leading many politicians and laypersons alike to call for a return to traditional values and beliefs and to emphasize the need for moral education. The implication is that somehow things were better in the "good old days," and that if we could somehow return to those times, things would be better again. Some things were better—for some people.

Fifty years ago, there were fewer divorces and less crime. AIDS and crack cocaine were unheard of, and violence in schools was almost nonexistent. At the same time, however, in 1950, the infant mortality rate was more than three times what it is today; racial and ethnic discrimination flourished in an atmosphere of bigotry and hate, and millions of Americans were routinely denied the right to vote because of the color of their skin; more than half of all Americans smoked cigarettes; and people over the age of 25 had completed a median of 6.8 years of school.

The social problems of today are the cumulative result of structural and cultural forces over time. Today's problems are not necessarily better or worse than those of generations ago—they are different and, perhaps, more diverse as a result of the increased complexity of social life. But, as surely as we brought the infant mortality rate down; passed laws to prohibit racial discrimination in education, housing, and employment; increased educational levels; and reduced the number of smokers, we can continue to meet the challenges of today's social problems. But how does positive social change occur? How does one alter something as amorphous as society? The answer is really quite simple. All social change takes place because of the acts of individuals. Every law, every regulation and policy, every social movement and media exposé, and every court decision began with one person.

The media attention surrounding the death of Rosa Parks in 2005 reminded the world how one person can make a difference. Rosa Parks was a seamstress in Montgomery, Alabama, in the 1950s. Like almost everything else in the South in the 1950s, public transportation was racially segregated. On December 1, 1955, Rosa Parks was on her way home from work when the "white section" of the bus she was riding became full. The bus driver told black passengers in the first row of the black section to relinquish their seats to the standing white passengers. Rosa Parks refused. She was arrested and put in jail, but her treatment so outraged the black community that a boycott of the bus system was organized by a new minister in town—Martin Luther King, Jr. The Montgomery bus boycott was a success. Just 11 months later, in November 1956, the U.S. Supreme Court ruled that racial segregation of public facilities was unconstitutional. Rosa Parks had begun a process that in time would echo her actions—the civil rights movement, the March on Washington, the 1963 Equal Pay Act, the 1964 Civil Rights Act, the 1965 Voting Rights Act, regulations against discrimination in housing, and affirmative action.

Was social change accomplished? In 1960, just 20 percent of blacks 25 years old and older had completed high school, compared to 40 percent of whites. By 2003, 80 percent of blacks 25 years old and older had completed high school, compared to 89 percent of whites (Stoops 2004). While many would point out that such changes and thousands like them have created other problems that need to be addressed, who among us would want to return to the "good old days" of the 1950s in Montgomery, Alabama?

Throughout this text we have mentioned just a few of the millions of individuals who make a difference daily, including:

- Nkosi Johnson, a South African boy born with HIV who publicly spoke out for the rights of HIV-infected individuals (see Chapter 2).
- Actor and football player Victor Rivas Rivers, a victim of child abuse who became a national spokesperson for the National Network to End Domestic Violence (see Chapter 5).
- A group of fifth-grade students in Denver who saved enough money to purchase the freedom of two Sudanese slaves (see Chapter 7).
- Dr. Kathleen Moltz, a pediatrician and professor at Wayne State University School of Medicine, who testified against the passage of the Federal Marriage Amendment before a U.S. Senate Judiciary Committee (see Chapter 11).
- 2004 Nobel Peace Prize recipient Wangari Maathai, who lead a grassroots environmental campaign to plant 30 million trees across Kenya (see Chapter 14).

While only a fraction of the readers of this text will occupy social roles that directly influence social policy, one need not be a politician or member of a social reform group to make a difference. We, the authors of this text, challenge you, the

"The only thing necessary for the triumph of evil is for good men [and women] to do nothing."

Edmund Burke
English political writer/orator

"Some people see things that are and say, 'Why?' He saw things that never were and said, 'Why not?'"

Ted Kennedy
Of his brother, Bobby Kennedy

reader, to make individual decisions and take individual actions to make the world a more humane, just, and peaceful place for all. If you feel that your own actions cannot alleviate problems that are bigger than you are, consider the perspective of former California state senator Tom Hayden:

> The action we take, successful or not, reminds people that progress is possible. If we don't take action, we give the appearance that it's impossible to change things. Any action, while not guaranteeing change, creates possibilities that weren't there before. (quoted by McKee 2006, p. 6).

So, where should you begin? Where Rosa Parks and others like her began—with a simple individual act of courage, commitment, and faith.

Appendix:
Methods of Data Analysis

Description, Correlation, Causation, Reliability and Validity, and Ethical Guidelines in Social Problems Research

There are three levels of data analysis: description, correlation, and causation. Data analysis also involves assessing reliability and validity.

Description

Qualitative research involves verbal descriptions of social phenomena. Having a homeless and single pregnant teenager describe her situation is an example of qualitative research.

Quantitative research often involves numerical descriptions of social phenomena. Quantitative descriptive analysis may involve computing the following: (1) means (averages), (2) frequencies, (3) mode (the most frequently occurring observation in the data), (4) median (the middle point in the data; half of the data points are above the median, and half are below), and (5) range (the highest and lowest values in a set of data).

Correlation

Researchers are often interested in the relationship between variables. *Correlation* refers to a relationship between or among two or more variables. The following are examples of correlational research questions: What is the relationship between poverty and educational achievement? What is the relationship between race and crime victimization? What is the relationship between religious affiliation and divorce?

If there is a correlation or relationship between two variables, then a change in one variable is associated with a change in the other variable. When both variables change in the same direction, the correlation is positive. For example, in general, the more sexual partners a person has, the greater the risk of contracting a sexually transmissible disease. As variable A (number of sexual partners) increases, variable

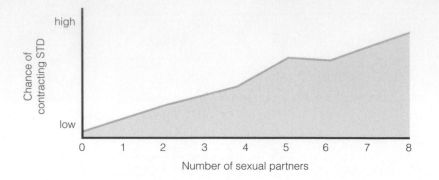

the more sexual partners a person has, the greater the risk of contracting a sexually transmissible disease. As variable A (number of sexual partners) increases, variable B (chance of contracting an STD) also increases. Similarly, as the number of sexual partners decreases, the chance of contracting an STD decreases. Notice that in both cases, the variables change in the same direction, suggesting a positive correlation (see Figure A.1).

When two variables change in opposite directions, the correlation is negative. For example, there is a negative correlation between condom use and contracting STDs. In other words, as condom use increases, the chance of contracting an STD decreases (see Figure A.2).

The relationship between two variables may also be curvilinear, which means that it varies in both the same and opposite directions. For example, suppose a researcher finds that after drinking one alcoholic beverage, research participants are more prone to violent behavior. After two drinks, violent behavior is even more likely, and this trend continues for three and four drinks. So far, the correlation between alcohol consumption and violent behavior is positive. After the research participants have five alcoholic drinks, however, they become less prone to violent behavior. After six and seven drinks, the likelihood of engaging in violent behavior decreases further. Now the correlation between alcohol consumption and violent behavior is negative. Because the correlation changed from positive to negative, we say that the correlation is curvilinear (the correlation may also change from negative to positive) (see Figure A.3).

A fourth type of correlation is called a spurious correlation. Such a correlation exists when two variables appear to be related, but the apparent relationship occurs only because each variable is related to a third variable. When the third variable is controlled through a statistical method in which the variable is held constant, the apparent relationship between the first two variables disappears. For example,

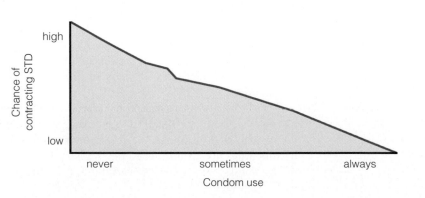

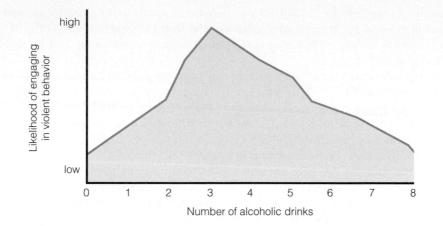

blacks have a lower average life expectancy than whites do. Thus, race and life expectancy appear to be related. This apparent correlation exists, however, because both race and life expectancy are related to socioeconomic status. Because blacks are more likely than whites to be impoverished, they are less likely to have adequate nutrition and medical care.

Causation

If the data analysis reveals that two variables are correlated, we know only that a change in one variable is associated with a change in another variable. We cannot assume, however, that a change in one variable causes a change in the other variable unless our data collection and analysis are specifically designed to assess causation. The research method that best assesses causality is the experimental method (discussed in Chapter 1).

To demonstrate causality, three conditions must be met. First, the data analysis must demonstrate that variable A is correlated with variable B. Second, the data analysis must demonstrate that the observed correlation is not spurious. Third, the analysis must demonstrate that the presumed cause (variable A) occurs or changes prior to the presumed effect (variable B). In other words, the cause must precede the effect.

It is extremely difficult to establish causality in social science research. Therefore, much social research is descriptive or correlative, rather than causative. Nevertheless, many people make the mistake of interpreting a correlation as a statement of causation. As you read correlative research findings, remember the following adage: "Correlation *does not* equal causation."

Reliability and Validity

Assessing reliability and validity is an important aspect of data analysis. *Reliability* refers to the consistency of the measuring instrument or technique; that is, the degree to which the way information is obtained produces the same results if repeated. Measures of reliability are made on scales and indexes (such as those in the *Self and Society* features in this text) and on information-gathering techniques, such as the survey methods described in Chapter 1.

Various statistical methods are used to determine reliability. A frequently used method is called the "test-retest method." The researcher gathers data on the same sample of people twice (usually one or two weeks apart) using a particular instrument or method and then correlates the results. To the degree that the results of the two tests are the same (or highly correlated), the instrument or method is considered reliable.

Measures that are perfectly reliable may be absolutely useless unless they also have a high validity. *Validity* refers to the extent to which an instrument or device measures what it intends to measure. For example, police officers administer "Breathalyzer" tests to determine the level of alcohol in a person's system. The Breathalyzer is a valid test for alcohol consumption.

Validity measures are important in research that uses scales or indexes as measuring instruments. Validity measures are also important in assessing the accuracy of self-report data that are obtained in survey research. For example, survey research on high-risk sexual behaviors associated with the spread of HIV relies heavily on self-report data on such topics as number of sexual partners, types of sexual activities, and condom use. Yet how valid are these data? Do survey respondents underreport the number of their sexual partners? Do people who say they use a condom every time they engage in intercourse really use a condom every time? Because of the difficulties in validating self-reports of number of sexual partners and condom use, we may not be able to answer these questions.

Ethical Guidelines in Social Problems Research

Social scientists are responsible for following ethical standards designed to protect the dignity and welfare of people who participate in research. These ethical guidelines include the following (Schutt 1999; Neuman 2000; American Sociological Association 2001):

1. *Freedom from coercion to participate.* Research participants have the right to decline to participate in a research study or to discontinue participation at any time during the study. For example, professors who are conducting research using college students should not require their students to participate in their research.
2. *Informed consent.* Researchers are required to inform potential participants of any aspect of the research that might influence a subject's willingness to participate. After informing potential participants about the nature of the research, researchers typically ask participants to sign a consent form indicating that the participants are informed about the research and agree to participate in it.
3. *Deception and debriefing.* Sometimes the researcher must disguise the purpose of the research in order to obtain valid data. Researchers may deceive participants as to the purpose or nature of a study only if there is no other way to study the problem. When deceit is used, participants should be informed of this deception (debriefed) as soon as possible. Participants should be given a complete and honest description of the study and why deception was necessary.
4. *Protection from harm.* Researchers must protect participants from any physical and psychological harm that might result from participating in a research study. This is both a moral and a legal obligation. It would not be ethical, for example, for a researcher studying drinking and driving behavior to observe an intoxicated individual leaving a bar, getting into the driver's seat of a car, and driving away.

    Researchers are also obligated to respect the privacy rights of research participants. If anonymity is promised, it should be kept. Anonymity is maintained in mail surveys by identifying questionnaires with a number coding system rather than with the participants' names. When such anonymity is not possible, as is the case with face-to-face interviews, researchers should tell participants that the information they provide will be treated as confidential. Although interviews may be summarized and excerpts quoted in published material, the identity of the individual participants is not revealed. If a research participant experiences either physical or psychological harm as a result of participation in a research study, the researcher is ethically obligated to provide remediation for the harm.

5. *Reporting of research.* Ethical guidelines also govern the reporting of research results. Researchers must make research reports freely available to the public. In these reports, a researcher should fully describe all evidence obtained in the study, regardless of whether the evidence supports the researcher's hypothesis. The raw data collected by the researcher should be made available to other researchers who might request it for purposes of analysis. Finally, published research reports should include a description of the sponsorship of the research study, its purpose, and all sources of financial support.

Glossary

abortion The intentional termination of a pregnancy.

absolute poverty The lack of resources that leads to hunger and physical deprivation.

acculturation Learning the culture of a group different from the one in which the learner was originally raised.

achieved status A status assigned on the basis of some characteristic or behavior over which the individual has some control.

acid rain The mixture of precipitation with air pollutants, such as sulfur dioxide and nitrogen oxide.

acquaintance rape Rape that is committed by someone known by the victim.

activity theory A theory claiming that the elderly disengage, in part, because they are structurally segregated and isolated with few opportunities to engage in active roles.

adaptive discrimination Discrimination that is based on the prejudice of others.

affirmative action A broad range of policies and practices to promote equal opportunity and diversity in the workplace and on campuses.

age grading The assignment of social roles to given chronological ages.

ageism The belief that age is associated with certain psychological, behavioral, and/or intellectual traits.

Aid to Families with Dependent Children (AFDC) Before 1996, a cash assistance program that provided single parents (primarily women) and their children with a minimum monthly income.

alienation The concept used by Karl Marx to describe the condition when workers feel powerlessness and meaninglessness as a result of performing repetitive, isolated work tasks. Alienation involves becoming estranged from one's work, the products one creates, other human beings, and/or one's self; it also refers to powerlessness and meaninglessness experienced by students in traditional, restrictive educational institutions.

alternative certification programs Programs that permit college graduates, without education degrees, to be certified to teach school based on job and/or life experiences.

amalgamation The physical blending of different racial and/or ethnic groups resulting in a new and distinct genetic and cultural population; results from the intermarriage of racial and ethnic groups over generations.

androgynous Having both feminine and masculine characteristics.

anomie A state of normlessness in which norms and values are weak or unclear; results from rapid social change and is linked to many social problems, including crime, drug addiction, and violence.

antimiscegenation laws Laws that prohibited interracial marriages.

arbitration Dispute settlement in which a neutral third party listens to evidence and arguments presented by conflicting groups and arrives at a decision or outcome that the two conflicting parties agree to accept.

ascribed status A status that society assigns to an individual on the basis of factors over which the individual has no control.

assimilation The process by which minority groups gradually adopt the cultural patterns of the dominant majority group.

automation The replacement of human labor with machinery and equipment.

aversive racism A subtle, often unintentional form of prejudice that involves (1) feelings of discomfort, uneasiness, disgust, and sometimes fear (as opposed to feelings of hate and hostility); and (2) the presence of pro-white attitudes as opposed to anti-black attitudes.

barrios In the United States, slums that are occupied primarily by Latinos.

behavior-based safety programs A controversial health and safety strategy used by business management that attributes health and safety problems in the workplace to workers' behavior, rather than to work processes and conditions.

beliefs Definitions and explanations about what is assumed to be true.

bias-motivated crime See *hate crime.*

bigamy In the United States, the criminal offense of marrying one person while still legally married to another.

bilingual education Educational instruction provided in two languages—the student's native language and another language. In the United States, bilingual education involves teaching individuals in both English and their non-English native language.

biodiversity The diversity of living organisms on earth.

biofeedback A process in which information that is fed back to the brain enables a person to change his or her biological functioning. Biofeedback treatment can teach a person to influence biological responses such as heart rate, nervous system arousal, muscle contractions, and even brain-wave functioning.

bioinvasion The emergence of organisms into regions where they are not native, usually as a result of being carried in the ballast water of ships, in packing material, and in shipments of crops and other goods that are traded around the world. Invasive species may compete with native species for food, start an epidemic, or prey on natives.

biomedicalization The view that medicine not only can control particular conditions, but also can transform bodies and lives.

biphobia Negative attitudes toward bisexuality and people who identify as bisexual.

biomonitoring A method of assessing a person's exposure to environmental chemicals or their metabolites (break-down products) by testing for the presence of these in blood or urine.

bisexuality A sexual orientation that involves emotional and sexual attraction to members of both sexes.

bonded labor The repayment of a debt through labor.

bourgeoisie The owners of the means of production.

Brady Bill A law that requires a five-day waiting period for handgun purchases so that sellers can screen buyers for criminal records or mental instability.

brain drain The phenomenon in developing countries of many individuals with the highest level of skill and education leaving the country in search of work abroad.

brownfields Abandoned or undeveloped sites that are located on contaminated land.

bullying Inherent in a relationship between individuals, groups, or individuals and groups, bullying entails an imbalance of power in which one individual or group intimidates or belittles another.

burden of disease The number of deaths in a population combined with the impact of premature death and disability on that population.

capitalism An economic system in which private individuals or groups invest capital to produce goods and services, for a profit, in a competitive market.

central city The largest city in a metropolitan area.

character education Education that emphasizes the moral and interpersonal aspects of an individual.

charter schools Public schools founded by parents, teachers, and communities, and maintained by school tax dollars.

chattel slavery An old form of slavery whereby slaves are considered property that can be bought and sold.

chemical dependency A condition in which drug use is compulsive, and users are unable to stop because of physical and/or psychological dependency.

child abuse The physical or mental injury, sexual abuse, negligent treatment, or maltreatment of a child under the age of 18 by a person who is responsible for the child's welfare.

child labor Involves children performing work that is hazardous, that interferes with a child's education, or that harms a child's health or physical, mental, spiritual, or moral development.

civil union A legal status that entitles same-sex couples who apply for and receive a civil union certificate to nearly all the benefits available to married couples.

classic rape A rape committed by a stranger, with the use of a weapon, resulting in serious bodily injury to the victim.

clearance rate The percentage of crimes in a jurisdiction for which the police report and arrest.

club drugs A general term for illicit, often synthetic drugs commonly used at nightclubs or all-night dances called "raves."

Code for Corporate Citizenship A proposed change to corporate law that would say: "The duty of directors henceforth shall be to make money for shareholders but not at the expense of the environment, human rights, public health and safety, dignity of employees, and the welfare of the communities in which the company operates."

Cold War The state of political tension and military rivalry that existed between the United States and the former Soviet Union from the 1950s through the late 1980s.

colonialism When a racial and/or ethnic group from one society takes over and dominates the racial and/or ethnic group(s) of another society.

common couple violence Occasional acts of violence that result from conflict that gets out of hand between persons in a relationship.

community policing A type of policing in which uniformed police officers patrol and are responsible for certain areas of the city as opposed to simply responding to crimes as they occur.

comparable worth The belief that individuals in occupations, even in different occupations, should be paid equally if the job requires "comparable" levels of education, training, and responsibility.

comprehensive primary health care An approach to health care that focuses on the broader social determinants of health, such as poverty and economic inequality, gender inequality, environment, and community development.

comprehensive sexuality education Sex education that includes topics such as abstinence, contraception, sexually transmitted diseases, HIV/AIDS, and disease-prevention methods

compressed workweek Workplace option in which employees work full-time, but in four 10-hour rather than five 8-hour days.

computer crime Any violation of the law in which a computer is the target or means of criminal activity.

conflict perspective A sociological perspective that views society as comprising different groups and interests competing for power and resources.

control theory A theory that a strong social bond between a person and society constrains some individuals from violating norms.

conversion therapy See *reparative therapy.*

cooperative learning Learning in which a heterogeneous group of students, of varying abilities, help one another with either individual or group assignments.

corporal punishment The intentional infliction of pain for a perceived misbehavior.

corporate downsizing The corporate practice of discharging large numbers of employees. Simply put, the term "downsizing" is a euphemism for mass firing of employees.

corporate violence The production of unsafe products and the failure of corporations to provide a safe working environment for their employees.

corporate welfare Laws and policies that favor corporations, such as low-interest government loans to failing businesses and special subsidies and tax breaks to corporations.

corporatocracy A system of government that involves ties between government and corporations and that serves the interests of corporations.

covenant marriage A type of marriage offered in Louisiana that permits divorce only under condition of fault or after a marital separation of more than two years.

crack A crystallized illegal drug product made by boiling a mixture of baking soda, water, and cocaine.

crime An act or the omission of an act that is a violation of a federal, state, or local law and for which the state can apply sanctions.

cultural imperialism The indoctrination into the dominant culture of a society; when cultural imperialism exists, the norms, values, traditions, and languages of minorities are systematically ignored.

cultural lag A condition in which the material part of the culture changes at a faster rate than the nonmaterial part.

cultural sexism The ways in which the culture of society perpetuates the subordination of individuals based on their sex classification.

culture of poverty The set of norms, values, and beliefs and self-concepts that contribute to the persistence of poverty among the underclass.

cumulative trauma disorders Also known as *repetitive motion disorders*, the most common type of workplace illness in the United States; includes muscle, tendon, vascular, and nerve injuries that result from repeated or sustained actions or exertions of different body parts.

cybernation The use of machines that control other machines in the production process; characteristic of post-industrial societies that emphasize service and information professions.

cyberwarfare Utilization of information technology to attack or manipulate the military and civilian infrastructure and information systems of an enemy.

cycle of abuse A pattern in abusive relationships whereby a violent or abusive episode is followed by a makeup period when the abuser expresses sorrow and asks for forgiveness and "one more chance." The "honeymoon period" may last for days, weeks, or even months before the next outburst of violence occurs.

date-rape drugs Drugs that are used to render victims incapable of resisting sexual assaults.

debt bondage See *bonded labor.*

de facto segregation Segregation that is not required by law but exists "in fact," often as a result of housing and socioeconomic patterns.

de jure segregation Segregation that is required by law.

deconcentration The redistribution of the population from cities to suburbs and surrounding areas.

decriminalization The removal of criminal penalties for a behavior, as in the decriminalization of drug use.

Defense of Marriage Act Federal legislation that states that marriage is a legal union between one man and one woman and denies federal recognition of same-sex marriage.

deforestation The conversion of forest land to nonforest land.

deindustrialization The loss and/or relocation of manufacturing industries.

deinstitutionalization The removal of individuals with psychiatric disorders from mental hospitals and large residential institutions to outpatient community mental health centers.

demographic transition theory A theory that attributes population growth patterns to changes in birthrates and death rates associated with the process of industrialization. In preindustrial societies, the population remains stable because, although the birthrate is high, the death rate is also high. As a society becomes industrialized, the birthrate remains high, but the death rate declines, causing rapid population growth. In societies with advanced industrialization, the birthrate declines, and this decline, in conjunction with the low death rate, slows population growth.

dependency ratio The number of societal members who are under 18 or are 65 and over compared to the number of people who are between 18 and 64.

dependent variable The variable that the researcher wants to explain. See also *independent variable*.

deregulation The reduction of government control of, for example, certain drugs.

desertification The degradation of semi-arid land, which results in the expansion of desert land that is unusable for agriculture. Desertification is caused by deforestation and overgrazing by cattle and other herd animals.

deskilling The tendency for workers in a post-industrial society to make fewer decisions and for labor to require less thought.

detente The philosophy of "negotiation rather than confrontation" in reference to relations between the United States and the former Soviet Union; put forth by Secretary of State Henry Kissinger's Declaration of Principles in 1972.

deterrence The use of harm or the threat of harm to prevent unwanted behaviors.

devaluation hypothesis The hypothesis that women are paid less because the work they perform is socially defined as less valuable than the work performed by men.

differential association A theory developed by Edwin Sutherland that holds that through interaction with others, individuals learn the values, attitudes, techniques, and motives for criminal behavior.

disability-adjusted life year (DALY) Years lost to premature death and years lived with illness or disability. More simply, 1 DALY equals 1 lost year of healthy life.

discrimination Actions or practices that result in differential treatment of categories of individuals.

disengagement theory A theory that the elderly disengage from productive social roles in order to relinquish these roles to younger members of society. As this process continues, each new group moves up and replaces another, which, according to disengagement theory, benefits society and all of its members.

distance learning Learning in which, by time or place, the student is separated from the teacher.

diversity training Workplace training programs designed to reduce prejudice and discrimination in the workplace, increase employees' awareness of cultural differences in the workplace and how these differences may affect job performance.

divorce law reform Policies and proposals designed to change divorce law. Usually, divorce law reform measures attempt to make divorce more difficult to obtain.

divorce mediation A process in which divorcing couples meet with a neutral third party (mediator) who assists the individuals in resolving such issues as property division, child custody, child support, and spousal support in a way that minimizes conflict and encourages cooperation.

divorce rate The number of divorces per 1,000 population.

domestic partners A status granted to unmarried couples, including gay and lesbian couples, by some states, counties, cities, and workplaces that conveys various rights and responsibilities. These may include coverage under a partner's health and pension plan, rights of inheritance and community property, tax benefits, access to married student housing, child custody and child and spousal support obligations, and mutual responsibility for debts.

domestic terrorism Domestic terrorism, sometimes called insurgent terrorism, occurs when the terrorist act involves victims, targets, institutions, governments, or citizens from one country.

double effect In reference to physician-assisted suicide (PAS), the use of medical interventions to relieve pain and suffering but that also may hasten death.

double jeopardy See *multiple jeopardy*.

doubling time The time required for a population to double in size from a given base year if the current rate of growth continues.

drug Any substance other than food that alters the structure and functioning of a living organism when it enters the bloodstream.

drug abuse The violation of social standards of acceptable drug use, resulting in adverse physiological, psychological, and/or social consequences.

drug addiction See *chemical dependency*.

dual-use technologies Defense-funded technological innovations with commercial and civilian use.

dumbing down The lowering of educational standards or expectations by students and/or teachers.

Earned Income Tax Credit (EITC) A refundable tax credit based on a working family's income and number of children. In addition to the federal EITC, many states also have a state EITC.

ecofeminism A perspective that views environmental problems as resulting from human domination of the environment and sees connections between the domination of women, people of color, children, and the poor and the domination of nature. In contrast to a male-oriented view of natural resources as a means to an end—a means to profit and power—ecofeminists often embrace a spiritual approach to addressing environmental problems and emphasize the close connection between women and nature.

e-commerce The buying and selling of goods and services over the Internet.

economic institution The structure and means by which a society produces, distributes, and consumes goods and services.

ecosystems The complex and dynamic relationships between forms of life and the environments they inhabit.

ecoterrorism Any crime intended to protect wildlife or the environment that is violent, puts human life at risk, or results in damages of $10,000 or more.

egalitarian relationships Relationships in which partners share decision making and assign family roles based on choice rather than on traditional beliefs about gender

elder abuse The physical or psychological abuse, financial exploitation, or medical abuse or neglect of the elderly.

emotion work Work that involves caring for and negotiating and empathizing with people.

Employment Nondiscrimination Act (ENDA) A bill that would make it illegal to discriminate based on sexual orientation. ENDA was introduced in Congress in 1994, and, as of 2003, had failed to pass the Senate.

environmental footprint The effect that each person makes on the environment

environmental injustice (See also *environmental racism*.) The tendency for socially and politically marginalized groups to bear the brunt of environmental ills.

environmental racism See *environmental injustice*.

environmental refugees Individuals who have migrated because they can no longer secure a livelihood because of deforestation, desertification, soil erosion, and other environmental problems.

epidemiological transition The shift from a society characterized by low life expectancy and parasitic and infectious diseases to one characterized by high life expectancy and chronic and degenerative diseases.

ergonomics The designing or redesigning of the workplace to prevent and reduce cumulative trauma disorders.

ethnicity A shared cultural heritage and/or national origin.

e-waste Waste from electronic equipment.

extended producer responsibility An environmental strategy whereby producers are responsible for taking back their old products, phasing out the toxic materials in manufacturing, and designing cleaner products with less waste generation.

experiment A research method that involves manipulating the independent variable to determine how it affects the dependent variable.

expulsion When a dominant group forces a subordinate group to leave the country or to live only in designated areas of the country.

Faith-Based and Community Initiative An initiative in the G. W. Bush administration that views faith-based (i.e., religious) organizations as effective social service providers and allows such organizations to compete for federal funding for programs that serve the needy.

familism A value system that encourages family members to put their family's well-being above their individual and personal needs.

family A kinship system of all relatives living together or recognized as a social unit, including adopted persons.

Family and Medical Leave Act (FMLA) A federal law passed in 1993 that requires all companies with 50 or more employees to provide eligible workers (who work at least 25 hours a week and have been working for at least a year) with up to 12 weeks of job-protected, unpaid leave so they can care for a seriously ill child, spouse, or parent; stay home to care for their newborn, newly adopted, or newly placed child; or take time off when they are seriously ill.

family household As defined by the U.S. Census Bureau, a family household consists of two or more persons related by birth, marriage, or adoption who reside together.

family preservation programs In-home interventions for families who are at risk of having a child removed from the home because of abuse or neglect.

feminism The belief that women and men should have equal rights and responsibilities.

feminization of poverty The disproportionate distribution of poverty among women.

fertility rate The average number of births per woman.

fetal alcohol syndrome A syndrome characterized by serious physical and mental handicaps as a result of maternal drinking during pregnancy.

field research A method of research that involves observing and studying social behavior in settings in which it naturally occurs; includes participant observation and nonparticipant observation.

flextime An option in work scheduling that allows employees to begin and end the workday at different times each day as long as they perform 40 hours of work per week.

folkways The customs and manners of society.

forced labor Also known as *slavery*, any work that is performed under the threat of punishment and is undertaken involuntarily.

free trade agreements Pacts between two countries, or a group of countries, that make it easier to trade goods across national boundaries. Free trade agreements reduce or eliminate foreign restrictions on exports, reduce or eliminate tariffs (or taxes) on imported goods, and prevent U.S. technology from being copied and used by competitors through protection of "intellectual property rights."

functionally illiterate Being unable to carry out many of the tasks required of an adult in today's society because of reading, writing, and/or quantitative deficiencies.

future shock The state of confusion resulting from rapid scientific and technological changes that challenge traditional values and beliefs.

gateway drug A drug (e.g., marijuana) that is believed to lead to the use of other drugs (such as cocaine and heroin).

gender The social definitions and expectations associated with being male or female.

gender tourism The recent tendency for definitions of masculinity and femininity to become less clear, resulting in individual exploration of the gender continuum.

gene monopolies Exclusive control over a particular gene as a result of government patents.

gene therapy The transplantation of a healthy gene to replace a defective or missing gene.

genetically modified food Also known as genetically engineered (GE) food and genetically modified organisms (GMOs); food products that result from the process of DNA recombination, in which scientists transfer genes from one plant into the genetic code of another plant. The most common genetically modified (GM) trait is herbicide tolerance (which produces crops that tolerate weed-killing chemicals), followed by insect resistance.

genetic engineering The manipulation of an organism's genes in such a way that the natural outcome is altered.

genetic screening The use of genetic maps to detect predispositions to human traits or disease(s).

genocide The deliberate, systematic annihilation of an entire nation or people.

gentrification A type of neighborhood revitalization in which middle- and upper-income persons receive tax incentives to buy and rehabilitate older homes in a depressed neighborhood.

gerontophobia Fear or dread of the elderly.

ghettos In the United States, slums that are occupied primarily by African Americans.

glass ceiling An invisible, socially created barrier that prevents some women and other minorities from being promoted into top corporate positions.

GLBT A term used to refer collectively to gays, lesbians, bisexuals, and transgendered individuals; see also "LGBT."

global economy An interconnected network of economic activity that transcends national borders.

global warming The increasing average global air temperature, caused mainly by the accumulation of various gases (see *greenhouse gases*) that collect in the atmosphere.

globalization The economic, political, and social interconnectedness among societies throughout the world.

greenhouse gases Gases (primarily carbon dioxide, methane, and nitrous oxide) that accumulate in the atmosphere and act like the glass in a greenhouse, holding heat from the sun close to the earth.

greenwashing The way in which environmentally and socially damaging companies portray their corporate image and products as being "environmentally friendly" or socially responsible.

guerrilla warfare Warfare in which organized groups oppose domestic or foreign governments and their military forces; often involves small groups of individuals who use camouflage and underground tunnels to hide until they are ready to execute a surprise attack.

harm reduction A recent public health position that advocates reducing the harmful consequences of drug use for the user as well as society as a whole.

hate crime Also known as a "bias-motivated crime"; and "ethnoviolence," an unlawful violent act motivated by prejudice or bias. Hate crimes include intimidation (e.g., threats), destruction/damage of property, physical assault, and murder.

Head Start A project begun in 1965 by the federal government to help preschool children from disadvantaged families.

health A state of complete physical, mental, and social well-being.

health maintenance organizations (HMOs) Prepaid group plans in which a person pays a monthly premium for comprehensive health care services.

heterosexism The institutional and societal reinforcement of heterosexuality as the privileged and powerful norm. Heterosexism is based on the belief that heterosexuality is superior to homosexuality and results in prejudice and discrimination against homosexuals and bisexuals.

heterosexuality A sexual orientation that involves the predominance of emotional and sexual attraction to persons of the other sex.

home schooling The education of children at home instead of in a public or private school; often part of a fundamentalist movement to protect children from perceived non-Christian values in the public schools.

homophobia Negative attitudes and emotions toward homosexuality and those who engage in same-sex sexual behavior.

homosexuality The predominance of cognitive, emotional, and sexual attraction to persons of the same sex.

HPI-1 The Human Poverty Index for developing countries.

HPI-2 The Human Poverty Index for industrialized countries.

human capital The skills, knowledge, and capabilities of the individual.

human capital hypothesis The hypothesis that female-male pay differences are a function of differences in women's and men's levels of education, skills, training, and work experience.

Human Poverty Index (HPI) A composite measure of poverty based on three measures of deprivation: (1) deprivation of a long, healthy life; (2) deprivation of knowledge; and (3) deprivation in decent living standards.

hypothesis A prediction or educated guess about how one variable is related to another variable.

identity theft The use of someone else's identification (e.g., social security number, birth date) to obtain credit or other economic rewards.

in-vitro fertilization (IVF) The union of an egg and a sperm in an artificial setting such as a laboratory dish.

incapacitation A criminal justice philosophy that views the primary purpose of the criminal justice system as preventing criminal offenders from committing further crimes against the public by putting them in prison.

incumbent upgrading Aid programs that help residents of depressed neighborhoods buy or improve their homes and stay in the community.

independent variable The variable that is expected to explain change in the dependent variable.

Index offenses Crimes identified by the FBI as the most serious, including personal crimes (homicide, rape, robbery, assault) and property crimes (burglary, larceny, car theft, arson).

individual discrimination The unfair or unequal treatment of individuals because of their group membership.

individualism The tendency to focus on one's individual self-interests rather than on the interests of one's family and community.

Industrial Revolution The period between the mid-eighteenth and the early nineteenth century when machines and factories became the primary means for producing goods. The Industrial Revolution led to profound social and economic changes.

infant mortality rate The number of deaths of infants under 1 year of age per 1,000 live births in a year.

infantilizing elders The portrayal of the elderly in the media as childlike in terms of clothes, facial expression, temperament, and activities.

infotech An abbreviation for "information technology"; any technology that carries information.

infowar The utilization of technology to manipulate or attack an enemy's military and civilian infrastructure and information systems.

infrastructure The underlying foundation that enables a city to function. Infrastructure includes water and sewer lines, phone lines, electricity cables, sidewalks, streets, curbs, lighting, and storm drainage systems.

institution An established and enduring pattern of social relationships. The five traditional social institutions are family, religion, politics, economics, and education. Institutions are the largest elements of social structure.

institutional discrimination Discrimination in which the normal operations and procedures of social institutions result in unequal treatment of minorities.

integration hypothesis A theory that the only way to achieve quality education for all racial and ethnic groups is to desegregate the schools.

intergenerational poverty Poverty that is transmitted from one generation to the next.

internalized homophobia A sense of personal failure and self-hatred among some lesbians and gay men due to social rejection and stigmatization.

Internet An international information infrastructure available through many universities, research institutes, government agencies, and businesses; developed in the 1970s as a Defense Department experiment.

intimate partner violence Actual or threatened violent crimes committed against persons by their current or former spouses, boyfriends, or girlfriends.

intimate terrorism Almost entirely perpetrated by men, a form of violence that is motivated by a wish to control one's partner and involves the systematic use of not only violence, but economic subordination, threats, isolation, verbal and emotional abuse, and other control tactics. This form of violence is more likely to escalate over time and to involve serious injury.

Jim Crow laws Laws that separated blacks from whites by prohibiting blacks from using "white" buses, hotels, restaurants, drinking fountains, and restrooms.

job exportation The relocation of U.S. jobs to other countries where products can be produced more cheaply.

job sharing A work option in which two people, often husband and wife, share and are paid for one job.

labeling theory A symbolic interactionist theory that is concerned with the effects of labeling on the definition of a social problem (e.g.,

a social condition or group is viewed as problematic if it is labeled as such) and with the effects of labeling on the self-concept and behavior of individuals (e.g., the label "juvenile delinquent" may contribute to the development of a self-concept and behavior consistent with the label).

labor unions Worker organizations that originally developed to protect workers and to represent them at negotiations between management and labor. Labor unions have played an important role in fighting for fair wages and benefits, healthy and safe work environments, and other forms of worker advocacy.

latent functions Consequences that are unintended and often hidden, or unrecognized; for example, a latent function of education is to provide schools that function as babysitters for employed parents.

law Norms that are formalized and backed by political authority.

legalization Making prohibited behavior legal; for example, legalizing marijuana or prostitution.

lesbigays A collective term sometimes used to refer to lesbians, gays, and bisexuals.

LGBT A term used to refer collectively to lesbians, gays, bisexuals, and transgendered individuals; see also "GLBT."

life chances A term used by Max Weber to describe the opportunity to obtain all that is valued in society, including happiness, health, income, and education.

life expectancy The average number of years that a person born in a given year can expect to live.

living wage laws Laws that require state or municipal contractors, recipients of public subsidies or tax breaks, or, in some cases, all businesses to pay employees wages significantly above the federal minimum, enabling families to live above the poverty line.

long-term unemployment rate The share of the unemployed who have been out of work for 27 weeks or more.

looking-glass self The idea that individuals develop their self-concept through social interaction.

macro sociology The study of large aspects of society, such as institutions and large social groups.

MADD (Mothers Against Drunk Driving) A social action group committed to reducing drunk driving.

mailbox economy The tendency for a substantial portion of local economies to be dependent on pension and social security checks received in the mail by older residents.

Malthusian theory The theory proposed by Thomas Malthus in which he predicted that population would grow faster than the food supply and that masses of people were destined to be poor and hungry. According to Malthus, food shortages would lead to war, disease, and starvation that would eventually slow population growth.

managed care A type of medical insurance plan that controls costs through monitoring and controlling the decisions of health care providers.

manifest functions Consequences that are intended and commonly recognized; for example, a manifest function of education is to transmit knowledge and skills to youth.

marital decline perspective A pessimistic view of the current state of marriage that includes the belief that (1) personal happiness has become more important than martial commitment and family obligations and (2) the decline in lifelong marriage and the increase in single-parent families have contributed to a variety of social problems, such as poverty, delinquency, substance abuse, violence, and the erosion of neighborhoods and communities.

marital resiliency perspective A view of the current state of marriage that includes the belief that (1) poverty, unemployment, poorly funded schools, discrimination, and the lack of basic services (such as health insurance and child care) represent more serious threats to the well-being of children and adults than does the decline in married two-parent families; and (2) divorce provides a second chance at happiness for adults and an escape from dysfunctional and aversive home environments for many children.

master status The status that is considered the most significant in a person's social identity.

maternal mortality rate A measure of deaths that result from complications associated with pregnancy, childbirth, and unsafe abortion.

McDonaldization The process by which the principles of the fast-food industry—efficiency, predictability, calculability, and control through technology—are being applied to more and more sectors of society.

means-tested programs Public assistance programs that have eligibility requirements based on income and assets.

mechanization The use of tools to accomplish tasks previously done by workers; characteristic of agricultural societies that emphasize the production of raw materials.

Medicaid A public assistance program designed to provide health care for the poor.

medicalization The tendency to label behaviors and conditions as medical problems in need of medical intervention.

Medicare A national public insurance program created by Title XVIII of the Social Security Act of 1965; originally designed to protect people 65 years of age and older from the rising costs of health care. In 1972, Medicare was extended to permanently disabled workers and their dependents and persons with end-stage renal disease.

Medigap The difference between Medicare benefits and the actual cost of medical care.

megacities Cities with 10 million residents or more.

melting pot The product of different groups coming together and contributing equally to a new, common culture.

mental health The successful performance of mental function, resulting in productive activities, fulfilling relationships with other people, and the ability to adapt to change and to cope with adversity.

mental illness A term that refers collectively to all mental disorders, which are health conditions that are characterized by alterations in thinking, mood, and/or behavior associated with distress and/or impaired functioning, and that meet specific criteria (such as level of intensity and duration) specified in the American Psychiatric Association's classification manual used to diagnose mental disorders: *Diagnostic and Statistical Manual of Mental Disorders*.

metropolis From the Greek meaning "mother city." See also *metropolitan area*.

metropolitan area A densely populated core area and any adjacent communities that have a high degree of social and economic integration with the core; a large city and its surrounding suburbs; also called a "metropolis."

micropolitan area A small city located beyond congested metropolitan areas.

micro-society school A simulation of the "real" or nonschool world where students design and run their own democratic, free-market society within the school.

micro sociology The study of the social psychological dynamics of individuals interacting in small groups.

military-industrial complex A term used by Dwight D. Eisenhower to connote the close association between the military and defense industries.

Millennium Development Goals A set of eight goals for reducing poverty and improving lives that leaders from 191 United Nations member countries pledged to achieve.

minority A category of people who are denied equal access to positions of power, prestige, and wealth because of their group membership.

mixed-use land use The location of homes, schools, shops, workplaces, and parks in close proximity to each other. A main purpose of mixed-use land use is to reduce driving distances.

modern racism A subtle form of racism that involves the belief that serious discrimination in America no longer exists, that any continuing racial inequality is the fault of minority group members, and the demands for affirmative action for minorities are unfair and unjustified.

modernization theory A theory claiming that as society becomes more technologically advanced, the position of the elderly declines.

monogamy Marriage between two partners; the only legal form of marriage in the United States.

morbidity Illnesses, symptoms, and impairments they produce.

mores Norms that have moral basis.

mortality Death.

multicultural education A broad range of programs and strategies designed to dispel myths, stereotypes, and ignorance about minorities; promote tolerance and appreciation of diversity; and include minority groups in the school curriculum.

multiple chemical sensitivity (MCS) Also known as "environmental illness," a condition whereby individuals experience adverse reactions when exposed to low levels of chemicals found in everyday substances (vehicle exhaust, fresh paint, house cleaning products, perfume and other fragrances, synthetic building materials, and numerous other petrochemical-based products). Symptoms of MCS include headache, burning eyes, difficulty breathing, stomach distress/nausea, loss of mental concentration, and dizziness. The onset of MCS is often linked to acute exposure to a high level of chemicals or to chronic long-term exposure.

multiple jeopardy The disadvantages associated with being a member of two or more minority groups.

mutual violent control A rare pattern of abuse when two intimate terrorists battle for control. See also *intimate terrorism*.

mutually assured destruction (MAD) A perspective that argues that if both sides in a conflict were to lose in a war, neither would initiate war.

naturalized citizen An immigrant who applied and met the requirements for U.S. citizenship.

needle exchange programs Programs designed to reduce transmission of HIV among injection drug users, their sex partners, and their children, by providing new, sterile syringes in exchange for used, contaminated syringes.

neglect A form of abuse involving the failure to provide adequate attention, supervision, nutrition, hygiene, health care, and a safe and clean living environment for a minor child or a dependent elderly individual.

New Urbanism A movement in urban planning, similar to *smart growth*, that approaches the idea of sustainable urban communities with the goal of raising the quality of life for all those in the community by creating compact communities with a sustainable infrastructure.

no-fault divorce A divorce that is granted based on the claim that there are irreconcilable differences within a marriage (as opposed to one spouse being legally at fault for the marital breakup).

non-family household A household that consists of one person who lives alone, two or more people as roommates, and cohabiting heterosexual or homosexual couples involved in a committed relationship.

norms Socially defined rules of behavior, including folkways, mores, and laws.

nuclear winter The predicted result of a thermonuclear war whereby dust storms and concentrations of particles would block out vital sunlight, lower temperatures in the Northern Hemisphere, and lead to the death of most living things on earth.

objective element of social problems Awareness of social conditions through one's own life experience and through reports in the media.

occupational sex segregation The concentration of women in certain occupations and of men in other occupations.

one-drop rule A rule that specified that even one drop of Negroid blood defined a person as black and, therefore, eligible for slavery.

operational definition In research, a definition of a variable that specifies how that variable is to be measured (or was measured) in the research.

organized crime Criminal activity conducted by members of a hierarchically arranged structure devoted primarily to making money through illegal means.

overt discrimination Discrimination that occurs because of an individual's own prejudicial attitudes.

Parental Alienation Syndrome An emotional and psychological disturbance in which children engage in exaggerated and unjustified denigration and criticism of a parent.

parity The concept of equality between mental health care insurance coverage and other health care insurance.

partial birth abortion Also called an intact dilation and extraction (D&X) abortion, the procedure may entail delivering the limbs and the torso of the fetus before it has expired.

patriarchal Literally, rule by father; today, connotes rule by males.

patriarchal terrorism A form of abuse in which husbands control their wives by the systematic use of not only violence, but also economic subordination, threats, isolation, and other control tactics.

patriarchy A tradition in which families are male dominated.

perinatal transmission The transmission of a virus (such as HIV) from an infected mother to a fetus or newborn.

Personal Responsibility and Work Opportunity Reconciliation Act (PRWORA) The 1996 legislation that affected numerous public assistance programs, primarily in the form of cutbacks and eligibility restrictions. This law ended Aid to Families with Dependent Children (AFDC) and replaced it with Temporary Assistance to Needy Families (TANF).

phased retirement Allows workers to ease into retirement by reducing hours worked a day, days worked a week, or months worked a year.

pink-collar jobs Jobs that offer few benefits, often have low prestige, and are disproportionately held by women.

planned obsolescence The manufacturing of products that are intended to become inoperative or outdated in a fairly short period of time.

pluralism A state in which racial and/or ethnic groups maintain their distinctness but respect each other and have equal access to social resources.

plural marriage The term used by Mormons/Latter Day Saints to refer to marriages involving one husband and two or more wives (also known as *polygyny*).

polyandry The concurrent marriage of one woman to two or more men.

polygamy A form of marriage in which one person may have two or more spouses.

polygyny The concurrent marriage of one man to two or more women.

population density The number of people per unit of land area.

population momentum Continued population growth that occurs even if a population achieves replacement-level fertility (2.1 births per woman) due to past high fertility rates which have resulted in large numbers of young women who are currently entering their childbearing years.

population transfer See *expulsion*.

post-industrialization The shift from an industrial economy dominated by manufacturing jobs to an economy dominated by service-oriented, information-intensive occupations.

postmodernism A worldview that questions the validity of rational thinking and the scientific enterprise.

post-traumatic stress disorder A set of symptoms that may result from any traumatic experience, including crime victimization, war, natural disasters, or abuse.

poverty The lack of resources necessary for material well-being—most importantly food and water, but also housing, land, and health care.

poverty line An annual level of personal or household income below which individuals or families are considered officially poor by the government.

preferred provider organizations (PPOs) Health care organizations in which employers who purchase group health insurance agree to send their employees to certain health care providers or hospitals in return for cost discounts.

prejudice Negative attitudes and feelings toward or about an entire category of people.

primary aging Biological changes associated with aging that are due to physiological variables such as cellular and molecular variation (e.g., gray hair).

primary assimilation The integration of different groups in personal, intimate associations such as friends, family, and spouses.

primary deviance Deviance committed before the person is caught and labeled as an offender; i.e., the act is defined as deviant.

primary group Small groups characterized by intimate and informal interaction.

primary prevention strategies Family violence prevention strategies that target the general population.

privatization A practice in which states hire businesses to provide services or operate local institutions, replacing government employees who formerly carried out these functions.

probation The conditional release of an offender who, for a specific time period and subject to certain conditions, remains under court supervision in the community.

proletariat Workers, often exploited by the bourgeoisie.

pronatalism A cultural value that promotes having children.

public assistance A general term referring to some form of support by the government to citizens who meet certain established criteria.

public housing An assistance program that provides federal subsidies for low-income housing units built, owned, and operated by local public housing authorities; also known as *subsidized housing.*

race A category of people who share distinct physical characteristics that are deemed socially significant.

racial profiling The law enforcement practice of targeting suspects based upon race.

racism The belief that certain groups of people are innately inferior to other groups of people based on their racial classification. Racism serves to justify discrimination against groups that are perceived as inferior.

refugees Immigrants who apply from abroad for admission on the basis of persecution or fear of persecution for their political or religious beliefs.

regionalism A form of collaboration among central cities and suburbs that encourages local governments to share responsibility for common problems.

registered partnerships Federally recognized same-sex relationships that convey most but not all of the rights of marriage (some countries also offer registered partnerships to opposite-sex couples).

regressive taxes Taxes that absorb a much higher proportion of the incomes of lower-income households than of higher-income households.

rehabilitation A criminal justice philosophy that views the primary purpose of the criminal justice system as changing the criminal offender through such programs as education and job training, individual and group therapy, substance abuse counseling, and behavior modification.

relative poverty A deficiency in material and economic resources compared with some other population.

reparative therapy Various therapies that are aimed at changing homosexuals' sexual orientation.

replacement level fertility The fertility rate that is needed to stabilize the population, which is 2.1 births per woman—slightly more than 2 because not all female children will live long enough to reach their reproductive years.

restorative justice A philosophy primarily concerned with reconciling conflict between the offender, the community, and the victim.

road rage Aggressive and violent driving behavior.

role A set of rights, obligations, and expectations associated with a status.

sample In survey research, the portion of the population selected to be questioned.

sanctions Social consequences for conforming to or violating norms. Types of sanctions include positive, negative, formal, and informal.

sandwich generation The generation that has the responsibility of simultaneously caring for their children and their aging parents.

school vouchers Tax credits that are transferred to the public or private school of a parent's choice.

science The process of discovering, explaining, and predicting natural or social phenomena.

scientific apartheid The growing gap between the industrial and developing countries in the rapidly evolving knowledge frontier.

second shift The household work and child care that employed parents (usually women) do when they return home from their jobs.

secondary aging Biological changes associated with aging that can be attributed to poor diet, lack of exercise, and increased stress.

secondary assimilation The integration of different groups in public areas and in social institutions, such as neighborhoods, schools, the workplace, and in government.

secondary deviance Deviance that results from being caught and labeled as an offender; i.e., the actor is defined as deviant.

secondary group A group characterized by impersonal and formal interaction.

secondary prevention strategies Family violence prevention strategies that target groups thought to be at high risk for family violence.

Section 8 housing A federal low-income housing program in which federal rent subsidies are provided either to tenants (in the form of certificates and vouchers) or to private landlords.

segregation The physical and social separation of categories of individuals, such as racial or ethnic groups.

selective primary health care An approach to health care that uses technocratic solutions to target a specific health problem, such as using immunization and oral rehydration therapy to promote child survival.

self-fulfilling prophecy A concept referring to the tendency for people to act in a manner consistent with the expectations of others.

senescence The biology of aging.

serial monogamy A succession of marriages in which a person has more than one spouse

over a lifetime but is legally married to only one person at a time.

sex A person's biological classification as male or female.

sexism The belief that there are innate psychological, behavioral, and/or intellectual differences between females and males and that these differences connote the superiority of one group and the inferiority of another.

sex slavery A form of forced labor in which girls are forced into prostitution by their own husbands, fathers, and brothers to earn money to pay family debts, or they are lured by offers of good jobs and then forced to work in brothels under the threat of violence.

sexual aggression Sexual interaction that occurs against one's will through the use of physical force, threat of force, pressure, use of alcohol/drugs, or use of position of authority.

sexual harassment An employer's requiring sexual favors in exchange for a promotion, salary increase, or any other employee benefit and/or the existence of a hostile environment that unreasonably interferes with job performance, as in the case of sexually explicit remarks or insults being made to an employee.

sexual orientation The identification of individuals as heterosexual, bisexual, or homosexual, based on their emotional and sexual attractions, relationships, self-identity, and lifestyle.

shaken baby syndrome A form of child abuse whereby the caretaker shakes a baby to the point of causing the child to experience brain or retinal hemorrhage. Shaken baby syndrome most often occurs in response to a baby, who typically is younger than 6 months, who won't stop crying.

single-payer system A single tax-financed public insurance program that replaces private insurance companies.

slavery The loss of free will, where a person is forced through violence or the threat of violence to give up the ability to sell freely his/her labor power.

slums Concentrated areas of poverty and poor housing in urban areas.

smart growth A strategy for managing urban sprawl that serves the economic, environmental, and social needs of communities.

social class Group of people who share a similar position or social status within the stratification system.

social group Two or more people who have a common identity, interact, and form a social relationship. Institutions are made up of social groups.

social movements An organized group with a common purpose to either promote or resist social change through collective action.

social problem A social condition that a segment of society views as harmful to members of society and in need of remedy.

social promotion The passing of students from grade to grade even if they are failing.

socialism An economic ideology that emphasizes public rather than private ownership.

Theoretically, goods and services are equitably distributed according to the needs of the citizens.

socialized medicine National health insurance systems in countries such as Canada, Great Britain, Sweden, Germany, and Italy.

sociological imagination A term coined by C. Wright Mills to refer to the ability to see the connections between our personal lives and the social world in which we live.

sodomy Oral and anal sexual acts.

soft money Money that flows through a loophole in the law to provide political parties, candidates, and contributors a means to evade federal limits on political contributions.

state The organization of the central government and government agencies such as the armed forces, police force, and regulatory agencies.

State Children's Health Insurance Program (SCHIP) A government health program funded by state and federal funds that was created to expand health coverage to uninsured children, many of whom come from families with incomes too high to qualify for Medicaid but too low to afford private health insurance.

status A position a person occupies within a social group.

stem cells Undifferentiated cells that can produce any type of cell in the human body.

stereotypes Exaggerations or generalizations about the characteristics and behavior of a particular group.

stigma Any personal characteristic associated with social disgrace, rejection, or discrediting.

strain theory A theory that when legitimate means of acquiring culturally defined goals are limited by the structure of society, the resulting strain may lead to crime or other deviance.

structural-functionalism A sociological perspective that views society as a system of interconnected parts that work together in harmony to maintain a state of balance and social equilibrium for the whole; focuses on how each part of society influences and is influenced by other parts.

structural sexism The ways in which the organization of society, and specifically its institutions, subordinate individuals and groups based on their sex classification.

structured choice Choices that are limited by the structure of society.

subcultural theories A set of theories arguing that certain groups or subcultures in society have values and attitudes that are conducive to crime and violence.

subculture The distinctive lifestyles, values, and norms of discrete population segments within a society.

subjective element of social problems The belief that a particular social condition is harmful to society, or to a segment of society, and that it should and can be changed.

suburbanization The development of urban areas outside central cities, and the movement of populations into these areas.

suburbs The urbanlike areas surrounding central cities.

survey research A method of research that involves eliciting information from respondents through questions; includes interviews (telephone or face-to-face) and written questionnaires.

sustainable development Societal development that meets the needs of current generations without threatening the future of subsequent generations.

sweatshops Work environments that are characterized by less than minimum wage pay, excessively long hours of work often without overtime pay, unsafe or inhumane working conditions, abusive treatment of workers by employers, and/or the lack of worker organizations aimed at negotiating better work conditions.

symbol Something that represents something else.

symbolic interactionism A sociological perspective emphasizing that human behavior is influenced by definitions and meanings that are created and maintained through symbolic interaction with others.

technological dualism A term referring to the tendency for technology to have both positive (e.g., time saving) and negative (e.g., unemployment) consequences.

technological fix The use of scientific principles and technology to solve social problems.

technology Activities that apply the principles of science and mechanics to the solution of specific problems.

technology-induced diseases Diseases that result from the use of technological devices, products, and/or chemicals.

telework A form of work that allows employees to work part- or full-time at home or at a satellite office.

Temporary Assistance to Needy Families (TANF) The welfare program that resulted from 1996 welfare reform legislation. Under the TANF program, which replaced Aid to Families with Dependent Children (AFDC), after two consecutive years of receiving aid, welfare recipients are required to work at least 20 hours per week or to participate in a state-approved work program (few exceptions are made). A lifetime limit of five years is set for families to receive benefits.

terrorism The premeditated use or threatened use of violence by an individual or group to gain a political objective.

tertiary prevention strategies Family violence prevention strategies that target families who have experienced family violence.

therapeutic cloning Use of stem cells from human embryos to produce body cells that can be used to grow needed organs or tissues.

therapeutic communities Organizations in which approximately 35 to 100 individuals reside for up to 15 months to abstain from drugs, develop marketable skills, and receive counseling.

total fertility rate The average lifetime number of births per woman in a population.

total immersion An educational program in which students, particularly elementary-age students, receive literacy and communication instruction entirely in a foreign language, usually Spanish.

tracking An educational practice in which students are grouped together on the basis of similar levels of academic achievement and abilities.

transgendered individuals Includes a range of people whose gender identities do not conform to traditional notions of masculinity and femininity. Transgendered individuals include homosexuals, bisexuals, cross-dressers, transvestites, and transsexuals.

transnational corporations Also known as *multinational corporations,* corporations that have their home base in one country and branches, or affiliates, in other countries.

transnational crime Crime that, directly or indirectly, involves more than one country.

transnational terrorism Terrorism that occurs when a terrorist act in one country involves victims, targets, institutions, governments, or citizens of another country.

triad dispute resolution Dispute resolution that involves two disputants and a negotiator.

triangulation The use of multiple methods and approaches to study a social phenomenon.

triple jeopardy See *multiple jeopardy.*

underclass A persistently poor and socially disadvantaged group that disproportionately experiences early sexual activity, unmarried parenthood, joblessness, reliance on public assistance, illegitimate income-producing activities (e.g., selling drugs), and substance use.

underemployment Unemployed workers as well as (1) those working part-time but who wish to work full-time ("involuntary" part-timers); (2) those who want to work but have been discouraged from searching by their lack of success ("discouraged" workers), and (3) others who are neither working nor seeking work but who indicate that they want and are available to work and have looked for employment in the last 12 months.

unemployment To be currently without employment, actively seeking employment, and available for employment, according to measures of U.S. unemployment. Unemployment figures do not include discouraged workers, who have given up on finding a job and are no longer looking for employment.

union density The percentage of workers who belong to unions.

universal health care See *socialized medicine.*

universal living wage A proposed wage policy whereby wages would be based on a single national formula that indexes the minimum wage to the local cost of housing as set each year by the U.S. Department of Housing and Urban Development (HUD). The proposed universal living wage is based on the premise that a person who works 40 hours a week should be able to afford basic housing, spending no more than 30 percent of income on housing.

upskilling The opposite of de-skilling; upskilling reduces employee alienation and increases decision-making powers.

urban area A spatial concentration of people whose lives are centered around nonagricultural activities.

urban sprawl The ever-increasing outward growth of urban areas.

urbanization The transformation of a society from a rural to an urban one.

values Social agreements about what is considered good and bad, right and wrong, desirable and undesirable.

variable Any measurable event, characteristic, or property that varies or is subject to change.

victimless crimes Illegal activities, such as prostitution or drug use, that have no complaining party; also called "vice crimes."

violent resistance Acts of violence by a partner that are committed in self-defense. Violent resistance is almost exclusively perpetrated by women against a male partner.

virtual reality Computer-generated three-dimensional worlds that change in response to the movements of the head or hand of the individual; a simulated experience of people, places, sounds, and sights.

war Organized armed violence aimed at a social group in pursuit of an objective.

wealth The total assets of an individual or household, minus liabilities.

wealthfare Governmental policies and regulations that economically favor the wealthy.

white-collar crime Includes both occupational crime, in which individuals commit crimes in the course of their employment, and corporate crime, in which corporations violate the law in the interest of maximizing profit.

WMD Weapons of mass destruction including chemical, biological, and nuclear weapons.

working poor Individuals who spend at least 27 weeks a year in the labor force (working or looking for work), but whose income falls below the official poverty line.

References

Chapter 1

Alvarado, Diana. 2000. "Student Activism Today." *Diversity Digest*. Available at http://www.inform.umd.edu/Diversity Web/Digest/sm99/activism.html

The American Freshman: National Norms for Fall 2004. 2005 (January). Los Angeles: Higher Education Research Institute, UCLA.

Batsell, Jake. 2002 "USA: Students Campaign for Coffee in Good Conscience." *The Seattle Times,* March 17.

Beeby, Dean. 2005. "Census Includes Same Sex Marriage Question." *Globe and Mail,* April 18. Available at http://globeand mail.com

Berkowitz, Elana. 2005. "They Say Tomato, Students say Justice." *Campus Progress,* March 18.

Blumer, Herbert. 1971. "Social Problems as Collective Behavior." *Social Problems* 8(3):298–306.

Boston College. 2003. *The Anti-Sweatshop Movement and Boston College.* Available at http://www.bc.edu/bc_org/cas/soc/justice

Caldas, Stephen, and Carl L. Bankston III. 1999. "Black and White TV: Race, Television Viewing, and Academic Achievement." *Sociological Spectrum* 19:39–61.

Canedy, Dana. 2003. "Lifting Veil for Photo ID Goes Too Far, Driver Says." *New York Times,* June 27. Available at http://www.nytimes.com/2003/06/27/national

Centers for Disease Control and Prevention. 1998 (September 18). *Trends in Sexual Risk Behaviors Among High School Students: United States, 1991–1997*. Morbidity and Mortality Weekly, Report 47. Available at http://www.cdc.gov/nchstp/dstd/MMWRs/Trends_Risk_Behaviors_HS_students.htm

Centers for Disease Control and Prevention. 1999. *Youth Risk Behavior Survey,* Tables 30 and 32. Atlanta: Centers for Disease Control.

Centers for Disease Control and Prevention. 2002. *Sexual Behaviors.* Youth Risk Behavior Surveillance System. Available at http://apps.nccde.cdc.gov/yrbss

Centers for Disease Control and Prevention. 2004 (May 21). *Youth Risk Behavior Surveillance: United States, 2003.* Morbidity and Mortality Weekly Report 53, no. SS-2. Available at http://www.cdc.gov/mmwr/PDF/SS/SS5302.pdf

Dordick, Gwendolyn A. 1997. *Something Left to Lose: Personal Relations and Survival Among New York's Homeless.* Philadelphia: Temple University Press.

Eitzen, Stanley, and Maxine Baca Zinn. 2000. *Social Problems.* Boston, MA: Allyn and Bacon.

Engel, Robin Shepard, and Robert E. Worden. 2003. "Police Officers' Attitudes, Behavior, and Supervisory Influences: An Analysis of Problem Solving." *Criminology* 41:131–166.

Fleming, Zachary. 2003. "The Thrill of It All." In *In Their Own Words,* Paul Cromwell, ed. (pp. 99–107). Los Angeles, CA: Roxbury.

Gallup Poll. 2005. *State of the Nation.* Available at http://www.gallup.com/poll/stateNation

Hewlett, Sylvia Ann. 1992. *When the Bough Breaks: The Cost of Neglecting Our Children.* New York: Harper Perennial.

Human Development Report. 2004. *2004 Human Development Report: Foreword.* Available at http://www.undp.org/hdr2004

Jacobs, Bruce A. 2003. "Researching Crack Dealers." In *In Their Own Words,* Paul Cromwell, ed. (pp. 1–11). Los Angeles, CA: Roxbury.

Jekielek, Susan M. 1998. "Parental Conflict, Marital Disruption, and Children's Emotional Well-Being." *Social Forces* 76: 905–935.

Killer Coke. 2005. Available at http://killer coke.org/grincamp.htm

Kmec, Julie A. 2003. "Minority Job Concentration and Wages." *Social Problems* 50:38–59.

Lewis, Barbara A. 1991. *The Kid's Guide to Social Action,* pp. 110–111. Minneapolis, MN: Free Spirit Publishing.

Merton, Robert K. 1968. *Social Theory and Social Structure.* New York: Free Press.

Mills, C. Wright. 1959. *The Sociological Imagination.* London: Oxford University Press.

Newport, Frank. 2003. *Public's Satisfaction at Lowest Point in Six Years.* Gallup Poll. Available at http://www.gallup.com

Newport, Frank. 2005. *Update: Americans Satisfaction with Aspects of Life in U.S.* Gallup Organization, January 17. Available at http://gallup.com/poll/content

Palacios, Wilson R., and Melissa E. Fenwick. 2003. "'E' is for Ecstasy." In *In Their Own Words,* Paul Cromwell, ed. (pp. 277–283). Los Angeles, CA: Roxbury.

Reiman, Jeffrey. 2004. *The Rich Get Richer and the Poor Get Prison.* Boston, MA: Allyn and Bacon.

Romer, D., R. Hornik, B. Stanton, M. Black, X. Li, I. Ricardo, and S. Feigelman. 1997. "Talking Computers: A Reliable and Private Method to Conduct Interviews on Sensitive Topics with Children." *Journal of Sex Research* 34:3–9.

---

*The authors and Wadsworth acknowledge that some of the Internet sources may have become unstable, that is, they are no longer hot links to the intended reference. In that case, the reader may want to access the article through the search engine or archives of the homepage cited (e.g., fbi.gov, cbsnews.com).

Sax, Linda, Sylvia Hurtado, Jennifer Lindholm, Alexander Astin, William Korn,
and Kathryn Mahoney. 2004. *The American Freshman: National Norms for Fall
2004*. Los Angeles, CA: Higher Education
Research Institute.

Schwalbe, Michael. 1998. *The Sociologically
Examined Life: Pieces of the Conversation*. Mountain View, CA: Mayfield.

Thomas, W. I. 1966 [1931]. "The Relation of
Research to the Social Process." In *W. I.
Thomas on Social Organization and So-
cial Personality*, Morris Janowitz, ed.
(pp. 289–305). Chicago: University of
Chicago Press.

Troyer, Ronald J., and Gerald E. Markle. 1984.
"Coffee Drinking: An Emerging Social
Problem." *Social Problems* 31:403–416.

Ukers, William H. 1935. *All About Tea*,
Vol. 1. New York: Tea and Coffee Trade
Journal Co.

Weir, Sara, and Constance Faulkner. 2004.
*Voices of a New Generation: A Feminist
Anthology*. Boston, MA: Pearson Educa-
tion Inc.

Wilson, John. 1983. *Social Theory*. Englewood Cliffs, NJ: Prentice Hall.

Chapter 2

"Access to Free Health Care Means Crime
Does Pay for Some Sick Inmates." 2001.
The Independent, April 20. Available at
http://theIndependent.com

Alan Guttmacher Institute. 2004. *New Evidence from Africa, Asia, and Latin America Underscores Impact of Violence
Against Women*. News release, December 10. Available at http://www
.guttmacher.org

Allen, P. L. 2000. *The Wages of Sin: Sex and
Disease, Past and Present*. Chicago: University of Chicago Press.

Altman, Dennis. 2003. "Understanding HIV/
AIDS as a Global Security Issue." In
Health Impacts of Globalization, Kelley
Lee, ed. (pp. 33–46). New York: Palgrave
MacMillan.

American Psychiatric Association. 2000. *Di-
agnostic and Statistical Manual of Mental
Disorders*, 4th ed. Text Revision DSM–TR.
Washington, DC: American Psychiatric
Association.

Ashford, Lori, and Donna Clifton. 2005 (February). *Women of Our World: 2005*. Population Reference Bureau. Available at
http://www.prb.org

Baker, Dean, and David Rosnick. 2005
(March 24). *The Burden of Social Security Taxes and the Burden of Excessive
Health Care Costs*. Center for Economic
and Policy Research. Available at http://
www.cepr.net

Barker, Kristin. 2002. "Self-Help Literature
and the Making of an Illness Identity: The
Case of Fibromyalgia Syndrome (FMS)."
Social Problems 49(3):279–300.

Bartley, Mel, Jane Ferrie, and Scott M.
Montgomery. 2001. "Living in a High-
Unemployment Economy: Understanding
the Health Consequences." In *Social De-
terminants of Health*, M. Marmot and
R. G. Wilkinson, eds. (pp. 81–104). New
York: Oxford University Press.

Bowman, S. A., S. L. Gortmaker, C. B. Ebbeling, M. A. Pereira, and D. S. Ludwig.
2004. "Effects of Fast-Food Consumption
on Energy Intake and Diet Quality Among
Children in a National Household Survey." *Pediatrics* 113:112–118.

Center for Mental Health Services. 2004.
Mental Health, United States, 2002.
Rockville, MD: Substance Abuse and
Mental Health Services Administration.

Centers for Disease Control and Prevention.
2004a. *Defining Overweight and Obesity*.
Available at http://www.cdc.gov

Centers for Disease Control and Prevention.
2004b. *HIV/AIDS Surveillance Report,
2003* (Vol. 15). Available at
http://www.cdc.gov/hiv/stats

Centers for Disease Control and Prevention.
2005a. *HIV/AIDS Among African Americans*. Fact sheet. Available at http://www
.cdc.gov

Centers for Disease Control and Prevention.
2005b. "HIV Prevalence, Unrecognized
Infection, and HIV Testing Among Men
Who Have Sex with Men: Five U.S. Cities,
June 2004–April 2005." *Morbidity and
Mortality Weekly Report* 54:597–601.

Centers for Disease Control and Prevention.
2005c (July 15). "Update: Syringe Exchange Programs—United States, 2002."
Morbidity and Mortality Weekly Report
54(27):673-676.

Children's Defense Fund. 2000. *The State of
America's Children Yearbook 2000*. Available at http://www.childrensdefense
.org/keyfacts.htm

Clarke, Adele E., Laura Mamo, Jennifer R.
Fishman, Janet K. Shim, and Jennifer Ruth
Fosket. 2003. "Biomedicalization: Tech-
noscientific Transformations of Health,
Illness, and U.S. Biomedicine." *American
Sociological Review* 68:161–194.

Cockerham, William C. 2004. *Medical Sociology*, 9th ed. Upper Saddle River, NJ: Prentice Hall.

Cockerham, William C. 2005. *Sociology of
Mental Disorder*, 7th ed. Upper Saddle
River, NJ: Prentice Hall.

Conyers, John. 2003. "A Fresh Approach to
Health Care in the United States: Improved and Expanded Medicare for All."
American Journal of Public Health
93(2):193.

Diamond, Catherine, and Susan Buskin. 2000.
"Continued Risky Behavior in HIV-
Infected Youth." *American Journal of
Public Health* 90(1):115–118.

Durkheim, Emile. 1951 [1897]. *Suicide: A
Study in Sociology*, J. A. Spaulding and
G. Simpson, trans.; G. Simpson, ed. New
York: Free Press.

Egwu, Igbo N. 2002. "Health Promotion Challenges in Nigeria: Globalization, Tobacco
Marketing, and Policies." In *Health Communication in Africa: Contexts, Constraints, and Lessons*, A. O. Alali and
B. A. Jinadu, eds. (pp. 40–55). Lanham,
MD: University Press of America.

Ehrenreich, Barbara. 2005. "A Society That
Throws the Sick Away." *Los Angeles
Times*, April 28. Available at http://
www.latimes.com

Ezzell, Carol. 2003. "Why? The Neuroscience
of Suicide." *Scientific American* 288(2):
45–51.

Families USA. 2005 (February). *Have Health
Insurance? Think You're Well Protected?
Think Again!* Fact sheet. Available at
http://www.familiesusa.org

Feachum, Richard G. A. 2000. "Poverty and
Inequality: A Proper Focus for the New
Century." *International Journal of Public
Health* 78:1–2.

Feldman, Debra S., Dennis H. Novack, and
Edward Gracely. 1998. "Effects of Managed Care on Physician Patient Relationships, Quality of Care, and the Ethical
Practice of Medicine: A Physician Survey." *Archives of Internal Medicine*
158:1626–1632.

Fisher, Edward, and Amerigo Farina. 1995.
"Attitudes Toward Seeking Professional
Psychological Help: A Shortened Form
and Considerations for Research." *Journal
of College Student Development* 36(4):
368–373.

Fox, Susannah. 2005. *Health Information
Online*. Pew Internet and American
Life Project. Available at http://www
.pewinternet.org.

Fox, Susannah, and Deborah Fallows. 2003.
Internet Health Resources. Pew Internet
and American Life Project. Available at
http://www.pewinternet.org

Fox, Susannah, and Lee Rainie. 2002. *Vital
Decisions*. Pew Internet and American
Life Project. Available at http://www.pew
internet.org

Freudenheim, M. H. 2005. "Digital Rx: Take
Two Aspirins and E-Mail Me in the Morn-
ing." *New York Times Online*, March 2.
Available at http://www.nytimes.com

Gallup Poll Social Series. 2002 (December).
Health and Healthcare. Princeton, NJ:
Gallup Organization.

Garfinkel, P. E., and D. S. Goldbloom. 2000.
"Mental Health: Getting Beyond Stigma
and Categories." *Bulletin of the World
Health Organization* 78(4):503–505.

Gilmore, Todd, and Richard Kronick. 2005.
"It's the Premiums, Stupid: Projections of
the Uninsured Through 2013." *Health Affairs*, April 5. Available at http://www
.healthaffairs.org

Goering, Laurie. 2005. "Beauty Tames Beast
of HIV Stigma." *Chicago Tribune*, February 28. Available at http://www.chicago
tribune.com

Goldberg, Lisa. 2005. "Educating Seniors on
Safe Sex." *Baltimore Sun*, March 1. Available at http://www.baltimoresun.com

Goldstein, Michael S. 1999. "The Origins of the Health Movement." In *Health, Illness, and Healing: Society, Social Context, and Self,* Kathy Charmaz and Debora A. Paterniti, eds. (pp. 31–41). Los Angeles: Roxbury.

Grunbaum, Jo Anne, Laura Kann, Steve Kinchen, James Ross, Joseph Hawkins, Richard Lowry, William A. Harris, Tim McManus, David Chyen, and Janet Collins. 2004. "Youth Risk Behavior Surveillance: United States, 2003 (Abridged)." *Journal of School Health* 74(8):307–324.

Gruttadaro, Darcy. 2005. "Federal Leaders Call on Schools to Help." *NAMI Advocate,* winter, pp. 7–9.

Halperin, D. T., M. J. Steiner, M. M. Cassell, E. C. Green, N. Hearst, D. Kirby, H. D. Gayle, and W. Cates. 2004. "The Time Has Come for Common Ground on Preventing Sexual Transmission of HIV." *The Lancet* 364:1913–1915.

Harrison, Mary M. 2002. "Silence That Promotes Stigma." *Teaching Tolerance* 21:52–53.

Himmelstein, D. U., E. Warren, D. Thorne, and S. Woolhandler. 2005. "MarketWatch: Illness And Injury As Contributors to Bankruptcy." *Health Affairs,* February 2. Available at http://www.healthaffairs.org

Hippert, Christine. 2002. "Multinational Corporations, the Politics of the World Economy, and Their Effects on Women's Health in the Developing World: A Review." *Health Care for Women International* 23:861–869.

HIV & AIDS Stigma and Discrimination. 2004 (November 30). Available at http://www.Avort.org

Hollander, D. 2003. "Having Two Parents Helps." *Perspectives on Sexual and Reproductive Health* 35(2):60.

Hollander, D. 2005. "Unprotected Anal Sex Is Not Uncommon Among Men with HIV Infection." *Perspectives on Sexual and Reproductive Health* 37(1). Available at http://www.agi-usa.org

Honberg, Ron. 2005. "Decriminalizing Mental Illness." *NAMI Advocate,* winter, pp. 4–5, 7.

Hoyert, D. L., H. Kung, and B. L. Smith. 2005. "Deaths: Preliminary Data for 2003." *National Vital Statistics Reports* 53(15), Table B. National Center for Health Statistics. Hyattsville, MD: U.S. Government Printing Office.

Hudson, C. G. 2005. "Socioeconomic Status and Mental Illness: Tests of the Social Causation and Selection Hypothesis." *American Journal of Orthopsychiatry* 75(1):3–18.

Hughes, Mary Elizabeth, and Linda J. Waite. 2002. "Health in Household Context: Living Arrangements and Health in Late Middle Age." *Journal of Health and Social Behavior* 43:1–21.

Human Rights Watch. 2003. *Ill-Equipped: U.S. Prisons and Offenders with Mental Illness.* Available at http://www.hrw.org

Human Rights Watch. 2005 (March 3). *U.S. Gag on Needle Exchange Harms U.N. AIDS Efforts.* Available at http://www.hrw.org

Institute of Medicine. 2004. *Insuring America's Health: Principles and Recommendations.* Washington, DC: National Academies Press.

Jha, A. K., E. S. Fisher, Z. Li, E. J. Orav, and A. M. Epstein. 2005. "Racial Trends in the Use of Major Procedures Among the Elderly." *New England Journal of Medicine* 353(7):683–691.

Johnson, Tracy L., and Elizabeth Fee. 1997. "Women's Health Research: An Introduction." In *Women's Health Research: A Medical and Policy Primer,* Florence P. Haseltine and Beverly Greenberg Jacobson, eds. (pp. 3–26). Washington, DC: Health Press International.

Joint United Nations Programme on HIV/ AIDS. 2004. *Global Summary of the HIV/ AIDS Epidemic: December 2004.* Available at http://www.unaids.org

Kaiser Commission on Medicaid and the Uninsured. 2004 (November). *The Uninsured: A Primer Key Findings About Americans Without Health Insurance.* Washington, DC: Kaiser Family Foundation.

Kaiser Family Foundation. 2000. *Preliminary Findings from a New National Survey of Teens on HIV/AIDS, 2000.* Available at http://www.kff.org

Kaiser Family Foundation. 2004a (June). *The HIV/AIDS Epidemic in the U.S. HIV/AIDS Policy Fact Sheet.* Available at http://www.kff.org

Kaiser Family Foundation. 2004b (August). *Survey of Americans on HIV/AIDS. Part 3. Experiences and Opinions by Race/ Ethnicity and Age.* Available at http://www.kff.org

Kaiser Family Foundation. 2004c. *Survey of Americans on HIV/AIDS. Part 2. HIV Testing.* Available at http://www.kff.org

Kaiser Family Foundation. 2005a (January). *November/December 2004 Health Poll Report Survey.* Available at http://www.kff.org

Kaiser Family Foundation. 2005b. *Trends and Indicators in the Changing Health Care Marketplace.* Available at http://www.kff.org

Kaiser Health Poll Report. 2004 (July-August). *Views of Managed Care and Customer Service.* Available at http://www.kff.org

Keith, Tamara. 2005. "Health Insurance for College Students: Burden or Blessing?" *Day to Day,* National Public Radio, March 8. Available at http://www.npr.org

Komiya, Noboru, Glenn E. Good, and Nancy B. Sherrod. 2000. "Emotional Openness as a Predictor of College Students' Attitudes Toward Seeking Psycho-

logical Help." *Journal of Counseling Psychology* 47(1)138–143.

Krugman, Paul. 2005. "Passing the Buck." *New York Times Online,* April 22. Available at http://www.nytimes.com

Ku, L., M. Broaddus, and V. Wachino. 2005 (January 31). *Medicaid and SCHIP Protected Insurance Coverage for Millions of Low-Income Americans.* Center on Budget and Policy Priorities. Available at http://www.cbpp.org

Lalasz, Robert. 2004 (November). *World AIDS Day 2004: The Vulnerability of Women and Girls.* Population Reference Bureau. Available at http://www.prb.org

Lantz, Paula M., James S. House, James M. Lepkowski, David R. Williams, Richard P. Mero, and Jieming Chen. 1998. "Socioeconomic Factors, Health Behaviors, and Mortality: Results from a Nationally Representative Prospective Study of U.S. Adults." *Journal of the American Medical Association* 279:1703–1708.

Lee, Kelley. 2003. "Introduction." In *Health Impacts of Globalization,* Kelley Lee, ed. (pp. 1–10). New York: Palgrave MacMillan.

Lehmann, Christine. 2003. "Parents Giving Up Custody of Mentally Ill Children." *Psychiatric News* 38(11):13–15.

Lerer, Leonard B., Alan D. Lopez, Tord Kjellstrom, and Derek Yach. 1998. "Health for All: Analyzing Health Status and Determinants." *World Health Statistics Quarterly* 51:7–20.

Lightfoot, M., D. Swendeman, M. J. Rotham-Borus, W. S. Comulada, and R. Weiss. 2005. "Risk Behaviors of Youth Living with HIV: Pre- and Post HAART." *American Journal of Health Behavior* 29(2):162–171.

Link, Bruce G., and Jo Phelan. 2001. "Social Conditions as Fundamental Causes of Disease." In *Readings in Medical Sociology,* 2nd ed., William C. Cockerham, Michael Glasser, and Linda S. Heuser, eds. (pp. 3–17). Upper Saddle River, NJ: Prentice Hall.

Mahapatra, Rajesh. 2005. "Groups Slam Indian Decision on Patent Law." *Business Week Online,* March 23. Available at http://www.businessweek.com

Malatu, Mesfin Samuel, and Carmi Schooler. 2002. "Causal Connections between Socioeconomic Status and Health: Reciprocal Effects and Mediating Mechanisms." *Journal of Health and Social Behavior* 43:22–41.

Mastny, L. and R. P. Cincotta. 2005. "Examining the Connection Between Population and Security." In *State of the World 2005: Redefining Global Security,* M. Renner, H. French, and E. Assadourian, eds. (pp. 22–39). New York: W. W. Norton.

McGinn, Anne P. 2003. "Combating Malaria." In *State of the World 2003,* L. Starke, ed. (pp. 62–83). New York: W. W. Norton.

Mercy, James A., Etienne G. Krug, Linda L. Dahlberg, and Anthony B. Zwi. 2003. "Violence and Health: The United States in a Global Perspective. *American Journal of Public Health* 93(2):256–261.

Millennium Ecosystem Assessment. 2005. *Millenium Ecosystem Assessment Synthesis Report.* Pre-Publication Final Draft. Available at http://www.millenium assessment.org

Miller, Joel E. 2005. "Assessing the Future Needs of Veterans with PTSD." *NAMI Advocate,* winter, pp. 14–15.

Mirowsky, John, and Catherine E. Ross. 2003. *Social Causes of Psychological Distress,* 2nd ed. New York: Walter de Gruyter.

Mui, Y. Q. 2005. "Md. Senate Backs School Snacks Limit." *Washington Post,* March 19, p. B10.

Murphy, Elaine M. 2003. "Being Born Female Is Dangerous for Your Health." *American Psychologist* 58(3):205–210.

Murray, C., and A. Lopez, eds. 1996. *The Global Burden of Disease.* Boston: Harvard University Press.

National Center for Chronic Disease Prevention and Health Promotion. 2004. *Physical Activity and Good Nutrition: Essential Elements to Prevent Chronic Diseases and Obesity.* Available at http://www.cdc.gov/nccdphp

National Center for Health Statistics. 2004a. *Health, United States, 2004 with Chartbook on Trends in the Health of Americans.* Hyattsville, MD: U.S. Government Printing Office.

National Center for Health Statistics. 2004b. *Obesity Still a Major Problem, New Data Show.* Fact sheet. Available at http://www.cdc.gov

National Center for Health Statistics. 2005 (March 23). *Early Release of Selected Estimates Based on Data from the January–September 2004 National Health Interview Survey.* Available at http://www.cdc.gov/nchs

National Council on Disability. 2002. *The Well-Being of Our Nation: An Inter-Generational Vision of Effective Mental Health Services and Supports.* Washington DC: National Council on Disability.

National Institute of Mental Health. 2003. *Real Men, Real Depression: Personal Stories.* Available at http://menand depression.nimh.nih.gov/personal.asp

Nelson, Lyle. 2003 (May). *How Many People Lack Health Insurance for How Long?* Congressional Budget Office. Available at http://www.cbo.gov

Newport, Frank. 2004 (December 2). *Assessing American's Mental Health.* Gallup Organization. Available at http://www.gallup.com

Ninan, Ann. 2003 (March). *"Without My Consent": Women and HIV-Related Stigma in India.* Population Reference Bureau. Available at http://www.prb.org

Olshansky, S. J., D. J. Pasaro, R. C. Hershow, J. Layden, B. A. Carnes, J. Brody, L. Hayflick, R. N. Butler, D. B. Allison, and D. S. Ludwig. 2005. "A Potential Decline in Life Expectancy in the United States in the 21st Century." *New England Journal of Medicine* 352(11):1138–1145.

Oxfam GB. 2004 (March 19). *The Cost of Childbirth.* Press release. Available at http://www.oxfam.org.uk/press/releases

Parsons, Talcott. 1951. *The Social System.* New York: Free Press.

Peeno, Linda. 2000. "Taking On the System." *Hope* (22):18–21.

Pereira, M. A., A. I. Kartashov, C. B. Ebbeling, L. Van Horn, M. I. Slattery, D. R. Jacobs, and D. S. Ludwig. 2005. "Fast-Food Habits, Weight Gain, and Insulin Resistance." *The Lancet* 365(9453):36–42.

Pirages, Dennis. 2005. "Containing Infectious Diseases." In *State of the World 2005: Redefining Global Security,* M. Renner, H. French, and E. Assadourian, eds. (pp. 42–59). New York: W. W. Norton.

PNHP (Physicians for a National Health Plan). 2005 (February 2). *Bankruptcy Study Highlights Need For National Health Insurance.* Available at http://www.pnhp.org/news

Quadagno, Jill. 2004. "Why the United States Has No National Health Insurance: Stakeholder Mobilization Against the Welfare State 1945–1996." *Journal of Health and Social Behavior* 45:25–44.

Rosenberg, Tina. 2005. "Think Again: AIDS." *Foreign Policy,* March-April. Available at http://www.foreignpolicy.com

Sachs, Jeffrey D. 2005. "A Practical Plan to End Poverty." *Washington Post,* January 17, p. A17.

Sager, Alan, and Deborah Socolar. 2004 (October 28). *2003 U.S. Prescription Drug Prices 81 Percent Higher Than in Other Wealthy Nations.* Data Brief No. 7. Boston University School of Public Health.

Sanders, David, and Mickey Chopra. 2003. "Globalization and the Challenge of Health for All. A View from Sub-Saharan Africa." In *Health Impacts of Globalization,* Kelley Lee, ed. (pp. 105–119). New York: Palgrave Macmillan.

Sanders, Jim. 2005. "Bill Would OK Condoms in Prisons." *The Sacramento Bee,* February 26. Available at http://www.sacbee.com

Santora, Marc. 2005. "Hold that Fat, New York Asks its restaurants." *New York Times,* August 11. Available at http://www.nytime.com.

Save the Children. 2002. *State of the World's Mothers, 2002.* Available at http://www.savethechildren.org

Schoenborn, Charlotte A. 2004 (December 15). "Marital Status and Health: United States, 1999–2002." Advance data from *Vital and Health Statistics,* no. 351. Hyattsville, MD: National Center for Health Statistics.

Shaffer, E. R., H. Waizkin, J. Brenner, and R. Jasso-Aguilar. 2005. "Global Trade and Public Health." *American Journal of Public Health* 95(1):23–34.

Sidel, Victor W., and Barry S. Levy. 2002. "The Health and Social Consequences of Diversion of Economic Resources to War and Preparation for War." In *War or Health: A Reader,* Ilkka Taipale, P. Helena Makela, Kati Juva, and Vappu Taipale, eds. (pp. 208–221). New York: Palgrave MacMillan.

SIECUS. 2004. *Public Support for Comprehensive Sexuality Education.* SIECUS Fact Sheet. Available at http://www.siecus.org

Skala, Nicholas. 2005 (spring). *PNHP Data Update.* Available at http://www.pnhp.org

SmokeFree Educational Services. 2003 (June 5). *5 Million Deaths a Year Worldwide from Smoking.* Corp-Watch. Available at http://www.corpwatch.org

"Stark Introduces Constitutional Amendment to Establish Right to Health Care." 2005 (March 3). Press release. Available at http://www.house.gov/stark

Stein, R., and C. Connolly. 2004. "Medicare Changes Policy on Obesity, Some Treatments May Be Covered." *Washington Post,* July 16. Available at http://www.washingtonpostonline

StigmaBusters Alert. 2004 (April 27). Available at http://www.nami.org

Stockard, Jean, and Robert M. O'Brien. 2002. "Cohort Effects on Suicide Rates: International Variations." *American Sociological Review* 67(6):854–872.

Sturm, Roland. 2005 (April). "Childhood Obesity: What We Can Learn From Existing Data on Social Trends, Part 2." *Preventing Chronic Disease* 2(2). Available at http://www.cdc.gov/pcd

Substance Abuse and Mental Health Services Administration. 2003 (July 21). *Treatment of Adults with Serious Mental Illness.* NHSDA (National Household Survey on Drug Abuse) Report. Available at http://www.samhsa.gov

Szasz, Thomas. 1970 [1961]. *The Myth of Mental Illness: Foundations of a Theory of Personal Conduct.* New York: Harper & Row.

Thomas, Caroline. 2003. "Trade Policy, the Politics of Access to Drugs, and Global Governance for Health." In *Health Impacts of Globalization,* Kelley Lee, ed. (pp. 177–191). New York: Palgrave MacMillan.

Timberg, C. 2005. "In S. Africa, Stigma Magnifies Pain of AIDS." *Washington Post,* January 14, p. A14.

UNICEF. 2004. *The State of the World's Children, 2004: Childhood Under Threat.* Available at http://www.unicef.org

United Nations Population Fund. 2002. *The State of World Population 2002: People, Poverty, and Possibilities.* Available at http://www.unfpa.org

United Nations Population Fund. 2004. *The State of World Population 2004.* Available at http://www.unfpa.org

UNOG (United Nations of Geneva). 2005. "Commission Adopts Five Resolutions and Three Decisions on Economic, Social, and Cultural Rights." *News and Media,* April 15. Available at http://www.unog.ch/unog/website/news_media.nsf

U.S. Department of Health and Human Services. 1999. *Mental Health: A Report of the Surgeon General—Executive Summary.* Rockville, MD: U.S. Government Printing Office.

U.S. Department of Health and Human Services. 2001. *Mental Health: Culture, Race, and Ethnicity—A Supplement to Mental Health: A Report of the Surgeon General.* Rockville, MD: U.S. Government Printing Office.

Vaccarino, V., S. S. Rathore, N. K. Wenger, P. D. Frederick, J. L. Abramson, H. V. Barron, A. Manhapra, S. Mallik, and H. M. Krumholz. 2005 (August 18). "Sex and Racial Differences in the Management of Acute Myocardial Infarction, 1994 through 2002." *New England Journal of Medicine* 353(7):671–682.

Veugelers, Paul J., and Angela L. Fitzgerald. 2005. "Effectiveness of School Programs in Preventing Childhood Obesity: A Multilevel Comparison." *American Journal of Public Health* 95(3):432–435.

Waxman, Henry A. 2004 (December). *The Context of Federally Funded Abstinence-Only Education Programs.* U.S. House of Representatives Committee on Government Reform. Minority Staff, Special Investigation Division. Available at http://www.democrats.reform.house.gov

Weitz, Rose. 2004. *The Sociology of Health, Illness, and Health Care: A Critical Approach,* 3rd ed. Belmont, CA: Wadsworth.

White, Frank. 2003. "Can International Public Health Law Help to Prevent War?" *Bulletin of the World Health Organization* 81(3):228.

WHO International Consortium in Psychiatric Epidemiology. 2000. "Cross-National Comparisons of the Prevalences and Correlates of Mental Disorders." *Bulletin of the World Health Organization* 78(4):413–426.

Williams, David R. 2003. "The Health of Men: Structured Inequalities and Opportunities." *American Journal of Public Health* 93(5):724–731.

Wooten, Jim. 2004. *We Are All The Same.* New York: Penguin Press.

World Health Organization. 1946. *Constitution of the World Health Organization.* New York: World Health Organization Interim Commission.

World Health Organization. 2000. *The World Health Report 2000.* Available at http://www.who.int

World Health Organization. 2002. *The World Health Report 2002: Reducing Risks, Promoting Healthy Life.* Available at http://www.who.int

World Health Organization. 2003. *The World Health Report 2003: Shaping the Future.* Available at http://www.who.int

World Health Organization. 2004. *The World Health Report 2004: Changing History.* Available at http://www.who.int

World Health Organization. 2005. *The World Health Report 2005: Make Every Mother and Child Count.* Available at http://www.who.int

Chapter 3

Abadinsky, Howard. 2004. *Drugs: An Introduction.* Belmont, CA: Wadsworth.

Affects Child. 2003. *How Does Alcohol Affect the World of a Child.* Available at http://www.alcoholfreechildren.org

Alcohol Epidemiology Program. 2002. *Public Support for Alcohol Policies: 1997–2001.* Youth Access to Alcohol Survey. Robert Wood Johnson Foundation, University of Minnesota.

Alcoholics Anonymous. 2005. Available at http://www.alcoholics-anonymous.org

American Heart Association (AHA). 2003a (July 30). *How Does the Tobacco Industry Target Youth?* Available at http://www.americanheart.org

American Heart Association. 2003b (August 7). *Cigarette Smoking and Cardiovascular Diseases.* Available at http://www.americanheart.org

American Heart Association. 2005 (May 7). *Cigarette Smoking and Cardiovascular Diseases.* Available at http://www.americanheart.org

Armstrong, Elizabeth, and Christina McCarroll. 2004. "Girls Lead in Teen Alcohol Use." *Seattle Times,* August 14. Available at http://seattletimes.nwsource.com

Associated Press. 1999 (December 30). *Alcoholism Touches Millions.* Available at http://www.abcnews.go.com

BBC (British Broadcasting Company). 2003 (July 29). *Know About Drugs So You Won't Make a Mistake About Drugs.* Available at http://www.bbc.uk/stoke/features

BBC. 2004a (February 22). *Blair Backs Drug Tests in Schools.* Available at http://newsvote.bbc.co.uk/mpapps/pagetools/print/news.bbc.co.uk/l/hi/uk_politics/3510605

BBC. 2004b. *Teens "Fake Drug Use to Fit In."* Available at http://newsvote.bbc.co.uk/mpapps/pagetoolsc

Becker, H. S. 1966. *Outsiders: Studies in the Sociology of Deviance.* New York: Free Press.

Belluck, Pam. 2003. "Methadone Grows as Killer Drug." *New York Times,* February 9. Available at http://www.nytimes.com

Brannigan, Martha. 2000. "Labs That Test Transportation Workers for Drugs Face Inquiry over Samples." *Wall Street Journal,* October 2, p. A4.

Brayton and Purcell Law Firm. 2005. *Recent Tobacco Law Suits.* Available at http://braytonlaw.com

Campaign for Tobacco-Free Kids. 2004 (September 17). *Special Report: Big Tobacco Still Targeting Kids.* Available at http://tobaccofreekids.org/reports/targeting/

Carroll, Joseph. 2005. *Crystal Meth, Child Molestation Top Crime Concerns.* Gallup Poll, May 3. Available at http://www.gallup.com

Centers for Disease Control and Prevention. 2004. *Alcohol Use and Pregnancy.* Available at http://www.cdc.gov/ncbddd/fas/factsheet

Charon, Joel. 2002. *Social Problems: Readings with Four Questions.* Belmont, CA: Wadsworth.

Cloud, John. 2000. "The Lure of Ecstasy." *Time,* June 5, pp. 63–72.

CNN. 2005a (April 10). "State Requiring Ignition Locks for DWI." Available at http://cnn.usnews

CNN. 2005b (May 10). *Alcohol Abandoned at Berkeley Frat Events.* Available at http://www.cnn.com

Columbia Tribune. 2005. "Suit Says State's Drug Testing Plan Is Illegal." *Columbia Tribune,* May 3. Available at http://columbiatribune.com

Conant, Eve. 2005. "A Possible New Role for a Banned Club Drug." *Newsweek,* May 2, p. 11.

Crittenden, Jules. 2002. "MIT Fraternity Settles Lawsuit in Freshman's Drinking Death." *Boston Herald.* Available at http://www.groups.yahoo.com/group/fraternalnews

DEA (Drug Enforcement Administration). 2000. *An Overview of Club Drugs.* Drug Intelligence Brief, February, pp. 1–10. Washington, DC: U.S. Department of Justice.

Drug Policy Alliance. 2003. *Drug Policy Around the World: The Netherlands.* Available at http://www.drugpolicy.org/global

Drug Policy Alliance. 2005a. *England.* Available at http://www.drugpolicy.org/global/drugpolicyby/westerneurop

Drug Policy Alliance. 2005b (January 6). *European Parliament on Record Admitting Drug War a Failure.* Available at http://www.dpf.org/news

Drug Policy Alliance. 2005c. *State by State.* Available at http://www.dpf.org/state bystate

Duke, Steven, and Albert C. Gross. 1994. *America's Longest War: Rethinking Our Tragic Crusade Against Drugs.* New York: G. P. Putnam.

Easley, Margaret, and Norman Epstein. 1991. "Coping with Stress in a Family with an Alcoholic Parent." *Family Relations* 40:218–224.

Eisner, Robin. 2005. "Marijuana Abuse: Age of Initiation, Pleasure of Response Foreshadow Young Adult Outcomes." *NIDA Notes* 9(5):1, 6–7.

Feagin, Joe R., and C. B. Feagin. 1994. *Social Problems*. Englewood Cliffs, NJ: Prentice Hall.

Firshein, Janet. 2003. "The Role of Biology and Genetics." *Moyers on Addiction*. Public Broadcasting System (PBS).

Fletcher, Michael A. 2000. "War on Drugs Sends More Blacks to Prison Than Whites." *Washington Post*, June 8, p. A10.

Gentry, Cynthia. 1995. "Crime Control Through Drug Control." In *Criminology*, 2d ed., Joseph F. Sheley, ed. (pp. 477–493). Belmont, CA: Wadsworth.

Goldstein, Adam, Rachel Sobel, and Glen Newman. 1999. "Tobacco and Alcohol Use in G-Rated Children's Animated Films." *Journal of the American Medical Association* 281:1121–1136.

Grant, Tavia. 2005. "Cannabis Spray Gets Go-Ahead." *The Globe and Mail,* April 19.

Greenhouse, Linda. 2002. "Justices Allow Schools Wider Use of Random Drug Tests for Pupils." *New York Times,* June 28. Available at http://www.nytimes.com

Gusfield, Joseph. 1963. *Symbolic Crusade: Status Politics and the American Temperance Movement*. Urbana: University of Illinois Press.

Hanson, Glen R. 2002. "New Vistas in Drug Abuse Prevention." *NIDA Notes* 16:3–4.

Healy, Patrick. 2005. "Sobriety Tests Are Becoming Part of the School Day." *New York Times,* March 3. Available at http://www.nytimes.com

Heroin Drug Conference. 1997. *Administrator's Message*. U.S. Department of Justice, Drug Enforcement Administration. Available at http://www.udsdoj.gov/dea/pubs/special/heroin.html

Jarvik, M. 1990. "The Drug Dilemma: Manipulating the Demand." *Science* 250:387–392.

Jefferson, David A. 2005. "America's Most Dangerous Drug." *Newsweek,* August 8, pp. 41–48.

Jernigan, David. 2005. "National Experiences, the U.S.A.: Alcohol and Young People Today." *Addiction* 100(3):271–273.

Johnson, Lynn M. 2003. *Alcohol, Drugs and Violence: Detrimental Effects on Children*. Available at http://www.babyparenting.about.com

Johnston, L. D., P. M. O'Malley, J. G. Bachman, and J. E. Schulenberg. 2004. "Overall Teen Drug Use Continues Overall Decline, But Use of Inhalants Rises." University of Michigan News Service, December 24.

Klutt, Edward C. 2000. *Pathology of Drug Abuse*. Available at http://www.medlib-utah.edu/webpath

Lee, Yoon, and M. Abdel-Ghany. 2004. "American Youth Consumption of Licit and Illicit Substances." *International Journal of Consumer Studies* 28(5):454–465.

Leonard, K. E., and H. T. Blane. 1992. "Alcohol and Marital Aggression in a National Sample of Young Men." *Journal of Interpersonal Violence* 7:19–30.

Leshner, Alan. 2003. "Using Science to Counter the Spread of Ecstasy Abuse." *NIDA Notes* 16:3–4.

Lyons, Linda. 2005. *Drug Use Still Among Americans' Top Worries*. Gallup Poll, April 5. Available at http://brain.gallup.com/content

MacCoun, Robert J., and Peter Reuter. 2001. "Does Europe Do It Better? Lessons from Holland, Britain and Switzerland." In *Solutions to Social Problems,* D. Stanley Eitzen and Craig S. Leedham, eds. (pp. 260–264). Boston: Allyn and Bacon.

MADD (Mothers Against Drunk Driving). 2003a. *Alcohol Advertising*. Available at http:www.madd.org/stats

MADD. 2003b. *The Limiting Factor: Economic Cost of Underage Drinking*. Available at http://www.madd.org/activism

MADD. 2005a. *Did you Know . . .* Available at http://www.madd.org/stats

MADD. 2005b *MADD Programs*. Available at http://www.madd.org/madd_programs

Mann, Judy. 2000. "Make War on the War on Drugs." *Washington Post,* July 26, p. C13.

Mayell, Hillary. 1999. "Tobacco on Course to Become World's Leading Cause of Death." *National Geographic News*. Available at http://ngnews/news/1999/12149

McCaffrey, Barry. 1998. *Remarks by Barry McCaffrey, Director, Office of National Drug Control Policy, to the United Nations General Assembly: Special Session on Drugs*. Office of National Drug Control Policy. Available at http://www.whitehousedrugpolicy.gov/news/speeches

Monitoring the Future (MTF). 2004. *National Results on Adolescent Drug Use*. Overview of Key Findings. Available at http://monitoringthefuture.org/pubs

Moore, Martha T. 1997. "Binge Drinking Stalks Campuses." *USA Today,* October 1, p. A3.

Morgan, Patricia A. 1978. "The Legislation of Drug Law: Economic Crisis and Social Control." *Journal of Drug Issues* 8:53–62.

National Institute on Alcohol Abuse and Alcoholism. 2000 (April). *Mechanisms of Addiction*. Alcohol Alert. National Institute on Alcohol Abuse and Alcoholism no. 47. Available at http://www.niaaa.nih.gov/publications/aa47.htm

NHSDA (National Household Survey on Drug Abuse). 2003. *Cigarette Brand Preference*. Available at http://www.samhsa.gov/oas/2k3/cigBrands

NIDA (National Institute on Drug Abuse). 2000. *Heroin Abuse and Addiction*. Research Report Series, NIH Publication 00-4165. Available at http://www.nida.nih/gov/ResearchReports/Heroin/Heroin.html

NIDA. 2003. *InfoFacts: Drug Addiction Treatment Methods*. Available at http://www.nida.org

NSDUH (National Survey on Drug Use and Health). 2004a (February 13). *Alcohol Dependence or Abuse Among Parents with Children Living at Home*. Office of Applied Studies, Substance Abuse and Mental Health Services Administration. Available at http://www.DrugAbuseStatistics.samhsa.gov

NSDUH. 2004b (December 31). *Driving Under the Influence (DUI) Among Young Persons*. Office of Applied Studies, Substance Abuse and Mental Health Services Administration. Available at http://www.DrugAbuseStatistics.samhsa.gov

NSDUH. 2005 (April 1). *Alcohol Use and Delinquent Behavior Among Youths*. Office of Applied Studies, Substance Abuse and Mental Health Services Administration. Available at http://www.DrugAbuseStatistics.samhsa.gov

ODCCP (Office of Drug Control and Crime Prevention). 2003. *Global Illicit Drug Trends 2002*. New York: United Nations.

ONDCP (Office of National Drug Control Policy). 2003a. *Budget Highlights*. Available at http://www.whitehousedrugpolicy.gov/publications/policy/ndsc03

ONDCP. 2003b. *Drug Data Summary*. Available at http://www.whitehousedrugpolicy.gov/publications/factsht/summary

ONDCP 2003c. *Drug Facts: Club Drugs*. Available at http://www.whitehousedrugpolicy.gov/club

ONDCP. 2003d. *Heroin*. Fact sheet. Available at http://www.whitehousedrugpolicy.gov/publications/factsht/heroin

ONDCP. 2003e. *Rohypnol*. Fact sheet. Available at http://www.whitehousedrugpolicy.gov/publications/factsht/rohypnol

ONDCP. 2004. *The Economic Costs of Drug Abuse in the United States: 1992–2002*. Section IV. Executive Office of the President. Available at http://www.whitehousedrugpolicy.gov

ONDCP. 2005a. *Club Drugs*. Available at http://www.whitehousedrugpolicy.gov/drugfact/club

ONDCP. 2005b. *Crack*. Available at http://www.whitehousedrugpolicy.gov/drugfact/crack

ONDCP. 2005c. *Inhalants*. Available at http://www.whitehousedrugpolicy.gov/inhalants/club

ONDCP. 2005d. *Marijuana*. Available at http://www.whitehousedrugpolicy.gov/drugfact/marijuana

ONDCP. 2005e. *Media Campaign: What's New*. Available at http:www.mediacampaign.org

ONDCP. 2005f. *The National Drug Control Strategy*. Available at http://www.whitehousedrugpolicy.gov

ONDCP. 2005g. *The President's National Drug Control Strategy: National Priorities*. Available at http://www.whitehousedrugpolicy.gov/publications/policiy/dncs05

ONDCP. 2005h (May 3). *White House Drug Czar, Research and Mental Health Communities Warn Parents That Marijuana Use Can Lead to Depression, Suicidal Thoughts, and Schizophrenia*. Available at http://www.whitehousedrugpolicy.gov/news/press05

Pew. 2002. *Illegal Drugs.* Pew Research Center for the People and Press Survey. Available at http://www.pollingreport.com/drugs

Reiman, Jeffrey. 2003. *The Rich Get Richer and the Poor Get Prison.* Boston: Allyn and Bacon.

Rorabaugh, W. J. 1979. *The Alcoholic Republic: An American Tradition.* New York: Oxford University Press.

Rychtarik, Robert G., Gerald J. Connors, Kurt H. Dermen, and Paul Stasiewicz. 2000. "Alcoholics Anonymous and the Use of Medications to Prevent Relapse." *Journal of Studies on Alcohol* 61:134–141.

Schemo, Diana Jean. 2002. "Study Calculates the Effects of College Drinking in the U.S." *New York Times,* April 10. Available at http://www.nytimes.com

Sheldon, Tony. 2000. "Cannabis Use Among Dutch Youth." *British Medical Journal* 321:655.

Siegel, Larry. 2006. *Criminology.* Belmont, CA: Wadsworth.

Singley, Devin. 2004. "Collegiate Alcohol Awareness Week Speaker Focuses on the Dangers of Hazing." *The Carolinian Online,* October 25. Available at http://carolinianonline.com/news/2004/10/25

State Legislatures. 2000. "All You Ever Wanted to Know About Drunk Drivers." *State Legislatures* 26:7.

Thio, Alex. 2004. *Deviant Behavior.* Boston, MA: Allyn and Bacon.

Thompson, Don. 2000. "States Ballot Questions Focus on Drug Rehab Instead of Prison." *Excite News.* Available at http://news.excite.com/news

Time/CNN. 2003. "The New Politics of Pot." *Time,* November 4, pp. 56–66.

UF News. 2002. *UF Researchers Link Increasing Prescription Methadone Use for Chronic Pain to Rise in Abuse Related Deaths.* University of Florida, October 3.

UNODC (United Nations Office on Drugs and Crime). 2004. *Drugs and Crime Trends in Europe and Beyond.* Available at http://unodc.org/pdf/factsheets

U.S. Department of Health and Human Services. 2002. *2001 National Household Survey on Drug Abuse.* Substance Abuse and Mental Health Service Administration. Washington, DC: U.S. Government Printing Office.

U.S. Department of Health and Human Services. 2004. *2003 National Survey on Drug Use and Health.* Substance Abuse and Mental Health Service Administration. Washington, DC: U.S. Government Printing Office.

Van Dyck, C., and R. Byck. 1982. "Cocaine." *Scientific American* 246:128–141.

Van Kammen, Welmoet B., and Rolf Loeber. 1994. "Are Fluctuations in Delinquent Activities Related to the Onset and Offset in Juvenile Illegal Drug Use and Drug Dealing?" *Journal of Drug Issues* 24:9–24.

Weitzman, Elissa, Henry Wechsler, and Toben F. Nelson. 2003. *Environment, Not Education, a Stronger Predictor of Binge Drinking Behavior Among College Freshman.* Press Release, Harvard School of Public Health, January 21.

Willing, Richard. 2002. "Study Shows Alcohol Is Main Problem for Addicts." *USA Today,* October 3, p. B4.

Witters, Weldon, Peter Venturelli, and Glen Hanson. 1992. *Drugs and Society,* 3rd ed. Boston: Jones & Bartlett.

World Drug Report. 2004. *Volume 1: Analysis.* Office of Drug Control and Crime Prevention. United Nations Publication E.04.X1.16. New York: United Nations.

Wysong, Earl, Richard Aniskiewicz, and David Wright. 1994. "Truth and Dare: Tracking Drug Education to Graduation and as Symbolic Politics." *Social Problems* 41:448–468.

Yamaguchi, Ryoko, Loyd Johnson, and Patrick O'Malley. 2003. *Student Drug Testing Not Effective in Reducing Drug Use.* News and Information Services. Ann Arbor: University of Michigan.

Zailckas, Koren. 2005. *Smashed: Story of a Drunken Girlhood.* New York: Viking.

Zickler, Patrick. 2003. "Study Demonstrates That Marijuana Smokers Experience Significant Withdrawal." *NIDA Notes* 17:7, 10.

Chapter 4

Albanese, Jay. 2000. *Criminal Justice.* Boston: Allyn and Bacon.

Amnesty International. 2005. *Amnesty International's Annual Death Penalty Report Finds Global Trend Toward Abolition.* Available at http://web.amnesty.org

Anderson, David. 1999. "The Aggregate Burden of Crime." *Journal of Law and Economics* 42:611–642.

Associated Press. 2003. "Video Games Get Updated Rating System." *New York Times,* June 20. Available at http://www.nytimes.com

Associated Press. 2005a. "Thief Takes Laptop with Berkeley Data." *New York Times,* March 29. Available at http://www.nytimes.com

Associated Press. 2005b (April 18). *UN Conference Takes Aim at Crime.* Available at http://theglobeandmail.com

Associated Press. 2005c (April 27). *Backlogs at Nation's Forensic Labs Undercut DNA's Crime Solving Value.* Available at http://www.cnn.com

Barkan, Steve. 2006. *Criminology.* Upper Saddle River, NJ: Pearson–Prentice Hall.

Becker, Howard S. 1963. *Outsiders: Studies in the Sociology of Deviance.* New York: Free Press.

BJS (Bureau of Justice Statistics). 2003. *Nation's Prison and Jail Population Exceeds 2 Million Inmates for the First Time.* Available at http://www.ojp.usdoj.gov/bjs/pub

BJS. 2004. *Probation and Parole Statistics.* Available at http://www.ojp.usdoj.gov/pandp

BJS. 2005a. *Capital Punishment Statistics.* Available at http://www.ojp.usdoj.gov/bjs/cp

BJS. 2005b. *Crime Characteristics.* Available at http://www.ojp.usdoj.gov/bjs/cvict_c

BJS. 2005c (April 24). *Nations Prison and Jail Population Grew by 932 Inmates per Week.* Available at http://www.ojp.usdoj.gov/bjs/pub/press

Brady Campaign. 2005. *Law Enforcement Opposition to HR. 800.* Available at http://www.bradycampaign.org

Butterfield, Fox. 2002a (June 3). *Study Shows Building Prisons Did Not Prevent Repeat Crimes.* Available at http://www.ojp.usdoj.gov/bjs/pub

Butterfield, Fox. 2002b. "Tight Budgets Force States to Reconsider Crime and Penalties." *New York Times,* January 21. Available at http://www.nytimes.com

Campo-Flores, Arian. 2005. "The Most Dangerous Gang in America." *Newsweek,* March 28, pp. 23–25.

Chesney-Lind, Meda, and Randall G. Shelden. 2004. *Girls, Delinquency, and Juvenile Justice.* Belmont, CA: Wadsworth.

Chicago Tribune. 2003. "Ford Fined, Ordered to Hand over Van Data." *Tribune News Services.* Available at http://www.lieffcabraser.com

Clear, Todd. 2006. "The Results of American Incarceration." In *Annual Editions: Criminal Justice,* Joseph Victor and Joanne Naughton, eds. (pp. 191–193). Dubuque, IA: Dushkin.

Coates, Sam. 2005. "Rader Gets 175 Years for BTK Slayings." *Washington Post,* August 19, p. A3. Available at http://www.washingtonpost.com

CODIS. 2005. *Mission Statement and Background.* Available at http://www.fbi.gov/hq/lab/codis

Cohen, Jon. 2005. *Poll: Identity Theft Concerns Rise.* ABC News, May 18. Available at http://abcnews.go.com/Business

Conklin, John E. 1998. *Criminology,* 6th ed. Boston: Allyn and Bacon.

Copeland, Larry. 2005. "I Could Tell He Was Going to Shoot Everybody." *USA Today,* March 13. Available at http://www.usatoday.com

COPS. 2005. *What Is Community Policing?* Available at http://www.cops.usdoj.gov

D'Alessio, David, and Lisa Stolzenberg. 2002. "A Multilevel Analysis of the Relationship Between Labor Surplus and Pretrial Incarceration." *Social Problems* 49:178–193.

D'Alessio, David, and Lisa Stolzenberg. 2003. "Race and the Probability of Arrest." *Social Forces* 81:1381–1397.

"Death Behind Bars." 2005. Editorial. *New York Times,* March 10. Available at http://www.nytimes.com.

Death Penalty Information Center. 2005. *DPIC Summary of The Innocence Protection Act of 2004.* Available at http://www.deathpenaltyinfo.org

Dickerson, Debra. 2000. "Racial Profiling: Are We All Really Equal in the Eyes of the Law?" *Los Angeles Times,* July 16. Available at http://www.latimes.com

Dixon, Travis L., and Daniel Linz. 2000. "Race and the Misrepresentation of Victimization on Local Television News." *Communication Research* 27:547–574.

Dobriansky, Paula. 2001. "The Explosive Growth of Globalized Crime." *Global Issues: Arresting Transnational Crime* 6(August):1–3.

Eisenberg, Daniel. 2002. "Jail to the Chiefs?" *Time,* August 12, pp. 23–26.

Eitle, David, Stewart D'Alessio, and Lisa Stolzenberg. 2002. "Racial Threat and Social Control: A Test of the Political, Economic, and Threat of Black Crime Hypothesis." *Social Forces* 81:557–576.

Erikson, Kai T. 1966. *Wayward Puritans.* New York: Wiley.

FBI (Federal Bureau of Investigation). 2004. *Crime in the United States, 2003.* Uniform Crime Reports. Washington, DC: U.S. Government Printing Office.

FBI. 2005. *Financial Crime Report to the Public.* Available at http://www.fbi.gov/publications/financial/fcs_report052005

Felson, Marcus. 2002. *Crime and Everyday Life,* 3rd ed. Thousand Oaks, CA: Sage.

Finckenauer, James O. 2000. "Meeting the Challenge of Transnational Crime." *National Institute of Justice Journal,* July, pp. 2–7.

Fletcher, Michael A. 2000. "War on Drugs Sends More Blacks to Prison than Whites." *Washington Post,* June 8, p. A10.

Florida Criminal Code. 2004. Available at http://www.leg.state.fl./statutes/index

Friedrichs, David O. 2004. *Trusted Criminals.* Belmont, CA: Wadsworth.

Gallup Poll. 2004. *Crime.* Gallup Poll News Service. Available at http://gallup.com/poll

Gallup Poll. 2005a (March 13). *Gallup Poll Social Series.* Available at http://gallup.com/documents

Gallup Poll. 2005b. *Guns.* Available at http://brain.gallup.com

Garey, M. 1985. "The Cost of Taking a Life: Dollars and Sense of the Death Penalty." *U.C. Davis Law Review* 18:1221–1273.

Gest, Ted, and Dorian Friedman. 1994. "The New Crime Wave." *U.S. News and World Report,* August 29, pp. 26–28.

Harrow, Robert O. 2005. "ID Data Conned from Firm." *Washington Post,* February 16. Available at http://www.washingtonpost.com

Harvard School of Public Health. 2002. *American Females at Highest Risk of Murder.* Press release. Harvard School of Public Health, April 17.

Heimer, Karen, and Stacy DeCoster. 1999. "The Gendering of Violent Delinquency." *Criminology,* May, pp. 377–389.

Herbert, Bob. 2002. "The Fatal Flaws." *New York Times,* February 11. Available at http://www.nytimes.com

Hirschi, Travis. 1969. *Causes of Delinquency.* Berkeley: University of California Press.

ICCC (Internet Crime Complaint Center). 2005. *IC3 2004 Internet Fraud-Crime Report.* Washington, DC: National White Collar Crime Center and the Federal Bureau of Investigation.

Innocence Project. 2005. *Post-Conviction DNA Testing Statutes.* Available at http://www.innocenceproject.org

Interpol. 1998. *Interpol Warning: Nigerian Crime Syndicate's Letter Scheme Fraud Takes on New Dimension.* Press Release. Available at http://www.kenpubs.co.uk/interpol.com/English/pres/nig.html

Irwin, John. 2005. *The Warehouse Prison.* Los Angeles, CA: Roxbury.

Jacobson, H., and R. Greca Green. 2002. "Computer Crimes." *American Criminal Law Review* 39:273–326.

Johnson, Dirk. 2005. "In Georgia, A Matter of Faith." *Newsweek,* March 28, p. 29.

Johnson, Robert. 1989. "'This Man Has Expired': Witness to an Execution." *Commonwealth,* January 13, pp. 9–15.

Kerley, Kent, Michael Benson, Matthew Lee, and Francis Cullen. 2004. "Race, Criminal Justice Contact, and Adult Position in the Social Stratification System." *Social Problems* 51(4):549–568.

Klapper, Bradley. 2005. "UN Told Governments Must Combat Internet Child Pornography." Associated Press, April 14. Available at http://informationweek.com

Kluger, Jeffery. 2002. "How Science Solves Crimes." *Time,* October 21, pp. 36–45.

Kong, Deborah, and Jon Swartz. 2000 "Experts See Rash of Hack Attacks Coming." *USA Today,* September 27, p. 1B.

Kubrin, Charis E. 2005. "Gangsters, Thugs, and Hustlas: Identity and the Code of the Street in Rap Music." *Social Problems* 52(3):360–378.

Kubrin, Charis, and Ronald Weitzer. 2003. "Retaliatory Homicide: Concentrated Disadvantage and Neighborhood Culture." *Social Problems* 50:157–180.

Lane, Charles. 2004. "Less Support for Death Sentence Cited for Decline in Executions." *Washington Post,* December 15. Available at http://www.washingtonpost.com

Laub, John, Daniel S. Nagan, and Robert Sampson. 1998. "Trajectories of Change in Criminal Offending: Good Marriages and the Desistance Process." *American Sociological Review* 63(April):225–238.

Lee, Jennifer. 2003. "Identity Theft Complaints Double in 2002." *New York Times,* January 22. Available at http://www.nytimes.com

Lichtblau, Eric. 2003. "Panel Clears Harsher Terms in Corporate Crime Cases." *New York Times,* January 9. Available at http://www.nytimes.com

Liptak, Adam. 2003. "Death Penalty Found More Likely When Victim Is White." *New York Times,* January 8. Available at http://www.nytimes.com

Lott, John R., Jr. 2003. "Guns Are an Effective Means of Self-Defense." In *Gun Control,* Helen Cothran, ed. (pp. 86–93). Farmington Hills, MI: Greenhaven Press.

MAD DADS. 2005. *About Us.* Available at http://www.maddads.com

Madriz, Esther. 2000. "Nothing Bad Happens to Good Girls." In *Social Problems of the Modern World,* Frances Moulder, ed. (pp. 293–297). Belmont, CA: Wadsworth.

"Major Rulings of the 2002–2003 Term." 2003. *USA Today,* June 27, p. 4A

Merkle, Daniel. 2004. "Crime Fears Linger: Public Still Concerned Despite Improvements." ABC News, October 8. Available at http://www.abcnews.go.com

Merton, Robert. 1957. *Social Theory and Social Structure.* Glencoe, IL: Free Press.

Moore, David. 2005. "Young Adults Most Likely to Be Crime Victims." Gallup Organization, March 21. Available at http://gallup.com/poll

Murray, Mary E., Nancy Guerra, and Kirk Williams. 1997. "Violence Prevention for the Twenty-First Century." In *Enhancing Children's Awareness,* Roger P. Weissberg, Thomas Gullota, Robert L. Hampton, Bruce Ryan, and Gerald Adams, eds. (pp. 105–128). Thousand Oaks, CA: Sage.

Myths and Facts About the Death Penalty. 1998. *Death Penalty: Focus on California.* Available at http://www.members.aol.com/Dpfocus/facts.htm

Napolitano, Jo. 2004. *Top Illinois Court Upholds Total Amnesty of Death Row.* Available at http://www.nytimes.com

National Night Out. 2005. *What is National Night Out?* Available at http://www.nationaltownwatch.org/nno/about

National Research Council. 1994. *Violence in Urban America: Mobilizing a Response.* Washington, DC: National Academy Press.

NCCPC (National Citizen's Crime Prevention Campaign). 2005. *McGruff the Crime Dog and the National Citizen's Crime Prevention Campaign.* Available at http://www.weprevent.com

Orator. 2005. *News and Information: 109 Congress.* Available at http://www.the orator.com/bills109

Pertossi, Mayra. 2000 (September 27). *Analysis: Argentine Crime Rate Soars.* Available at http://www.news.excite.com

Pew Research Center. 2000. "Respondents' Perception of Safety." Pew Research Center for the People and the Press, May 12. Available at http://www.peoplepress.org/april00rpt.htm

Philips, Julie. 2002. "White, Black, and Latino Homicide Rates: Why the Difference?" *Social Problems* 49:349–374.

Pickler, Nedra. 2000 (September 6). *Documents Point to Tire Problem.* Available at http://www.news.excite.com

Reid, Sue Titus. 2003. *Crime and Criminology,* 10th ed. Boston: McGraw-Hill.

Reiman, Jeffrey. 2004. *The Rich Get Richer and the Poor Get Prison.* Boston: Allyn and Bacon.

Reisig, Michael, and Roger Parks. 2004. "Community Policing and Quality of Life." In *Community Policing,* Wesley Skogan, ed. (pp. 207–227). Belmont, CA: Wadsworth.

Richtel, Matt. 2002. "Credit Card Theft Thrives Online as Global Market." *New York Times.* Available at http://www.nytimes.com

Richtel, Matt. 2003. "Mayhem, and Far from the Nicest Kind." *New York Times,* February 10. Available at http://www.nytimes.com

Ripley, Amanda. 2003. "The Night Detective." *Time,* January 6, pp. 45–50.

Ripley, Amanda. 2005. "The DNA Dragnet." *Time,* January 24, pp. 39–40.

Romano, Lois. 2005. "More Complete Portrait of BTK Suspect Is Emerging." *Washington Post,* March 5. Available at http://www.washingtonpost.com.

Rosoff, Stephen, Henry Pontell, and Robert Tillman. 2002. "White Collar Crime." In *Social Problems: Readings with Four Questions,* Joel M. Charon, ed. (pp. 339–350). Belmont, CA: Wadsworth.

Rubin, Paul H. 2002. "The Death Penalty and Deterrence." *Forum,* winter pp. 10–12.

Schiesel, Seth. 2005. Growth of Wireless Internet Opens New Paths for Thieves." *New York Times,* March 19. Available at http://www.nytimes.com

Shelden, Randall, Sharon Tracy, and William Brown. 2004. *Youth Gangs in American Society.* Belmont, CA: Wadsworth

Sherman, Lawrence. 2003. "Reasons for Emotion." *Criminology* 42:1–37.

Siegel, Larry. 2006. *Criminology.* Belmont, CA: Wadsworth.

The Situation in the Netherlands. 2003. *Prostitution in Holland.* Available at http://www.ex.ac.uk/politics/por_data

Skogan, Wesley, and Jeffery Roth. 2004. "Introduction." In *Community Policing,* Wesley Skogan, ed. (pp. xvii–xxiv). Belmont, CA: Wadsworth.

Steffensmeier, Darryl, and Stephen Demuth. 2000. "Ethnicity and Sentencing Outcomes in U.S. Federal Courts: Who Is Punished More Harshly?" *American Sociological Review* 65:705–729.

Stephens, Gene. 2006. "Global Trends in Crime." In *Annual Editions: Criminal Justice,* Joseph Victor and Joanne Naughton, eds. (pp. 16–20). Dubuque, IA: Dushkin.

Surgeon General. 2002. "Cost-Effectiveness." In *Youth Violence: A Report of the Surgeon General.* Available at http://www.mentalhealth.org/youthviolence/surgeongeneral

Sutherland, Edwin H. 1939. *Criminology.* Philadelphia: Lippincott.

Thio, Alex. 2004. *Deviant Behavior.* Boston: Allyn and Bacon.

Thomas, Landon. 2005. "On Wall Street, a Rise in Dismissals over Ethics." *New York Times,* March 29. Available at http://www.nytimes.com/2005/03/29/business

Travis, Jeremy, and Michelle Waul. 2002. *Reflections on the Crime Decline: Lessons for the Future.* Washington, DC: Urban Institute, Justice Policy Center.

Tucker, Abigail, and Stephen Kiehl. 2005. "A Youth Drifted from Deep Family Ties to Deep Trouble." *Baltimore Sun,* March 18. Available at http://newsday.com/news/nationworld/nation

United Nations. 1997. *Crime Goes Global.* Document DPI/1518/SOC/CON/30M. New York: United Nations.

United Nations. 2003. *Creating Guidelines for Restorative Justice Programmes.* Available at http://www.restorativejustice.org/rjs/feature/2003

U.S. Bureau of the Census. 2004. *Statistical Abstract of the United States,* 124th edition. Washington, DC: U.S. Government Printing Office.

U.S. Department of Justice. 2003. "Global Crime Issues." In *International Center Global Crimes Issues,* National Institute of Justice. Washington, DC: U.S. Government Printing Office.

U.S. Department of Justice. 2004. *Crime in the United States, 2003.* Washington, D.C.

Vander Ven, Thomas, Francis Cullen, Mark Carrozza, and John Wright. 2001. "Home Alone: The Impact of Maternal Employment on Delinquency." *Social Problems* 48:236–257.

Victim-Offender Reconciliation Program. 2005. *About Victim-Offender Mediation and Reconciliation.* Available at http://www.vorp.com

Wallis, Claudia. 2005. "Too Young to Die." *Time,* March 14, p. 40.

Warner, Barbara, and Pamela Wilcox Rountree. 1997. "Local Social Ties in a Community and Crime Model." *Social Problems* 4(4):520–536.

Weed and Seed. 2005. *The Weed and Seed Strategy.* Available at http://www.weedandseeddatacenter.org

Weitzer, Ronald, and Steven Tuch. 2004. "Race and Perception of Police Misconduct." *Social Problems* 51(3):305–325.

White House. 2005 (July 15). *Remarks by President at Signing of Identity Theft Penalty Enhancement Act.* Available at http://www.whitehouse.gov/news/release

Williams, Linda. 1984. "The Classic Rape: When Do Victims Report?" *Social Problems* 31:459–467.

Wilgoren, Jodi. 2003. "Governor Empties Illinois Death Row." *New York Times,* January 12. Available at http://www.nytimes.com

Chapter 5

Aassve, A. 2003. "The Impact of Economic Resources on Premarital Childbearing and Subsequent Marriage Among Young American Women." *Demography* 40:105–126.

Adams, Bert N. 2004. "Families and Family Study in International Perspective." *Journal of Marriage and Family* 66(5):1076–1088.

Administration for Children and Families. 2003. *Prevention Pays: The Costs of Not Preventing Child Abuse and Neglect.* U.S. Dept. of Health and Human Services. Available at http://www.acf.hhs.gov

Ahrons, C. 2004. *We're Still Family: What Grown Children Have to Say About Their Parents' Divorce.* New York: Harper-Collins

Amato, Paul. 1999. "The Postdivorce Society: How Divorce Is Shaping the Family and Other Forms of Social Organization." In *The Postdivorce Family: Children, Parenting, and Society,* R. A. Thompson and P. R. Amato, eds. (pp. 161–190). Thousand Oaks, CA: Sage.

Amato, Paul. 2003. "The Consequences of Divorce for Adults and Children." In *Family in Transition,* 12th ed., Arlene S. Skolnick and Jerome H. Skolnick, eds. (pp. 190–213). Boston: Allyn and Bacon.

Amato, Paul. 2004. "Tension Between Institutional and Individual Views of Marriage." *Journal of Marriage and Family* 66:959–965.

Amato, P. R. and J. Cheadle. 2005 "The Long Reach of Divorce: Divorce and Child Well-Being Across Three Generations." *Journal of Marriage and the Family* 67:191–206.

Amato, P. R., D. R. Johnson, A. Booth, and S. J. Rogers. 2003. "Stability and Change in Marital Quality Between 1980 and 2000." *Journal of Marriage and Family* 65:1–22.

American Association of Blood Banks. 2002 (October). *Annual Report Summary for Testing in 2001.* Bethesda, MD: American Association of Blood Banks.

American Council on Education and University of California. 2004–2005. *The American Freshman: National Norms for Fall, 2004.* Los Angeles: Los Angeles Higher Education Research Institute.

Anderson, Kristin L. 1997. "Gender, Status, and Domestic Violence: An Integration of Feminist and Family Violence Approaches." *Journal of Marriage and the Family* 59:655–669.

Applewhite, Ashton. 2003. "Covenant Marriage Would Not Benefit the Family." In *The Family: Opposing Viewpoints,* Auriana Ojeda, ed. (pp. 189–195). Farmington Hill, MI: Greenhaven Press.

As-Sanie, S., A. Gantt, and M. S. Rosenthal. 2004. "Pregnancy Prevention in Adolescents." *American Family Physician* 70:1517–1519.

Babcock, J. C., S. A. Miller, and C. Siard. 2003. "Toward a Typology of Abusive Women: Differences Between Partner-Only and Generally Violent Women in the Use of Violence." *Psychology of Women Quarterly* 27:153–161.

Bachu, Amara. 1999. "Trends in Premarital Childbearing." *Current Population Reports*, pp. 23–197. Washington, DC: U.S. Bureau of the Census.

Beitchman, J. H., K. J. Zuker, J. E. Hood, G. A. daCosta, D. Akman, and E. Cassavia. 1992. "A Review of the Long-Term Effects of Child Sexual Abuse." *Child Abuse and Neglect* 16:101–119.

Block, Nadine. 2003. "Disciplinary Spanking Should Be Banned." In *Child Abuse: Opposing Viewpoints,* L. I. Gerdes, ed. (pp. 182–190). Farmington Hills, MI: Greenhaven Press.

Browning, Christopher R., and Edward O. Laumann. 1997. "Sexual Contact Between Children and Adults: A Life Course Perspective." *American Sociological Review* 62:540–560.

Browning, Don S. 2003. *Marriage and Modernization: How Globalization Threatens Marriage and What to Do About It.* Grand Rapids, MI: William B. Eerdmans.

Carrington, Victoria. 2002. *New Times: New Families.* Dordrecht, Netherlands: Kluwer Academic.

Centers for Disease Control and Prevention. 2004. "Youth Risk Behavior Surveillance, United States, 2003." *Morbidity and Mortality Weekly Report* 53, no. SS-2.

Cherlin, Andrew J., Linda M. Burtin, Tera R. Hurt, and Diane M. Purvin. 2005. "The Influence of Physical and Sexual Abuse on Marriage and Cohabitation." *American Sociological Review* 69:768–789.

Cole, Charles L., Anna L. Cole, and Jessica G. Gandolfo. 2000. "Marriage Enrichment for Newlyweds: Models for Strengthening Marriages in the New Millennium." Poster Presentation at the 62nd Annual Conference of the National Council on Family Relations, Minneapolis, November 10–13.

Coontz, Stephanie. 1992. *The Way We Never Were: American Families and the Nostalgia Trap.* New York: Basic.

Coontz, Stephanie. 1997. *The Way We Really Are.* New York: Perseus

Coontz, Stephanie. 2000. "Marriage: Then and Now." *Phi Kappa Phi Journal* 80:10–15.

Coontz, Stephanie. 2004. "The World Historical Transformation of Marriage." *Journal of Marriage and Family* 66(4):974–979.

Coontz, Stephanie. 2005. "For Better, For Worse." *Washington Post,* May 1. Available at http://www.washingtonpost.com

Daniel, Elycia. 2005. "Sexual Abuse of Males." In *Sexual Assault: The Victims,*

the Perpetrators, and the Criminal Justice System,* Frances P. Reddington and Betsy Wright Kreisel, eds. (pp. 133–140). Durham, NC: Carolina Academic Press.

Davis, Shannon N., and Theodore N. Greenstein. 2004. "Cross-National Variations in the Division of Household Labor." *Journal of Marriage and Family* 66:1260–1271.

Decuzzi, A., D. Knox, and M. Zusman. 2004. "The Effect of Parental Divorce on Relationships with Parents and Romantic Partners of College Students." Roundtable discussion, Southern Sociological Society, Atlanta, April 17.

Demian. 2004 (December 12). *Civil Solidarity Pact.* Partners Task Force for Gay and Lesbian Couples. Available at http://www.buddybuddy.com

Demo, David H., Mark A. Fine, and Lawrence H. Ganong. 2000. "Divorce as a Family Stressor." In *Families and Change: Coping with Stressful Events and Transitions,* 2nd ed., P. C. McKenry and S. J. Price, eds. (pp. 279–302). Thousand Oaks, CA: Sage.

DiLillo, D., G. C. Tremblay, and L. Peterson. 2000. "Linking Childhood Sexual Abuse and Abusive Parenting: The Mediating Role of Maternal Anger." *Child Abuse and Neglect* 24:767–769.

Drummond, Tammerlin. 2000. "Mom on Her Own." *Time,* August 28, pp. 54–55.

Edin, Kathryn. 2000. "What Do Low-Income Single Mothers Say About Marriage?" *Social Problems* 47(1):112–133.

Edin, Kathryn, and Laura Lein. 1997. *Making Ends Meet: How Single Mothers Survive Welfare and Low-Wage Work.* New York: Russell Sage Foundation.

Edleson, J. L., L. F. Mbilinyi, S. K. Beeman, and A. K. Hagemeister. 2003. "How Children Are Involved in Adult Domestic Violence." *Journal of Interpersonal Violence* 18:18–32.

Edwards, Tamala M. 2000. "Flying Solo." *Time,* August 28, pp. 49–53.

Emery, Robert E. 1999. "Postdivorce Family Life for Children: An Overview of Research and Some Implications for Policy." In *The Postdivorce Family: Children, Parenting, and Society,* R. A. Thompson and P. R. Amato, eds. (pp. 3–27). Thousand Oaks, CA: Sage.

Emery, Robert E., David Sbarra, and Tara Grover. 2005. "Divorce Mediation: Research and Reflections." *Family Court Review* 43(1):22–37.

Family Court Reform Council of America. 2000. *Parental Alienation Syndrome.* Rancho Santa Margarita, CA: Family Court Reform Council of America.

Family Violence Prevention Fund. 2005 (February 24). *Testimony of the Family Violence Prevention Fund on Welfare Reform and Marriage Promotion Initiatives: Submitted to the House Ways and Means Committee.* Available at http://endabuse.org

FBI (Federal Bureau of Investigation). 2004. *Crime in the United States: 2003.* Available at http://www.fbi.gov

Federal Interagency Forum on Child and Family Statistics. 2004. *America's Children in Brief: Key National Indicators.* Washington, DC: U.S. Government Printing Office.

Few, April L., and Karen H. Rosen. 2005. "Victims of Chronic Dating Violence: How Women's Vulnerabilities Link to Their Decision to Stay." *Family Relations* 54(2):265–279.

Fields, Jason. 2004. "America's Families and Living Arrangements: 2003." *Current Population Reports.* Washington, DC: U.S. Census Bureau.

Fisher, Bonnie S., Francis T. Cullen, and Michael G. Turner. 2000. The *Sexual Victimization of College Women.* National Institute of Justice and Bureau of Justice Statistics. Washington, DC: U.S. Department of Justice.

Fogle, Jean M. 2003. "Domestic Violence Hurts Dogs, Too." *Dog Fancy,* April, p. 12.

Forum on Child and Family Statistics. 2000. *America's Children: Key National Indicators of Well-Being, 2000.* Available at http://www.childstats.gov

Gardner, Richard A. 1998. *The Parental Alienation Syndrome,* 2nd ed. Cresskill, NJ: Creative Therapeutics.

Gelles, Richard J. 1993. "Family Violence." In *Family Violence: Prevention and Treatment,* Robert L. Hampton, Thomas P. Gullotta, Gerald R. Adams, Earl H. Potter III, and Roger P. Weissberg, eds. (pp. 1–24). Newbury Park, CA: Sage.

Gelles, Richard J. 2000. "Violence, Abuse, and Neglect in Families." In *Families and Change: Coping with Stressful Events and Transitions,* 2nd ed., P. C. McKenry and S. J. Price, eds. (pp. 183–207). Thousand Oaks, CA: Sage.

Gilbert, Neil. 2003. "Working Families: Hearth to Market." In *All Our Families,* 2nd ed., M. A. Mason, A. Skolnick, and S. D. Sugarman, eds. (pp. 220–243). New York: Oxford University Press.

Global Initiative to End All Corporal Punishment of Children. 2005 (March). *Legality of Corporal Punishment Worldwide.* Available at http://www.endcorporalpunishment.org

Goldstein, Joshua R., and Catherine T. Kenney. 2001. "Marriage Delayed or Marriage Forgone? New Cohort Forecasts of First Marriage for U.S. Women." *American Sociological Review* 66:506–519.

Gore, Al, and Tipper Gore. 2002. *Joined at the Heart.* New York: Henry Holt.

Grych, John H. 2005. "Interparental Conflict as a Risk Factor for Child Maladjustment: Implications for the Development of Prevention Programs." *Family Court Review* 43(1):97–108.

Hacker, Andrew. 2003. *Mismatch: The Growing Gulf Between Women and Men.* New York: Scribner.

Hackstaff, Karla B. 2003. "Divorce Culture: A Quest for Relational Equality in Marriage." In *Family in Transition*, 12th ed., Arlene S. Skolnick and Jerome H. Skolnick, eds. (pp. 178–190). Boston: Allyn and Bacon.

Hamilton, Brandy E., Joyce A. Martin, and Paul D. Sutton. 2004 (November 23). "Births: Preliminary Data for 2003." *National Vital Statistics Reports*, 53(9). Available at http://www.cdc.gov/nchs

Harrington, Donna, and Howard Dubowitz. 1993. "What Can Be Done to Prevent Child Maltreatment?" In *Family Violence: Prevention and Treatment*, Robert L. Hampton, Thomas P. Gullotta, Gerald R. Adams, Earl H. Potter III, and Roger P. Weissberg, eds. (pp. 258–280). Newbury Park, CA: Sage.

Hawkins, Alan J., Jason S. Carroll, William J. Doherty, and Brian Willoughby. 2004. "A Comprehensive Framework for Marriage Education." *Family Relations* 53(5):547–558.

Hendy, H. M., D. Eggen, C. Gustitus, K. C. McLeod, and P. Ng. 2003. "Decision to Leave Scale: Perceived Reasons to Stay in or Leave Violent Relationships." *Psychology of Women Quarterly* 27:162–173.

Hewlett, Sylvia Ann, and Cornel West. 1998. *The War Against Parents: What We Can Do for Beleaguered Moms and Dads*. Boston: Houghton Mifflin.

Heyman, R. E., and A. M. S. Slep. 2002. "Do Child Abuse and Interpersonal Violence Lead to Adult Family Violence?" *Journal of Marriage and Family* 64:864–870.

Hochschild, Arlie Russell. 1989. *The Second Shift: Working Parents and the Revolution at Home*. New York: Viking.

Hochschild, Arlie Russell. 1997. *The Time Bind: When Work Becomes Home and Home Becomes Work*. New York: Henry Holt.

Hogan, D. P., R. Sun, and G. T. Cornwell. 2000. "Sexual and Fertility Behaviors of American Females Aged 15–19 Years: 1985, 1990, and 1995. *American Journal of Public Health* 90:1421–1425.

Jackson, Shelly, Lynette Feder, David R. Forde, Robert C. Davis, Christopher D. Maxwell, and Bruce G. Taylor. 2003 (June). *Batterer Intervention Programs: Where Do We Go From Here?* U.S. Department of Justice. Available at http://www.usdoj.gov

Jalovaara, M. 2003. "The Joint Effects of Marriage Partners' Socioeconomic Positions on the Risk of Divorce." *Demography* 40:67–81.

Jasinski, J. L., L. M. Williams, and J. Siegel. 2000. "Childhood Physical and Sexual Abuse as Risk Factors for Heavy Drinking Among African-American Women: A Prospective Study." *Child Abuse and Neglect* 24:1061–1071.

Jekielek, Susan M. 1998. "Parental Conflict, Marital Disruption, and Children's Emotional Well-Being." *Social Forces* 76:905–935.

Johnson, Michael P. 2001. "Patriarchal Terrorism and Common Couple Violence: Two Forms of Violence Against Women." In *Men and Masculinity: A Text Reader*, T. F. Cohen, ed. (pp. 248–260). Belmont, CA: Wadsworth.

Johnson, Michael P., and Kathleen Ferraro. 2003. "Research on Domestic Violence in the 1990s: Making Distinctions." In *Family in Transition*, 12th ed., A. S. Skolnick and J. H. Skolnick, eds. (pp. 493–514). Boston: Allyn and Bacon.

Jones, C. 2005. "California's Gay-Marriage Ban Tossed Out." *USA Today*, March 15, p. A1.

Jones, Rachel K., Alison Purcell, Sushella Singh, and Lawrence B. Finer. 2005. "Adolescents' Reports of Parental Knowledge of Adolescents' Use of Sexual Health Services and Their Reactions to Mandated Parental Notification for Prescription Contraception." *JAMA* 293(3):340–348.

Jorgensen, Stephen R. 2000. "Adolescent Pregnancy Prevention: Prospects for 2000 and Beyond." Presidential address at the National Council on Family Relations, 62nd Annual Conference, Minneapolis, November 11.

Kaiser Family Foundation. 2005 (January). *U.S. Teen Sexual Activity*. Available at http://www.kff.org

Kasindorf, Martin. 2002. "Men Wage Battle on 'Paternity Fraud.'" *USA Today*, December 3, p. A3.

Kaufman, Joan, and Edward Zigler. 1992. "The Prevention of Child Maltreatment: Programming, Research, and Policy," In *Prevention of Child Maltreatment: Developmental and Ecological Perspectives*, Diane J. Willis, E. Wayne Holden, and Mindy Rosenberg, eds. (pp. 269–295). New York: Wiley.

Kimmel, Michael S. 2004. *The Gendered Society*, 2nd ed. New York: Oxford University Press.

Kitzmann, K. M., N. K. Gaylord, A. R. Holt, and E. D. Kenny. 2003. "Child Witnesses to Domestic Violence: A Meta-Analytic Review." *Journal of Clinical and Consulting Psychology* 71:339–352.

Knox, David (with Kermit Leggett). 1998. *The Divorced Dad's Survival Book: How to Stay Connected with Your Kids*. New York: Insight Books.

Knox, D., L. Sturdivant, M. E. Zusman, and A. P. Sandie. 2000. "Single Motherhood: College Student Views." *College Student Journal* 34:585–588.

Knutson, John F., and Mary Beth Selner. 1994. "Punitive Childhood Experiences Reported by Young Adults over a 10-Year Period." *Child Abuse and Neglect* 18:155–166.

LaFraniere, Sharon. 2005. "Entrenched Epidemic: Wife-Beatings in Africa." *New York Times*, August 11, pp. A1 and A8.

Laungani, P. 2005. "Changing Patterns of Family Life in India" In *Families in Global Perspective*, J. L. Roopnarine and U. P. Gielen eds. (pp. 85–103). Boston: Pearson, Allyn & Bacon.

Lewin, Tamar. 2000. "Fears for Children's Well-Being Complicates a Debate over Marriage." *New York Times Online*, November 4. Available at http://www.nytimes.com

Lindsey, Linda L. 2005. *Gender Roles: A Sociological Perspective*, 4th ed. Upper Saddle River, NJ: Pearson Prentice Hall.

Lloyd, Sally A. 2000. "Intimate Violence: Paradoxes of Romance, Conflict, and Control." *National Forum* 80(4):19–22.

Lloyd, Sally A., and Beth C. Emery. 2000. *The Dark Side of Courtship: Physical and Sexual Aggression*. Thousand Oaks, CA: Sage.

Loy, E., L. Machen, M. Beaulieu, and G. L. Greif. 2005. "Common Themes in Clinical Work with Women Who Are Domestically Violent." *American Journal of Family Therapy* 33:33–42.

Luker, Kristin. 1996. *Dubious Conceptions: The Politics of Teenage Pregnancy*. Cambridge, MA: Harvard University Press.

Magdol, L., T. E. Moffitt, A. Caspi, and P. A. Silva. 1998. "Hitting Without a License: Testing Explanations for Differences in Partner Abuse Between Young Adult Daters and Cohabitors." *Journal of Marriage and the Family* 60:41–55.

Mason, Mary Ann. 2003. "The Modern American Step-Family: Problems and Possibilities." In *All Our Families*, 2nd ed., Mary Ann Mason, Arlene Skolnick, and Stephen D. Sugarman, eds. (pp. 96–116). New York: Oxford University Press.

Mason, Mary Ann, Arlene Skolnick, and Stephen D. Sugarman. 2003. "Introduction." In *All Our Families*, 2nd ed., Mary Ann Mason, Arlene Skolnick, and Stephen D. Sugarman, eds. (pp. 1–13). New York: Oxford University Press.

Mauldon, Jane. 2003. "Families Started by Teenagers." In *All Our Families*, 2nd ed., Mary Ann Mason, Arlene Skolnick, and Stephen D. Sugarman, eds. (pp. 40–65). New York: Oxford University Press.

McFraser, Peter. 2005 (February 15). *Paternity Testing: Are You Raising Someone Else's Child?* Available at http://www.articlecity.com/articles/health/article_1351.shtml

McKenry, P. C., J. M. Serovich, T. L. Mason, and K. E. Mosack. 2004. "Perpetration of Gay and Lesbian Violence: A Disempowerment Perspective." Paper presented at the Annual Conference of the National Council on Family Relations, Orlando, Florida. November.

Mercy, James A., Etienne G. Krug, Linda L. Dahlberg, and Anthony B. Zwi. 2003. "Violence and Health: The United States in a Global Perspective." *American Journal of Public Health* 93(2):256–261.

Mindel, Charles H., Robert W. Habenstein, and Roosevelt Wright Jr. 1998. *Ethnic Families in America: Patterns and Variations.* Upper Saddle River, NJ: Prentice Hall.

Monson, C. M., G. R. Byrd, and J. Langhinrichsen-Rohling. 1996. "To Have and to Hold: Perceptions of Marital Rape." *Journal of Interpersonal Violence* 11:410–424.

Moore, David W. 2003. *Family, Health Most Important Aspects of Life.* Gallup News Service, January 3. Available at http://www.gallup.org

Munson, M. L., and P. D. Sutton. 2004 (June 10). "Births, Marriages, Divorces, and Deaths: Provisional Data for 2003." *National Vital Statistics Reports* 52(22). Hyattsville, MD: National Center for Health Statistics.

National Campaign to Prevent Teen Pregnancy. 2004 (February). *Teen Birth Rates in the United States, 1940–2003.* Available at http://www.teenpregnancy.org

National Center for Injury Prevention and Control. 2004. *Intimate Partner Violence: Fact Sheet.* Available at http://www.cdc.gov/ncipc

National Center for Injury Prevention and Control. 2005. *Child Maltreatment: Fact Sheet.* Available at http://www.cdc.gov/ncipc

National Gay and Lesbian Task Force. 2005 (January). *Second-Parent Adoption in the United States.* Available at http://www.ngltf.org

National Mental Health Association. 2003. *Effective Discipline Techniques for Parents: Alternatives to Spanking.* Strengthening Families Fact Sheet. Available at http://www.nmha.org

National Parenting Association. 1996. *What Will Parents Vote For? Findings of the First National Survey of Parent Priorities.* New York: National Parenting Association.

Nelson, B. S., and K. S. Wampler. 2000. "Systemic Effects of Trauma in Clinic Couples: An Exploratory Study of Secondary Trauma Resulting from Childhood Abuse." *Journal of Marriage and Family Counseling* 26:171–184.

Newton, James P. 2005. "Abuse in the Elderly: A Perennial Problem." *Gerontology* 22(1):1.

Nock, Steven L. 1995. "Commitment and Dependency in Marriage." *Journal of Marriage and the Family* 57:503–514.

O'Reilly, S., D. Knox, M. Zusman, and H. R. Thompson. 2005. "What Women Want." Poster, Annual Meeting of the Eastern Sociological Society, Washington, DC, March 18.

Parker, Marcie R., Edward Bergmark, Mark Attridge, and Jude Miller-Burke. 2000. "Domestic Violence and Its Effect on Children." *National Council on Family Relations Report* 45(4):F6–F7.

Pasley, K. 2000. "Stepfamilies Doing Well Despite Challenges." *National Council on Family Relations* 45:6–7.

Pasley, Kay, and Carmelle Minton. 2001. "Generative Fathering After Divorce and Remarriage: Beyond the 'Disappearing Dad.'" In *Men and Masculinity: A Text Reader,* T. F. Cohen, ed. (pp. 239–248). Belmont CA: Wadsworth.

Popenoe, David. 1996. *Life Without Father.* New York: Free Press.

Rand, M. R. 2003. "The Nature and Extent of Recurring Intimate Partner Violence Against Women in the United States." *Journal of Comparative Family Studies* 34:137–146.

Realini, J. P. 2004. "Teenage Pregnancy Prevention: What Can We Do?" *American Family Physician* 70:1457–1458.

Rennison, Callie M. 2003. *Intimate Partner Violence, 1993–2001.* Bureau of Justice Statistics, Crime Data Brief. Washington, DC: U.S. Department of Justice.

Ricci, L., A. Giantris, P. Merriam, S. Hodge, and T. Doyle. 2003. "Abusive Head Trauma in Maine Infants: Medical, Child Protective, and Law Enforcement Analysis." *Child Abuse and Neglect* 27:271–283.

Roopnarine, J. L., and U. P. Gielen. 2005. "Families in Global Perspective: An Introduction" In *Families in Global Perspective,* J. L. Roopnarine and U. P. Gielen, eds. (pp. 3–13). Boston: Pearson, Allyn & Bacon.

Rubin, D. M., C. W. Christian, L. T. Bilaniuk, K. A. Zaxyczny, and D. R. Durbin. 2003. "Occult Head Injury in High-Risk Abused Children." *Pediatrics* 111:1382–1386.

Russell, D. E. 1990. *Rape in Marriage.* Bloomington: Indiana University Press.

Schacht, Thomas E. 2000. "Protection Strategies to Protect Professionals and Families Involved in High-Conflict Divorce." *UALR Law Review* 22(3):565–592.

Scott, K. L., and D. A. Wolfe. 2000. "Change Among Batterers: Examining Men's Success Stories." *Journal of Interpersonal Violence* 15:827–842.

Shapiro, Joseph P., and Joannie M. Schrof. 1995. "Honor Thy Children." *U.S. News and World Report,* February 27, pp. 39–49.

Shepard, Melanie F., and James A. Campbell. 1992. "The Abusive Behavior Inventory: A Measure of Psychological and Physical Abuse." *Journal of Interpersonal Violence* 7(3):291–305.

SIECUS (Sex Information and Education Council of the United States). 2004. *Public Support for Comprehensive Sexuality Education.* Fact sheet. Available at http://www.siecus.org

Sigle-Rushton, W., and S. McLanahan. 2002. "The Living Arrangements of New Unmarried Mothers." *Demography* 39:415–433.

Simmons, T., and M. O'Connell. 2003. *Married-Couple and Unmarried Partner Households: 2000.* Census 2000 Special Reports. Available at http://www.census.gov

Simonelli, C. J., T. Mullis, A. N. Elliott, and T. W. Pierce. 2002. "Abuse by Siblings and Subsequent Experiences of Violence Within the Dating Relationship." *Journal of Interpersonal Violence* 17:103–121.

Singh, Susheela, and Jacqueline E. Darroch. 2000. "Adolescent Pregnancy and Childbearing: Levels and Trends in Developed Countries." *Family Planning Perspectives* 32(1):14–23.

Smith, J. 2003. "Shaken Baby Syndrome." *Orthopaedic Nursing.* 22:196–205.

Sorensen, Elaine and Helen Oliver. 2002 (April). *Policy Reforms Are Needed to Increase Child Support from Poor Families.* Washington, DC: Urban Institute.

Spiegel, D. 2000. "Suffer the Children: Long-Term Effects of Sexual Abuse." *Society* 37:18–20.

Stanley, Scott M., Howard J. Markman, Michelle St. Peters, and B. Douglas Leber. 1995. "Strengthening Marriage and Preventing Divorce: New Directions in Prevention Research." *Family Relations* 44:392–401.

Steimle, Brynn M., and Stephen F. Duncan. 2004. "Formative Evaluation of a Family Life Education Web Site." *Family Relations* 53(4):367–376.

Stein, Rob. 2005. "Pharmacists' Rights at Front of New Debate." *Washington Post,* March 28, p. A01. Available at http://www.washingtonpost.com

Stock, J. L., M. A. Bell, D. K. Boyer, and F. A. Connell. 1997. "Adolescent Pregnancy and Sexual Risk-Taking Among Sexually Abused Girls." *Family Planning Perspectives* 29:200–203.

Stone, R. D. 2004. *No Secrets, No Lies: How Black Families Can Heal from Sexual Abuse.* New York: Broadway Books

Straus, Murray. 2000. "Corporal Punishment and Primary Prevention of Physical Abuse." *Child Abuse and Neglect* 24:1109–1114.

Terry-Humen, Elizabeth, Jennifer Manlove, and Kristin A. Moore. 2005. *Playing Catch-Up: How Children Born to Teen Mothers Fare.* Washington, DC: National Campaign to Prevent Teen Pregnancy.

Thakkar, R. R., P. M. Gutierrez, C. L. Kuczen, and T. R. McCanne. 2000. "History of Physical and/or Sexual Abuse, and Current Suicidality in College Women." *Child Abuse and Neglect* 24:1345–1354.

Ulman, A. 2003. "Violence by Children Against Mothers in Relation to Violence Between Parents and Corporal Punishment by Parents." *Journal of Comparative Family Studies* 34:41–56.

Umberson, D., K. L. Anderson, K. Williams, and M. D. Chen. 2003. "Relationship

Dynamics, Emotion State, and Domestic Violence: A Stress and Masculine Perspective." *Journal of Marriage and the Family* 65:233–247.

United Nations Development Programme. 2000. *Human Development Report 2000.* Cary, NC: Oxford University Press.

U.S. Bureau of Labor Statistics. 2005. *Employment Characteristics of Families in 2004.* Available at http://www.bls.gov

U.S. Bureau of the Census. 2000. *America's Families and Living Arrangements.* Current Population Reports, Series P20-537. Washington, DC: U.S. Government Printing Office.

U.S. Bureau of the Census. 2005. *Statistical Abstract of the United States: 2004–2005,* 124th ed. Washington, DC: U.S. Government Printing Office.

U.S. Census Bureau. 2004. *Annual Social and Economic Supplement.* Available at http://www.census.gov

U.S. Conference of Mayors. 2004. *USCM-Sodexho USA Hunger and Homelessness Survey.* Available at http://www.usmayors.org

U.S. Department of Health and Human Services. 2000 (June 17). *HHS Fatherhood Initiative.* Available at http://www.hhs.gov/news/press/2000pres/20000617.html

U.S. Department of Health and Human Services, Administration on Children, Youth, and Families. 2005. *Child Maltreatment 2003.* Washington, DC: U.S. Government Printing Office.

Ventura, Stephanie. 1995 (June). *Births to Unmarried Mothers: United States, 1980–92.* Vital and Health Statistics, Series 21(no. 53). Available at http://www.cdc.gov/nchs

Ventura, S. J., and Christine A. Bachrach. 2000 (October 18). *Nonmarital Childbearing in the United States, 1940–99.* National Vital Statistics Report 48(16).

Ventura, Stephanie J., Sally C. Curtin, and T. J. Mathews. 2000 (April 24). *Variations in Teenage Birth Rates, 1991–1998: National and State Trends.* National Vital Statistics Report 48(6).

Viano, C. Emilio. 1992. "Violence Among Intimates: Major Issues and Approaches." In *Intimate Violence: Interdisciplinary Perspectives,* C. E. Viano, ed. (pp. 3–12). Washington, DC: Hemisphere.

Walker, Alexis J. 2001. "Refracted Knowledge: Viewing Families Through the Prism of Social Science." In *Understanding Families into the New Millennium: A Decade in Review,* Robert M. Milardo, ed. (pp. 52–65). Minneapolis: National Council on Family Relations.

Wallerstein, Judith S. 2003. "Children of Divorce: A Society in Search of Policy." In *All Our Families,* 2nd ed., Mary Ann Mason, Arlene Skolnick, and Stephen D. Sugarman, eds. (pp. 66–95). New York: Oxford University Press.

Waxman, Henry A. 2004 (December). *The Context of Federally Funded Abstinence-*

Only Education Programs. U.S. House of Representatives Committee on Government Reform, Minority Staff, Special Investigation Division. Available at http://www.democrats.reform.house.gov

Whiffen, V. E., J. M. Thompson, and J. A. Aube. 2000. "Mediators of the Link Between Childhood Sexual Abuse and Adult Depressive Symptoms." *Journal of Interpersonal Violence* 15:1100–1120.

Whitehead, Barbara Dafoe, and David Popenoe. 2005. *The State of Our Unions: The Social Health of Marriage in America.* National Marriage Project, Rutgers University. Available at http://marriage.rutgers.edu

Chapter 6

Administration for Children and Families. 2002. "Early Head Start Benefits Children and Families." U.S. Department of Health and Human Services. Available at http://www.acf.hhs.gov

Albelda, Randy, and Chris Tilly. 1997. *Glass Ceilings and Bottomless Pits: Women's Work, Women's Poverty.* Boston: South End Press.

Alex-Assensoh, Yvette. 1995. "Myths About Race and the Underclass." *Urban Affairs Review* 31:3–19.

Altieri, Miguel. 2003 (June 10). *The Case Against Agricultural Biotechnology: Why Are Transgenic Crops Incompatible with Sustainable Agriculture in the Third World?* Corpwatch. Available at http://www.corpwatch.org

Anderson, Sarah, John Cavanagh, Chuck Collins, Chris Hartman, and Felice Yeskel. 2000. *Executive Excess: Seventh Annual CEO Compensation Survey.* Boston: Institute for Policy Studies and United for a Fair Economy.

Anderson, Sarah, John Cavanagh, Chris Hartman, Scott Klinger, and Stacey Chan. 2004. *Executive Excess: 11th Annual CEO Compensation Survey.* Boston: Institute for Policy Studies and United for a Fair Economy.

Andrews, Edmund. 2003. "Economic Inequality Grew in 90's Boom, Fed Reports." *New York Times Online,* January 23. Available at http://www.nytimes.com

Associated Press. 2005. "Judge Halts Grant Over Religion." *New York Times,* January 16. Available at http://www.nytimes.com

Bailey, Ronald. 2004. "Scientific Arguments Against Biotechnology Are Fallacious." In *Genetically Engineered Foods,* Nancy Harris, ed. (pp. 80–93). Farmington Hills, MI: Greenhaven Press.

Benbrook, Charles M. 2004. *Impacts of Genetically Engineered Crops on Pesticide Use in the United States: The First Nine Years.* BioTech InfoNet, Technical Paper 7. Available at http://www.biotech-info.net

Bickel, G., M. Nord, C. Price, W. Hamilton, and J. Cook. 2000. *United States Depart-*

ment of Agriculture Guide to Measuring Household Food Security. Alexandria, VA: U.S. Department of Agriculture, Food and Nutrition Service.

Boston, Rob. 2005 (February 15). *"Faith-Based" Flim-Flam Initiative Didn't Have a Prayer Says Former White House Aid.* Americans United for a Separation of Church and State. Available at http://blog.au.org

Briggs, Vernon M., Jr. 1998. "American-Style Capitalism and Income Disparity: The Challenge of Social Anarchy." *Journal of Economic Issues* 32(2):473–481.

Brown, Lester R. 2001. "Eradicating Hunger: A Growing Challenge." In *State of the World 2001,* Lester R. Brown, Christopher Flavin, and Hilary French, eds. (pp. 43–62). New York: W.W. Norton.

Bureau of Labor Statistics. 2005. *Characteristics of Minimum Wage Workers: 2004.* Available at http://stats.bls.gov

Center for Food Safety and International Forum on Globalization. 2005. *Worldwide Regulation, Prohibition, and Production of Genetically Modified Crops and Foods.* Available at http://www.centerforfoodsafety.org

Center on Budget and Policy Priorities. 2005 (May 31). *New Study Finds Poor Medicaid Beneficiaries Face Growing Out-of-Pocket Medical Costs.* Available at http://www.cbpp.org

Chasanov, Amy, and Jeff Chapman. 2005. "Minimum Wage: State Success, Federal Failure." *EPI Journal,* winter, pp. 1 and 7.

Children's Defense Fund. 2002. *Low-Income Families Bear the Burden of State Child Care Cuts.* Available at http://www.childrensdefense.org

Children's Defense Fund. 2003. *Children in the United States.* Available at http://www.childrensdefense.org

Citizens for Tax Justice. 2002 (April 17). *Surge in Corporate Tax Welfare Drives Corporate Tax Payments Down to Near Record Low.* Available at http://www.ctj.org

Citizens for Tax Justice. 2005 (April 21). *International Tax Comparisons, 1965–2003 (Federal, State, and Local).* Available at http://www.ctj.org

Corcoran, Mary, and Terry Adams. 1997. "Race, Sex, and the Intergenerational Transmission of Poverty." In *Consequences of Growing Up Poor,* Greg J. Duncan and Jeanne Brooks-Gunn, eds. (pp. 461–517). New York: Russell Sage Foundation.

Davis, Kingsley, and Wilbert Moore. 1945. "Some Principles of Stratification." *American Sociological Review* 10:242–249.

DeNavas-Walt, Carmen, Bernadette D. Proctor, and Robert J. Mills. 2004. *Income, Poverty, and Health Insurance in the United States: 2003.* U.S. Census Bureau, Current Population Reports P60-226. Washington, DC: U.S. Government Printing Office.

DeNavas-Walt, Carmen, Bernadette D. Proctor, and Cheryl Hill Lee. 2005. *Income, Poverty, and Health Insurance in the United States: 2004.* U.S. Census Bureau, Current Population Reports P60-229. Washington, DC: U.S. Government Printing Office.

Deng, Francis M. 1998. "The Cow and the Thing Called 'What': Dinka Cultural Perspectives on Wealth and Poverty." *Journal of International Affairs* 52(1):101–115.

Dordick, Gwendolyn. 1997. *Something Left to Lose: Personal Relations and Survival Among New York's Homeless.* Philadelphia: Temple University Press.

Dowd, Maureen. 2005. "United States of Shame." *New York Times,* September 3. Available at http://www.nytimes.com

Duncan, Greg J., and Jeanne Brooks-Gunn. 1997. "Income Effects Across the Life Span: Integration and Interpretation." In *Consequences of Growing Up Poor,* Greg J. Duncan and Jeanne Brooks-Gunn, eds. (pp. 596–610). New York: Russell Sage Foundation.

Duncan, Greg J., and P. Lindsay Chase-Lansdale. 2001. "Welfare Reform and Child Well-Being." Paper presented at the Blank/Haskins conference, "The New World of Welfare Reform," Washington, D.C., February 1–2.

Ebaugh, Helen Rose, Janet Saltzman Chafetz, and Paula F. Pipes. 2005. "Faith-Based Social Service Organizations and Government Funding: Data From a National Survey." *Social Science Quarterly* 86(2): 273–292.

Economic Policy Institute. 2000. *Issue Guide to the Minimum Wage.* Available at http://www.epinet.org

Edin, Kathryn, and Laura Lein. 1977. *Making Ends Meet.* New York: Russell Sage Foundation.

Epstein, William M. 2004. "Cleavage in American Attitudes Toward Social Welfare." *Journal of Sociology and Social Welfare* 31(4):177–201.

ETC Group. 2003a. *Broken Promise? Monsanto Promotes Terminator Seed Technology.* News release, April 23. Available at http://www.etcgroup.org

ETC Group. 2003b. *Contamination by Genetically Modified Maize in Mexico Much Worse Than Feared.* Available at http://www.etcgroup.org

Fields, Jason. 2003. *Children's Living Arrangements and Characteristics: March 2002.* U.S. Census Bureau, Current Population Reports, P20–547. Washington, DC: U.S. Census Bureau.

Fletcher, Michael A. 2005. "Two Fronts on the War on Poverty." *Washington Post,* May 17, p. A01.

Food Research and Action Center. 2004. "Food Stamp Participation Increases in October 2003 to More Than 23.3 Million Persons." *Current News and Analysis,* January 6. Available at http://www.frac.org

Fremstad, Sean. 2004 (January 30). *Recent Welfare Reform Research Findings: Implications for TANF Reauthorization and State TANF Policies.* Center on Budget and Policy Priorities. Available at http://www.cbpp.org

Global Knowledge Center on Crop Biotechnology. 2005. *Summary Report 1996–2004.* Available at http://www.isaaa.org

Goesling, Brian. 2001. "Changing Income Inequalities Within and Between Nations: New Evidence." *American Sociological Review* 66:745–761.

Gonzalez, David. 2005. "From Margins of Society to Center of Tragedy." *New York Times,* September 2. Available at http://www.nytimes.com

Hallman, William K., W. Carl Hebden, Cara L. Cuite, Helen L. Aquino, and John T. Lang. 2004. *Americans and GM Food: Knowledge, Opinion, and Interest in 2004.* Food Policy Institute. Rutgers University, New Brunswick, NJ. Available at http://www.foodpolicyinstitute.org

Halweil, Brian. 2005. "Grain Harvest and Hunger Both Grow." In *Vital Signs 2005,* Linda Starke, ed. (pp. 22–23). New York: W. W. Norton.

The Hatcher Group. 2005. *Greener Pastures for Working Families.* Resource Brief No. 2. Bethesda MD: The Hatcher Group.

Haygood, Wil. 2005. "Living Paycheck to Paycheck Made Leaving Impossible." *Washington Post,* September 4, p. A33. Available at http://www.washingtonpost.com

Hickey, Ellen, and Anuradha Mittal. 2003. *Voices from the South: The Third World Debunks Corporate Myths on Genetically Engineered Crops.* Food First Institute for Food and Development Policy and Pesticide Action Network North America. Available at http://www.foodfirst.org

Hill, Lewis E. 1998. "The Institutional Economics of Poverty: An Inquiry into the Causes and Effects of Poverty." *Journal of Economic Issues* 32(2):279–286.

hooks, bell. 2000. *Where We Stand: Class Matters.* New York: Routledge.

Jargowsky, Paul A. 1997. *Poverty and Place: Ghettos, Barrios, and the American City.* New York: Russell Sage Foundation.

Kennedy, Bruce P., Ichiro Kawachi, Roberta Glass, and Deborah Prothrow-Stith. 1998. "Income Distribution, Socioeconomic Status, and Self-Rated Health in the U.S.: Multilevel Analysis." *British Medical Journal* 317(7163):917–921.

Kingsley, G. Thomas, and Kathryn L. S. Pettit. 2003 (May). *Concentrated Poverty: A Change in Course.* Urban Institute. Available at http://www.urban.org/nnip

Kraut, Karen, Scott Klinger, and Chuck Collins. 2000. *Choosing the High Road: Businesses That Pay a Living Wage and Prosper.* Boston: United for a Fair Economy.

Leventhal, Tama, and Jeanne Brooks-Gunn. 2003. "Moving to Opportunity: An Exper-imental Study of Neighborhood Effects on Mental Health." *American Journal of Public Health* 93(9):1576–1585.

Lewis, Oscar. 1966. "The Culture of Poverty." *Scientific American* 2(5):19–25.

Lewis, Oscar. 1998. "The Culture of Poverty: Resolving Common Social Problems." *Society* 35(2):7–10.

Living Wage Resource Center. 2005 (May 27). Personal Communication. Living Wage Resource Center, Boston, MA. Available at http://www.livingwagecampaign.org

Llobrera, Joseph, and Bob Zahradnik. 2004. *A HAND UP: How State Earned Income Tax Credits Helped Working Families Escape Poverty in 2004.* Center on Budget and Policy Priorities. Available at http://www.cbpp.org

Luker, Kristin. 1996. *Dubious Conceptions: The Politics of Teenage Pregnancy.* Cambridge, MA: Harvard University Press.

Malatu, Mesfin Samuel, and Carmi Schooler. 2002. "Causal Connections Between Socioeconomic Status and Health: Reciprocal Effects and Mediating Mechanisms." *Journal of Health and Social Behavior* 43:22–41.

Mann, Judy. 2000 (May 15). "Demonstrators at the Barricades Aren't Very Subtle, But They Sometimes Win." *Washington Spectator* 26(10):1–3.

Massey, D. S. 1991. "American Apartheid: Segregation and the Making of the American Underclass." *American Journal of Sociology* 96:329–357.

Mayer, Susan E. 1997. *What Money Can't Buy: Family Income and Children's Life Chances.* Cambridge, MA: Harvard University Press.

McIntyre, Robert. 2005 (April 12). *Tax Cheats and Their Enablers.* Citizens for Tax Justice. Available at http://www.ctj.org

Mendelson, Joseph. 2002. *Why Biotechnology Will Not Feed the World.* Center for Food Safety. Available at http://www.centerforfoodsafety.org

Michel, Sonya. 1998. "Childcare and Welfare (In)justice." *Feminist Studies* 24:44–54.

Mishel, Lawrence. 2005. "The Economy *Is* a Values Issue." *EPI Journal,* winter, pp. 1 and 6.

Mishel, Lawrence, Jared Bernstein, and Sylvia Allegretto. 2005. *The State of Working America 2004/2005.* New York: Cornell University Press.

Narayan, Deepa. 2000. *Voices of the Poor: Can Anyone Hear Us?* New York: Oxford University Press.

National Center for Children in Poverty. 2004. *The Effects of Parental Education on Income.* Available at http://www.nccp.org

National Law Center on Homelessness and Poverty in America. 2004. *Homelessness in the United States and the Human Right to Housing.* Available at http://www.nlchp.org

Newman, Katherine S. 1999. *No Shame in My Game: The Working Poor in the Inner*

City. New York: Alfred A. Knopf and The Russell Sage Foundation.

Nord, Mark. 2005. "Measuring U.S. Household Food Insecurity." *Amber Waves* 3(2):10–11.

Office of Family Assistance. 2004. *TANF Sixth Annual Report to Congress.* Available at http://www.acf.hhs.gov

Oxfam. 2005. *Paying the Price: Why Rich Countries Must Invest Now in a War on Poverty.* Available at http://www.oxfam.org

Parisi, Domenico, Steven Michael Grice, and Michael Taquino. 2003. "Poverty and Inequality in the Context of Welfare Reform." *Social Problems Forum: The SSSP Newsletter,* 34(2):19–21.

Purdum, Todd S. 2005. "Across U.S., Outrage at Response." *New York Times,* September 3. Available at http://www.nytimes.com

Roberts, John. 2005 (January 17). *Thai Government Puts Tourism Ahead of the Poor in Tsunami Relief Effort.* World Socialist Web Site. Available at http://www.wsws.org

Rothstein, Richard. 2004. *Class and Schools.* Washington DC: Economic Policy Institute.

Ruiz-Marrero, Carmelo. 2002. *Genetic Pollution: Starlink Corn Invades Mexico.* CorpWatch. Available at http://www.corpwatch.org

Sanderson, Stephen K., and Arthur S. Alderson. 2005. *World Societies: The Evolution of Human Social Life.* Boston: Pearson Education.

Schifferes, Steve. 2004. "Can Globalization Be Tamed?" *BBC News Online,* February 24. Available at http://www.bbc.co.uk

Schiller, Bradley R. 2004. *The Economics of Poverty and Discrimination,* 9th ed. Upper Saddle River, NJ: Pearson Prentice Hall.

Seccombe, Karen. 2001. "Families in Poverty in the 1990s: Trends, Causes, Consequences, and Lessons Learned." In *Understanding Families into the New Millennium: A Decade in Review,* Robert M. Milardo, ed. (pp. 313–332). Minneapolis: National Council on Family Relations.

Shapiro, Isaac. 2005 (March 7). *What New CBO Data Indicate About Long-Term Income Distribution Trends.* Center on Budget and Policy Priorities. Available at http://www.cbpp.org

Sobolewski, Juliana M., and Paul R. Amato. 2005. "Economic Hardship in the Family of Origin and Children's Psychological Well-Being in Adulthood." *Journal of Marriage and Family* 67(1):141–156.

Sorensen, Elaine. 2003. *Child Support Gains Some Ground.* Urban Institute. Available at http://www.urbaninstitute.org

Stocking, Barbara. 2005 (January 5). *The Tsunami and the Bigger Picture.* Oxfam. Available at http://www.oxfam.org

Streeten, Paul. 1998. "Beyond the Six Veils: Conceptualizing and Measuring Poverty." *Journal of International Affairs* 52(1):1–8.

Susskind, Yifat. 2005 (May). *Ending Poverty, Promoting Development: MADRE Criticizes the United Nations Millennium Development Goals.* Available at http://www.madre.org

Turner, Margery Austin, Susan J. Popkin, G. Thomas Kingsley, and Deborah Kaye. 2005 (April). *Distressed Public Housing: What it Costs to Do Nothing.*" The Urban Institute. Available at http://www.urban.org

UNDP (United Nations Development Programme). 1997. *Human Development Report 1997.* New York: Oxford University Press.

UNDP. 2000. *Human Development Report 2000.* New York: Oxford University Press.

UNDP. 2003. *Human Development Report 2003.* New York: Oxford University Press.

United Nations. 1997. *Report on the World Social Situation, 1997.* New York: United Nations.

United Nations Population Fund. 2002. *State of World Population 2002: People, Poverty, and Possibilities.* New York: United Nations.

United Nations Population Fund. 2004. *State of World Population 2004: The Cairo Consensus at Ten—Population, Reproductive Health, and the Global Effort to End Poverty.* New York: United Nations.

Universal Living Wage. 2005. *Universal Living Wage.* Available at http://www.universallivingwage.org

U.S. Census Bureau. 2002 (March). *Annual Demographic Survey, March Supplement. Current Population Survey.* Available at http://ferret.bls.census.gov

U.S. Census Bureau. 2004. *Historical Poverty Tables: People.* Available at http://www.census.gov/hhes/poverty/histpov

U.S. Census Bureau. 2005a. *Current Population Survey: Annual Social and Economic Supplement.* Available at http://www.census.gov

U.S. Census Bureau. 2005b. *Poverty Thresholds 2004.* Available at http://www.census.gov

U.S. Conference of Mayors. 2004. *USCM–Sodexo USA Hunger and Homelessness Survey: A Status Report on Hunger and Homelessness in America's Cities.* Washington, DC: U.S. Conference of Mayors.

USDA (U.S. Department of Agriculture). 2004. *Characteristics of Food Stamp Households: Fiscal Year 2003.* Available at http://www.fns.usda.gov

U.S. Department of Health and Human Services. 2004 (March 30). *Welfare Rolls Drop Again.* News release. Available at http://www.hhs.gov

Van Kempen, Eva T. 1997. "Poverty Pockets and Life Chances: On the Role of Place in Shaping Social Inequality." *American Behavioural Scientist* 41(3):430–450.

Wagstaff, Adam. 2003. "Child Health on a Dollar a Day: Some Tentative Cross-Country Comparisons." *Social Science and Medicine* 57:1529–1538.

Washburn, Jennifer. 2004. "The Tuition Crunch." *Atlantic Monthly* 293(1):140.

Whitehead, Barbara Dafoe, and David Popenoe. 2004. *The State of Our Unions: The Social Health of Marriage in America.* The National Marriage Project, Rutgers University. Available at http://marriage.rutgers.edu

Wilson, William J. 1987. *The Truly Disadvantaged: The Inner City, the Underclass, and Public Policy.* Chicago: University of Chicago Press.

Wilson, William J. 1996. *When Work Disappears: The World of the New Urban Poor.* New York: Knopf.

World Bank. 2001. *World Development Report: Attacking Poverty, 2000/2001.* Herndon, VA: World Bank and Oxford University Press.

World Bank. 2005. *Global Monitoring Report 2005.* Available at http://www.worldbank.org

World Health Organization. 2002. *The World Health Report 2002.* Available at http://www.who.int/pub/en

World Population News Service. 2003. "Reducing Poverty Is Key to Global Stability." *Popline,* May–June, p. 4.

Zedlewski, Sheila R. 2003. *Work and Barriers to Work Among Welfare Recipients in 2002.* Urban Institute. Available at http://www.urban.org

Zedlewski, Sheila R., Sandi Nelson, Kathryn Edin, Heather L. Koball, and Kate Roberts. 2003. *Families Coping Without Earnings or Government Cash Assistance.* Urban Institute. Available at http://www.urbaninstitute.org

Zedlewski, Sheila R., and Kelly Rader. 2005 (March 31). *Feeding America's Low Income Children.* New Federalism: National Survey of America's Families Series No. B-65. Urban Institute. Available at http://www.urban.org

Chapter 7

AFL-CIO. 2004. *Ask a Working Woman Survey Report.* Available at http://www.aflcio.org

AFL-CIO. 2005. *Death on the Job: The Toll of Neglect,* 14th ed. Available at http://www.aflcio.org

Austin, Colin. 2002. "The Struggle for Health in Times of Plenty." In *The Human Cost of Food: Farmworkers' Lives, Labor, and Advocacy,* C. D. Thompson Jr. and M. F. Wiggins, eds. (pp. 198–217). Austin: University of Texas Press.

Bales, Kevin. 1999. *Disposable People: New Slavery in the Global Economy.* Berkeley: University of California Press.

Barlett, Donald L., and James B. Steele. 1998. "Corporate Welfare: First in a Series." *Time* 152(19):36–39.

Barstow, David, and Lowell Bergman. 2003. "Deaths on the Job, Slaps on the Wrist." *New York Times Online,* January 10. Available at http://www.nytimes.com

Bassi, Laurie J., and Jens Ludwig. 2000. "School-to-Work Programs in the United States: A Multi-Firm Case Study of Training, Benefits, and Costs." *Industrial and Labor Relations Review* 53(2):219–239.

Bello, Walden. 2001. "Lilliputians Rising: 2000: The Year of Global Protest Against Corporate Globalization." *Multinational Monitor* 22(1–2). Available at http://www.essential.org/monitor

Benjamin, Medea. 1998. "What's Fair About Fair Labor Association (FLA)?" *Sweatshop Watch.* Available at http://www.sweatshopwatch.org

"Big Business for Reform." 2000. *Multinational Monitor* 21(11). Available at http://www.essential.org/monitor

Bond, James T., Cindy Thompson, Ellen Galinsky, and David Prottas. 2002. *Highlights of the National Study of the Changing Workforce.* Families and Work Institute. Available at http://www.familiesandwork.org

Bowles, Diane O. 2000. "Growth in Telework." Paper presented at the symposium "Telework and the New Workplace of the 21st Century," Xavier University, New Orleans, October 16. U.S. Department of Labor. Available at http://www.dol.gov/dol/asp/public/telework/htm

Bronfenbrenner, Kate. 2000. "Raw Power: Plant-Closing Threats and the Threat to Union by Organizing." *Multinational Monitor* 21(12). Available at http://www.essential.org/monitor

Bureau of Labor Statistics. 2004. *Workplace Injuries and Illnesses in 2003.* Available at http://www.bls.gov

Bureau of Labor Statistics. 2005a. "Employment Characteristics of Families in 2004." Available at http://www.bls.gov

Bureau of Labor Statistics. 2005b (March 30). *Lost-Worktime Injuries and Illnesses: Characteristics and Resulting Days Away From Work, 2003.* Available at http://www.bls.gov

Bureau of Labor Statistics. 2005c. *Manufacturing.* Available at http://www.bls.gov

Bureau of Labor Statistics. 2005d (August 25). *National Census of Fatal Occupational Injuries in 2004.* Available at http://www.bls.gov

Bureau of Labor Statistics. 2005e. *Union Members in 2004.* Available at http://www.bls.gov

Bureau of Labor Statistics. 2005f. *Women in the Labor Force: A Databook.* Available at http://www.bls.gov

Cantor, David, Jane Waldfogel, Jeffrey Kerwin, Mareena McKinley Wright, Kerry Levin, John Rauch, Tracey Hagerty, and Martha Stapelton Kudela. 2001. *Balancing the Needs of Families and Employers: The Family and Medical Leave Surveys, 2000 Update.* U.S. Department of Labor. Available at http://www.dol.gov/dol

Caston, Richard J. 1998. *Life in a Business-Oriented Society: A Sociological Perspective.* Boston: Allyn and Bacon.

Center for Responsive Politics. 2005a. *The Bush Administration Corporate Connections.* Available at http://www.opensecrets.org

Center for Responsive Politics. 2005b. *Tracking the Payback.* Available at http://www.opensecrets.org

Cernasky, Rachel. 2002 (December). *Slavery: Alive and Thriving in the World Today—The Satya Interview with Kevin Bales.* Available at http://www.satyamag.com

Clarke, Tony. 2002. "Twilight of the Corporation." In *Social Problems, Annual Editions 02/03,* 30th ed., Kurt Finster-Busch, ed. (pp. 41–45). Guilford, CT: McGraw-Hill/Dushkin.

Cockburn, Andrew. 2003. "21st Century Slaves." *National Geographic,* September, pp. 2–11, 18–24.

Durkheim, Emile. 1966 [1893]. *On the Division of Labor in Society,* G. Simpson, trans. New York: Free Press.

"Editorial: What Is Society Willing to Spend on Human Beings?" 2000. *Multinational Monitor* 21(11). Available at http://www.essential.org

Frederick, James, and Nancy Lessin. 2000. "Blame the Worker: The Rise of Behavior-Based Safety Programs." *Multinational Monitor* 21(11). Available at http://www.essential.org/monitor

Galinsky, Ellen, and James T. Bond. 1998. *The 1998 Business Work-Life Study.* New York: Families and Work Institute.

Galinsky, Ellen, James T. Bond, Stacy S. Kim, Lois Backon, Erin Brownfield, and Kelly Sakai. 2005. *Overwork in America: When the Way We Work Becomes Too Much.* New York: Families and Work Institute.

Galinsky, Ellen, and Stacy S. Kim. 2000. "Navigating Work and Parenting by Working at Home: Perspectives of Workers and Children Whose Parents Work at Home." Paper presented at the symposium "Telework and the New Workplace of the 21st Century," Xavier University, New Orleans, October 16. U.S. Department of Labor. Available at http://www.dol.gov/dol/asp/public/telework/htm

Gallup Organization. 2005. "Race Relations." *The Gallup Brain.* Available at http://www.gallup.com

"The Garment Industry." 2001. *Sweatshop Watch.* Available at http://www.sweatshopwatch.org

George, Kathy. 2003 (December 1). "Myanmar: Unocal Faces Landmark Trial over Slavery." *CorpWatch.* Available at http://www.corpwatch.org

Gordon, David M. 1996. *Fat and Mean: The Corporate Squeeze of Working Americans and the Myth of Managerial "Downsizing."* New York: Free Press.

Greenhouse, Steven. 2000. "Anti-Sweatshop Movement Is Achieving Gains Overseas." *New York Times,* January 26. Available at http://www.nytimes.com

Greenhouse, Steven. 2005. "Wal-Mart Workers Are Finding a Voice Without a Union." *New York Times,* September 3. Available at http://www.nytimes.com

Hargis, Michael J. 2001. "Bangladesh: Garment Workers Burned to Death." *Industrial Worker* 1630, 98(1). Available at http://www.parsons.www.org

Hochschild, Arlie Russell. 1997. *The Time Bind: When Work Becomes Home and Home Becomes Work.* New York: Holt.

Human Rights Watch. 2000. *Unfair Advantage: Workers' Freedom of Association in the United States Under International Human Rights Standards.* Available at http://www.hrw.org

Human Rights Watch. 2001. *World Report 2001.* Available at http://www.hrw.org

Inglehart, Ronald. 2000. "Globalization and Postmodern Values." *Washington Quarterly,* winter, pp. 215–228.

International Confederation of Free Trade Unions. 2005. *2004 Annual Survey of Violations of Trade Union Rights.* Available at http://www.icftu.org

International Labour Organization. 2003. *Global Employment Trends: 2003.* Geneva: International Labour Organization.

International Labour Organization. 2005a. *A Global Alliance Against Forced Labour.* Geneva: International Labour Office.

International Labour Organization. 2005b. *Global Employment Trends 2004.* Geneva: International Labour Office.

International Labour Organization. 2005c. *World Employment Report 2004–2005.* Geneva: International Labour Office.

International Telework Association and Council. 2004 (September 2). *Work at Home Grows in Past Year by 7.5% in U.S.* Available at http://www.workingfromanywhere.org

Jensen, Derrick. 2002. "The Disenchanted Kingdom: George Ritzer on the Disappearance of Authentic Culture." *The Sun* (June):38–53.

Kenworthy, Lane. 1995. *In Search of National Economic Success.* Thousand Oaks, CA: Sage.

Kiefer, Heather M. 2003 (November 4). *Stressed Out? Not U.S. Workers.* The Gallup Organization. Available at http://www.gallup.com

Koch, Kathy. 1998. "High-Tech Labor Shortage." *CQ Researcher* 8(16):361–384.

Kukreja, Anil, and George M. Neely Sr. 2000. "Strategies for Preventing the Digital Divide." Paper presented at the symposium "Telework and the New Workplace of the 21st Century," Xavier University, New Orleans, October 16. U.S. Department of Labor. Available at http://www.dol.gov/dol/asp/public/telework/htm

Labor's "Female Friendly" Agenda. 1998. Labor Relations Bulletin no. 690, p. 2.

Lenski, Gerard, and J. Lenski. 1987. *Human Societies: An Introduction to Macrosociology,* 5th ed. New York: McGraw-Hill.

Leonard, Bill. 1996 (July). "From School to Work: Partnerships Smooth the Transition." *HR Magazine* (Society for Human Resource Management). Available at http://www.shrm.org

Levitan, Sar A., Garth L. Mangum, and Stephen L. Mangum. 1998. *Programs in Aid of the Poor,* 7th ed. Baltimore: Johns Hopkins University Press.

Lovelace, Glenn. 2000. "The Nuts and Bolts of Telework." Paper presented at the symposium "Telework and the New Workplace of the 21st Century," Xavier University, New Orleans, October 16. U.S. Department of Labor. Available at http://www.dol.gov/dol/asp/public/telework/htm

Lovell, Vicky. 2005 (April). *Valuing Good Health: An Estimate of Costs and Savings for the Healthy Families Act.* Institute for Women's Policy Research. Available at http://www.iwpr.org

MacEnulty, Pat. 2005 (September). "An Offer They Can't Refuse: John Perkins on His Former Life as an Economic Hit Man." *The Sun* 357:4–13.

Miers, Suzanne. 2003. *Slavery in the Twentieth Century: The Evolution of a Global Problem.* Walnut Creek, CA: AltaMira Press.

Mishel, Lawrence, Jared Bernstein, and Sylvia Allegretto. 2005. *The State of Working America, 2004/2005.* Ithaca, NY: ILR Press.

Moen, Phyllis. 2003. "Epilogue: Toward a Policy Agenda." In *It's About Time: Couples and Careers,* P. Moen, ed. (pp. 333–337). Ithaca, NY: Cornell University Press.

Mokhiber, Russell, and Robert Weissman. 1998. "Focus on the Corporation. *Multinational Monitor.* Available at http://www.essential.org

Mokhiber, Russell, and Robert Weissman. 2001 (January 4). "The Corporate Conservative Administration." *Focus on the Corporation.* Available at http://www.essential.org

Moloney, Anastasia. 2005. "Terror as Anti-Union Strategy: The Violent Suppression of Labor Rights in Colombia." *Multinational Monitor* 26(3–4).

Moore, David W. 2002. *Public Support for Unions Remains Strong.* Gallup News Service. Available at http://www.gallup.com

National Labor Committee. 2001 (January 16). *Nightmare at J.C. Penney Contractor.* Available at http://www.nlcnet.org

National League of Cities. 2004. *The American Dream in 2004: A Survey of the American People.* Washington, DC: National League of Cities.

"New OSHA Policy Relieves Employees." 1998. *Labor Relations Bulletin* no. 687, p. 8.

OECD. 2004 (October). *Clocking in and Clocking Out: Recent Trends in Working Hours.* OECD Policy Brief. Available at http://www.oecd.org

O'Rourke, Dara. 2003. "Outsourcing Regulation: Analyzing Nongovernmental Systems of Labor Standards and Monitoring." *Policy Studies Journal* 31(1):1–29.

Passaro, Jamie. 2003 (January). "Fingers to the Bone: Barbara Ehrenreich on the Plight of the Working Poor." *The Sun,* pp. 4–10.

Perkins, John. 2004. *Confessions of an Economic Hit Man.* San Francisco: Berrett-Koehler Publishers, Inc.

Pratt, Joanne H. 2000. "Telework and Society: Implications for Corporate and Societal Cultures." Paper presented at the symposium "Telework and the New Workplace of the 21st Century," Xavier University, New Orleans, October 16. U.S. Department of Labor. Available at http://www.dol.gov/dol/asp/public/telework/htm

Riley, Patricia, Anu Mandavilli, and Rebecca Heino. 2000. "Observing the Impact of Communication and Information Technology on 'NetWork.'" Paper presented at the symposium "Telework and the New Workplace of the 21st Century," Xavier University, New Orleans, October 16. U.S. Department of Labor. Available at http://www.dol.gov/dol/asp/public/telework/htm

Ritzer, George. 1995. *The McDonaldization of Society: An Investigation into the Changing Character of Contemporary Social Life.* Thousand Oaks, CA: Pine Forge Press.

Schaeffer, Robert K. 2003. *Understanding Globalization: The Social Consequences of Political, Economic, and Environmental Change,* 2nd ed. Lanham, MD: Rowman & Littlefield.

Scott, Robert E. 2003 (November). *The High Price of Free Trade.* Briefing Paper 147. Economic Policy Institute. Available at http://www.epinet.org

Scott, Robert E., and David Ratner. 2005 (July 20). *NAFTA'S Cautionary Tale.* Economic Policy Institute Briefing Paper 214. Available at http://www.epinet.org

"Sex Trade Enslaves Millions of Women, Youth." 2003. *Popline* 25:6.

Shipler, David K. 2005. *The Working Poor.* New York: Vintage Books.

Stettner, Andrew, and Sylvia A. Allegretto. 2005 (May 26). *The Rising Stakes of Job Loss: Stubborn Long-Term Joblessness Amid Falling Unemployment Rates.* Briefing Paper 162. Washington, DC: Economic Policy Institute.

Still, Mary C., and David Strang. 2003. "Institutionalizing Family-Friendly Policies." In *It's About Time: Couples and Careers,* Phyllis Moen, ed. (pp. 288–309). Ithaca, NY: Cornell University Press.

Students and Scholars Against Corporate Misbehavior. 2005 (August 12). *Looking for Mickey Mouse's Conscience: A Survey of the Working Conditions of Disney Factories in China.* Available at http://www.nlcnet.org

Thompson, Charles D., Jr. 2002. "Introduction." In *The Human Cost of Food: Farmworkers' Lives, Labor, and Advocacy,* C. D. Thompson Jr. and M. F. Wiggins, eds. (pp. 2–19). Austin: University of Texas Press.

Uchitelle, Louis. 2003. "A Missing Statistic: U.S. Jobs that Went Overseas." *New York Times,* October 5. Available at http://www.nytimes.com

"Unions Forge Global Network." 2002 (April 22). *CorpWatch.* Available at http://www.corpwtch.org

United Nations. 2005. *The Millennium Development Goals Report.* New York: United Nations.

U.S. Bureau of the Census. 2004. *Statistical Abstract of the United States: 2004–2005,* 124th ed. Washington, DC: U.S. Government Printing Office.

U.S. Department of Labor. 2001 (January 16). *OSHA Statement: Statement of Assistant Secretary Charles N. Jeffress on Effective Date of OSHA Ergonomic Standard.* Available at http://www.osha.gov

Vandivere, Sharon, Kathryn Tout, Martha Zaslow, Julia Calkins, and Jeffrey Cappizzano. 2003. *Unsupervised Time: Family and Child Factors Associated with Self-Care.* Assessing the New Federalism Occasional Paper No. 71. Urban Institute. Available at http://www.urbaninstitute.org

Watkins, Marilyn. 2002. *Building Winnable Strategies for Paid Family Leave in the United States.* Economic Opportunity Institute. Available at http://www.econop.org

Went, Robert. 2000. *Globalization: Neoliberal Challenge, Radical Responses.* Sterling, VA: Pluto Press.

Werhane, Patricia H., Tara J. Radin, and Norman E. Bowie. 2004. *Employment and Employee Rights.* Malden, MA: Blackwell.

"Workers at Risk." 2003. *Multinational Monitor* 24(6). Available at http://www.essential.org/monitor

Wright, Carter. 2001. "A Clean Sweep: Justice for Janitors." *Multinational Monitor* 22(1–2). Available at http://www.essential.org

Chapter 8

AAUP (Association of University Professors). 2003. *Distance Education.* Available at http://aaup.org/issues/DistanceEd

Addington, Lynne, Sally Ruddy, Amanda Miller, Jill Devoe, and Kathryn Chandler. 2004. "Are America's Schools Safe? Students Speak Out." In *Schools and Society,* Jeanne Ballantine and Joan Spade, eds. (pp. 161–164). Belmont, CA: Thomson Wadsworth.

AERA (American Educational Research Association). 2004. *English Language Learners: Boosting Academic Achievement.* Research Points. Available at http://www.aera.net

Archer, Jeff. 2005a. "Connecticut Files Court Challenge to NCLB." *Education Week,* August 31. Available at http://www.edweek.org

Archer, Jeff. 2005b. "Connecticut Pledges First State Legal Challenge to NCLB Law." *Education Week,* April 13. Available at http://www.edweek.org

Arenson, Laren. 1998. "More Colleges Plunging into the Uncharted Waters of On-Line Courses." *New York Times,* November 2, p. A14.

ASCE (American Society of Civil Engineers). 2005. "Schools." In *America's Crumbling Infrastructure Eroding the Quality of Life.* Available at http://www.asce.org/report card/2005

Associated Press. 2004. "Poll: Student Cheating Prevalent." *Detroit News,* April 30. Available at http://www.detnews.com

Baker, David P., and Deborah P. Jones. 1993. "Creating Gender Equality: Cross-National Gender Stratification and Mathematical Performance." *Sociology of Education* 66:91–103.

Bankston, Carl, and Stephen Caldas. 1997. "The American School Dilemma: Race and Scholastic Performance." *Sociological Quarterly* 8(3):423–429.

Barton, Paul E. 2005 (February). *One Third of a Nation: Rising Drop out Rates and Declining Opportunities.* Princeton, NJ: Educational Testing Service.

BBC (British Broadcasting Company). 2003. "School Drop Outs 'Global Problem.'" *BBC News,* November 21. Available at http://www.news.bbs.co.uk

Benson, Lorna. 2005 (March 25). "Jeff Weise's Enigmatic Internet Persona." Minnesota's Public Radio. Available at http://www.minnesota.publicradio.org/features

Boaz, David, and Morris Barrett. 2004. *What Would a School Voucher Buy? The Real Cost of Private Schools.* Cato Institute Briefing Paper No. 25. Available at http://www.cato.org/pubs

Bowman, Darcia. 2002. "Lethal School Shootings Resemble Workplace Rampages, Report Says." *Education Week,* May 29. Available at http://www.edweek.com

Bushweller, Kevin. 1995. "Turning Our Backs on Boys." *Education Digest,* January, pp. 9–12.

Call, Kathleen, Lorie Grabowski, Jeylan Mortimer, Katherine Nash, and Chaimun Lee. 1997. "Impoverished Youth and the Attainment Process." Paper presented at the annual meeting of the American Sociological Association, Toronto, August.

CAPE (Council for American Private Education). 2005. *Facts and Studies: Private School Statistics at a Glance.* Available at http://www.capenet.org/facts.html

Carnevale, Dan. 2001. "Commission Says Federal Rules on Distance Education Must Be Updated." *Chronicle of Higher Education,* January 5, p. A46.

Carvin, Andy. 1997. *EdWeb: Exploring Technology and School Reform.* Available at http://edweb.gsn.org

CBS. 2003. *Cracking Down on Diploma Mills.* Available at http://www.cbsnews.com/stories/2003

CBS. 2005 (March 11). *South Beach Diet Goes to Schools.* Available at http://www.cbsnews.com/stories

CDF (Children's Defense Fund). 2003a. *Administration Proposal to Test Head Start Children.* Available at http://www.childrensdefense.org/hs_test_ proposal

CDF. 2003b. *New TV Ad Tells Congress: "Head Start's Not Broken, Don't Break It."* Available at http://www.childrens defense.org/release

CDF. 2004a. *Educational Resource Disparities for Minority and Low Income Children.* Available at http://www.childrensdefense.org/eduation/resources-disparities

CDF. 2004b. *The Road to Dropping Out.* Available at http://www.childrens defense.org/education/dropping-out

CNN. 2005 (May 3). *Utah Governor Snubs "No Child" Requirements.* Available at http://www.cnn.com

Coleman, James S., J. E. Campbell, L. Hobson, J. McPartland, A. Mood, F. Weinfield, and R. York. 1966. *Equality of Educational Opportunity.* Washington, DC: U.S. Government Printing Office.

Cronin, John Gage Kingbury, Martha McCall, and Branin Bowe. 2005 (April). *The Impact of No Child Left Behind Act on Student Achievement and Growth.* Northwest Evaluation Association. Research Brief. Available at http://www.nwea.org

Day, Sherri. 2003. "Sizing Up Snapple's Drink Deal with New York City." *New York Times,* September 12, p. C2.

Dillon, Sam. 2003. "Cameras Watching Students, Especially in Biloxi." *New York Times,* September 24. Available at http://www.nytimes.com

Dobbs, Michael. 2005a. "National Teacher's Union, Three School Districts File 'No Child' Law Suit." *Washington Post,* April 20. Available at http://www.washingtonpost.com

Dobbs, Michael. 2005b. "Youngest Students Most Likely to Be Expelled." *Washington Post,* May 16. Available at http://www.washingtonpost.com

EFA Global Monitoring Report Team. 2003. " Why Are Girls Still Held Back?" In *Gender and Education for All: The Leap to Equality.* Global Monitoring Report 2003/4. UNESCO. Available at http://portal.unesco.org/education/en/ev.php-url_id=24152&url_do=do_topic&url_section=201.html

Elam, Stanley M., Lowell C. Rose, and Alec M. Gallup. 1994. "The 26th Annual Phi Delta Kappa/Gallup Poll of the Public's Attitudes Toward the Public Schools." *Phi Delta Kappa,* September, 41–56.

Entwisle, Doris, Carl Alexander, and Linda Olson. 2004. "Temporary as Compared to Permanent High School Drop-Out." *Social Forces* 82(3):1181–1205.

ETS (Educational Testing Service). 2005. *The Praxis Series: Professional Assessment for Beginning Teachers.* Available at http://www.ets.org/praxis

Evans, Lorraine, and Kimberly Davies. 2000. "No Sissy Boys Here." *Sex Roles,* February, pp. 255–271.

Fact Sheet on the Major Provisions of the Conference Report to H.R. 1, the No Child Left Behind Act. 2003. Available at http://www.ed/gov/print/nclb

Fletcher, Robert S. 1943. *History of Oberlin College to the Civil War.* Oberlin, OH: Oberlin College Press.

Flexner, Eleanor. 1972. *Century of Struggle: The Women's Rights Movement in the United States.* New York: Atheneum.

Frankenberg, Erika, and Chungmei Lee. 2002. *Race in American Public Schools: Rapidly Resegregating School Districts.* Cambridge, MA: Harvard University, The Civil Rights Project.

Frankenberg, Erika, Chungmei Lee, and Gary Orfield. 2003. *Multiracial Society with Segregated Schools: Are We Losing the Dream?* Cambridge, MA: Harvard University, The Civil Rights Project. Available at http://www.civilrightsproject.harvard.edu/research/reseg03

Gallup. 2005a. *Education.* Available at http://www.brain.gallup.com

Gallup. 2005b. *Most Important Problem.* Available at http://www.brain.gallup.com

Goldberg, Carey. 1999. "After Girls Get Attention, Focus Is on Boys' Woes." In *Themes of the Times: New York Times,* p. 6. Upper Saddle River, NJ: Prentice-Hall.

Guernsey, Lisa. 2000. "Education: Web's New Come-On." *New York Times,* March 16, pp. C1 and C7.

Haga, Chuck, Howie Padilla, and Richard Meryhew. 2005. "Jeff Weise: A Mystery in a Life of Hardship." *Star Tribune,* March 23. Available at http://www.startribune.com/dynamic

Harris Poll. 2003. *As Economy Grows the Public's Priorities for Growth Are Health, Education and Defense.* Available at http://www.harrisinteractive.com

Head Start. 2004. *Head Start Program Fact Sheet.* Available at http://www.policy almanac.org/eduation/archive

Hurst, Marianne. 2005. "When It Comes to Bullying, There Are No Boundaries." *Education Week,* February 8. Available at http://www.edweek.org

Independent Sector. 2002. "Engaging Youth in Lifelong Service." *Newsroom.* Available at http://www.independentsector.org

Jacobs, Joanne. 1999. "Gov. Davis to Make Volunteering Mandatory for Students." *Jose Mercury News,* April 23.

Jencks, Christopher, and Meredith Phillips. 1998. "America's Next Achievement Test:

Closing the Black-White Test Score Gap." *The American Prospect*, September–October, pp. 44–53.

Jenkins, Robert F. 2006. "The Pendulum of Change." Unpublished essay.

Kanter, Rosabeth Moss. 1972. "The Organization Child: Experience Management in a Nursery School." *Sociology of Education* 45:186–211.

Koch, James V. 1998. "How Women Actually Perform in Distance Education." *Chronicle of Higher Education* 45:A60.

Kozol, Jonathan. 1991. *Savage Inequalities: Children in America's Schools*. New York: Crown.

Lareau, Annette. 1989. *Home Advantage: Social Class and Parental Intervention in Elementary Education*. Philadelphia: Falmer Press.

Lareau, Annette, and Erin Horvat. 2004. "Moments of Social Inclusion and Exclusion." In *Schools and Society*, Jeanne Ballantine and Joan Spade, eds. (pp. 276–286). Belmont, CA: Thomson Wadsworth.

Leandro v. State, 346 N.C. 336, 488 S.E.2d 249 (1997); aff'd. Hoke Cty Bd. of Educ. v. State, 530 PA02, Filed July 30, 2004.

Leonhardt, David. 2005. "The College Drop Out Boom." *New York Times*, May 24. Available at http://www.nytimes.com

Levinson, Arlene. 2000. *Study Evaluates Higher Education*. Excite News. Available at http://www.excite.com/news/ap/00130/08/grading

Lewin, Tamar. 2003. "Writing in Schools Is Found Both Dismal and Neglected." *New York Times*, April 26. Available at http://www.nytimes.com

Literacy Volunteers of America. 2003. *Facts on Literacy in America*. Available at http://www.literacyvolunteers.org

Manzo, Kathleen K. 2005. "College-Based High Schools Fill Growing Need." *Education Week*, May 25. Available at http://www.edweek.org

Mason, Heather. 2005 (April 5). *Public: Society Powerless to Stop School Shootings*. Available at http://www.brain.gallup.com

Man, Rosalind Y. 1992. "The Validity and Devolution of a Concept: Student Alienation." *Adolescence* 27(107):739–740.

Merton, Robert K. 1968. *Social Theory and Social Structure*. New York: Free Press.

MetLife. 2005. *The MetLife Survey of the American Teacher*. Available at http://www.metlife.com/WPSAssets

Mollison, Andrew. 2001. "Boom in Charter Schools Continues Despite Mixed Results." *Atlanta Journal Constitution*, September 24. Available at http://www.accessatlanta.com/partners/ajc/epaper

Morse, Jodie. 2002a. "Flunking Lunch." *Time*, December 2, pp. 74–75.

Morse, Jodie. 2002b. "Learning While Black." *Time*, May 27, pp. 50–52.

Morse, Jodie. 2002c. "A Victory for Vouchers." *Time*, July 8, pp. 52–53.

Muller, Chandra, and Katherine Schiller. 2000. "Leveling the Playing Field?" *Sociology of Education* 73:196–218.

Murnane, Richard J. 1994. "Education and the Well-Being of the Next Generation." In *Confronting Poverty: Prescriptions for Change*, Sheldon H. Danziger, Gary D. Sandefur, and Daniel H. Weinberg, eds. (pp. 289–307). New York: Russell Sage Foundation.

NAAL (National Assessment of Adult Literacy). 2005. *FAQ*. Available at http://nces.ed.gov/naal/faq

NABE (National Association of Bilingual Education). 2005. *Head Start Tests Neither "Valid" or "Reliable," GAO Study Finds*. Available at http://www.nabe.org

Nash, Madeleine. 2003. "Obesity Goes Global." *Time*, August 25, p. 53.

National Science Board. 2003. "Elementary and Secondary Education: IT in Schools." *Indicators 2002*, chapter 1. Available at http://www.nsf.gov/sbe/srs/seind02/c1/c1s8.htm

Natriello, Gary. 1995. "Dropouts: Definitions, Causes, Consequences, and Remedies." In *Transforming Schools*, Peter W. Cookson Jr. and Barbara Schneider, eds. (pp. 107–128). New York: Garland.

NBPTS (National Board for Professional Teaching Standards). 2005. *About NBPTS*. Available at http://www.nbpts.org/about

NCCAI (North Carolina Child Advocacy Institute). 2005. *What the Courts Decided in the Leandro Case: A Brief History*. Available at http://ncchild.org/leandro_index files

NCD (National Council on Disability). 2000. *Executive Summary*. Available at http://www.ncd.gov/newsroom/publications

NCES (National Center for Education Statistics). 2000. *Teacher Trends*. Available at http://www.nces.ed.gov/fastfacts

NCES. 2002a. *Drop Out Rates in the United States*. Available at http://www.nces.ed.gov/pubs2002

NCES. 2002b. *Student Participation in Distance Education*. Available at http://nces.ed.gov/programs.coe

NCES. 2003a. *Academic Background of College Graduates Who Enter and Leave Teaching*. Contexts of Elementary and Secondary Education. Available at http://nces.ed.gov/programs

NCES. 2003b. "Distance Learning." http://nces.ed.gov/fastfacts

NCES. 2003c. *Fast Fact: Family Reading*. Available at http://www.nces.ed.gov/fastfacts

NCES. 2003d. *Special Analysis 2002: Private Schools—A Brief Portrait*. Available at http://www.nces.ed.gov/programs

NCES. 2003e. *Young Children's Access to Computers in the Home and at School in 1999 and 2000: Executive Summary*. Available at http://www.nces.ed.gov

NCES. 2004a. *Condition of Education*. U.S. Department of Education, Washington, D.C.

NCES. 2004b. *Digest of Education Statistics 2003*. U.S. Department of Education, Washington, D.C.

NCES. 2004c. *1.1 Million Home Schooled Students in the U.S. in 2003*. Issue Brief. Available at http://www.nces.ed.gov

NCES. 2005a. *Comparative Indicators of Education in the United States and Other G8 Countries: 2004*. Available at http://nces.ed.gov/pubsearch

NCES. 2005b. *Indicators of School Safety and Crime: Executive Summary*. U.S. Department of Education, Washington, D.C.

NCJRS (National Criminal Justice Reference Service). 2003. *School Safety Resources: Facts and Figures*. Available at http://www.ncjrs.org/school_safety

NPR (National Public Radio). 2003. *NPR/Kaiser/Kennedy School Education Survey*. Available at http://www.npr.org/programs/specials

Official Press Release. 2005 (August 22). *State Sues Federal Government over Illegal Unfunded Mandates Under No Child Left Behind Act*. Connecticut Attorney General's Office. Available at http://www.cslib.org/attygenl/press/2005

Orfield, Gary, and Chungmei Lee. 2005. *Why Segregation Matters: Poverty and Educational Inequality*. Cambridge, MA: Harvard University: The Civil Rights Project. Available at http://www.civilrightsproject.harvard.edu/research/deseg05

PEQIS. 2003. *Distance Education at Degree Granting Postsecondary Institutions: 2000–2001*. Available at http://www.nces.ed.gov/surveys/peqis

Pollock, William. 2000. *Real Boys' Voices*. New York: Random House.

A Quality Teacher in Every Classroom. 2002. White House Fact Sheet. Available at http://www.ed.gov/print/news/press release

Ramierz-Valles, and Amanda Brown. 2003. "Latinos' Community Involvement in HIV/AIDS: Organizational and Individual Perspectives on Volunteering." *AIDS Education and Prevention* 15:90–104.

Rand. 2003. *Rand Study Finds California Charter Schools Produce Achievement Gains Similar to Conventional Public Schools*. Available at http://www.rand.org

Richard, Alan. 2005. "Court Showdown over Fla. Vouchers Near." *Education Week*, May 25. Available at http://www.edweek.org

Riehl, Carolyn. 2004. "Bridges to the Future: Contributions of Qualitative Research to the Sociology of Education." In *Schools and Society*, Jeanne Ballantine and Joan Spade, eds. (pp. 56–72). Belmont, CA: Thomson Wadsworth.

Ripley, Amanda, and Sonja Steptoe. 2005. "Inside the Revolt over Bush's School Rules." *Time*, May 9, pp. 30–33.

Rodriguez, Richard. 1990. "Searching for Roots in a Changing World." In *Social Problems Today,* James M. Henslin, ed. (pp. 202–213). Englewood Cliffs, NJ: Prentice-Hall.

Roscigno, Vincent. 1998. "Race and the Reproduction of Educational Disadvantage." *Social Forces* 76(3):1033–1060.

Rosenbaum, James. 2002. "Beyond College for All: Career Paths for the Forgotten Half." In *Schools and Society,* Jeanne Ballantine and Joan Spade, eds. (pp. 485–490). Belmont, CA: Thomson Wadsworth.

Rosenthal, Robert, and Lenore Jacobson. 1968. *Pygmalion in the Classroom: Teacher Expectations and Pupils' Intellectual Development.* New York: Holt, Rinehart & Winston.

Roundtree, Pamela Wilcox. 2000. "Weapons at School: Are the Predictors Generalizable Across Context?" *Sociological Spectrum* 20:291–324.

Rumberger, Russell W. 1987. "High School Dropouts: A Review of Issues and Evidence." *Review of Educational Research* 57:101–121.

Saad, Lydia. 2005 (May 17). *Math Problematic for U.S. Teens.* Available at http://www.gallup.com

Sack, Joetta. 2000. "States Lack in Enforcing IDEA, Study Asserts." *Education Week,* February 2. Available at http://www.edweek.org

Sack, Joetta. 2005. "Utah Passes Bill to Trump 'No Child' Law." *Education Week,* April 27. Available at http://www.edweek.org

Sadovnik, Alan. 2004. "Theories in the Sociology of Education." In *Schools and Society,* Jeanne Ballantine and Joan Spade, eds. (pp. 7–26). Belmont, CA: Thomson Wadsworth.

Safe School Initiative. 2004. *The Final Report and Findings of the Safe School Initiative.* Washington, DC: U.S. Secret Service and U.S. Department of Education.

Saporito, Salvatore. 2003. "Private Choices, Public Consequences: Magnet School Choice and Segregation by Race and Poverty." *Social Problems* 50:181–203.

Schemo, Diana. 2003. "Neediest Schools Receive Less Money, Report Finds." *New York Times,* August 9. Available at http://www.nytimes.com

Schneider, Mike. 2005 (March 6). *Florida School Testing South Beach Diet.* Available at http://apnews.myway.com

Selingo, Jeffery. 2003. "What Americans Think About Higher Education." *Chronicle of Higher Education,* May, pp. A10–A17.

Setzer, J., and L. Lewis. 2005. *Distance Education Courses for Public Elementary and Secondary School Students: 2002–2003.* NCES 2005-010. U.S. Department of Education, Washington, D.C.

Slobogin, Kathy. 2002. *Survey: Many Students Say Cheating's OK.* CNN News. Available at http://www.cnn.com

Sommers, Christina. 2000. *The War Against Boys.* New York: Simon and Schuster.

Spade, Joan. 2004. "Gender and Education in the United States." In *Schools and Society,* Jeanne Ballantine and Joan Spade, eds. (pp. 287–295). Belmont, CA: Thomson Wadsworth.

Steptoe, Sonja. 2003. "Taking the Alternative Route." *Time,* January 13, pp. 50–51.

Toppo, Greg, 2004. "Low-Income College Students Are Increasingly Left Behind." *USA Today,* January 14. Available at http://www.usatoday.com

UCLA. 2005. *"Williams v. California."* Brochure. Institute for Democracy, Education and Access. Los Angeles: University of California.

UNESCO. 2005. *United Nations Literacy Decade: 2003–2012.* Available at http://portal.unesco.org/education/en

U.S. Census Bureau. 2003. *Facts and Figures: Back to School.* Available at http://www.census.gov/Press-Release

U.S. Census Bureau. 2004. *Statistical Abstracts of the United States.* Washington, DC: U.S. Government Printing Office.

U.S. Census Bureau. 2005. *Facts for Features.* Press release. Available at http://www.census.gov/PressRelease

U.S. Charter Schools. 2005. *FAQ.* Available at http://www.uscharterschool.org

U.S. Department of Education. 2000. *The Baby Boom Echo: No End in Sight,* Figure 2. Available at http://www.ed.gov/pubs/bbecho00/figure2

U.S. Department of Education. 2003. *International Education Report: U.S. Students Are Average.* Press release. Available at http://www.ed.gov/news/pressreleases/2003/09/09152003b.html

U.S. Department of Education. 2004 (March 15). *New, Flexible Policies Help Teachers Become Highly Qualified.* Available at http://www.ed.gov/news/pressreleases

U.S. Department of Health and Human Services. 2001. *Building Their Futures.* Summary Report. Available at http://www2.acf.dhhs.gov/programs/hsb/EHS

Viadero, Debra. 2005. "Study Sees Positive Effects of Teacher Certification." *Education Week,* April 27. Available at http://www.edweek.org

Wakefield, Julie, 2002. "Learning the Hard Way." *Environmental Health Perspectives* 110(6):1.

Webb, Julie. 1989. "The Outcomes of Home-Based Education: Employment and Other Issues." *Educational Review* 41:121–133.

Weiner, Rebecca. 2000. "Industry Group's Education Study Draws Conclusions and Critics." *New York Times.* Available at http://www.nytimes.com

WEP (Wake Education Partnership). 2005. *The Issues.* Raleigh, NC: Wake County Schools.

What Is Character Education? 2003. Boston College: Center for the Advancement of Ethics and Character.

Wildavsky, Ben. 2001. "Teach Your Children Well—On the Web?" *U.S. News and World Report,* January 8. Available at http://www.usnews.com

Wilgoren, Jodi. 2001. "Calls for Change in the Scheduling of the School Day." *New York Times,* January 10. Available at http://www.nytimes.com

Winter, Greg. 2005. "Study Finds Poor Performance by Nations' Education Schools." *New York Times,* March 15. Available at http://www.nytimes.com

Winters, Rebecca. 2001. "From Home to Harvard." *Time,* September 11. Available at http://www.time.com/time/magazine/articles

Zigler, Edward, Sally Styfco, and Elizabeth Gilman. 2004. "The National Head Start Program for Disadvantaged Preschoolers." In *Schools and Society,* Jeanne Ballantine and Joan Spade, eds. (pp. 341–346). Belmont, CA: Thomson Wadsworth.

Zimmerman, Frederick, Gwen Glew, Dimitri Christakis and Wayne Katon. 2005. "Early Cognitive Stimulation, Emotional Support, and Television Watching as Predictors of Subsequent Bullying Among Grade School Children." *Archives of Pediatric and Adolescent Medicine* 159(4):384–388.

Chapter 9

Allport, G. W. 1954. *The Nature of Prejudice.* Cambridge, MA: Addison-Wesley.

Amer, Mildred L. 2005. *Membership of the 109th Congress: A Profile.* Congressional Research Service. Available at http://www.senate.gov

American Council on Education and American Association of University Professors. 2000. *Does Diversity Make a Difference? Three Research Studies on Diversity in College Classrooms.* Washington, DC: American Council on Education and American Association of University Professors.

BBC. 2005 (July 15). *UK Muslim Leader Barred from US.* BBC News. Available at http://newsvote.bbc.co.uk

Beeman, Mark, Geeta Chowdhry, and Karmen Todd. 2000. "Educating Students About Affirmative Action: An Analysis of University Sociology Texts." *Teaching Sociology* 28(2):98–115.

Brace, C. Loring. 2005. *"Race" Is a Four-Letter Word.* New York: Oxford University Press.

Capps, Randolph, Michael E. Fix, Jason Ost, Jane Reardon-Anderson, and Jeffrey S. Passel. 2005 (February 8). *The Health and Well-Being of Young Children of Im-*

migrants. Urban Institute. Available at http://www.urban.org

Carroll, James. 2005a (July 19). "The Border Mentality." *Boston Globe.* Available at http://www.boston.com

Carroll, Joseph. 2005b (July 12). *Who Are the People in Your Neighborhood?* Gallup Organization. Available at http://www.gallup.com

Chu, Jeff. 2005. "Who Gets the Break?" *Time,* July 11, p. 60.

Cohen, Mark Nathan. 1998. "Culture, Not Race, Explains Human Diversity." *Chronicle of Higher Education* 44(32):B4–B5.

Conley, Dalton. 1999. *Being Black, Living in the Red: Race, Wealth, and Social Policy in America.* Berkeley: University of California Press.

Conley, Dalton. 2002. "The Importance of Being White." *Newsday,* October 13. Available at http://www.newsday.com

Day, Jennifer Cheeseman, and Eric Newburger, 2002. *The Big Payoff: Educational Attainment and Synthetic Estimates of Work-Life Earnings.* Washington, DC: U.S. Census Bureau.

Dees, Morris. 2000 (December 28). Personal correspondence. Morris Dees, co-founder of the Southern Poverty Law Center, 400 Washington Avenue, Montgomery, AL 36104.

EEOC (Equal Employment Opportunity Commission). 2003 (July 17). *Muslim Pilot Fired Due to Religion and Appearance, EEOC Says in Post-9/11 Backlash Discrimination Suit.* Press release. U.S. Equal Employment Opportunity Commission. Available at http://www.EEOC.gov

EEOC. 2004. *Retail Distribution Centers: How New Business Processes Impact Minority Labor Markets.* U.S. Equal Employment Opportunity Commission. Available at http://www.EEOC.gov

EEOC. 2005a. *EEOC Agrees to Landmark Resolution of Discrimination Case Against Abercrombie & Fitch.* Press release. Available at http://www.eeoc.gov

EEOC. 2005b. *Georgetowne Place to Pay $650,000 to Settle EEOC Race Discrimination Lawsuit.* Press release. Available at http://www.eeoc.gov

Etzioni, Amitai. 1997. "New Issues: Re-Thinking Race." *Public Perspective,* June–July, pp. 39–40. Available at http://www.roper center.unconn.edu

FBI (Federal Bureau of Investigation). 2004. *Hate Crime Statistics 2003.* Available at http://www.fbi.gov

Fields, Jason. 2004 (November). *America's Families and Living Arrangements: 2003.* Current Population Reports P20-553. U.S. Census Bureau. Available at http://www.census.gov

Fitrakis, Bob, and Harvey Wasserman. 2004. "Hearings on Ohio Voting Put 2004 Election in Doubt." *The Free Press,* November 18. Available at http://www.thefreepress.org

Fix, Michael E., and Randolph Capps. 2002. *The Dispersal of Immigrants in the 1990s.* Washington DC: Urban Institute.

Foust, Dean, Brian Grow, and Aixa M. Pascual. 2002. "The Changing Heartland." *Business Week,* September 9, pp. 80–84.

Gaertner, Samuel L., and John F. Dovidio. 2000. *Reducing Intergroup Bias: The Common Ingroup Identity Model.* Philadelphia: Taylor & Francis.

Gallup Organization. 2005. *Race Relations.* Gallup Brain. Available at http://www.gallup.org

Glaser, Jack, Jay Dixit, and Donald P. Green. 2002. "Studying Hate Crime with the Internet: What Makes Racists Advocate Racial Violence?" *Journal of Social Issues* 58(1):177–193.

Glaubke, Christina Roman, and Katharine Heintz-Krowles. 2004. *Fall Colors: Prime Time Diversity Report 2003–04.* Oakland, CA: Children Now & the Media Program.

Goldstein, Joseph. 1999. "Sunbeams." *The Sun* 277(January):48.

Goodnough, Abby. 2001. "New York City Is Short-Changed in School Aid, State Judge Rules." *New York Times,* January 11. Available at http://www.nytimes.com

Greenberg, Daniel S. 2003. "Supreme Court Sets Showdown on Affirmative Action." *The Lancet* 361(March 1):762. Available at http://www.thelancet.com

Greenhouse, Steven. 2003. "Suit Claims Discrimination Against Hispanics on Job." *New York Times,* February 9. Available at http://www.nytimes.com

Greenhouse, Steven. 2005. "Wal-Mart to Pay U.S. $11 Million in Lawsuit on Illegal Workers." *New York Times,* March 19. Available at http://www.nytimes.com

Grieco, Elizabeth M., and Rachel C. Cassidy. 2001 (March). *Overview of Race and Hispanic Origin: Census 2000 Brief.* Washington, DC: U.S. Census Bureau. Available at http://www.census.gov

Gurin, Patricia. 1999. "New Research on the Benefits of Diversity in College and Beyond: An Empirical Analysis." *Diversity Digest,* spring, pp. 5–15.

"Hate on Campus." 2000. *Intelligence Report* 98(spring):6–15.

Healey, Joseph F. 1997. *Race, Ethnicity, and Gender in the United States: Inequality, Group Conflict, and Power.* Thousand Oaks, CA: Pine Forge Press.

"Hear and Now." 2000 (fall). *Teaching Tolerance,* p. 5.

Hodgkinson, Harold L. 1995. "What Should We Call People? Race, Class, and the Census for 2000." *Phi Delta Kappa,* October, pp. 173–179.

Holzer, Harry. 2005 (May 14). *New Jobs in Recession and Recovery: Who Are Getting Them and Who Are Not?* Urban Institute. Available at http://www.urban.org

Holzer, Harry, and David Neumark. 2000. "Assessing Affirmative Action." *Journal of Economic Literature* 38(3):483–568.

hooks, bell. 2000. *Where We Stand: Class Matters.* New York: Routledge.

Humphreys, Debra. 1999. "Diversity and the College Curriculum: How Colleges and Universities Are Preparing Students for a Changing World." *Diversity-Web.* Available at http://www.inform.umd.edu

Humphreys, Debra. 2000. "National Survey Finds Diversity Requirements Common Around the Country." *Diversity Digest,* fall. Available at http://www.diversity web.org

Jackson, Camille. 2005 (September 19). "Katrina: Decoding the Language of Race and Class." *Tolerance in the News.* Teaching Tolerance. Available at http://tolerance .org.news

Jay, Gregory. 2005 (March 17). *Introduction to Whiteness Studies.* Available at http://www.uwm.edu/~gjay

Jensen, Derrick. 2001. "Saving the Indigenous Soul: An Interview with Martin Prechtel." *The Sun,* 304(April):4–15.

Kaplan, David E., and Lucian Kim. 2000. "Nazism's New Global Threat." *U.S. News Online,* September 25. Available at http://www.usnews.com

Keita, S. O. Y., and Rick A. Kittles. 1997. "The Persistence of Racial Thinking and the Myth of Racial Divergence." *American Anthropologist* 99(3):534–544.

King, Joyce E. 2000. "A Moral Choice." *Teaching Tolerance* 18(fall):14–15.

Kozol, Jonathan. 1991. *Savage Inequalities: Children in America's Schools.* New York: Crown.

Larsen, Luke J. 2004 (August). *The Foreign-Born Population in the United States: 2003.* Current Population Reports P20-551. U.S. Census Bureau. Available at http://www.census.gov

Lawrence, Sandra M. 1997. "Beyond Race Awareness: White Racial Identity and Multicultural Teaching." *Journal of Teacher Education* 48(2):108–117.

Lovin, Jack, and Jack McDevitt. 1995. "Landmark Study Reveals Hate Crimes Vary Significantly by Offender Motivation." *Klanwatch Intelligence Report,* August, pp. 7–9.

Lollock, Lisa. 2001 (March). *The Foreign-Born Population in the United States: March 2000.* Current Population Reports P20-534. Washington DC: U.S. Bureau of the Census.

Maril, Robert Lee. 2004. *Patrolling Chaos: The U.S. Border Patrol in Deep South Texas.* Lubbock, TX: Texas Tech University Press.

Massey, Douglas, and Nancy Denton. 1993. *American Apartheid: Segregation and the Making of an American Under-Class.* Cambridge, MA: Harvard University Press.

Massey, Douglas S., and Garvey Lundy. 2001. "Use of Black English and Racial Discrimination in Urban Housing Markets: New

Methods and Findings." *Urban Affairs Review* 36(4):452–469.

Morrison, Pat. 2002. "September 11: A Year Later—American Muslims Are Determined Not to Let Hostility Win." *National Catholic Reporter,* September 6, 38(38): 9–10.

Moser, Bob. 2004. "The Battle of 'Georgiafornia.'" *Intelligence Report* 116:40–50.

NAACP. 2000 (November 11). *NAACP Voting Irregularities Public Hearing.* Press release. National Association for the Advancement of Colored People. Available at http://www.NAACP.org

NAACP. 2001 (January 10). *NAACP National Civil Rights Groups File Florida Voting Rights Lawsuit to Eliminate Unfair Voting Practices.* Press release. National Association for the Advancement of Colored People. Available at http://www.NAACP.org

Nash, Manning. 1962. "Race and the Ideology of Race." *Current Anthropology* 3:258–288.

"National Briefing." 2005. *New York Times,* August 11, p. A19.

National Immigration Law Center. 2005a. *Basic Facts About In-State Tuition for Undocumented Immigrant Students.* Available at http://www.nilc.org

National Immigration Law Center. 2005b. *Overview of Immigrant Eligibility for Federal Programs.* Available at http://www.nilc.org

Navarro, Mireya. 2003. "For New York's Black Latinos, a Growing Racial Awareness." *New York Times,* April 28. Available at http://www.nytimes.com

Nemes, Irene. 2002. "Regulating Hate Speech in Cyberspace: Issues of Desirability and Efficacy." *Information and Communication Technology Law* 11(3):193–220.

"Neo-Nazi Label Woos Teens With Hate-Music Sampler." 2004. *Intelligence Report* 116(winter):5.

Nichols, Bruce. 2005. "Truck Driver Avoids Death Penalty." *Dallas Morning News,* March 23. Available at http://www.dallasnews.com

NPR, Kaiser Family Foundation, and Kennedy School of Government. 2005. *Immigration: Summary of Findings.* NPR/Kaiser/Kennedy School Poll. Available at http://www.npr.org/news/specials/polls/2004/immigration/summary.pdf

Office of Immigration Statistics. 2005. *Southwest Border Apprehensions.* Available at http://www.uscis.gov

Olson, James S. 2003. *Equality Deferred: Race, Ethnicity, and Immigration in America Since 1945.* Belmont CA: Wadsworth/Thomson.

Orfield, Gary. 2001 (July). *Schools More Separate: Consequences of a Decade of Resegregation.* Cambridge, MA: Harvard University, Civil Rights Project.

Ossorio, Pilar, and Troy Duster. 2005. "Race and Genetics." *American Psychologist* 60(1):115–128.

Pager, Devah. 2003. "The Mark of a Criminal Record." *American Journal of Sociology* 108(5):937–975.

Parsons, Sharon, William Simmons, Frankie Shinhoster, and John Kilburn. 1999. "A Test of the Grapevine: An Empirical Examination of Conspiracy Theories Among African-Americans." *Sociological Spectrum* 19(2):201–222.

Passel, Jeffrey S., Randolph Capps, and Michael Fix. 2004 (January 12). *Undocumented Immigrants: Facts and Figures.* Urban Institute. Available at http://www.urban.org

Paul, Pamela. 2003. "Attitudes Toward Affirmative Action." *American Demographics* 25(4):18–19.

Pew Research Center. 2003 (May 13). *Conflicted Views of Affirmative Action.* Available at http://www.people-press.org

Pew Research Center. 2005 (September 8). *Huge Racial Divide over Katrina and Its Consequences.* Available at http://www.people-press.org

Plous, S. 2003. "Ten Myths About Affirmative Action." In *Understanding Prejudice and Discrimination,* S. Plous, ed. (pp. 206–212). New York: McGraw-Hill.

Potok, Mark. 2004. "The Year in Hate." *Intelligence Report* 113(spring):29–54.

Potok, Mark. 2005 (September 20). *In Katrina's Wake, White Supremacists Spew Hatred.* Southern Poverty Law Center. Available at http://www.splcenter.org/news/new.jsp

Ramirez, Roberto R., and G. Patricia de la Cruz. 2003 (June). *The Hispanic Population in the United States: March 2002.* Current Population Reports P20-545. Washington, DC: U.S. Census Bureau.

Rosenfeld, Michael J., and Byung-Soo Kim. 2005. "The Independence of Young Adults and the Rise of Interracial and Same-Sex Unions." *American Sociological Review* 70(August):541–562.

Rothenberg, Paula S. 2002. *White Privilege.* New York: Worth.

Sax, L. J., S. Hurtado, J. A. Lindholm, A. W. Astin, W. S. Korn, and K. M. Mahoney. 2004. *The American Freshman: National Norms for Fall 2004.* Los Angeles: Higher Education Research Institute, UCLA.

Schaefer, Richard T. 1998. *Racial and Ethnic Groups,* 7th ed. New York: Harper-Collins.

Schiller, Bradley R. 2004. *The Economics of Poverty and Discrimination,* 9th ed. Upper Saddle River, NJ: Pearson Education.

Schmidt, Peter. 2004 (January 30). "New Pressure Put on Colleges to End Legacies in Admissions." *Chronicle of Higher Education* 50(21):A1.

Schmitt, Eric. 2001a. "Analysis of Census Finds Segregation Along with Diversity."

New York Times, April 4. Available at http://www.nytimes.com

Schmitt, Eric. 2001b. "Blacks Split on Disclosing Multiracial Roots." *New York Times,* March 31. Available at http://www.nytimes.com

Schuman, Howard, and Maria Krysan. 1999. "A Historical Note on Whites' Beliefs About Racial Inequality." *American Sociological Review* 64:847–855.

Shipler, David K. 1998. "Subtle vs. Overt Racism." *Washington Spectator* 24(6):1–3.

SPLC (Southern Poverty Law Center). 2005a (June). "Center Wins $1.35 Million Judgment Against Violent Border Vigilantes." *SPLC Report* 35(2):3.

SPLC. 2005b (March). "Hate Group Numbers Up Slightly in 2004." *SPLC Report* 35(1):3.

SPLC. 2005c (June). "New Lawsuits Seek Reform of Abusive Labor Practices." *SPLC Report* 35(2):1.

Tolbert, Caroline J., and John A. Grummel. 2003. "Revisiting the Racial Threat Hypothesis: White Voter Support for California's Proposition 209. *State Politics and Policy Quarterly* 3(2):183–202 and 215–216.

Turner, Margery Austin, Stephen L. Ross, George Galster, and John Yinger. 2002. *Discrimination in Metropolitan Housing Markets.* Washington, DC: Urban Institute.

Turner, Margery Austin, and Felicity Skidmore. 1999. *Mortgage Lending Discrimination: A Review of Existing Evidence.* Washington, DC: Urban Institute.

U.S. Citizenship and Immigration Services. 2003. *General Naturalization Requirements.* Available at http://www.uscis.gov

U.S. Department of Labor. 2002. *Facts on Executive Order 11246 Affirmative Action.* Available at http://www.dol.gov

Van Ausdale, Debra, and Joe R. Feagin. 2001. *The First R: How Children Learn Race and Racism.* Lanham, MD: Rowman & Littlefield.

"White Power Bands." 2002. *Hate in the News,* January. Southern Poverty Law Center. Available at http://www.tolerance.org

Williams, Eddie N., and Milton D. Morris. 1993. "Racism and Our Future." In *Race in America: The Struggle for Equality,* Herbert Hill and James E. Jones Jr., eds. (pp. 417–424). Madison: University of Wisconsin Press.

Williams, Richard, Reynold Nesiba, and Eileen Diaz McConnell. 2005. "The Changing Face of Inequality in Home Mortgage Lending." *Social Problems* 52(2):181–208.

Willoughby, Brian. 2003. "Hate on Campus." *Tolerance in the News,* June 13. Available at http://www.tolerance.org

Wilson, William J. 1987. *The Truly Disadvantaged: The Inner City, the Underclass and Public Policy.* Chicago: University of Chicago Press.

Winter, Greg. 2003a. "Schools Resegregate, Study Finds." *New York Times,* January 21. Available at http://www.nytimes.com.

Winter, Greg. 2003b. "U. of Michigan Alters Admissions Use of Race." *New York Times,* August 29. Available at http://www.nytimes.com

Zack, Naomi. 1998. *Thinking About Race.* Belmont, CA: Wadsworth.

Zinn, Howard. 1993. "Columbus and the Doctrine of Discovery." In *Systemic Crisis: Problems in Society, Politics, and World Order,* William D. Perdue, ed. (pp. 351–357). Fort Worth: Harcourt Brace Jovanovich.

Chapter 10

Abernathy, Michael. 2003. "Male Bashing on TV." *Tolerance in the News.* Available at http://www.tolerance.org/news

Alley, Thomas R., and Catherine M. Hicks. 2005. "Peer Attitudes Towards Adolescent Participants in Male- and Female-Oriented Sports." *Adolescence* 40(158): 273–280.

American College of Obstetricians and Gynecologists Committee on Ethics. 2004. *Ethics in Obstetrics and Gynecology,* 2nd ed. Washington, DC: American College of Obstetricians and Gynecologists.

American Sociological Association. 2004 (July). *The Best Time to Have a Baby: Institutional Resources and Family Strategies Among Early Career Sociologists.* ASA Research Brief.

Amnesty International. 2005 (April 26). *Afghanistan: Stoned to Death—Human Rights Scandal,* Press release. Available at http://www.amnesty.org

Anderson, David A., and Mykol Hamilton. 2005. "Gender Role Stereotyping of Parents in Children's Picture Books: The Invisible Father." *Sex Roles* 52:145–151.

Anderson, Margaret L. 1997. *Thinking About Women,* 4th ed. New York: Macmillan.

Athreya, Bama. 2003. "Trade Is a Women's Issue." *ATTAC,* February 20. Available at http://www.globalpolicy.org/socecon

Baker, Robin, Gary Kriger, and Pamela Riley. 1996. "Time, Dirt, and Money: The Effects of Gender, Gender Ideology, and Type of Earner Marriage on Time, Household Task, and Economic Satisfaction Among Couples with Children." *Journal of Social Behavior and Personality* 11:161–177.

Banister, Judith. 2003. *Shortage of Girls in China: Causes, Consequences, International Comparisons, and Solutions.* Population Reference Bureau online. Available at http://www.prb.org

Bannon, Lisa. 2000. "Why Girls and Boys Get Different Toys." *Wall Street Journal,* February 14, p. B1.

Barak, Azy. 2005. "Sexual Harassment on the Internet." *Social Science Computer Review* 23(1):77–92.

Barko, Naomi. 2003. "Equal Pay for Equal Work." In *Women's Rights,* Shasta Gaughen, ed. (pp. 43–48). Farmington, MA: Greenhaven Press.

Basow, Susan A. 1992. *Gender: Stereotypes and Roles,* 3rd ed. Pacific Grove, CA: Brooks/Cole.

Beeman, Mark, Geeta Chowdhry, and Karmen Todd. 2000. "Educating Students About Affirmative Action: An Analysis of University Sociology Texts." *Teaching Sociology* 28(2):98–115.

Begley, Sharon. 2000. "The Stereotype Trap." *Newsweek,* November 6, pp. 66–68.

Beutel, Ann M., and Margaret Mooney Marini. 1995. "Gender and Values." *American Sociological Review* 60:436–448.

Bianchi, Susanne M., Melissa A. Milkie, Liana C. Sayer, and John Robinson. 2000. "Is Anyone Doing the Housework? Trends in the Gender Division of Household Labor." *Social Forces* 79:191–228.

Bittman, Michael, and Judy Wajcman. 2000. "The Rush Hour: The Character of Leisure Time and Gender Equity." *Social Forces* 79:165–189.

Bly, Robert. 1990. *Iron John: A Book About Men.* Boston: Addison-Wesley.

Budig, Michelle. 2003. "Male Advantage and the Gender Composition of Jobs: Who Rides the Glass Escalator?" *Social Problems* 49:258–277.

Bureau of Labor Statistics. 2000. *Report on the Youth Labor Force.* Washington, DC: U.S. Department of Labor.

Bureau of Labor Statistics. 2005. *Kids' Careers.* Available at http:///www.bls.gov/k12/index.htm

Burger, Jerry M., and Cecilia H. Solano. 1994. "Changes in Desire for Control over Time: Gender Differences in a Ten-Year Longitudinal Study," *Sex Roles* 31:465–472.

Burk, Martha. 2005 (July 27). "Reaching the Top—and Still Finding a Pay Gap." *Common Dreams.* Available at http://www.commondreams.org

CBS News. 2003a (February 17). *Air Force Cadets Claim Rape Cover Up.* Available at http://www.cbsnews.com/stories/2003

CEDAW (Committee on the Elimination of Discrimination Against Women). 2003 (August 7). *Welcome New Legislation to Foster Gender Equality.* Press release.

Cejka, Mary Ann, and Alice Eagly. 1999. "Gender Stereotypic Images of Occupations Correspond to the Sex Segregation of Employment." *Personality and Social Psychology Bulletin* 25:413–423.

Center for American Women in Politics. 2005. *Women in Elected Offices, 2005.* Available at http://rci.rutgers.edu/cawp

Chavez, Linda. 2000. *The Color Bind.* Berkeley: University of California Press.

"Chinese Look to Their Daughters." 2004. *Toronto Star,* August 15, p. F5.

CNN. 2003 (March 7). *More Air Force Cadets Speak Out.* Available at http://www.cnn.com

Cohen, Philip, and Matt Huffman. 2003. "Individuals, Jobs, and Labor Markets: The Devaluation of Women's Work." *American Sociological Review* 68:443–463.

Cohen, Theodore. 2001. *Men and Masculinity.* Belmont, CA: Wadsworth.

Common Dreams. 2005 (March 24). *Media Advisory: Women's Opinions Also Missing on Television.* Available at http://www.commondreams.org/news2005/0324-07.htm

Connolly, Ceci. 2005. "Access to Abortion Pared at State Level." *Washington Post,* August 29, p. A1.

Cullen, Lisa Takeuchi. 2003. "I Want Your Job, Lady!" *Time,* May 12, pp. 52–56.

Cunningham, Mick. 2001. "Parental Influences on the Gendered Division of Housework." *American Sociological Review* 66:184–203.

Darnovsky, Marcy. 2004. "Revisiting Sex Selection." *Gene Watch* 17(1). Available at http://www.gene-watch.org

Davis, James A., Tom W. Smith, and Peter V. Marsden. 2002. *General Social Surveys, 1972–2002: 2nd ICPSR Version.* Chicago: National Opinion Research Center.

Dittrich, Liz. 2002. *About-Face Facts on the Media.* Available at http://www.about-face.org/r/facts/childrenmedia

Dumais, Susan A. 2002. "Cultural Capital, Gender, and School Success: The Role of Habitus." *Sociology of Education* 75: 44–68.

Equal Employment Opportunity Commission. 2003 (July 31). *Women of Color Make Gains in Employment and Job Status.* Available at http://www.eeoc.org/pres

Equal Employment Opportunity Commission. 2005. *Sexual Harassment.* Available at http://www.eeoc.gov

Evans, Lorraine, and Kimberly Davies. 2000. "No Sissy Boys Here." *Sex Roles,* February, pp. 255–271.

Faludi, Susan. 1991. *Backlash: The Undeclared War Against American Women.* New York: Crown.

Fitzpatrick, Catherine. 2000. "Modern Image of Masculinity Changes with Rise of New Celebrities." *Detroit News,* June 24. Available at http://www.detnews.com/2000/religion/0006/24

Gallagher, Sally K. 2004. "The Marginalization of Evangelical Feminism." *Sociology of Religion* 65:215–237.

Goffman, Erving. 1963. *Stigma.* Englewood Cliffs, NJ: Prentice Hall.

Gupta, Sanjay. 2003. "Why Men Die Young." *Time,* May 12, p. 84.

Guy, Mary Ellen, and Meredith A. Newman. 2004. "Women's Jobs, Men's Jobs: Sex Segregation and Emotional Labor." *Public Administration Review.* 64:289–299.

Hawkesworth, Mary, Kathleen J. Casey, Krista Jenkins, and Kathleen E. Kleeman. 2001. *Legislating By and For Women: A Comparison of the 103rd and 104th Congress.* Center for American Women and Politics, Rutgers University.

Hewlett, Sylvia Ann. 2002. "Executive Women and the Myth of Having It All." *Harvard Business Review,* April, pp. 66–73.

Heyzer, Noeleen. 2003. "Enlisting African Women to Fight AIDS." *Washington Post,* July 8. Available at http://www.global policy.org/socecon/inequal

Hochschild, Arlie. 1989. *The Second Shift.* London: Penguin.

Hochschild, Arlie. 2001. *The Time Bind: When Work Becomes Home and Home Becomes Work.* New York: Owl Books.

Hoffnung, Michele. 2004. "Wanting It All: Career, Marriage, and Motherhood During College-Educated Women's 20s." *Sex Roles* 50:711–723.

Hollingsworth, Leslie D. 2005. "Ethical Considerations in Prenatal Sex Selection." *Health and Social Work* 30(2):126–134.

Human Rights Watch. 2005. *CEDAW: The Women's Treaty.* Available at http://hrw .org/campaigns/cedaw/

"India's Disappearing Females." 2004. *Futurist* 38(2):8.

Institute for Women's Policy Research. 2004. *The Status of Women in the States.* Available at http://www.iwpr.org

International Women's Right's Project. 2000. *The First CEDAW Impact Study.* Available at http://www.yorku.ca/iwrp/ cedawReport

Ison, Elizabeth R. 2005 (August). "California Supreme Court Says Workers Who Are Not Involved in Office Romance Can Sue Over the Affair." *HR California.* Available at http://www.hrcalifornia.com

Jackson, Robert, and Meredith A. Newman. 2004. "Sexual Harassment in the Federal Workplace Revisited: Influences on Sexual Harassment." *Public Administration Review* 64(6):705–717.

Jones, Jeffrey M. 2005 (March 10). *Public Believes Men, Women Have Equal Abilities in Math, Science.* Gallup News Services, Gallup Organization.

Kilbourne, Barbara S., Georg Farkas, Kurt Beron, Dorothea Weir, and Paula England. 1994. "Returns to Skill, Compensating Differentials, and Gender Bias: Effects of Occupational Characteristics on the Wages of White Women and Men." *American Journal of Sociology* 100:689–719.

Kilbourne, Jean. 2000. *Killing Us Softly 3: Advertising's Image of Women.* Northampton, MA: Media Education Foundation.

Lawler, Andrew. 1999. "Tenured Women Battle to Make It Less Lonely at the Top." *Science* 286:1272–1279.

Long, J. Scott, Paul D. Allison, and Robert McGinnis. 1993. "Rank Advancement in Academic Careers: Sex Differences and the Effects of Productivity." *American Sociological Review* 58:703–722.

Lopez-Claros, Augusto, and Saadia Zahidi. 2005. *Women's Empowerment: Measuring the Global Gender Gap.* World Economic Forum.

Lorber, Judith. 1998. "Night to His Day." In *Reading Between the Lines,* Amanda Konradi and Martha Schmidt, eds. (pp. 213–220). Mountain View, CA: Mayfield.

"Major Rulings of the 2002–2003 Term." 2003. *Daily Reflector,* June 27, p. 4A

Martin, Patricia Yancey. 1992. "Gender, Interaction, and Inequality in Organizations." In *Gender, Interaction, and Inequality,* Cecilia Ridgeway, ed. (pp. 208–231). New York: Springer-Verlag.

Mattingly, Marybeth J., and Suzanne M. Bianchi. 2003. "Gender Differences in the Quantity and Quality of Free Time: The U.S. Experience." *Social Forces.* 81(3):999–1030.

Mazure, C. M., G. P. Keita, and M. Blehar. 2002. *Summit on Women and Depression: Proceedings and Recommendations.* Washington, DC: American Psychological Association.

McBrier, Debra Branch. 2003. "Gender and Career Dynamics Within a Segmented Professional Labor Market: The Case of Law Academic." *Social Forces* 81:1201–1266.

McCammon, Susan, David Knox, and Caroline Schacht. 1998. *Making Choices in Sexuality.* Pacific Grove, CA: Brooks/Cole.

McGregor, Liz. 2003. *Women Bear Brunt of AIDS Toll.* Available at http://www .globalpolicy.org/socecon/develop

Mensch, Barbara, and Cynthia Lloyd. 1997. *Gender Differences in the Schooling Experiences of Adolescents in Low-Income Countries: The Case of Kenya.* Policy Research Working Paper 95. New York: Population Council.

Messner, Michael A., and Jeffrey Montez de Oca. 2005. "The Male Consumer as Loser: Beer and Liquor Ads in Mega Sports Media Events." *Signs* 30:1879–1909.

Miller, Carol D. 2004. "Participating but Not Leading: Women's Underrepresentation in Student Government Leadership Positions." *College Student Journal,* September, pp. 423–428.

Moen, Phyllis, and Yan Yu. 2000. "Effective Work/Life Strategies: Working Couples, Working Conditions, Gender, and Life Quality." *Social Problems* 47:291–326.

Morin, Richard, and Megan Rosenfeld. 2000. "The Politics of Fatigue." In *Annual Editions: Social Problems,* Kurt Finsterbusch, ed. (pp. 152–154). Guilford, CT: Dushkin/McGraw-Hill.

Nanda, Serena. 2000. *Gender Diversity: Cross-Cultural Variations.* Long Grove, IL: Wavelane Press.

National Council of Women's Organizations. 2005. *National Council of Women's Organizations Homepage.* Available at http://www.women'sorganizations.org

National Organization for Women. 2004. *Civil Rights Groups Protect Affirmative Action in Two Key States.* Available at http://www.now.org/issues/affirm/040204affirm.html

National Science Foundation. 2003. *Science and Engineering Doctorate Awards: 2003.* National Science Foundation, Division of Science Resource Studies. Available at http://www.nsf.gov

National Women's Law Center. 2004. "Commission Finds University of Colorado Knew of Alleged Sexual Harassment in Football Recruiting Program and Failed to Take Appropriate Action." *NWLC News Room.* Available at http://www.nwlc.org/details.cfm?id'1884&section'newsroom

Nelson, Robert, and William Bridges. 1999. *Legalizing Gender Inequality:* New York: Cambridge University Press.

Nichols-Casebolt, Ann, and Judy Krysik. 1997. "The Economic Well-Being of Never and Ever-Married Mother Families." *Journal of Social Service Research* 23(1):19–40.

Orecklin, Michelle. 2003. "Now She's Got Game." *Time,* March 3, pp. 57–59.

Orenstein, Peggy. 2001. *Flux: Women on Sex, Work, Love, Kids, and Life in a Half-Changed World.* New York: Anchor.

Padavic, Irene, and Barbara Reskin. 2002. *Men and Women at Work,* 2nd ed. Thousand Oaks, CA: Pine Forge Press.

Palmberg, Elizabeth. 2005. "Teach a Women to Fish and Everyone Eats: Why Women Are the Key to Global Poverty." *Sojourner's Magazine,* June, pp. 1–4. Available at http://www.sojo.net

Parker, Kathleen. 2000. "It's Time for Women to Get Angry but Not at Men." *Greensboro News Record,* March 14, p. F4.

Poe, Marshall. 2004. "The Other Gender GAP." *The Atlantic* (online), January-February. Available at http://www .theatlantic.com

Pollack, William. 1998. *Real Boys: Rescuing Our Sons from the Myths of Boyhood.* New York: Random House.

Pollack, William. 2000. "The Columbine Syndrome." *National Forum* 80:39–42.

Population Reference Bureau. 2005. *Women of Our World: 2005.* Washington, DC: Population Reference Bureau.

Purcell, Piper, and Lara Stewart. 1990. "Dick and Jane in 1989." *Sex Roles* 22:177–185.

Quist-Areton, Ofeibea. 2003. "Fighting Prejudice and Sexual Harassment of Girls in Schools." *All Africa,* June 12. Available at http://www.globalpolicy.org/socecon

Rasmussen Reports. 2005 (April 8). *72% Say They're Willing to Vote for Woman President.* Available at http://www.rasmussenreports.com/2005/Woman%20President.htm

Reid, Pamela T., and Lillian Comas-Diaz. 1990. "Gender and Ethnicity: Perspectives on Dual Status." *Sex Roles* 22:397–408.

Renzetti, Claire, and Daniel Curran. 2003. *Women, Men and Society.* Boston: Allyn and Bacon.

Reskin, Barbara, and Debra McBrier. 2000. "Why Not Ascription? Organizations' Employment of Male and Female Managers." *American Sociological Review* 65:210–233.

Robinson, John P., and Suzanne Bianchi. 1997. "The Children's Hours." *American Demographics,* December, pp. 1–6.

Ross, Catherine E., and Marieke Van Willigen. 1996. "Gender, Parenthood, and Anger." *Journal of Marriage and the Family* 58(3):572–584.

Sadker, Myra, and David Sadker. 1990. "Confronting Sexism in the College Classroom." In *Gender in the Classroom: Power and Pedagogy,* S. L. Gabriel and I. Smithson, eds. (pp. 176–187). Chicago: University of Illinois Press.

Sanchez, Diana T., and Jennifer Crocker. 2005. "How Investment in Gender Ideals Affects Well-Being: The Role of External Contingencies of Self-Worth." *Psychology of Women Quarterly* 29:63–77.

Sapiro, Virginia. 1994. *Women in American Society.* Mountain View, CA: Mayfield.

Sax, Linda, Sylvia Hurtado, Jennifer Lindholm, Alexander Astin, William Korn, and Kathryn Mahoney. 2004. *The American Freshman: National Norms for Fall 2004.* Los Angeles: Higher Education Research Institute.

Schwalbe, Michael. 1996. *Unlocking the Iron Cage: The Men's Movement, Gender Politics, and American Culture.* New York: Oxford University Press.

Sheehan, Molly. 2000. "Women Slowly Gain Ground in Politics." In *Vital Signs: The Environmental Trends That Are Shaping Our Future,* Linda Starke, ed. (pp. 152–153). New York: W. W. Norton.

Simpson, Ruth. 2005. "Men in Non-Traditional Occupations: Career Entry, Career Orientation, and Experience of Role Strain." *Gender Work and Organization* 12(4):363–380.

Skuratowicz, Eva, and Larry W. Hunter. 2004. "Where Do Women's Jobs Come From? Job Resegregation in an American Bank." *Work and Occupations* 31:73–110.

Smith, Shellee. 2003 (August 29). "New Report on Academy Sex Assault." *MSNBC .com.* Available at http://www.msnbc .com/m/pt/printms

Smolken, Rachael. 2000. "Girls' SAT Scores Still Lag Boys'." *Post Gazette.* Available at http://www.post-gazette.com/headlines/ 20000830sat2.asp

Snell, William E., Jr. 1997. *The Beliefs About Women Scale.* Department of Psychology, College of Liberal Arts, Southeast Missouri State University.

Sommers, Christina. 2000. *The War Against Boys.* New York: Simon and Schuster.

Sondhaus, Elizabeth L., Richard M. Kurtz, and Michael J. Strube. 2001. "Body Attitude, Gender, and Self-Concept: A 30-Year Perspective." *Journal of Psychology* 135: 413–429.

Stein, Rob. 2004. "A Boy for You, A Girl for Me: Technology Allows Choice." *Washington Post,* December 14, p. A1.

Thomas, Cathy Booth. 2003. "Conduct Unbecoming." *Time,* March 10, pp. 46–47.

Tinklin, Teresa, Linda Croxford, Alan Ducklin, and Barbara Frame. 2005. "Gender and Attitudes to Work and Family Roles: The Views of Young People at the Millennium." *Gender and Education* 17: 129–143.

Uggen, C., and A. Blackstone. 2004. "Sexual Harassment as a Gendered Expression of Power." *American Sociological Review* 69:64–92.

Ulick, Josh. 2004. "The Science of Sex Selection." *Newsweek,* January 26, 143(4): 48–49.

United Nations. 2000. *The World's Women 2000: Trends and Statistics.* New York: United Nations Statistics Division.

United Nations. 2005. *Final Report and Appraisal of the Beijing Declaration and Platform for Action and the Outcome Document of the 23rd Special Session of the General Assembly.* Available at http://un.org/womenwatch/daw/Review

United Nations Population Fund. 2003a. *State of the World Population, 2003.* Available at http://www.unfpa.org

United Nations Population Fund. 2003b. "Women and Gender Inequality." In *State of the World Population, 2002.* Available at http://www.unfpa.org/swp

University of Michigan News Service. 2005. *Working Moms Need to Negotiate Better Terms on Childcare Burden.* Available at http://www.umich.edu/news/ ?Releases/2005/Feb05

Urban Institute. 2003 (September 2). *Gender Gap in Higher Education Focus of New Urban Institute Research.* Available at http://www.urban.org

U.S. Bureau of the Census. 2005. *Statistical Abstract of the United States: 2004–2005,* 124th ed. Washington, DC: U.S. Government Printing Office.

U.S. Census Bureau. 2004. *Historical Poverty Tables: People.* Available at http://www .census.gov/hhes/poverty/histpov

U.S. Department of Defense. 2005. *Special Department of Defense Briefing on New Sexual Assault Policy.* Available at http:// www.defenselink.mil/transcripts/2005/ tr20050104-1922.html

U.S. Department of Labor. 2005. *Women in the Labor Force: A Databook.* Report 985. Washington, DC: U.S. Government Printing Service.

White, Gayle. 2000. "Carter Cuts Ties to 'Rigid' Southern Baptists." *Atlanta Constitution-Journal,* October 20. Available at http://www.ajc.com

Williams, Christine L. 1995. *Still a Man's World: Men Who Do Women's Work.* Berkeley: University of California Press.

Williams, Joan. 2000. *Unbending Gender: Why Family and Work Conflict and What to Do About It.* Oxford: Oxford University Press.

Williams, John E., and Deborah L. Best. 1990. *Sex and Psyche: Gender and Self Viewed Cross-Culturally.* London: Sage.

Williams, Kristi, and Debra Umberson. 2004. "Marital Status, Marital Transitions, and Health: A Gendered Life Course Perspective." *Journal of Health and Social Behavior* 45:81–98.

Winfield, Nicole. 2000. "Activists Give U.S. Mixed Rating for Efforts on Gender." *Boston Globe,* June 8, p. A21.

Winters, Rebecca. 2005. "Harvard's Crimson Face." *Time,* January 31, p. 52.

Witt, S. D. 1996. "Traditional or Androgynous: An Analysis to Determine Gender Role Orientation of Basal Readers." *Child Study Journal* 26:303–318.

Women's International Network. 2000. "Reports from Around the World." *WIN News,* Autumn, pp. 50–58.

World Bank. 2003 (April). *Gender Equality and the Millennium Development Goals.* Gender and Development Group, World Bank.

World Economic Forum. 2005. *Women's Empowerment: Measuring the Global Gender Gap.* Geneva: World Economic Forum.

World Global Issues. 2003. *Women.* Available at http://www.osearth.cpm/resources

Yoder, Janice D., and Patricia Aniakudo. 1997. "Outsiders Within the Firehouse: Subordination and Difference in the Social Interactions of African American Women Firefighters." *Gender and Society* 11(3): 324–341.

Yoder, P. Stanley, N. Abderrahim, and A. Zhuzhuni. 2004. *Female Genital Cutting in the Demographic and Health Surveys: A Critical and Comparative Analysis.* DHS Comparative Reports 7. Calverton, MD: ORC Macro.

Yumiko, Ehara. 2000. "Feminism's Growing Pains." *Japan Quarterly* 47:41–48.

Zakrzewski, Paul. 2005. "Daddy, What Did You Do in the Men's Movement?" *Boston Globe,* June 19. Available at http://www .bostonglobe.com/news/globe

Zimmerman, Marc A., Laurel Copeland, Jean Shope, and T. E. Dielman. 1997. "A Longitudinal Study of Self-Esteem: Implications for Adolescent Development." *Journal of Youth and Adolescence* 26(2): 117–141.

Chapter 11

Allport, G. W. 1954. *The Nature of Prejudice.* Cambridge, MA: Addison-Wesley.

Amato, Paul R. 2004. "Tension Between Institutional and Individual Views of Marriage." *Journal of Marriage and Family* 66:959–965.

Amnesty International. 2005. *Stonewalled: Police Abuse and Misconduct Against Lesbian, Gay, Bisexual, and Transgender People in the United States.* Available at http://www.amnesty.org

Aravosis, John. 1999. "Privacy: The Impact on Lesbian, Gay, Bisexual, and Transgender Community." In *Access Denied Version 2.0: The Continuing Threat Against Internet Access and Privacy and Its Impact on the Lesbian, Gay, Bisexual, and Trangender Community,* Gay and Lesbian Alliance Against Defamation, ed. (pp. 30–33). New York: Gay and Lesbian Alliance Against Defamation.

Bayer, Ronald. 1987. *Homosexuality and American Psychiatry: The Politics of Diagnosis,* 2nd ed. Princeton, NJ: Princeton University Press.

Besen, Wayne. 2000. "Introduction." In *Feeling Free: Personal Stories—How Love and Self-Acceptance Saved Us from "Ex-Gay" Ministries,* Human Rights Campaign, ed. (p. 7). Washington, DC: Human Rights Campaign Foundation.

Black, Dan, Gary Gates, Seth Sanders, and Lowell Taylor. 2000 (May). "Demographics of the Gay and Lesbian Population in the United States: Evidence from Available Systematic Data Sources." *Demography* 37(2):139–154.

Bobbe, Judith. 2002. "Treatment with Lesbian Alcoholics: Healing Shame and Internalized Homophobia for Ongoing Sobriety." *Health and Social Work* 27(3):218–223.

Bontempo, Daniel E., and Anthony R. D'Augelli. 2002. "Effects of At-School Victimization and Sexual Orientation on Lesbian, Gay, or Bisexual Youths' Health Risk Behavior." *Journal of Adolescent Health* 30:364–374.

Bowes, John. 1999. "Conclusions." In *Access Denied Version 2.0: The Continuing Threat Against Internet Access and Privacy and Its Impact on the Lesbian, Gay, Bisexual, and Trangender Community,* Gay and Lesbian Alliance Against Defamation, ed. (pp. 38–44). New York: Gay and Lesbian Alliance Against Defamation.

Bradford, Judith, Kirsten Barrett, and Julie A. Honnold. 2002. *The 2000 Census and Same-Sex Households: A User's Guide.* New York: National Gay and Lesbian Task Force Policy Institute, Survey and Evaluation Research Laboratory, and Fenway Institute. Available at http://www .thetaskforce.org

Buhl, Larry. 2005 (June 16). "Youth's Blog Stirs Uproar Over 'Ex-Gay' Camp." *Planet Out.* Available at http://www.planetout .com

Button, James W., Barbara A. Rienzo, and Kenneth D. Wald. 1997. *Private Lives, Public Conflicts: Battles over Gay Rights in American Communities.* Washington, DC: CQ Press.

Cahill, Sean. 2005. *The Glass Nearly Half Full: 47% of U.S. Population Lives in Jurisdiction with Sexual Orientation Nondiscrimination Law.* National Gay and Lesbian Task Force. Available at http:// www.thetaskforce.org

Cahill, Sean, Mitra Ellen, and Sarah Tobias. 2002. *Family Policy: Issues Affecting Gay, Lesbian, Bisexual, and Transgender Families.* National Gay and Lesbian Task Force Policy Institute. Available at http://www .thetaskforce.org

Cahill, Sean, and Kenneth T. Jones. 2003. *Child Sexual Abuse and Homosexuality: The Long History of the 'Gays as Pedophiles' Fallacy.* National Gay and Lesbian Task Force. Available at http://www .thetaskforce.org

Cahill, Sean, and Samuel Slater. 2004. *Marriage: Legal Protections for Families and Children.* Policy brief. Washington, DC: National Gay and Lesbian Task Force Policy Institute.

Chase, Bob. 2000. *NEA President Bob Chase's Historic Speech from 2000 GLSEN Conference.* Available at http://www.glsen.org

Cianciotto, Jason, and Sean Cahill. 2003. *Educational Policy: Issues Affecting Lesbian, Gay, Bisexual, and Transgender Youth.* Washington, DC: National Gay and Lesbian Task Force Policy Institute.

Colvin, Roddrick. 2004. *The Extent of Sexual Orientation Discrimination in Topeka, Kansas.* National Gay and Lesbian Task Force Policy Institute. Available at http:// www.thetaskforce.org

"Constitutional Protection." 1999. *GayLawNet.* Available at http://www.nexus .net.au/~dba/news.html#top

Curtis, Christopher. 2003 (October 7). "Poll: U.S. Public is 50–50 on Gay Marriage." *PlanetOut.* Available at http://www .planetout.com

Curtis, C. 2004. "Poll: 1 in 20 High School Students Is Gay." *PlanetOut.* Available at http://www.planetout.com

Curtis, Christopher. 2005 (August 31). "Group Links Katrina to Annual Gay Party." *PlanetOut.* Available at http://www .planetout.com

"Custody and Visitation." 2000. Human Rights Campaign FamilyNet. Available at http://familynet.hrc.org

D'Emilio, John. 1990. "The Campus Environment for Gay and Lesbian Life." *Academe* 76(1):16–19.

Diamond, Lisa M. 2003. "What Does Sexual Orientation Orient? A Biobehavioral Model Distinguishing Romantic Love and Sexual Desire." *Psychological Review* 110(1):173–192.

Dozetos, Barbara. 2001a (March 7). "School Shooter Taunted as 'Gay.'" *PlanetOut.* Available at http://www.planetout.com

Durkheim, Emile. 1993. "The Normal and the Pathological." In *Social Deviance,* Henry N. Pontell, ed. (pp. 33–63). Englewood Cliffs, NJ: Prentice-Hall. (Originally published in *The Rules of Sociological Method,* 1938.)

Esterberg, K. 1997. *Lesbian and Bisexual Identities: Constructing Communities, Constructing Selves.* Philadelphia: Temple University Press.

FBI (Federal Bureau of Investigation). 2004. *Hate Crime Statistics 2003.* Available at http://www.fbi.gov

Firestein, B. A. 1996. "Bisexuality as Paradigm Shift: Transforming Our Disciplines." In *Bisexuality: The Psychology and Politics of an Invisible Minority,* B. A. Firestein, ed. (pp. 263–291). Thousand Oaks, CA: Sage.

Fone, Byrne. 2000. *Homophobia: A History.* New York: Henry Holt.

Frank, Barney. 1997. "Foreword." In *Private Lives, Public Conflicts: Battles over Gay Rights in American Communities,* by J. W. Button, B. A. Rienzo, and K. D. Wald (pp. i–xi). Washington, DC: CQ Press.

Frank, David John, and Elizabeth H. McEneaney. 1999. "The Individualization of Society and the Liberalization of State Policies on Same-Sex Relations, 1984–1995." *Social Forces* 77(3):911–944.

Franklin, Karen. 2000. "Antigay Behaviors Among Young Adults." *Journal of Interpersonal Violence* 15(4):339–362.

Garnets, L., G. M. Herek, and B. Levy. 1990. "Violence and Victimization of Lesbians and Gay Men: Mental Health Consequences." *Journal of Interpersonal Violence* 5:366–383.

Gay and Lesbian Alliance Against Defamation. 1999. Appendix A of the "Frequently Asked Questions." In *Access Denied Version 2.0: The Continuing Threat Against Internet Access and Privacy and Its Impact on the Lesbian, Gay, Bisexual, and Transgender Community,* Gay and Lesbian Alliance Against Defamation, ed. (pp. 45–48). New York: Gay and Lesbian Alliance Against Defamation.

Gay, Lesbian, and Straight Education Network. 2000. *Homophobia 101: Teaching Respect for All.* Gay, Lesbian, and Straight Education Network. Available at http:// www.glsen.org

Gilman, Stephen E., Susan D. Cochran, Vickie M. Mays, Michael Hughes, David Ostrow, and Ronald C. Kessler. 2001. "Risk of Psychiatric Disorders Among Individuals Reporting Same-Sex Sexual Partners in the National Comorbidity Survey." *American Journal of Public Health* 91(6):933–939.

Goode, Erica E., and Betsy Wagner. 1993. "Intimate Friendships." *U.S. News and World Report,* July 5, pp. 49–52.

Green, John C. 2004. *The American Religious Landscape and Political Attitudes: A Baseline for 2004.* Pew Forum on Religion and Public Life. Available at http:// pewforum.org

Held, Myka. 2005 (March 16). *Mix It Up: T-Shirts and Activism.* Available at http://www.tolerance.org/teens

Holthouse, David. 2005. "Curious Cures." *Intelligence Report* 117(spring):14.

Human Rights Campaign. 2000. *Feeling Free: Personal Stories—How Love and Self-*

Acceptance Saved Us from "Ex-Gay" Ministries. Washington, DC: Human Rights Campaign Foundation.

Human Rights Campaign. 2005a (March 22). *Recruiting Age Bump Highlights Consequences of "Don't Ask, Don't Tell."* HRC press release. Available at http://www.hrc.org

Human Rights Campaign. 2005b. *The State of the Workplace for Lesbian, Gay, Bisexual, and Transgendered Americans, 2004*. Washington, DC: Human Rights Campaign. Available at http://www.hrc.org

Human Rights Watch. 2001. *World Report 2001*. Available at http://www.hrw.org

International Gay and Lesbian Human Rights Commission. 1999. *Antidiscrimination Legislation*. Available at http://www.iglhrc.org/news/factsheets/990604-antidis.html

International Gay and Lesbian Human Rights Commission. 2003a. *Where Having Sex Is a Crime: Criminalization and Decriminalization of Homosexual Acts*. Available at http://www.iglhrc.org

International Gay and Lesbian Human Rights Commission. 2003b. *Where You Can Marry: Global Summary of Registered Partnership, Domestic Partnership, and Marriage Laws*. Available at http://www.iglhrc.org

International Gay and Lesbian Human Rights Commission. 2004 (January 30). *IGLHRC Calls for Global Mobilization to Help Pass the United Nations Resolution on Sexual Orientation and Human Rights*. Available at http://www.iglhrc.org

Israel, Tania, and Jonathan J. Mohr. 2004. "Attitudes Toward Bisexual Women and Men: Current Research, Future Directions." In *Current Research on Bisexuality*, Ronald C. Fox, ed. (pp. 117–134). New York: Harrington Park Press.

"Jail, Death Sentences in Africa." 2001 (February 21). *PlanetOut*. Available at http://www.planetout.com

Javier, Loren. 1999. "The World Since Access Denied." In *Access Denied Version 2.0: The Continuing Threat Against Internet Access and Privacy and Its Impact on the Lesbian, Gay, Bisexual, and Transgender Community*, Gay and Lesbian Alliance Against Defamation, ed. (pp. 6–9). New York: Gay and Lesbian Alliance Against Defamation.

Johnston, Eric. 2003 (October 14). "Clinton Blasts Bush's Gay Marriage Attack." *PlanetOut*. Available at http://www.planetoutcom

Johnston, Eric. 2005 (May 5). "Mass. Releases Data on Same-Sex Marriages." *PlanetOut*. Available at http://www.planetout.com

Kaiser Family Foundation. 2000. *Sex Education in America: A View from Inside the Nation's Classrooms*. Menlo Park, CA: Henry J. Kaiser Family Foundation.

Kinsey, A. C., W. B. Pomeroy, and C. E. Martin. 1948. *Sexual Behavior in the Human Male*. Philadelphia: W. B. Saunders.

Kinsey, A. C., W. B. Pomeroy, C. E. Martin, and P. H. Gebhard. 1953. *Sexual Behavior in the Human Female*. Philadelphia: W. B. Saunders.

Kirkpatrick, R. C. 2000. "The Evolution of Human Sexual Behavior." *Current Anthropology* 41(3):385–414.

Kosciw, Joseph G., and M. K. Cullen. 2002. *The 2001 National School Climate Survey*. Gay, Lesbian, and Straight Education Network. Available at http://www.glsen.org

Landis, Dan. 1999 (February 17). "Mississippi Supreme Court Made a Tragic Mistake in Denying Custody to Gay Father, Experts Say." *American Civil Liberties Union News*. Available at http://www.aclu.org

LAWbriefs. 2000 (fall). "Recent Developments in Sexual Orientation and Gender Identity Law." *LAWbriefs*, v. 3, no. 3.

LAWbriefs. 2003 (spring). "Recent Developments in Sexual Orientation and Gender Identity Law." *LAWbriefs*, v. 6, no. 1.

LAWbriefs. 2005a (April). "Recent Developments in Sexual Orientation and Gender Identity Law." *LAWbriefs*, v. 7, no. 1.

LAWbriefs. 2005b (July). "Recent Developments in Sexual Orientation and Gender Identity Law." *LAWbriefs*, v. 7, no. 2.

Lever, Janet. 1994. "The 1994 Advocate Survey of Sexuality and Relationships: The Men." *The Advocate*, August 23, pp. 16–24.

Loftus, Jeni. 2001. "America's Liberalization in Attitudes Toward Homosexuality, 1973 to 1998." *American Sociological Review* 66:762–782.

Louderback, L. A., and B. E. Whitley. 1997. "Perceived Erotic Value of Homosexuality and Sex-Role Attitudes as Mediators of Sex Differences in Heterosexual College Students' Attitudes Toward Lesbians and Gay Men." *Journal of Sex Research* 34:175–182.

Lyall, Sarah. 2005. "New Course by Royal Navy: A Campaign to Recruit Gays." *New York Times*, February 22. Available at http://www.nytimes.com

Mackay, J. 2001. "Global Sex: Sexuality and Sexual Practices Around the World." *Sexuality and Relationship Therapy* 16:71–82.

Mathison, Carla. 1998. "The Invisible Minority: Preparing Teachers to Meet the Needs of Gay and Lesbian Youth." *Journal of Teacher Education* 49:151–155.

McCammon, Susan, David Knox, and Caroline Schacht. 2004. *Choices in Sexuality*, 2nd ed. Cincinnati: Atomic Dog.

McKinney, Jack. 2004 (February 8). *The Christian Case for Gay Marriage*. Pullen Memorial Baptist Church. Available at http://www.pullen.org

Michael, Robert T., John H. Gagnon, Edward O. Laumann, and Gina Kolata. 1994. *Sex in America: A Definitive Survey*. Boston: Little, Brown.

Miller, Patti, McCrae A. Parker, Eileen Espejo, and Sarah Grossman-Swenson. 2002. *Fall Colors: Prime Time Diversity Report 2001–02*. Oakland CA: Children Now and The Media Program.

Mohipp, C., and M. M. Morry. 2004. "Relationship of Symbolic Beliefs and Prior Contact to Heterosexuals' Attitudes Toward Gay Men and Lesbian Women." *Canadian Journal of Behavioral Science* 36(1):36–44.

Moltz, Kathleen. 2005 (April 13). *Testimony of Kathleen Moltz*. Testimony given before the United States Senate Judiciary Committee, Subcommittee on the Constitution, Civil Rights and Property Rights. Human Rights Campaign. Available at http://www.hrc.org

Moore, David W. 1993. "Public Polarized on Gay Issue." *Gallup Poll Monthly* 331(April):30–34.

Moser, Bob. 2005. "Holy War." *Intelligence Report* 117(spring):8–21.

National Coalition of Anti-Violence Programs. 2005. *2004 National Hate Crimes Report: Anti-Lesbian, Gay, Bisexual and Transgender Violence in 2004*. New York: National Coalition of Anti-Violence Programs.

National Gay and Lesbian Task Force. 2004 (June). *Anti-Gay Parenting Laws in the U.S*. National Gay and Lesbian Task Force. Available at http://www.thetaskforce.org

National Gay and Lesbian Task Force. 2005a (May). *Anti-Gay Marriage Measures in the U.S*. National Gay and Lesbian Task Force. Available at http://www.thetaskforce.org

National Gay and Lesbian Task Force. 2005b (May). *Hate Crime Laws in the U.S*. National Gay and Lesbian Task Force. Available at http://www.thetaskforce.org

National Gay and Lesbian Task Force. 2005c. *The Issues: Seniors*. National Gay and Lesbian Task Force. Available at http://www.thetaskforce.org

National Gay and Lesbian Task Force. 2005d. *Second-Parent Adoption in the U.S*. National Gay and Lesbian Task Force. Available at http://www.thetaskforce.org

National Gay and Lesbian Task Force. 2005e. *States Banning Sexual Orientation Discrimination*. Available at http://www.thetaskforce.org

"NCLR Wins Equal Tax Benefits for Nonbiological Lesbian Mother." 2000 (fall). *NCLR Newsletter* (National Center for Lesbian Rights), pp. 1 and 10. Available at http://www.nclrights.org/index.html

Newport, Frank. 2002 (September). *In-Depth Analysis: Homosexuality*. Gallup Organization. Available at http://www.gallup.com

Newport, Frank. 2005 (April 19). *Americans Turn More Negative Toward Same-Sex Marriage*. Gallup Poll News Service. Available at http://www.gallup.com

Page, Susan. 2003. "Gay Rights Tough to Sharpen into Political 'Wedge Issue.'" *USA Today*, July 28, p. 10A.

Patel, S. 1989. *Homophobia: Personality, Emotional, and Behavioral Correlates*. Master's thesis, East Carolina University, Greenville, N.C.

Patel, S., T. E. Long, S. L. McCammon, and K. L. Wuensch. 1995. "Personality and Emotional Correlates of Self-Reported Antigay Behaviors." *Journal of Interpersonal Violence* 10:354–366.

Patterson, Charlotte J. 2001. "Family Relationships of Lesbians and Gay Men." In *Understanding Families into the New Millennium: A Decade in Review*, Robert M. Milardo, ed. (pp. 271–288). Minneapolis: National Council on Family Relations.

Paul, J. P. 1996. "Bisexuality: Exploring/Exploding the Boundaries." In *The Lives of Lesbians, Gays, and Bisexuals: Children to Adults*, R. Savin-Williams and K. M. Cohen, eds. (pp. 436–461). Fort Worth: Harcourt Brace.

Platt, Leah. 2001. "Not Your Father's High School Club." *American Prospect* 12(1):A37–A39.

Plugge-Foust, C., and George Strickland. 2000. "Homophobia, Irrationality, and Christian Ideology: Does a Relationship Exist? *Journal of Sex Education and Therapy* 25:240–244.

Potok, Mark. 2005 (spring). "Vilification and Violence." *Intelligence Report* 117:1.

Price, Jammie, and Michael G. Dalecki. 1998. "The Social Basis of Homophobia: An Empirical Illustration." *Sociological Spectrum* 18:143–159.

Puccinelli, Mike. 2005 (April 19). "Students Support, Decry Gays with T-Shirts." *CBS 2 Chicago*. Available at http://cbs2 chicago.com

Rankin, Susan R. 2003. *Campus Climate for Gay, Lesbian, Bisexual, and Transgendered People: A National Perspective*. New York: National Gay and Lesbian Task Force Policy Institute. Available at http://www.thetaskforce.org

Rideout, Victoria, Caroline Richardson, and Paul Resnick. 2002. *See No Evil: How Internet Filters Affect the Search for Online Health Information*. Washington, DC: Kaiser Family Foundation. Available at http://www.kff.org

Roderick, T. 1994. *Homonegativity: An Analysis of the SBS-R*. Master's thesis, East Carolina University, Greenville, N.C.

Roderick, T., S. L. McCammon, T. E. Long, and L. J. Allred. 1998. "Behavioral Aspects of Homonegativity." *Journal of Homosexuality* 36:79–88.

Rosin, Hanna, and Richard Morin. 1999 (January 11). "In One Area, Americans Still Draw a Line on Acceptability." *Wash-ington Post National Weekly Edition* 16(11):8.

Saad, Lydia. 2005 (May 20). *Gay Rights Attitudes a Mixed Bag*. Gallup Organization. Available at http://www.gallup.com

Safe Zone Resources. 2003. *National Consortium of Directors of LGBT Resources in Higher Education*. Available at http://www.lgbtcampus.org

Sanday, Peggy. R. 1995. "Pulling Train." In *Race, Class, and Gender in the United States*, 3rd ed., P. S. Rothenberg, ed. (pp. 396–402). New York: St. Martin's Press.

Sapp, Jeff. 2005 (January 24). "Internet Filters Block 'Gay' and 'Lesbian'." *Tolerance in the News*. Available at http://www.tolerance.org

Schellenberg, E. Glenn, Jessie Hirt, and Alan Sears. 1999. "Attitudes Toward Homosexuals Among Students at a Canadian University." *Sex Roles* 40(1/2):139–152.

Schiappa, E., P. B. Gregg, and D. E. Hewes. 2005. "The Parasocial Contact Hypothesis." *Communication Monographs* 72(1):92–115.

SIECUS (Sexuality Information and Education Council of the United States). 2000. *Fact Sheets: Sexual Orientation and Identity*. New York: SIECUS.

Simmons, Tavia, and Martin O'Connell. 2003 (February). *Married-Couple and Unmarried-Partner Households: 2000*. Washington, DC: U.S. Census Bureau.

Singh, Daniel P., and Heather D. Wathington. 2003. "Valuing Equity: Recognizing the Rights of the LGBT Community." *Diversity Digest* 7 (1–2):8–9.

Smith, D. M., and G. J. Gates. 2001. *Gay and Lesbian Families in the United States: Same-Sex Unmarried Partner Households—Preliminary Analysis of 2000 U.S. Census Data*. Washington, DC: Human Rights Campaign. Available at http://www.hrc.org

Snorton, Riley. 2005 (April 1). *GLSEN's 2004 State of the States Report Is the First Objective Analysis of Statewide Safe Schools Policies*. Available at http://www.glsen.org

Sullivan, A. 1997. "The Conservative Case." In *Same-Sex Marriage: Pro and Con*, A. Sullivan, ed. (pp. 146–154). New York: Vintage Books.

Sullivan, A. 2003. "Legalizing Same-Sex Marriage Would Strengthen Marriage." In *The Family: Opposing Viewpoints*, A. Ojeda, ed. (pp. 196–200). Farmington Hill, MI: Greenhaven Press.

Sylivant, S. 1992. *The Cognitive, Affective, and Behavioral Components of Adolescent Homonegativity*. Master's thesis, East Carolina University, Greenville, N.C.

Thompson, Cooper. 1995. "A New Vision of Masculinity." In *Race, Class, and Gender in the United States*, 3rd ed., P. S. Rothenberg, ed. (pp. 475–481). New York: St. Martin's Press.

Tobias, Sarah, and Sean Cahill. 2003. *School Lunches, the Wright Brothers, and Gay Families*. National Gay and Lesbian Task Force. Available at http://www.thetask force.org

The United Methodist Church and Homosexuality. 1999. Available at www.religious tolerance.org/hom_umc.htm

White, Mel. 2005 (spring). "A Thorn In Their Side." *Intelligence Report* 117:27–30.

Wilcox, Clyde, and Robin Wolpert. 2000. "Gay Rights in the Public Sphere: Public Opinion on Gay and Lesbian Equality." In *The Politics of Gay Rights*, Craig A. Rimmerman, Kenneth D. Wald, and Clyde Wilcox, eds. (pp. 409–432). Chicago: University of Chicago Press.

Yang, Alan. 1999. *From Wrongs to Rights 1973 to 1999: Public Opinion on Gay and Lesbian Americans Moves Toward Equality*. New York: Policy Institute of the National Gay and Lesbian Task Force.

Yvonne. 2004 (December 1). *Devout Christian Finds a Reason to Stand Up for Equality*. Human Rights Campaign. Available at http://www.hrc.org

Chapter 12

AARP. 2003. "Fighting Ageism." *AARP Magazine*, July-August. Available at http://www.aarpmagazine.org

AHA (American Heart Association). 2005. *Childhood Obesity: Early Prevention Offers Best Solution*. Available at http://www.americanheart.org

"AIDS Creating Global 'Orphans Crisis.'" 2003. *CBS News*, July 10. Available at http://www.cbsnews.com/stories/2002

Allen, Claude A. 2002. *Speech to Trafficking Convention*. Speech presented in Honolulu, November 8. Available at http://www.iabolish.com/today/experience/health

Allen, C. L. 1998. "Euthanasia: Why Torture Dying People When We Have Sick Animals Put Down?" *Australian Psychologist* 33:12–15.

AOA (Administration on Aging). 2001. *The Older Americans Act*. Available at http://www.aoa.gov/may2001/factsheets/OAA.htm

AOA. 2003. *Older Americans Act*. Fact sheets. Available at http://www.aoa.gov/press/fact

AOA. 2004. *A Profile of Older Americans: 2004*. Washington, DC: U.S. Department of Health and Human Services. Available at http://www.aoa.dhhs.gov/aoa/stats

APA (American Psychological Association). 2005. *Elder Abuse and Neglect: In Search of Solutions*. Available at http://www.apa.org/pi/aging/eldabuse

Associated Press. 2005. "Supreme Court to Review Assisted Suicide Law in Oregon." *New York Times*, February 22. Available at http://www.nytimes.com

Atchley, Robert C. 2000. *Social Forces and Aging*. Belmont, CA: Wadsworth.

Becker, J. 2004. *Children as Weapons of War*. Human Rights Watch. Available at http://www.hrw.org/wrzk4

Binstock, Robert H. 1986. "Public Policy and the Elderly." *Journal of Geriatric Psychiatry* 19:115–143.

Birnbaum, Jeffrey. 2005. "AARP Leads with Wallet in Fight over Social Security." *Washington Post,* March 29. Available at http://www.washingtonpost.com

Bradley, Ed. 2005. "Can a Video Game Lead to Murder?" *CBS News,* March 6. Available at http://www.cbsnews.com/stories/2005

Brazzini, D. G., W. D. McIntosh, S. M. Smith, S. Cook, and C. Harris. 1997. "The Aging Woman in Popular Film: Underrepresented, Unattractive, Unfriendly, and Unintelligent." *Sex Roles* 36:531–543.

Brooks-Gunn, Jeanne, and Greg Duncan. 1997. "The Effects of Poverty on Children." *Future of Children* 7(2):55–70

Carpenter, MacKenzie, and Ginny Kopas. 1999. "Casualties of Custody Wars: Special Report." *Pittsburgh Post-Gazette*. Available at http://www.post-gazette.com/custody

Carroll, Joseph. 2005 (May 17). *Retirement Age Expectations and Realities*. Gallup Organization. Available at http://gallup.com/poll

CASA Alianza. 2003. *Living in the Streets*. Available at http://www.casa-alianza.org

Casey Family Programs. 2003. *Assessing the Effects of Foster Care*. Casey National Alumni Study. Available at http://www.casey.org

CBS and New York Times. 2000. *Priorities* CBS/New York Times Poll, September 9–11. Available at http://www.pollingreport.com/priorit1/htm

CDF (Children's Defense Fund). 2000. *Child Care Now!* Available at http://www.childrensdefense.org/childcare/cc_polls.html

CDF. 2003a. *Among Industrialized Countries, the United States Ranks . . .* Available at http://www.childrensdefense.org/factsfigures

CDF. 2003b. *Basic Facts on Poverty*. Available at http://www.childrensdefense.org

CDF. 2003c. *Media Violence*. Fact sheet. Available at http://www.childrensdefense.org

CDF. 2004a (October). *2003 Facts on Child Poverty in America*. Available at http://www.childrensdefense.org/family/income/childpoverty

CDF. 2004b (November 24). *Children's Health Jeopardized to Subsidize Powerful Special Interests*. Available at http://www.childrensdefense.org/childwatch

CDF. 2005. *Each Day in America*. Available at http://www.childrensdefense.org/data/eachday

Center for Health Statistics, Oregon Health Division. 2003. *Oregon's Death with Dignity Act: The Second Year Experience*. Oregon's Death with Dignity Act Annual Report, 2002. Salem: Center for Heath Statistics, Oregon Health Division.

CWLA (Children's Welfare League of America). 2005. *Quick Facts About Foster Care*. Available at http://www.cwla.org

DeAngelis, Tori. 1997. "Elderly May Be Less Depressed Than the Young." *APA Monitor,* October. Available at http://www.apa.org/monitor/oct97/elderly.html

Dewan, Shaila. 2005. "Abused Children Are Found Entitled to Legal Aid." *New York Times,* February 9. Available at http://www.nytimes.com

Donlan, Maggie M., Ori Ashman, and Becca R. Levy. 2005. "Re-Vision of Older Television Characters: A Stereotype-Awareness Intervention." *Journal of Social Issues* 61(2):307–319.

Dorman, Peter. 2001. *Child Labour in the Developed Economies*. Geneva: International Labour Office.

Duncan, Greg, W. Jean Yeung, Jeanne Brooks-Gunn, and Judith Smith. 1998. "How Much Does Childhood Poverty Affect the Life Chance of Children?" *American Sociological Review* 63:402–423.

"Elderly Become 'Muti' Targets." 2003. *African Eye News,* April 15.

EurekAlert. 2003 (April 22). *Children's Stereotypes of Aging Start Early*. Available at http://www.globalaging.org

FBI (Federal Bureau of Investigation). 2004. *Crime in the United States, 2003*. Uniform Crime Reports. Washington, DC: U.S. Government Printing Office.

Federal Elections Commission. 2005. *Voter Registration and Turnout by Age, Gender, and Race*. Available at http://www.fec.gov

Finsterbusch, Kurt. 2001. *Clashing Views on Controversial Social Issues*. Guilford, CN: Dushkin.

Flaherty, Kristina. 2004. "Abused Seniors Turn to Alameda Court for Help." *California Bar Journal,* October. Available at http://www.calbar.ca.gov/state/calbar

Gallup. 2004. *Children and Violence*. Gallup Poll. Available at http://www.gallup.com/poll/content

Gallup. 2005. *Social Security*. Gallup Poll. Available at http://brain.gallup.com/poll/content

Generations United. 2003. *About GU*. Available at http://www.gu.org

Goldberg, Beverly. 2000. *Age Works*. New York: Free Press.

Goozner, Merrill. 2005. "Don't Mess with Success." *AARP Bulletin,* January, pp. 12–15.

Green, Shane. 2003. "Hidden Abuse of Elderly Emerging Problem for Japan." *Sydney Morning Herald,* June 21. Available at http://www.globalaging.org/elderrights

Greenhouse, Linda. 2005. "Supreme Court Removes Hurdle to Age Bias Suits." *New York Times,* March 31. Available at http://www.nytimes.com

Harris, Diana K. 1990. *Sociology of Aging*. New York: Harper and Row.

Harris, Kathleen, and Jeremy Marmer. 1996. "Poverty, Paternal Involvement, and Adolescent Well-Being." *Journal of Family Issues* 17(5):614–640.

Hooyman, Nancy R., and H. Asuman Kiyak. 1999. *Social Gerontology: A Multidisciplinary Perspective,* 2nd ed. Boston: Allyn and Bacon.

Hounsell, Cindy, and Pat Humphlett. 2003. "Older Minority Women Need Retirement Help Now." *WomenENews,* May 29. Available at http://www.globalaging.org

Human Rights Watch. 2001 (January). "Underage and Unprotected: Child Labor in Egypt's Cotton Fields." *Human Rights Watch* 13(1). Available at http://www.hrw.org/reports/2001/Egypt

Human Rights Watch. 2003. *Child Farmworkers*. Available at http://www.hrw.org/campaigns

Human Rights Watch. 2005a. *Child Labor*. Available at http://www.hrw.org/about/projects/crd/child-labor

Human Rights Watch. 2005b. *Orphans and Abandoned Children*. Available at http://www.hrw.org/about/projects/crd/child-orphan

Jackson, Maggie. 2003. "More Sons Are Juggling Jobs and Care for Parents." *New York Times,* June 15. Available at http://www.globalaging.org/elderrights

Jacobson, Linda. 2000. "Children's Early Needs Seen as Going Unmet." *Washington Post,* October 3. Available at http://washingtonpost.com/wp-dyn/articles/A4014-2000Oct5.html

Jones, Jeffrey M. 2005. "Retirement Expectations, Reality Not Entirely Consistent." Gallup Organization, May 2. Available at http://gallup.com/poll

Kaiser Family Foundation. 2005a (March 9). *Media Multi-Tasking Changing the Amount and Nature of Young People's Media Use*. Available at http://www.kff.org/entmedia

Kaiser Family Foundation. 2005b (March). *Medicare: The Medicare Prescription Drug Benefit.* Available at http://www.kff.org

Kakuchi, Suvendrini. 2003. "Japan's Elderly Refuse to Fade Away." *Asian Times,* June 22. Available at http://www.atimes.com/japan-econ

Kotlowitz, Alex. 1991. *There Are No Children Here*. New York: Doubleday.

Legislative Agenda. 2005. *Vote Kids Legislative Agenda*. Available at http://votekids.everychildmatters.org

Leichtentritt, R. D., and K. D. Rettig. 2000. *Conflicting Value Considerations for End-of-Life Decisions*. Poster presented at 62nd Annual Meeting of National Council on Family Relations, Minneapolis, November 12.

Livni, Ephrat. 2000. "Exercise, the Anti-Drug." *ABC News,* September 21. Available at http://abcnews.go.com/sections/living/Daily/News/Depression

Luce, J. M. 1997. "Withholding and Withdrawal of Life Support: Ethical, Legal, and Clinical Aspects." *New Horizons* 5:30–37.

Mack, Brandy, and Kathi Jones. 2003 (August 5). "Elder Abuse: Identification and Prevention." *ASHA Leader* 8:10–12A.

Matras, Judah. 1990. *Dependency, Obligations, and Entitlements: A New Sociology of Aging, the Life Course, and the Elderly.* Englewood Cliffs, NJ: Prentice-Hall.

McCoy, Kevin. 2003. "Study Offers First Assisted-Living Guidelines." *USA Today,* April 29. Available at http://www.usatoday.com

Meier, Diane, Carol-Ann Emmons, Ann Litke, Sylvan Wallenstein, and Sean Morrison. 2003. "Characteristics of Patients Requesting and Receiving Physician Assisted Suicide." *Archives of Internal Medicine* 163:1537–1542.

Miner, Sonia, John Logan, and Glenna Spitze. 1993. "Predicting Frequency of Senior Center Attendance." *The Gerontologist* 33:650–657.

Moore, David W. 2005 (May 17). *Three in Four Americans Support Euthanasia.* Gallup Organization. Available at http://www.gallup.com

Morin, Richard, and Jim VandeHei. 2005. "Social Security Plan's Support Dwindling." *Washington Post,* June 9. Available at http://www.washingtonpost.com

National Consumers League. 2005. *Clocking In for Trouble: Teens and Unsafe Work.* Available at http://nclnet.org/labor/childlabor/jobreport

NCHS (National Center for Health Statistics). 2003. *Fast Stats: Health of the Elderly.* Available at http://www.cdc.gov/nchn

NCSC (National Council of Senior Citizens). 2000. *Our Issues: Poverty.* Available at http://www.ncscinc.org/issues/poverty.htm

Nicholson, Trish. 2003. "Age Complaints Surge as Midlife Workers Find the Going Harder." *AARP Bulletin,* March. Available at http://www.globalaging.org/elderrights/us

NIH (National Institutes of Health). 2003. *New Prevalence Study Suggests Dramatically Rising Numbers of People with Alzheimer's Disease.* Available at http://www.nih.gov/news

NIMH (National Institute of Mental Health). 2003. *Older Adults: Depression and Suicide Facts.* Available at http://www.mental-health-matters.com/articles

NMHA (National Mental Health Association). 2003. *Children's Mental Health Statistics.* Available at http://www.nimh.org/children/prevent

OLPA (Office of Legislative Policy and Analysis). 2005. *Elder Justice Act.* Available at http://olpa.od.hih.gov/legislation/108/pendinglegislation

Palmore, Erdman B. 1984. "The Retired." In *Handbook on the Aged in the United States,* Erdman B. Palmore, ed. (pp. 63–75). Westport, CT: Greenwood Press.

Palmore, Erdman B. 1999. *Ageism: Negative and Positive.* New York: Springer.

PBS. 2005. *The Last Hope.* Available at http://www.pbs.org/society/fosterres

Pear, Robert. 2002. "9 out of 10 Nursing Homes Lack Adequate Staff, Study Finds." *New York Times,* February 18. Available at http://www.globalaging.org/elderrights

Peterson, Karen. 2003. "Kids Are Better Off Than Adults Believe." *USA Today,* July 2. Available at http://www.usatoday.com

Porter, Eduardo, and Mary Walsh. 2005. "Retirement Turns into a Rest Stop as Benefits Dwindle." *New York Times,* February 9. Available at http://www.nytimes.com

Quadagno, Jill. 2005. *Aging and the Life Course.* Boston: McGraw-Hill.

Ramirez, Eddy. 2002. "Ageism in the Media Is Seen as Harmful to Health of Elderly." *Los Angeles Times,* September 5. Available at http://www.globalaging.org

Randolph, A. G., M. B. Zollo, R. S. Wigton, and T. S. Yeh. 1997. "Factors Explaining Variability Among Caregivers in the Intent to Restrict Life-Support Interventions in a Pediatric Intensive Care Unit." *Critical Care Medicine* 25:435–439.

Reio, Thomas G., and Joanne Sanders-Reto. 1999. "Combating Workplace Ageism." *Adult Learning* 11:10–13.

Riley, Matilda White. 1987. "On the Significance of Age in Sociology." *American Sociological Review* 52(February): 1–14.

Riley, Matilda W., and John W. Riley. 1992. "The Lives of Older People and Changing Social Roles." In *Issues in Society,* Hugh Lena, William Helmreich, and William McCord, eds. (pp. 220–231). New York: McGraw-Hill.

Seeman, Teresa E., and Nancy Adler. 1998. "Older Americans: Who Will They Be?" *National Forum,* spring, pp. 22–25.

Sloan, Allan, 2005. "Social Security: A Daring Leap." *Newsweek,* February 14, pp. 41–45.

"Snapshots of the Elderly." 2000. *Washington Post.* Available at http://www.washingtonpost.com/wp-srv/health/images/elderly.html

SOTWC (State of the World's Children). 2005. *United Nations Children's Fund.* Available at http://www.unicef.org/publications

Stolberg, Sheryl. 2005. "Schiavo's Case May Reshape American Law." *New York Times,* April 1. Available at http://www.nytimes.com

Street Children. 2005. "The Facts." *The New Internationalist,* no. 377. Available at http://www.newint.org/issue377/facts

Taylor, Humphrey. 2002 (January 9). *2-to-1 Majorities Continue to Support Rights to Both Euthanasia and Doctor Assisted Suicide.* Harris Poll 2. Available at http://www.harrisinteractive.com

Thurow, Lester C. 1996. "The Birth of a Revolutionary Class." *New York Times Magazine,* May 19, pp. 46–47.

Toner, Robin. 1999. "A Majority over 45 Say Sex Lives Are Just Fine." *New York Times,* August 4, p. A10.

Torres-Gil, Fernando. 1990. "Seniors React to Medicare Catastrophic Bill: Equity or Selfishness?" *Journal of Aging and Social Policy* 2(1):1–8.

Totenberg, Nina. 2005. "Supreme Court Ends Death Penalty for Juveniles." *National Public Radio,* March 22. Available at http://www.npr.org/templates/story

UNICEF. 2005a. *Child Labour and Exploitation.* Available at http://www.unicef.org/girlseducation

UNICEF. 2005b. *Conventions on the Rights of the Child.* FAQ. Available at http://www.unicef.org.crc/faq

United Nations. 1998. "The First Nearly Universally Ratified Human Rights Treaty in History." *Status.* Washington, DC: UNICEF. Available at http://www.unicef.org/crc/status

United Nations. 2003a. *Facts on Children: Child Protection.* Available at http://www.unicef.org/media

United Nations. 2003b. *Indicators of Youth and Elderly Populations.* New York: United Nations Statistics Division. Available at http://www.unstat.un.org/unsd/demographics/social

United Nations. 2003c. *Introduction to Convention on the Rights of the Child.* Available at http://www.unicef.org/crc/introduction

United Nations. 2003d. *The State of the World's Children.* New York: United Nations Children's Fund.

U.S. Bureau of the Census. 1980. *1980 Census of Population, General Population Characteristics, United States Summary.* Publication PC80-1-B1. Washington, DC: U.S. Government Printing Office.

U.S. Bureau of the Census. 2000a. *2000 Census of Population, Profiles of General Demographic Characteristics, United States.* Available at http://www.census.gov/prod/cen2000/dp1/2kh00.pdf (accessed on September 27, 2001).

U.S. Bureau of the Census. 2000b. *Projections of the Total Resident Population by 5-Year Age Groups, and Sex with Special Age Categories: Middle Series, 2050 to 2070.* Available at http://www.census.gov/population/projections/nation/summary/np-t3-g.txt (accessed on September 27, 2001).

U.S. Bureau of the Census. 2002. *International Data Base.* Washington, DC: U.S. Department of Commerce.

U.S. Bureau of the Census. 2004. *Statistical Abstract of the United States, 2004,* 124th ed. Washington, DC: U.S. Government Printing Office.

U.S. Department of Health and Human Services. 1999. "Children's Health Insurance Program National Back-to-School Kick Off." *HHS News,* September 22. Available at http://www.hcfa.gov/init/9909922wh.htm

U.S. Department of Labor. 1995. *By the Sweat and Toil of Children.* v. 2, *The Use of Child Labor in U.S. Agricultural Imports and Forced and Bonded Child Labor.* Washington, DC: U.S. Department of Labor, Bureau of International Labor Affairs.

Vollmar, Valeria. 2003. *Recent Developments in Physician-Assisted Suicide.* Available at http://www.willamette.edu/wucl/pas

World Youth Report. 2005. *General Assembly Economic and Social Council: Report of the Secretary-General.* New York: United Nations.

YRBSS (Youth Risk Behavior Surveillance System). 2004. *Morbidity and Mortality Weekly Report,* May 21, v. 53, no. SS2.

Zhong Xin Net. 2003. *China Becomes an Aging Society.* Available at http://www.globalaging.org/elderrights/world

Chapter 13

American Community Survey. 2004. Washington, DC: Belden Russonello & Stewart.

Americans' Attitudes Toward Walking and Creating Better Walking Communities. 2003 (April), Washington, DC: Belden Russonello & Stewart Research and Communications.

Blatt, Harvey. 2005. *America's Environmental Report Card: Are We Making the Grade?* Cambridge, MA: MIT Press.

Bongaarts, John, and Susan Cotts Watkins. 1996. "Social Interactions and Contemporary Fertility Transitions." *Population and Development Review* 22(4):639–682.

"The Bridge to the 21st Century Leads to Gridlock In and Around Decaying Cities." 1997. *Washington Spectator* 23(12).

Brown, Lester. 2000. "Overview: The Acceleration of Change." In *Vital Signs: The Environmental Trends That Are Shaping Our Future,* Linda Starke, ed. (pp. 17–29). New York: W. W. Norton.

Butz, Bill. 2005. *The World's Next "Population Problem."* Population Reference Bureau. Available at http://www.prb.org

Catley-Carlson, Margaret, and Judith A. M. Outlaw. 1998. "Poverty and Population Issues: Clarifying the Connections." *Journal of International Affairs* 52(1):233–243.

Cincotta, Richard, Robert Engelman, and Daniele Anastasion. 2003. *The Security Demographic: Population and Civil Conflict After the Cold War.* Washington, DC: Population Action International.

Clark, David. 1998. "Interdependent Urbanization in an Urban World: An Historical Overview." *Geographical Journal* 164(1):85–96.

Cowherd, Phil. 2001. "What Is the Business Case for Investing in Inner-City Neighborhoods?" *Public Management* 83(1):12–14.

DaVanzo, Julie, David M. Adamson, Nancy Belden, and Sally Patterson. 2000. *How Americans View World Population Issues: A Survey of Public Opinion.* Santa Monica, CA: Rand Corporation.

DeNavas-Walt, Carmen, Bernadette D. Proctor, and Robert J. Mills. 2004. *Income, Poverty, and Health Insurance Coverage in the United States: 2003.* U.S. Census Bureau, Current Population Reports P60-226. Washington, DC: U.S. Government Printing Office.

Douthwaite, M., and P. Ward. 2005. "Increasing Contraceptive Use in Rural Pakistan: An Evaluation of the Lady Health Worker Programme." *Health Policy and Planning* 20(2):117–123.

Durning, Alan. 1996. *The City and the Car.* Northwest Environment Watch. Seattle: Sasquatch Books.

Egan, Timothy. 2005. "Vibrant Cities Find One Thing Missing: Children." *New York Times,* March 24. Available at http://www.nytimes.com

Ewing, Reid, Tom Schmid, Richard Killingsworth, Amy Zlot, and Stephen Raudenbush. 2003. "Relationship Between Urban Sprawl and Physical Activity, Obesity, and Morbidity." *American Journal of Health Promotion* 18(1):47–57.

Fisher, Claude. 1982. *To Dwell Among Friends: Personal Networks in Town and City.* Chicago: University of Chicago Press.

Froehlich, Maryann. 1998. "Smart Growth: Why Local Governments Are Taking a New Approach to Managing Growth in Their Communities." *Public Management* 80(5):5–9.

Galster, George, Diane Levy, Noah Sawyer, Kenneth Temkin, and Christopher Walker. 2005 (June 30). *The Impact of Community Development Corporations on Urban Neighborhoods.* Urban Institute. Available at http://www.urban.org

Gans, Herbert. 1984. *The Urban Villagers,* 2nd ed. New York: Free Press.

Gardner, Gary. 1999. "Cities Turning to Bicycles to Cut Costs, Pollution, and Crime." *Public Management* 81(1):23.

Gardner, Gary. 2003. "Bicycle Production Seesaws." In *Vital Signs,* M. Renner and M. O. Sheehan, eds. (pp. 58–59). New York: W. W. Norton.

Gardner, Gary. 2005. "Bicycle Production Recovers." In *Vital Signs,* L. Starke, ed. (pp. 58–59). New York: W. W. Norton.

Goodrich, Catherine. 2004. "Women's Empowerment, Population, and Development: An Update on the Cairo Conference." *The Reporter,* winter, pp. 1–15. Population Connection. Available at http://www.popconnect.org

Jensen, Derrick. 2001. "Road to Ruin: An Interview with Jan Lundberg." *The Sun* 302:4–13.

Johnson, William C. 1997. *Urban Planning and Politics.* Chicago: American Planning Association and Planners Press.

Kimuna, Sitawa R., and Donald J. Adamchak. 2001. "Gender Relations: Husband-Wife Fertility and Family Planning Decisions in Kenya." *Journal of Biosocial Science* 33:13–23.

Lalasz, Robert. 2005 (July). *Baby Bonus Credited with Boosting Australia's Fertility Rate.* Population Reference Bureau. Available at http://www.prb.org

Leahy, Elizabeth. 2003. "As Contraceptive Use Rises, Abortions Decline." *Popline,* November-December, pp. 3 and 8.

Lindstrom, Matthew J., and Hugh Bartling. 2003. "Introduction." In *Suburban Sprawl: Culture, Theory, and Politics,* M. J. Lindstrom and H. Bartling, eds. (pp. xi–xxvii). Lanham, MD: Rowman & Littlefield.

Livernash, Robert, and Eric Rodenburg. 1998. "Population Change, Resources, and the Environment." *Population Bulletin* 53(1):1–36.

Millennium Ecosystem Assessment. 2005. *Ecosystems and Human Well-Being: Synthesis.* Washington, DC: Island Press.

National Research Council. 2003. *Cities Transformed: Demographic Change and Its Implications in the Developing World —Panel on Urban Population Dynamics,* Mark R. Montgomery, Richard Stren, Barney Cohen, and Holly Reed, eds. Committee on Population. Washington, DC: National Academy Press.

Nelson, Mara. 2004. "Silence and Myths: A Response to the 'Birth Dearth.'" *The Reporter,* winter, pp. 17–21 and 24. Population Connection. Available at http://www.popconnect.org

Orfield, Myron. 1997. *Metropolitics: A Regional Agenda for Community and Stability.* Washington, DC: Brookings Institution Press, and Cambridge, MA: Lincoln Institute of Land Policy.

Pelley, Janet. 1999. "Building Smart-Growth Communities." *Environmental Science and Technology News* 33(1):28A–32A.

Population Reference Bureau. 2004a. *Human Population: Fundamentals of Growth Patterns of World Urbanization.* Available at http://www.prb.org

Population Reference Bureau. 2004b. "Transitions in World Population." *Population Bulletin* 59(1) (entire issue).

Population Reference Bureau. 2005. *2005 World Population Data Sheet.* Washington, DC: Population Reference Bureau. Available at http://www.prb.org

Rees, Amanda. 2003. "New Urbanism: Visionary Landscapes in the Twenty-First Century." In *Suburban Sprawl: Culture, Theory, and Politics,* M. J. Lindstrom and H. Bartling, eds. (pp. 93–114). Lanham, MD: Rowman & Littlefield.

Rodriquez, Luis. 2000. "Urban Renewal: The Resurrection of an Ex-Gang Member" (in an interview with Derrick Jensen). *The Sun*, April, pp. 4–13.

Schrank, David, and Tim Lomax. 2005 (May). *The 2005 Urban Mobility Report.* Texas Transportation Institute and Texas A&M University System. Available at http://mobility.tamu.edu

Schueller, Jane. 2005. "Boys and Changing Gender Roles." *YouthNet.* YouthLens 16 (August). Available at http://www.fhi.org.

Sheehan, M. O. 2001. "Making Better Transportation Choices." In *State of the World 2001,* L. Starke, ed. (pp. 103–122). New York: W. W. Norton.

Sheehan, M. O. 2003. "Uniting Divided Cities." In *State of the World 2003,* L. Starke, ed. (pp. 130–151). New York: W. W. Norton.

Shevis, Jim. 1999. "More Affluent Than Their Inner-City Neighbors, Suburbanites Still Have Growth Problems." *Washington Spectator* 25(19):1–3.

Smart Growth Network. 2004. *About Smart Growth.* Available at http://www.smartgrowth.org

Tittle, Charles. 1989. "Influences on Urbanism: A Test of Predictions from Three Perspectives." *Social Problems* 36(3):270–288.

UNEP (United Nations Environment Programme). 2005. *GEO Yearbook: An Overview of Our Changing Environment 2004/5.* Available at http://www.unep.org

Union of Concerned Scientists. 2003. *Cars and Trucks and Air Pollution.* Available at http://www.ucsusa.org

United Nations. 2005a. *The Millennium Development Goals Report.* New York: United Nations.

United Nations. 2005b. "World Population Prospects: The 2004 Revision." *Population Newsletter* 79.

United Nations Population Division. 2004. *World Urbanization Prospects: The 2003 Revision.* New York: United Nations.

United Nations Population Fund. 1997. *1997 State of the World Population.* New York: United Nations.

United Nations Population Fund. 1999. *The State of World Population 1999.* New York: United Nations.

United Nations Population Fund. 2004. *The State of World Population 2004.* New York: United Nations

U.S. Conference of Mayors. 2001. *Traffic Congestion and Rail Investment.* Washington, DC: Global Strategy Group.

U.S. Department of Housing and Urban Development. 2004. "Revitalizing Neighborhoods by Redeveloping Brownfields: The HUD Brownfields Economic Development Initiative." *Research Works* 1(3):2.

Warren, Roxanne. 1998. *The Urban Oasis: Guideways and Greenways in the Human Environment.* New York: McGraw Hill.

Weeks, John R. 2005. *Population: An Introduction to Concepts and Issues,* 9th ed. Belmont, CA: Wadsworth/Thomson.

Weiland, Katherine. 2005. *Breeding Insecurity: Global Security Implications of Rapid Population Growth.* Washington, DC: Population Institute.

Wirth, Louis. 1938. "Urbanism as a Way of Life." *American Journal of Sociology* 44:8–20.

Wiseman, Paul. 2005. "Towns Hope Cash-for-Babies Incentives Boost Populations." *USA Today,* July 28. Available at http://www.usatoday.com

Wolff, Kurt H. 1978. *The Sociology of George Simmel.* Toronto: Free Press.

Women's Studies Project. 2003. *Women's Voices, Women's Lives: The Impact of Family Planning.* Family Health International. Available at http://www.fhi.org

World Health Organization. 2003. *World Health Report 2003: Shaping the Future.* Geneva: World Health Organization.

World Watch Institute. 2004 (January 7). *State of the World 2004: Consumption by the Numbers.* Press release. Available at http://www.worldwatch.org

Zabin, L. S., and K. Kiragu. 1998. "The Health Consequences of Adolescent Sexual and Fertility Behavior in Sub-Saharan Africa." *Studies in Family Planning* 2(June 29):210–232.

Chapter 14

Aeck, Molly. 2005. "Biofuel Use Growing Rapidly." In *Vital Signs 2005,* Linda Starke, ed. (pp. 38–39). New York: W. W. Norton.

American Lung Association. 2005. *Lung Disease Data in Culturally Diverse Communities: 2005.* Available at http://www.lungusa.org

Aslam, Abid. 2005 (October 9). "Planet Faces Nightmare Forecasts as Chinese Consumption Grows and Grows." *Common Dreams News Center.* Available at http://www.commondreams.org

Associated Press. 2005 (April 13). "Wal-Mart Pledges $35 Million for Wildlife." MSNBC.com. Available at http://www.msnbc.msn.com.

Ayres, Ed. 2004. "The Hidden Shame of the Global Industrial Economy." *World Watch Magazine,* January-February, pp. 18–29.

Baillie, J. E. M., C. Hilton-Taylor, and S. N. Stuart, eds. 2004. *2004 IUCN Red List of Threatened Species: A Global Species Assessment.* Cambridge, England: International Union for Conservation of Nature and Natural Resources.

Blatt, Harvey. 2005. *America's Environmental Report Card: Are We Making the Grade?* Cambridge, MA: MIT Press.

Bohan, Caren. 2005 (August 23). "The Illulissat Glacier, a Wonder of the World Melting Away." *Common Dreams News Center.* Available at http://www.commondreams.org

Bright, Chris. 2003. "A History of Our Future." In *State of the World 2003,* Linda Starke, ed. (pp. 3–13). New York: W. W. Norton.

Brown, Lester R. 2003. *Plan B: Rescuing a Planet Under Stress and a Civilization in Trouble.* New York: W. W. Norton.

Brown, Lester R., and Jennifer Mitchell. 1998. "Building a New Economy." In *State of the World 1998,* Lester R. Brown, Christopher Flavin, and Hilary French, eds. (pp. 168–187). New York: W. W. Norton.

Bruce, Nigel, Rogelio Perez-Padilla, and Rachel Albalak. 2000. "Indoor Air Pollution in Developing Countries: A Major Environmental and Public Health Challenge." *Bulletin of the World Health Organization* 78(9):1078–1092.

Bruno, Kenny, and Joshua Karliner. 2002. *Earthsummit.biz: The Corporate Takeover of Sustainable Development.* CorpWatch and Food First Books. Available at http://www.corpwatch.org

"Bulgarian Green Leader Threatened with Death." 2005 (March 8). *Common Dreams News Center.* Available at http://www.commondreams.org

Bullard, Robert D. 2000. *Dumping in Dixie: Race, Class, and Environmental Quality,* 3rd ed. Boulder, CO: Westview Press.

Bullard, Robert D., and Glenn S. Johnson. 1997. "Just Transportation." In *Just Transportation: Dismantling Race and Class Barriers to Mobility,* Robert D. Bullard and Glenn S. Johnson, eds. (pp. 1–21). Stony Creek, CT: New Society.

Cappiello, Dina. 2003. "Corporations Co-opt Earth Day." *Houston Chronicle,* April 22. Available at http://www.houstonchronicle.com

Caress, Stanley, Anne C. Steinemann, and Caitlin Waddick. 2002. "Symptomatology and Etiology of Multiple Chemical Sensitivities in the Southeastern United States." *Archives of Environmental Health* 57(5):429–436.

Centers for Disease Control and Prevention, National Center for Environmental Health. 2003. *Second National Report on Human Exposure to Environmental Chemicals.* Atlanta, GA: Centers for Disease Control and Prevention.

Centers for Disease Control and Prevention, National Center for Environmental Health. 2005. *Third National Report on Human Exposure to Environmental Chemicals.* Atlanta, GA: Centers for Disease Control and Prevention.

Chafe, Zoe. 2005. "Bioinvasions." In *State of the World 2005,* L. Starke, ed. (pp. 60–61). New York: W. W. Norton.

Chivian, Eric, and Aaron S. Bernstein. 2004. "Embedded in Nature: Human Health and Biodiversity." *Environmental Health Perspectives,* 112(1). Available at http://ehp.niehs.nih.gov

Cincotta, Richard P., and Robert Engelman. 2000. *Human Population and the Future of Biological Diversity.* Washington, DC: Population Action International.

Clarke, Tony. 2002. "Twilight of the Corporation." In *Social Problems, Annual Editions 02/03,* 30th ed., Kurt Finster-Busch, ed. (pp. 41–45). Guilford, CT: McGraw-Hill/Dushkin.

Clarren, Rebecca. 2005. "Dirty Politics, Foul Air." *The Nation,* March 14, pp. 6 and 8.

Clayton, Mark. 2005. "States Take on Feds Over Environment." *Christian Science Monitor,* October 6. Available at http://www.csmonitor.com

Computer Take Back Campaign. 2005. *Fifth Annual Computer Report Card.* Available at http://www.computertakeback.com

Cooper, Arnie. 2004 (September). "Twenty-Eight Words That Could Change the World: Robert Hinkley's Plan to Tame Corporate Power." *The Sun* 345:4–11.

Denson, Bryan. 2000. "Shadowy Saboteurs." *IRE Journal* 23(May-June):12–14.

DesJardins, Andrea. 1997. *Sweet Poison: What Your Nose Can't Tell You About the Dangers of Perfume.* Available at http://members.aol.com/enviroknow/perfume/sweet-poison.htm

Edwards, Bob, and Anthony Ladd. 2000. "Environmental Justice, Swine Production, and Farm Loss in North Carolina." *Sociological Spectrum* 20(3):263–290.

Energy Information Administration. 2005. *International Energy Annual 2003.* U.S. Department of Energy. Available at http://www.eia.doe.gov

Environmental Defense. 2004. *2004 Annual Report.* Available at http://www.environmentaldefense.org

Environmental Working Group. 2005 (July 14). *Body Burden: The Pollution in Newborns.* Available at http://www.ewg.org

EPA (U.S. Environmental Protection Agency). 2004a. (August 8). *EPA Releases 12th Annual National Listing of Fish Advisories.* Available at http://www.epa.gov

EPA. 2004b. *What You Need to Know About Mercury in Fish and Shellfish.* Available at http://www.epa.gov

Faber, Daniel. 2005. "Building a Transnational Environmental Justice Movement: Obstacles and Opportunities in the Age of Globalization." In *Coalitions Across Borders,* Joe Bandy and Jackie Smith, eds. (pp. 43–68). Lanham, MA: Rowman & Littlefield.

Fisher, Brandy E. 1998. "Scents and Sensitivity." *Environmental Health Perspectives* 106(12). Available at http://ehpnet1.niehs.nih.gov/docs/1998/106-12/focus-abs.html

Fisher, Brandy E. 1999. "Focus: Most Unwanted." *Environmental Health Perspectives* 107(1). Available at http://ehpnet1.niehs.nih.gov/docs/1999/107-1/focus-abs.html

Flavin, Chris, and Molly Hull Aeck. 2005 (September 15). "Cleaner, Greener, and Richer." *Tom.Paine.com.* Available at http://wu.oneworld.net

Food and Drug Administration. 2005 (May). *Pesticide Program Residue Monitoring 2003.* Available at http://www.cfsan.fda.gov

Fornos, Werner. 2005. "*Homo Sapiens:* An Endangered Species?" *Popline* 27:4.

French, Hilary. 2000. *Vanishing Borders: Protecting the Planet in the Age of Globalization.* New York: W. W. Norton.

Gaier, Rodrigo. 2005 (February 24). "Brazil Environmentalist Shot in Rain Forest." *Common Dreams News Center.* Available at http://www.commondreams.org

Gallup Organization. 2005. *Energy.* The Gallup Brain. Available at http://www.gallup.org

Gardner, Gary. 2005. "Forest Loss Continues." In *Vital Signs 2005,* Linda Starke, ed. (pp. 92–93). New York: W. W. Norton.

Gartner, John. 2005 (September 13). "Sierra Club Gets Behind the Wheel." *Wired News.* Available at http://www.wired.com/news

Goodstein, Laurie. 2005. "Evangelical Leaders Swing Influence Behind Effort to Combat Global Warming." *New York Times,* March 10. Available at http://www.nytimes.com

Gottlieb, Roger S. 2003a. "Saving the World: Religion and Politics in the Environmental Movement." In *Liberating Faith,* Roger S. Gottlieb, ed. (pp. 491–512). Lanham, MD: Rowman & Littlefield.

Gottlieb, Roger S. 2003b. "This Sacred Earth: Religion and Environmentalism." In *Liberating Faith,* Roger S. Gottlieb, ed. (pp. 489–490). Lanham, MD: Rowman & Littlefield.

Hager, Nicky, and Bob Burton. 2000. *Secrets and Lies: The Anatomy of an Anti-Environmental PR Campaign.* Monroe, ME: Common Courage Press.

Hunter, Lori M. 2001. *The Environmental Implications of Population Dynamics.* Santa Monica, CA: Rand Corporation.

Intergovernmental Panel on Climate Change. 2000. *Land Use, Land-Use Change, and Forestry.* Available at http://www.ipcc.ch

Intergovernmental Panel on Climate Change. 2001a. *Climate Change 2001: Impacts, Adaptation, and Vulnerability.* United Nations Environmental Programme and the World Meteorological Organization. Available at http://www.ipcc.ch

Intergovernmental Panel on Climate Change. 2001b. *Climate Change 2001: The Scientific Basis.* United Nations Environmental Programme and the World Meteorological Organization. Available at http://www.ipcc.ch

International Fund for Animal Welfare. 2005. *Caught in the Web: Wildlife Trade on the Internet.* Available at http://www.ifaw.org

Jan, George P. 1995. "Environmental Protection in China." In *Environmental Policies in the Third World: A Comparative Analysis,* O. P. Dwivedi and Dhirendra K. Vajpeyi, eds. (pp. 71–84). Westport, CT: Greenwood Press.

Janofsky, Michael. 2005. "Pentagon Is Asking Congress to Loosen Environmental Laws." *New York Times,* May 11. Available at http://www.nytimes.com

Jarboe, James F. 2002 (February 12). *The Threat of Eco-Terrorism.* Congressional Statement. Federal Bureau of Investigation. Available at http://www.fbi.gov

Jensen, Derrick. 2001. "A Weakened World Cannot Forgive Us: An Interview with Kathleen Dean Moore." *The Sun* 303 (March):4–13.

Johnson, Geoffrey. 2004 (April). *Don't Be Fooled: America's Ten Worst Greenwashers.* The Green Life. Available at http://www.thegreenlife.org

Kaplan, Sheila, and Jim Morris. 2000. "Kids at Risk." *U.S. News and World Report,* June 19, pp. 47–53.

Koenig, Dieter. 1995. "Sustainable Development: Linking Global Environmental Change to Technology Cooperation." In *Environmental Policies in the Third World: A Comparative Analysis,* O. P. Dwivedi, and Dhirendra K. Vajpeyi, eds. (pp. 1–21). Westport, CT: Greenwood Press.

Lenssen, Nicholas. 2005. "Nuclear Power Rises Once More." In *Vital Signs 2005,* Linda Starke, ed. (pp. 32–33). New York. W. W. Norton.

Leutwyler, Kristin. 2001. "The Poor Face More Environmental Hazards." *Scientific American,* January 8. Available at http://www.sciam.com/news

Levin, Phillip S., and Donald A. Levin. 2002. "The Real Biodiversity Crisis." *American Scientist* 90(1):6.

Little, Amanda Griscom. 2005. "Maathai on the Prize: An Interview with Nobel Peace Prize Winner Wangari Maathai." *Grist Magazine,* February 15. Available at http://www.grist.org

Lyons, Linda. 2005 (April 19). *Daily Concerns Overshadow Environment Worries.* Gallup Organization. Available at http://www.gallup.com

MacDonald, Mia, and Danielle Nierenberg. 2003. "Linking Population, Women, and Biodiversity." In *State of the World 2003,* Linda Starke, ed. (pp. 38–61). New York: W. W. Norton.

Makower, Joel, Ron Pernick, and Clint Wilder. 2005. *Clean Energy Trends 2005.* Clean Edge. Available at http://www.cleanedge.com

Margesson, Rhoda. 2005. "Environmental Refugees." In *State of the World 2005,* Linda Starke, ed. (pp. 40–41). New York: W. W. Norton.

Massachusetts Department of Environmental Protection. 2001. *TV and Computer Reuse and Recycling.* Available at http://www.state.ma.us/dep/recycle/crt/crthome.htm

Mastny, Lisa. 2003. "State of the World: A Year in Review." In *State of the World 2003,* Linda Starke, ed. (pp. xix–xxiii). New York: W. W. Norton.

McDaniel, Carl N. 2005. *Wisdom for a Livable Planet.* San Antonio: Trinity University Press.

McGinn, Anne Platt. 2000. "Endocrine Disrupters Raise Concern." In *Vital Signs 2000,* Lester R. Brown, Michael Renner, and Brian Halweil, eds. (pp. 130–131). New York: W. W. Norton.

McMichael, Anthony J., Kirk R. Smith, and Carlos F. Corvalan. 2000. "The Sustainability Transition: A New Challenge." *Bulletin of the World Health Organization* 78(9):1067.

Mead, Leila. 1998. "Radioactive Wastelands." *The Green Guide* 53(April 14):1–3.

Millennium Ecosystem Assessment. 2005. *Ecosystems and Human Well-Being: Synthesis.* Washington, DC: Island Press.

Mishra, Vinod, Robert D. Retherford, and Kirk R. Smith. 2002. "Indoor Air Pollution: The Quiet Killer." *Asia Pacific Issues* 63:1–8.

Mooney, Chris. 2005. "Some Like It Hot." *Mother Jones,* May-June. Available at http://www.MotherJones.com

Motavalli, Jim. 2001. "Is There an Afterlife for Your Computer? Grappling with America's Techno-Trash Dilemma." *Environmental Defense* 32(2):6.

National Assessment Synthesis Team. 2000. *Climate Change Impacts on the United States: The Potential Consequences of Climate Variability and Change.* Washington, DC: U.S. Global Change Research Program.

National Environmental Education and Training Foundation, and Roper Starch Worldwide. 1999. *1999 NEETF/Roper Report Card.* Washington, DC: National Environmental Education and Training Foundation.

National Oceanic and Atmospheric Administration. 2005 (January 13). *Climate of 2004 Annual Review.* Available at http://www.ncdc.noaa.gov

Niihuis, Michelle. 2003. "Coal-Miner's Slaughter." *Grist Magazine,* April 14. Available at http://www.gristmagazine.com

Northwest Product Stewardship Council (Computer Subcommittee). 2000 (October). *A Guide to Environmentally Preferable Computer Purchasing.* Available at http://www.govlink.org/nwpsc

"Nukes Rebuked." 2000 (July 1). *Washington Spectator* 26(13):4.

PBS. 2001. *Trade Secrets: A Moyers Report.* Available at http://www.pbs.org/trade secrets/program/program.html

Pearce, Fred. 2005. "Climate Warming as Siberia Melts." *New Scientist,* August 11. Available at http://www.NewScientist.com

Pimentel, David, Maria Tort, Linda D'Anna, Anne Krawic, Joshua Berger, Jessica Rossman, Fridah Mugo, Nancy Don, Michael Shriberg, Erica Howard, Susan Lee, and Jonathan Talbot. 1998. "Ecology of Increasing Disease: Population Growth and Environmental Degradation." *BioScience* 48(October):817–827.

Pope, Carl. 2005 (July 7). *America Needs Real Energy Solutions.* Sierra Club press release. Available at http://www.sierraclub.org

"A Prescription for Reducing the Damage Caused by Dams." 2001 (March). *Environmental Defense* 32(2).

Public Citizen. 2005 (February). *NAFTA Chapter 11 Investor-State Cases: Lessons for the Central America Free Trade Agreement.* Public Citizens Global Trade Watch Publication E9014. Available at http://www.citizen.org

Reel, Monte. 2005. "Murder Galvanizes Nun's Cause." *Washington Post,* February 21, p. A1.

Reese, April. 2001 (February). *Africa's Struggle with Desertification.* Population Reference Bureau. Available at http://www.prb.org/regions/africa/africadesertification.html

Renner, Michael. 1996. *Fighting for Survival: Environmental Decline, Social Conflict, and the New Age of Insecurity.* New York: W. W. Norton.

Renner, Michael. 2004. "Moving Toward a Less Consumptive Economy." In *State of the World 2004,* Linda Starke, ed. (pp. 96–119). New York: W. W. Norton.

Reuters. 2005 (August 23). *Antarctic Ozone Hole Grows from Last Year: WMO.* Available at http://today.reuters.com

Rifkin, Jeremy. 2004. *The European Dream: How Europe's Vision of the Future Is Quietly Eclipsing the American Dream.* New York: Tarcher/Penguin.

Rogers, Sherry A. 2002. *Detoxify or Die.* Sarasota, FL: Sand Key.

Saad, Lydia. 2002. *Americans Sharply Divided on Seriousness of Global Warming.* Gallup News Service. Available at http://www.gallup.org

Sampat, Payal. 2003. "Scrapping Mining Dependence." In *State of the World 2003,* Linda Starke, ed. (pp. 110–129). New York: W. W. Norton.

Sawin, Janet. 2004. "Making Better Energy Choices." In *State of the World 2004,* Linda Starke, ed. (pp. 24–43). New York: W. W. Norton.

Sawin, Janet. L. 2005a. "Global Wind Growth Continues." In *Vital Signs 2005,* Linda Starke, ed. (pp. 34–35). New York: W. W. Norton.

Sawin, Janet L. 2005b. "Solar Energy Markets Booming." In *Vital Signs 2005,* Linda Starke, ed. (pp. 36–37). New York: W. W. Norton.

Sax, L. J., S. Hurtado, J. A. Lindholm, A. W. Astin, W. S. Korn, and K. M. Mahoney. 2004. *The American Freshman: National Norms for Fall 2004.* Los Angeles: Higher Education Research Institute, UCLA.

Schaeffer, Robert K. 2003. *Understanding Globalization: The Social Consequences of Political, Economic, and Environmental Change,* 2nd ed. Lanham, MD: Rowman & Littlefield.

Schapiro, Mark. 2004. "New Power for 'Old Europe.'" *The Nation,* December 27, pp. 11–16.

Schmidt, Charles W. 2004. "Environmental Crimes: Profiting at the Earth's Expense." *Environmental Health Perspectives* 112(2):A96–A103.

Silicon Valley Toxics Coalition. 2001. *Just Say No to E-Waste: Background Document on Hazards and Waste from Computers.* Available at http://www.svtc.org

Silicon Valley Toxics Coalition. 2002. *Exporting Harm: The High-Tech Trashing of Asia.* Available at http://www.svtc.org

Silicon Valley Toxics Coalition. 2005 (March 7). *Toxic Sentence: E-Waste, Prisons, and Environmental Justice.* Available at http://www.svtc.org

Smith, Velma, John Coequyt, and Richard Wiles. 2000. *Clean Water Report Card.* Washington, DC: Environmental Working Group.

Sterritt, Angela. 2005. "Indigenous Youth Challenge Corporate Mining." Cultural Survival, *Weekly Indigenous News,* July 15. Available at http://209.200.101.189/home.cfm

Stoll, Michael. 2005 (September-October). "A Green Agenda for Cities." *E Magazine* 16(5). Available at http://www.emagazine.com

Stretesky, Paul. 2003. "The Distribution of Air Lead Levels Across U.S. Counties: Implications for the Production of Racial Inequality." *Sociological Spectrum* 23:91–118.

Thibodeau, Patrick. 2002. "Handling E-Waste." *Computerworld* 36(47):46.

UNDP (United Nations Development Programme). 2005. *Human Development Report 2005.* Available at http://hdr.undp.org

UNEP (United Nations Environment Programme). 2002. *North America's Environment: A Thirty-Year State of the Environment and Policy Retrospectives.* Available at http://www.na.unep.net

UNEP. 2003 (August 3). *Basic Facts and Data on the Science and Politics of Ozone Protection.* Available at http://www.unep.org/ozone

UNEP. 2005. *GEO Yearbook 2004/5.* Available at http://www.unep.org

United Nations Population Fund. 2004. *The State of World Population 2004.* New York: United Nations.

U.S. Bureau of the Census. 2004. *Statistical Abstract of the United States: 2004–2005,* 124th ed. Washington, DC: U.S. Government Printing Office.

U.S. Department of Energy. 2005. *Alternative Fuel Vehicles.* Available at http://www.eere.energy.gov/afdc

U.S. Department of Health and Human Services. 2004. *11th Report on Carcinogens.* Washington, DC: Public Health Service.

Vajpeyi, Dhirendra K. 1995. "External Factors Influencing Environmental Policymaking: Role of Multilateral Development Aid Agencies." In *Environmental Policies in the Third World: A Comparative Analysis,* O. P. Dwivedi and Dhirendra K. Vajpeyi, eds. (pp. 24–45). Westport, CT: Greenwood Press.

Vartan, Starre. 2005. "Money Matters: Buying Cleaner Energy." *E Magazine* 16(5). Available at http://www.emagazine.com

Vedantam, Shankar. 2005a. "Nuclear Plants Not Keeping Track of Waste." *Washington Post,* April 19. Available at http://www.washingtonpost.com

Vedantam, Shankar. 2005b. "Storage Plan Approved for Nuclear Waste." *Washington Post,* September 10. Available at http://www.washingtonpost.com

Wald, Matthew L. 2005. "Casks Gain Favor as Method for Storing Nuclear Waste." *New York Times,* June 5. Available at http://www.nytimes.com

Wal-Mart Watch. 2005 (April 6). *Wal-Mart Watch.* Press release. Available at http://walmartwatch.com

Warren, Karen J. 2000. *Ecofeminist Philosophy: A Western Perspective on What It Is and Why It Matters.* Lanham, MD: Rowman & Littlefield.

Weiner, Tim. 2001. "Terrific News in Mexico City: Air Is Sometimes Breathable." *New York Times* (online), January 5. Available at http://www.nytimes.com

World Resources Institute. 2000. *World Resources 2000–2001: People and Ecosystems—The Fraying Web of Life.* Washington, DC: World Resources Institute.

Youth, Howard. 2003. "Watching Birds Disappear." In *State of the World 2003,* Linda Starke, ed. (pp. 85–109). New York: W. W. Norton.

Chapter 15

ACLU (American Civil Liberties Union). 2005. *Utah Businesses, Free Speech Groups, and Individuals Challenge Restrictions on Internet Speech.* Privacy and Technology. Available at http://www.aclu.org/Privacy

Addison, John T., Douglas Fox, and Christopher Ruhm. 2000. "Technology, Trade Sensitivity, and Labor Displacement." *Southern Economic Journal* 66:682–699.

AMA (American Medical Association). 2001. "Should Genetic Information Be Treated Separately?" *Genethics,* January. Available at http://www.ama-assn.org/ama/pub

ANOL (A Nation Online). 2004. *A Nation Online: Entering the Broadband Age.* Washington, DC: National Telecommunications and Information Administration.

Associated Press. 2005a. "Italian Scientists Say They Have Cloned a Second Horse." *Globe and Mail,* April 16. Available at http://www/theglobeandmail.com

Associated Press. 2005b. "Judge Allows Microsoft Suit to Proceed." *Washington Post,* June 14. Available at http://www.washingtonpost.com

Attewell, Paul, Belkis Suazo-Garcia, and Juan Battle. 2003. "Computers and Young Children: Social Benefit or Social Problem?" *Social Forces* 82:277–296.

Barbash, Fred. 2005. "High Court Takes Up Abortion Consent Case." *Washington Post,* May 25. Available at http://www.washingtonppost.com

BBC. 2003. "Hi-Tech Workplaces No Better than Factories." *BBC News,* November 27. Available at http://www.bbc.co.uk

Bell, Daniel. 1973. *The Coming of Post-Industrial Society: A Venture in Social Forecasting.* New York: Basic Books.

Beniger, James R. 1993. "The Control Revolution." In *Technology and the Future,* Albert H. Teich, ed. (pp. 40–65). New York: St. Martin's Press.

Boles, Margaret, and Brenda Sunoo. 1998. "Do Your Employees Suffer from Techno-Phobia?" *Workforce* 77(1):21.

Brasher, Philip. 2000. "Government Probes Biotech Corn Allegations." *Excite News.* Available at http://www.news.excite.com/news/ap/000918

Brin, David. 1998. *The Transparent Society: Will Technology Force Us to Choose Between Privacy and Freedom?* Reading, MA: Addison Wesley.

Buchanan, Allen, Dan Brock, Norman Daniels, and Daniel Wikler. 2000. *From Chance to Choice: Genetics and Justice.* New York: Cambridge University Press.

Buechner, Maryanne, Lev Grossman, and Anita Hamilton. 2002. "The Coolest Invention of 2002." *Time,* November 18, pp. 73–81.

Burkholder, Richard. 2005. *Internet Use: Behind the Great Firewall of China.* Available at http://www.gallup.com/poll

Bush, Corlann G. 1993. "Women and the Assessment of Technology." In *Technology and the Future,* Albert H. Teich, ed. (pp. 192–214). New York: St. Martin's Press.

CAW (Center for the Advancement of Women). 2003. *Women in Science, Engineering, and Technology.* Available at http://www.advancewomen.org

Ceruzzi, Paul. 1993. "An Unforeseen Revolution." In *Technology and the Future,* Albert H. Teich, ed. (pp. 160–174). New York: St. Martin's Press.

Chepesiuk, Ron. 1999. "Where the Chips Fall: Environmental Health in the Semiconductor Industry." *Environmental Health Perspectives* 107(9). Available at http://www.junkscience.com/aug99/chips.htm

Clarke, Adele E. 1990. "Controversy and the Development of Reproductive Sciences." *Social Problems* 37(1):18–37.

CNN. 2004 (February 5). *Spam Invasion Targets Mobile Phones..* Available at http://www.cnn.com/2004

CNN. 2005a (June 6). *Info on 3.9 Million Citigroup Customers Lost.* Available at http://www.money.cnn.com

CNN. 2005b (June 15). *Poll: Most Want Congress to Make Sure Internet Safe.* Available at http://cnn.technology

Conrad, Peter. 1997. "Public Eyes and Private Genes: Historical Frames, New Constructions, and Social Problems." *Social Problems* 44:139–154.

D'Agnese, Joseph. 2000. "The 11th Annual Discover Awards." *Discover* 21(7). Available at http://www.discover.com/jul_00featawards.html.

DeNoon, Daniel. 2005 (March 8). *Study: Computer Design Flaws May Create Dangerous Hospital Errors.* Available at http://my.webmd.com

Deutsche Welle. 2003. *German Kids Go to Camp for Internet Addiction.* Available at http://www.dw-world.de/dw/article

Durkheim, Emile. 1973 [1925]. *Moral Education.* New York: Free Press.

Ehrenfeld, David. 1998. "A Techno-Pox upon the Land." *Harper's,* October, pp. 13–17.

Eibert, Mark D. 1998. "Clone Wars." *Reason* 30(2):52–54.

Eilperin, Juliet, and Rick Weiss. 2003. "House Votes to Prohibit All Human Cloning." *Washington Post,* February 28. Available at http://www.washingtonpost.com

FBI (Federal Bureau of Investigation). 2002. *Online Child Pornography: Innocent Images National Initiative.* Available at http://www.fbi.gov

Freudenheim, Milt. 2005. "Digital Rx: Take Two Aspirins and E-mail Me in the Morning." *New York Times,* March 2. Available at http://www.nytimes.com

Gallup. 2000. *Abortion Issues.* Available at http://www.gallup.com/poll/indicators

Gallup. 2003. *Abortion Issues.* Available at http://www.brain.gallup.com

Gallup. 2005. *Abortion.* Available at http://www.brain.gallup.com

Gibbs, Nancy. 2003. "The Secret of Life." *Time,* February 17, pp. 42–45.

Glendinning, Chellis. 1990. *When Technology Wounds: The Human Consequences of Progress.* New York: William Morrow.

Global Internet Project. 1998. *The Workplace.* Available at http://www.gip.org/gip2g.html

Global Reach. 2004. *Global Internet Statistics.* Available at http://global-reacg.biz/globalstats

Goodman, Paul. 1993. "Can Technology Be Humane?" In *Technology and the Future,* Albert H. Teich, ed. (pp. 239–255). New York: St. Martin's Press.

Gottlieb, Scott. 2000. "Abortion Pill Is Approved for Sale in United States." *British Medical Journal* 321:851.

Hamilton, Anita, Maryanne Murray Beuchner, Lev Grossman, Simon Crittle, and Sora Song. 2004. "Coolest Inventions: Zoom Zoom." *Time*, November 29, pp. 6–15.

Hancock, LynNell. 1995. "The Haves and the Have-Nots." *Newsweek*, February 27, pp. 50–53.

Hasson, Judi, and Graeme Browning. 2002. *The Federal Government Has Become One of the Biggest Online Retailers*. Pew Internet and American Life. Available at http://www.pewinternet/reports

Hayday, Graham. 2003. *Technology in the Workplace: More Important Than Managers?* Available at http://news.zdnet.co.uk

Hormats, Robert D. 2001. "Asian Connection." *Across the Board* 38:47–50.

Human Genome Project. 2005a. *Gene Therapy*. Available at http://www.ornl.gov/sci/techresources/Human_Genome/medicine/genetherapy

Human Genome Project. 2005b. *Genetics and Patenting*. Available at http://www.ornl.gov/sci/techresources/Human_Genome/elsi/patents

Human Genome Project. 2005c. *Medicine and the New Genetics*. Available at http://www.ornl.gov/sci/techresources/Human_Genome/medicine

International Labor Organization. 2001. *Digital Divide Is Wide and Getting Wider*. World Employment Report. Geneva: International Labor Organization. Available at http://www.ilo.org/public/english/bureau/inf/pkits/wer2001

Internet Pornography Statistics. 2005. Available at http://www.internet-filter-review.toptenreviews.com

Internet World Stats. 2005. *Internet Usage Statistics: The Big Picture*. Available at http://www.internetworldstats.com

Jones, Caroline. 2004. *Teleworking: The Quiet Revolution*. Gartner Group. Available at http://www.gartner.com

Jordan, Tim, and Paul Taylor. 1998. "A Sociology of Hackers." *Sociological Review*, November, pp. 757–778.

Kahin, Brian. 1993. "Information Technology and Information Infrastructure." In *Empowering Technology: Implementing a U.S. Strategy*, Lewis M. Branscomb, ed. (pp. 135–166). Cambridge, MA: MIT Press.

Kahn, A. 1997. "Clone Mammals . . . Clone Man." *Nature*, March 13, p. 119.

Kalb, Claudia. 2005. "Ethics, Eggs, and Embryos." *Newsweek*, June 20, pp. 52–53.

Kaplan, Carl S. 2000 (December 22). *The Year in Technology Law*. Available at http://www.nytimes.com/2000/12/22/technology/22CYBERLAW

Kilen, Mike. 2003. "She's a Little Fighter." *Des Moines Register*, February 23. Available at http://www.desmoinesregister.com/news

Klein, Matthew. 1998. "From Luxury to Necessity." *American Demographics* 20(8): 8–12.

Klotz, Joseph. 2004. *The Politics of Internet Communication*. Lanham, MD: Rowman and Littlefield.

Krebs, Brian. 2005. "Uncle Sam Gets 'D-Plus' on Cyber-Security." *Washington Post*, February 16. Available at http://www.washingtonpost.com

Krim, Jonathon. 2005. "Net Aids Access to Sensitive ID Data." *Washington Post*, April 3. Available at http://www.washingtonpost.com

Kuhn, Thomas. 1973. *The Structure of Scientific Revolutions*. Chicago: University of Chicago Press.

Lee, Jennifer. 2003. "Identity Theft Complaints Double in 2002." *New York Times*, January 22. Available at http://www.nytimes.com

Lemonick, Michael, and Dick Thompson. 1999. "Racing to Map Our DNA." *Time Daily* 153:1–6. Available at http://www.time.com

Lynch, Colum. 2005. "U.N. Backs Human Cloning Ban." *Washington Post*, March 8. Available at http://www.washingtonpost.com

Macklin, Ruth. 1991. "Artificial Means of Reproduction and Our Understanding of the Family." *Hastings Center Report*, January-February, pp. 5–11.

MacMillan, Robert. 2004. "Primer: Children, the Internet and Pornography." *Washington Post*, June 29. Available at http://www.washingtonpost.com

MacMillan, Robert. 2005. "My Cubicle, My Cell." *Washington Post*, May 19. Available at http://www.washingtonpost.com

March of Dimes. 2003. *What We Know and What We Don't*. Available at http://www.marchofdimes.com/prematurity

Markoff, John. 2000. "Report Questions a Number in Microsoft Trial." *New York Times* (online), August 28. Available at http://www.nytimes.com/library/tech/00/08

Markoff, John. 2002. "Protesting the Big Brother Lens, Little Brother Turns an Eye Blind." *New York Times*, October 7. Available at http://www.nytimes.com

Mayer, Sue. 2002. "Are Gene Patents in the Public Interest?" *BIO-IT World*, November 12. Available at http://www.bio-itworld.com

McCormick, S. J., and A. Richard. 1994. "Blastomere Separation." *Hastings Center Report*, March-April, pp. 14–16.

McDermott, John. 1993. "Technology: The Opiate of the Intellectuals." In *Technology and the Future*, Albert H. Teich, ed. (pp. 89–107). New York: St. Martin's Press.

McFarling, Usha L. 1998. "Bioethicists Warn Human Cloning Will Be Difficult to Stop." *Raleigh News and Observer*, November 18, p. A5.

McGann, Rob. 2005a (January 6). *Most Active Web Users Are Young, Affluent*. Jupiter Research. Available at http://www.clickz.com/stats/sector/demographics

McGann, Rob. 2005b (February 9). *Online Banking Increased 47 Percent Since 2002*. Jupiter Research. Available at http://www.clickz.com/stats/sector/demographics

McNair, James. 2003. "Rising Unemployment Affects Even Tech Sector." *Cincinnati Enquirer*, March 26, p. 1.

Merton, Robert K. 1973. "The Normative Structure of Science." In *The Sociology of Science*, Robert K. Merton, ed. Chicago: University of Chicago Press.

Mesthene, Emmanuel G. 1993. "The Role of Technology in Society." In *Technology and the Future*, Albert H. Teich, ed. (pp. 73–88). New York: St. Martin's Press.

Murphie, Andrew, and John Potts. 2003. *Culture and Technology*. New York: Palgrave Macmillan.

National Science Foundation. 2002. *Science and Engineering Indicators*. Available at http://www.nsf.gov

National Science Foundation. 2004. *Science and Engineering Indicators*. Available at http://www.nsf.gov

NCSL (National Conference of State Legislatures). 2005a. *State Genetic Privacy Laws*. Available at http://ncls.org/programs/health

NCSL. 2005b. *State Human Cloning Laws*. Available at http://ncls.org/programs/health

NIC (National Intelligence Council). 2003. *The Global Technology Revolution*, preface and summary. Rand Corporation. Available at http://www.rand.org

Nie, Norman, Alberto Simpser, Irena Stepanikova, and Lu Zheng. 2004 (December). *Ten Years After the Birth of the Internet: How Do Americans Use the Internet in Their Daily Lives?* Stanford Institute for the Quantitative Study of Society.

1999 Emerging Technology Finalist. 2000. Available at http://www.discover.com/awards/awards_emerging.html

Nowell, Paul. 2005a. "Bank of America Loses Tape with Federal Workers Data." *Washington Post*, February 26. Available at http://www.washingtonpost.com

Nowell, Paul. 2005b. "Banks Notify Customers of Data Theft." *Washington Post*, May 23. Available at http://www.washingtonpost.com

Ogburn, William F. 1957. "Cultural Lag as Theory." *Sociology and Social Research* 41:167–174.

O'Meara, Kelly Patricia. 2003. "Cheap Labor at America's Expense." *Insight on the News*, May 27, p. 32.

Papadakis, Maria. 2000 (March 31). *Complex Picture of Computer Use in Home Emerges*. National Science Foundation NSF000-314.

Perrolle, Judith A. 1990. "Computers and Capitalism." In *Social Problems Today*,

James M. Henslin, ed. (pp. 336–342). Englewood Cliffs, NJ: Prentice-Hall.

Pew. 2003. *Summary of Findings.* Pew Internet and American Life. Available at http://www.pewinternet.org/reports

Pew Research Center. 2005 (May 23). *More See Benefits of Stem Cell Research.* Available at http://people-press.org/commentary

Pollack, Andrew. 2000. "Nations Agree on Safety Rules for Biotech Food." *New York Times,* January 30. Available at http://www.nytimes.com/library/national/science

Postman, Neil. 1992. *Technopoly: The Surrender of Culture to Technology.* New York: Alfred A. Knopf.

Powers, Richard. 1998. "Too Many Breakthroughs." *New York Times,* November 19, p. 35.

Rabino, Isaac. 1998. "The Biotech Future." *American Scientist* 86(2):110–112.

Reuters. 2005. *Report: China's New Bid to Gag Web.* Available at http://www.cnn.technology

Rosenberg, Debra. 2003. "The War over Fetal Rights." *Newsweek,* June 9, pp. 40–44.

Rosenberg, Jim. 1998. "Troubles and Technologies." *Editor and Publisher* 131(6):4.

Sampat, Payal. 2000. "Internet Use Accelerates." In *Vital Signs: The Environmental Trends That Are Shaping Our Future,* Linda Starke, ed. (pp. 94–104). New York: W. W. Norton.

Schaefer, Naomi. 2001. "The Coming Internet Privacy Scrum." *American Enterprise* 12:50–51.

Schneider, Greg. 2004. "Slowdown Forces Many to Wander for Work." *Washington Post,* November 9, p. A01.

Shand, Hope. 1998. "An Owner's Guide." *Mother Jones,* May-June, p. 46.

Sloan, L. A. 1983. "Abortion Attitude Scale." *Journal of Health Education* 14(3).

Tanaka, Jennifer. 2000. "An Extreme Reaction." *Newsweek,* September 25, p. 75.

Taylor, Chris. 2003. "Spam's Big Bang." *Time,* June 16, pp. 50–53.

Tesler, Pearl. 2003. *Universal Robots: The History and Workings of Robots.* Tech Museum of Innovation. Available at http://www.thetech.org/robotics

Toffler, Alvin. 1970. *Future Shock.* New York: Random House

Tumulty, Karen, 2005. "Why Bush's Ban Could Be Reversed." *Time,* May 23, pp. 26–30.

U.S. Bureau of the Census. 2004. *Statistical Abstract of the United States, 2004–2005,* 124th ed. Washington, DC: U.S. Government Printing Office.

U.S. Congress. 2005. *Anti Phishing Act of 2005: Legislation Details.* Available at http://www.congress.org/congressorg/bill

U.S. Department of Commerce. 2004. *A Nation Online.* Washington, DC: Department of Commerce.

U.S. Department of Justice. 1999. *1999 Report on Cyberstalking: A New Challenge for Law Enforcement and Industry.* Washington, DC: Department of Justice.

U.S. Newswire. 2003 (October 2). *Feminists Condemn House Passage of Deceptive Abortion Ban, Urge Activists to March on Washington.* Available at http://www.usnewswire.com

Weinberg, Alvin. 1966. "Can Technology Replace Social Engineering?" *University of Chicago Magazine* 59(October):6–10.

Weiss, Rick. 2003. "First Cloned Horse Announced." *Washington Post,* August 6. Available at http://www.washingtonpost.com

Weiss, Rick. 2005a. "House Leaders Agree to Vote on Relaxing Stem Cell Limits." *Washington Post,* March 24. Available at http://www.washingtonpost.com

Weiss, Rick. 2005b. "Stem Cell Injections Repair Spinal Cord Injuries in Mice." *Washington Post,* September 20. Available at http://www.washingtonpost.com

Welter, Cole H. 1997. "Technological Segregation: A Peek Through the Looking Glass at the Rich and Poor in an Information Age." *Arts Education Policy Review* 99(2):1–6.

Whine, Michael. 1997. "The Far Right on the Internet." In *The Governance of Cyberspace,* Brian D. Loader, ed. (pp. 209–227). London: Routledge.

White House. 2003 (November 5). *President Bush Signs Partial Birth Abortion Ban Act of 2003.* Available at http://www.whitehouse.gov/news/release

Willing, Richard. 2005. "Federal Judge Blocks 'Partial Birth Abortion' Ban." *USA Today,* June 1. Available at http://www.usatoday.com

Winner, Langdon. 1993. "Artifact/Ideas as Political Culture." In *Technology and the Future,* Albert H. Teich, ed. (pp. 283–294). New York: St. Martin's Press.

Chapter 16

American Jewish Committee. 2005. *2004 Annual Survey of American Jewish Opinion.* Available at http://www.ajc.org

Arms Control Association. 2003a (January). *Start II and Its Extension at a Glance.* Available at http://www.armscontrol.org/factsheets

Arms Control Association. 2003b (May). *The Nuclear Proliferation Treaty at a Glance.* Available at http://www.armscontrol.org/factsheets

Associated Press. 2005. *Pentagon Tests Negative for Anthrax.* MSNBC, March 15. Available at http://www.msnbc.msn.com

Baliunas, Sallie. 2004. "Anthrax Is a Serious Threat." In *Biological Warfare,* William Dudley, ed. (pp. 53–58). Farmington Hills, MA: Greenhaven Press.

Barkan, Steven, and Lynne Snowden. 2001. *Collective Violence.* Boston: Allyn and Bacon.

Barnes, Steve. 2004. "No Cameras in Bombing Trial." *New York Times,* January 29, p. 24.

Barry, John, Richard Wolffe, and Evan Thomas. 2005. "War of Nerves." *Time,* July 4, pp. 20–26.

BBC. 2005. *French Say Firm "No" to EU Treaty.* Available at http://news.bbc.co.uk

Bellamy, Carol. 2003. *Children Are War's Greatest Victims.* Organization for Economic Cooperation and Development. Available at http://www.oecdobserver.org

Berrigan, Frida, and William Hartung. 2005. *U.S. Weapons at War: Promoting Freedom or Fueling Conflict?* World Policy Institute Report. Available at http://www.worldpolicy.org/projects

Bjorgo, Tore. 2003. "Root Causes of Terrorism." Paper presented at the International Expert Meeting, June 9–11. Oslo: Norwegian Institute of International Affairs.

Bonn International Center for Conversion. 1998. *Conversion Survey, 1998,* ch. 6. Bonn, Germany: Bonn International Center for Conversion.

Borum, Randy. 2003. "Understanding the Terrorist Mind-Set." *FBI Law Enforcement Bulletin,* July, pp. 7–10.

Brauer, Jurgen. 2003. "On the Economics of Terrorism." *Phi Kappa Phi Forum,* spring, pp. 38–41.

Brinkley, Joel, and David E. Sanger. 2005. "North Koreans Agree to Resume Nuclear Talks." *New York Times,* July 10. Available at http://www.nytimes.com

Carneiro, Robert L. 1994. "War and Peace: Alternating Realities in Human History." In *Studying War: Anthropological Perspectives,* S. P. Reyna and R. E. Downs, eds. (pp. 3–27). Langhorne, PA: Gordon & Breach.

Cauffman, Elizabeth, Shirley Feldman, Jaime Waterman, and Hans Steiner. 1998. "Post-Traumatic Stress Disorder Among Female Juvenile Offenders." *Journal of the American Academy of Child and Adolescent Psychiatry* 37:1209–1217.

Center for Defense Information. 2003. *Military Almanac.* Available at http://www.cdi.org

Center for Public Integrity. 2005. *Military Professional Resources, Inc. Background.* Available at http://www.publicintegrity.org

Cha, Ariana. 2005. "Computers Simulate Terrorism's Extremes." *Washington Post,* July 4, p. A1.

CNN. 2001a (January 31). *Libyan Bomber Sentenced to Life.* Available at http://www.europe.cnn.com/2001/LAW

CNN. 2001b (February 23). *Rape War Crime Verdict Welcomed.* Available at http://www.cnn.com/2001/WORLD/europe

CNN. 2003 (February 26). *Poll: Muslims Call U.S. "Ruthless, Arrogant."* Available at http://www.cnn.com/usnews

Cohen, Ronald. 1986. "War and Peace Proneness in Pre- and Post-industrial States." In *Peace and War: Cross-Cultural Perspectives,* M. L. Foster and R. A. Rubinstein, eds. (pp. 253–267). New Brunswick, NJ: Transaction Books.

Conflict Research Consortium. 2003. *Mediation.* Available at http://www.colorado.edu/conflict/peace

Congressional Research Service. 2005. *USA Patriot Act Sunset.* CRS Report RS21704. Washington, DC: Library of Congress.

Crary, David. 2005. "Army Marriages Dying in the War on Terror." *Salt Lake Tribune,* June 30. Available at http://www.sltrib.com

Davenport, Christian. 2005. "Guard's New Pitch: Fighting Words." *Washington Post,* April 28. Available at http://www.washingtonpost.com

Deen, Thalif. 2000 (September 9). *Inequality Primary Cause of Wars, Says Annan.* Available at http://www.hartford-hwp.com/archives

De Jong, J., W. Scholte, M. Koeter, and A. Hart. 2000. "The Prevalence of Mental Health Problems in Rwandan and Burundese Refugee Camps." *Acta Psychiatrica Scandinavica* 102:171–177.

Dickey, Christopher. 2005. "Iran's Nuclear Lies." *Newsweek,* July 11, pp. 36–40.

Dixon, William J. 1994. "Democracy and the Peaceful Settlement of International Conflict." *American Political Science Review* 88(1):14–32.

Dodds, Paisley. 2005. "Amnesty International Takes Aim at U.S." *Washington Post,* May 25. Available at http://www.washingtonpost.com

Doyle, Michael. 1986. "Liberalism and World Politics." *American Political Science Review* 80(December):1151–1169.

Environmental Media Services. 2002 (October 7). *Environmental Impacts of War.* Available at http://www.ems.org

Feder, Don. 2003. "Islamic Beliefs Led to the Attack on America." In *The Terrorist Attack on America,* Mary E. Williams, ed. (pp. 20–23). Farmington Hills, MA: Greenhaven Press.

"Foreign Conservationists Under Siege." 2000. Editorial. *New York Times,* April 1, p. A14.

Franke, Volker C. 2001. "Generation X and the Military: A Comparison of Attitudes and Values Between West Point Cadets and College Students." *Journal of Political and Military Sociology* 29:92–119.

Funke, Odelia. 1994. "National Security and the Environment." In *Environmental Policy in the 1990s: Toward a New Agenda,* 2nd ed., Norman J. Vig and Michael E. Kraft, eds. (pp. 323–345). Washington, DC: Congressional Quarterly.

Gallup. 2005a. *Military and National Defense.* Available at http://www.gallup.com/poll

Gallup. 2005b. *War on Terrorism.* Available at http://www.brain.gallup.com

Garrett, Laurie. 2001. "The Nightmare of Bioterrorism." *Foreign Affairs* 80:76.

Ghosh, Aparisim. 2005. "Inside the Mind of an Iraqi Suicide Bomber." *Time,* July 4, pp. 22–29.

Gioseffi, Daniela. 1993. "Introduction." In *On Prejudice: A Global Perspective,* Daniela Gioseffi, ed. (pp. xi–l). New York: Anchor Books, Doubleday.

Global Security. 2005. *Liberian Conflict.* Available at http://www.globalsecurity.org

Goure, Don. 2003. "First Casualties? NATO, the U.N." *MSNBC News,* March 20. Available at http://www.msnbc.com/news

Haavisto, Pekka. 2005. "Environmental Impacts of War." In *State of the World: 2005,* Linda Starke, ed. (pp. 158–159). New York: W. W. Norton.

Hayman, Peter, and Douglas Scaturo. 1993. "Psychological Debriefing of Returning Military Personnel: A Protocol for Post-Combat Intervention." *Journal of Social Behavior and Personality* 8(5):117–130.

Hooks, Gregory, and Leonard E. Bloomquist. 1992. "The Legacy of World War II for Regional Growth and Decline: The Effects of Wartime Investments on U.S. Manufacturing, 1947–72." *Social Forces* 71(2):303–337.

Howe, Kevin. 2003 (April 18). *War Games at Navy School.* Available at http://www.montereyherald.com

Huntington Samuel. 1996. *The Clash of Civilizations and the Remaking of World Order.* New York: Simon and Schuster.

Inglesby, Thomas, Donald Henderson, John Bartlett, Michael Archer et al. 1999. "Anthrax as a Biological Weapon: Medical and Public Health Management." *Journal of the American Medical Association* 281:1735–1745.

International Campaign to Ban Landmines. 2005. *What is the Mine Ban Treaty?* Available at http://www.icbl.org

Interpol. 1998. *Frequently Asked Questions About Terrorism.* Available at http://www.kenpubs.co.uk/interpol.com/English/faq

"Iraq: Status Report." 2005. *Time,* January 31, p. 29.

Jones, Jeffrey M. 2005 (June 13). *Nearly 6 in 10 Americans Support Troop Reductions in Iraq.* Gallup Organization. Available at http://www.gallup.com

Kemper, Bob. 2003. "Agency Wages Media Battle." *Chicago Tribune,* April 7. Available at http://www.chicagotribune.com

Klare, Michael. 2001. *Resource Wars: The New Landscape of Global Conflict.* New York: Metropolitan Books.

Klingman, Avigdor, and Zehara Goldstein. 1994. "Adolescents' Response to Unconventional War Threat Prior to the Gulf War." *Death Studies* 18:75–82.

Knickerbocker, Brad. 2002. "Return of the Military-Industrial Complex?" *Christian Science Monitor,* February 13. Available at http://www.csmonitor.com/2002/0213

Kyl, Jon, and Morton Halperin. 1997. "Q: Is the White House's Nuclear-Arms Policy on the Wrong Track?" *Insight on the News* 42:24–28.

Lamont, Beth. 2001. "The New Mandate for UN Peacekeeping." *The Humanist* 61:39–41.

Landmines. 2003. *The Problem: Impact of Landmines.* Available at http://www.landmines.org

Laqueur, Walter. 2003. *No End to War: Terrorism in the 21st Century.* New York: Continuum International.

Laqueur, Walter. 2006. "The Terrorism to Come." In *Annual Editions 05–06,* Kurt Finsterbusch, ed. (pp. 169–176). Dubuque, IA: McGraw-Hill/Dushkin.

Lederer, Edith. 2005. "Annan Lays out Sweeping Changes to U.N." *Associated Press,* May 20. Available at http://www.apnews.myway.com

Li, Xigen, and Ralph Izard. 2003. "Media in a Crisis Situation Involving National Interest: A Content Analysis of Major U.S. Newspapers' and TV Network' Coverage of the 9/11 Tragedy." *Newspaper Research Journal* 24:1–16.

Lifsher, Marc. 1999. "Critics Assail Plans to Dispose of Paint Cans." *Wall Street Journal,* June 16, p. CA1.

Lloyd, John. 1998. "The Dream of Global Justice." *New Statesman* 127:28–30.

Lorch, Donatella, and Preston Mendenhall. 2000. "A War's Hidden Tragedy." *Newsweek.* Available at http://www.msnbc.com/news

Lumpkin, John J. 2005. *U.S. Spending More per Soldier Than Ever.* Associated Press, February 16. Available at http://www.apnews.myway.com

McDonald, Tim. 2002. "U.S.: Cyber Strike Could Earn Military Response." *News Factor Network,* February 14. Available at http://www.newsfactor.com

McElroy, Wendy. 2003. *Iraq War May Kill Feminism as We Know It.* Fox News Channel, March 18. Available at http://www.fox news.com

McKenna, Corey. 2005. "Bill to Create Assistant Secretary for Cybersecurity at DHS Delivered to Full House." *Government Technology,* April 22. Available at http://www.govtech.net

Messmer, Ellen. 2000 (November 20). *U.S. Army Kick-Starts Cyberwar Machine.* Available at http://www.nwfusion.com

Military Professional Resources Inc. 2005. *Our Mission.* Available at http://www.mpri.com

Miller, Susan. 1993. "A Human Horror Story." *Newsweek,* December 27, p. 17

Monbiot, George. 2005. "The Victims of the Tsunami Pay the Price of War on Iraq." *The Guardian,* January 4. Available at http://www.monbiot.com/archives

Morahan, Lawrence. 2003. "Hussein Threatens to Become Symbol of Nationalism in

Iraq." *Townhall,* July 22. Available at http://www.townhall.com/news

Myers-Brown, Karen, Kathleen Walker, and Judith A. Myers-Walls. 2000. "Children's Reactions to International Conflict: A Cross-Cultural Analysis." Paper presented at the National Council of Family Relations, Minneapolis, November 20.

NASP (National Association of School Psychologists). 2003. "Children and Fear of War and Terrorism." Available at http://www.nasponline.org/NEAT/children

National Center for Post Traumatic Stress Disorder. 2003. *What Is Post Traumatic Stress Disorder?* National Center for Post Traumatic Stress Disorder. Available at http://www.ncptsd.org/facts

National Center for Post Traumatic Stress Disorder. 2005. *What Is Posttraumatic Stress Disorder?* National Center for Post Traumatic Stress Disorder. Available at http://www.ncptsd.va.gov

National Priorities Project. 2005. *Better Security for Less Money in the United States.* Available at http://www.nationalpriorities.org

National Security. 2003. *Special Focus: Cyberwarfare.* Available at http://www.tecsoc.org/natsec

Nelson, Murry R. 1999. "An Alternative Medium of Social Education: The 'Horrors of War' Picture Cards." *Social Studies* 88: 100–108.

Office of Management and Budget. 2005. *Department of Homeland Security.* Available at http://www.whitehouse.gov/omb/budget

O'Prey, Kevin P. 1995. *The Arms Export Challenge: Cooperative Approaches to Export Management and Defence Conversion.* Washington, DC: Brookings Institution.

Pape, Robert A. 2005. *Dying to Win. The Strategic Logic of Suicide Bombers.* New York: Random House.

Paul, Annie Murphy. 1998. "Psychology's Own Peace Corps." *Psychology Today* 31:56–60.

PBS (Public Broadcasting System). 2003. *FAQ: Cybersecurity.* Available at http://www.pbs.org/wgbh

Perera, David. 2005. "Bush Backs Boost for Cybersecurity." *Federal Computer Week,* February 7. Available at http://www.fcw.com

Perl, Raphael. 2003. *Terrorism, the Future, and U.S. Policy.* Congressional Research Service. Washington, DC: Library of Congress.

Pew Research Center. 2005 (July 13). *Guantanamo Prisoner Mistreatment Seen as an Isolated Incident.* Available at http://pewtrust.org

Pfefferbaum, Betty. 1997. "Post-Traumatic Stress Disorder in Children: A Review of the Last Ten Years." *Journal of the American Academy of Child and Adolescent Psychiatry* 36:1503–1512.

Porter, Bruce D. 1994. *War and the Rise of the State: The Military Foundations of Modern Politics.* New York: Free Press.

Powell, Bill, and Tim McGirk. 2005. "The Man Who Sold the Bomb." *Time,* February 14, pp. 22–31.

Priest, Dana, and Spenser Hsu. 2005. "U.S. Sees Drop in Terrorist Threats." *Washington Post,* April 30. Available at http://www.washingtonpost.com.

Renner, Michael. 1993. "Environmental Dimensions of Disarmament and Conversion." In *Real Security: Converting the Defense Economy and Building Peace,* Karl Cassady and Gregory A. Bischak, eds. (pp. 88–132). Albany: State University of New York Press.

Renner, Michael. 2000. "Number of Wars on Upswing." In *Vital Signs: The Environmental Trends That Are Shaping Our Future,* Linda Starke, ed. (pp. 110–111). New York: W. W. Norton.

Renner, Michael. 2005. "Disarming Postwar Societies." In *State of the World: 2005,* Linda Starke, ed. (pp. 122–123). New York: W. W. Norton.

Reuters. 2001 (November 26). *Death Toll Climbs Despite Lull in Sri Lanka Conflict.* Available at http://www.in.news.yahoo.com.

Romero, Anthony. 2003. "Civil Liberties Should Not Be Restricted During Wartime." In *The Terrorist Attack on America,* Mary Williams, ed. (pp. 27–34). Farmington Hills, MA: Greenhaven Press.

Rosenberg, Tina. 2000. "The Unbearable Memories of a U.N. Peacekeeper." *New York Times,* October 8, pp. 4 and 14.

Salladay, Robert. 2003 (April 7). *Anti-War Patriots Find They Need to Reclaim Words, Symbols, Even U.S. Flag from Conservatives.* Available at http://www.commondreams.org

Save the Children. 2003. "Report: Women, Children Bear Brunt of War." *USA Today,* May 6. Available at http://www.usatoday.com

Save the Children. 2005. *One World, One Wish: The Campaign to Help Children and Women Affected by War.* Available at http://www.savethechildren.org

Scheff, Thomas. 1994. *Bloody Revenge.* Boulder, CO: Westview Press.

Schmitt, Eric. 2005. "Pentagon Seeks to Shut Down Bases Across Nation." *New York Times,* May 14. Available at http://www.nytimes.com

Sengupta, Somini. 2003. "Taylor Steps Down as President of Liberia." *New York Times,* August 11. Available at http://www.nytimes.com

Shanahan, John J. 1995. "Director's Letter." *Defense Monitor* 24(6):8.

Shrader, Katherin. 2005. "WMD Commission Releases Scathing Report." *Washington Post,* March 31. Available at http://www.washingtonpost.com

Sietzen, Frank, Jr. 2000. *"Battlelabs" Beef Up Space Defense.* Available at http://www.msnbc.com/ews/4/2426

Smith, Craig. 2004. "Libya to Pay More to French in '89 Bombing." *New York Times,* January 9, p. 6.

Smith, Lamar. 2003. "Restricting Civil Liberties During Wartime Is Justifiable." In *The Terrorist Attack on America,* Mary Williams, ed. (pp. 23–26). Farmington Hills, MA: Greenhaven Press.

Social Science Research Council. 2003. *Introduction to New World Order.* Available at http://www.ssrc.org/sept11/essays

Starr, J. R., and D. C. Stoll. 1989. *U.S. Foreign Policy on Water Resources in the Middle East.* Washington, DC: Center for Strategic and International Studies.

State of Hawaii. 2000. *Dual-Use Technologies.* Available at http://www.state.hi.us/dbedt/ert/key

Stipp, David. 2004. "The United States Must Spend More on High-Tech Defenses Against Biological Warfare." In *Biological Warfare,* William Dudley, ed. (pp. 109–116). Farmington Hills, MA: Greenhaven Press.

Stockholm International Peace Research Institute. 2005. *SIPRI Yearbook 2005: Armaments, Disarmament, and International Security.* Oxford: Oxford University Press.

Strobel, Warren, David Kaplan, Richard Newman, Kevin Whitelaw, and Thomas Grose. 2001. "A War in the Shadows." *U.S. News and World Report* 130:22.

Sutker, Patricia B., Madeline Uddo, Karen Brailey, and Albert N. Allain Jr. 1993. "War Zone Trauma and Stress-Related Symptoms in Operation Desert Storm (ODS) Returnees." *Journal of Social Issues* 40(4):33–50.

Sweet, Alec Stone, and Thomas L. Brunell. 1998. "Constructing a Supranational Constitution. Dispute Resolution and Governance in the European Community." *American Political Science Review* 92: 63–82.

Tavernise, Sabrina. 2005. "Many Iraqis See Sectarian Roots in New Killing." *New York Times,* May 27. Available at http://www.nytimes.com

United Nations. 2003. *Some Questions and Answers.* Available at http://www.unicef.org

United Nations. 2005. *United Nations Peacekeeping Operations.* Available at http://www.un.org/Depts/dpko

U.S. Bureau of the Census. 2004. *Statistical Abstract of the United States: 2004–2005,* 124th ed. Washington, DC: U.S. Government Printing Office.

U.S. Department of Homeland Security. 2003. *Building a Secure Homeland.* Available at http://www.dhs.gov

U.S. State Department. 2004. *Patterns of Global Terrorism.* Available at http://www.state.gov

Vesely, Milan. 2001. "UN Peacekeepers: Warriors or Victims?" *African Business* 261: 8–10.

Viner, Katharine. 2002. "Feminism as Imperialism." *The Guardian*, September 21. Available at http://www.guardian.co.uk

Vlahos, Kelley Beaucar. 2005. "How Secure Is Federal 'Cybersecurity'?" Fox News. Available at http://www.foxnews.com

Waters, Rob. 2005. "The Psychic Costs of War." *Psychotherapy Networker*, March-April, pp. 1–3.

Wax, Emily. 2003. "War Horror: Rape Ruining Women's Health." *Miami Herald*, November 3. Available at http://www.miami.com

Weiner, Tim. 2005. "Air Force Seeks Bush's Approval for Space Weapons Program." *New York Times*, May 18. Available at http://www.nytimes.com

Williams, Mary E., ed. 2003. *The Terrorist Attack on America*. Farmington Hills, MA: Greenhaven Press.

Worldwatch Institute. 2003. *Vital Signs: The Trends That Are Shaping Our Future.* New York: W. W. Norton.

Zakaria, Fareed. 2000. "The New Twilight Struggle." *Newsweek*, October 12. Available at http://www.msnbc.com/news

Epilogue

McKee, Tim. 2006 (January). "A More Perfect Union: Rom Hayden on Democracy and Redemption." *The Sun* 361:4–11.

Stoops, Nicole. 2004 (June). *Educational Attainment in the United States: 2003*. Current Population Reports P20-550. Washington D.C.: U.S. Census Bureau.

Credits

This page constitutes an extension of the copyright page. We have made every effort to trace the ownership of all copyrighted material and to secure permission from copyright holders. In the event of any question arising as to the use of any material, we will be pleased to make the necessary corrections in future printings. Thanks are due to the following authors, publishers, and agents for permission to use the material indicated.

Chapter 1 **2:** AP/Wide World Photos; **3:** AP/Wide World Photos; **23:** © Paul Fusco/Magnum Photos.

Chapter 2 **30:** AP/Wide World Photos; **41:** Digital Vision/Getty Images; **44, 45:** Photos courtesy of the National Institute of Mental Health; **56:** AP/Wide World Photos; **63:** © Per-Anders Pettersson/Getty Images; **68:** © Bill Aron/PhotoEdit, Inc.

Chapter 3 **74:** AP/Wide World Photos; **78:** © Partnership For a Drug-Free America®; **83:** Campaign for Tobacco-Free Kids®/www.tobaccofreekids.org; **94:** Photo by Dennis Oda, June 2005; **96:** AP/Wide World Photos.

Chapter 4 **111:** © Michael Newman/PhotoEdit, Inc.; **117:** © William Campbell/Corbis; **119:** © A. Ramey /PhotoEdit, Inc.; **130:** AP/Wide World Photos.

Chapter 5 **139 (top to bottom):** © AFP/Getty Images; © Caroline Schacht; **154:** © Kevin Winter/Getty Images; **176:** © Getty Images; **179:** Courtesy of The East Los Angeles Men's Health Center, a project of Bienvenidos Children's Center, Inc., www.bienvenidos.org.

Chapter 6 **184 (top to bottom):** © Reuters/Amit Dave/Landov; AP/Wide World Photos; **192:** Photo courtesy of the author; **195:** Courtesy of Texas A&M, photograph by Damian Medina; **198:** © Elena Rooraid/PhotoEdit, Inc.; **203:** USDA Food Stamp Program.

Chapter 7 **228:** © Mark Peterson/Corbis; **230:** © Guenter Stand/VISUM/The Image Works; **233:** © Courtesy of Caroline Schacht and David Knox; **234:** © Kimberly Butler/Time Life Pictures/Getty Images; **250:** Kirschner/United Students Against Sweatshops; **252:** © UPI Photo/Roger L. Wollenberg/Landov.

Chapter 8 **258:** © Karen Kasmauski/Woodfin Camp & Associates **261 (left to right):** © James D. Wilson/Woodfin Camp & Associates; © Dan Habib; **267:** © Michael Newman/PhotoEdit, Inc.; **279:** AP/Wide World Photos.

Chapter 9 **290:** AP/Wide World Photos; **293:** © David Turnley/Corbis; **296:** © Reuters NewMedia, Inc./Corbis; **297:** AP/Wide World Photos; **303:** Photo courtesy of Caroline Schacht and David Knox; **320:** AP/Wide World Photos; **327** © Southern Poverty Law Center.

Chapter 10 **335:** © David Turnley/Corbis; **338:** © Marjorie Farrell/The Image Works; **353:** Source, Bureau of Labor Statistics; **355:** Bangor Daily News photo by Denise Henhoffer; **360:** © Reuters/NASA/HO/Landov; **365 (left to right):** AP/Wide World Photos; © Bob Sacha.

Chapter 11 **369:** Courtesy of Fine By Me, photo by Michael Yaksich; **371:** AP/Wide World Photos; **373:** Palms of Manasota; **381:** AP/Wide World Photos; **384:** © Reuters NewMedia/Corbis; **404 (clockwise from top left:** Courtesy of University of Pennsylvania; Courtesy of University of North Carolina at Chapel Hill; Courtesy of Cornell College; Courtesy of Purdue University; **405:** AP/Wide World Photos.

Chapter 12 **412:** © Tom Miner/The Image Works; **413:** © Reuters NewMedia/Corbis; **419:** © Tomas Bravo/Reuters/Corbis; **426:** © Tony Freeman/PhotoEdit, Inc.; **434:** Courtesy of National Center on Elder Abuse; **437:** © Frank Fournier/Contact Press Images.

Chapter 13 **456:** © David Turnley/Corbis; **457:** © Peter Johnson/Corbis; **458:** 2005 James Gathany; William Nicholson/CDC; **463:** © Dave G. Houser/Corbis; **465:** Courtesy of Wheeling National Heritage Area.

Chapter 14 **473:** AP/Wide World Photos; **482:** © Ted Spiegel/Corbis; **404:** © RII Production/Robert Harding World Imagery/Getty Images; **488:** AP/Wide World Photos; **500:** AP/Wide World Photos; **502:** © Morton Beebe, S.F./Corbis; **504:** Ed Hancock, Courtesy of DOE/NREL; **506:** © Micheline Pelletier/Corbis.

Chapter 15 **516 (left to right):** © Bettmann/Corbis; © Sonda Dawes/The Image Works; **518:** © Tom Grill/Corbis; **522:** © Adam Lubroth/Stone/Getty Images; **530:** © Robert Visser/Corbis Sygma; **534:** Becky Levine; **536:** © David Sams/Stock, Boston.

Chapter 16 **547:** © Bob Mahoney/The Image Works; **550:** © David Pollack/Corbis; **554:** Photo courtesy of the author; **559:** © Sion Touhig/Corbis; **563:** © AFP/Getty Images; **570:** © Michel Lipschitz/AP/Wide World Photos; **573:** © Majid Saeedi/Getty Images.

Name Index

Page numbers in italics refer to information in tables, figures, illustrations, and captions.

Subject Index

Page numbers in italics refer to tables, figures, illustrations and captions.